Free Ideal Rings and Localization in General Rings

Proving that a polynomial ring in one variable over a field is a principal ideal domain can be done by means of the Euclidean algorithm, but this does not extend to more variables. However, if the variables are not allowed to commute, giving a free associative algebra, then there is a generalization, the weak algorithm, which can be used to prove that all one-sided ideals are free.

This book presents the theory of free ideal rings (firs) in detail. Particular emphasis is placed on rings with a weak algorithm, exemplified by free associative algebras. There is also a full account of localization, which is treated for general rings, but the features arising in firs are given special attention. Each section has a number of exercises, including some open problems, and each chapter ends in a historical note.

PAUL COHN is Emeritus Professor of Mathematics at the University of London and Honorary Research Fellow at University College London.

NEW MATHEMATICAL MONOGRAPHS

Editorial Board

Béla Bollobás
William Fulton
Frances Kirwan
Peter Sarnak
Barry Simon

For information about Cambridge University Press mathematics publications visit
http://publishing.cambridge.org/stm/mathematics

Already published in New Mathematical Monographs:

Representation Theory of Finite Reductive Groups
Marc Cabanes, Michel Enguehard

Harmonic Measure
John B. Garnett, Donald E. Marshall

Heights in Diophantine Geometry
Enrico Bombieri, Walter Gubler

Free Ideal Rings and Localization in General Rings

P. M. COHN
Department of Mathematics
University College London

CAMBRIDGE UNIVERSITY PRESS
Cambridge, New York, Melbourne, Madrid, Cape Town, Singapore, São Paulo

Cambridge University Press
The Edinburgh Building, Cambridge CB2 2RU, UK

Published in the United States of America by Cambridge University Press, New York

www.cambridge.org
Information on this title: www.cambridge.org/9780521853378

© Cambridge University Press 2006

This book is in copyright. Subject to statutory exception
and to the provisions of relevant collective licensing agreements,
no reproduction of any part may take place without
the written permission of Cambridge University Press.

First published 2006

Printed in the United Kingdom at the University Press, Cambridge

A catalogue record for this book is available from the British Library

ISBN-13 978-0-521-85337-8 hardback
ISBN-10 0-521-85337-0 hardback

Cambridge University Press has no responsibility for the persistence or accuracy of URLs
for external or third-party internet websites referred to in this book, and does not guarantee
that any content on such websites is, or will remain, accurate or appropriate.

To my granddaughters Chasya and Ayala

Contents

Preface		*page* xi
Note to the reader		xiv
Terminology, notation and conventions used		xvi
List of special notation		xx

0 Generalities on rings and modules — 1
  0.1 Rank conditions on free modules — 1
  0.2 Matrix rings and the matrix reduction functor — 7
  0.3 Projective modules — 12
  0.4 Hermite rings — 19
  0.5 The matrix of definition of a module — 25
  0.6 Eigenrings and centralizers — 33
  0.7 Rings of fractions — 37
  0.8 Modules over Ore domains — 47
  0.9 Factorization in commutative integral domains — 52
  Notes and comments on Chapter 0 — 58

1 Principal ideal domains — 60
  1.1 Skew polynomial rings — 60
  1.2 The division algorithm — 66
  1.3 Principal ideal domains — 73
  1.4 Modules over principal ideal domains — 77
  1.5 Skew Laurent polynomials and Laurent series — 86
  1.6 Iterated skew polynomial rings — 98
  Notes and comments on Chapter 1 — 105

2 Firs, semifirs and the weak algorithm — 107
  2.1 Hereditary rings — 107

2.2	Firs and α-firs	110
2.3	Semifirs and n-firs	113
2.4	The weak algorithm	124
2.5	Monomial K-bases in filtered rings and free algebras	131
2.6	The Hilbert series of a filtered ring	141
2.7	Generators and relations for $GE_2(R)$	145
2.8	The 2-term weak algorithm	153
2.9	The inverse weak algorithm	156
2.10	The transfinite weak algorithm	171
2.11	Estimate of the dependence number	176
	Notes and comments on Chapter 2	183

3 Factorization in semifirs — 186

3.1	Similarity in semifirs	186
3.2	Factorization in matrix rings over semifirs	192
3.3	Rigid factorizations	199
3.4	Factorization in semifirs: a closer look	207
3.5	Analogues of the primary decomposition	214
	Notes and comments on Chapter 3	223

4 Rings with a distributive factor lattice — 225

4.1	Distributive modules	225
4.2	Distributive factor lattices	231
4.3	Conditions for a distributive factor lattice	237
4.4	Finite distributive lattices	243
4.5	More on the factor lattice	247
4.6	Eigenrings	251
	Notes and comments on Chapter 4	261

5 Modules over firs and semifirs — 263

5.1	Bound and unbound modules	264
5.2	Duality	269
5.3	Positive and negative modules over semifirs	272
5.4	The ranks of matrices	281
5.5	Sylvester domains	290
5.6	Pseudo-Sylvester domains	300
5.7	The factorization of matrices over semifirs	304
5.8	A normal form for matrices over a free algebra	311
5.9	Ascending chain conditions	320

5.10	The intersection theorem for firs	326
	Notes and comments on Chapter 5	329

6 Centralizers and subalgebras — 331

6.1	Commutative subrings and central elements in 2-firs	331
6.2	Bounded elements in 2-firs	340
6.3	2-Firs with prescribed centre	351
6.4	The centre of a fir	355
6.5	Free monoids	357
6.6	Subalgebras and ideals of free algebras	367
6.7	Centralizers in power series rings and in free algebras	374
6.8	Invariants in free algebras	379
6.9	Galois theory of free algebras	387
6.10	Automorphisms of free algebras	396
	Notes and comments on Chapter 6	407

7 Skew fields of fractions — 410

7.1	The rational closure of a homomorphism	411
7.2	The category of R-fields and specializations	418
7.3	Matrix ideals	428
7.4	Constructing the localization	437
7.5	Fields of fractions	444
7.6	Numerators and denominators	455
7.7	The depth	466
7.8	Free fields and the specialization lemma	474
7.9	Centralizers in the universal field of fractions of a fir	482
7.10	Determinants and valuations	491
7.11	Localization of firs and semifirs	500
7.12	Reversible rings	511
	Notes and comments on Chapter 7	515

Appendix — 519
A. Lattice theory — 519
B. Categories and homological algebra — 524
C. Ultrafilters and the ultraproduct theorem — 538
Bibliography and author index — 540
Subject index — 566

Preface

It is not your duty to complete the work,
But neither are you free to desist from it.
R. Tarphon, Sayings of the Fathers.

One of the questions that intrigued me in the 1950s was to find conditions for an embedding of a non-commutative ring in a skew field to be possible. I felt that such an embedding should exist for a free product of skew fields, but there seemed no obvious route. My search eventually led to the notion of a *free ideal ring, fir* for short, and I was able to prove (i) the free product of skew fields (amalgamating a skew subfield) is a fir and (ii) every fir is embeddable in a skew field. Firs may be regarded as the natural generalization (in the non-commutative case) of principal domains, to which they reduce when commutativity is imposed. The proof of (i) involved an algorithm, which when stated in simple terms, resembled the Euclidean algorithm but depended on a condition of linear dependence. In this form it could be used to characterize free associative algebras, and this 'weak' algorithm enables one to develop a theory of free algebras similar to that of a polynomial ring in one variable. Of course free algebras are a special case of firs, and other facts about firs came to light, which were set forth in my book *Free Rings and their Relations* (a pun and a paradox). It appeared in 1971 and in a second edition in 1985. A Russian translation appeared in 1975.

More recently there has been a surprising increase of interest, in many fields of mathematics, in non-commutative theories. In functional analysis there has been a greater emphasis on non-commutative function algebras and quantum groups have been introduced in the study of non-commutative geometry, while quantum physics uses non-commutative probability theory, in which even free associative algebras have made their appearance. The localization developed in *Free Rings* has also found a use by topologists. All this, and the fact that many proofs have been simplified, has encouraged me to write a book based on the earlier work, but addressed to a wider audience. Since skew fields play a prominent role, the prefix 'skew' will often be left out, so fields are generally assumed to be not necessarily commutative.

The central part is Chapter 7, in which non-commutative localization is studied. For any ring R the various homomorphisms into fields are described by their singular kernels, the matrices with a singular image, which form a resemblance to prime ideals and so are called *prime matrix ideals*. Various classes of rings, such as firs and semifirs, are shown to be embeddable in fields, and an explicit criterion is given for such an embedding of a general ring to be possible, as well as conditions for a universal field of fractions to exist. This is the case for firs, while for free algebras the universal field of fractions can be shown to be 'free'. The existence of the localization now has a simpler and more direct proof, which is described in Sections 7.1–7.4. It makes only occasional reference to earlier chapters (mainly parts of Chapter 0) and so can be read at any stage.

In the remaining chapters the theory of firs is developed. Their similarity to principal ideal domains is stressed; the theory of the latter is recalled in Chapter 1, while Chapter 0 brings essential facts about presentations of modules over general rings, particularly projective modules, facts that are perhaps not as well known as they should be. Chapter 2 introduces firs and semifirs and deals with the most important example, a ring possessing a generalized form of the division algorithm called the *weak algorithm*. The unique factorization property of principal ideal domains has an analogue in firs, which applies to square matrices as well; this result and its consequences for modules are discussed in Chapter 3. It turns out that the factors of any element form a modular lattice (as in principal ideal domains), which in the case of free algebras is even distributive; this result is the subject of Chapter 4. In Chapter 5 the module theory of firs and semifirs is studied; this leads to a wider class of rings, the Sylvester domains (characterized by Sylvester's law of nullity), which share with semifirs the property of possessing a universal field of fractions. Chapter 6 examines centres, centralizers and subalgebras of firs and semifirs.

Results from lattice theory, homological algebra and logic that are used in the book are recalled in an Appendix. Thus the only prerequisites needed are a basic knowledge of algebra: rings and fields, up to about degree level. Although much of the work already occurs in *Free Rings*, the whole text has been reorganized to form a better motivated introduction and there have been many improvements that allow a smoother development. On the other hand, the theory of skew field extensions has been omitted as a fuller account is now available in my book on skew fields (SF; see p. xv). The rather technical section on the work of

Gerasimov, leading to information on the localization of n-firs, has also been omitted.

I have had the help of many correspondents in improving this edition, and would like to express my appreciation. Foremost among them is G. M. Bergman, who in 1999–2000 ran a seminar at the University of California at Berkeley on the second edition of *Free Rings*, and provided me with over 300 pages of comments, correcting mistakes, outlining further developments and raising interesting questions. As a result the text has been greatly improved. I am also indebted to V. O. Ferreira for his comments on *Free Rings*.

My thanks go also to the staff of the Cambridge University Press for the efficient way they have carried out their task.

University College London
October 2005 P. M. Cohn

Note to the reader

Chapter 0 consists of background material from ring theory that may not be entirely standard, whereas the Appendix gives a summary of results from lattice theory, homological algebra and logic, with reference to proofs, or in many cases, sketch proofs. Chapter 1 deals with principal ideal domains, and so may well be familiar to the reader, but it is included as a preparation for what is to follow. The main subject matter of the book is introduced in Chapter 2 and the reader may wish to start here, referring back to Chapters 1 or 0 only when necessary. In any case Chapter 2 as well as Chapter 3 are used throughout the book (at least the earlier parts, Sections 2.1–2.7 and 3.1–3.4), as is Chapter 5, while Chapters 4 and 6 are largely independent of the rest. The first half of Chapter 7 (Sections 7.1–7.5) is quite independent of the preceding chapters, except for some applications in Section 7.5, and it can also be read at any stage.

All theorems, propositions, lemmas and corollaries are numbered consecutively in a single series in each section, thus Theorem 4.2.5 is followed by Corollary 4.2.6, and this is followed by Lemma 4.2.7, in Section 4.2, and in that section they are referred to as Theorem 2.5, Corollary 2.6, Lemma 2.7 (except in the enunciation). The end or absence of a proof is indicated by ∎. A few theorems are quoted without proof. They are distinguished by letters, e.g. Theorem 7.8.A. There are exercises at the end of each section; the harder ones are marked * and open-ended (or open) problems are marked °, though sometimes this may refer only to the last part; the meaning will usually be clear.

References to the bibliography are by author's name and the year of publication, though 19 is omitted for publications between 1920 and 1999. Publications by the same author in a given year are distinguished by letters. The following books by the author, which are frequently referred to, are indicated by abbreviations:

CA. *Classic Algebra*. John Wiley & Sons, Chichester 2000.
BA. *Basic Algebra, Groups, Rings and Fields*. Springer-Verlag, London 2002.
FA. *Further Algebra and Applications*. Springer-Verlag, London 2003.
SF. *Skew Fields, Theory of General Division Rings*. Encyclopedia of Mathematics and its Applications, 57. Cambridge University Press, Cambridge 1995.
UA. *Universal Algebra*, rev. edn. Mathematics and its Applications, Vol. 6. D. Reidel, Publ. Co., Dordrecht 1981.
IRT. *Introduction to Ring Theory*. Springer Undergraduate Mathematics Series. Springer-Verlag, London 2000.
FR.1. *Free Rings and their Relations*. London Math. Soc. Monographs No. 2. Academic Press, New York 1971.
FR.2. *Free Rings and their Relations*, 2nd edn. London Math. Soc. Monographs No. 19. Academic Press, New York 1985.

Terminology, notation and conventions used

For any set X, the number of its elements, or more generally, its cardinality is denoted by $|X|$. If a condition holds for all elements of X except a finite number, we say that the condition holds for *almost all* members of X.

All rings occurring are associative, but not generally commutative (in fact, much of the book reduces to well-known facts in the commutative case). Every ring has a unit-element or one, denoted by 1, which is inherited by subrings, preserved by homomorphisms and acts as the identity operator on modules. The same convention applies to monoids (i.e. semigroups with one). A ring may consist of 0 alone; this is so precisely when $1 = 0$ and R is then called the *zero ring*. Given any ring R, the opposite ring R^o is defined as having the same additive group as R and multiplication $a.b = ba$ $(a, b \in R)$. With any property of a ring we associate its left–right dual, which is the corresponding property of the opposite ring. Left–right duals of theorems, etc. will not usually be stated explicitly.

We shall adopt the convention of writing (as far as practicable) homomorphisms of left modules on the right and vice versa. Mappings will be composed accordingly, although we shall usually give preference to writing mappings on the right, so that fg means 'first f, then g'. If R is any ring, then for any left R-module M, its *dual* is $M^* = \text{Hom}_R(M, R)$, a right R-module; similarly on the other side. The space of $m \times n$ matrices over M is written ${}^m M^n$, and we shall also write ${}^m M$ for ${}^m M^1$ (column vectors) and M^n for ${}^1 M^n$ (row vectors). A similar notation is used for rings.

In any ring R the set of non-zero elements is denoted by $R^\times$, but this notation is mostly used for integral domains, where $R^\times$ contains 1 and is closed under multiplication. Thus an integral domain need not be commutative. If $R^\times$ is a group under multiplication, R is called a *field*; occasionally the prefix 'skew' is used, to emphasize the fact that our fields need not be commutative. An element u in a ring or monoid is *invertible* or a *unit* if it has an *inverse* u^{-1} satisfying

$uu^{-1} = u^{-1}u = 1$. Such an inverse is unique if it exists at all. The units of a ring (or monoid) R form a group, denoted by $U(R)$. The ring of all $n \times n$ matrices over R is written $\mathfrak{M}_n(R)$ or R_n. The set of all square matrices over R is denoted by $\mathfrak{M}(R)$. Instead of $U(R_n)$ we also write $GL_n(R)$, the general linear group. The matrix with (i, j)-entry 1 and the rest zero is denoted by e_{ij} and is called a *matrix unit* (see Section 0.2). An *elementary* matrix is a matrix of the form $B_{ij}(a) = I + ae_{ij}$, where $i \neq j$; these matrices generate a subgroup $E_n(R)$ of $GL_n(R)$, the *elementary group*. By a *permutation matrix* we understand the matrix obtained by applying a permutation to the columns of the unit matrix. It is a member of the *extended* elementary group $E_n^*(R)$, the group generated by $E_n(R)$ and the matrix $I - 2e_{11}$. If in a permutation matrix the sign of one column is changed whenever the permutation applied was odd, we obtain a *signed* permutation matrix; these matrices generate a subgroup $P_n(R)$ of $E_n(R)$.

An element u of a ring is called a *left zero-divisor* if $u \neq 0$ and $uv = 0$ for some $v \neq 0$; of u is neither 0 nor a left zero-divisor, it is called *right regular*. Thus u is right regular whenever $uv = 0$ implies $v = 0$. Corresponding definitions hold with left and right interchanged. A left or right zero-divisor is called a *zero-divisor*, and an element that is neither 0 nor a zero-divisor is called *regular*. These terms are also used for matrices, not necessarily square. Over a field a square matrix that is a zero-divisor or 0 is also called *singular*, but this term will not be used for general rings.

An element u of a monoid is called *regular* if it can be cancelled, i.e. if $ua = ub$ or $au = bu$ implies $a = b$. If every element of a monoid S can be cancelled, S is called a *cancellation monoid*. A monoid is called *conical* if $ab = 1$ implies $a = 1$ (and so also $b = 1$).

An element of a ring is called an *atom* if it is a regular non-unit and cannot be written as a product of two non-units. A factorization is called *proper* if all its factors are non-units; if all its factors are atoms, it is called a *complete* factorization. An integral domain is said to be *atomic* if every element other than zero or a unit has a complete factorization. If a, b are elements of a commutative monoid, we say that *a divides b* and write $a|b$ if $b = ac$ for some element c.

The maximum condition or ascending chain condition on a module or the left or right ideals of a ring or a monoid is abbreviated as ACC. If a module satisfies ACC on submodules on at most n generators, we shall say that it satisfies ACC_n. In particular, left (right) ACC_n for a ring R is the ACC on a n-generator left (right) ideals of R. A module (or ring) satisfying ACC_n for all n is said to satisfy pan-ACC. Similar definitions apply to the minimum condition or descending chain condition. DCC for short.

Two elements a, b of a ring (or monoid) R (or matrices) are *associated* if $a = ubv$ for some $u, v \in U(R)$. If $u = 1$ ($v = 1$), they are *right* (*left*) associated; if $u = v^{-1}$, they are *conjugate* under $U(R)$. A polynomial in one variable (over any ring) is said to be *monic* if the coefficient of the highest power is 1. Two elements a, b of a ring R are *left coprime* if they have no common left factor apart from units; they are *right comaximal* if $aR + bR = R$. Clearly two right comaximal elements are left coprime, but not necessarily conversely. Two elements a, b are said to be *right commensurable* if there exist a', b' such that $ab' = ba' \neq 0$. Again, corresponding definitions apply on the other side. A row $(a_1, \ldots, a_n)$ of elements in R is said to be *unimodular* if the right ideal generated by the a_i is R; thus a pair is unimodular precisely when it is right comaximal. Similarly, a column is unimodular if the left ideal generated by its components is R.

Let A be a commutative ring; by an *A-algebra* we understand a ring R which is an A-module such that the multiplication is bilinear. Sometimes we shall want a non-commutative coefficient ring A; this means that our ring R is an A-bimodule such that $x(yz) = (xy)z$ for any x, y, z from R or A; this will be called an *A-ring*. To rephrase the definitions, a A-ring is a ring R with a homomorphism $\alpha \mapsto \alpha.1$ of A into R, while an A-algebra is a ring R with a homomorphism of A into the centre of R. Moreover, the use of the term 'A-algebra' implies that A is commutative. Frequently our coefficient ring will be a skew field, usually written K, or also k when it is assumed to be commutative.

Let R be an A-ring. A family (u_i) of elements of R is *right linearly dependent* over A or *right A-dependent* if there exist $\lambda_i \in A$ almost all but not all zero, such that $\sum u_i \lambda_i = 0$. In the contrary case (u_i) is *right A-independent*. Occasionally we speak of *a set* being linearly dependent; this is to be understood as a family indexed by itself. For example, two elements of an integral domain R are *right commensurable* if and only if they are right linearly R-dependent and both non-zero.

If A, B are matrices, we write $\begin{pmatrix} A & 0 \\ 0 & B \end{pmatrix}$ as $A \oplus B$ or $\mathrm{diag}(A, B)$. We shall also sometimes write columns as rows, with a superscript T to indicate transposition (reflexion in the main diagonal). In such cases the blocks are to be transposed as a whole, thus $(A, B)^{\mathrm{T}}$ means $\begin{pmatrix} A \\ B \end{pmatrix}$, not $\begin{pmatrix} A^{\mathrm{T}} \\ B^{\mathrm{T}} \end{pmatrix}$. For any $m \times n$ matrix A its *index* $i(A)$ is defined as $n - m$, and $m \times n$, or n in case $m = n$, is described as its *size* or *order*.

The letters $\mathbb{N}, \mathbb{N}_{>0}, \mathbb{Z}, \mathbb{F}_p, \mathbb{Q}, \mathbb{R}, \mathbb{C}$ stand as usual for the set (respectively ring) of non-negative integers, all positive integers, all integers, all integers mod p, rational, real and complex numbers, respectively. If $T \subseteq S$, the complement of T in S is written $S \backslash T$.

In a few places in Chapter 7 and the Appendix some terms from logic are used. We recall that a *formula* is a statement involving elements of a ring or group. Formulae can be combined by forming a *conjunction* $P \wedge Q$ (P and Q), a *disjunction* $P \vee Q$ (P or Q) or a *negation* $\neg P$ (not P). A formula that is not formed by conjunction, disjunction or negation from others is called *atomic*.

List of special notation

(Notation that is either standard or only used locally has not always been included.)

$\lvert I \rvert$	cardinality of the set I, xvi
$R^\times$	set of non-zero elements in a ring R, xvi
$U(R)$	group of units in a ring R, xvii
R°	opposite ring, xvi
$\mathfrak{M}_n(R), R_n$	ring of all $n \times n$ matrices over R, xvii
M^n	set of all n-component row vectors over a module M, xvi, 7
M^I	direct power of M indexed by a set I, 1
$M^{(I)}$	direct sum of $\lvert I \rvert$ copies of M, 1
$^m M$	set of all m-component column vectors over M, xvi, 7
$^m M^n$	set of all $m \times n$ matrices over M, xvi, 7
$i(A)$	index of a matrix A, xviii
M^*	dual of M, xvi, 2
$\rho(A)$	inner rank of a matrix A, 3
$\rho^*(A)$	stable rank of A, 5
$V_{m,n}$	canonical non-IBN ring, 7
e_{ij}	matrix unit, xvii, 8
e_i, e_i^T	row, resp. column vector, 9f.
Rg, Rg_n	category of rings, $n \times n$ matrix rings, 9
I_0	0×0 matrix, 9
$\mathfrak{W}_n$	n-matrix reduction functor, 11
$_R\mathrm{Mod}$	category of left R-modules
$_R\mathrm{proj}$	category of finitely generated projective left R-modules, 14
$S(R)$	monoid of finitely generated projective left R-modules, 14
$K_0(R)$	Grothendieck group, 14
$J(R)$	Jacobson radical of R, 15

List of special notation

$GL_n(R)$	general linear group of $n \times n$ invertible matrices over R, xvii
$E_n(R)$	subgroup generated by the elementary matrices, xvii, 117
$E_n^*(R)$	extended elementary group, xvii, 147
$P_n(R)$	subgroup of signed permutation matrices, xvii, 159
$Tr_n(R)$	subgroup of upper unitriangular matrices, 159
$\chi(M)$	characteristic of a module with a finite-free resolution of finite length, 26
$I(-)$	idealizer, 33
$E(-)$	eigenring, 33
M_t	submodule of torsion elements, 48
$\Gamma_{\geq 0}$	set of all elements ≥ 0 in an ordered additive group Γ, 337
$\Gamma_{> 0}$	set of all elements > 0 in Γ
$a\|b$	a divides b, 52
$l(c)$ or $\|c\|$	length of an element c of a UF-monoid, 56
$a \parallel b$	a is a total divisor of b, 79
$A_1(k)$	Weyl algebra over k, 64
$R_{(n)}$	component of a filtered ring, 125
$R_{\lceil n \rceil}$	component of an inversely filtered ring, 157
$k\langle X \rangle$	free k-algebra on X, 135
$D_K \langle X \rangle$	free tensor D-ring centralizing K, 136
$k\langle\langle X \rangle\rangle$	formal power series ring on X over k, 161
$k\langle\langle X \rangle\rangle_{\text{rat}}$	subring of all rational power series, 167
$k\langle\langle X \rangle\rangle_{\text{alg}}$	subring of all algebraic power series, 167
c^*	bound of an element c (in a PID), 342
$\iota(c)$	inner automorphism defined by c
$d(f)$	degree of a polynomial f, 60f.
$o(f)$	order of a power series f, 88
$L(cR, R)$	lattice of principal right ideals containing cR, 116
$H(X), H(R:k)$	Hilbert series, 142
Tor	category of torsion modules (over a semifir), 193
$Tr(-)$	transpose of a module, 269
Pos, Neg	category of positive, negative modules over a semifir (in Section 4.4 $\mathscr{P}os$ is also used for the category of all finite partially ordered sets), 273
M_b	bound component of a module M, 264f.
$B(\Sigma)$	set of displays, 437
Inv(R)	set of all invariant elements of R, 57, 333
X^*	free monoid on an alphabet X, 358
$\|f\|$	length of a word f, 358
f^λ, A^λ	terms of highest degree in f, A, 312

$\mathfrak{M}(R)$	set of all square matrices over R, 411
Ker φ	singular kernel of homomorphism φ, 423
$A \nabla B$	determinantal sum of A and B, 429
$\sqrt{\mathfrak{a}}$	radical of a matrix ideal $\mathfrak{a}$, 434
$\mathcal{F}_R$	category of all R-fields and specializations, 420
ε_R	full subcategory of epic R-fields, 421
$\mathcal{Z}$ ($\mathcal{Z}_0$)	least matrix (pre-)ideal, 444
$\mathcal{N} = \sqrt{\mathcal{Z}}$	matrix nilradical, 444
$o(A)$	order of an admissible matrix A, 414
$X(R)$	field spectrum of R, 442
$\mathcal{D}(A)$	singularity support of the matrix A, 442
$\Phi(R)$	set of all full matrices over R, 447
$K_1(U)$	Whitehead group of a field U, 495
G^{ab}	universal abelianization of a group G, 494
$D(R)$	divisor group of R, 497

0
Generalities on rings and modules

This chapter collects some facts on rings and modules, which form neither part of our subject proper, nor part of the general background (described in the Appendix). By its nature the content is rather mixed, and the reader may well wish to begin with Chapter 1 or even Chapter 2, and only turn back when necessary.

In Section 0.1 we describe the conditions usually imposed on the ranks of free modules. The formation of matrix rings is discussed in Section 0.2; Section 0.3 is devoted to projective modules and the special class of Hermite rings is considered in Section 0.4.

Section 0.5 deals with the relation between a module and its defining matrix, and in particular the condition for two matrices to define isomorphic modules. This and the results on eigenrings and centralizers in Section 0.6 are mainly used in Chapters 4 and 6.

The Ore construction of rings of fractions is behind much of the later development, even when this does not appear explicitly. In Section 0.7 we recall the details and apply it in Section 0.8 to modules over Ore domains; it turns out that the (left or right) Ore condition has some unexpected consequences. In Section 0.9 we recall some well-known facts on factorization in commutative rings, often stated in terms of monoids, in a form needed later.

0.1 Rank conditions on free modules

Let R be any ring, M an R-module and I a set. The direct power of M with index set I is denoted by M^I, while the direct sum is written $M^{(I)}$. When I is finite, with n elements, these two modules agree and are written as M^n, as usual. More precisely, M^n denotes the set of rows and ^nM the set of columns of length n.

With every left R-module M we can associate its dual
$$M^* = \mathrm{Hom}_R(M, {}_RR),$$
consisting of all linear functionals on M with the natural right R-module structure defined by $(\alpha c, x) = (\alpha, cx)$, where $x \in M, \alpha \in M^*$ and $c \in R$. Similarly, every right R-module N has as dual the left R-module $N^* = \mathrm{Hom}_R(N, R_R)$. In particular, $(R^n)^* \cong {}^nR, ({}^nR)^* \cong R^n$; more generally, if P is a finitely generated projective left R-module, then P^* is a finitely generated projective right R-module and $P^{**} \cong P$. For if $P \oplus Q \cong R^n$, then $P^* \oplus Q^* \cong {}^nR$ and $P^{**} \oplus Q^{**} \cong R^n$. Now the obvious map $\delta_P : P \to P^{**}$, which maps $x \in P$ to $\hat{x} : f \mapsto \langle f, x \rangle$ is an isomorphism, because $\delta_P \oplus \delta_Q = 1$.

Let R be any ring and M a left R-module with a minimal generating set X. If X is infinite, then any generating set of M has at least $|X|$ elements, and in particular, any two minimal generating sets of M have the same cardinality. However, when X is finite, this need not be so, thus a free module on a finite free generating set may have minimal generating sets of different sizes. We shall say that R^n has *unique rank* if it is not isomorphic to R^m for any $m \neq n$. Using the pairing provided by * we see that R^n has unique rank if and only if ${}^nR$ has unique rank. For any free module F of unique rank n we write $n = \mathrm{rk}(F)$.

A ring R is said to have the *invariant basis property* or *invariant basis number* (IBN) if every free R-module has unique rank. Most rings commonly encountered have IBN, but we shall meet examples of non-zero rings where this property fails to hold.

Occasionally we shall need stronger properties than IBN. A ring R is said to have *unbounded generating number* (UGN) if for every n there is a finitely generated R-module that cannot be generated by n elements. Since any n-generator module is a homomorphic image of a free module of rank n, it follows that in a ring with UGN a free module of rank n cannot be generated by fewer than n elements, and this condition characterizes rings with UGN. It also shows that UGN implies IBN.

A ring R is said to be *weakly n-finite* if every generating set of n elements in R^n is free; if this holds for all n, R is called *weakly finite* (WF). Weakly 1-finite rings are sometimes called 'directly finite', 'von Neumann finite' or 'inverse symmetric'. As an example of weakly finite rings we have projective-free rings, where a ring is called *projective-free* if every finitely generated projective module is free, of unique rank.

Let R be any non-zero ring and suppose that R^n has a generating set of m elements, for some $m, n \geq 1$. Then we have a surjection $R^m \to R^n$, giving rise to an exact sequence
$$0 \to K \to R^m \to R^n \to 0.$$

Since R^n is free, the sequence splits and so $R^m \cong R^n \oplus K$. This shows that the three properties defined here may be stated as follows:

IBN. *For all m, n, $R^m \cong R^n$ implies $m = n$.*
UGN. *For all m, n, $R^m \cong R^n \oplus K$ implies $m \geq n$.*
WF. *For all n, $R^n \cong R^n \oplus K$ implies $K = 0$.*

By describing the change of basis, we can express these conditions in matrix form:

IBN. *For any $A \in {}^m R^n, B \in {}^n R^m$, if $AB = I_m$, $BA = I_n$, then $m = n$.*
UGN. *For any $A \in {}^m R^n, B \in {}^n R^m$, if $AB = I_m$, then $n \geq m$.*
WF. *For any $A, B \in R^n$, if $AB = I$, then $BA = I$.*

We see that a ring has IBN if and only if every invertible matrix has index zero; it has UGN if and only if every matrix with a right inverse has non-negative index, and it is weakly finite if and only if all inverses of square matrices are two-sided. The UGN condition can also be defined in terms of the rank of a matrix, which over general rings is defined as follows. Given any matrix A, of the different ways of writing A as a product, $A = PQ$, we choose one for which the number of rows of Q is least. This number is called the *inner rank* of A, written $\rho(A)$ or ρA, and the corresponding factorization of A is called a *rank factorization*. For matrices over a field this notion of rank reduces to the familiar rank; now we observe that a ring has UGN if and only if the inner rank of any $n \times n$ unit matrix is n. Such a matrix is said to be *full*. Thus a matrix is full if and only if it is square, say $n \times n$, and cannot be written as a product of an $n \times r$ by an $r \times n$ matrix, where $r < n$. We note that every non-zero element (in any ring) is full as a 1×1 matrix, and the unit matrix of every size is full precisely if the ring has UGN. Over a field the full matrices are just the regular matrices (see Section 5.4), but in general there is no relation between full and regular matrices.

Either set of the above conditions makes it clear that the zero ring is weakly finite, but has neither IBN nor UGN. For a non-zero ring,

$$\text{WF} \Rightarrow \text{UGN} \Rightarrow \text{IBN},$$

and if a ring R has any of these properties, then so does its opposite R°. Moreover, if $R \to S$ is a homomorphism and S has IBN or UGN, then so does R. Clearly any field (even skew) has all properties; more generally this holds for any Noetherian ring (see BA, theorem 4.6.7 or Exercise 5 below), as well as any subring of a field. Using determinants, we see that every non-zero commutative ring also has all three properties. Examples of rings having IBN but not UGN, and rings having UGN but not weakly finite, may be found in Cohn [66a] or in SF, Section 5.7 (see also Exercise 2 and Section 2.11).

For a non-zero ring without IBN there exist positive integers h, k such that

$$R^h \cong R^{h+k}, \quad h, k \geq 1. \tag{1}$$

The first such pair (h, k) in the lexicographic ordering is called the *type* of the ring R. We observe that for a ring R of type (h, k) $R^m \cong R^n$ holds if and only if $m = n$ or $m, n \geq h$ and $m \equiv n \pmod{k}$ (see, e.g. UA, Theorem X.3.2, p. 340).

Proposition 0.1.1. *Let $f: R \to S$ be a homomorphism between non-zero rings. If R does not have IBN and its type is (h, k), then S does not have IBN and if its type is (h', k'), then $h' \leq h, k'|k$.*

Here it is important to bear in mind that all our rings have a unit element, which is preserved by homomorphisms and inherited by subrings.

Proof. By hypothesis (1) holds, hence there exist $A \in {}^m R^n, B \in {}^n R^m$ satisfying $AB = I, BA = I$, with $m = h, n = h + k$. Applying f we get such matrices over S, whence it follows that $S^h \cong S^{h+k}$, so S cannot have IBN and $h' \leq h, k'|k$. ∎

The next result elucidates the connexion between weak finiteness and UGN.

Proposition 0.1.2. *A ring R has UGN if and only if some non-zero homomorphic image of R is weakly finite.*

Proof. If a non-zero homomorphic image S of R is weakly finite, then S has UGN, hence so does R. Conversely, assume that the zero ring is the only weakly finite homomorphic image of R. By adjoining the relations $YX = I$, for all pairs of square matrices X, Y satisfying $XY = I$, we obtain a weakly finite ring S. For suppose that $I - AB = \Sigma_1^r U_i(I - Y_i X_i) V_i$, where $X_i Y_i = I$. By taking $X = X_1 \oplus \cdots \oplus X_r, Y = Y_1 \oplus \cdots \oplus Y_r, U = (U_1, \ldots, U_r), V = (V_1, \ldots, V_r)^T$, we can write this as

$$I - AB = U(I - YX)V, \tag{2}$$

and $XY = I$. If A, B are $n \times n$ and X, Y are $m \times m$, then U is $n \times m$ and V is $m \times n$. Suppose that $n \geq m$; on replacing X, Y by $X \oplus I, Y \oplus I$, respectively, where I is the unit matrix of order $n - m$, and completing U, V to square matrices by adding columns, respectively rows of zeros, we obtain an equation (2), where all matrices are square of order n. Similarly, if $n \leq m$, we can achieve the same result by taking diagonal sums of A, B with I. Writing $Z = AX + U(I - YX), W = YB + (I - YX)V$, we have $ZW = I$ and $ZY = A, XW = B$, hence $I - BA = X(I - WZ)Y$. Therefore S is weakly finite and so must be the zero ring. It follows that R becomes zero by adjoining a finite number of such matrix equations, and by taking diagonal sums we obtain a single pair

X, Y, each $s \times s$, say, for which this happens. Thus $XY = I$, while the ideal generated by the entries of $I - YX$ is the whole ring. Replacing each of X, Y by a diagonal sum of an appropriate number of copies, we may assume that there exist $p \in R^s, q \in {}^sR$ such that $p(I - YX)q = 1$. Therefore we have

$$I_{s+1} = \begin{pmatrix} X \\ p(I - YX) \end{pmatrix} (Y \quad (I - YX)q),$$

and this equation shows that UGN fails for R. ∎

For another characterization of weak finiteness we shall need to refine the notion of inner rank. Given a non-zero ring R, let A be any matrix over R and consider $A \oplus I_r$, the diagonal sum of A and the $r \times r$ unit matrix. Since any factorization of A can also be used to factorize $A \oplus I_r$, it follows that $\rho(A \oplus I_r) \leq \rho(A) + r$. Therefore we have

$$\rho(A) \geq \rho(A \oplus I_1) - 1 \geq \rho(A \oplus I_2) - 2 \geq \ldots \tag{3}$$

If this sequence has a finite limit, we denote it by $\rho^*(A)$ and call it the *stable rank* of A. An $n \times n$ matrix of stable rank n is said to be *stably full*. Thus a square matrix A is stably full if and only if $A \oplus I_r$ is full for all $r \geq 1$. Hence every stably full matrix is full, but the converse need not hold.

For a ring without UGN the unit matrix of some size is not full, say the $n \times n$ unit matrix I_n has rank $n - 1$. Then the sequence (3) is unbounded below and we formally put $\rho^*(A) = -\infty$ for every matrix A. If R has UGN, we have $\rho(A \oplus I_r) \geq r$, so in this case the sequence (3) is bounded below by 0 and hence has a limit; thus in any ring with UGN the stable rank of every matrix exists as a non-negative integer. Conversely, if the stable rank exists for some matrix A, say $\rho^*(A) = r$, then for some n and all $s \geq n$, $\rho(A \oplus I_s) = r + s$. Hence for any $t \geq 0, r + s + t = \rho(A \oplus I_s \oplus I_t) \leq \rho(A \oplus I_s) + \rho(I_t) = r + s + \rho(I_t)$. Thus $\rho(I_t) \geq t$ and this proves that R has UGN. We now have the following connexion with weak finiteness.

Proposition 0.1.3. *For any non-zero ring R the following are equivalent:*

(a) R is weakly finite,
(b) every non-zero matrix over R has a stable rank, which is positive,
(c) every non-zero idempotent matrix over R has a stable rank, which is positive.

Proof. We note that in each case the stable rank is finite. Now let A be any $m \times n$ matrix over R, of stable rank t, say; then for some $s \geq 0$, $\rho(A \oplus I_s) = t + s = r$, say. So we can write

$$\begin{pmatrix} A & 0 \\ 0 & I_s \end{pmatrix} = \begin{pmatrix} B \\ B' \end{pmatrix} (C \quad C'), \tag{4}$$

where $B \in {}^m R^r, B' \in {}^s R^r, C \in {}^r R^n, C' \in {}^r R^s$. Thus we have $B'C' = I, BC' = 0 = B'C, BC = A$.

To prove (a) $\Rightarrow$ (b), assume that $\rho^*(A) = 0$; then $r = s$, so by weak finiteness, $C'B' = I$, hence $B = 0 = C$ and therefore $A = BC = 0$.

(b) $\Rightarrow$ (c) is clear; to prove (c) $\Rightarrow$ (a), assume that R is not weakly finite. Then there exist $B', C' \in R_s$ with $B'C' = I, C'B' \neq I$, so (4) holds with $m = n = r = s, A = B = C = I - C'B'$, and this is a non-zero idempotent matrix of zero stable rank. ∎

In conclusion we note another consequence of weak finiteness.

Proposition 0.1.4. *Let R be a weakly n-finite ring and let $A \in {}^r R^n, A' \in {}^n R^r, B \in {}^n R^s, B' \in {}^s R^n$ be such that $AB = 0, AA' = I_r, B'B = I_s$, where $r + s = n$. Then there exists $P \in GL_n(R)$ such that*

$$A = (I_r \ 0)P, \quad B = P^{-1} \begin{pmatrix} 0 \\ I_s \end{pmatrix}.$$

Proof. These equations just state that A constitutes the first r rows of P, while B forms the last s columns of P^{-1}. To prove this result, we have by hypothesis

$$\begin{pmatrix} A \\ B' \end{pmatrix} (A' \ B) = \begin{pmatrix} I_r & 0 \\ B'A' & I_s \end{pmatrix},$$

where all the matrices are $n \times n$. By subtracting $B'A'$ times the first r rows from the last s we reduce the right-hand side to I, so the result follows by taking $P = (A, B'E)^T, P^{-1} = (A', B)$, where $E = I - A'A$. ∎

Exercises 0.1

1. Show that over a ring of type $(h, k)(k \geq 1)$ every finitely generated module can be generated by h elements. Find a bound for the least number of elements in a basis of a finitely generated free module.
2. If K is a non-zero ring and I an infinite set, show that $R = \text{End}(K^{(I)})$ does not have IBN and determine its type.
3. If every finitely generated R-module is cyclic, show that R cannot be an integral domain; in particular, obtain this conclusion for a ring of type $(1, k)$.
4. A ring R is said to have *bounded decomposition type* (BDT), if there is a function $r(n)$ such that R^n can be written as a direct sum of at most $r(n)$ terms. Show that any ring with BDT is weakly finite.
5. Show that a ring with left ACC_n for some $n \geq 1$ is weakly n-finite. Deduce that a left (or right) Noetherian ring, or more generally, a ring with left (or right) pan-ACC is weakly finite. (Recall that 'pan-ACC' stands for 'ACC_n for all n'.) Obtain the same conclusion for DCC_n. (*Hint*: See Exercise 7.10.)

6. Let R be a non-zero ring without IBN and for fixed $m, n(m \neq n)$ consider pairs of mutually inverse matrices $A \in {}^m R^n$, $B \in {}^n R^m$. Show that if A', B' is another such pair, then $P = A'B$ is an invertible matrix such that $PA = A'$, $BP^{-1} = B'$. What is P^{-1}?
7. Let R be a weakly n-finite ring. Given maps $\alpha : R^r \to R^n$ and $\beta : R^n \to R^s (r + s = n)$ such that $\alpha\beta = 0$, α has a right inverse and β has a left inverse, then there exists an automorphism μ of R^n such that $\alpha\mu : R^r \to R^n$ is the natural inclusion and $\mu\pi = \beta$, where $\pi : R^n \to R^s$ is the natural projection. Show that conversely, every ring with this property is weakly n-finite. (*Hint*: Imitate the proof of Proposition 1.4.)
8. Show that a ring R is weakly n-finite if and only if (F): *Every surjective endomorphism of R^n is an automorphism.* If a non-zero ring R has the property (F), show that every free homomorphic image of R^n has rank at most n. Deduce that every non-zero weakly finite ring has UGN.
9*. Which of IBN, UGN, weak finiteness (if any) are Morita invariants?
10°. Characterize the rings all of whose homomorphic images are weakly finite.
11. (Leavitt [57]) Show that if a ring R has a non-zero free module F with no infinite linearly independent subset, then F has unique rank.
12*. (Montgomery [83]) Let A be an algebra over the real numbers with generators a_0, a_1, b_0, b_1 and defining relations $a_0 b_0 - a_1 b_1 = 1$, $a_1 b_0 + a_0 b_1 = 0$. Show (by using a normal form for the elements of A) that A is an integral domain, hence weakly 1-finite, but not weakly 2-finite. Show also that $A \otimes_R C$ is not weakly 1-finite (see also Exercise 2.11.8).
13°. Is the tensor product of two weakly finite k-algebras again weakly finite?
14°. Is every weakly 1-finite von Neumann regular ring weakly finite?
15. Let $V_{m,n}$ be a k-algebra with $2mn$ generators, arranged as an $m \times n$ matrix A and an $n \times m$ matrix B and defining relations (in matrix form) $AB = I$, $BA = I$ (the 'canonical non-IBN ring' for $m \neq n$). Show that $V_{1,n}$ is a simple ring for $n > 1$; what is $V_{1,1}$?
16. (M. Kirezci) If $V_{m,n}$ is defined as in Exercise 15 and $m < n$, show that there is a homomorphism $V_{m,n+r(n-m)} \to V_{m,n}$, for any $r > 0$. [*Hint*: If in $V_{m,n}$, $A = (A_1, A_2)$, $B = (B_1, B_2)^T$, where A_1, B_1 are square, verify that $(A_1^r, A_1^{r-1} A_2, \ldots, A_2)$ and $(B_1^r, B_2 B_1^{r-1}, \ldots, B_2)^T$ are mutually inverse.] Deduce that $V_{1,n}$, for $n > 1$, can be embedded in $V_{1,2}$.

0.2 Matrix rings and the matrix reduction functor

Given a ring R, consider a left R-module which is expressed as a direct sum of certain submodules:

$$M = U_1 \oplus \cdots \oplus U_n. \tag{1}$$

Let $\pi_i : M \to U_i$ be the canonical projections and $\mu_i : U_i \to M$ the canonical injections, for $i = 1, \ldots, n$. Thus $(x_1, \ldots, x_n)\pi_i = x_i$, $x\mu_i = (0, \ldots, x, \ldots, 0)$ with x in the ith place and 0 elsewhere. Clearly we have

$$\mu_i \pi_j = \delta_{ij}, \tag{2}$$

where δ is the Kronecker delta: $\delta_{ij} = 1$ if $i = j$ and 0 otherwise. Further,

$$\sum \pi_i \mu_i = 1. \tag{3}$$

With each endomorphism f of M we can associate a matrix (f_{ij}), where $f_{ij} : U_i \to U_j$ is defined by

$$f_{ij} = \mu_i f \pi_j. \tag{4}$$

Similarly, any family of homomorphisms $f_{ij} : U_i \to U_j$ gives rise to an endomorphism f of M defined by

$$f = \sum \pi_i f_{ij} \mu_j. \tag{5}$$

These two processes are easily seen to be mutually inverse and if we add and multiply two families (f_{ij}) and (g_{ij}) 'matrix fashion': $(f+g)_{ij} = f_{ij} + g_{ij}, (fg)_{ik} = \Sigma f_{ij} g_{jk}$, the correspondence is an isomorphism, so that we have

Theorem 0.2.1. *Let R be any ring. If M is a left R-module, expressed as a direct sum as in (1), then each element f of $End_R(M)$ can be written as a matrix (f_{ij}) where $f_{ij} : U_i \to U_j$, is obtained by (4) and in turn gives rise to an endomorphism of M by (5), and this correspondence is an isomorphism.* ■

In the particular case where all summands are isomorphic, we have $M \cong U^n$ and so we find

Corollary 0.2.2. *Let A be a ring, U a left A-module and $R = End_A(U)$. Then for any $n \geq 1$ we have*

$$End_A(U^n) \cong \mathfrak{M}_n(R). \quad \blacksquare \tag{6}$$

The matrix ring $\mathfrak{M}_n(R)$ in (6) is also denoted by R_n. Let us consider it more closely. Writing $e_{ij} = \pi_i \mu_j$, we obtain from (2) and (3) the equations

$$e_{ij} e_{kl} = \delta_{jk} e_{il}, \quad \Sigma e_{ii} = 1. \tag{7}$$

The e_{ij} are just the matrix units and the matrix ring R_n may be defined as the ring generated by R and n^2 elements $e_{ij}(i, j = 1, \ldots, n)$ satisfying the conditions (7) and $ae_{ij} = e_{ij}a$ for all $a \in R$. The general element of R_n is then uniquely expressible as $\sum a_{ij} e_{ij}$ ($a_{ij} \in R$). In fact matrix rings are characterized by (7), which gives a decomposition of 1 in R into n idempotents: $1 = e_{11} + \cdots + e_{nn}$.

Theorem 0.2.3. *Let S be any ring with n^2 elements e_{ij} satisfying the equations (7). Then $S \cong \mathfrak{M}_n(R)$, where R is the centralizer of all the e_{ij}.*

Proof. For each $a \in S$ we define $a_{ij} = \sum_\nu e_{\nu i} a e_{j\nu}$; then it is easily checked that $a_{ij} \in R$ and $a = \sum_{ij} a_{ij} e_{ij}$. Now the correspondence $a \leftrightarrow (a_{ij})$ is seen to be an isomorphism: $S \cong R_n$. ∎

Using the language of categories, we can say that the process of forming the $n \times n$ matrix ring is a functor from Rg, the category of rings, to Rg_n, the category of $n \times n$ matrix rings: to each ring R corresponds the matrix ring R_n and to each ring homomorphism $f : R \to S$ there corresponds the homomorphism from R_n to S_n obtained by applying f to the separate matrix entries; conversely, any homomorphism $R_n \to S_n$ arises in this way from a homomorphism $R \to S$, because R is characterized within R_n as the centralizer of the e_{ij}. Moreover, every object T in Rg_n is of the form $\mathfrak{M}_n(C)$, where C is the centralizer of the e_{ij} in T. This shows the functor $\mathfrak{M}_n$ to be a category equivalence (BA, Proposition 3.3.1 or Appendix B below). Thus we have proved

Theorem 0.2.4. *The matrix functor $\mathfrak{M}_n$ establishes an equivalence between the categories Rg and Rg_n, for any $n \geq 1$.* ∎

Of course this is just an instance of the well-known Morita equivalence (see Appendix B). Given a left A-module U with endomorphism ring $R = \text{End}_A(U)$, when we considered U^n as an A-module, its endomorphism ring turned out to be R_n. But we can also consider U^n as an A_n-module; in that case its endomorphism ring, i.e. the centralizer of A_n in $\text{End}(U^n)$, is the centralizer of the matrix basis $\{e_{ij}\}$ in R_n, i.e. R itself. Thus we have

$$\text{End}_{A_n}(U^n) \cong R. \tag{8}$$

In the two cases (6) and (8), U^n may be visualized as consisting of row vectors and column vectors, respectively, over U. We shall distinguish these cases by writing the set of column vectors as ${}^n U$ and the set of row vectors as U^n. More generally, we denote by ${}^m U^n$ the set of all $m \times n$ matrices with entries in U, and omit reference to either of m or n equal to 1. For a ring R, R_n is just ${}^n R^n$, considered as a ring. We shall also allow m or n to be 0. Thus ${}^0 U^n$ is the set of matrices with no rows and n columns; there is one such matrix for each n (including $n = 0$). Similarly for ${}^m U^0$; of course R_0 is the zero ring. The unique 0×0 matrix over R will be written 1_0, and an $m \times n$ matrix where $mn = 0$ will be called a *null matrix*.

If M is an (R, S)-bimodule, then ${}^m M^n$ is an (R_m, S_n)-bimodule in a natural way. As an example, take R itself, considered as an R-bimodule; the set of row vectors R^n has a natural (R, R_n)-bimodule structure and the set of column vectors ${}^n R$ a natural (R_n, R)-bimodule structure. Writing R_R, ${}_R R$ for R as right, respectively left R-module and $\rho_a : x \mapsto xa$, $\lambda_a : x \mapsto ax$ for the right, respectively left multiplication by a, we have $\operatorname{End}_R({}_RR) \cong R$ via the map $a \mapsto \rho_a$ and $\operatorname{End}_R(R_R) \cong R^\circ$ via the map $a \mapsto \lambda_a$, where the opposite ring R° means that the R-endomorphisms of R_R form a ring anti-isomorphic to R (because $\lambda_{ab} = \lambda_b \lambda_a$). In this case equations (6) and (8) become

$$\operatorname{End}_R(R^n) \cong R_n, \quad \operatorname{End}_{R_n}(R^n) \cong R^\circ. \tag{9}$$

The row vectors $e_1 = (1, 0, \ldots, 0), e_2 = (0, 1, 0, \ldots, 0), \ldots$ and the corresponding column vectors e_i^T form bases for R^n, ${}^nR$ respectively, as R-modules, called the *standard bases*.

Returning to the case of a general R-module M, we can summarize the relation between M and ${}^nM$ as follows.

Theorem 0.2.5. *Let R be a ring and M a left R-module with endomorphism ring E. Then ${}^nM$ may be regarded as an R_n-module in a natural way, with endomorphism ring E, and there is a lattice-isomorphism between the lattice of R-submodules of M and the lattice of R_n-submodules of ${}^nM$, in which (R, E)-bimodules correspond to (R_n, E)-bimodules.*

Proof. The first part is just a restatement of (8). To establish the isomorphism we recall that ${}^nM$ consists of columns of vectors over M; any submodule N of M corresponds to a submodule ${}^nN$ of ${}^nM$ and the correspondence

$$N \mapsto {}^nN \tag{10}$$

is order-preserving. Conversely, if P is an R_n-submodule of ${}^nM$, then the n projections $\pi_i : P \to M$ ($i = 1, \ldots, n$) all have the same image and associate with P a submodule of M. The correspondence $P \mapsto P\pi_1$ easily seen to be an order-preserving map inverse to (10), hence (10) is an order-isomorphism between lattices, and so a lattice-isomorphism. The rest follows because the E-action on M and on ${}^nM$ is compatible with the R-action. ∎

The equivalence between R and R_n may be used to reduce any categorical question concerning a finitely generated module to a question for a cyclic module, over an appropriate ring. For, given M, generated as left R-module by $u_1, \ldots, u_n$, say, we apply the functor

$$M \mapsto {}^nM = \operatorname{Hom}_R({}^nR, M) = {}^nR \otimes_R M, \tag{11}$$

and pass to the left R_n-module ${}^nM$, which is generated by the single element $(u_1, \ldots, u_n)^T$.

Thus we have proved

Theorem 0.2.6. *Any R-module M with an n-element generating set corresponds to a cyclic R_n-module under the equivalence (11).* ∎

For example, if R is a principal ideal ring, then as is well known (see Proposition 1.4.5 for the case of a principal ideal domain), any submodule of an n-generator module over R can be generated by n elements. Applying Theorem 2.6 we see that any submodule of a cyclic R_n-module is cyclic, in particular, R_n is again a principal ideal ring. In the other direction, if R_n is a principal ideal ring, then any submodule of a cyclic R_n-module is cyclic, whence it follows that any submodule of an n-generator R-module can be generated by n elements. This can happen for some $n > 1$ in rings that are not principal ideal rings (see Webber [70]).

Another functor of importance in what follows is the *matrix reduction functor* $\mathfrak{W}_n$, which is defined as the left adjoint of the $n \times n$ matrix functor. We note the rule for its construction: given any ring R or more generally, a K-algebra for some base ring K, we form a ring $\mathfrak{F}_n(R; K)$ by adjoining a set of matrix units e_{ij} to R that centralize K. Since this ring contains a set of matrix units, it has, by Theorem 2.3, the form S_n, where S is the centralizer of the e_{ij}. Now S is the *n-matrix reduction* of R, as K-algebra. When $R = K$, S_n becomes K_n; in general, this ring S contains, for each $a \in R$, n^2 elements a_{ij} and these elements centralize K and the e_{ij} in $\mathfrak{F}_n(R; K)$. In other words, we take the elements of R and interpret them as $n \times n$ matrices, with the elements of K as scalars. In terms of the coproduct $*$ the definitions of $\mathfrak{F}_n$ and of $\mathfrak{W}_n$ may be written

$$\mathfrak{F}_n(R; K) = R *_K \mathfrak{M}_n(K) \cong \mathfrak{W}_n(R; K) \otimes_K \mathfrak{M}_n(K). \qquad (12)$$

Examples:

1. $R = k[x]$, the polynomial ring in x over a field k. To obtain $\mathfrak{W}_n(R; k)$, we write x as an $n \times n$ matrix, thus we have the free algebra on n^2 indeterminates (see Section 2.5).

2. R is the k-algebra generated by a, b with defining relation $ab = 1$. Here $\mathfrak{W}_n(R; k)$ is the algebra on $2n^2$ generators a_{ij}, b_{ij} with defining relations in matrix form $AB = I$, where $A = (a_{ij})$, $B = (b_{ij})$. Thus we obtain the universal ring that is not weakly n-finite.

3. R is the k-algebra generated by a, b with defining relation $ab = 0$. Now $\mathfrak{W}_n(R; k)$ again has generators as in example 2, with defining relation (in matrix form) $AB = 0$. Whereas R has zero-divisors, $\mathfrak{W}_n(R; k)$ is an integral domain for $n > 1$ (see Section 2.11 or SF, theorem 5.7.6).

Exercises 0.2

1. (Palmer [94]) Let R be any ring; if there exist $e, w \in R$ satisfying $ew^{i-1}e = \delta_{i1}e$ (Kronecker delta), $\Sigma_1^n w^{i-1}ew^{1-i} = w^n = 1$, then $R = \mathfrak{M}_n(S)$, where S is the centralizer of e, w in R. (*Hint*: Calculate $w^{i-1}ew^{1-j}$.)

2. If R satisfies left ACC_n, show that R_r satisfies left ACC_k, where $k = [n/r]$ is the greatest integer below n/r.
3. If every left ideal of a ring R can be generated by r elements, show that for any n-generator left R-module M, every submodule of M can be generated by nr elements.
4. Show that if R satisfies left pan-ACC, then so does every finitely generated free left R-module.
5. If a ring R is injective, as left R-module over itself, show that R_n ($n > 1$) has the same property.
6. For any ring R, show that R and R_n ($n > 1$) have isomorphic centres. Prove this fact by characterizing the centre of R as the set of all natural transformations of the identity functor on $_R\text{Mod}$.
7. Let $R = K_n$ be a full matrix ring and $f : R \to S$ any ring homomorphism. Show that S is a full $n \times n$ matrix ring, say $S = L_n$ and there is a homomorphism $\phi : K \to L$ inducing f.
8. (G. M. Bergman) Let $n \geq 1$ be an integer and R a ring in which every right ideal that is not finitely generated has a finitely generated direct summand that cannot be generated by n elements. Show that R satisfies right ACC_n.
9. (Jacobson [50]) Let R be a non-zero ring that is not weakly 1-finite, say $ab = 1 \neq ba$. Writing $e_{ij} = b^{i-1}a^{j-1} - b^i a^j$, show that the e_{ij} satisfy the first set of equations (7) for matrix units and the universal weakly 1-finite image of R is $R/\mathfrak{e}$, where $\mathfrak{e}$ is the ideal generated by e_{11}.
10. Let R be a projective-free ring. Show that the only rings Morita-equivalent to R are the full matrix rings R_n ($n = 1, 2, \ldots$).

0.3 Projective modules

Let R be any ring; if P is a finitely generated projective left R-module, generated by n elements, say, then we have $P \oplus P' \cong R^n$ for some projective R-module P'. The projection of R^n on P is given by an idempotent $n \times n$ matrix E and we may write $P = R^n E$; in fact, P is the left R-module generated by the rows of E. We record conditions for two idempotent matrices to define isomorphic projective modules.

Proposition 0.3.1. *Let R be any ring and let $E \in R_n$, $F \in R_r$ be idempotent matrices. Then the following conditions are equivalent:*

(a) $E = XY, F = YX$ for some $X \in {}^n R^r, Y \in {}^r R^n$;
(b) $E = AB, F = BA$, for some $A \in {}^n R^r, B \in {}^r R^n$, where $EA = AF = A, FB = BE = B$;
(c) *the projective left R-modules defined by E and F are isomorphic:* $R^n.E \cong R^r.F$;
(d) *the projective right R-modules defined by E and F are isomorphic:* $E.{}^n R \cong F.{}^r R$.

Proof. (a) ⇒ (b). Assume (a) and put $A = XYX, B = YXY$; then (b) follows by the idempotence of XY and YX. (b) ⇒ (c). Let P, Q be the left R-modules generated by the rows of E, F, respectively. Since $EA = A = AF$, the right multiplication by A maps E to F; similarly right multiplication by B maps F to E, and these maps are mutually inverse, because $EAB = E, FBA = F$. (c) ⇒ (a). Let $\theta : R^n E \to R^r F$ be an isomorphism and suppose that θ maps E to X while θ^{-1} maps F to Y. Since $X \in R^r F$, we have $XF = X$; hence $E = E\theta\theta^{-1} = X\theta^{-1} = (XF)\theta^{-1} = X(F\theta^{-1}) = XY$. Thus $E = XY$, and similarly, $F = YX$. Thus (a), (b), (c) are equivalent; by the symmetry of (a) they are also equivalent to (d). ∎

Two idempotent matrices E, F that are related as in Proposition 3.1 are said to be *isomorphic*. They will be called *conjugate* if there is an invertible matrix U such that $F = U^{-1}EU$. When R has IBN, this is only possible when E and F are of the same order, but we shall not make this restriction. If P and Q are the projective modules defined by $E \in R_n, F \in R_r$, respectively, and $P \oplus P' \cong R^n, Q \oplus Q' \cong R^r$, then conjugacy of E and F just means that $P \cong Q$ and $P' \cong Q'$. The following is a condition for conjugacy:

Proposition 0.3.2. *Let R be a ring and E, F idempotent matrices over R. Then E and F are conjugate if and only if E is isomorphic to F and $I - E$ is isomorphic to $I - F$.*

Proof. If E, F are conjugate, say $F = U^{-1}EU$, then we can take $X = EU, Y = U^{-1}E$; it follows that $XY = E, YX = F$, so E and F are isomorphic. Since we also have $I - F = U^{-1}(I - E)U, I - E$ and $I - F$ are also isomorphic. Conversely, when the isomorphism holds, we have $E = AB, F = BA, EAF = A, FBE = B$ and $I - E = A'B', I - F = B'A', (I - E)A'(I - F) = A', (I - F)B'(I - E) = B'$. Let us put $X = A + A', Y = B + B'$; then $B'(I - E) = B'$, hence $B'E = 0$ and similarly $EA' = 0$; hence $YEX = BEA + B'EA + BEA' + B'EA' = BEA = F$. For the same reason, $Y(I - E)X = I - F$, hence $YX = I$. By symmetry, $XY = I$, and it follows that E and F are conjugate. ∎

We also note the following result, relating E to the number of generators of P.

Lemma 0.3.3. *For any ring R, let P be a projective R-module defined by an idempotent $n \times n$ matrix, $P = R^n E$. Then the minimal number of generators of P is the inner rank of E; thus if E has inner rank r, then P can be generated by r but no fewer elements.*

Proof. Given any factorization

$$E = XY, \text{ where } X \text{ is } n \times r \text{ and } Y \text{ is } r \times n, \tag{1}$$

the matrix $F = YXYX$ is an idempotent isomorphic to E and so defines a module isomorphic to P; since F is $r \times r$, it follows that P can be generated by r elements. Conversely, if P can be generated by r elements, then it can be represented as the image of an idempotent $r \times r$ matrix F, hence $E = XFY$ and $\rho E \leq r$. Thus the minimal number of generators of P equals the least value of r in (1), i.e. ρE. ∎

For any ring R denote by $_R$proj (proj$_R$) the category of all finitely generated projective left (right) R-modules and all homomorphisms between them. We have seen in Section 0.1 that the correspondence $P \mapsto P^*$ defines a duality (i.e. anti-equivalence) between $_R$proj and proj$_R$ such that $P^{**} \cong P$.

We shall denote by $\mathcal{S}(R)$ the monoid whose elements are the isomorphism classes of objects in $_R$proj; thus each $P \in {}_R$proj defines an element $[P]$ of $\mathcal{S}(R)$, where $[P] = [P']$ if and only if $P \cong P'$. The operation on $\mathcal{S}(R)$ is given by

$$[P] + [Q] = [P \oplus Q].$$

Clearly this is well-defined, i.e. the right-hand side depends only on $[P], [Q]$ and not on P, Q themselves. We see that $\mathcal{S}(R)$ is a commutative monoid, in which we may regard $[R]$ as a distinguished element. It is *conical*, i.e. $\alpha + \beta = 0$ implies $\alpha = 0$ and hence $\beta = 0$. Its universal group (see Section 0.7), often called the *Grothendieck group*, is the projective module group $K_o(R)$ (see e.g. Milnor [71]). By the duality between $_R$proj and proj$_R$ we have $\mathcal{S}(R^o) \cong \mathcal{S}(R), K_o(R^o) \cong K_o(R)$. The element of $K_o(R)$ corresponding to P may be written (P), so the general element has the form $(P) - (Q)$ and we have $(P) = (P')$ in $K_o(R)$ if and only if $P \oplus S \cong P' \oplus S$ for some $S \in {}_R$proj. Here S may be taken to be free of finite rank, so we have

$$(P) = (P') \text{ in } K_o \text{ if and only if } P \oplus R^n \cong P' \oplus R^n \text{ for some } n \geq 0. \tag{2}$$

We can equally well define $\mathcal{S}(R)$ in terms of idempotent matrices. For any ring R a finitely generated projective left R-module is generated by the rows of an idempotent matrix; thus if P is generated by n elements, it has the form $R^n E$, where E is an idempotent $n \times n$ matrix. By Proposition 3.1, two projective R-modules $R^n E$ and $R^m F$ are isomorphic if and only if there exist matrices $X (n \times m), Y (m \times n)$, such that the matrices $XY = E$ and $YX = F$ are idempotent. Moreover, if E corresponds to P and F to Q, then the diagonal sum $E \oplus F$ corresponds to the direct sum $P \oplus Q$, so $\mathcal{S}(R)$ may be defined as the set of isomorphism classes of idempotent matrices with the operation $E \oplus F$.

The structure of $\mathcal{S}(R)$ is closely related to certain properties of the ring R, while $K_0(R)$ reflects the corresponding stable properties. This is illustrated in the next result, where by a *stably free* module we understand a module P such that $P \oplus R^m \cong R^n$, for some integers $m, n \geq 0$.

Proposition 0.3.4. *Let R be any ring and denote by $\lambda : \mathbb{Z} \to K_0(R)$ the homomorphism mapping 1 to (R). Then (i) R has IBN if and only if λ is injective and (ii) every finitely generated projective module is stably free if and only if λ is surjective.*

Further, R is projective-free if and only if the natural homomorphism $\mathbb{N} \to \mathcal{S}(R)$, is an isomorphism.

Proof. Clearly λ fails to be injective if and only if $n(R) = 0$ in $K_0(R)$ for some $n \neq 0$, say $n > 0$. Then $R^n \oplus P \cong P$ for some P, and if $P \oplus P' \cong R^r$, then $R^{n+r} \cong R^r$, so either $r = 0$ and R is the zero ring, or $r > 0$ and IBN fails in R. The converse is clear.

If every finitely generated projective module P is stably free, then for any P there are integers m, n such that $P \oplus R^m \cong R^n$, hence $[P] = n[R] - m[R] = (n-m)[R] = \lambda(n-m)$, and conversely, if $[P] = \lambda(r)$, then $P \oplus R^m \cong R^{m+r}$ for some $m \geq 0$ and P is stably free. The final assertion follows because R is projective-free if and only if every P satisfies $[P] = n[R]$ in $\mathcal{S}(R)$, for a unique $n \in \mathbb{N}$, depending on P. ∎

It will be useful to relate the monoid $\mathcal{S}(R)$ to $\mathcal{S}(R/\mathfrak{a})$, where $\mathfrak{a}$ is an ideal contained in $J(R)$, the Jacobson radical of R. We recall that $J(R)$ is defined as the intersection of all maximal left (or equivalently, all maximal right) ideals of R (equivalently, the set of all $a \in R$ such that $1 - xa$ is a unit, for all $x \in R$, see BA, Section 5.3).

We shall be particularly concerned with rings R for which $R/J(R)$ is a field. They are the *local* rings, characterized in

Proposition 0.3.5. *For any non-zero ring R the following conditions are equivalent:*

(a) R is a local ring, i.e. $R/J(R)$ is a (skew) field,
(b) the non-units in R form an ideal,
(c) for any $a \in R$, either a or $1 - a$ has a one-sided (at least) inverse, not necessarily on the same side.

Proof. (a) $\Rightarrow$ (b). Put $J = J(R)$. If R/J is a field, then J is the unique maximal ideal and hence consists of non-units. For $u \notin J$ we have $uv \equiv vu \equiv 1 \pmod{}$

J) for some $v \in R$, hence $uv = 1 + t$ ($t \in J$) is a unit, so u has a right inverse that is two-sided, by symmetry. Thus J is the set of all non-units.

(b) $\Rightarrow$ (c) is clear. To prove (c) $\Rightarrow$ (a), we first note that an idempotent $e \neq 1$ cannot have even a one-sided inverse, for if $eu = 1$, then $e = e^2 u = eu = 1$. Thus when (c) holds, R has no idempotents $\neq 0, 1$. Next, if $ab = 1$, then ba is a non-zero idempotent, hence $ba = 1$, so all one-sided inverses are two-sided.

Now let $u \in R$; if u has no inverse, then neither does ux, for any $x \in R$, hence $1 - ux$ always has an inverse, so $u \in J$. Thus all non-units are in J and (a) follows. ∎

Here (a) or (b) is the usual form of the definition of a local ring, while (c) is the easiest to verify.

Let R be a ring with an ideal $\mathfrak{a} \subseteq J(R)$; write $\bar{R} = R/\mathfrak{a}$ and for any left R-module M, put $\bar{M} = M/\mathfrak{a}M$. Then $\bar{M}$ is an $\bar{R}$-module in a natural way and if M is finitely generated non-zero, then $M/JM \neq 0$ by Nakayama's lemma, hence $\bar{M} \neq 0$, because $\mathfrak{a} \subseteq J$. Suppose now that $f : R \to S$ is a surjective homomorphism. The next result gives conditions for ker f to be contained in $J(R)$. Here a ring homomorphism is called *local* if it maps non-units to non-units.

Lemma 0.3.6. *Let $f : R \to S$ be a surjective ring homomorphism. Then the following conditions are equivalent:*

(a) ker $f \subseteq J(R)$,
(b) the homomorphism induced by f on $n \times n$ matrix rings (for any $n \geq 1$) is local,
(c) f is a local homomorphism.

Proof. (a) $\Rightarrow$ (b) Let A be a matrix over R and suppose that A^f is a unit, say $A^f B = I$. Take B_0 over R such that $B_0^f = B$; then $AB_0 = I + C$, where $C^f = 0$ and so C has entries in $J(R)$. It is easily checked that $J(R_n) = J(R)_n$; so it follows that $I + C$ is a unit, hence $AB_0(I + C)^{-1} = I$. Thus A has a right inverse; by symmetry it has a left inverse, and so is a unit. Now (b) $\Rightarrow$ (c) is clear, and to prove (c) $\Rightarrow$ (a), let $a \in \ker f$. Then $a^f = 0$, hence $(1 + ax)^f = 1$, so $1 + ax$ has an inverse for all $x \in R$ and it follows that $a \in J(R)$, thus (a) holds. ∎

Any ring homomorphism $f : R \to S$ induces a monoid homomorphism $\mathcal{S}(f) : \mathcal{S}(R) \to \mathcal{S}(S)$ that preserves the distinguished element: $\mathcal{S}(f)[P] = [S \otimes_R P]$. This homomorphism need be neither injective nor surjective, even when f is, but we have

Theorem 0.3.7. *Let R, S be any rings and $\varphi : R \to S$ a homomorphism that is surjective and local; in particular S may be $R/J(R)$, with the natural map φ.*

Then the induced homomorphism $\mathcal{S}(\varphi) : \mathcal{S}(R) \to \mathcal{S}(S)$ *is an embedding. Thus if S is projective-free, then so is R. If S has IBN, UGN or is weakly finite, then the same is true of R.*

Proof. Write $\bar{R} = R/\ker \varphi$ and $\bar{M} = M/(\ker \varphi)M$. Given any finitely generated projective R-modules P, Q such that $\bar{P} \cong \bar{Q}$, we have the following diagram, where f is an isomorphism:

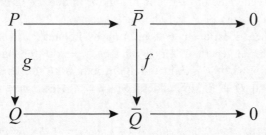

The map g to make the diagram commutative exists since P is projective. If coker $g = L$ then $\bar{L} = 0$ implies $L = 0$, by Nakayama's lemma, hence g is surjective. Therefore Q splits P, i.e. $P \cong Q \oplus M$, where $M = \ker g$. By the diagram, $\bar{M} = 0$, hence $M = 0$ and this proves g to be an isomorphism, as claimed. Further, when $R/\ker \varphi$ is projective-free, then $\mathcal{S}(S) \cong \mathbb{N}$ and the natural homomorphism preserves the generator of $\mathbb{N}$, whence $\mathcal{S}(R) \cong \mathbb{N}$. The assertions for IBN and UGN are evident; for weak finiteness we take square matrices A, B over R and suppose that $AB = I$; then their images in S are mutually inverse, hence $BA = I + C$, where C has entries in $J(R)$ and so $I + C$ is invertible. This shows that A is invertible. By Lemma 3.6 all these results hold when $S = R/J(R)$. ∎

Of course the embedding obtained here will not in general be an isomorphism. We remark that the result also follows by considering the corresponding idempotent matrices (see the remarks before Proposition 3.1). Since every field is projective-free, and any projective-free ring is clearly weakly finite, we obtain from Theorem 3.7,

Corollary 0.3.8. *Every local ring is projective-free and hence weakly finite.* ∎

Sometimes a more general notion of local ring is needed. Let us call R a *matrix local ring* if $R/J(R)$ is simple Artinian. By Wedderburn's theorem this means that $R/J(R) \cong K_n$, where K is a field and $n \geq 1$; n is sometimes called the *capacity* of R. When $n = 1$, we are back in the case of a local ring. By contrast this is sometimes called a *scalar* local ring, but we generally omit the qualifier, so 'local ring' will mean as usual 'scalar local ring'.

We conclude with some conditions for a ring to be projective-free.

Theorem 0.3.9. *Let R be a non-zero ring such that the set of all full matrices admits diagonal sums. Then R is projective-free.*

Proof. Since $R \neq 0$, the 1×1 unit matrix is full, hence the unit matrix of any order is full, so R has IBN (it even has UGN). It remains to prove that every finitely generated projective R-module P is free. Let n be the minimal number of generators of P and choose Q such that $P \oplus Q = R^n$. We have $P = R^n E$, where E is idempotent and full, by Lemma 3.3; similarly, if the minimal number of generators for Q is m, then $Q = R^m F$ for some idempotent $m \times m$ matrix F, where F is again full. By hypothesis, $E \oplus F$ is again full, and its image is $P \oplus Q \cong R^n$; by fullness, $m + n = n$. Hence $m = 0, Q = 0$ and $P \cong R^n$. ■

Theorem 0.3.10. *Let R be a ring. If for every $n \geq 1$ the product of two full $n \times n$ matrices is again full, then every finitely generated projective module is stably free.*

Proof. Suppose there is a finitely generated projective module that is not stably free; we choose such a module P with the least number of generators, n say. We have $P = R^n E$, where E is an idempotent matrix, which must be full, by Lemma 3.3 and the minimality of n. The module $Q = R^n(I - E)$ is such that $P \oplus Q = R^n$, and since $E(I - E) = 0, I - E$ cannot be full, so Q can be generated by fewer than n elements and hence is stably free. Thus $Q \oplus R^s \cong R^r$ and so $P \oplus R^r \cong P \oplus Q \oplus R^s \cong R^{n+s}$, which shows P to be stably free. ■

If the hypothesis of Theorem 3.10 holds and in addition every full matrix is stably full, i.e., whenever a matrix A is full, then so is $A \oplus 1$, then each projective is actually free, as we can see by verifying the hypothesis of Theorem 3.9. Thus let A, B be full matrices; then $A \oplus I$, $I \oplus B$ are full (for unit matrices of any size), hence so is their product $A \oplus B$. Moreover, since I is full, R has IBN. This proves

Corollary 0.3.11. *Let R be a ring. If the product of any two full matrices of the same size is full and any full matrix is stably full, then R is projective-free.* ■

Exercises 0.3

1. Verify the equivalence of the two definitions of $\mathcal{S}(R)$, in terms of projective modules and idempotent matrices.

2. Let R be a ring and $J = J(R)$. By considering the kernel of the homomorphism $R_n \to (R/J)_n$ induced by the natural homomorphism $R \to R/J$, show that $J(R_n) \cong J_n$.
3°. Show that if R is a matrix local ring and $n > 1$, then so is R_n. Does the converse hold?
4. For any ring R, show that $R/J(R)$ is weakly finite if and only if R is weakly finite. Deduce that any matrix local ring is weakly finite. Show that if R has IBN or UGN, then so does $R/J(R)$.
5. Let K be a field and R a subring such that for any $x \in K$, either x or x^{-1} lies in R. Show that for any non-unit $a \in R$, $a(1-a)^{-1} \in R$; deduce that R is a local ring.
6. In any local ring show that the additive order of 1 is 0 or a prime power.
7. Let R be a local ring with residue-class field $K = R/J(R)$. If M is a finitely generated left R-module such that $K \otimes M = 0$, show that $M = 0$.
8. Show that any Artinian matrix local ring is a full matrix ring over a scalar local ring. (*Hint*: Recall that in an Artinian ring idempotents can be lifted from $R/J(R)$ to R, see e.g. FA, Lemma 4.3.2.)
9. Let R be the ring of rational quaternions with denominator prime to p, an odd prime. Show that the Jacobson radical of R is pR and R/pR is the ring of quaternions over $\mathbb{F}_p$. Deduce that R is a matrix local ring which is not a matrix ring over a scalar local ring.
10. Show that for any ring R the following are equivalent (see Lorimer [92]):
    (a) R is local and any finitely generated left ideal is principal,
    (b) the principal left ideals of R are totally ordered by inclusion,
    (c) all left ideals of R are totally ordered by inclusion.
11. (Beck [72]) Let P be a finitely generated projective left R-module. If P/JP is free over R/J, where J is the Jacobson radical of R, show that P is free over R. (This holds even if P is not finitely generated, see Beck [72].)
12*. (Kaplansky [58]) Let P be a projective module over a local ring. Show that any element of P can be embedded in a free direct summand of P; deduce that *every* projective module over a local ring is free.

0.4 Hermite rings

The conditions of IBN, UGN and weak finiteness discussed in Section 0.1 hold in most rings normally encountered, and counter-examples belong to the pathology of the subject. By contrast, the property defined below forms a significant restriction on the ring.

Clearly any stably free module is finitely generated projective. If $P \oplus R^m$ is free but not finitely generated, then P is necessarily free (see Exercise 9). In any case we shall mainly be concerned with finitely generated modules.

A ring R is called an *Hermite ring* if it has IBN and any stably free module is free. More specifically, if n-generator free modules have unique rank and any left R-module P is free whenever $P \oplus R^r \cong R^m (r \leq m \leq n)$, then R is called an *n*-Hermite-ring. Thus, denoting by $\mathcal{H}(\mathcal{H}_n)$ the class of all (*n*-)Hermite rings, we have

$$\mathcal{H}_1 \supseteq \mathcal{H}_2 \supseteq \ldots \quad \text{and} \quad \cap \mathcal{H}_n = \mathcal{H}. \tag{1}$$

From results in Section 2.11 (or also SF, 5.7) it follows that all the inclusions in (1) are proper. The class of Hermite rings clearly includes all projective-free rings, and is contained in the class of all weakly finite rings. More generally, it is easily seen that every *n*-Hermite ring is weakly *n*-finite.

To describe Hermite rings in terms of matrices, let us call an $m \times n$ matrix A over a ring R *completable* in R, if either $m = n$ and A is invertible, or A can be completed to an invertible matrix by adjoining $n - m$ rows, if $m < n$, or $m - n$ columns, if $m > n$. We note that in a non-zero ring, if a matrix with a right inverse is completable, then its index must be non-negative. For suppose that $AB = I$, where A is $m \times n$ and $m > n$. Since A is completable, there is an $m \times m - n$ matrix A' such that (A, A') is invertible. Let $(C, C')^T$ be the inverse; then we have

$$\begin{pmatrix} C \\ C' \end{pmatrix} (A \quad A') = I.$$

Hence $CA = I$, which together with $AB = I$ shows that $C = B$. Further, $CA' = 0$, $C'A' = I$, but C is the inverse of A, hence $A' = 0$, which contradicts the fact that $C'A' = I$. So this case cannot occur.

We now have the following description of Hermite rings in terms of matrices, where a *unimodular row* is defined as a $1 \times n$ matrix with a right inverse.

Theorem 0.4.1. *For any non-zero ring R and any $n \geq 1$ the following conditions are equivalent:*

(a) R is n-Hermite.
(b) Every $r \times m$ matrix over R, where $r, m \leq n$, with a right inverse is completable, more precisely, if $AB = I$, where $A \in {}^rR^m, r \leq m \leq n$, then there is an invertible matrix with A as its first r rows, whose inverse has B for the first r columns.
(c) Every unimodular row over R of length at most n is completable.
Moreover, R is n-Hermite if and only if its opposite $R°$ is.

Proof. (a) $\Rightarrow$ (b). Let R be *n*-Hermite and take $A \in {}^rR^m, B \in {}^mR^r$ such that $AB = 1$. Interpreting A, B as mappings α, β between R^r and R^m, we have a split exact sequence

$$0 \to R^r \xrightarrow{\alpha} R^m \xrightarrow{\beta'} P \to 0, \tag{2}$$

where $P = \operatorname{coker} \alpha$. It follows that $R^m \cong R^r \oplus P$; thus P is stably free and hence free: $P \cong R^s$, where $s = m - r$. Since the sequence (2) is split, β' has a left inverse $\alpha' : R^s \to R^m$ and if A', B' are the matrices corresponding to α', β', then $AB = I$, $A'B' = I$, $AB' = 0$, and so

$$\begin{pmatrix} A \\ A' \end{pmatrix} (B \quad B') = \begin{pmatrix} I & 0 \\ A'B & I \end{pmatrix}.$$

The matrix on the right is invertible and multiplying by its inverse on the left, we obtain

$$I = \begin{pmatrix} I & 0 \\ -A'B & I \end{pmatrix} \begin{pmatrix} A \\ A' \end{pmatrix} (B \quad B') = \begin{pmatrix} A \\ A'(I - BA) \end{pmatrix} \cdot (B \quad B'),$$

and since the ring R, being n-Hermite, is weakly n-finite, the matrices on the right are inverses of each other and are of the required form.

(b) $\Rightarrow$ (c) is clear. To prove (c) $\Rightarrow$ (a) we must show that when (c) holds, then $P \oplus R^r \cong R^m$ implies $P \cong R^{m-r}$; consider first the case $r = 1$. Thus let $P \oplus R \cong R^m$. Then we have again a split exact sequence

$$0 \to R \xrightarrow{\alpha} R^m \xrightarrow{\beta'} P \to 0.$$

Let a be the $1 \times m$ matrix corresponding to α; since the sequence is split, there is a map $\beta : R^m \to R$ such that $\alpha\beta = 1$. If β is represented by b, then $ab = I$. By hypothesis, there exists $A' \in {}^{m-1}R^m$ such that $(a, A')^T$ is invertible. Let the inverse, correspondingly partitioned, be (c, C'); then $\beta' : R^m \to P$ is represented by C', hence $P \cong R^{m-1}$. Suppose now that $P \oplus R^r \cong R^m$; we claim that $r \leq m$, for if $r > m$, then by successively cancelling R we obtain $P \oplus R^{r-m} = 0$, a contradiction, since $R \neq 0$. Taking $P = 0$, we also see that R has IBN up to rank n. Thus $r < m$, and by successive cancelling we find that $P \cong R^{m-r}$; in particular, when $r = m$, it follows that $P = 0$. Hence R is n-Hermite. Now the rest follows from (a) by duality. ∎

The proof shows that we have the following characterizations of Hermite rings:

Corollary 0.4.2. *For any non-zero ring the following conditions are equivalent:*

(a) R is an Hermite ring,
(b) if $P \oplus R \cong R^m$, then $P \cong R^{m-1}$,
(c) if $P \oplus R^r \cong R^m$, then $r \leq m$ and $P \cong R^{m-r}$. ∎

For 2-Hermite rings there is a simple criterion that is sometimes useful:

Proposition 0.4.3. *An integral domain R is 2-Hermite if and only if, for any right comaximal pair a, b, $aR \cap bR$ is principal.*

Proof. If a, b are right comaximal, then the mapping $\mu : (x, y)^T \mapsto ax - by$ is a surjective homomorphism of right R-modules ${}^2R \to R$, giving rise to the exact sequence

$$0 \to P \longrightarrow {}^2R \xrightarrow{\mu} R \to 0 . \qquad (3)$$

where $P = \ker \mu$. Thus any right comaximal pair a, b leads to a sequence (3), and such a sequence always splits to give $R \oplus P \cong {}^2R$. Conversely, if $R \oplus P \cong {}^2R$, we have a split exact sequence (3), and if $(a, -b)$ is the matrix of μ, then a, b are right comaximal. Now $(x, y)^T \in P$ if and only if $ax = by \in aR \cap bR$; thus $P \cong aR \cap bR$ and P is free if and only if $aR \cap bR$ is principal. ∎

The following characterization of Hermite rings in terms of the notion of rank is also of interest. We recall from the definitions, that whenever the stable rank $\rho^*(A)$ is defined, then for any matrix A we have

$$\rho(A) \geq \rho^*(A). \qquad (4)$$

Proposition 0.4.4. *A non-zero ring is Hermite if and only if the stable rank of every matrix exists and equals its inner rank.*

Proof. We shall use Proposition 1.3 and the factorization of $A \oplus I_s$ given there. Suppose that R is Hermite, hence with UGN, and let A be a matrix with the factorization

$$\begin{pmatrix} A & 0 \\ 0 & I_s \end{pmatrix} = \begin{pmatrix} B \\ B' \end{pmatrix} (C \quad C'),$$

where as before, B' is $s \times r$. Then $r \geq s$ and if $r = s$, then $A = 0$ by Proposition 1.3, so we may assume that $r > s$. Since $B'C' = I$, there exist (by Theorem 4.1) $B'' \in {}^{r-s}R^r, C'' \in {}^rR^{r-s}$ such that $(B', B'')^T$ and (C', C'') are mutually inverse. Hence

$$B = B(C' \quad C'')\begin{pmatrix} B' \\ B'' \end{pmatrix} = BC''B'',$$

because $BC' = 0$. It follows that $A = BC = BC''.B''C$, hence $\rho(A) \leq r - s \leq \rho^*(A)$, and so by (4) we have equality of ranks. Conversely, if the stable rank equals the inner rank, then by Proposition 1.3, R is weakly finite. To show that R is Hermite it is enough, by Theorem 4.1, to show that every unimodular row is completable. Let a be a unimodular row of length n, say $ab = 1$, and put $F = I - ba$. Then $F^2 = F, Fb = 0 = aF$, a set of equations summed up in

$$\begin{pmatrix} F & 0 \\ 0 & I \end{pmatrix} = \begin{pmatrix} F \\ a \end{pmatrix} (F \quad b).$$

0.4 Hermite rings

This shows that $\rho^*(F) = r < n$, so $\rho(F) = r$, say $F = B'A'$, where B' is $n \times r$ and A' is $r \times n$. We now have $(B', b)(A', a)^T = I$; since I is full, $r + 1 = n$ and by weak finiteness these matrices are mutually inverse, so R is Hermite. ∎

Over an Hermite ring we have the following stability property of matrix factorizations.

Proposition 0.4.5. *Let R be an $(n + r)$-Hermite ring, $C \in R_n$ and suppose that there is a factorization into matrices (not necessarily square)*

$$C \oplus I_r = P_1 \cdots P_t.$$

Then there are invertible matrices $U_i (i = 0, 1, \ldots, t)$, $U_0 = U_t = I$, such that

$$U_{i-1}^{-1} P_i U_i = P_i' \oplus I \quad \text{and} \quad C = P_1' \ldots P_t'.$$

Proof. By induction it will be enough to treat the case of two factors; thus we have

$$\begin{pmatrix} C & 0 \\ 0 & I \end{pmatrix} = AB = \begin{pmatrix} A' \\ A'' \end{pmatrix} (B' \quad B''),$$

with an appropriate block decomposition. We have $A''B'' = I$, and since R is $(n + r)$-Hermite, A'' forms the last r rows of an invertible matrix P say, and B'' the last r columns of the inverse, i.e. $A'' = (0, I)P$, $B'' = P^{-1}(0, I)^T$. If we replace A, B by AP^{-1}, PB, we obtain

$$\begin{pmatrix} C & 0 \\ 0 & I \end{pmatrix} = \begin{pmatrix} A_1 & A_2 \\ 0 & I \end{pmatrix} \begin{pmatrix} B_1 & 0 \\ B_2 & I \end{pmatrix}.$$

On multiplying out, we find that $A_2 = 0, B_2 = 0, C = A_1 B_1$ and now the conclusion follows by induction. ∎

A square matrix C will be called a *stable matrix atom* if $C \oplus I_r$ is an atom for all $r \geq 1$. From Proposition 4.5 we obtain

Corollary 0.4.6. *Over an Hermite ring every matrix atom is stable.* ∎

We can specialize our ring still further. A ring R is called *cancellable* if for any projective modules $P, Q, P \oplus R \cong Q \oplus R$ implies $P \cong Q$. It is clear that every projective-free ring is cancellable and every cancellable ring is Hermite. Thus for non-zero rings the following classes of rings become smaller as we go down the list:

1. Rings with invariant basis number.
2. Rings with unbounded generating number.
3. Weakly finite rings.

4. Hermite rings.
5. Cancellable rings.
6. Projective-free rings.

We have seen in Section 0.3 that a finitely generated projective left R-module P is given by an idempotent matrix E, representing the projection of R^n on P. Further, by Proposition 3.1, P is free, of rank r say, if and only if $P \cong R^r$, i.e. if $E = AB$, where A is $n \times r$ and B is $r \times n$ such that $BA = I$. An idempotent matrix with this property is said to be *split*. Thus we obtain the following criterion for a ring to be projective-free:

Proposition 0.4.7. *A ring is projective-free if and only if it has IBN and each idempotent matrix is split.* ∎

Unlike some of the other properties, projective-freeness is not a Morita invariant. We therefore define a ring R to be *projective-trivial* if there exists a projective left R-module P, called the *minimal projective* of R, such that every finitely generated projective left R-module M has the form P^n, for an integer n that is uniquely determined by M. Clearly being projective-trivial is a Morita invariant, and a projective-trivial ring R is projective-free precisely when its minimal projective is R. The precise relationship between these two concepts is elucidated in

Theorem 0.4.8. *For any ring R the following properties are equivalent:*

(a) R is a full matrix ring over a projective-free ring,
(b) there exists n such that for every finitely generated projective module P, P^n is free of unique rank,
(c) R is Morita equivalent to a projective-free ring,
(d) R is projective-trivial.

$(a^o)-(d^o)$ *the corresponding properties for the opposite ring.*

Proof. Clearly (a) ⇒ (b) ⇒ (c) ⇒ (d). Now assume (d): R is projective-trivial, with minimal projective P, say. Since R is finitely generated projective, we have

$$R \cong P^n, \qquad (5)$$

for some positive integer n. Write $E = \mathrm{End}_R(P)$; then by Corollary 2.2 we find, on taking endomorphism rings in (5), that $R \cong E_n$. Here E is again projective-trivial, and (5) shows that its minimal projective is $\mathrm{End}_R(P) = E$, hence E is projective-free, i.e. (a) holds. Now the final assertion follows by the obvious symmetry of (a). ∎

Exercises 0.4

1. Show that a 1-Hermite ring is the same as a weakly 1-finite ring.
2. Let A, B be matrices whose indices have the same sign. Show that $A \oplus B$ is completable whenever A and B are. Prove the converse when R is Hermite.
3. Which of the properties 1.–6. are Morita invariant? For the others describe the rings that are Morita invariant to them.
4. If in an Hermite ring, $AB = I$ and B is completed to an invertible matrix (B, B'), show that for suitably chosen A', $(A, A')^T$ has the inverse $(B, B' - BAB')$.
5. Given $A \in {}^m R^n$, $B \in {}^n R^m$, where $m < n$, over any ring R, such that $AB = I_m$, show that A is completable if and only if $A{:}0 = \{x \in {}^n R | Ax = 0\}$ is free of rank $n - m$ (Kazimirskii and Lunik [72]).
6. Define an *n-projective-free* ring as a ring over which every n-generator projective module is free of unique rank. State and prove an analogue of Theorem 4.1 for such rings.
7°. Find examples of Hermite rings that are not cancellable.
8. If R is any commutative ring and $P \oplus R^{n-1} \cong R^n$, show that $P \cong R$. [*Hint*: In the exterior algebra on P show that $\Lambda^k P = 0$ for $k > 1$ and calculate $\Lambda^n(P \oplus R^{n-1})$.]
9*. (M. R. Gabel, see Lam [78]) If P is not finitely generated but $P \oplus R^m = F$, where F is free, show (by writing this as a split exact sequence with F as middle term) that $F = F_0 \oplus F_1$, where each F_i is free, F_0 is finitely generated and $F = P + F_0$, $F_0 \cong (P \cap F_0) \oplus R^m$. Deduce that $P/(P \cap F_0) \cong F_1$ and hence show that P is free.
10. (Lam [76]) Let R be any ring and P a projective module that has R as a direct summand. If $P \oplus R^m \cong R^n$, where $n > m$, show that P^{m+1} is free. (*Hint*: If $P \cong Q \oplus R$, compute P^{m+1} and use R^m to 'liberate' P and the resulting R^n to 'liberate' Q^m.)
11. (Ojanguren and Sridharan [71]). Show that the polynomial ring $D[x, y]$ over a non-commutative field D is not 2-Hermite, by verifying that for suitable $a, b \in D$, the pair $(x + a, y + b)$ is a unimodular row, but is not completable. (*Hint*: Choose non-commuting a and b; for the last part apply Proposition 4.3. See also Exercise 1.1.11.)
12. Show that a ring is Hermite if and only if it has IBN and for every idempotent matrix E that splits, $I - E$ also splits.

0.5 The matrix of definition of a module

Given a ring R, we have, for any R-module M, a presentation

$$G \to F \to M \to 0,$$

where F, G are free. If F may be taken of finite rank, M is *finitely generated;* this holds even if F is merely projective (and finitely generated), for on replacing F, G by $F \oplus P$, $G \oplus P$ for a suitable finitely generated projective P, we obtain a free module $F \oplus P$ of finite rank mapping onto M. If G may be taken to be of finite rank, M is *finitely related*, and if both F and G can be taken to be of finite rank, M is *finitely presented*. Thus a finitely presented left R-module M has a presentation

$$R^m \xrightarrow{\alpha} R^n \longrightarrow M \longrightarrow 0. \tag{1}$$

Here M is determined up to isomorphism by a *presenting matrix* for α. Conversely, every $m \times n$ matrix A defines a finitely presented left R-module M in this way, as $M = \operatorname{coker} \alpha$, where α is the mapping from R^m to R^n described by A. We note the following property of modules that are finitely related but not finitely generated.

Proposition 0.5.1. *Over an arbitrary ring R, any finitely related R-module is the direct sum of a finitely presented module and a free module.*

Proof. We have $M = F/G$, where G is finitely generated. Write $F = F' + F''$, where F' is free on the generators occurring in elements of G and F'' is free on the remaining generators of F. Then $G \subseteq F'$, hence $M \cong (F'/G) \oplus F''$, which is the required decomposition. ∎

Returning to (1), we see that α is injective if and only if A is left regular. In that case M has a 'finite free resolution' of length 1:

$$0 \to R^m \xrightarrow{\alpha} R^n \longrightarrow M \longrightarrow 0, \tag{2}$$

and we define the *characteristic* $\chi(M)$ of M as the index of the presenting matrix, thus $\chi(M) = n - m$. In a ring with IBN this is well-defined and independent of the choice of presentation, by Schanuel's lemma (Appendix Lemma B.5, or also Theorem 0.5.3 below).

It should be observed that for general rings, modules with a finite free resolution of length at most 1 are very special; however, for the rings discussed in later chapters they include all finitely presented modules, which is why we treat them in more detail. In particular, we can show that for such modules, short exact sequences correspond to matrix equations.

Proposition 0.5.2. *Let R be a ring with IBN. Given any left R-modules M, M', M'' with finite free resolutions of length 1 and a left regular matrix C presenting M, there exists a short exact sequence*

$$0 \to M' \to M \to M' \to 0 \tag{3}$$

if and only if there exists a factorization

$$C = AB, \tag{4}$$

where A, B are left regular matrices presenting M', M'', respectively, and

$$\chi(M) = \chi(M') + \chi(M''). \tag{5}$$

Conversely, any equation (4) between left regular matrices corresponds to a short exact sequence (3).

Proof. Given (3) with the stated properties, there exists a free module F mapping onto M, with free kernel H, both of finite rank. We also have a surjection $F \to M''$, and if the kernel is denoted by K, then $K \supseteq H$ and we have the commutative diagram with exact rows and columns:

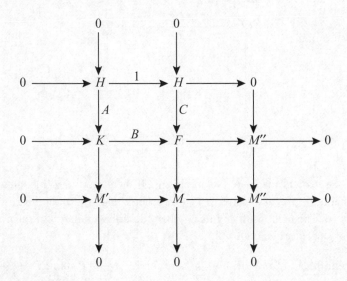

Since M'' has a finite free resolution of length 1, there are free modules $F_1 \supseteq K_1$ of finite rank such that $M'' \cong F_1/K_1$. By Schanuel's lemma we have $K \oplus F_1 \cong K_1 \oplus F$, therefore on replacing F by $F \oplus F_1$, K by $K \oplus F_1$ and H by $H \oplus F_1$ we can ensure that K is also free. If the matrices defining M', M'', M are A, B, C, respectively, then by the commutativity of the diagram, $C = AB$. Now $\chi(M) = \text{rk } F - \text{rk } H$, $\chi(M') = \text{rk } K - \text{rk } H$, $\chi(M'') = \text{rk } F - \text{rk } K$, and (5) follows. Conversely, given left regular matrices A, B, C satisfying (4), we obtain the first two columns of the above diagram, hence the third follows by the dual of the 3×3 lemma (see Mac Lane [63], p. 49, or Appendix Lemma B.3). ∎

We shall call two matrices over R *left similar* if the left modules they define are isomorphic; *right similar* matrices are defined correspondingly, and two matrices are called *similar* if they are left and right similar. Thus in an integral domain R two elements a and b are similar if and only if $R/aR \cong R/bR$, or equivalently, $R/Ra \cong R/Rb$.

The precise relationship between similar matrices was found by Fitting [36]. This relation can be simplified by restricting attention to matrices that are left regular, corresponding to the case where α in (1) is injective. We shall give an explicit description of similarity in this case; in essence this is just a formulation of Schanuel's lemma (Mac Lane [63], p. 101, or also below).

Two maps between R-modules, $\alpha : Q \to P, \alpha' : Q' \to P'$ are said to be *associated* if there is a commutative square

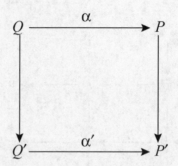

where the vertical maps are isomorphisms. If there are two R-modules S, T such that $\alpha \oplus 1_S$ is associated to $1_T \oplus \alpha'$, then α and α' are said to be *stably* associated. The next result and its corollary describe similarity of matrices in terms of stable association.

Theorem 0.5.3. *Let R be any ring and let $\alpha : Q \to P$ and $\alpha' : Q' \to P'$ be two homomorphisms of left R-modules. Then the following conditions are equivalent:*

(a) there is an isomorphism $\mu : Q \oplus P' \to P \oplus Q'$ of the form

$$\mu = \begin{pmatrix} \alpha & \beta \\ \gamma & \delta \end{pmatrix} \quad \text{with inverse} \quad \mu^{-1} = \begin{pmatrix} \delta' & \beta' \\ \gamma' & \alpha' \end{pmatrix},$$

(b) α is stably associated to α'.

Further, these conditions imply

(c) $\operatorname{coker} \alpha \cong \operatorname{coker} \alpha'$,

and if P, P' are projective modules and α, α' are injections, then the converse holds, so (a), (b), (c) are then equivalent.

Proof. (a) ⇒ (b). If we take $S = P', T = P$, we obtain the commutative square

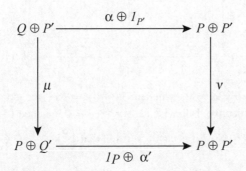

with the vertical isomorphisms

$$\mu = \begin{pmatrix} \alpha & \beta \\ \gamma & \delta \end{pmatrix} \quad \text{and} \quad \nu = \begin{pmatrix} 1 & 0 \\ \gamma & 1 \end{pmatrix} \begin{pmatrix} 1 & -\beta' \\ 0 & 1 \end{pmatrix}.$$

This result is also proved more simply by the equation

$$\begin{pmatrix} \alpha & \beta \\ 0 & 1 \end{pmatrix} = \begin{pmatrix} 1 & 0 \\ \gamma' & \alpha' \end{pmatrix} \begin{pmatrix} \alpha & \beta \\ \gamma & \delta \end{pmatrix}. \tag{6}$$

(b) ⇒ (a). If α is stably associated to α', we have a commutative square, which is expressed by an equation

$$\begin{pmatrix} \alpha & 0 \\ 0 & 1 \end{pmatrix} \begin{pmatrix} p & q \\ r & s \end{pmatrix} = \begin{pmatrix} x & y \\ z & t \end{pmatrix} \begin{pmatrix} 1 & 0 \\ 0 & \alpha' \end{pmatrix},$$

where

$$\begin{pmatrix} p & q \\ r & s \end{pmatrix}^{-1} = \begin{pmatrix} s' & q' \\ r' & p' \end{pmatrix}, \quad \begin{pmatrix} x & y \\ z & t \end{pmatrix}^{-1} = \begin{pmatrix} t' & y' \\ z' & x' \end{pmatrix}.$$

Now (a) follows with

$$\mu = \begin{pmatrix} \alpha & y \\ -r' & p't \end{pmatrix}, \quad \mu^{-1} = \begin{pmatrix} pt' & -q \\ z' & \alpha' \end{pmatrix},$$

as is easily checked.

(b) ⇒ (c) is clear. Now let P, P' be projective, α, α' injective and assume (c). Then there exist maps $\gamma : P' \to P$ and $\beta' : P \to P'$ making the following diagram commutative, and γ induces $-\gamma' : Q' \to Q$, while β' induces $-\beta : Q \to Q'$.

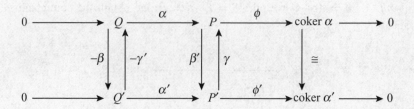

Further, $(1_P - \beta'\gamma)\phi = 0$, whence $1 - \beta'\gamma = \delta'\alpha$ for some $\delta': P \to Q$, because P is projective. Likewise $(1 - \gamma\beta')\phi' = 0$, whence $1 - \gamma\beta' = \delta\alpha'$ for some $\delta: P' \to Q'$. Now it is easily verified that $\begin{pmatrix} \alpha & \beta \\ \gamma & \delta \end{pmatrix}: Q \oplus P' \to P \oplus Q'$ has inverse $\begin{pmatrix} \delta' & \beta' \\ \gamma' & \alpha' \end{pmatrix}: P \oplus Q' \to Q \oplus P'$, which proves (a). ∎

The implication (c) ⇒ (a) (under the given conditions) is just the assertion of Schanuel's lemma. The proof of the equivalence (a) ⇔ (b) shows that the definition of stable association can be made a little more precise.

Corollary 0.5.4. *If $\alpha: Q \to P$ is stably associated to $\alpha': Q' \to P'$, then $\alpha \oplus 1_{P'}$ is associated to $1_P \oplus \alpha'$. Hence two matrices $A \in {}^r R^m$ and $A' \in {}^s R^n$ are stably associated, qua maps, if and only if $A \oplus I_n$ is associated to $I_m \oplus A'$.* ∎

In terms of matrices we obtain the following criteria by taking P, P', Q, Q' to be free.

Corollary 0.5.5. *Let $A \in {}^r R^m$, $A' \in {}^s R^n$ be any two matrices. Then of the following, (a) and (b) are equivalent and imply (c):*

(a) there exists an $(r+n) \times (s+m)$ matrix $\begin{pmatrix} A & * \\ * & * \end{pmatrix}$ with an inverse of the form $\begin{pmatrix} * & * \\ * & A' \end{pmatrix}$,

(b) A and A' are stably associated,

(c) A and A' are left similar.

If A, A' are left regular, all three conditions are equivalent. Moreover, two regular matrices are left similar if and only if they are right similar.

Proof. The equivalence follows from Theorem 5.3, while the left–right symmetry is a consequence of the evident symmetry of (a) or (b). ∎

0.5 The matrix of definition of a module

Two matrices A, A' standing in the relation (a) are said to be *GL-related*. By Corollary 5.5 this means the same as 'stably associated'. We also note that an invertible (square) matrix is stably associated to the unique 0×0 matrix 1_0.

From (a) we see that if two matrices A, A' over R are stably associated, then their images under any ring homomorphism are again stably associated. In particular, if A maps to a unit under some homomorphism, then so does A'. We further note that for any ring with IBN any two *GL*-related matrices have the same index, which is also the characteristic of the corresponding left R-modules, assuming the matrices to be left regular.

Over a weakly finite ring the notion of similarity of matrices can still be simplified. Consider a relation

$$AB' = BA' \qquad (7)$$

between matrices. This can also be written

$$(A \quad B)\begin{pmatrix} -B' \\ A' \end{pmatrix} = 0.$$

We shall call A, B *right comaximal* if the matrix $(A \quad B)$ has a right inverse, and A', B' *left comaximal* if $(A' \quad B')^T$ has a left inverse. Now (7) is called a *comaximal* relation if A, B are right comaximal and A', B' left comaximal. We shall find that in a weakly finite ring stable association can be described in terms of comaximal relations.

Proposition 0.5.6. *Let R be any ring and let $A \in {}^rR^m, A' \in {}^sR^n$. Then the following two relations are equivalent:*

(a) A, A' satisfy a comaximal relation (7),
(b) there is an $(r+n) \times (s+m)$ matrix with first row block $(A \quad B)$, with a right inverse whose last column block has the form $(-B' \quad A')^T$.

In particular, (a) and (b) hold whenever

(c) A and A' are stably associated,

and in a weakly finite ring (a)–(c) are equivalent for two matrices of the same index.

Proof. Suppose that A, A' satisfy a comaximal relation (7), say

$$AD' - BC' = I, \quad DA' - CB' = I. \qquad (8)$$

Then on writing

$$M = \begin{pmatrix} A & B \\ C & D \end{pmatrix} \quad \text{and} \quad N = \begin{pmatrix} D' & -B' \\ -C' & A' \end{pmatrix}, \qquad (9)$$

we have $MN = \begin{pmatrix} I & 0 \\ P & I \end{pmatrix}$, where $P = CD' - DC'$. Hence M has the right inverse

$$\begin{pmatrix} D' & -B' \\ -C' & A' \end{pmatrix} \begin{pmatrix} I & 0 \\ -P & I \end{pmatrix} = \begin{pmatrix} * & -B' \\ * & A' \end{pmatrix},$$

and (b) follows. Conversely, if N in (9) is a right inverse of M, then (7) and (8) hold, hence (7) is then a comaximal relation. This shows that (a) $\Leftrightarrow$ (b).

Now (c) $\Rightarrow$ (b) by Corollary 5.5, and (b) $\Rightarrow$ (c) under the given conditions, because when $m - r = n - s$, then $r + n = s + m$, and for a square matrix over a weakly finite ring any right inverse is an inverse. ∎

For later use we note the explicit form of the relation of stable association between A and A':

$$\begin{pmatrix} A & 0 \\ 0 & I \end{pmatrix} \begin{pmatrix} D' & -B' \\ -C' & A' \end{pmatrix} = \begin{pmatrix} I + BC' & -B \\ -C' & I \end{pmatrix} \begin{pmatrix} I & 0 \\ 0 & A' \end{pmatrix}; \qquad (10)$$

another form of such a relation is given by (6). We also restate the criterion for stable association derived in Proposition 5.6.

Proposition 0.5.7. *In a weakly finite ring R, two matrices A and A' are stably associated if and only if they have the same index and satisfy a comaximal relation $AB' = BA'$.* ∎

Finally we note a remark on the invertibility of endomorphisms that will be useful later.

Lemma 0.5.8. *Given modules M, N over any ring, let $s: M \to N$, $t: N \to M$ be module homomorphisms, so that st, ts are endomorphisms of M, N, respectively; further denote the identity mappings on M, N by e, f, respectively. Then $e - st$ is an automorphism of M if and only if $f - ts$ is an automorphism of N and the inverses are related by the equations*

$$(f - ts)^{-1} = f + t(e - st)^{-1} s, \qquad (11)$$
$$(e - st)^{-1} = e + s(f - ts)^{-1} t. \qquad (12)$$

Proof. Assume that $e - st$ is invertible. Then

$$(f - ts)(f + t(e - st)^{-1} s) = f - ts + t(e - st)^{-1} s - tst(e - st)^{-1} s$$
$$= f - ts + t(e - st)(e - st)^{-1} s$$
$$= f.$$

Thus $f - ts$ has the right inverse given by (11); a similar calculation shows that this is also a left inverse, so $f - ts$ is an automorphism of N. The reverse implication follows by symmetry. ∎

Exercises 0.5

1. Show that a matrix is stably associated to I if and only if it is a unit; if it is stably associated to an $m \times n$ zero matrix, where $m, n > 0$, then it is a zero-divisor.
2. Let A be a matrix over any ring R. Show that the left R-module presented by A is zero if and only if A has a right inverse.
3. Let R be a ring and $A \in {}^m R^n, B \in {}^n R^m$. Show that $I + AB$ is stably associated to $I + BA$. Deduce that $I + AB$ is a unit if and only if $I + BA$ is; prove this directly by evaluating $I - B(I + AB)^{-1}A$.
4°. Under what circumstances is AB stably associated to BA?
5. Let R be a ring with UGN. If A, A' satisfy a comaximal relation (7), show that $i(A) \geq i(A')$. Deduce that if A, A' satisfy a comaximal relation and A', A likewise, then A and A' are stably associated.
6. Show that the condition on the index cannot be omitted from Proposition 5.7. Hence find examples of pairs of matrices (over a weakly finite ring, say) that satisfy a comaximal relation but are not stably associated.
7*. Let R be a non-zero ring and $S = \text{End}_R(R^{(\mathbb{N})})$. Show that $S_2 \cong S$; is it the case that any two stably associated 1×1 matrices are associated?
8*. Since the relation of stable association is clearly transitive, it follows by Corollary 5.5 that being GL-related is transitive. Give a direct proof of this fact. (*Hint*: Take the case of elements first.)

0.6 Eigenrings and centralizers

Let R be a ring, M a left R-module and N a submodule of M. We define the *idealizer* of N in M over R as the set

$$I(N) = \{\beta \in \text{End}_R(M) | N\beta \subseteq N\}.$$

Clearly $I(N)$ is a subring of $\text{End}_R(M)$ and if we put $\mathfrak{a} = \text{Hom}_R(M, N)$, then $\mathfrak{a}$ is a left ideal in $\text{End}_R(M)$ and a two-sided ideal in $I(N)$. The quotient ring $E(N) = I(N)/\mathfrak{a}$ is called the *eigenring* of N in M over R. Writing $Q = M/N$, we have a natural ring homomorphism $I(N) \to \text{End}_R(Q)$; the kernel is easily seen to be $\mathfrak{a}$, so we obtain an injection

$$E(N) \to \text{End}_R(Q). \tag{1}$$

Suppose now that M is projective. Then any endomorphism ϕ of Q can be lifted to an endomorphism β of M such that $N\beta \subseteq N$; this shows the map (1) to be surjective, and so an isomorphism. We state the result as

Proposition 0.6.1. *Given any ring R, if P is a projective left R-module and N a submodule of P with eigenring E(N), then there is a natural isomorphism*

$$E(N) \cong \mathrm{End}_R(P/N).$$ ∎

In particular, if $P = R$, then $\mathfrak{a}$ is a left ideal of R and we have $I(\mathfrak{a}) = \{x \in R \mid \mathfrak{a}x \subseteq \mathfrak{a}\}$ and $E(\mathfrak{a}) = I(\mathfrak{a})/\mathfrak{a} \cong \mathrm{End}_R(R/\mathfrak{a})$.

Now let $A \in {}^mR^n$ and let $\alpha: R^m \to R^n$ be the corresponding map. Taking $P = R^n, N = \mathrm{im}\,\alpha$, we have $\mathrm{End}_R(R^n) \cong R_n$, hence $I(N) = \{\beta \in R_n \mid N\beta \subseteq N\}$ and $\mathfrak{a} = \mathrm{Hom}_R(R^n, N) \cong {}^nN$, as left R_n-module. We define the left *idealizer* of a matrix A over R as the corresponding set of matrices

$$I(A) = \{B \in R_n \mid AB = B'A \text{ for some } B' \in R_m\},$$

and the *left eigenring* of A as the quotient ring $E(A) = I(A)/({}^nR^m)A$. By Proposition 6.1, $E(A) \cong \mathrm{End}_R(M)$, where M is the left R-module defined by A. The right eigenring of A is defined similarly, and it is clear that for a regular matrix A the left and right eigenrings are isomorphic, the isomorphism being induced by the mapping

$$B \mapsto B', \text{ where } AB = B'A.$$

In the particular case where $m = n = 1$, the matrix becomes an element a of R and we have $E(a) = I(a)/Ra \cong \mathrm{End}_R(R/Ra)$.

Given any matrices A, B over R, if M, N are the left R-modules defined by them, then each R-homomorphism $f: M \to N$ is completely specified by a matrix P over R such that

$$AP = P'B \tag{2}$$

for some matrix P'. If $I(A, B)$ denotes the set of all such P and $\mathfrak{b}$ is the left R-module spanned by the rows of B, then as before,

$$I(A, B)/\mathfrak{b} \cong \mathrm{Hom}_R(M, N) \tag{3}$$

is an isomorphism of $(E(A), E(B))$-bimodules.

For later use we record the effect of a change of base field on $\mathrm{Hom}_R(M, N)$:

Proposition 0.6.2. *Let R be a k-algebra, where k is a commutative field. Given a field extension E/k, write $R_E = R \otimes_k E$ and for any R-module M denote the extension $M \otimes_k E$ by M_E. If M, N are R-modules such that $\mathrm{Hom}_R(M, N)$ is finite-dimensional over k, then*

$$\mathrm{Hom}_R(M, N) \otimes_k E \cong \mathrm{Hom}_{R_E}(M_E, N_E). \tag{4}$$

0.6 Eigenrings and centralizers

Proof. There is a natural map from the left- to the right-hand side in (4), which is clearly injective, so it will be enough to show that both sides have the same dimension.

Let (e_i), (f_λ) be bases for M, N as k-spaces (possibly infinite-dimensional); then the action of R is given by

$$e_i x = \sum \rho_{ij}(x) e_j, \quad f_\lambda x = \sum \sigma_{\lambda\mu}(x) f_\mu \quad (x \in R),$$

and $\mathrm{Hom}_R(M, N)$ is the space of all solutions α over k of the system

$$\sum_j \rho_{ij}(x) \alpha_{j\mu} = \sum_\lambda \alpha_{i\lambda} \sigma_{\lambda\mu}(x). \tag{5}$$

Let $V = \mathrm{Hom}_k(M, N)$ as k-space and C be the subspace of solutions $(\alpha_{i\lambda})$ of (5). Then $V = C \oplus D$ for some k-space D and by hypothesis C is finite-dimensional, say $\dim_k C = n$. Let $p_1, \ldots, p_n$ be linearly independent functionals on V such that $\cap \ker p_\nu = D$. Then (5) together with

$$p_\nu(\alpha) = 0 \quad (\nu = 1, \ldots, n),$$

has only the trivial solution over k and the same holds over E. It follows that the solution of (5) is again n-dimensional over E, and so (4) holds. ∎

The eigenring of a ring element is closely related to its centralizer and to some extent both may be treated by the same method, by the device of adjoining an indeterminate. The basic result is:

Theorem 0.6.3. *Let R be a ring and $S = R[t]$ the ring obtained by adjoining a central indeterminate t to R. Given $a, b \in R$, write $C = C(a, b) = \{x \in R | ax = xb\}$. Then there is a natural isomorphism of abelian groups:*

$$C(a, b) \cong \mathrm{Hom}_S(S/S(t - a), S/S(t - b)).$$

Proof. Let R_a denote R viewed as left S-module with t acting by right multiplication by a. By mapping $t \mapsto a$ we define a left S-module homomorphism $S \to R_a$ with kernel $S(t - a)$, and so

$$\mathrm{Hom}_S(S/S(t - a), S/S(t - b)) \cong \mathrm{Hom}_S(R_a, R_b).$$

For any $f : R_a \to R_b$ we have $a(1f) = af = (t1)f = t(1f) = (1f)b$; therefore the rule $f \mapsto 1f$ defines a homomorphism from $\mathrm{Hom}_S(R_a, R_b)$ to $C(a, b)$. Conversely, for any $x \in C(a, b)$, right multiplication by x defines a left S-linear map $R_a \to R_b$; so we obtain a homomorphism $C(a, b) \to \mathrm{Hom}_S(R_a, R_b)$, clearly inverse to the previous map. ∎

By putting $b = a$ we can express the centralizer of a as an eigenring:

Corollary 0.6.4. *The centralizer of an element $a \in R$ is isomorphic to the eigenring of $t - a$ in the polynomial ring $R[t]$.* ∎

The following result is well known in the special case of matrix rings over a field, where it is used to obtain the canonical form of a matrix (see CA, p. 355).

Proposition 0.6.5. *Let R be a ring and t a central indeterminate. Then two elements a, b of R are conjugate under $U(R)$ if and only if $t - a$ and $t - b$ satisfy a comaximal relation*

$$f.(t - a) = (t - b).g \tag{6}$$

in $R[t]$, and in any such comaximal relation (6), f and g can be found to lie in $U(R)$.

Proof. If a, b are conjugate, say $ua = bu$, where $u \in U(R)$, then clearly $u(t - a) = (t - b)u$ is a comaximal relation. Conversely, assume a comaximal relation (6). By subtracting an expression $(t - b)h(t - a)$ from both sides, we obtain the equation

$$u(t - a) = (t - b)v, \tag{7}$$

where $u = f - (t - b)h$, $v = g - h(t - a)$. Here we may choose h so that u has degree 0 in t, i.e. $u \in R$. Then on comparing degrees in (7) we find that $v \in R$, while a comparison of highest terms shows that $v = u$ and so

$$ua = bu. \tag{8}$$

Further, since $u \equiv f \pmod{(t - b)R[t]}$, u and $t - b$ are still right comaximal, say

$$up + (t - b)q = 1, \text{ where } p, q \in R[t]. \tag{9}$$

Replacing p by $p - (t - a)k$ for suitable $k \in R[t]$ and using (7), we can reduce (9) to the case where p has degree 0. Then $q = 0$, by comparing degrees, and now (9) shows p to be a right inverse to u. By the symmetry of (8), u also has a left inverse and so is a unit. Now (8) shows a and b to be conjugate, as claimed. ∎

Exercises 0.6

1. In any ring R, if $ab' = ba'$, show that $a'b$ lies in the idealizer of $Rb'b$ and that of $a'aR$.

0.7 Rings of fractions

2. Let R be a ring and t a central indeterminate. Given $a, b \in R^\times$, show that the elements $a^n b (n = 0, 1, \ldots)$ are right linearly dependent over R if and only if $t - a$ and b are right commensurable in $R[t]$.
3. In any ring R, show that $\operatorname{Hom}_R(R/aR, R/bR) \cong R/dR$ where d is the largest element (in terms of divisibility) similar to a left factor of a and a right factor of b.
4. Let A be a matrix with eigenring E. Show that $A \oplus \cdots \oplus A$ (r terms A) has eigenring E_r.
5. Show that a unit has zero eigenring and conversely, an element with zero eigenring over an integral domain is a unit.
6. Let R be a ring and t a central indeterminate. If $t - a$ and $h \in R[t]$ satisfy a comaximal relation $f.(t - a) = h.g$, show that g can be taken to lie in R, but not in general f. (*Hint*: Use a nilpotent element of R to construct f as an invertible element of degree 2.)
7*. (Robson [72]) (a) In a ring R, let $\mathfrak{a} = \mathfrak{m}_1 \cap \ldots \cap \mathfrak{m}_k$, where the $\mathfrak{m}_i$ are maximal left ideals (such an $\mathfrak{a}$ is called *semimaximal*). If $B = \{b \in R | \mathfrak{a}b \subseteq \mathfrak{m}_1\}$ and A is the idealizer of $\mathfrak{a}$, show that $B/\mathfrak{m}_1$ is a simple left A-module.

   (b) If $\mathfrak{a}$ and A are as before, show that any simple left R-module is either simple as left A-module or is a homomorphic image of $R/\mathfrak{a}$.

   (c) With the notation as before, let M be a simple left R-module. Then M is simple as left A-module, unless for some i, $M \cong R/\mathfrak{m}_i$ and $\mathfrak{a}R \not\subseteq \mathfrak{m}_i$. In that case M has a unique composition series $R \supset A + \mathfrak{m}_i \supset \mathfrak{m}_i$.
8*. (G. M. Bergman) Prove Proposition 6.2 under the hypothesis that M is finitely generated, as R-module.
9. Given two left R-modules with finite free resolutions of length 1, $U = R^n/R^m X$, $V = R^s/R^r Y$, show that $\operatorname{Ext}^1_R(V, U) = {}^r R^n/(Y^s R^n + {}^r R^m X)$. Similarly if $W = {}^h R/Z({}^k R)$ is a right R-module, show that $\operatorname{Tor}^R_1(W, V) = (Z^k R^s \cap {}^h R^r Y)/Z^k R^r Y$.
10. Let R be any ring and A an $m \times n$ matrix over R, which is not right full. Given $B \in I(A)$, if B represents zero in $E(A)$, show that B is not full.

0.7 Rings of fractions

As is well known, a commutative ring has a field of fractions if and only if it is an integral domain. In the general case this condition is still necessary, but not sufficient, as Malcev [37] has shown. Malcev then gave a set of necessary and sufficient conditions for a semigroup to be embeddable in a group (Malcev [39],[40]; see also UA, p. 268), but for rings the problem of finding an embeddability criterion remained open until 1971 (see Cohn [71a] and Chapter 7 below). However, a simpler set of sufficient conditions was found by Ore [31] and after some generalities we briefly recall the details.

Let R, R' be any rings and S a subset of R. A homomorphism $f : R \to R'$ is said to be *S-inverting* if f maps S into $U(R')$. It is clear that there always exists a universal S-inverting ring R_S, obtained by adjoining for each $x \in S$ a new element x' and adding the relations $xx' = x'x = 1$. The construction shows that the natural map $\lambda: R \to R_S$ is an S-inverting homomorphism and it is easily checked that any S-inverting homomorphism can be factored uniquely by λ (see FA, Section 7.1). However, it is not easy to decide when λ is injective, or indeed when it is non-zero. The same construction can be used to find, for a given monoid M with a subset S, a universal S-inverting monoid M_S with a natural homomorphism $\lambda: M \to M_S$. In particular, taking $S = M$, we obtain a group $\mathcal{G}(M)$, with a homomorphism $M \to \mathcal{G}(M)$, which is universal for all homomorphisms from M to a group. $\mathcal{G}(M)$ is known as the *universal group* of M, also called Grothendieck group, see Section 0.3.

Ore's construction asks under what conditions the elements of the universal S-inverting ring R_S can be written in the form as^{-1}, where $s \in S$. Clearly it is necessary for $s^{-1}a$ to be expressible in this form, say $s^{-1}a = a_1 s_1^{-1}$. On multiplying up, we find the condition $as_1 = sa_1$; this may also be stated as

O.1   For any $a \in R, s \in S, aS \cap sR \neq \emptyset$.

In addition we shall also need a cancellation condition:

O.2   For each $a \in R, s \in S$, if $sa = 0$, then $at = 0$ for some $t \in S$, and $0 \notin S$.

Further, it is convenient to assume S to be *multiplicative*, i.e. to contain 1 and be closed under multiplication. A multiplicative subset of a ring satisfying O.1 and O.2 will be called a *right Ore set*. If R has such a subset, then by 0.2 and multiplicativity, $1 \neq 0$, so R must be non-zero. In the expression as^{-1} of an element, a is called the *numerator* and s the *denominator*. Now the basic result may be stated as

Theorem 0.7.1.   *Let R be a ring and S a right Ore set in R. Then all the elements of the universal S-inverting ring R_S can be expressed in the form as^{-1}, where $a \in R, s \in S$. When the right Ore set S consists of regular elements, then the natural map $\lambda: R \to R_S$ into the universal S-inverting ring is an embedding. Conversely, when $RS^{-1} = \{as^{-1} | a \in R, s \in S\}$ is a ring, then S is a right Ore set.*

Proof.   We shall give a sketch of the proof, referring to FA, 7.1 for the details. Define a relation on $R \times S$ by setting $(a_1, s_1) \sim (a_2, s_2)$ whenever there exist $t_1, t_2 \in R$ such that $s_1 t_1 = s_2 t_2 \in S$ and $a_1 t_1 = a_2 t_2$; this is easily verified to be an equivalence. Denoting the equivalence class of (a, s) by a/s, where a is the numerator and s the denominator, we observe first that any pair of elements can be brought to a common denominator: if $a_i \in R, s_i \in S$ are given and $t_i \in R$ is chosen as before (by the Ore condition), so that $s_1 t_1 = s_2 t_2 \in S$, then $a_1/s_1 = a_1 t_1/s_1 t_1, a_2/s_2 = a_2 t_2/s_2 t_2$ and $s_1 t_1 = s_2 t_2$. To define addition, we first bring the elements to a common denominator and then put $a_1/s + a_2/s =$

0.7 Rings of fractions

$(a_1 + a_2)/s$. To multiply a_1/s_1 and a_2/s_2, we determine $a_3 \in R, s_3 \in S$ such that $s_1 a_3 = a_2 s_3$ and then put $a_1/s_1 . a_2/s_2 = (a_1 a_3)/(s_2 s_3)$. Of course it needs to be checked that the results do not depend on the choice of a_1, s_1 and a_2, s_2 within their equivalence classes. This is a routine verification that may be left to the reader, as well as the verification of the ring laws. The mapping $a \mapsto a/1$ is easily seen to be a homomorphism from R to R_S, which is injective when S consists of regular elements. The converse is also a straightforward verification. ■

The ring R_S is also called the *localization* of R at S; when S consists of regular elements, so that λ is injective, R_S is called the *ring of fractions* of R by S. When R is an integral domain and $S = R^\times$, R is called a *right Ore domain*; its localization $R_{R^\times}$ is the field of fractions of R. The condition for an integral domain R to be right Ore is therefore

$$aR \cap bR \neq 0 \text{ for all } a, b \in R^\times. \tag{1}$$

Corresponding definitions apply on the left and we speak of an *Ore domain* when the side is not specified (rather in the way one speaks of a module).

For a right Ore domain R the localization at $R^\times$ is also called the field of right fractions of R. By symmetry every left Ore domain can be embedded in a field of left fractions, and for a two-sided Ore domain the fields of left and of right fractions coincide, by the uniqueness of the latter (see below). We also note that every commutative integral domain is both a left and right Ore domain.

As a special case of Theorem 7.1 we obtain

Corollary 0.7.2. *Any right (or left) Ore domain can be embedded in a field, and the least such field, unique up to isomorphism, is the universal $R^\times$-inverting ring. Hence any (left or right) Ore domain is weakly finite.*

Proof. To establish the uniqueness, suppose that there exist two embeddings into fields $R \to K$, $R \to K'$. The identity map on R can be extended to elements ab^{-1} and this shows K, K' to be isomorphic. Suppose there are two embeddings of R in the field of fractions K; then we have an automorphism of K, which reduces to the identity on R. For if $ab^{-1} \leftrightarrow a'b'^{-1}$, we can find a common denominator and so obtain $cd^{-1} \leftrightarrow c'd^{-1}$; multiplying by d we find that $c = c'$ and so the automorphism reduces to the identity on R, as claimed. The last statement is clear, since R is a subring of a field. ■

In general, when λ is not injective, its kernel has the form

$$\ker \lambda = \{a \in R | at = 0 \text{ for some } t \in S\}. \tag{2}$$

We remark that any finite set of elements of R_S may be brought to a common denominator, which is a right multiple of the given denominators. The case of two elements was dealt with in the proof of Theorem 7.1; now let $c_1, \ldots, c_n \in R_S$ and use induction on n. We first bring c_1, c_2 to a common denominator b and then bring $b^{-1}, c_3, \ldots, c_n$ to a common denominator b'. This is a right multiple of the denominators of $b^{-1}, c_3, \ldots, c_n$, hence also of that of c_1, c_2 and so it is the desired common denominator. Thus we have

Proposition 0.7.3. *Let R be a ring and S a right Ore set in R. Then any finite set of elements of the localization R_S can be brought to a common denominator, which is a right multiple of the denominators of all the given elements.* ∎

To find Ore sets in a ring one looks for its 'large' elements. An element c of an integral domain R is said to be *right large* if $cR \cap aR \neq 0$ for all $a \in R^\times$. The set L of all right large elements is always multiplicative. For clearly $1 \in L$ and if $a, b \in L$ and $c \in R^\times$, then there exist $x, y \in R^\times$ such that $ax = cy$ and there exist $u, v \in R^\times$ such that $bu = xv$, hence $abu = axv = cyv$ and this shows that $ab \in L$. Further, if $aR \cap bR \neq 0$ implies $ab' = ba'$ with either a' or b' in L, then it follows that L is a right Ore set. For if $a \in L$ and $b \in R^\times$, then $aR \cap bR \neq 0$, say $ab' = ba' \neq 0$. If $a' \in L$, then $aR \cap bL \neq \emptyset$, as claimed; otherwise $b' \in L$ and then $ab' \in L$, hence $ba' \in L$. Now for any $c \in R^\times$ there exist $x, y \in R^\times$ with $ba'x = bcy$, hence $a'x = cy$ and this shows again that $a' \in L$. Thus we have proved

Proposition 0.7.4. *In any integral domain R the set L of right large elements is a submonoid. If $aR \cap bR \neq 0$ for $a, b \in L$ implies that $ab' = ba'$, where either a' or b' is right large, then the set L of all right large elements in R is a right Ore set and the natural map $R \to R_S$ is an embedding.* ∎

If R is a ring and T is a right Ore set in R, then any T-inverting homomorphism to a ring S, $f: R \to S$ extends in a unique fashion to a homomorphism of R_T into S, by the universal property of R_T. Sometimes we shall need this result for R-subrings of R_T; the proof is quite similar to that of Theorem 7.1, though it does not actually follow as a special case.

Proposition 0.7.5. *Let R, S be rings and $f: R \to S$ an injective homomorphism. If T is a right Ore set in R such that Tf is regular and R' is an R-subring of R_T such that*

$$a \in R'b \, (a \in R, b \in T) \text{ implies } af \in S.bf, \tag{3}$$

then f extends to a unique homomorphism $f': R' \to S$ and f' is again injective.

Proof. Given $r \in R'$, we can write $r = ab^{-1} (a \in R, b \in T)$, thus $a = rb$ and so $af = s.bf$ for some $s \in S$, by (3). We define $rf' = s$ and note that if instead of a, b we had used au, bu, where $bu \in T$, then $(au)f = s.(bu)f$ with the same s, so any expression $r = au.(bu)^{-1}$ leads to the same value of rf'. Since any two representations of r can be brought to a common denominator, they lead to the same value for rf' and this shows f' to be well-defined. The homomorphism property and injectivity follow as in the proof of Theorem 7.1. ∎

An integral domain R that is not a right Ore domain must contain two non-zero elements a, b that are right incommensurable: $aR \cap bR = 0$. It follows that the right ideal $aR + bR$ is a free right R-module of rank 2. Moreover, the elements $a^n b (n = 0, 1, 2, \ldots)$ are right linearly independent; for if $\sum a^i bc_i = 0 (c_i \in R$, not all 0), then by cancelling on the left as many factors a as possible, we can write this equation as

$$bc_0 + abc_1 + \cdots + a^r bc_r = 0 \quad (c_0 \neq 0),$$

hence $bc_0 \in aR \cap bR$, a contradiction. This proves

Proposition 0.7.6. *An integral domain that is not a right Ore domain contains free right ideals of any finite or countable rank.* ∎

Since a free right ideal of countable rank is not finitely generated, we obtain

Corollary 0.7.7. *Any right Noetherian domain is a right Ore domain.* ∎

Examples of non-Ore domains are free associative algebras of rank at least 2 (to be defined in Section 2.5).

Let R, A, B be any rings, $\alpha: R \to A, \beta: R \to B$ two homomorphisms and M an (A, B)-bimodule. Then an (α, β)-*derivation* from R to M is a map $\delta: R \to M$ that is additive and satisfies

$$(xy)^\delta = x^\alpha y^\delta + x^\delta y^\beta. \tag{4}$$

In particular, if $A = R$ and $\alpha = 1$, we speak of a (right) β-derivation. Putting $x = y = 1$ in (4) and observing that $1^\alpha = 1^\beta = 1$, we see that any (α, β)-derivation δ satisfies $1^\delta = 0$. It is easily verified that ker δ is a subring of R, called the *ring of constants* (with respect to δ). Moreover, any element of ker δ that is invertible in R is also invertible in ker δ, as follows by the formula (itself easily checked):

$$(x^{-1})^\delta = -(x^{-1})^\alpha . x^\delta . (x^{-1})^\beta.$$

We list some examples of derivations.

1. Let $M = R = A = B = k(t)$ be the field of rational functions in t over some field k, and let f' be the usual derivative of f. Then on taking $\alpha = \beta = 1$, we obtain the familiar formula

$$(fg)' = fg' + f'g,$$

as a special case of (4).

2. Let $M = R = A = B$ and let α be any automorphism of R. Then $f^\delta = f^\alpha - f$ is a right α-derivation. In particular, when $R = k[t]$ and $\alpha: f(t) \mapsto f(t+1)$, then δ is the differencing operator $f(t) \to f(t+1) - f(t)$.

3. For any R, A, B, M take $m \in M$ and define $\delta_m : R \to M$ by the rule

$$\delta_m : x \mapsto x^\alpha m - m x^\beta. \tag{5}$$

This is easily seen to be an (α, β)-derivation; it is called the *inner* (α, β)-*derivation* induced by m. Thus the differencing operator in example 2 is the inner α-derivation induced by 1, where α is the translation operator. A derivation that is not inner is called *outer*.

With any (A, B)-module M we can associate the ring $\begin{pmatrix} A & M \\ 0 & B \end{pmatrix}$ consisting of all matrices

$$\begin{pmatrix} a & m \\ 0 & b \end{pmatrix} \quad (a \in A, b \in B, m \in M),$$

with the usual matrix addition and multiplication. The (A, B)-bimodule property just ensures that we get a ring in this way:

$$\begin{pmatrix} a & m \\ 0 & b \end{pmatrix} \begin{pmatrix} a' & m' \\ 0 & b' \end{pmatrix} = \begin{pmatrix} aa' & am' + mb' \\ 0 & bb' \end{pmatrix}.$$

Given maps $\alpha : R \to A, \beta : R \to B, \delta : R \to M$, we can define a map from R to $\begin{pmatrix} A & M \\ 0 & B \end{pmatrix}$ by the rule

$$x \mapsto \begin{pmatrix} x^\alpha & x^\delta \\ 0 & x^\beta \end{pmatrix}, \tag{6}$$

and it is easily checked that this is a ring homomorphism if and only if α, β are homomorphisms and δ is an (α, β)-derivation. This alternative method of defining derivations is often useful, for example in proving

Theorem 0.7.8. *Let R, A, B be rings, T a right Ore set in R and M an (A, B)-bimodule. Then any T-inverting homomorphism $\alpha : R \to A$ extends to a unique homomorphism $\alpha' : R_T \to A$, and given T-inverting homomorphisms*

$\alpha: R \to A$, $\beta: R \to B$, any (α, β)-derivation $\delta: R \to M$ extends to an (α', β')-derivation δ' of R_T into M.

Proof. The existence and uniqueness of α' follow because R_T is universal T-inverting. Now δ defines a homomorphism (6) from R to $\begin{pmatrix} A & M \\ 0 & B \end{pmatrix}$, which is T-inverting and therefore extends to a homomorphism of $R_T : x \mapsto \begin{pmatrix} x^{\alpha'} & x^{\delta'} \\ 0 & x^{\beta'} \end{pmatrix}$. It follows that δ' is an (α', β')-derivation. ∎

We conclude this section by briefly discussing a special class of monoids that are embeddable in groups. Let S be a cancellation monoid; an element c of S is said to be *rigid* (sometimes called *equidivisible*), if

$$ab' = ba' = c \tag{7}$$

implies $a = bu$ or $b = au$, for some $u \in S$. Thus c is rigid if the left factors of c form a chain under the ordering by divisibility. When (7) holds, and $a = bu$, then $bub' = ba'$, hence $a' = ub'$; this shows the condition to be left–right symmetric. A monoid is said to be *rigid* if it admits cancellation and all its elements are rigid. Thus S is rigid if it is a cancellation monoid such that

$$aS \cap bS \neq \emptyset \Rightarrow aS \subseteq bS \text{ or } bS \subseteq aS. \tag{8}$$

Theorem 0.7.9. *Every rigid monoid is embeddable in a group.*

Proof. Let S be a rigid monoid and for $a, b \in S$ denote by $\{a.b\}$ the set of all elements of the form $a'b''$, where $a = a'a''$, $b = b'b''$. We first establish the following assertions:

(α) $u \in \{ac.b\}$, $u \notin \{a.b\} \Rightarrow u \in aS$,

(β) $uc \in \{a.bc\} \Rightarrow u \in \{a.b\}$.

To prove (α) we have by definition, $u = pq$, $ac = px$, $b = yq$. By rigidity, $a \in pS$ or $p \in aS$. In the first case $u \in \{a.b\}$, which is excluded, so $p \in aS$ and hence $u \in aS$. To prove (β), let $uc = pq$, $a = px$, $bc = yq$. Now either $p = uz$; then $a = uzx$ and so $u \in \{a.b\}$, or $u = pz$; then $q = zc$, $b = yz$ and again $u \in \{a.b\}$. Thus (α) and (β) are established.

We next consider the set of all expressions

$$p = a_0 b_1^{-1} a_1 b_2^{-1} \ldots a_{n-1} b_n^{-1} a_n, \quad \text{where } a_i, b_i \in S. \tag{9}$$

The expression (9) is said to have *length* n; it is said to be *reduced* if

$$a_i \notin \{b_i.b_{i+1}\} \quad (i = 1, \ldots, n-1), \quad b_i \notin \{a_i.a_{i-1}\} \quad (i = 1, \ldots, n). \tag{10}$$

Clearly if (9) is reduced, none of the a's or b's except possibly a_0 or a_n can be 1. We define the following elementary transformations on (9), for any $x \in S$:

$$R_i : a_{i-1}b_i^{-1} \mapsto a_{i-1}x(b_ix)^{-1} \quad (i = 1, \ldots, n),$$
$$L_i : b_i^{-1}a_i \mapsto (xb_i)^{-1}xa_i \quad (i = 1, \ldots, n).$$

Two expressions are called *equivalent* if we can pass from one to the other by a finite chain of elementary transformations and their inverses; clearly this is an equivalence. We note in particular that every element of S forms an expression of length 0, which is reduced and admits no elementary transformations.

We claim that a reduced expression stays reduced under elementary transformation. Consider the effect of R_j on (9) and the first condition (10). It is clear that this will not be affected unless $j = i$ or $i + 1$. We take these cases in turn.

R_i. Suppose that $a_i \in \{b_ix.b_{i+1}\}$. Since $a_i \notin \{b_i.b_{i+1}\}$, we have $a_i \in b_iS$ by (α), but this contradicts the fact that $b_i \notin \{a_i.a_{i-1}\}$.

R_{i+1}. Since $a_i \notin \{b_i.b_{i+1}\}$, we have $a_ix \notin \{b_i.b_{i+1}x\}$ by (β).

R_i^{-1}. Let $a_{i-1} = a'_{i-1}x, b_i = b'_ix$; if $a_i \in \{b'_i.b_{i+1}\}$, then $a_i \in \{b'_ix.b_{i+1}\} = \{b_i.b_{i+1}\}$, a contradiction.

R_{i+1}^{-1}. Let $a_i = a'_ix, b_{i+1} = b'_{i+1}x$ and $a'_i \in \{b_i.b'_{i+1}\}$; then $a_i = a'_ix \in \{b_i.b'_{i+1}x\} = \{b_i.b_{i+1}\}$, which is again a contradiction.

By symmetry L_j leaves the first condition (10) unaffected, and we can deal similarly with the second condition (10) by considering the formal inverse of (9).

Thus the conditions (10) are unaffected by elementary transformations, so for any reduced expression (9) the length is an invariant of the equivalence class. In particular, two expressions of length 0 are equivalent if and only if they are equal, as elements of S.

We now define $\mathcal{G}(S)$ as a group of permutations on the set of equivalence classes of reduced expressions and verify that S acts faithfully; this will show that S is embedded in $\mathcal{G}(S)$. Given $c \in S$ and a reduced expression p as in (9), we define

$$pc = \begin{cases} a_0b_1^{-1}a_1 \ldots a_{n-1}b_n^{-1}a_nc & \text{if } b_n \notin \{a_nc.a_{n-1}\}, \\ a_0b_1^{-1}a_1 \ldots b_{n-1}^{-1}a'c'' & \text{if } c = c'c'', a_{n-1} = a'a'', b_n = a_nc'a''. \end{cases} \quad (11)$$

Clearly the first form is reduced; when it does not apply, we have $b_n \in \{a_nc.a_{n-1}\}$, but $b_n \notin \{a_n.a_{n-1}\}$, hence by ($\alpha$), $b_n = a_nu$ and so $a_nu \in \{a_nc.a_{n-1}\}$. By the left–right dual of (β), $u \in \{c.a_{n-1}\}$, so $c = c'c'', a_{n-1} = a'a'', u = c'a''$ and $b_n = a_nc'a''$, which is the second alternative in (11). It is

0.7 Rings of fractions

reduced, for if $b_{n-1} \in \{a'c''.a_{n-2}\}$, then since $b_{n-1} \notin \{a_{n-1}.a_{n-2}\}$, we have $b_{n-1} \notin \{a'.a_{n-2}\}$ and so by (α), $b_{n-1} = a'v$, but then $a_{n-1} \in \{a'v.a_nc'a''\}$, which contradicts the condition $a_{n-1} \notin \{b_{n-1}.b_n\}$.

A routine verification shows that this action is compatible with the elementary transformations R_i, L_i and their inverses, so that (11) defines an action on the equivalence classes. Next we define for $c \in S$ and p as in (9),

$$pc^{-1} = \begin{cases} a_0b_1^{-1}a_1 \ldots b_n^{-1}a_nc^{-1}1 & \text{if } c \notin \{1.a_n\} \text{ and } a_n \notin \{b_n.c\}, \\ a_0b_1^{-1}a_1 \ldots b_n^{-1}u & \text{if } a_n = uc, \\ a_0b_1^{-1}a_1 \ldots b_{n-1}^{-1}a_{n-1}(c'b'')^{-1}1 & \text{if } a_n = b'c'', b_n = b'b'', c = c'c''. \end{cases} \quad (12)$$

In the case $n = 0$, the centre line applies, but a_0 is then omitted, i.e. $pc^{-1} = u$. The first form is clearly reduced. If it does not hold, suppose that $c \in \{1.a_n\}$, say $a_n = uc$. By hypothesis, $b_n \notin \{a_n.a_{n-1}\} = \{uc.a_{n-1}\}$, hence $b_n \notin \{u.a_{n-1}\}$ and this shows the second form to be reduced. Finally, if $c \notin \{1.a_n\}$ but $a_n \in \{b_n.c\}$, let $c = c'c'', b_n = b'b'', a_n = b'c''$; then we are in the third case and it will be reduced, provided that $a_{n-1} \notin \{b_{n-1}.c'b''\}$. So suppose that $a_{n-1} \in \{b_{n-1}.c'b''\}$; since $a_{n-1} \notin \{b_{n-1}.b_n\}$, we have $a_{n-1} \in Sb''$ by the left–right dual of (α), say $a_{n-1} = vb''$; but then $b_n \in \{a_n.a_{n-1}\}$, which is a contradiction. Again it is straightforward to show that the action is compatible with L_i, R_i.

To verify that we have a representation, we shall use $\sim$ to indicate equivalence.

(i) $cc^{-1} = 1$. If pc has the first form (11), it is clear that $pcc^{-1} = p$. For the second form we have

$$pcc^{-1} = \begin{cases} a_0b_1^{-1}a_1 \ldots b_{n-1}^{-1}a'c'^{-1}1, \\ \sim a_0b_1^{-1}a_1 \ldots b_{n-1}^{-1}a_{n-1}(c'a'')^{-1}1 \sim p. \end{cases}$$

(ii) $c^{-1}c = 1$. If pc^{-1} has the first form (12), all is clear. If the second applies, we have

$$pc^{-1}c = a_0b_1^{-1}a_1 \ldots b_n^{-1}uc \sim p,$$

and for the third,

$$pc^{-1}c = a_0b_1^{-1} \ldots b_{n-1}^{-1}a_{n-1}(c'b'')^{-1}c$$
$$\sim a_0b_1^{-1} \ldots b_{n-1}^{-1}a_{n-1}b''^{-1}c'' \sim p.$$

(iii) $(pc)d = p(cd)$. If $p(cd)$ has the first form (11), then $b_n \notin \{a_ncd.a_{n-1}\}$, hence $b_n \notin \{a_nc.a_{n-1}\}$, and so

$$(pc)d = a_0b_1^{-1} \ldots b_n^{-1}a_ncd = p(cd).$$

If $p(cd)$ has the second form (11) and $b_{n-1} \notin \{a'c''d.a_{n-2}\}$, then
$$(pc)d = a_0 b_1^{-1} \ldots b_{n-1}^{-1} a'c''d = p(cd).$$

If $b_{n-1} \in \{a'c''d.a_{n-2}\}$, say $a'c''d = e'e''$, $a_{n-2} = f'f''$, $b_{n-1} = e'f''$, then
$$(pc)d = a_0 b_1^{-1} \cdots b_{n-2}^{-1} f'e'' = p(cd).$$

It is clear that $p1 = p$, so we have a representation of S by permutations of the classes of reduced expressions (9). Further, for any $x, y \in S$, if x and y have the same action, then $x = 1.x = 1.y = y$; this shows that S acts faithfully, so S is embedded in $\mathcal{G}(S)$, as claimed. ■

Exercises 0.7

1. In a monoid S, if aba is invertible, show that a and b are both invertible. Show also that it is not enough for ab to be invertible. What is the generalization to n elements?
2. Let R be any ring; show that any $R^\times$-inverting homomorphism into a non-zero ring must be injective.
3. Verify the formula (2) for the kernel of λ.
4. Let R be an integral domain. Show that any Ore subring of R is contained in a maximal Ore subring.
5. Show that a direct limit of Ore domains is again an Ore domain.
6. Let R be a ring and T a left and right Ore set in R. If R is (left or right) Ore, Noetherian or Artinian, show that the same is true of R_T.
7. In any ring show that any left factor of a right large element is again right large. In an integral domain, is the same true of any right factor?
8. Let R be a right Ore domain with right ACC_1 and $\mathfrak{a}$ an ideal of R that is principal as left ideal. If $R/\mathfrak{a}$ is an integral domain, show that it is again a right Ore domain.
9. If R is an ordered ring that is a right Ore domain, show that the ordering can be extended in a unique way to the field of fractions of R.
10*. Let R be an integral domain that is not right Ore and let $n \geq 1$. Show that $^n R$ can be embedded in R as a right ideal, and if $^n R$ does not have unique rank, show that $^n R$ contains a strictly descending chain of direct summands that are free of rank n. Deduce that if an integral domain satisfies right pan-ACC then R has IBN.
11. Let R be a right Ore domain and K its field of fractions. If $A \in R_n$ is right regular in R_n show that it is right regular in K_n and hence invertible, with an inverse of the form Bd^{-1}, $B \in R_n$, $d \in R^\times$. Deduce that every right zero-divisor in R_n is a left zero-divisor. Does the reverse implication hold generally?
12. Let $E \supset F$ be a skew field extension of finite right dimension. Show that in the polynomial ring $E[x]$ the monic polynomials with coefficients in F form a

right Ore set. (*Hint*: In the relation $au' = ua'$, where $a, a' \in E[x]$, $u, u' \in F[x]$, u monic, equate coefficients and eliminate the coefficients of a'.)

13. (L. G. Makar-Limanov) Let S be a cancellation monoid. Given $a, b \in S$, denote by T the submonoid generated by a, b. Show that if $aT \cap bT = \varnothing$, then T is free on a, b. Deduce that a cancellation monoid containing no free submonoid on more than one element can be embedded in a group.

14. Let R be a ring with IBN and S a right Ore set; show that the localization R_S need not have IBN. (*Hint*: Take a ring generated by the entries of rectangular matrices A, B with defining relations $AB = \lambda I, BA = \lambda I$, where λ is another generator, which is central.)

15*. Let R be an Hermite ring and T a right Ore set; show that the localization R_T need not be Hermite. (*Hint*: See Exercise 14; use the completion with respect to the powers of λ.)

16°. Let R be a right hereditary right Ore domain. Can every right ideal be generated by two elements? (This is true in the commutative case, but as we shall see later, false in the non-Ore case.)

17*. (S. Rosset) Let G be a group and A a torsion-free abelian normal subgroup of G. Show that in the group algebra kG (over a field k) the set $(kA)^\times$ is a left and right Ore set consisting of regular elements.

18. Show that the kernel of a derivation acting on a local ring is again a local ring.

19. Prove Leibniz's formula for derivations:

$$(ab)\delta^n = \sum_i \binom{n}{i} (a\delta^i)(b\delta^{n-i}).$$

More generally, if δ is an α-derivation, show that

$$(ab)\delta^n = \sum_i a\delta^i . bf_i^n(\alpha, \delta),$$

where $f_i^n(\alpha, \delta)$ is the coefficient of t^i in the formal expansion of $(t\alpha + \delta)^n$.

20. If δ is a derivation on an integral domain of prime characteristic p, show that δ^p is again a derivation.

21. If δ is a nilpotent derivation of exponent r on an integral domain K (i.e. $\delta^r = 0 \neq \delta^{r-1}$) and $r > 1$, show that K has prime characteristic p and $r = p^t$. (*Hint*: Apply δ^r to ab, where $b\delta \neq 0 = b\delta^2$ and use Leibniz's formula to show that $p|r$; now repeat the argument with δ replaced by δ^p.)

22. Let D be a skew field with centre F and let R be the F-algebra generated by all multiplicative commutators in $D: R = FD'$. Show that R is a (left and right) Ore domain with field of fractions D.

0.8 Modules over Ore domains

Many results on modules over commutative integral domains hold more generally either for right modules or for left modules over right Ore domains. For convenience we shall deal with left modules over left or right Ore domains in this section and leave the reader to make the necessary modifications.

Let R be an integral domain and M a left R-module. An element $x \in M$ is called a *torsion element* if $ax = 0$ for some $a \in R^\times$. When R is a left Ore domain, the set M_t of all torsion elements of M is easily seen to be a submodule. If $M_t = 0$, we say that M is *torsion-free*; if $M_t = M$, one calls M a *torsion module*. This definition is the customary one, at least over Ore domains; later, in Sections 3.2 and 5.3 we shall use this term in a different sense, so we shall reserve the term for later and call M_t the submodule of torsion elements. It is clear that for any module M over an Ore domain, M_t is a module of torsion elements and M/M_t is torsion-free; moreover, these two properties serve to determine M_t.

Let R be a ring and T a left Ore set in R; then the localization R_T may be expressed as a direct limit

$$R_T = \varinjlim \{t^{-1}R \,|\, t \in T\}.$$

For, given $t_1, t_2 \in T$, there exists $t \in Tt_1 \cap Rt_2$ and so $t_1^{-1}R \cup t_2^{-1}R \subseteq t^{-1}R$. This process can be applied to modules as well as rings; for simplicity we state the result only for Ore domains, the case of principal interest:

Proposition 0.8.1. *Let R be a left Ore domain, K its field of fractions and M a left R-module. Then $K \otimes_R M$ can be described as the set of all formal products $b^{-1}x (x \in M, b \in R^\times)$ subject to the relations: $b^{-1}x = b'^{-1}x'$ if and only if there exist $u, v \in R^\times$ such that $ux = vx'$, $ub = vb'$. Moreover, the kernel of the canonical map*

$$M \to K \otimes M \qquad (1)$$

is M_t, so (1) is an embedding if and only if M is torsion-free.

Proof. Any element of $K \otimes M$ has the form $x = \sum b_i^{-1} a_i \otimes x_i$. If b is a common left multiple for the $b_i : c_i b_i = b$, then

$$x = \sum b^{-1} c_i a_i \otimes x_i = b^{-1} \left(\sum c_i a_i x_i \right).$$

Thus every element of $K \otimes M$ has the form $b^{-1}x$, $x \in M$, $b \in R^\times$. Given $p = b^{-1}x$ and $p' = b'^{-1}x'$, there exist $u, v \in R^\times$ such that $ub = vb' = c$, and we have $cp = ux$, $cp' = vx'$. If $p = p'$, then $cp = cp'$, i.e. $ux = vx'$ in M; conversely, if $cp = cp'$, then $ux = vx'$ and so $p = c^{-1}.cp = c^{-1}.cp' = p'$. Now it follows that $b^{-1}x = 0$ if and only if $ux = 0$ for some $u \in R^\times$, i.e. precisely if $x \in M_t$. Hence the kernel of (1) is M_t and the rest is clear. ∎

For a right R-module there is no such convenient description, but in that case there are two ways of describing the linear functionals on M, using the dual $M^* = \mathrm{Hom}_R(M, R)$.

0.8 Modules over Ore domains

Proposition 0.8.2. *Let R be a right Ore domain with field of fractions K and let M be a finitely generated left R-module. Then there is a natural isomorphism of right K-modules*

$$M^* \otimes_R K \cong \operatorname{Hom}_K(K \otimes M, K). \qquad (2)$$

Proof. By adjoint associativity, applied to $({}_K K_R, {}_R M, {}_K K)$, we have

$$\operatorname{Hom}_K(K \otimes_R M, K) \cong \operatorname{Hom}_R(M, K)$$
$$\cong \operatorname{Hom}_R(M, \varinjlim Rb^{-1}).$$

Since M is finitely generated, we can find a common denominator for the images of elements of M, so we can replace $\varinjlim Rb^{-1}$ by Rb^{-1} for any given homomorphism; thus we have

$$\operatorname{Hom}_R(M, \varinjlim Rb^{-1}) \cong \varinjlim \operatorname{Hom}_R(M, R)b^{-1} \cong M^* \otimes_R K.$$

■

We have seen that the field of fractions K of a left Ore domain R has the form $K = \varinjlim b^{-1}R$; here each $b^{-1}R$ is a free right R-module. Let us call a module *semifree*[*] if every finite subset is contained in a finitely generated free submodule. Then we can say that K is semifree as right R-module, hence flat, therefore, if a family of elements in a left R-module M is linearly independent, then so is its image in $K \otimes M$. Hence the dimension of $K \otimes M$ as a vector space over K equals the cardinality of a maximal linearly independent subset of M. This number is an invariant of M, which we shall call the *rank* of M and denote by rk M. In particular, rk $M = 0$ precisely when M consists of torsion elements. On free modules the rank clearly agrees with our previous definition of rank, and since tensoring preserves exactness, we have

Proposition 0.8.3. *Let R be a left Ore domain. If $0 \to M' \to M \to M'' \to 0$ is an exact sequence of left R-modules, then*

$$\operatorname{rk} M = \operatorname{rk} M' + \operatorname{rk} M''.$$

In particular, if N is a submodule or a homomorphic image of M, then $rk N \leq rk M$. ■

The last assertion, relating to homomorphic images, holds (under an appropriate definition of rank) for a large class of rings, including all that can be

---

[*] This is sometimes called 'locally free', but we shall avoid that term, as it has quite a different meaning in commutative algebra.

embedded in fields, and hence most of the rings considered later. However, apart from this, none of the other assertions holds with 'right' in place of 'left' Ore domain. Thus let R be any right Ore domain (or indeed, any integral domain) that is not left Ore and let $x, y \in R^\times$ be such that $Rx \cap Ry = 0$; then R contains the left ideal $Rx + Ry$, which is isomorphic to R^2; this shows that the first part of Proposition 8.3 cannot be extended to such rings. For an example showing that K need not be semifree as right R-module, take x and y as before; then the submodule $x^{-1}R + y^{-1}R$ of the right R-module K contains $a, b \neq 0$ such that $ax = by$ (namely x^{-1} and y^{-1}), but such elements do not exist in R and hence do not exist in general domains; however, it remains true for right Bezout domains (see Proposition 2.3.19).

The following property of right Ore domains is not in general shared by left Ore domains (see Exercise 3).

Proposition 0.8.4. *Let R be a right Ore domain and K its field of fractions. Then any left K-module, considered as left R-module, is semifree, and in particular, torsion-free.*

Proof. Let M be a finitely generated R-submodule of a left K-module, which may without loss of generality be taken to be K^n, for some n. We can choose a common right denominator $c \in R^\times$ for the components of the finite generating set of M. Then $M \subseteq R^n.c^{-1}$ and the latter is a free R-module. ∎

By combining this result with Proposition 8.1, we obtain

Corollary 0.8.5. *If R is a left and right Ore domain, then every finitely generated torsion-free R-module is embeddable in a free R-module.* ∎

Finally we note that the flatness of the ring of fractions, well known in the commutative case, continues to hold in the Ore case.

Proposition 0.8.6. *Let R be a ring and T a right Ore set in R. Then R_T is left R-flat. If R is any integral domain, then $R_{R^\times}$ is non-zero and left R-flat if and only if R is a right Ore domain.*

Proof. We have $R_T = \varinjlim Rc^{-1} (c \in T)$, therefore R_T is a direct limit of free left R-modules Rc^{-1} and hence is flat, in particular, $K = R_{R^\times}$ is so when R is right Ore.

Conversely, if $R_{R^\times}$ is left R-flat and non-zero, take $a, b \in R^\times$; then $a.a^{-1} - b.b^{-1} = 0$, hence there exist $u_i \in K$, $p_i, q_i \in R$ such that $a^{-1} = \sum p_i u_i$, $b^{-1} = \sum q_i u_i$, $ap_i - bq_i = 0$. Not all the p_i, q_i can vanish, say $p_1, q_1 \neq 0$; then $ap_1 = bq_1$ is the desired right multiple. ∎

The last part of this proposition shows in effect that if R is a right Ore domain with field of fractions K, then K is left R-flat but not right R-flat unless R is also left Ore.

Exercises 0.8

1. Let K be a field and E a commutative field with a subring A isomorphic to a subring of K. Show that $K \otimes_A E$ is an Ore domain, provided that it is an integral domain. (*Hint*: Note that every element of $K \otimes_A E$ has a right multiple of the form $a \otimes 1$, where $a \in K$.)
2. Let F be a commutative field, E an algebraic commutative field extension and A an F-algebra that is a right Ore domain with field of fractions K. If $A \otimes_F E$ is an integral domain, show that it is a right Ore domain with field of fractions $K \otimes_F E$.
3. Let R be a left but not right Ore domain and K its field of fractions. Show that K, as left R-module, has rank 1 but is not semifree (see Exercise 5.1.7).
4. (Gentile [60]) Let R be a subring of a field. If every finitely generated torsion-free left R-module can be embedded in a free left R-module, show that R is right Ore. Note that this is a converse to Corollary 8.5. Investigate the truth of other possible converses.
5°. Does Exercise 4 remain true when R is merely assumed to be an integral domain, not necessarily contained in a skew field?
6. Show that a projective left ideal $\mathfrak{a}$ of a left Ore domain is finitely generated. (*Hint*: Use a projective coordinate system to show that $\mathfrak{a}$ is invertible or 0, see BA, Proposition 10.5.1.) Deduce that every projective left R-module that is uniform (i.e. any two non-zero submodules have a non-zero intersection) is finitely generated.
7*. Let R be a right Ore domain, K its field of fractions and $\mathfrak{a}$ any non-zero right ideal of R. Show that $\mathfrak{a} \otimes_R K \cong K$ (as right R-modules). Show that in $K \otimes_R K$ any element $s \in K$ satisfies $1 \otimes s = s \otimes 1$ and deduce that $K \otimes_R K \cong K$. [This is equivalent to the assertion that the embedding $R \to K$ is an epimorphism in the category of rings (see Theorem 7.2.1); this equivalence actually holds for any ring R with a homomorphism to a field K (Corollary 7.2.2).]
8. (Bergman [67]) Let R be a right Ore domain and K its field of fractions. Prove that the following conditions on a finitely generated left R-module M are equivalent:
    (a) the canonical map $M \to K \otimes M$ is an embedding,
    (b) M is embeddable in a K-module (*qua* left R-module),
    (c) M is embeddable in a free left R-module,
    (d) $\text{Hom}_R(M, R)$ distinguishes elements of M,
    (e) $\text{Hom}_R(M, K)$ distinguishes elements of M.
9°. Find the relations between (a)–(e) of Exercise 8 when (i) K is a field and R a subring generating K as a field, (ii) K is any ring and R a subring. Find conditions on the finitely generated R-module M for (a)–(e) to be equivalent.
10. Show that for any finitely generated left R-module M over a (left and right) Ore domain R with field of fractions K, $K \otimes M \cong K \otimes M^{**}$.

0.9 Factorization in commutative integral domains

As is well known, a commutative integral domain is called a *unique factorization domain* (UFD for short) if every element not zero or a unit can be expressed as a product of atoms and any such expression is unique except for the order of the factors and up to associates. This definition makes it clear that unique factorization is a property of the multiplicative monoid of the ring, even though other aspects of the ring are usually needed to establish it. We therefore restate the definition in terms of monoids.

Any commutative monoid S has a preordering by divisibility:

$$a|b \text{ if and only if } b = ac \text{ for some } c \in S. \tag{1}$$

If $a|b$ and $b|a$ in a cancellation monoid S, then

$$a = bu \text{ for some unit } u \in S, \tag{2}$$

i.e. a and b are *associated*. We recall that a monoid is called *conical* if $xy = 1$ implies $x = y = 1$; for a cancellation monoid this just means that 1 is the only unit. Clearly (1) is a partial ordering of S precisely when S is conical. With every commutative cancellation monoid S, having a group of units U, we can associate a conical monoid S/U whose elements are the classes of associated elements of S. Since the relation (2) between a and b clearly defines a congruence on S, the set of these classes forms a monoid in a natural way.

A commutative cancellation monoid S will be called a *UF-monoid* if the associated conical quotient monoid S/U is free commutative. With this definition it is clear that a commutative ring R is a UFD if and only if $R^\times$ forms a UF-monoid under multiplication. In studying unique factorization in commutative rings we can therefore limit ourselves to UF-monoids.

To state the conditions for unique factorization in monoids succinctly, let us define a *prime* in a commutative monoid S as an element p of S that is a non-unit and such that

for any $a, b \in S$, $p|ab$ implies $p|a$ or $p|b$.

Clearly any associate of a prime is again a prime. Further, a prime in a cancellation monoid is necessarily an atom. For any prime p is a non-unit and if $p = ab$, then $p|a$ or $p|b$, say $a = pq$; hence $p = ab = pqb$ and by cancelling p we have $qb = 1$, so b is a unit. The converse is false: an atom need not be prime, e.g. consider the monoid generated by a, b with the defining relation $a^2 = b^2$; here a is an atom but not prime. In fact for a commutative cancellation monoid, the converse, together with a finiteness condition, is easily seen to ensure that we have a UF-monoid.

0.9 Factorization in commutative integral domains

For later applications it is useful to consider a slightly more general case. An element c of a monoid S (not necessarily commutative) is said to be *right invariant*, if c is regular and for any $x \in S$ there exists $x' \in S$ such that $xc = cx'$. Since c is regular, x' is uniquely determined by x. *Left invariant* elements are defined similarly and c is called *invariant* if it is left and right invariant, i.e. c is regular and

$$cS = Sc.$$

Lemma 0.9.1. *In any monoid S the set $\mathrm{Inv}(S)$ of all invariant elements is a submonoid containing all units of S. More generally, if two elements of the equation*

$$c = ab$$

are invariant, then so is the third.

Proof. Clearly every unit of S is invariant. If $c = ab$, where a, b are invariant and $cx = cy$, then $abx = aby$, hence $bx = by$ and so $x = y$, which shows c to be right regular; left regularity follows similarly. Further, $cS = abS = aSb = Sab = Sc$, hence c is invariant. Suppose now that a and c are invariant. If $bx = by$, then $cx = abx = aby = cy$, and it follows that $x = y$, therefore b is right regular. Suppose next that $xb = yb$ and let $ax = x_1 a, ay = y_1 a$; then $x_1 c = x_1 ab = axb = ayb = y_1 ab = y_1 c$; hence $x_1 = y_1$, so $ax = x_1 a = y_1 a = ay$, and hence $x = y$. This shows b to be left regular. Now $abS = cS = Sc = Sab = aSb$, and it follows that $bS = Sb$, so b is invariant. Similarly, when b and c are invariant, then so is a. ∎

If every element of S is invariant, we say that S is *invariant*. Since every element is then regular, an invariant monoid always has cancellation. Moreover, in any invariant monoid S, $xy = 1$ implies $yx = 1$. For if $xy = 1$, then $y'x = 1$ for some $y' \in S$, hence $y' = y'xy = y$, so y is a two-sided inverse of x, as claimed. Invariant monoids clearly include all commutative cancellation monoids and they share with the latter the property that right associates are the same as left associates; more generally, the preordering by left divisibility (which is defined in any monoid) and that by right divisibility coincide. For if $a = bc$, then also $a = cb'$ and $a = c'b$ for some $b', c' \in S$; thus the relation $a|b$ is unambiguous in an invariant monoid. Further we can define primes as in commutative monoids and we can again associate a conical monoid S/U with S, whose elements are the classes of associated elements. An invariant monoid S with group of units U will be called a *UF-monoid* if its associated conical quotient monoid S/U is free commutative. This clearly generalizes the previous definition.

Let S be a commutative cancellation monoid; for any finite family (a_i) we define the *highest common factor* (HCF for short; also called GCD = greatest common divisor) as an element d such that $d|a_i$ for all i and any d' with the same property divides d. Similarly the least common multiple (LCM for short) is defined as an element m such that $a_i|m$ for all i and any m' with the same property is a multiple of m. We note that the HCF and LCM are each defined only up to a unit factor but they are unique elements of S/U. Since left and right divisibility in an invariant monoid coincide, it is clear that the notions of HCF and LCM can also be defined in that case. By contrast, in a general monoid (or ring) we need to speak of highest common left (or right) factor and least common right (or left) multiple, a case that will be considered later (in Section 2.8).

The relation between HCF and LCM is elucidated in

Proposition 0.9.2. *In any invariant monoid S two elements a and b have an HCF whenever they have an LCM, and the HCF d and LCM m are then related (in a localization of S) by the equations*

$$m = bd^{-1}a, \quad d = am^{-1}b. \tag{3}$$

Moreover, if in a commutative integral domain a and b have an HCF of the form $d = au + bv$, then they have an LCM m and (3) holds.

Proof. Suppose that a, b have an LCM m. Then $a|m$, hence $ba|bm$, so $bam^{-1}|b$. Thus $b = bam^{-1}c$ for some c; by cancellation $am^{-1}c = 1$, hence $cam^{-1} = 1$, and so $cam^{-1}b = b$, i.e. $am^{-1}b|b$. By symmetry, $am^{-1}b|a$ and it follows that $am^{-1}b$ is a common factor of a and b. Now suppose that $u|a, u|b$; then $bu^{-1}a$ is divisible by a and b, hence also by m, so we have $bu^{-1}a = fm$ for some f. Thus $bu^{-1}am^{-1}b = fb = bf'$ for some f', and so $am^{-1}b = uf'$. This shows that $u|am^{-1}b$ and it shows $am^{-1}b$ to be an HCF of a, b. Writing $am^{-1}b = d$, we clearly have $m = bd^{-1}a$ and (3) holds.

Suppose now that a, b in a commutative integral domain have an HCF d such that $d = au + bv$ and put $m = bd^{-1}a$. Clearly $a, b|m$; if $n \in S$ is such that $a, b|n$, say $n = rb = sa$, then $b|nu + sbv = s(au + bv) = sd$, and so $bd^{-1}a|sa = n$. It follows that $m = bd^{-1}a$ is indeed the LCM of a and b and (3) is satisfied. ∎

The relation $d = au + bv$ is known as *Bezout's relation*. As Exercise 6 shows, without it the LCM may not exist.

For the study of factorizations ACC_1 is particularly important. Thus let S be a cancellation monoid with left and right ACC_1 and take any $c \in S$. Then $cS \neq S$ if and only if c is a non-unit and in that case, by right ACC_1, there is a maximal principal right ideal p_1S such that $cS \subseteq p_1S \subset S$. This means that

$c = p_1c_1$ and p_1 is an atom. Repeating the procedure on c_1 we see that unless it is a unit, we can write $c_1 = p_2c_2$, where p_2 is an atom. Continuing in this fashion, we get a strictly ascending sequence of principal left ideals

$$Sc \subset Sc_1 \subset Sc_2 \subset \ldots,$$

which must terminate by left ACC_1. It follows that every non-zero element of S is either a unit or a product of atoms. A cancellation monoid (or an integral domain) with this property is said to be *atomic*, and what we have proved can be stated as

Proposition 0.9.3. *Any cancellation monoid, in particular any integral domain, with left and right ACC_1 is atomic.* ∎

It is clear that in an invariant monoid left and right ACC_1 coincide, and as we have just seen, such a monoid is atomic, but the converse is not generally true, even for commutative integral domains (see Grams [74], Zaks [82] and Exercise 9 below). We have

Theorem 0.9.4. *In any invariant monoid S the following conditions are equivalent:*

(a) *S is a UF-monoid, i.e. S/U is free commutative, where U is the group of units,*
(b) *S satisfies ACC_1 and any two elements have an HCF,*
(c) *S satisfies ACC_1 and any two elements have an LCM,*
(d) *S satisfies ACC_1 and the intersection of any two principal ideals is principal,*
(e) *S is atomic and every atom of S is prime.*

Here the assertion obtained by replacing the intersection in (d) by the union (or even by the sum) is not equivalent to the others.

Proof. None of the conditions is affected if we pass to the associated conical quotient monoid $T = S/U$, and 1 is the only unit in T. It is clear that a free commutative monoid is conical and satisfies (b), so (a) ⇒ (b). To prove (b) ⇒ (c), assume (b). Given $a, b \in T$, there is a common multiple, namely ab. Let m be a common multiple of a, b for which mS is maximal; if m' is another common multiple of a, b, we claim that $m|m'$; for otherwise the HCF, d say, of m and m' is again a common multiple and has the property that $mT \subset dT$. Thus m is in fact the *least* common multiple, as claimed.

It is clear that (c) ⇔ (d), because two elements a, b have an LCM m if and only if $aT \cap bT = mT$. To prove (c) ⇒ (e), let $a, b \in T$ and let p be an atom such that $p|ab$. Denote the LCM of p and a by m; then $m = ap_1 = pa_1$ say, and since ap is a common multiple, we have $ap = md = ap_1d$, say. Thus $p = p_1d$, but p is an atom, so either (i) $d = 1$ and $p = p_1$ or (ii) $p_1 = 1$ and $d = p$ (because 1 is the only unit). Case (i): $m = ap$ is an LCM; since $p|ab$, $a|ab$, we have $ab = me = ape$ for some $e \in T$, therefore $b = pe$, $p|b$. Case (ii): $m = a$ is an LCM, hence $p|a$. Thus p is prime and T is atomic, by ACC$_1$, so (e) holds.

Finally, to prove (e) $\Rightarrow$ (a), take distinct atoms a, b and let $ab = ba_1$, say. Then $a|ba_1$ but a does not divide b. Since a is prime by (e), it follows that $a|a_1$, say $a_1 = au$. Thus $ab = bau$; by symmetry $ba = abv = bauv$, hence $uv = 1$ and so $u = v = 1$ and $ab = ba$. This shows that the monoid generated by the atoms is commutative, but this is the whole of T, since T is atomic. This also shows $T = S/U$ to be commutative. Now the uniqueness proof follows a well-known pattern. If

$$c = p_1^{\alpha_1} \cdots p_n^{\alpha_n} = p_1^{\beta_1} \cdots p_n^{\beta_n}, \tag{4}$$

where the p_i are distinct atoms and $0 \leq \alpha_1 < \beta_1$ say, then p_1 divides c but not $p_2^{\alpha_2} \cdots p_n^{\alpha_n}$, hence $p_1|p_1^{\alpha_1}$, therefore $\alpha_1 > 0$. So we can cancel p_1 in (4) and use induction on $\sum \alpha_i$ to obtain (a). ∎

This result shows in particular that in a UF-monoid the number of prime factors in a complete factorization of an element c is the same for all factorizations; it will be denoted by $l(c)$ and called the *length* of c.

Frequently one needs to know the effect of localizing on unique factorization. Again we begin by setting out the problem in terms of monoids. If S is an invariant monoid, then any submonoid T is a right Ore set and we can form the localization S_T. The following criterion for S to be a UF-monoid generalizes a theorem of Nagata [57].

Theorem 0.9.5. *Let S be an invariant monoid and T a submonoid generated by certain primes of S. Further, assume that S is atomic and that the localization S_T is a UF-monoid. Then S is itself a UF-monoid.*

Proof. By Theorem 9.4 we need only verify that every atom of S is prime. Denote the canonical map $S \to S_T$ by $x \mapsto x'$ and let p be an atom of S. We claim that p' is an atom or a unit; for if p' is a product of non-units, say, then $p' = af^{-1}bg^{-1}$, where a, b are non-units and $f, g \in T$ are products of primes. Each such prime divides either a or b and cancelling them in turn, we find that $p = a_1b_1$, where a_1 or b_1 must be a unit and it follows that p' is an atom or a unit. We treat these two cases separately.

(i) p' is an atom and hence a prime in S_T. If $p|ab$, say $pc = ab$, then $p'c' = a'b'$, hence $p'|a'$ or $p'|b'$, say the former. This means that

$$pe = ar, \quad \text{where } e \in S, r \in T.$$

No prime factor of r can be an associate of p, for otherwise p' would be a unit. Hence the prime factors of r divide e, and cancelling them one by one, we obtain an equation $pe_1 = a$, so $p|a$, which shows p to be a prime.

(ii) p' is a unit in S_T. Then $pq = r$, where $r \in T$ and $q \in S$. Again we can cancel the prime factors of r one by one, but this time we find that one of them is associated to p, for otherwise p would be a unit. Hence p is a prime. ∎

Applied to rings, Theorem 9.5 yields

Theorem 0.9.6. *In any ring R let T be a submonoid of $Inv(R)$, the monoid of all invariant elements in R. Further, assume that $Inv(R)$ is generated by certain elements that are primes in $Inv(R)$. If $Inv(R)$ is atomic and its image in the localization R_T is a UF-monoid, then $Inv(R)$ is itself a UF-monoid.* ∎

Taking R to be a commutative integral domain, we find that $Inv(R) = R^\times$ and we obtain the following slight generalization of Nagata's theorem:

Corollary 0.9.7. *Let R be a commutative atomic integral domain. If T is a submonoid of $R^\times$ generated by certain primes and the localization R_T is a UFD, then R is a UFD.* ∎

Exercises 0.9

1. Show that every Noetherian integral domain is atomic.
2. Let k be a commutative field and $R = k[x, y, z, t]$, $\mathfrak{a} = (xt - yz)R$. Show that the ring $R/\mathfrak{a}$ (the coordinate ring of a quadric) is an atomic integral domain, but not a UFD.
3. If an invariant monoid S satisfies ACC_1 and the join of any two principal ideals is principal, show that S is a UF-monoid. Show that the converse is false, by considering the multiplicative monoid of a suitable UFD.
4. If S is an invariant monoid and Q a normal submonoid, i.e. $aQ = Qa$ for all $a \in S$, define the quotient monoid S/Q and show that it is again invariant.
5. Show that any invariant element in a simple ring is a unit.
6. Let R be the subring of $\mathbb{Z}[x]$ consisting of all polynomials with even coefficient of x. Show that two elements of R may have an HCF without having an LCM.
7*. (Novikov [84]) Let S be an invariant monoid generated by two elements. Show that the associated conical monoid S/U is commutative. What happens for more than two generators?
8. Let S be a monoid (in multiplicative notation) and let kS be its monoid algebra (over a commutative field k). Show that kS is atomic if and only if S is. Likewise for ACC_1.
9*. (G. M. Bergman) Let $\alpha_1, \alpha_2, \ldots$ be a sequence of real numbers, linearly independent over $\mathbb{Q}$, such that $0 < \alpha_n < 1/n$ and denote by S the submonoid of the additive group of real numbers generated by all elements $\alpha_n, 1/n - \alpha_n (n = 1, 2, \ldots)$. Verify that S contains all positive rational numbers and so does not satisfy ACC_1.

Show further that for any $s \in S$, if $s \neq 0$, then for some $n \geq 1$, either $s \geq \alpha_n$ or $s \geq 1/n - \alpha_n$. Deduce that S is atomic.

10°. Is every invariant conical monoid necessarily commutative?

Notes and comments on Chapter 0

Much of this material is part of the folklore and the citations given below are probably far from exhaustive. The first thorough discussion of IBN (treated in Section 0.1) occurs in Leavitt [57]; Shepherdson [51] gives an example of a ring that is weakly 1- but not 2-finite. For a connected account of IBN, UGN and weak finiteness (showing that these classes are distinct), see Cohn [66a]. Proposition 1.2 is proved for regular rings by Goodearl [79], Theorem 18.3, and generally by Malcolmson [80a]. Proposition 1.3 is taken from Cohn and Schofield [82], where Lemma 3.3 is also proved. The notion of inner rank was defined in Bergman [67], but goes back much further; almost any pre-1914 book on matrix theory defines the rank of a matrix A as the least number of terms in the expression of A as a sum of *dyads*, i.e. products of a column by a row, which is a matrix of inner rank 1. Stephenson [66] has shown that the injective hull of a non-Ore domain is a ring Q satisfying $Q^{\mathbb{N}} \cong Q$; see also O'Neill [91].

The matrix theory of Section 0.2 is fairly standard and occurred in FR.0.1, but is not always stated in this explicit form. The matrix reduction functor was first used in Bergman [74b] and Cohn [72c, 79]; for a fuller account see SF, 1.7. The projective module group $K_0(R)$, taken from K-theory, contains much of the 'stabilized' information about the category $_R$proj, though for best results one needs to take the affine structure into account; see an instructive study by Goodearl and Warfield [81]. The (unstabilized) facts about $S(R)$ are well known, e.g. Theorem 3.7 generalizes a result of Bass [68], p. 90. Lemma 3.6 is due to Bergman, who also helped with Theorems 3.9, 3.10.

The term Hermite-ring (or H-ring) was introduced by Lissner [65] (the term was used earlier by Kaplansky [49] for a special type of Bezout ring). The symmetry of the condition occurs repeatedly, e.g. in Drogomizhska [70] and Kazimirskii and Lunik [72]; see also Lam [78] for the commutative case. Theorem 4.1 occurs in Lam [78]; the present proof is based on Cohn [2000a], while Corollary 4.2 is new. Proposition 4.3 is due to Ojanguren and Sridharan [71]. Proposition 4.4 is essentially Proposition 5.6.2 of FR.2 and goes back to Cohn and Schofield [82]. The stability properties (Proposition 4.5, Corollary 4.6) were new in FR.2, while the notion 'projective-free' is defined (as 'p-free') in Cohn [66c], where Theorem 4.8 is also proved.

The notion of stable association and its connexion with the matrix of definition of a module is implicit in Fitting's work [36]; in the form given in Section 0.5 it is developed from the factorization theory in Cohn [63a], see also Cohn [82a]. Lemma 5.8 was suggested by Bergman in connexion with Proposition 7.2.6 (see also Exercises 3 and 2.7.9 below). The concepts of idealizer and eigenring seem to have been first used by Ore [32] (though also implicit in Levitzki's work) in his papers on the formal theory of differential equations. Proposition 6.1 (for the case $P = R$) is due to Fitting [35], and special cases of Theorem 6.3 are well known (see e.g. Amitsur [58]); they are stated in this form by Cohn [70a].

Since Ore's original construction (Corollary 7.2; see Ore [31]) there have been innumerable papers dealing with extensions, analogues for monoids, etc. We have here concentrated on the cases used later in the book. For a comprehensive survey see Elizarov [69]. Corollary 7.7 is due to Goldie [58]; this is an easy special case of his main theorem (the customary proof actually gives Proposition 7.6). Theorem 7.9 is due to Doss [48], who proves it by applying the Malcev conditions (itself a non-trivial verification). The present proof, presented in FR.2, corrects an error in FR.1 (pointed out by L. A. Bokut). The discussion in Section 0.8 is based on Bergman [67]; Proposition 8.2 occurs in Cohn and Schofield [82] and Corollary 8.5 in Gentile [60].

The results of Section 0.9 are for the most part well known, though their formulation for invariant (rather than commutative cancellation) monoids was new in Section 3.1 of FR.2. Theorem 9.5 was proved by Nagata [57] for commutative Noetherian domains.

1
Principal ideal domains

Since the main classes of rings considered in this work generalize principal ideal domains, it seems reasonable to start by recalling the properties of the latter. We begin in Section 1.1 by looking at examples that will be important to us later, the skew polynomial rings, and in Section 1.2 discuss the division algorithm, which forms a paradigm for later concepts. Sections 1.3 and 1.4 recall well-known properties of principal ideal domains and their modules, while Section 1.5 describes the Malcev–Neumann construction of the ordered series field of an ordered group, and the Bergman conjugacy theorem. The concluding Section 1.6 deals with Jategaonkar's iterated skew polynomial rings, leading to one-sided PIDs with a transfinite-valued division algorithm. The later parts of Sections 1.5 and 1.6 are not essential for an understanding of the rest and so may be omitted on a first reading.

1.1 Skew polynomial rings

Polynomial rings are familiar to the reader as the rings obtained from commutative rings by adjoining one or more indeterminates. Here we want to discuss a generalization that is often useful in providing examples and counter-examples. It differs from the usual polynomial ring $k[x]$ in one indeterminate x over a field k in that k need not be commutative, nor commute with x. However, it resembles the classical case in that every element is unique of the form

$$f = a_0 + xa_1 + \cdots + x^n a_n, \tag{1}$$

where the a_i lie in the ground ring (which need not be a field). Thus let R be any non-zero ring and consider a ring containing R as subring, as well as an element x such that every element f of the subring A generated by R and x is uniquely expressible in the form (1). If $a_n \neq 0$, we define the *degree* of

1.1 Skew polynomial rings

f as $d(f) = n$, as in the commutative case. This function d has the following properties:

D.1. For $a \in A^\times$, $d(a) \in \mathbb{N}$, while $d(0) = -\infty$,
D.2. $d(a - b) \leq \max\{d(a), d(b)\}$.

We shall assume further, that it satisfies

D.3. $d(ab) = d(a) + d(b)$.

A function d on a non-zero ring A will be called a *degree-function* if it satisfies D.1–D.3; it will be called *trivial* if it is 0 on all of $A^\times$.

For any degree-function we have, by D.3, $d(1) = 0$. By D.2 we have, as for valuations, $d(a) = d(-a)$ and

$$d(a + b) \leq \max\{d(a), d(b)\}, \qquad (2)$$

with equality holding whenever $d(a) \neq d(b)$ ('every triangle is isosceles'). We note that in our case the elements of degree zero are just the non-zero elements of R, showing that R is an integral domain. More generally, as a consequence of D.3, the set $A^\times$ is closed under multiplication and contains 1, so A is also an integral domain. Further, for any $a \in R^\times$, the product ax has degree 1 and so is of the form

$$ax = xa^\alpha + a^\delta \quad (a \in R^\times). \qquad (3)$$

In the first place we note that a^α, a^δ, are uniquely determined by a and moreover α is injective; by comparing the expressions for $(a+b)x$ and $(ab)x$ we see that α is an endomorphism, while δ is such that

$$(a + b)^\delta = a^\delta + b^\delta, \quad (ab)^\delta = a^\delta b^\alpha + ab^\delta. \qquad (4)$$

Thus δ is an α-*derivation*, as defined in Section 0.7. We observe that the additive structure of A is determined by (1), while the multiplicative structure follows from (3): by the distributive law it is enough to know $x^m a.x^n b$, and by (3) we have

$$x^m a.x^n b = x^{m+1} a^\alpha x^{n-1} b + x^m a^\delta x^{n-1} b,$$

which allows us to compute $x^m a.x^n b$ in all cases, by induction on n. Thus A is completely fixed when R, α, δ are given. We shall write

$$A = R[x; \alpha, \delta], \qquad (5)$$

and call A the *skew polynomial ring* in x over R determined by α, δ. Instead of $R[x; \alpha, 0]$ one also writes $R[x; a]$ and $R[x; 1, 0]$ is just $R[x]$, the usual polynomial ring in a central indeterminate x.

It remains to show that for any integral domain R with an injective endomorphism α and an α-derivation δ, there always exists a skew polynomial ring $R[x; \alpha, \delta]$. As in the commutative case we define it by its action on infinite sequences. Consider the direct power $M = R^{\mathbb{N}}$ as right R-module and define an additive group endomorphism of M by the rule

$$x : (a_i) \mapsto \left(a_i^\delta + a_{i-1}^\alpha\right) \quad (a_{-1} = 0). \tag{6}$$

Clearly right multiplication of M by an element of $R^\times$ is injective, so we may identify R with its image in End(M). Now the action of the endomorphism x defined by (6) satisfies the rule

$$\begin{aligned}(c_i)ax &= (c_i a)x = ((c_i a)^\delta + (c_{i-1} a)^\alpha) \\ &= \left(c_i^\delta a^\alpha + c_i a^\delta + c_{i-1}^\alpha a^\alpha\right) \\ &= \left(c_i^\delta + c_{i-1}^\alpha\right) a^\alpha + (c_i) a^\delta \\ &= (c_i)(x.a^\alpha + a^\delta).\end{aligned}$$

This proves that (3) holds and, moreover, that every element of the subring of End(M) generated by R and x can be brought to the form (1). This form is unique, because when $f = a_0 + xa_1 + \cdots + x^n a_n$ is applied to $(1, 0, 0, \ldots)$, it produces $(a_0, a_1, \ldots, a_n, 0, \ldots)$. Further, the function $d(f)$, defined as the subscript of the highest non-zero coefficient a_i, is easily seen to satisfy D.1–D.3, using the fact that R is an integral domain and α is injective. So the polynomial ring is again an integral domain; in all we have proved

Theorem 1.1.1. *Let R be an integral domain with an injective endomorphism α and an α-derivation δ. Then there is a skew polynomial ring $R[x; \alpha, \delta]$ that is an integral domain, and every skew polynomial ring arises in this way.* ∎

The result is not left–right symmetric because in (3) the coefficients were written on the right. One therefore sometimes introduces the *left* skew polynomial ring, in which the coefficients are on the left and the commutation rule instead of (3) is

$$xa = a^\alpha x + a^\delta. \tag{7}$$

In general $R[x; \alpha, \delta]$ will not be a left skew polynomial ring, but when α is an automorphism of R, with inverse β say, then on replacing a by a^β in (3) and rearranging the terms, we obtain

$$xa = a^\beta x - a^{\beta\delta}. \tag{8}$$

Thus we have

Proposition 1.1.2. *The ring $R[x; \alpha, \delta]$ is a left skew polynomial ring provided that α is an automorphism.* ∎

We have an analogue of the Hilbert basis theorem for skew polynomial rings, with a similar proof, but we shall need to restrict α to be an automorphism, as Exercise 3 below shows.

Proposition 1.1.3. *Let R be a right Noetherian domain, α an automorphism and δ an α-derivation of R. Then the skew polynomial ring $A = R[x; \alpha, \delta]$ is again right Noetherian.*

Proof. This is essentially as in the commutative case. Let us consider A as a left skew polynomial ring, i.e. with coefficients on the left, as we may, by Proposition 1.2, since α is an automorphism. If $\mathfrak{a}$ is a right ideal of A, let $\mathfrak{a}_0 = \sum_1^k c_i R$ be the ideal of its leading coefficients and for $i = 1, \ldots, k$ take a polynomial f_i in $\mathfrak{a}$ with leading coefficient c_i. If $n = \max \{d(f_i)\}$, we can reduce every element of $\mathfrak{a}$ to a polynomial of degree less than n and a linear combination of $f_1, \ldots, f_k$. For each degree $i < n$ there is a finite basis B_i for the polynomials in $\mathfrak{a}$ of degree i and the union of all the B_i and $\{f_1, \ldots, f_k\}$ forms a finite basis for $\mathfrak{a}$. ∎

If K is any field, with an endomorphism α and an α-derivation δ, then $K[x; \alpha, \delta]$ is a right Noetherian domain (it is even a principal right ideal domain, as we shall see in Section 1.3), hence right Ore, by Corollary 0.7.7, and so has a field of fractions, which we shall call the *skew rational function field* and denote by $K(x; \alpha, \delta)$. More generally, let R be a right Ore domain with field of fractions K. If α is an injective endomorphism of R and δ an α-derivation, they can be extended to K, by Theorem 0.7.8, and we have the inclusions

$$R[x; \alpha, \delta] \subseteq K[x; \alpha, \delta] \subseteq K(x; \alpha, \delta).$$

Any element u of $K(x; \alpha, \delta)$ has the form fg^{-1}, where $f, g \in K[x; \alpha, \delta]$. On bringing the coefficients of f and g to a common right denominator we can write $f = f_1 c^{-1}, g = g_1 c^{-1}$, where $f_1, g_1 \in R[x; \alpha, \delta]$ and $c \in R^\times$. Hence $u = fg^{-1} = f_1 g_1^{-1}$ and we have proved

Proposition 1.1.4. *Any skew polynomial ring over a right Ore domain is again a right Ore domain.* ∎

Here we localized at the set of all non-zero polynomials; so we obtain by Theorem 0.7.1,

Corollary 1.1.5. *If R is a right Ore domain with an injective endomorphism α and an α-derivation δ, then the non-zero polynomials in $R[x; \alpha, \delta]$ form a right Ore set.* ∎

We now give some examples of skew polynomial rings, both as illustration and for later use.

1. The *complex-skew* polynomial ring $\mathbb{C}[x;^-]$ consists of all polynomials with complex coefficients and commutation rule

$$ax = x\bar{a}, \text{ where } \bar{a} \text{ is the complex conjugate of } a.$$

We observe that the centre of this ring is $\mathbb{R}[x^2]$, the ring of all real polynomials in x^2. The residue class ring mod $x^2 + 1$ is the field of real quaternions.

2. Let $k = F(t)$ be the rational function field in an indeterminate t over a commutative field F. The usual derivative $f \mapsto f'$ defines a derivation on k, and this gives rise to a skew polynomial ring $R = k[x; 1, ']$, the ring of differential operators.

3. Let F be a commutative field of characteristic $p \neq 0$ and E/F a separable field extension of degree p, say $E = F(\xi)$, where $\xi^p - \xi \in F$. The mapping $\alpha : f(\xi) \mapsto f(\xi + 1)$ defines an automorphism of order p and we have the skew polynomial ring $E[t; \alpha]$.

4. Let k be any field with an endomorphism α and let $c \in k^\times$. Then the mapping $\delta : \alpha \mapsto ac - ca^\alpha$ defines an α-derivation, called the *inner* α-derivation determined by c. The skew polynomial ring $R = k[x; \alpha, \delta]$ can then be written as $k[y; \alpha]$, where $y = x - c$, as is easily verified. Similarly, if α is an inner automorphism, say $a^\alpha = b^{-1}ab$, then the skew polynomial ring $k[x; \alpha]$ can be written as $k[y]$, where $y = xb^{-1}$.

5. Let K be any field and denote by $A_1(K)$ the K-ring generated by x, y centralizing K, with the defining relation $xy - yx = 1$. This ring $A_1(K)$, called the *Weyl algebra* on x, y over K, may also be defined as the skew polynomial ring $A[y; 1, ']$, where $A = K[x]$ and $'$ is the derivation with respect to x (as in Example 2). Example 2 above is obtained by localizing at the set of all monic polynomials on x over k, and Example 3 by putting $\xi = xy, t = y$ and then localizing at the set of all monic polynomials in ξ over F.

The Weyl algebra is useful as an example of a finitely generated infinite-dimensional algebra which in characteristic 0 is simple. For in any non-zero ideal we can pick an element $f(x, y) \neq 0$ of least possible x-degree. The ideal still contains $\partial f/\partial x = fy - yf$, which is of lower degree and so must be zero. Therefore $f = f(y)$ is a polynomial in y alone. If its y-degree is minimal, then $\partial f/\partial y = xf - fx = 0$, hence $f = c \in K^\times$; therefore the ideal contains a non-zero element of the ground field and so must be the whole ring. This shows $A_1(K)$ to be simple. Further, $A_1(K)$ is Noetherian, by Proposition 1.3.

6. The k-algebra generated by x, y with the defining relation $xy = y(x+1)$ may be defined as $R = A[y; \tau]$, where $A = k[x]$ is the polynomial ring with the shift automorphism $\tau : x \mapsto x + 1$; R is called the *translation ring*.

7. Let p be a prime, $q = p^n$ a prime power, $\mathbb{F}_q$ the field of q elements and T the endomorphism $f \mapsto f^p$ of $\mathbb{F}_q[x]$. If the operation of multiplying by $a \in \mathbb{F}_q$ is simply denoted by a, then each polynomial $\sum a_i T^i$ defines an endomorphism of $\mathbb{F}_q[x]$, and it is easily verified that (applying endomorphisms on the right) we have $aT = Ta^p$; hence the endomorphisms of $\mathbb{F}_q[x]$ form a skew polynomial ring $\mathbb{F}_q[T; \sigma]$, where $\sigma : a \mapsto a^p$. (This has an application to finite fields; see Ore [33]).

Exercises 1.1

1. Let $R = K[x; \alpha, \delta]$ be a skew polynomial ring over a field K. Show that K may be regarded as a right R-module by letting each $a \in K$ act by right multiplication by a and letting x correspond to the action by δ. When is this representation faithful?
2. Let $R = K[x; \alpha, \delta]$ be a skew polynomial ring. If $\alpha\delta = \delta\alpha$, show that α may be extended to an endomorphism of R by taking $x^\alpha = x$; what value could be assigned to x^δ in this case?
3. Let $R = K[x; \alpha, \delta]$ be a skew polynomial ring, where K is an integral domain and α an endomorphism such that $K^\alpha a \cap K^\alpha = 0$ for some $a \in K^\times$. Show that R is not left Ore. If K is a field, show that R is left Ore if and only if α is an automorphism.
4. (Bergman) Let $A = k[t_i \,|\, i \in \mathbb{Z}]$ and α the shift automorphism $t_i \mapsto t_{i+1}$. Show that in $R = A[x; \alpha]$ the monic polynomials do not form a right Ore set. [*Hint*: Consider the equation $(x-1)f = cg$.]
5. Prove the existence of the skew polynomial ring (Theorem 1.1) by means of the diamond lemma (see Bergman [78a] or FA, Lemma 1.4.1).
6. Let R be an integral domain with an injective endomorphism α. If S is a right Ore set in R admitting α, show that S is also a right Ore set for $R[x; \alpha]$ and that the localization of $R[x; \alpha]$ at S can be obtained by localizing R at S, i.e. $R[x; \alpha]_S \cong R_S[x; \alpha]$.
7. (Ore [32]) Let R be a skew polynomial ring over a field K and let f, g be polynomials of degrees m, n, respectively. Denote by K_0 the centralizer in K of f, and let r be the dimension of K as right K_0-space. Show that $\mathrm{Hom}_R(R/fR, R/gR)$ is a right K_0-space of dimension at most rm. [*Hint*: Use (3) of 0.6.]
8*. (D. A. Jordan) Let k be a field, $K = k(x_i \,|\, i \in \mathbb{Z})$, $E = k(x_i \,|\, i > 0)$ and α the automorphism $x_i \mapsto x_{i+1}$ of K. Let S be the set of all monic polynomials in a central indeterminate t over E, put $A = K[t]_S$ and extend α to A by the rule $t^\alpha = t$. Show that $A[y; \alpha]$ is right Noetherian and left Ore but not left Noetherian. [*Hint*: Use the fact that if a ring is right (left) Ore or right (left) Noetherian, then its localization at a right Ore set has the same property.]
9*. (L. Lesieur) Let R be a right Noetherian domain and α an endomorphism of R. Show that $R[x; \alpha]$ is right Noetherian if and only if, for any sequence of right ideals $\mathfrak{a}_i$ such that $\mathfrak{a}_i \subseteq \mathfrak{a}_{i+1}$, there exists n_0 such that $\mathfrak{a}_{n+1} = \mathfrak{a}_n^\alpha R$ for all $n \geq n_0$.

10. (Ince [27], Section 5.5) In the ring of differential polynomials $R = k(x)[D; 1, ']$ show that $P = D^2 - 2x^{-2}$ and $Q = D^3 - D.3x^{-2} + 3x^{-3}$ commute but cannot be written as polynomials in the same element of R. (*Hint*: Verify that $P^3 = Q^2$ and note that P, Q are obtained by conjugating D^2, D^3, respectively, by $D + x^{-1}$ in the field of fractions of R.)

11. Let K be a non-commutative field and $R = K[x, y]$ the polynomial ring in two central indeterminates. If $[a, b] = c \neq 0$ for $a, b \in K$, verify that $[x + a, y + b] = c$; deduce that $(x + a)R \cap (y + b)R$ is isomorphic to a stably free but non-free right ideal of R (thus R is not Hermite).

12. (Bergman and Dicks [78]) Let $\phi : R \to S$ be a k-algebra homomorphism. The *multiplication mapping* $\mu : S \otimes_R S \to S$ is given by $x \otimes y \mapsto xy (x, y \in S)$, and its kernel is denoted by $\Omega_{S/R}$, while the *universal derivation* of S relative to R, $d : S \to \Omega_{S/R}$ is defined by $s \mapsto s \otimes 1 - 1 \otimes s$ (Eilenberg). Show that there exists an exact sequence

$$0 \to \mathrm{Tor}_1^R(S, S) \to S \otimes_R \Omega \otimes_R S \xrightarrow{d\phi} \Omega_{S/k} \to \Omega_{S/R} \to 0$$

(the mapping $d\phi$ is called the *derivative* of ϕ relative to the category of k-algebras). Show further that (a)–(d) below are equivalent:
 (a) $d\phi$ is injective,
 (b) $\Omega_{S/R} = 0$,
 (c) $S \otimes_R S \to S$ is surjective,
 (d) ϕ is an epimorphism (in the category of k-algebras).
(See also Proposition 7.2.1).

13. Let $A = k[t]$ and $\alpha : f(t) \mapsto f(t + 1)$. Show that for suitable elements $a \in A$, $c \in R = A[x, x^{-1}; \alpha]$, a and c are right comaximal but $ca^\alpha \notin aR$; deduce that R is not 2-Hermite.

14. Let K be a right Noetherian domain and $A = K[x; \alpha, \delta]$ a skew polynomial ring with an automorphism α. If $c \in K$ is a non-unit such that $\sum c^{\delta^i}.K = K$, show that $cA \cap xA$ is stably free but not free. Deduce that for any right Noetherian domain K, $A_1(K)$ has stably free right ideals that are not free. Conclude that the Weyl algebra $A_1(k)$ is not 2-Hermite (see also Corollary 5.3 below).

1.2 The division algorithm

In the study of rings of algebraic integers as well as rings of polynomials the division algorithm is an important tool, leading to the familiar Euclidean algorithm. Here we make a general study, but much of this section is not essential for later work.

A ring R is said to satisfy the *division algorithm* relative to a function δ on R taking values in a well-ordered set, if
DA. Given $a, b \in R$, if $a \neq 0$, then there exist $q, r \in R$ such that

$$b = aq + r, \quad \text{where } \delta(r) < \delta(a). \tag{1}$$

1.2 The division algorithm

Here q is the *quotient* and r the *remainder*. We note that if a is chosen in R so as to minimize $\delta(a)$, we must have $r = 0$ in (1), so that $b = aq$. Since this holds for all $b \in R$, a must then have a right inverse. This also shows that $\delta(0)$ has the least value, usually taken to be 0, or also sometimes -1 or $-\infty$. Strictly speaking, DA should be called the *right* division algorithm, since it is not left–right symmetric, but we shall usually omit the qualifying adjective.

It is often convenient to replace DA by the following condition, which demands less, but is easily seen to be equivalent (see Exercise 1):

A. If $a, b \in R$ and $\delta(a) \leq \delta(b)$, then there exists $c \in R$ such that

$$\delta(b - ac) < \delta(b). \tag{2}$$

We note that any condition such as DA or A is relative to a function δ, but to investigate the existence of an algorithm we need not presuppose that δ is given. For any ring R and subsets S, T of R let us put $S + T = \{s + t \mid s \in S, t \in T\}$ and define the *derived set* of S as

$$S' = \{x \in R \mid S + xR = R\}. \tag{3}$$

Thus S' is the set of divisors for which we can always perform the division with a remainder in S. Now define a sequence of subsets $\{S_n\}$ of R recursively by putting

$$S_0 = \{0\}, \quad S_{n+1} = S_n \cup S_n'. \tag{4}$$

For example, S_1 consists of 0 and all right invertible elements of R. These sets form an ascending chain

$$\{0\} = S_0 \subseteq S_1 \subseteq \ldots . \tag{5}$$

If their union is the whole ring,

$$\bigcup S_n = R, \tag{6}$$

we shall say that R is *Euclidean* and define a function $\phi : R \to \mathbb{N}$ by

$$\phi(x) = \min\{n \mid x \in S_n\}. \tag{7}$$

Thus S_n consists of all $x \in R$ such that $\phi(x) \leq n$. The concepts of Euclidean ring and division algorithm are related in the following way:

Theorem 1.2.1. *If R is a Euclidean ring, then R satisfies the division algorithm relative to the $\mathbb{N}$-valued function ϕ on R defined by (7). Here*

$$\phi(0) = 0, \quad \phi(1) = 1, \tag{8}$$

and R is an integral domain if and only if

$$\phi(ab) \geq \phi(a) \quad \text{for all } a, b \in R^\times. \tag{9}$$

Conversely, if R satisfies the division algorithm relative to a function $\delta: R \to \mathbb{N}$, then R is Euclidean and

$$\phi(x) \leq \delta(x) \quad \text{for all } x \in R.$$

Proof. Suppose that R is Euclidean; then it is clear that (8) holds. If $a, b \in R$, $a \neq 0$, are given, say $\phi(a) = n > 0$, then $S_{n-1} + aR = R$, so there exist $r \in S_{n-1}, c \in R$ such that $r + ac = b$, hence $\phi(b - ac) = \phi(r) \leq n - 1 < \phi(a)$; this proves A and hence DA. If R is an integral domain, then for $a, b \in R^\times$ we have $ab \neq 0$, so $\phi(ab) = n > 0$ for some $n \in \mathbb{N}$, hence $R = S_{n-1} + abR \subseteq S_{n-1} + aR$, therefore $a \in S_n$ and $\phi(a) \leq n = \phi(ab)$, i.e. (9). Of course, when (9) holds, then $\phi(ab) > 0$ for $a, b \in R^\times$ and by (8), $1 \neq 0$, so R is an integral domain.

Conversely, assume that R satisfies the division algorithm relative to some $\mathbb{N}$-valued function δ. For $n \in \mathbb{N}$, put $T_n = \{x \in R \mid \delta(x) \leq n\}$; we shall show by induction on n that $T_n \subseteq S_n$ for all n. For $n = 0$ this is clear, so assume that $n \geq 0$, $T_n \subseteq S_n$ and consider $a \in R$ such that $\delta(a) = n + 1$. By DA, for each $b \in R$ there exists $q \in R$ such that $\delta(b - aq) < \delta(a) = n + 1$, so $b - aq \in T_n$ and $b \in T_n + aR \subseteq S_n + aR$. This holds for all $b \in R$, so $S_n + aR = R$ and $a \in S_{n+1}$. Hence $T_{n+1} \subseteq S_{n+1}$ and by induction, $T_n \subseteq S_n$ for all n. Since $\cup T_n = R$, we see that (6) holds, so R is Euclidean and $\phi(b) \leq \delta(b)$ for all $b \in R$. ∎

If a ring satisfies the division algorithm relative to some $\mathbb{N}$-valued function δ on R, then by Theorem 2.1, ϕ (given by (7)) is defined and is the smallest $\mathbb{N}$-valued function for which the algorithm holds. By repeated application of the division algorithm to a and b we obtain a series of equations, known as the *Euclidean algorithm*, leading to the HCLF of a and b; this is certainly familiar to the reader and we shall encounter it (in a generalized form) in Section 2.8. Later, in Proposition 2.4, we shall see that for any $a \in R$, $\delta(a)$ is an upper bound to the number of steps in the Euclidean algorithm for any pair a, b, so in a sense ϕ gives the 'fastest' Euclidean algorithm.

Even when (6) fails, we can continue the sequence transfinitely, putting

$$S_{\alpha+1} = S_\alpha \cup S'_\alpha, \quad S_\lambda = \cup_{\alpha<\lambda} S_\alpha \text{ at a limit ordinal } \lambda. \tag{10}$$

If $S_\tau = R$ for some ordinal τ, we say that R is *transfinitely Euclidean* and define an ordinal function ϕ on R by

$$\phi(x) = \min\{\alpha \mid x \in S_\alpha\}. \tag{11}$$

1.2 The division algorithm

We note that if $x \in S_\lambda$, where λ is a limit ordinal, then $x \in S_\alpha$ for some $\alpha < \lambda$, hence $\phi(x)$ is never a limit ordinal.

Clearly R is Euclidean if and only if we can take $\tau = \omega$. In any case, to check whether R is transfinitely Euclidean, we need only consider ordinals not exceeding $|R|$.

As before we have (with the same proof)

Theorem 1.2.2. *If R is a transfinitely Euclidean ring, then R satisfies the division algorithm relative to the ordinal-valued function ϕ defined on R by (11). This function again satisfies (8), and (9) holds if and only if R is an integral domain.*

Conversely, if R satisfies the division algorithm relative to an ordinal-valued function δ on R, then R is transfinitely Euclidean and $\delta(x) \geq \phi(x)$ for all $x \in R$. ∎

In many Euclidean rings the remainder r in the division algorithm (1) is uniquely determined and we record the conditions for this to happen.

Proposition 1.2.3. *Let R be a Euclidean ring relative to the function δ. Then the remainder in the division (1) is unique if and only if*

(i) $\delta(a - b) \leq \max\{\delta(a), \delta(b)\}$ for all $a, b \in R$, and
(ii) $\delta(a) \leq \delta(ab)$ for all $a, b \in R$ such that $ab \neq 0$.

Proof. Suppose that remainders are unique. Given $a, b \in R$, put $x = a$, $y = a - b$; then $x = y.0 + a = y.1 + b$, hence by uniqueness, $\delta(a) \geq \delta(y)$ or $\delta(b) \geq \delta(y)$, i.e. (i). To establish (ii), if $\delta(ab) < \delta(a)$ but $ab \neq 0$, then $0 = ab - ab = a.0 + 0$, which contradicts uniqueness.

Conversely, when (i) and (ii) hold and $x = yq_1 + r = yq_2 + s$, where $\delta(r), \delta(s) < \delta(y)$, then on writing $q = q_2 - q_1$, if $yq = r - s \neq 0$, we have $\delta(yq) = \delta(r - s) < \delta(y) \leq \delta(yq)$, which is a contradiction, hence $r = s$. ∎

In fact, the uniqueness of the remainder ensures that we have the fastest algorithm:

Proposition 1.2.4. *Let R be a Euclidean ring relative to δ. If the remainder in the division algorithm is unique and the values of δ and their limits form an initial segment of the ordinals, then δ is the least function for which the algorithm holds.*

Proof. Let ϕ be the least function for which the algorithm holds. Then $\phi(x) \leq \delta(x)$ for all $x \in R$ and we must show that equality holds. Assume the contrary; then

$$\phi(a) < \delta(a) \quad \text{for some } a \in R, \tag{12}$$

and we can choose a so that $\phi(a)$ is as small as possible. If we take $b \in R$ such that $\delta(b) = \phi(a)$, then there is a unique element $r = b - aq$ such that $\delta(r) < \delta(a)$. In fact, since $\delta(b) < \delta(a)$, we have $r = b$ by uniqueness, and for any q such that $aq \neq 0$, $\delta(b - aq) \geq \delta(a) > \phi(a)$. Now take $c = b - aq$ such that $\phi(c) < \phi(a)$; then $\phi(c) < \phi(a) < \delta(a) \leq \delta(c)$, and this contradicts the choice of a. Hence (12) cannot be satisfied and the result follows. ∎

Sometimes it is convenient to put $\phi(0) = -\infty$, $\phi(1) = 0$. With this definition we can show that for Euclidean domains with unique remainder the least function defining the algorithm is in fact a degree-function (satisfying D.1–D.3 of Section 1.1).

Proposition 1.2.5. *Let R be a Euclidean domain and let ϕ be the least function with values in $\mathbb{N} \cup \{-\infty\}$ defining the algorithm. Then*

$$\phi(ab) \geq \phi(a) + \phi(b) \quad \text{for } a, b \neq 0, \tag{13}$$

with equality provided that remainders are unique.

Proof. By Theorem 2.1, $\phi(ab) \geq \phi(a)$. Now fix $c \neq 0$; clearly $\phi(cx)$ still has values in $\mathbb{N} \cup \{-\infty\}$, as does $\delta(x) = \phi(cx) - \phi(c)$. We shall show that R is Euclidean relative to the function $\delta(x)$. Given $a, b \in R$, where $a \neq 0$, we have $ca \neq 0$ and by the division algorithm,

$$cb = ca.q + s, \quad \phi(s) < \phi(ca). \tag{14}$$

Here $s = cr$, where $r = b - aq$. We can now cancel c in (14) and obtain $b = aq + r$, and by subtracting $\phi(c)$ from the inequality we have $\delta(r) < \delta(a)$. Thus R is Euclidean relative to δ. By Theorem 2.1 we have $\delta(x) \geq \phi(x)$, which is (13). If the remainders for ϕ are unique, then they are also unique for δ, so $\delta = \phi$ by Proposition 2.4, and it then follows that equality holds in (13). ∎

The principal applications of the division algorithm are to two classes of rings:

(i) rings of algebraic integers,
(ii) polynomial rings over fields.

In (i) the role of δ is played by the norm; in (ii) one uses the degree of the polynomial. Of these only the latter is a degree-function, and since we shall mainly be concerned with generalizations of (ii), we shall concentrate on rings satisfying the division algorithm relative to a degree-function.

Let R be any ring with a degree-function d. Then as we have seen in Section 1.1, R is an integral domain; moreover, any unit of R has degree 0, for if $ab = 1$, then $d(a) + d(b) = d(ab) = d(1) = 0$, hence $d(a) = d(b) = 0$. In particular, if

1.2 The division algorithm

R is a field, then every non-zero element has degree 0, so d is trivial and the division algorithm holds relative to d.

Next consider the polynomial ring in an indeterminate x over a field K. Of course this satisfies the division algorithm relative to the usual degree-function, and as we have seen in Section 1.1, this holds even for the skew polynomial ring $K[x; \alpha, \delta]$ relative to an endomorphism α and an α-derivation δ. However, the left division algorithm is not satisfied unless α is an automorphism of K.

We now show that the examples just given exhaust the rings with a right division algorithm relative to a degree-function.

Theorem 1.2.6. *Let R be a ring with a degree-function $d : R \to \mathbb{N} \cup \{-\infty\}$. Then R satisfies the right division algorithm relative to d if and only if either (i) R is a field or (ii) R is of the form $K[x; \alpha, \delta]$ for some field K, with endomorphism α and α-derivation δ, where $d(x) > 0$.*

Here d is trivial in (i), and in (ii) is a multiple of the usual degree.

Proof. Suppose that R satisfies condition A relative to d and write $K = \{a \in R \mid d(a) \leq 0\}$. By the properties of the degree-function, K is a subring of R. Given $a \in K^\times$, we have $d(a) = 0$, hence there exists $b \in R$ such that $d(ab - 1) < d(a) = 0$, so $ab = 1$, and $d(b) = 0$, i.e. $b \in K^\times$. Thus every non-zero element of K has a right inverse in K, whence it follows that K is a field. If d is trivial then R has no elements of positive degree, so $R = K$ and case (i) follows. Otherwise we take an element x, say, of least positive degree in R and assert that every element of R is of the form

$$a_0 + xa_1 + \cdots + x^n a_n, \quad \text{where } a_i \in K, n \geq 0. \tag{15}$$

For if this were not so, let b be an element of least degree that is not of the form (15). By DA, there exists $q \in R$ such that $d(b - xq) < d(x)$ and by the definition of x it follows that $b - xq \in K$. Thus for some $a \in K$ we have

$$b = xq + a, \tag{16}$$

and $d(q) < d(x) + d(q) = d(xq) \leq \max\{d(b), d(a)\} = d(b)$. Therefore, by the choice of b, q must have the form $q = \sum x^i a_i (a_i \in K)$. Inserting this expression in (16), we obtain

$$b = \sum x^{i+1} a_i + a,$$

which contradicts the assumption that b is not of the form (15). Moreover, the form (15) for any element of R is unique, for otherwise we should have a relation

$$c_0 + xc_1 + \cdots + x^n c_n = 0,$$

say, where $c_n \neq 0$. Hence

$$d(x^n) = d(x^n c_n) \leq \max\{d(x^i c_i) \mid i = 0, 1, \ldots, n-1\},$$

i.e. $nd(x) \leq (n-1)d(x)$, in contradiction to the assumption that $d(x) > 0$. Finally we have, for any $a \in K$,

$$ax = xa_1 + a_2,$$

where $d(a_2) < d(x)$ and $d(a_1) \leq d(ax) - d(x) = 0$, hence $a_1, a_2 \in K$. By the uniqueness of the form (15) it follows as in Section 1.1 that $a \mapsto a_1$ is an endomorphism α, say, of K and $a \mapsto a_2$ is an α-derivation δ, thus $R = K[x; \alpha, \delta]$ as asserted.

To prove the converse, take the skew polynomial ring $K[x; \alpha, \delta]$ with a degree-function d. We have already seen that the elements of $K^\times$ must have degree 0, and by hypothesis, $d(x) = \lambda > 0$, hence the degrees $d(x^n) = n\lambda$ are all different for different n. It follows that for $c_n \neq 0$,

$$d(c_0 + xc_1 + \cdots + x^n c_n) = \max\{d(x^i c_i) \mid i = 0, 1, \ldots, n\} = n\lambda.$$

Thus all degrees are multiples of λ. We may therefore divide the degrees by λ and so obtain the usual degree-function on $K[x; \alpha, \delta]$. As we saw earlier, the right division algorithm holds for this degree-function. ∎

Exercises 1.2

1. Show, by induction on $\delta(b - ac)$, that condition A is equivalent to DA.
2. Show that if a ring R has a division algorithm relative to a function that is constant on $R^\times$, then R is a field.
3. Show that the ring of integral quaternions over $\mathbb{Z}$ (the rational integers) is Euclidean relative to the norm function [q is called *integral* if it is a linear combination of $1, i, j, k$ and $(1 + i + j + k)/2$)]. Does this still hold for the ring of quaternions with integer coefficients? (Recall that every commutative principal ideal domain is integrally closed in its field of fractions. See BA, 9.4)
4. (Sanov [67]) Let R be a commutative Euclidean domain relative to an $\mathbb{N}$-valued function ϕ and on the matrix ring R_n define $|A| = \phi(\det A)$. Show that for any $A, B \in R_n$ with $|A| \neq 0$, there exist $P, Q \in R_n$ such that

$$B = AQ + P, \quad 0 < |P| < |A| \text{ or } P = 0.$$

  Use this result to obtain a reduction to triangular form for matrices over R.
5°. (P. Samuel) Determine all Euclidean rings in which the number of atomic factors in the complete factorization of an element serves as a value function.
6. Let R be a Euclidean domain with a fastest algorithm given by ϕ. Given $a \in R^\times$, show that $\phi(xa) = \phi(x)$ for some $x \in R^\times$ if and only if a is a unit.

7. If R is a Euclidean domain and T is a right Ore set in R, show that the localization R_T is again Euclidean.
8*. (Lenstra [74]) Show that for a commutative ring R with unique remainder algorithm, $U(R) \cup \{0\} = k$ is a field. Deduce that a commutative ring has a unique remainder algorithm relative to a function δ if and only if δ is a degree-function or $R = \mathbb{F}_2 \times \mathbb{F}_2$.
9. (Lemmlein [54]) Let R be an integral domain with an $\mathbb{N}$-valued unbounded function ϕ such that $\phi(0) = 0$, $\phi(x) > 0$ for $x \neq 0$, and there exists n_0 such that for any $x \in R$ satisfying $\phi(x) > n_0$ there exists $y \in R$ such that $\phi(y - qx) \geq \phi(x)$ for all $q \in R$. Show that R is not Euclidean.
10. (S. Singh) Find an integral domain that is Euclidean and whose least value function does not satisfy D.3 of Section 1.1.

1.3 Principal ideal domains

The reader will be familiar with the notion of a principal ideal domain, as an integral domain, usually commutative, in which every ideal is *principal*, i.e. it can be generated by a single element. Here we shall be interested in the non-commutative case, where one has to distinguish between left and right ideals. Thus a *principal right ideal domain* is an integral domain in which every right ideal is principal, and a principal left ideal domain has a corresponding definition, while a domain that is both left and right principal will be called a *principal ideal domain*, often abbreviated to PID; a principal right ideal domain is briefly referred to as a right PID and similarly for a left PID. An integral domain in which every finitely generated right ideal is principal is called a *right Bezout domain;* left Bezout domains are defined similarly, and when both conditions hold we speak of a *Bezout domain*.

Suppose R is a ring with a division algorithm relative to a degree-function d; we claim that R is then a right PID. The degree-function shows it to be an integral domain; now let $\mathfrak{a}$ be any right ideal of R; we have to show that $\mathfrak{a}$ is principal. When $\mathfrak{a} = 0$, there is nothing to prove, so assume that $\mathfrak{a} \neq 0$ and let $a \in \mathfrak{a}$ be a non-zero element of least degree. We claim that $\mathfrak{a} = aR$; for clearly $aR \subseteq \mathfrak{a}$ and if $b \in \mathfrak{a}$, then $b = aq + r$, where $d(r) < d(a)$. Hence $r = b - aq \in \mathfrak{a}$; by the minimality of $d(a)$, $r = 0$ and this shows that $b \in aR$. Thus $\mathfrak{a}$ is indeed a principal right ideal, and we obtain

Proposition 1.3.1. *Let R be any ring with a degree-function d satisfying the division algorithm DA. Then R is a principal right ideal domain.* ∎

As an example consider a skew polynomial ring $R = K[x; \alpha, \delta]$, where K is a field with an endomorphism α and α-derivation δ. Given $f = x^m a + \ldots, g = x^n b + \ldots \in R$, where only the leading terms are indicated, if $d(f) \leq d(g)$, then

$m \leq n$ and so $g - fa^{-1}x^{n-m}b$ has degree less than n and this shows that the division algorithm holds in R. It follows that R is a right PID. When α is an automorphism, R also satisfies the left-hand analogue of DA and it follows that R is a PID. So we have proved

Theorem 1.3.2. *Let K be any skew field with an endomorphism α and an α-derivation δ. Then the skew polynomial ring $K[x; \alpha, \delta]$ is a principal right ideal domain; it is a principal ideal domain whenever α is an automorphism.* ∎

Sometimes a slight refinement of this result is useful:

Proposition 1.3.3. *Let A be any principal right ideal domain with an endomorphism α mapping $A^\times$ into $U(A)$ and an α-derivation δ. Then the skew polynomial ring $A[x; \alpha, \delta]$ is again a principal right ideal domain.*

Proof. The ring $R = A[x; \alpha, \delta]$ may no longer satisfy DA, but under the given condition we can proceed as follows. Given a non-zero right ideal $\mathfrak{a}$ in R, let m be the least degree of elements of $\mathfrak{a}$, and consider the leading coefficients of elements of degree m in $\mathfrak{a}$. Together with 0 they clearly form a right ideal in A; let a be a generator of this right ideal and $f = x^m a + \ldots$ a polynomial with this leading coefficient. We claim that $\mathfrak{a} = fR$; in one direction this inclusion is again clear, so let $g \in \mathfrak{a}$, say $g = x^n b + \ldots$. By the definition of m we have $n \geq m$; if $n = m$, then $b \in aA$, say $b = ac$ and so $d(g - fc) < d(g)$. Otherwise $n > m$ and now $fx = x^{m+1} a^\alpha + \ldots$ has a unit as leading coefficient and so fxa' is monic for some $a' \in A$; now $d(g - fxa'x^{n-m-1}b) < d(g)$. Thus we have in all cases found an h such that $g - fh$ has degree less than $n = d(g)$, so it follows as before that $\mathfrak{a} = fR$, as claimed. ∎

In the commutative theory an important result states that every commutative PID is a unique factorization domain. For general PIDs there is a corresponding result, though inevitably rather more complicated. In any ring we recall that an *atom* is a non-unit that cannot be written as a product of two non-units. If every element other than 0 or a unit is a product of atoms, the ring is said to be *atomic*. Two elements a, b are said to be *associated* if there exist units u, v such that $a = ubv$. In Section 0.5 two elements a, b of an integral domain R were called similar if $R/aR \cong R/bR$ or equivalently, if a, b are stably associated; in particular, two associated (regular) elements are always similar. Given two non-zero elements a, b of an integral domain R, consider any factorizations of a and b:

$$a = u_1 u_2 \cdots u_r, \qquad (1)$$

$$b = v_1 v_2 \cdots v_s. \qquad (2)$$

1.3 Principal ideal domains

The number of atomic factors of a is called its *length*, thus $l(a) = r$. These factorizations of a and b are said to be *isomorphic* if $r = s$ and there is a permutation σ of $(1, \ldots, r)$ such that u_i is similar to $v_{i\sigma}$. Our first observation is that similar elements have isomorphic factorizations.

Proposition 1.3.4. *Let R be an integral domain and let a, b be non-zero elements of R that are similar. Then any factorization of a gives rise to an isomorphic factorization of b.*

Proof. Any factorization of a may be regarded as a chain of cyclic submodules from R to aR, and by the isomorphism $R/aR \cong R/bR$ we obtain a chain from R to bR in which corresponding factors are isomorphic. ∎

Here the permutation involved in the isomorphism is the identity, but we shall soon meet cases where this is not so. Let us define a *unique factorization domain* (UFD) as an atomic integral domain R such that any two complete factorizations of a given element are isomorphic. It is easily seen that this reduces to the usual definition in the commutative case; the following result provides a source of non-commutative UFDs.

Theorem 1.3.5. *Every principal ideal domain is a unique factorization domain.*

Proof. Let R be a PID; then left and right ACC_1 holds in R, and as in Section 0.9 we see that every element of R, not zero or a unit, has a complete factorization. Suppose $a \in R$ has the complete factorization (1). This corresponds to a chain of submodules

$$R \supset u_1 R \supset u_1 u_2 R \supset \ldots \supset u_1 \ldots u_r R = aR. \tag{3}$$

Let $a = v_1 v_2 \ldots v_s$ be a second factorization of a into atoms and consider the corresponding chain

$$R \supset v_1 R \supset \ldots \supset v_1 \ldots v_s R = aR. \tag{4}$$

Since R is a PID, every right ideal containing aR has the form cR, where c is a left factor of a, and (3), (4) are actually composition series from R to aR, because the u_i and v_j are atoms. So we can apply the Jordan–Hölder theorem to conclude that $r = s$ and for some permutation σ of $(1, \ldots, r)$, $R/u_i R \cong R/v_{i\sigma} R$, i.e. u_i is similar to $v_{i\sigma}$. ∎

Later, in Section 3.2, we shall find a far-reaching generalization of this result. Let us now take a general ring R and look at cyclic modules $R/\mathfrak{a}$, where $\mathfrak{a}$ is a right ideal of R, and examine when two such modules are isomorphic. We shall say that two right ideals $\mathfrak{a}$, $\mathfrak{a}'$ of a ring R are *similar* if

$$R/\mathfrak{a} \cong R/\mathfrak{a}'. \tag{5}$$

When these right ideals are principal, say $\mathfrak{a} = aR$, $\mathfrak{a}' = a'R$, this reduces to the notion defined in Section 0.5. The following criterion for similarity is often useful:

Proposition 1.3.6. *In any ring R, two right ideals $\mathfrak{a}$, $\mathfrak{a}'$ are similar if and only if there exists $c \in R$ such that (i) $\mathfrak{a} + cR = R$ and (ii) $\{x \in R | cx \in \mathfrak{a}\} = \mathfrak{a}'$.*

In an abbreviated notation, (ii) may be expressed as $\mathfrak{a}' = c^{-1}\mathfrak{a}$.

Proof. Suppose that $\mathfrak{a}$ is similar to $\mathfrak{a}'$ and let 1 (mod $\mathfrak{a}'$) correspond to c (mod $\mathfrak{a}$) in the isomorphism (5). Then c generates R (mod $\mathfrak{a}$), so (i) holds. Further x (mod $\mathfrak{a}'$) corresponds to cx (mod $\mathfrak{a}$), so $cx \in \mathfrak{a}$ if and only if $x \in \mathfrak{a}'$, which is (ii). Conversely, when (i) and (ii) hold, then $R/\mathfrak{a} \cong (\mathfrak{a} + cR)/\mathfrak{a} \cong cR/(\mathfrak{a} \cap cR) \cong R/\mathfrak{a}'$. ∎

Note that whereas the relation of similarity is clearly symmetric, the criterion of Proposition 3.6 is not, so there are two ways of applying the result, once as it stands and once with $\mathfrak{a}$ and $\mathfrak{a}'$ interchanged. If $c \in R$ is right regular, (ii) takes on the form

$$\mathfrak{a} \cap cR = c\mathfrak{a}'. \tag{6}$$

In particular, when R is an integral domain and $\mathfrak{a}$ is a proper right ideal, then $c \neq 0$ by (i), so (ii) can then be put in the form (6).

We also see from (ii) that $\mathfrak{a}'$ is determined in terms of $\mathfrak{a}$ and c. For example, if $\mathfrak{a}$ is a maximal right ideal, then (i) holds provided that $c \notin \mathfrak{a}$ and it follows that the right ideal $\mathfrak{a}'$ determined by (ii) is also maximal, because of (5). The result may be stated as

Corollary 1.3.7. *If $\mathfrak{a}$ is a maximal (proper) right ideal of a ring R and $c \in R \setminus \mathfrak{a}$, then the set $c^{-1}\mathfrak{a} = \{x \in R \mid cx \in \mathfrak{a}\}$ is a right ideal similar to $\mathfrak{a}$ and hence is also maximal.* ∎

Exercises 1.3

1. Let R be a commutative Bezout domain and F its field of fractions. Show that every element of F can be written in the form a/b, where a and b are coprime (i.e. without a common factor). To what extent is this representation unique?
2. Show that a principal ideal domain R is primitive if and only if it has an unbounded atom (i.e. it has an atom a such that Ra contains no non-zero ideal).

3. Show that a Bezout domain with right ACC_1 is right principal.
4. Let R be an integral domain that is weakly finite. Show that if R has an infinite centre, then the polynomial ring $R[t]$ is again weakly finite.
5. What happens in Exercise 4 when the centre of R is finite?
6°. (A. Hausknecht) Let R be a principal ideal domain; if the units together with 0 form a field k, is R necessarily a polynomial ring over k?
7. Let R be an integral domain in which every right ideal generated by two elements is principal. Show that R is a right Bezout domain.
8. (Hasse [28]) Show that a commutative integral domain R is a principal ideal domain if and only if there is an $\mathbb{N}$-valued function ϕ on $R^\times$ such that (i) $a|b$ imples $\phi(a) \leq \phi(b)$, with equality if and only if $aR = bR$ and (ii) if neither of a, b divides the other, then there exist $p, q, r \in R$ such that $pa + qb = r$, $\phi(r) < \min\{\phi(a), \phi(b)\}$.
9°. Generalize Exercise 8 to obtain a characterization of Bezout domains.
10°. Investigate rings with a positive *real*-valued function satisfying the conditions (i) and (ii) of Exercise 8.
11. Show that over the field of real quaternions the equation $x^2 + 1 = 0$ has infinitely many roots. (*Hint*: Observe that any conjugate of a root is again a root.)
12*. Show that Bezout domains form an elementary class (i.e. they can be defined by elementary sentences; the class of PIDs is not closed under ultraproducts and so cannot be elementary). (*Hint*: Use Exercise 7.)
13*. Show that a PID may be characterized as a Bezout domain such that (a) every non-unit ($\neq 0$) has an atomic left factor, and (b) left ACC_1 holds. Deduce that any right PID that is elementarily equivalent to a PID is itself a PID. [*Hint*: (a) is elementary, but not (b); see Cohn [87a].]
14*. Let K be a field with an automorphism α, no power of which is an inner automorphism of K, and let δ be an α-derivation. Show that in the skew polynomial ring $R = K[x; \alpha, \delta]$ the monic right invariant elements form a monoid M and either (i) M is generated by a single element $d \neq 1$ and all ideals of R are of the form $d^\nu R (\nu = 0, 1, \ldots)$, or (ii) $M = \{1\}$ and R is simple (Cohn [77a]). [*Hint*: Choose a monic element u of least degree subject to the condition $cu = uc'$ for all $c \in K$, and apply the division algorithm.]
15. Let $R = K[x; \alpha, \delta]$ be as in Exercise 14. Show that in case (i) M is a right Ore set and R_M is a simple PID while in case (ii) R is a simple right PID that is left principal if and only if α is an automorphism of K.

1.4 Modules over principal ideal domains

Let R be a principal ideal domain; then R is in particular a (left and right) Ore domain, so every R-module M has a submodule M_t of torsion elements with the torsion-free quotient M/M_t (see Section 0.8). Suppose now that M is finitely generated torsion-free left R-module and let K be the field of fractions of R. Then M can be embedded in $K \otimes M$ by Proposition 0.8.1 and the latter is semifree as left R-module (Proposition 0.8.4); this shows that M is free and we obtain

Proposition 1.4.1. *Let R be a principal ideal domain. Then any finitely generated torsion-free left (or right) R-module is free.* ∎

This result still holds for left modules over right PIDs (even over right Bezout domains, see Proposition 2.3.19), though not for left modules over left PIDs, as the remark after Proposition 0.8.3 shows. In that case a stronger hypothesis is needed to ensure freeness (see the remark after Corollary 2.1.3).

Let R be a PID or more generally, a Bezout domain; then R clearly has IBN (in fact it is weakly finite, but this is not needed yet). Any finitely generated projective module is clearly torsion-free and hence free; thus R is projective-free and we obtain

Corollary 1.4.2. *Every Bezout domain, in particular, every principal ideal domain is projective-free, hence it is an Hermite ring.* ∎

We shall also need a description of Ore sets in PIDs; as long as 0 is excluded, the set will be regular, so we need not worry about property O.2 of Section 0.7. More generally we shall consider right PIDs:

Proposition 1.4.3. *Let R be a principal right ideal domain and S a submonoid of $R^\times$ such that (i) $ab \in S$ implies $b \in S$, and (ii) if $a \in S$, then any element similar to a is in S. Then S is a right Ore set in R.*

Proof. Given $a \in R, u \in S$, we have to show that $aS \cap uR \neq \emptyset$. For $a = 0$ this is clear, so let $a \neq 0$; then $aR + uR = dR$, say $a = da_1, u = du_1$. Here $u_1 \in S$ by (i) and (a_1, u_1) is unimodular, hence there is an invertible matrix P such that $(a_1, u_1)P = (1, 0)$. Moreover, we have a relation $a_1 u' = u_1 a' \neq 0$. If

$$P^{-1}(u', -a')^{\mathrm{T}} = (v, -b)^{\mathrm{T}}, \tag{1}$$

then $0 = (a_1, u_1)(u', -a')^{\mathrm{T}} = (1, 0)(v, -b)^{\mathrm{T}} = v$; hence $P(0, -b)^{\mathrm{T}} = (u', -a')^{\mathrm{T}}$ and this shows that b is a common right factor of u' and a', say $u' = u''b, a' = a''b$. By cancelling b and equating the last components in (1) we find that u'' and a'' are left comaximal. Now $a_1 u'' = u_1 a''$ is a comaximal relation, so by Corollary 0.5.5 and Proposition 0.5.6, u'' is similar to u_1 and by (ii), $u'' \in S$. ∎

Let us return to an arbitrary finitely generated module M over a PID R. Then M/M_t is finitely generated torsion-free, hence free, and so it can be lifted to a free submodule F of M complementing M_t; since $F \cong M/M_t$, it is unique up to isomorphism and we have

1.4 Modules over principal ideal domains

Corollary 1.4.4. *Let R be a principal ideal domain. For any finitely generated R-module M there exists a free submodule F such that*

$$M = M_t \oplus F.$$

Here M_t is uniquely determined as the submodule of torsion elements of M, while F is unique up to isomorphism. ∎

The principal ideal property also translates to a property of the modules:

Proposition 1.4.5. *Let R be a principal ideal domain and $n \in \mathbb{N}$. Given an n-generator left R-module M, any submodule of M has an n-element generating set. If moreover, M is free, then so is the submodule.*

Proof. By writing $M = F/L$, where F is free of rank n, we see that it is enough to prove the result for free modules. So take F to be free of rank n and N any submodule. Let π be the projection of F on the first coordinate; then $N\pi$ is a left ideal of R and so is principal. Hence we have the exact sequence

$$0 \to N \cap \ker \pi \to N \to N\pi \to 0.$$

Clearly $\ker \pi$ is free of rank $n - 1$, hence by induction $N \cap \ker \pi$ can be generated by $n - 1$ elements, while $N\pi$ is generated by a single element. It follows that N can be generated by n elements. When M is free, N is torsion-free and hence free by Corollary 4.4. ∎

Now every module over a PID R has a finite free resolution of length at most 1, so it has a characteristic, as defined in Section 0.5, and by Proposition 4.5 this is non-negative for any finitely generated R-module. In Chapter 2 we shall meet a class of rings, over which every finitely presented module has a finite free resolution of length at most 1, but where the characteristic can assume any integer values, negative as well as positive.

Proposition 4.5 also shows that over a PID, every finitely generated module is finitely presented. Let M be a finitely generated left R-module with the presentation

$$R^m \xrightarrow{\alpha} R^n \longrightarrow M \to 0, \qquad (2)$$

and let A be the $m \times n$ matrix over R that represents the homomorphism $\alpha : R^m \to R^n$ relative to the standard bases in R^m, R^n. Then M is completely specified by A, and if we change the bases in R^m and R^n, this amounts to replacing A by PAQ^{-1}, where $P \in GL_m(R)$, $Q \in GL_n(R)$. Thus our next task will be to find a normal form for matrices under association. To state the result, we need another definition. In an integral domain, an element a is said to be a *total divisor* of b, written $a \| b$, if there exists an invariant element c such that $a|c|b$.

We observe that an element is not generally a total divisor of itself; in fact $a \| a$ if and only if a is invariant. The invariant element of shortest length divisible by a is also called its *bound*; clearly it is unique up to associates. If the ring R is simple, it has no non-unit invariant elements and $a \| b$ implies that either a is a unit or b is 0. In a PID the invariant elements serve as the generators of ideals.

Proposition 1.4.6. *(i) In any ring R, an element c is invariant if and only if c is regular and the left and right ideals of R generated by c coincide.*

(ii) If R is an integral domain, then any non-zero ideal that is principal both as left and as right ideal has an invariant generator.

Proof. (i) is clear, since $cR = Rc$ for any invariant element c. For (ii) take $aR = Ra'$ and let $a = ua'$, $a' = av$; then $a = uav$. Now $ua \in aR$, say $ua = aw$, so $a = awv$, hence $wv = 1$. In an integral domain this shows v to be a unit, similarly for u, and so $aR = Ra$. ∎

The notation $\mathrm{diag}(a_1, \ldots, a_r)$ for a matrix with $a_1, \ldots, a_r$ on the main diagonal and 0s elsewhere will be used here even for matrices that are not square; the exact size will be indicated explicitly, unless it is clear from the context, as in (3) below. For any matrix the maximum number of left linearly independent rows is called its *row rank*; *column rank* is defined similarly as the maximum number of right linearly independent columns. We now have the following reduction for matrices over a PID, known (in the commutative case) as the *Smith normal form*.

Theorem 1.4.7. *Let R be a principal ideal domain and $A \in {}^m R^n$. Then the row and column rank of A are the same; denoting the common value by r, we can find $P \in GL_m(R)$, $Q \in GL_n(R)$ such that*

$$PAQ^{-1} = \mathrm{diag}(e_1, \ldots, e_r, 0, \ldots, 0), \quad e_i \| e_{i+1}, e_r \neq 0. \qquad (3)$$

Proof. We have the following four types of operations on the columns of A, of which the first three are the well-known elementary operations:

(i) *interchange two columns and change the sign of one,*
(ii) *add a right multiple of one column to another,*
(iii) *multiply a column on the right by a unit factor,*
(iv) *multiply two columns on the right by an invertible 2×2 matrix.*

As is well known, (i) and (ii) correspond to right multiplication by an elementary matrix, while (iii) corresponds to multiplying by a diagonal matrix. The object of using (iv) is to replace the first two elements in the columns by their highest common left factor and 0, respectively. Thus if these elements are

a, b, not both zero, we have $aR + bR = kR$, say $a = ka_1, b = kb_1$. This means that (a_1, b_1) has a right inverse, and (since a PID is evidently an Hermite ring), it can be completed to an invertible 2×2 matrix C say. Hence $(k, 0)C = (a, b)$ and $(a, b)C^{-1} = (k, 0)$, as required. Corresponding operations can of course be carried out on the rows, acting on the left.

We can now proceed with the reduction. If $A = 0$, there is nothing to prove; otherwise we bring a non-zero element to the $(1, 1)$-position in A, by permuting rows and permuting columns, using (i). Next we use (iv) to replace a_{11} successively by the HCLF of a_{11} and a_{12}, then by the HCLF of the new a_{11} and a_{13}, etc. After $n - 1$ steps we have transformed A to a form where the first row is zero except for a_{11}. By symmetry the same process can be applied to the first column of A; in this reduction the first row of A may again become non-zero, but this can happen only if the length (i.e. the number of factors) of a_{11} is reduced; therefore by induction on the length of a_{11} we transform A to $P_0 A Q_0^{-1} = a_{11} \oplus A_1$. By another induction, on $\max(m, n)$, we reach the form

$$P_1 A Q_1^{-1} = \mathrm{diag}(a_1, a_2, \ldots, a_r, 0, \ldots, 0),$$

where P_1, Q_1 are invertible matrices and the a_i are non-zero. Here r, the number of non-zero a_i is the row rank and the column rank. Consider a_1 and a_2; for any $d \in R$ we have

$$\begin{pmatrix} 1 & d \\ 0 & 1 \end{pmatrix} \begin{pmatrix} a_1 & 0 \\ 0 & a_2 \end{pmatrix} = \begin{pmatrix} a_1 & da_2 \\ 0 & a_2 \end{pmatrix},$$

and now we can again diminish the length of a_1 unless a_1 is a left factor of da_2 for all $d \in R$, i.e. unless $a_1 R \supseteq R a_2$. But in that case $a_1 R \supseteq R a_2 R \supseteq R a_2$; thus $a_1 | c | a_2$, where c is the invariant generator of the ideal $R a_2 R$. Hence $a_1 \| a_2$, and by repeating the argument we obtain the form

$$P A Q^{-1} = \mathrm{diag}(e_1, e_2, \ldots, e_r, 0, \ldots, 0), \quad \text{where}$$

P, Q are invertible, $e_i \| e_{i+1}$ and $e_r \neq 0$.

We see that this matrix has row and column rank r. Clearly A and PAQ^{-1} have the same column rank; similarly for the row rank and so the assertion follows. ∎

We remark that if R is a Euclidean domain (hence a PID), we can instead of (iv) use the Euclidean algorithm, with an induction on the degree instead of the length, to accomplish the reduction in Theorem 4.7. Most of the PIDs we encounter will in fact be Euclidean.

We record two consequences of Theorem 4.7.

Corollary 1.4.8. *Let R be a principal ideal domain that is simple. Then any matrix over R is associated to $I \oplus a \oplus 0$, and hence stably associated to $a \oplus 0$, where $a \in R$.*

Proof. If $a \| b$, then either $b = 0$ or a is a unit. Now any unit can be transformed to 1 by applying (iii), so there can be only one diagonal element not 0 or 1. ∎

In the case of a field every non-zero element is a unit, so in this case every matrix A is associated to $I_r \oplus 0$, where r is the rank of A, a fact well known from linear algebra (see CA, Section 4.7).

As a further application of Theorem 4.7 we describe the rank of a matrix over $K(t)$, where K is a field. We recall that a polynomial of degree n in t over a commutative field k cannot have more than n zeros in k. Over a skew field this is no longer true, as the example of $t^2 + 1$ over the quaternions shows (see Exercise 3.11, also SF, Section 3.4). However, a polynomial of degree n over a skew field K has at most n zeros in the centre of K; this follows as in the commutative case:

Lemma 1.4.9. *Let K be a field with infinite centre C and consider the polynomial ring K[t], in a central indeterminate t, with field of fractions K(t). If $A = A(t)$ is a matrix over K[t], then the rank of A over K(t) is the supremum of the ranks of $A(\alpha)$, $\alpha \in C$. In fact this supremum is assumed for all but a finite number of values of α.*

Proof. By Theorem 4.7 we can find invertible matrices P, Q over $K[t]$ such that

$$PAQ^{-1} = \mathrm{diag}(f_1, \ldots, f_r, 0, \ldots, 0), \quad \text{where } f_i \in K[t]. \tag{4}$$

The product of the non-zero diagonal terms on the right gives us a polynomial f whose zeros in C are the only points of C at which $A = A(t)$ falls short of its maximum rank, and the number of these cannot exceed $\deg f$. ∎

We now come to the application of Theorem 4.7 to describe modules over a PID, in a result that generalizes the fundamental theorem of abelian groups.

Theorem 1.4.10. *Let R be a principal ideal domain. Then any finitely generated left R-module M is a direct sum of cyclic modules*

$$M \cong R/Re_1 \oplus \cdots \oplus R/Re_r \oplus R^{n-r}, \tag{5}$$

where $e_i \| e_{i+1}$, and this condition determines the e_i up to similarity.

Proof. Let M be defined by a presentation (2) with matrix A. By Theorem 4.7, A is associated to $\mathrm{diag}(e_1, \ldots, e_r, 0, \ldots, 0)$ with $e_i \| e_{i+1}$, and since this

1.4 Modules over principal ideal domains

change does not affect the module, we obtain (5). It only remains to prove the uniqueness.

We begin with the remark that modules of finite length can be cancelled. Thus if

$$M \oplus N \cong M \oplus N', \tag{6}$$

where M, N, N' are of finite length, then $N \cong N'$. This follows from the Krull–Schmidt theorem (see FA, Section 4.1 or IRT, Chapter 2).

Now let us write $R/Ra_1 \oplus \cdots \oplus R/Ra_r$ as $[a_1, \ldots, a_r]$ for short. Then $[a] \cong [b]$ if and only if a is similar to b, and by what has been said above, $[a, b_1, \ldots, b_r] \cong [a, c_1, \ldots, c_s]$ implies $[b_1, \ldots, b_r] \cong [c_1, \ldots, c_s]$. We take two representations of M as direct sums of cyclic modules:

$$M \cong [d_1, \ldots, d_r] \cong [e_1, \ldots, e_r], \quad d_i \| d_{i+1}, e_i \| e_{i+1}. \tag{7}$$

It is no loss of generality to assume the same number of summands on both sides, since we can always add zero summands, represented by unit factors: $R/R = 0$. Further we may suppose that the torsion-free part has been split off, so that the d_i, e_i are all different from 0. If $r = 1$, the result is clear, by what has been said, so let $r > 1$ and use induction on r. We shall write $l(a)$ for the length of a and assume that $l(d_1) \geq l(e_1)$; further, let $d_1|c|d_2$, where c is invariant. If N is any left R-module, cN is a submodule; more specifically, if $N = R/Ra$ and c is invariant, then $N/cN \cong R/(Ra + Rc)$. Now consider M/cM; writing $Re_i + Rc = Rf_i$, we have by (7),

$$M/cM \cong [d_1, c, \ldots, c] \cong [f_1, f_2, \ldots, f_r], \tag{8}$$

and $l(f_i) \leq l(c) (i = 1, \ldots, r), l(f_1) \leq l(e_1) \leq l(d_1)$. Comparing lengths in (8) (which must be equal, as the length of a composition series for M/cM), we find that $l(d_1) + (r-1)l(c) = \sum l(f_i)$, i.e.

$$l(d_1) - l(f_1) + \sum_{2}^{r}(l(c) - l(f_i)) = 0.$$

Since each term is non-negative, all are zero and $l(f_1) = l(e_1) = l(d_1)$, $l(f_i) = l(c)$. It follows that $[f_1] \cong [e_1]$, $[f_i] \cong [c] (i > 1)$, and now (8) reads

$$[d_1, c, \ldots, c] \cong [e_1, c, \ldots, c].$$

By cancellation we find that e_1 is similar to d_1; so we may cancel the first term on both sides of (7) and obtain $[d_2, \ldots, d_r] \cong [e_2, \ldots, e_r]$. Now an induction on r gives the result. ∎

If M consists of torsion elements, the last term in (5) is absent. If moreover, R is simple, then there are no non-unit invariant elements, so at most one of the

e_i in (5) can be a non-unit. But units can in any case be omitted, so (5) reduces to a single term; hence over a simple PID any finitely generated module of torsion elements is cyclic. In Proposition 4.12 below we shall obtain this result in a somewhat more general context.

Theorem 4.10 shows that the e_i in Theorem 4.7 are determined up to similarity; we shall generally omit the units among them, since they do not contribute to M. The e_i are called the *invariant factors* of the matrix A or also of the module M. The condition imposed on the e_i (that each e_i be a total divisor of the next) ensures that (5) is a decomposition of M into cyclic modules with as *few* terms as possible. At the other extreme we have a decomposition into as *many* terms as possible, i.e. a complete direct decomposition into indecomposable modules. The indecomposable parts must then be cyclic, by Theorem 4.10 and they are unique up to order and isomorphism, by the Krull–Schmidt theorem. The factors in this case are called the *elementary divisors* of the module (or the matrix). For example, over the integers, $M = [3, 15, 750]$ has the invariant factors 3, 15, 750 and the elementary divisors 3; 3, 5; 2, 3, 5^3.

It is of interest to extend the decomposition of Theorem 4.10 beyond the principal case. Here the following lemma is useful:

Lemma 1.4.11. *Let R be a non-Artinian simple ring. Then any R-module of finite length is cyclic.*

Proof. Let M be a left R-module of finite length and suppose first that M has a simple submodule that is a direct summand:

$$M = M' \oplus S, \quad \text{where } S \text{ is simple.} \tag{9}$$

By induction on the length, M' is cyclic, say $M' = Ru$. We denote by $\mathfrak{a}$ the left annihilator of u in R; it is a left ideal and since R is non-Artinian, $\mathfrak{a} \neq 0$. If $\mathfrak{a}S = 0$, then $S = RS = \mathfrak{a}RS = 0$, a contradiction, so there exists $v \in S$ such that $\mathfrak{a} \not\subseteq \mathrm{Ann}(v)$. It is clear that $v \neq 0$, so $Rv = S$ by simplicity; we claim that $u + v$ generates M. Consider $\mathrm{Ann}(u + v)$; if $x(u + v) = 0$, then $xu = xv = 0$ by (9), so $x \in \mathfrak{a} \cap \mathrm{Ann}(v)$. Thus the map $x \mapsto x(u + v)$ gives an isomorphism

$$R/(\mathfrak{a} \cap \mathrm{Ann}(v)) \cong R(u + v).$$

On the left we have a module of length $> l(M')$, hence of length $l(M)$, so $R(u + v) = M$ and M is cyclic, as claimed.

There remains the case when no simple submodule is a direct summand. Of course, M has a simple submodule (unless $M = 0$), S say. By induction hypothesis, M/S is cyclic, generated by $u + S$, say. Hence $M = Ru + S$ and by hypothesis, $Ru \cap S \neq 0$, so $Ru \supseteq S$ by the simplicity of S, therefore $Ru = M$ and M is cyclic. ∎

1.4 Modules over principal ideal domains

As a consequence we have

Proposition 1.4.12. *Let R be an atomic simple principal left ideal domain, not a field. Then any finitely generated left R-module of torsion elements is cyclic.*

Proof. Since R is non-Artinian simple, we need only verify that a finitely generated left R-module of torsion elements is of finite length. By induction it is enough to check this for cyclic modules and this clearly holds by unique factorization and the fact that all left ideals are principal. ∎

This result then shows that over an atomic simple principal left ideal domain (e.g. any simple PID) R, every regular matrix A is stably associated to an element of R.

Exercises 1.4

1. Show that (iv) is not needed in the proof of Theorem 4.7 if every invertible matrix is a product of elementary and diagonal matrices. What simplifications are possible when R is commutative? Prove the uniqueness in this case (Theorem 4.10).
2. Verify that the bound of an element, if it exists, is unique up to associates.
3°. What kind of reduction theorem can be proved for R when R is (i) an atomic principal left ideal domain or (ii) a Bezout domain?
4. (Kaplansky [49]) By an *elementary divisor ring* is meant a ring over which every matrix admits a diagonal reduction as in Theorem 4.7. Show that a ring over which every $m \times n$ matrix, where $m, n \leq 2$, admits a diagonal reduction is an elementary divisor ring.
5. (Kaplansky [49]) Show that an elementary divisor ring that is an integral domain is weakly finite.
6°. Is every commutative Bezout domain an elementary divisor ring?
7. Let R be a finitely generated module over a PID R. Given a decomposition (5) into cyclic modules that are as 'short' as possible, show directly that the e_i can be numbered so that each is a total divisor of the next.
8*. Let R be a PID and M a finitely generated R-module. Prove directly that the bounds of the elementary divisors of M are independent of the choice of decomposition of M.
9. Use Lemma 4.11 to prove that any left or right ideal in a simple Noetherian domain can be generated by two elements.
10. A ring R is called *semi-Euclidean* (D. Goldschmidt) if there is a function $\phi : R^\times \to \mathbb{N}$ such that for any $a, b \in R^\times$ either $a = bq + r, \phi(r) < \phi(b)$, or $\phi(a) = \phi(b)$ and $b = aq + r, \phi(r) < \phi(a)$. Prove a triangular reduction of matrices over semi-Euclidean rings. Show that every valuation ring is semi-Euclidean.
11. Let A be a right Bezout domain and K its field of fractions. Show that every finitely generated right A-submodule of K is cyclic. (*Hint*: Use Proposition 0.7.3.)
12°. Which of the results of this section go over to principal left ideal domains, or to Bezout domains?

1.5 Skew Laurent polynomials and Laurent series

Let A be an integral domain and α an automorphism of A, and consider the skew polynomial ring $A[x; \alpha]$. We can localize this ring at the powers of x; the resulting ring is denoted by $A[x, x^{-1}; \alpha]$ and is called the *skew Laurent polynomial ring*; it reduces to the Laurent polynomial ring $A[x, x^{-1}]$ when $\alpha = 1$. The latter may be thought of as the group algebra of the infinite cyclic group and many of the concepts introduced apply to general group algebras. This is true particularly of derivations.

Thus let G be any group, kG its group algebra over a field k and $\varepsilon : kG \to k$ the augmentation mapping, defined as a k-linear mapping such that $g\varepsilon = 1$ for all $g \in G$. The $(\varepsilon, 1)$-derivations of kF, where F is the free group, are called the *Fox derivatives* (Fox [53]). We shall have no more to say on this topic, except to remark that if F is free on $X = \{x_i\}$, each mapping $d : X \to kF$ defines a unique Fox derivative. Each such d can be written as a linear combination of the d_i, where d_i maps x_j to δ_{ij}. For if $u \in kF$ has the form $u = u\varepsilon + \sum(x_i - 1)d_i u$, and d is any Fox derivative, then $df = \sum dx_i . d_i f$.

Rings of skew (Laurent) polynomials are often useful in constructing counter-examples; as an illustration we shall obtain conditions for such a ring to be non-Hermite. We shall need a lemma on Ore sets in skew polynomial rings.

Lemma 1.5.1. *Let A be a right Noetherian domain with an automorphism α, an α-derivation δ and put $R = A[x; \alpha, \delta]$. Then the set S of all monic polynomials in R is a right Ore set.*

Proof. By Proposition 1.3, R is again right Noetherian. Now take $f \in S$, $g \in R^\times$ and put $\deg f = d$. Then R/fR is a free right A-module of rank d and hence Noetherian. Therefore the submodule generated by the images of $g, gx, gx^2, \ldots$ is finitely generated over A, say $gx^n \equiv \sum gx^i h_i \pmod{fR}$; hence $gu = fv$ for some $u \in S, v \in R$, as we had to show. ∎

Proposition 1.5.2. *Let A be a right Noetherian domain, α an automorphism and δ an α-derivation of A and write (i) $R = A[x; \alpha, \delta]$ or (ii) $R = A[x, x^{-1}; \alpha]$. If $a, c \in A$ are such that a is a non-unit and $a, x + c$ are right comaximal in R, and in case (ii) $c a^\alpha \notin aA$, then $aR \cap (x + c)R$ is stably free but non-principal, hence R is not 2-Hermite.*

Proof. By Proposition 0.4.3 it is enough to show that $aR \cap (x + c)R$ is not principal. In case (i) every element of R is a polynomial in x, while in case (ii) every element is associated to a polynomial. By Lemma 5.1 the monic polynomials form a right Ore set, so there exists a monic polynomial f such that $af = (x + c)u$.

1.5 Skew Laurent polynomials and Laurent series

Secondly, we have

$$ax = xa^\alpha + a^\delta = (x+c)a^\alpha + (a^\delta - ca^\alpha), \tag{1}$$

where δ is taken to be 0 in case (ii). By hypothesis there exist $a_1, a_2 \in A$, $a_1 \neq 0$, such that $(a^\delta - ca^\alpha)a_1 = aa_2$, hence on multiplying (1) by a_1 on the right and simplifying, we find

$$a(xa_1 - a_2) = (x+c)a^\alpha a_1.$$

We put $g = xa_1 - a_2$. If $aR \cap (x+c)R$ were principal, equal to ahR for some polynomial h say, then $g \in hR$, so $d(h) \leq 1$. Thus h (after multiplication by a suitable unit in case (ii)) has the form $h = xb_1 + b_2$, where $b_i \in A$. Since $ah \in (x+c)R$, $d(ah) \geq 1$, so $d(h) \geq 1$ and $b_1 \neq 0$. Now $f \in hR$, say $f = hd$; comparing highest terms and bearing in mind that f is monic, of degree r, say, we find $x^r = xb_1 x^{r-1} d_1$. Since α is an automorphism, this shows b_1 to be a unit, and dividing h by b_1 we may take it to be of the form $h = x + b$ ($b \in A$). By the definition of h we have $a(x+b) = (x+c)k$, i.e.

$$a(x+b) = xa^\alpha + a^\delta + ab = (x+c)k \quad \text{for some } k \in R. \tag{2}$$

It follows that $k = a^\alpha$, $ck = ab + a^\delta$, and so

$$ca^\alpha = ab + a^\delta. \tag{3}$$

In case (ii) $a^\delta = 0$ and so (3) is excluded by hypothesis. When (i) holds, we have

$$(x+c)p + aq = 1, \quad \text{for some } p, q \in R. \tag{4}$$

If we write $q = (x+b)q_1 + r$, where $q_1 \in R, r \in A$, then by (2), $aq = a(x+b)q_1 + ar = (x+c)kq_1 + ar$, and so

$$1 = (x+c)(p+kq_1) + ar.$$

A comparison of degrees shows that $p + kq_1 = 0$, $ar = 1$, which contradicts the fact that a is a non-unit. Thus $aR \cap (x+c)R$ is not principal, even though it is stably free, as we see by considering the short exact sequence

$$0 \to aR \cap (x+c)R \to R^2 \to R \to 0. \qquad \blacksquare$$

As an example consider the Weyl algebra; this may be written $A[x; 1, ']$, where $A = k[y]$ and $'$ is d/dy. Here y is a non-unit and x, y are comaximal; this answers Exercise 1.1.14.

Corollary 1.5.3. *The Weyl algebra $A_1(k)$ is not a 2-Hermite ring.* $\blacksquare$

Explicitly this means that $R = A_1(k)$ contains a non-principal right ideal $\mathfrak{a}$ such that $\mathfrak{a} \oplus R \cong R^2$. Moreover, the precise form of $\mathfrak{a}$ is given by Proposition

5.2 (see also Exercise 1.11). In fact, every finitely generated projecive module over a Weyl algebra is stably free (see Stafford [77]).

Besides polynomial rings we shall also need formal power series rings. Taking first the case of a zero derivation, we can describe the *formal power series ring* in x over A with endomorphism α, denoted by $R = A[[x; \alpha]]$, as the ring of all infinite series

$$f = a_0 + xa_1 + x^2 a_2 + \ldots, \tag{5}$$

with componentwise addition and multiplication based on the commutation rule $ax = xa^\alpha$. With each power series f we associate its *order* $o(f)$, defined as the suffix of the first non-zero coefficient in (5). When A is an integral domain and α is injective, this satisfies the conditions for an order-function analogous to D.1–D.3:

O.1. for $a \in A^\times$, $o(a) = 0$, while $o(0) = \infty$,
O.2. $o(a - b) \geq \min\{o(a), o(b)\}$,
O.3. $o(ab) = o(a) + o(b)$.

We can localize this ring at the set of all positive powers of x and so obtain the ring of *formal Laurent series*

$$f = \sum_{n=-k}^{\infty} x^n a_n. \tag{6}$$

We shall examine this ring more closely in the case where $A = K$ is a field. To express the multiplication we shall at first assume that α is an automorphism; putting $\beta = \alpha^{-1}$, we can write the commutation rule in the form

$$ax^{-1} = x^{-1} a^\beta.$$

Now it is an easy matter to show that the set of all series of the form (6) forms a field. If $o(f) = -k$, so that $a_{-k} \neq 0$, then we can write $f = x^{-k} a_{-k}(1 - g)$, where $o(g) > 0$, and so f has the inverse $f^{-1} = (\sum g^i) a_{-k}^{-1} x^k$. The resulting field is denoted by $K((x; \alpha))$ and may be obtained from the power series ring $K[[x; \alpha]]$ by formally inverting x. Since $K[x; \alpha]$ is embedded in the power series ring, it is also embedded in $K((x; \alpha))$, therefore, by the uniqueness of the field of fractions, so is $K(x; \alpha)$.

From O.3 above we see that $o(f)$ will not be an order-function unless $\delta = 0$, so when $\delta \neq 0$, the above method cannot be used, essentially because left multiplication by non-zero elements of K is not continuous in the x-adic topology, as the equation

$$ax = xa^\alpha + a^\delta \tag{7}$$

1.5 Skew Laurent polynomials and Laurent series

shows. To overcome this difficulty we introduce $y = x^{-1}$ and rewrite the commutation formula (7) in terms of y as

$$ay^{-1} = y^{-1}a^\alpha + a^\delta; \tag{8}$$

multiplying up, we obtain

$$ya = a^\alpha y + ya^\delta y. \tag{9}$$

In operator form (9) may be written as $L_y = \alpha R_y + \delta L_y R_y$, where L_y (R_y) stands for left (right) multiplication by y. Solving this equation for L_y, we obtain

$$ya = a^\alpha y + a^{\delta\alpha} y^2 + a^{\delta^2\alpha} y^3 + \ldots . \tag{10}$$

The result may be summed up as

Proposition 1.5.4. *Let K be a field with endomorphism α and α-derivation δ. Then the skew function field $K(x; \alpha, \delta)$ may be embedded in the field of skew Laurent series in $y = x^{-1}$ with commutation rules (8), (9).* ∎

When α is an automorphism, we can write every Laurent series in the form $\sum y^i a_i$, but this is no longer possible if α is merely an endomorphism. However, the same normal form can be achieved as follows. In any case α must be injective, as a field homomorphism, so we have an isomorphism between K and its image K^α. Let us write K_n for the image of K under α^n ($n \geq 0$); since α^n, like α, is injective, it provides an isomorphism between $K = K_0$ and K_n. For each $m = 1, 2, \ldots$ we take an isomorphic copy K_{-m} of K and embed K_{-m} in K_{-m-1} by identifying K_{-m} with the image of K_{-m-1} under α. In this way we obtain a filtration

$$\ldots K_n \subset K_{n-1} \subset \ldots \subset K_1 \subset K_0 \subset K_{-1} \subset \ldots ,$$

whose union is again a field, which we shall write as $K^{[\alpha]}$. Since $\alpha : K_n \to K_{n+1}$ is an isomorphism for all $n \in \mathbb{Z}$, α is an automorphism of $K^{[\alpha]}$.

We remark that $K^{[\alpha]}$ may also be obtained more directly as follows. In the skew function field $K(x; \alpha)$ we have the inner automorphism induced by x, which agrees with α on K (because $a^\alpha = x^{-1}ax$ for all $a \in K$); hence $K_{-m} = x^m K x^{-m}$ and $K^{[\alpha]} = \cup x^m K x^{-m}$. So in order to form Laurent series when α is not surjective (and $\delta = 0$), we may take $K^{[\alpha]}((x; \alpha))$.

When α is not surjective and $\delta \neq 0$, we can still form $K^{[\alpha]}$ but now δ will not be defined on all of $K^{[\alpha]}$ and there is no natural way of doing so unless we have a commutation relation between α and δ, such as $\alpha\delta = \delta\alpha$.

The power series representation is often useful for rational functions, e.g. for determining the centre of a rational function field. For simplicity we assume that $\delta = 0$.

Proposition 1.5.5. *Let K be a field and α an endomorphism of K, no positive power of which is inner. If C is the centre of K and C_0 the subfield of C fixed by α, then the centre of $K(x; \alpha)$ is C_0.*

Proof. If α^{-r} is inner, then so is α^r, hence no power $\alpha^i (i \neq 0)$ is inner. Consider a power series $f = \sum x^i a_i$; if this centralizes $K(x; \alpha)$, then $bf = fb$ for all $b \in K$, hence $\sum x^i (b^{\alpha^i} a_i - a_i b) = 0$, so $a_i b = b^{\alpha^i} a_i$, but α^i is not inner for $i \neq 0$; hence $a_i = 0$ for $i \neq 0$ and so $f = a_0 \in C$. Further we have $fx = xf$ and it follows that $a_0^\alpha = a_0$, i.e. $f \in C_0$. ∎

Sometimes it is useful to have a criterion for the rationality of a power series. Such a criterion is familiar for complex series, and this carries over to the general case, to provide the following rationality criterion:

Theorem 1.5.6. *A power series $\sum x^i a_i \in K((x; \alpha))$ is a rational function of x if and only if there exist integers r, n_0 and elements $c_1, \ldots, c_r \in K$ such that*

$$a_n = a_{n-1}^\alpha c_1 + a_{n-2}^{\alpha^2} c_2 + \cdots + a_{n-r}^{\alpha^r} c_r \quad \text{for all } n > n_0. \tag{11}$$

Proof. This is just the condition that

$$\left(\sum x^i a_i\right)\left(1 - \sum_1^r x^j c_j\right)$$

should be a polynomial, except for a factor x^{-k}. ∎

As an illustration of this result we have

Corollary 1.5.7. *Let $K \subseteq L$ be fields and α an automorphism of L mapping K into itself. Then*

$$K((x; \alpha)) \cap L(x; \alpha) = K(x; \alpha). \tag{12}$$

Proof. Clearly the field on the right is contained in the left-hand side. Conversely, any element of $L(x; \alpha)$ may be written uniquely as a Laurent series $f = \sum x^i a_i$ with coefficients a_i in L, and if it is a Laurent series over K, it follows that $a_i \in K$. If f belongs to the left-hand side of (12), it is rational over L and so its coefficients satisfy the above criterion. Thus the equations (11) have a solution for the c_i in L. They are linear equations with coefficients in K and hence have a solution in K. This means that $f \in K(x; \alpha)$ and the equality (12) is established. ∎

In the commutation relation for a formal power series ring over K we may from the beginning allow all higher powers; the most general relation is then of the form

$$ax = xa^{\delta_0} + x^2 a^{\delta_1} + x^3 a^{\delta_2} + \ldots, \tag{13}$$

where $\delta_0, \delta_1, \delta_2, \ldots$ is a sequence of mappings of K into itself. As in the case of skew polynomial rings (Section 1.1) we can show that the δs are additive and δ_0 preserves 1, while δ_i for $i > 0$ maps 1 to 0. Moreover, δ_0 must be an endomorphism of K, while the δs satisfy

$$(ab)^{\delta_n} = \sum_{i=0}^{n} a^{\Delta_i^n} b^{\delta_i}, \tag{14}$$

where Δ_i^n is the coefficient of t^{n+1} in $(\sum t^{k+1} \delta_k)^{i+1}$. Such a sequence $(\delta_0, \delta_1, \ldots)$ with $\delta_0 = \alpha$ is sometimes called a *higher α-derivation*. We shall not need a special notation for the ring defined by such a higher derivation.

As an example of a ring with a commutation formula (13) let us again take a skew polynomial ring $K[x; \alpha, \delta]$, where K is a field. In the skew function field $K[x; \alpha, \delta]$ consider the subring generated by K and x^{-1}; here it will be convenient to write the coefficients on the left. Writing $y = x^{-1}$, we have as before, $ya = a^\alpha y + y a^\delta y$, or in operator form, $L_y = \alpha R_y + \delta L_y R_y$, we again obtain (10), and this is indeed of the form (13), with $\delta_n = \delta^n \alpha$, except for a change of sides. In particular, if δ is nilpotent, say $\delta^{r+1} = 0$, then (10) reduces to the polynomial formula

$$ya = a^\alpha y + a^{\delta \alpha} y^2 + \cdots + a^{\delta^r \alpha} y^{r+1}. \tag{15}$$

Of course not every higher α-derivation is of the special form $\delta_n = \delta^n \alpha$, but it is a remarkable result, due to T. H. M. Smits [67], that if in (13) $\delta_i = 0$ for $i > r$, then (with another mild restriction) we do indeed have $\delta_n = \delta^n \alpha$. The rest of this section will not be needed later and so can be omitted without loss of continuity.

Theorem 1.5.8. *Let A be the ring of polynomials in an indeterminate y with coefficients in a field K, with the normal form*

$$f = a_0 + a_1 y + \cdots + a_n y^n \quad (a_i \in K), \tag{16}$$

such that $o(f) = \min\{i \mid a_i \neq 0\}$ is an order-function, and for any $a \in K$,

$$ya = a^\alpha y + a^{\delta_1} y^2 + \cdots + a^{\delta_r} y^{r+1}. \tag{17}$$

Assume further that (i) r is independent of a, (ii) α is an automorphism of K and (iii) $\alpha, \delta_1, \ldots, \delta_r$ are right linearly independent over K, in the sense that for all $b_1, \ldots, b_{r+1}$ and all $a \in K$,

$$a^\alpha b_1 + a^{\delta_1} b_2 + \cdots + a^{\delta_r} b_{r+1} = 0 \text{ implies } b_1 = b_2 = \ldots = b_{r+1} = 0.$$

Then A is obtained from a skew polynomial ring $R = K[x; \alpha, \delta]$, where δ is a nilpotent α-derivation: $\delta^{r+1} = 0$, by adjoining $y = x^{-1}$ and passing to the subring generated over K by y. Conversely, every skew polynomial ring with a nilpotent α-derivation δ leads to a ring satisfying (15), with $\delta_i = \delta^i \alpha$.

Proof. The converse has already been established. To prove the direct part, we have from (17), by induction on n,

$$y^n a = a_{nn} y^n + a_{nn+1} y^{n+1} + \cdots + a_{nk} y^k, \qquad (18)$$

where k may depend on n, but not on a. We shall write (18) as

$$y^n a = \sum a_{ni} y^i, \qquad (19)$$

where the summation is over all i and $a_{ni} = 0$ for $i < n$ or $i > k$. Clearly $a \mapsto (a_{ni})$ is a matrix representation of K over itself. From (17) we find, by induction on n,

$$a_{nn} = a^{\alpha^n}. \qquad (20)$$

For $n = 1$ we have from (19),

$$y(ab) = \sum (ab)_{1i} y^i,$$

$$(ya)b = \left(\sum a_{1i} y^i\right) b = \sum a_{1i} b_{ij} y^j.$$

Hence for $j > r + 1$,

$$\sum a_{1i} b_{ij} = 0 \quad (j = r + 2, \ldots).$$

Now $a_{11} = a^\alpha$, $a_{1i} = a^{\delta_{i-1}}$, and all these elements are right linearly independent over K, by (iii) above, so we obtain

$$b_{ij} = 0 \quad \text{for } i = 1, \ldots, r + 1; j = r + 2, \ldots.$$

Thus (18) takes the form

$$y^i a = a_{ii} y^i + a_{ii+1} y^{i+1} + \cdots + a_{ir+1} y^{r+1}. \qquad (21)$$

In particular, taking $i = r + 1$ and remembering (20), we find

$$y^{r+1} a = a^{\alpha^{r+1}} y^{r+1}. \qquad (22)$$

Similarly, for $i = r$, (21) becomes

$$y^r a = a^{\alpha^r} y^r + a_{rr+1} y^{r+1}.$$

Let us put $\alpha^{-1} = \beta$ and write a^γ for a_{rr+1}; using (22), we may write this relation formally as

$$y^{-1} a = y^r a^{\beta^{r+1}} y^{-r-1} = a^\beta y^{-1} + a^{\beta^{r+1} \gamma}.$$

1.5 Skew Laurent polynomials and Laurent series

If we define δ by setting $a^\delta = -a^{\beta^r \gamma}$, this relation takes on the form

$$y^{-1}a = a^\beta y^{-1} - a^{\beta \delta}. \tag{23}$$

We now replace a^β by a and recall that $\beta^{-1} = \alpha$, $y^{-1} = x$, and obtain

$$xa^\alpha = ax - a^\delta,$$

which by rearrangement yields

$$ax = xa^\alpha + a^\delta.$$

Thus A is obtained from the skew polynomial ring $K[x; \alpha, \delta]$ by taking the subring generated by K and $y = x^{-1}$. To find the relation between δ and the δ_i in (17) we apply y^{-1} to (17) and use (23):

$$a = y^{-1}(a^\alpha y + a^{\delta_1} y^2 + \cdots + a^{\delta_r} y^{r+1})$$
$$= a + a^{\delta_1 \beta} y + \cdots + a^{\delta_r \beta} y^r - a^\delta y - a^{\delta_1 \beta \delta} y^2 - \cdots - a^{\delta_r \beta \delta} y^{r+1}.$$

Equating coefficients, we find that $\delta_1 \beta = \delta$, $\delta_i \beta = \delta_{i-1} \beta \delta$, $\delta_r \beta \delta = 0$, hence by induction, $\delta_i \beta = \delta^i$, and so we obtain the desired relations

$$\delta_i = \delta^i \alpha, \quad i = 1, \ldots, r, \quad \delta^{r+1} = 0. \qquad \blacksquare$$

The power series ring and the Laurent series ring are special cases of the following construction, which allows the group algebra of any ordered group to be embedded in a field.

Let M be an ordered monoid, i.e. a monoid with a total ordering '$<$' such that $a_i < b_i (i = 1, 2)$ implies $a_1 a_2 < b_1 b_2$. By a *convex* submonoid we shall understand a submonoid S such that $a, b \in S$ and $a \leq x \leq b$ implies $x \in S$. Let K be any ring and consider the direct power K^M, regarded as a K-module. With each $f \in K^M$ we associate its *support*, defined as

$$D(f) = \{a \in M \mid f(a) \neq 0\}.$$

The elements of finite support may be written as finite sums $\sum f(a)a$ and just constitute the monoid ring KM of M over K, with the multiplication rule

$$fg = h, \quad \text{where } h(c) = \sum_{ab=c} f(a)g(b). \tag{24}$$

When M is a group, the latter sum may also be written $\sum f(x)g(x^{-1}c)$. Now let $R = K((M))$ be the set of elements of K^M with well-ordered support; here the definition (24) for the product still makes sense, for if the terms $f(a)g(b)$ are ordered so that $a_1 < a_2 < \ldots$ for the arguments of f, then for $c = a_i b_i$ we have $b_1 > b_2 > \ldots$ and this must break off, by the well-ordering of $D(g)$. Thus each $h(c)$ is defined; let us show further that h again has well-ordered support and that R is in fact a ring. Clearly we may suppose that the supports of f and g are both infinite. If $c_n \in D(h)$, say $c_n = a_n b_n$, consider a sequence

$$c_1 > c_2 > \ldots .$$

By the well-ordering of $D(f)$ the sequence (a_n) has a subsequence that is increasing, hence the corresponding subsequence of b's is decreasing and so must terminate. Thus $D(h)$ is well-ordered and this shows that $h = fg \in R$; clearly $f + g \in R$ and it is easily checked that R is a K-ring. We shall call $R = K((M))$ the *ordered series ring* of M over K. We remark that R has an order function defined on it, with values in M, by

$$o(f) = \min\{a \in M \mid f(a) \neq 0\}.$$

If $o(f) = a$, then the term $f(a)a$ is called the *leading term* of f. We claim that f is invertible if and only if the leading term of f is invertible. For this condition is clearly necessary for invertibility. Conversely, assume that it holds, let $f = f_0 a_0 + \ldots$, and write $g = a_0^{-1} f_0^{-1} f$; then g has the form $1 - h$, where $h = \sum h(x) x$ has support consisting entirely of elements > 1. Formally we can write

$$p = 1 + h + h^2 + \ldots . \qquad (25)$$

If we can show that $p \in R$, then it is clear that p is indeed an inverse of g, and this will prove f to be invertible.

We shall show that $D(p)$ is well-ordered; for if not, then we would have an infinite descending chain

$$z_1 = u_{11} \ldots u_{1n_1} > \ldots > z_i = u_{i1} \ldots u_{in_i} > \ldots, \qquad (26)$$

where $u_{ij} \in D(h)$. By omitting some of the z_k we may assume that $n_1 \leq n_2 \leq \ldots$. Let v_i be the least of $u_{i1}, \ldots, u_{in_i}$; then $v_1 > v_2 > \ldots$, and this contradicts the fact that $D(h)$ is well-ordered. The same argument shows that any element of M occurs in at most a finite number of the $D(h^n)$. For otherwise we would have an infinite chain as in (26), but with equality signs; now the same argument as before shows that the chain breaks off. Thus p is well-defined and we have proved

Theorem 1.5.9. *Let K be a ring and M an ordered monoid. Then the set $K((M))$ of power series with well-ordered support is a ring, the ordered series ring of M over K, and an element of this ring is invertible if and only if its leading term is invertible.* ∎

When M is a group and K is a field, every non-zero element of $K((M))$ has an inverse and we obtain the *Malcev–Neumann construction*:

1.5 Skew Laurent polynomials and Laurent series

Corollary 1.5.10. *Let K be a field and G an ordered group. Then the ordered series ring K((G)) is a field.* ∎

We shall want to apply this result to free groups, so we shall need to prove that every free group G can be totally ordered. This follows for example, by writing the elements of G as infinite power products of basic commutators and taking the lexicographic ordering of the exponents (see Hall [59], Chapter 11, or also Exercise 10).

We conclude this section with an interesting result due to G. M. Bergman, on normal forms under conjugation in ordered series rings.

Theorem 1.5.11. *Let K be a commutative ring and M an ordered monoid, and let f be an element of $K((M))$ with invertible leading term $f_u.u$. Then there exists an element q in $K((M))$ with leading term 1 such that $q^{-1}fq$ has support entirely in the centralizer of u in M.*

Further, q may be chosen so that no element of its support except the leading term 1 commutes with u. Under this hypothesis q is unique.

Proof. If $u = 1$, there is nothing to prove, so we may assume that $u \neq 1$ and on replacing f by f^{-1} if necessary (bearing in mind that f is invertible, by Theorem 5.9), we may suppose that $u > 1$. Further, we may assume without loss of generality that $f_u = 1$.

We shall denote the centralizer of u in M by C_u. Our aim will be to show that any term αt of f such that $t \notin C_u$ can be got rid of by conjugating by an element $(1 + \alpha tu^{-1})$ or $(1 - \alpha u^{-1}t)$, at the expense of adding higher terms. The process is then repeated on the new leading term. We shall need to set up some machinery to show that the construction of q can be made to 'converge'.

We shall use the customary notation $[u, v]$, $[u, v)$, etc. for closed, half-open, etc. intervals in M and let ∞ be such that $s < \infty$ for all $s \in M$. This will allow us to use the phrase: 'the leading term of f is αt', even when $f = 0$, in which case t is taken to be ∞ and α undefined. If $u, s \in M$, where u is invertible, we put $s/u = \max\{su^{-1}, u^{-1}s\}$; it is easy to verify that $x < y \Rightarrow x/u < y/u$.

Let X be the set of triples (t, g, p), where $t \in (u, \infty]$, $g, p \in K((M))$ and g has leading term u and support in $C_u \cap [u, t)$, while p has leading term 1 and support in $[1, t/u)$ such that $pgp^{-1} - f$ has a leading term of form αt. By our convention this means that $pgp^{-1} = f$ if and only if $t = \infty$. We partially order X by writing $(t, g, p) \leq (t', g', p')$ if $t \leq t'$, $D(g' - g) \subseteq [t, t')$ and $D(p' - p) \subseteq [t/u, t'/u)$; these conventions just mean that the series g' and p' 'extend' g and p, respectively.

If the leading term of $f - u$ is αt, then $(t, u, 1) \in X$; this shows that $X \neq \emptyset$. Our aim is to show that X is inductive, so that Zorn's lemma can be applied.

Suppose then that we have a chain $\{(t_\lambda, g_\lambda, p_\lambda) | \lambda \in I\}$ in X. We can 'piece together' the g_λ to get a common extension g, with support in $\cup_\lambda [u, t_\lambda)$, and similarly from the p_λ form a common extension p with support in $\cup_\lambda [1, (t_\lambda/u))$. It is clear that g and p again have well-ordered supports; we also see that $pgp^{-1} - f$ will have support in $[u, \infty) \setminus \cup_\lambda [u, t_\lambda)$. If we write the leading term of $pgp^{-1} - f$ as αt, we see that (t, g, p) majorizes the given chain; hence X is inductive and so by Zorn's lemma, X has a maximal element. To complete the proof we show that if $(t, g, p) \in X$ and $t < \infty$, then we can construct $(t', g', p') > (t, g, p)$. It will follow that a maximal element of X must have the form (∞, h, q), so that $qhq^{-1} - f = 0$, i.e. $h = q^{-1}fq$, as claimed.

Let the leading term of $pgp^{-1} - f$ be αt, say. We shall find an element g' extending g and having support in $[u, t]$ (recall that $D(g) \subseteq [u, t))$ and p' extending p with support in $[1, t/u]$, and show that $D(p'g'p'^{-1} - f) \subseteq (t, \infty)$. Hence if $\alpha t'$ is the leading term of $p'g'p'^{-1} - f$, then $(t, g, p) < (t', g', p') \in X$. We distinguish three cases.

(i) $tu^{-1} > u^{-1}t$. Then $t/u = tu^{-1}$ and we write $p' = p - \alpha t u^{-1}$; since p has leading term 1, we see that $p'^{-1} = p^{-1} + \alpha t u^{-1} +$ higher terms. Take $g' = g$; on multiplying out $p'g'p'^{-1}$ we find that the new terms introduced are $(-\alpha t u^{-1}).u.1 = -\alpha t$, $1.u.(\alpha t u^{-1}) = \alpha u t u^{-1}$, and higher terms. Since $tu^{-1} > u^{-1}t$, we have $utu^{-1} > t$, so the lowest term introduced is $-\alpha t$; this cancels the leading term αt of $pgp^{-1} - f$, hence $D(p'g'p'^{-1} - f) \subseteq (t, \infty)$ as claimed.

(ii) $tu^{-1} < u^{-1}t$. Now $t/u = u^{-1}t$; we put $p' = p + \alpha u^{-1}t$ and again take $g' = g$. The lowest terms introduced are $(\alpha u^{-1}t).u.1 = \alpha u^{-1}tu$ and $1.u.(-\alpha u^{-1}t) = -\alpha t$. Here the latter is the lower and again this cancels the leading term αt of $pgp^{-1} - f$.

(iii) $tu^{-1} = u^{-1}t$. Now t commutes with u; in this case the terms arising under (i), (ii) would cancel and so be of no help in eliminating the leading term of $pgp^{-1} - f$. So we set $p' = p$ and $g' = g - \alpha t$, which is permissible because $t \in C_u$. Now the lowest term by which $pgp^{-1} - f$ has changed is $1.(-\alpha t).1 = -\alpha t$, so here too, $p'g'p'^{-1} - f$ has support in (t, ∞).

This then proves the existence of q; since in the above construction we never added a term from C_u to our p's, we can clearly take q so that

$$D(q) \cap C_u = \{1\}. \tag{27}$$

Let q be so chosen and suppose that $q' \neq q$ is another element with leading term 1 such that $D(q'^{-1}fq') \subseteq C_u$. Write $q' = q(1 + h)$, $g = q^{-1}fq$, so that $g' = q'^{-1}fq' = (1 + h)^{-1}g(1 + h)$. If αt is the leading term of h, suppose that $t \notin C_u$; then g' would have a term ut or tu (whichever is the smaller), but by hypothesis, $D(g') \subseteq C_u$, so $t \in C_u$. This means that $q' = q(1 + h)$ will have in

1.5 Skew Laurent polynomials and Laurent series

its support a term $t \neq 1$, which commutes with u, a contradiction. Hence q is the unique element with the desired property and satisfying (27). ∎

We note that the assertion of Theorem 5.11 is strongest when the centralizer C_u is small. In particular, in a free group the centralizer of any element $\neq 1$ is a cyclic group. It follows that every element f with leading term αu, $u \neq 1$, is conjugate to a series in a single variable, and this even holds when the leading term is α, by applying the argument to $f - \alpha$. Thus we obtain

Corollary 1.5.12. *Let G be a free group and k a commutative field. In the ordered series ring $k((G))$ every element is conjugate to a Laurent series in a single variable.* ∎

Exercises 1.5

1. Show that if R is weakly finite, then so is $R[x]$. What happens for $R[x; \alpha, \delta]$ or $R[[x]]$?
2°. Obtain an analogue of Proposition 5.4 when $\delta \neq 0$ and α is not surjective (try the cases $\alpha\delta = \pm\delta\alpha$ first).
3°. To what extent are the conditions (i)–(iii) of Theorem 5.8 necessary?
4. (P. Samuel) If R is a Euclidean domain, show that $R[[x]][x^{-1}]$ is also Euclidean. [*Hint*: Define $\phi(\sum x^i a_i) = \phi(a_s)$, where $s = \min\{i \mid a_i \neq 0\}$.]
5. (Dress [71]) Let R be an integral domain. If $R[[x]][x^{-1}]$ is a Euclidean domain relative to a function ϕ, show that R is Euclidean relative to ϕ_R, where $\phi_R(a) = \min\{\phi(f) \mid d(f) \in Ra\}$, where $d(f)$ is the coefficient of the least power occurring in f.
6. Show that in an ordered group G any inner automorphism is order-preserving. Deduce that an element f of $k((G))$ commutes with $u \in G$ if and only if the support of f lies in the centralizer of u.
7*. (L. G. Makar–Limanov) Show that for any commutator $[a, b] = ab - ba$ in an ordered series ring $k((G))$ the coefficient of 1 is 0. Deduce that $[[a, b], b] = 0$ implies $[a, b] = 0$.
8. In the ordered series ring $k((F))$ of a free group F over a field k show: if f commutes with $\sum \lambda_i u^i \notin k(u \in F)$, then f commutes with u. Deduce that two elements of $k((F))$ commuting with the same element of $k((F))\backslash k$ commute with each other, i.e. commutativity is transitive on $k((F))\backslash k$. (*Hint*: Use the fact that commutativity is transitive on $F\backslash\{1\}$.)
9. Let M be an ordered monoid for which commutativity is transitive on $M\backslash\{1\}$. Show that f commutes with a conjugate of g if and only if the leading term of f commutes with a conjugate of the leading term of g.
10. Let F be a free group and F_n the nth term of the lower central series (defined as the subgroup generated by all repeated commutators of weight n). Given that F_n/F_{n+1} is free abelian and $\cap F_n = 1$, show that F can be ordered.
11*. (G. M. Bergman) Let $\mathbb{R}\langle\langle X \rangle\rangle$ be the free power series ring over the real numbers. Verify that if $|X| = r$, the elements $1 - x(x \in X)$ with inverses $\sum x^n$ form the free

generators of a free group of rank r. Show that $R\langle\langle X\rangle\rangle$ and hence the free group can be totally ordered by taking any ordering on X, extending it to the lexicographic ordering of the free monoid on X and then ordering $R\langle\langle X\rangle\rangle$ by the sign of its lowest term.

12. Let R be a principal ideal domain with endomorphism α and α-derivation δ and let $S = R[x, x^{-1}; \alpha, \delta]$ be the skew Laurent polynomial ring over R. Writing $y = x^{-1}$, show that the y-adic completion of S is again a principal ideal domain (see Cohn [87a]). What can be said when R is only left or right principal?

13°. A group is called *right-ordered* if it has a total ordering such that $a < b \Rightarrow ac < bc$. Given a right-ordered group G, is its group algebra embeddable in a field?

1.6 Iterated skew polynomial rings

From any ring R with an automorphism α we can form the skew polynomial ring $R[x; \alpha]$, but this will not be a principal ideal domain unless R was a field. However, when α is only required to be an endomorphism, we get a PID under wider conditions, as we saw in Proposition 3.3. The exact class was determined by Jategaonkar [69a], who used it to give an ingenious construction of 'iterated skew polynomial rings', which form a useful source of counter-examples. We shall here follow Lenstra's presentation in showing more generally that the iterated skew polynomial rings of Jategaonkar type ('J-rings') form the precise class of integral domains with a unique remainder algorithm.

Let R be a ring, α an endomorphism and δ an α-derivation of R. Then the skew polynomial ring $R[x; \alpha, \delta]$ is called a *J-skew polynomial ring* over R or simply a *J-ring* if α is injective and maps $R^\times$ into $U(R)$.

Given a ring R, a subring K and an ordinal number τ, we shall say that R is an *iterated skew polynomial ring* of type τ over K if R contains an ascending chain of K-rings $R_\lambda (\lambda \leq \tau)$, such that

J.1. $R_0 = K$,
J.2. $R_{\lambda+1}$ is a skew polynomial ring over $R_\lambda (\lambda < \tau)$,
J.3. $R_\mu = \cup_{\lambda < \mu} R_\lambda$ for any limit ordinal $\mu \leq \tau$,
J.4. $R_\tau = R$.

From the definition it is clear that every element of R can be written uniquely as

$$\sum x_{\lambda_1} \ldots x_{\lambda_r} a_{\lambda_1 \ldots \lambda_r}, \quad a_{\lambda_1 \ldots \lambda_r} \in K, \lambda_1 \geq \ldots \geq \lambda_r. \tag{1}$$

If K is a field and each $R_{\lambda+1}$ is a J-skew polynomial ring over R_λ, R will be called a *J-ring of type τ* over K.

1.6 Iterated skew polynomial rings

From Proposition 1.4 we know that any skew polynomial ring over a right Ore domain (with an injective endomorphism) is again right Ore. Jategaonkar's basic observation was that a J-skew polynomial ring over a right PID is again right principal (Proposition 3.3). We now show that a J-ring over a Euclidean domain is again Euclidean. In a Euclidean ring we shall denote the function giving the least algorithm by θ, where $\theta(0) = -\infty$. We also recall the multiplication of ordinals (see BA, Section 1.2): $2.\omega = \omega, \omega.2 = \omega + \omega$.

Proposition 1.6.1. *Let R be an integral domain with unique remainder algorithm, defined by θ_R and suppose that $S = R[x; \alpha, \delta]$ is a J-ring. Then S has again a unique remainder algorithm, with the function*

$$\theta\left(\sum_0^n x^i a_i\right) = \lambda.n + \theta_R(a_n) \quad (a_n \neq 0), \tag{2}$$

for some ordinal λ.

Proof. Let $a = \sum_0^n x^i a_i$, $b = \sum_0^m x^i b_i (a_n, b_m \neq 0)$, and note that $\theta = \theta_R$ on R. If $\theta(a) \geq \theta(b)$, then $n \geq m$. Either $n > m$; then $a' = a - bx(b_m^\alpha)^{-1} x^{n-m-1} a_n$ has degree $< n$ and so $\theta(a') < \theta(a)$; or $n = m$, $\theta(a_n) \geq \theta(b_n)$ and so by the algorithm in R, $\theta(a_n - b_n c) < \theta(a_n)$ for suitable $c \in R$, hence $\theta(a - bc) < \theta(a)$. This shows S to be Euclidean; now the conditions of Proposition 2.3 are easily checked, so we have a unique remainder algorithm on S. ∎

Since any field has a unique remainder algorithm, we obtain by transfinite induction

Corollary 1.6.2. *For any ordinal τ, a J-ring of type τ over a field is an integral domain with a unique remainder algorithm.* ∎

It is of interest to note that the converse also holds:

Theorem 1.6.3. *A ring R is a J-ring of type τ over a field (for some ordinal τ) if and only if R is an integral domain with unique remainder algorithm.*

Proof. The direct part was proved in Corollary 6.2. For the converse, let R be an integral domain with unique remainder algorithm $\theta = \theta_R$; by Proposition 2.5, θ is a degree-function. Take an ordinal τ bounding θ and define Λ as the set of ordinals $\lambda \leq \tau$ in the range of θ such that

$$\lambda > 0 \text{ and } \beta, \gamma < \lambda \Rightarrow \beta + \gamma < \lambda. \tag{3}$$

We claim

$$\lambda \in \Lambda, \beta < \lambda \Rightarrow \beta + \lambda = \lambda. \tag{4}$$

For the unique solution γ of $\beta + \gamma = \lambda$ cannot be $< \lambda$ by (3), but clearly $\gamma \leq \lambda$, hence $\gamma = \lambda$.

We index Λ by an initial segment of the ordinals such that $\lambda_\beta < \lambda_\gamma \Rightarrow \beta < \gamma$, and define

$$R_\beta = \{a \in R \mid \theta(a) < \lambda_\beta\} \quad (\beta < \tau).$$

Since θ is a degree-function, R_β is a subring of R. Now $\lambda_0 = 1$, so $R_0 = \{a \in R \mid \theta(a) \leq 0\}$, and this is a field, K say. For any limit ordinal γ we have $\lambda_\gamma = \lim_{\beta<\gamma} \lambda_\beta$ and hence

$$R_\gamma = \bigcup_{\beta<\gamma} R_\beta,$$

while $R_\tau = R$, so it only remains to show that for any $\beta < \tau$, $R_{\beta+1}$ is a J-skew polynomial ring over R_β. Choose $x \in R$ such that $\theta(x) = \lambda_\beta$ and denote by $R_\beta[x]$ the set of all polynomials $f = \sum x^i a_i$ ($a_i \in R_\beta$). We claim that $R_\beta[x]$ is a J-skew polynomial ring. Given $f \in R_\beta[x]$ of degree n with leading coefficient a_n, we have $\theta(f) = \lambda_\beta.n + \theta(a_n)$, hence $f \neq 0$ and the expression for it is unique. Given $a \in R_\beta^\times$, we have $ax = xa^\alpha + a^\delta$ for unique $a^\alpha, a^\delta \in R$ and $\theta(a^\delta) < \theta(x) = \lambda_\beta$, so $a^\delta \in R_\beta$. Next we have $\theta(ax) = \theta(a) + \theta(x) = \theta(x)$ by (4), so $\theta(xa^\alpha) = \theta(x)$, whence $\theta(a^\alpha) = 0$ and it follows that $a^\alpha \in R_0 \subset R_\beta$. Therefore α, δ map R_β into itself (if we define $0^\alpha = 0^\delta = 0$) and $R_\beta[x]$ is indeed a J-skew polynomial ring.

To establish the equality $R_{\beta+1} = R_\beta[x]$ we note that $\lambda = \lambda_\beta.\omega$ is the smallest ordinal of Λ that is $> \lambda_\beta$, hence $\lambda_{\beta+1} = \lambda_\beta.\omega$. We take $f \in R_{\beta+1} \backslash R_\beta[x]$ such that $\theta(f)$ has its least value. By the Euclidean algorithm, $f = xa + b$ with $b \in R_\beta$. As before, $\theta(f) = \theta(x) + \theta(a) = \lambda_\beta + \theta(a)$, since $a \neq 0$. But $\theta(f) < \lambda_{\beta+1}$, so we have $\theta(f) = \lambda_\beta.n + \gamma$ with $\gamma < \lambda_\beta$ and $1 \leq n < \omega$. Hence $\theta(a) = \lambda_\beta(n-1) + \gamma < \lambda_\beta n \leq \theta(f)$, therefore $a \in R_\beta[x]$, by the minimality of $\theta(f)$, and so $f = xa + b \in R_\beta[x]$, a contradiction. This shows that $R_{\beta+1} = R_\beta[x]$, as claimed. ∎

There remains the problem of constructing J-rings of a given type over a field. The J-rings of type 0 are fields, J-rings of type 1 are the usual skew polynomial rings over fields, J-rings of type 2 may be obtained by an ad hoc construction (see Exercise 8), but beyond that it is no easier to construct a J-ring of finite type than one of arbitrary type. In particular it is not possible to give a recursive construction, because the set $K^\times$ must contain an isomorphic copy of $R^\times$, so that K depends very much on τ. To construct K we shall use the field containing a free group, constructed in Section 1.5.

1.6 Iterated skew polynomial rings

Let $X = \{x_\alpha\}(\alpha < \tau)$ be a sequence of indeterminates and denote by E the ordered series field constructed from the free group on X over k by Corollary 5.10. Denote by K the subfield generated over k by the elements

$$u_{\beta\alpha_1\ldots\alpha_r} = (x_{\alpha_1}\ldots x_{\alpha_r})^{-1}x_\beta x_{\alpha_1}\ldots x_{\alpha_r}, \quad \alpha_1 \geq \ldots \geq \alpha_r, \alpha_1 > \beta, r \geq 1. \quad (5)$$

Lemma 1.6.4. *In the field K defined above, the centralizer of x_γ, for any γ is k.*

Proof. Let F be the free group on X and G the subgroup of F generated by the right-hand sides of (5). Each of these generators has odd length in the x's, and in any expression of an element of G as a power product of the u's, cancellation cannot affect the central factor of any u. It follows that each such expression must begin with a factor x_α^{-1} and end in a factor x_β, even after all the cancellations have been made, for this is true of the u's and their inverses. In particular it follows that G does not contain x_γ^n, for any $n \neq 0$.

Consider any $a \in K$; this is a power series: $a = \sum a_u u$, where u runs over G.

Conjugation by x_γ maps G into itself; explicitly we have

$$x_\gamma^{-1} u_{\beta\alpha_1\ldots\alpha_r} x_\gamma = u_{\alpha_r\gamma}^{-1} \ldots u_{\alpha_i\gamma}^{-1} u_{\beta\alpha_1\ldots\alpha_{i-1}\gamma} u_{\alpha_i\gamma} \ldots u_{\alpha_r\gamma},$$

where i is such that $\alpha_{i-1} \geq \gamma > \alpha_i$.

Since x_γ commutes only with the powers x_γ^n, it follows that conjugation by x_γ fixes only $1 \in G$ and moves all other elements in infinite orbits; to be precise, the elements in each orbit arise from the positive powers of x_γ, since G admits conjugation by x_γ, but not by x_γ^{-1}. Hence $x_\gamma^{-1} a x_\gamma = a$ is possible only if $a_u = 0$ for $u \neq 1$, i.e. $a = a_1 \in k$. ∎

To prove the existence of a J-ring of given type, let G be as before and denote by ϕ_γ the endomorphism of G induced by conjugation with x_γ; this can clearly be extended to an endomorphism of K, again denoted by ϕ_γ. Thus we have

$$ax_\gamma = x_\gamma a^{\phi_\gamma} \quad (a \in K). \quad (6)$$

Let E again be the ordered series ring on G over K and R the subring of E generated by K and the x_β ($\beta < \tau$). By (6), each element of R can be written as a finite sum

$$\sum x_{\alpha_1}\ldots x_{\alpha_r} a_{\alpha_1\ldots\alpha_r}, \quad \text{where } a_{\alpha_1\ldots\alpha_r} \in K. \quad (7)$$

Fix any term in (7) and let α_i be the last suffix such that $\alpha_i < \alpha_{i+1} \geq \cdots \geq \alpha_r$. Then we can use (5) to pull x_{α_i} through to the right; this will only change the coefficient. Repeating the process if necessary we can ensure that $\alpha_1 \geq \cdots \geq \alpha_r$ in each term of (7). We assert that under this condition the expression (7) is unique. Thus assume that the expression (7) vanishes, where now $\alpha_1 \geq \cdots \geq \alpha_r$, and assume that not all coefficients are 0. Let α be the highest suffix such that x_α occurs in (7); then we have the equation

$$\sum x_\alpha^i c_i = 0,$$

where each c_i is a polynomial in the x_β ($\beta < \alpha$), with right coefficients in K, and not all c_i vanishing. On conjugating by x_α we get

$$\sum x_\alpha^i c_i^{\phi_\alpha} = 0,$$

where the coefficients now lie in K; hence x_α is right algebraic over K. If we take a monic equation of least degree satisfied by x_α and conjugate with x_α, we get another monic equation, which must equal the first, by uniqueness. By Lemma 6.4 it follows that the coefficients lie in k, so x_α is algebraic over k, which is a contradiction. This proves the uniqueness of (7), so we obtain

Theorem 1.6.5. *Let k be any field and τ an ordinal. Then there exists a J-ring of type τ over k (with zero derivation).* ■

By Corollary 6.2 the resulting J-ring has a Euclidean algorithm and hence is a principal right ideal domain. However, it is not atomic if $\tau > 1$, for then we have

$$x_2 = x_1 x_2 u_{12}^{-1} = x_1^n x_2 u_{12}^{-n},$$

so x_2 has factorizations of arbitrary length.

A closely related class of rings has been studied by Brungs [69b], namely rings in which the set of all right ideals is well-ordered by inclusion, i.e. they are totally ordered and every set has a greatest member. Actually it is enough to take the set of all principal right ideals. Such a ring contains a unique maximal principal proper right ideal that is clearly also the unique maximal proper right ideal and hence is the Jacobson radical $J = J(R)$ of R. Hence J is two-sided and R/J is a field, because it has no non-trivial right ideals. Thus R is a local ring and we have the following structure theorem:

Theorem 1.6.6. *Let R be a ring in which the set of all principal right ideals is well-ordered by inclusion. Then all right ideals of R are principal and are in fact two-sided; thus all regular elements of R are right invariant.*

Proof. Let $\mathfrak{a}$ be a non-zero right ideal, bR the maximal principal right ideal properly contained in $\mathfrak{a}$ and $c \in \mathfrak{a} \setminus bR$. Then $cR \subseteq \mathfrak{a}$ but $cR \not\subseteq bR$, hence $bR \subset cR \subseteq \mathfrak{a}$, and so $cR = \mathfrak{a}$. Thus all right ideals are principal. Further, R is right invariant, for otherwise take a maximal right but not left ideal aR; there exists $b \in R$ such that $baR \supset aR$, hence $a = bac$, where c is a non-unit and so lies

1.6 Iterated skew polynomial rings

in J. But baR is two-sided, by the maximality of aR, so $b.ba = ba.b'$, hence $ba = b.bac = bab'c$. This shows that $ba(1 - b'c) = 0$, where $c \in J$, and so $1 - b'c$ is a unit. It follows that $ba = 0$, which is a contradiction, and this shows each right ideal of R to be two-sided. ∎

In order to examine these rings more closely, we note that they have the following property:

Lemma 1.6.7. *Let R be a ring as in Theorem 6.6, and $a, b, c \in R$. If $ab \in cR, b \notin cR$, then $a^n \in cR$ for some $n \geq 1$.*

Proof. If $\cap a^n R = dR$, then $adR = dR$, hence $d = adu$ for some $u \in R$. Now if $bR \subseteq dR$, say $b = dv$, then $b = dv = aduv = advu' = abu' \in cR$, which contradicts the hypothesis; thus $dR \subset bR$, whence $a^n R \subseteq bR$ for some n, and so $a^{n+1} \in abR \subseteq cR$. ∎

We now define elements $p_\alpha (\alpha \geq 0)$ of R as follows: $p_0 R$ is the maximal right ideal of R, and for any $\alpha > 0$, p_α is defined (up to a right unit factor) by

$$p_\alpha R = \begin{cases} \bigcap_{\beta < \alpha} p_\beta R & \text{if } \alpha \text{ is a limit ordinal,} \\ \cap p_{\alpha-1}^n R & \text{otherwise.} \end{cases}$$

Clearly, if $\alpha > \beta$, then $p_\alpha R \subset p_\beta R$; moreover, $p_\beta p_\alpha R \subseteq p_\alpha R$, and by definition of p_α, $p_\alpha = p_\beta c$ for some non-unit c. Since $p_\beta^n \notin p_\alpha R$ for all n, we have $c \in p_\alpha R$, so $p_\alpha \in p_\beta p_\alpha R$, and therefore $p_\alpha R = p_\beta p_\alpha R$, i.e.

$$p_\beta p_\alpha = p_\alpha u_{\alpha \beta} \quad \text{for } \beta < \alpha, \text{ where } u_{\alpha \beta} \in U(R). \tag{8}$$

We claim that each $a \in R^\times$ is expressible uniquely as

$$a = p_{\alpha_1} \ldots p_{\alpha_r} u, \quad \text{where } u \in U(R), \alpha_1 \geq \cdots \geq \alpha_r. \tag{9}$$

For let α be the least ordinal such that $a \notin p_\alpha R$; then α cannot be a limit ordinal, say $\alpha = \alpha' + 1$ and $a \in p_{\alpha'}^n R$ for some n. Taking n as large as possible, we have

$$a = p_{\alpha'}^{n'} a', \quad \text{where } a' \notin p_{\alpha'} R.$$

By repeating this process on a' we find by induction that a has the form (9) as claimed, and this expression is unique, from the way it was found.

We denote by σ the order type of the sequence of right ideals of R and express σ in the form

$$\sigma = \omega^{\tau_1} n_1 + \omega^{\tau_2} n_2 + \cdots + \omega n_{k-1} + n_k + 1, \tag{10}$$

where $\tau_1 > \tau_2 > \cdots > \tau_{k-2} > 1$. Then the p_α are indexed by all $\alpha < \tau_1$, and

(10) corresponds to the relation

$$p_{\tau_1}^{n_1} p_{\tau_2}^{n_2} \cdots p_1^{n_{k-1}} p_0^{n_k} = 0. \tag{11}$$

It only remains to construct a ring with these properties. To do this we take a J-ring of type τ_1 over any field and localize at the set of all polynomials with non-zero constant term, using Proposition 4.3. Now add the relations (11), where the p_α correspond to the x_α.

Exercises 1.6

1°. Show in Proposition 3.3 that the sufficient condition is also necessary when $\delta = 0$. What happens in general?

2. Show that a left Ore domain whose right ideals are well-ordered (under inclusion) is a principal ideal domain.

3. (Jategaonkar [69a]) Let P be the localization of a J-ring of type τ at the set of all polynomials with non-zero constant term. Show that the Jacobson radical of P is $J = x_1 P$. If transfinite powers of J are defined by the equations $J^{\alpha+1} = J^\alpha J$ and $J^\alpha = \cap_{\beta<\alpha} J^\beta$ at a limit ordinal α, show that $J^\alpha \supseteq x_\alpha P$. Deduce that $J^\alpha \neq 0$ if and only if $\alpha < \tau$. Show also that J^α is a two-sided ideal.

4. (Jategaonkar [69a]) Let R be a J-ring of type τ. Show that the elements $1 + x_\alpha$ ($\alpha < \tau$) are left linearly independent over R. If τ is a limit ordinal, show that every non-zero right ideal contains an ideal of the form $x_\alpha R$. Show also that $\mathfrak{a} = \sum R(1 + x_\alpha)$ is a proper left ideal and that no maximal left ideal containing $\mathfrak{a}$ can contain a non-zero ideal. Deduce that in this case R is left but not right primitive.

5. By a *strong prime ideal* in a ring R is meant an ideal $\mathfrak{p}$ such that $R/\mathfrak{p}$ is an integral domain. Determine the strong prime ideals in a J-ring of given type.

6. Show that a *reduced ring* (i.e. without non-zero nilpotent elements), whose right ideals are well-ordered is an integral domain.

7. (Brungs [69b]) Let R be a ring whose right ideals are well-ordered of type σ. Show that R is an integral domain if and only if $\sigma = \omega^\tau$, and R is left Noetherian if and only if $\sigma \leq \omega$.

8*. Let k be a commutative field with an endomorphism α such that k contains an element t transcendental over k^α and denote by K the subring of $k(y)$ consisting of all fractions $f.(1 + yg)^{-1}$, where $f, g \in k[y]$. Show that α can be extended to K by letting $y \mapsto t$ and verify that the resulting endomorphism maps $K^\times$ into $U(K)$. Show that the power series ring $R = K[[x; \alpha]]$ is a principal right ideal domain in which the right ideals are well-ordered, and determine the order type of its chain of right ideals (see Cohn [67]).

9. Let R be a principal right ideal domain that is also left Ore. If all atoms of R are right associated to a single one, p say, show that $J(R) = pR$; deduce that R is a right principal valuation ring.

10. Let F be the free group on x, y, with the ordering as defined in the text, and let S be the monoid of elements ≥ 1. Show that S is conical and rigid, but not atomic; deduce that S is locally free, i.e. every finitely generated submonoid is contained in a free submonoid of S.

Notes and comments on Chapter 1

Skew polynomial rings were first studied systematically by Ore [33a]. The proof of Proposition 1.3 is modelled on the commutative case and Proposition 1.4 is due to Curtis [52]; the analogue for power series does not hold, see Kerr [82].

The familiar Euclidean algorithm occurs in Euclid, Book VII, Propositions 1 and 2, as a method of finding the highest common factor of two integers. The extension to polynomials was not undertaken until the 16th century, when Simon Stevin in his *Arithmetic* (1585), Book II, Problem LIII, uses it to find the HCF of two polynomials. He remarks that this application is probably new, since Pedro Nuñez, writing only a few years earlier (*Libro de Algebra*, 1567) attempts to treat the same problem, but does not get beyond a few generalities; this was possibly because he considered polynomials with integer coefficients, which present a harder problem.

There is a very extensive literature dealing with the Euclidean algorithm in algebraic number fields; most of this does not concern us, but Motzkin [49], who determines the imaginary quadratic extensions admitting a Euclidean algorithm with respect to any function, introduces the notion of derived set and proves most of Theorem 2.1. As in FR.2 we have followed Samuel's ([71]) definition of derived set, see also Rodosski [80]. Propositions 2.3–2.5, giving conditions for the algorithm to be defined by a degree-function, are taken from Lenstra [74], who also determines all rings with a transfinite unique remainder algorithm; see also Section 1.6. The first commutative examples of a (genuinely) transfinite Euclidean algorithm were found by Hiblot [75]. Theorem 2.6 is due to Jacobson [34] and was found again (independently) by Cohn [61a].

An interesting generalization of the Euclidean algorithm is considered by Leutbecher [78], who defines a ring R to be *quasi-Euclidean* if there is a function of two arguments $\theta : R^2 \to \mathbb{N}$ such that for $(a, b) \in R \times R^\times$ there exists $q \in R$ such that $\theta(-b, a - bq) < \theta(a, b)$. He shows that this is sufficient to derive many of the usual consequences of the Euclidean algorithm and in particular he proves that a ring R is quasi-Euclidean if and only if R is a GE_2-ring and every matrix $A \in R_2$ is right associated to a matrix with $(1, 2)$-entry 0.

The first non-commutative UFD to have been studied is the ring $K[D; 1, ']$ of linear differential operators. It is discussed at some length by Schlesinger [1897], who proves that it is an integral domain. Landau [1902] shows that all complete factorizations of a given operator have the same length, and corresponding irreducible factors have the same order (= degree in D). Loewy [1903] shows that corresponding factors are 'equivalent' operators, in a sense introduced by Poincaré, and this turns out to correspond to the notion of similarity. A large number of papers on the subject appeared at this time. The first abstract account of this ring was given by Ore [32], who also introduced the notion of 'eigenring'. A further generalization, to PIDs, is undertaken by Asano [38]. This and much other work is summarized in chapter 3 of Jacobson [43], where the criterion of Proposition 3.6 is proved for PIDs. Proposition 3.3 is due to Jategaonkar

[69a]. The general notion of non-commutative UFD is defined in Cohn [63a], where it is shown to include a wider class of rings, which will be considered in Chapter 3.

The general results on PIDs, 4.1–4.6, are part of the folklore. Theorem 4.7 was proved in a weak form, for (non-commutative) Euclidean domains, by Wedderburn [32], and the full form by Jacobson [37]. This was generalized to PIDs by Teichmüller [37] and a uniqueness statement added by Nakayama [38]. A final touch, describing exact conditions for the equivalence of two reductions, was given by Guralnick, Levy and Odenthal [88]. If elements a, b in an integral domain R satisfy $aR \supseteq cR \supseteq bR$ for a right invariant element c, a is called a *total right divisor* of b. Now Theorem 4.7 can be generalized to right principal Bezout domains but with 'total divisor' replaced by 'total right divisor'; this is proved in Cohn [87a]. In fact the result holds for right principal Ore domains, for as we shall see in Section 2.2, such a ring is a semifir, and being Ore, has to be Bezout. Our account in Section 1.4 follows Jacobson [43] with some simplifications (see Amitsur [63]). For a general study of elementary divisor rings see Kaplansky [49]. An example of a PID for which the non-elementary operation (iv) is needed is the ring of integers in $\mathbb{Q}(\sqrt{-19})$, which is therefore not Euclidean, though principal, see Cohn [66b]. Lemma 4.11 appears to be folklore, communicated to the author by Stafford, who also (in Stafford [85]) provided the source for Proposition 5.2.

Polynomial rings with the commutation rule (15) have been studied by Smits [68a], to whom Theorem 5.8 is due. The use of generalized Laurent series has a long history; infinite series with support other than $\mathbb{N}$ were considered by Levi-Civita [1892] and skew Laurent series were used by Hilbert [1899] to construct an ordered field that is not commutative. Proposition 5.4 is implicit in Schur [1904]: Hahn [1907] showed that every totally ordered abelian group Γ can be embedded in a (lexicographically ordered) ordinal power of $\mathbb{R}$, and in the same paper introduced the ring $\mathbb{R}((\Gamma))$. Theorem 5.9 (for groups) was proved independently by Malcev [48] and Neumann [49]; our proof follows the former source, but is stated for monoids. Another proof, based on properties of algebras with a divisibility ordering was given by Higman [52]. Theorem 5.11 and Corollary 5.12 are due to Bergman [78b].

The notion of unique remainder algorithm is described by Lenstra [74], to whom Proposition 6.1 is due. Jategaonkar [69a] constructed his J-rings by transfinite induction; the more direct proof given here was new in FR.1. Theorem 6.3 was proved by Lenstra [74]; see also Korotkov [76]. Theorem 6.6 (for rings with well-ordered set of right ideals) is due to Brungs [69b], who also describes their structure.

2
Firs, semifirs and the weak algorithm

After a brief preamble on hereditary rings (Section 2.1), this chapter introduces our main topic, free ideal rings (firs) which form a generalization of principal ideal domains (Section 2.2 and 2.3); frequently they satisfy a weak algorithm relative to a filtration (Section 2.4), which generalizes the division algorithm (relative to a degree-function), to which it reduces in the commutative case. The most important example is the free associative algebra over a field k, characterized as a filtered k-algebra with weak algorithm in Section 2.5, while a useful invariant, the Hilbert series, is described in Section 2.6. Some consequences of the weak algorithm are traced out in Section 2.7 and 2.8; the inverse weak algorithm, using a generalization of the order-function, is used to describe power series rings in Section 2.9 and a transfinite form of the weak algorithm is applied in Section 2.10 to construct one-sided examples. In Section 2.11 a method is described which in many cases allows one to read off from the presentation of a ring whether the n-term weak algorithm holds. This enables one to construct quite naturally n-firs that are not $(n + 1)$-firs.

2.1 Hereditary rings

Homological algebra classifies rings according to their global dimension, i.e. the length of projective resolutions of modules. The case of zero dimension (semisimple rings) is fairly well known, and we shall mainly be concerned with the next case; a ring has global dimension 1 precisely when all submodules of projective modules are projective but the ring is not semisimple. As is well known, this holds for left modules, say, if all left ideals are projective. By taking a little care in the proof it is possible to derive a more precise result, which will be needed later. Given a cardinal α (finite or infinite), a module is said to be *α-generated* or an *α-generator module* if it has a generating set of cardinal not exceeding α. A ring R is called *left α-hereditary* if every α-generated left ideal is projective. Thus to say that R is left $|R|$-hereditary is to say that every left ideal of R is projective, i.e. R is *left hereditary*. A ring is *left semihereditary* if it is left n-hereditary for all natural numbers n. Corresponding definitions apply on the right.

Theorem 2.1.1. *Let α, κ be any cardinals and let R be a left α-hereditary ring. If F is a free left R-module of rank κ, then every α-generated submodule N of F is isomorphic to a direct sum of at most κ left ideals of R, each α-generated projective. More precisely, this direct sum has finitely many terms if α is finite and at most $min(\alpha, \kappa)$ terms for infinite α.*

Proof. We shall identify cardinals with their least ordinals. Let $\{e_\iota | \iota < \kappa\}$ be a basis of F, put $F_\lambda = \sum_{\iota < \lambda} Re_\iota$ and for each $\iota < \kappa$ write $p_\iota : F \to R$ for the ιth coordinate projection. If $N \cap F_{\iota+1}$ is α-generated, then so is its projection on R via p_ι, hence projective; now $\ker(p_\iota | N \cap F_{\iota+1}) = N \cap F_\iota$, so we have

$$N \cap F_{\iota+1} = (N \cap F_\iota) \oplus P_{\iota+1}, \tag{1}$$

where $P_{\iota+1} \cong (N \cap F_{\iota+1}) p_\iota$. If this holds for all $\iota < \kappa$, it follows that $N = \oplus_{\iota < \kappa} P_{\iota+1}$ as claimed; it remains to show that $N \cap F_{\iota+1}$ is α-generated for all $\iota < \kappa$, and for infinite α there are no more than α terms. There are two cases.

(i) α is finite. Then N lies in a submodule of F generated by finitely many e's, so we may assume $\kappa = k$ finite. Then $F = F_{k+1}$, $N \cap F_{k+1} = N$ is α-generated and by (1) for $\iota + 1 = k$, $N \cap F_k$ is α-generated. Now a downward induction on k shows that each $N \cap F_i$ is α-generated, as we had to prove.

(ii) α is infinite. Take a generating set $\{n_\beta | \beta < \alpha\}$ for N and for $\beta < \alpha$ denote by N_β the submodule of N generated by all $n_\gamma (\gamma < \beta)$. Then $N \cap F_{\iota+1} = \cup_{\beta < \alpha} (N_\beta \cap F_{\iota+1})$ is a union of α submodules, each β-generated (by induction over α) and hence α-generated; since $\alpha^2 = \alpha$, it is α-generated, as claimed. ∎

We note the special cases of hereditary and semihereditary rings; when R is hereditary, $\alpha = |R|$ and we can omit the hypothesis on the submodule.

Corollary 2.1.2. *Let R be a left hereditary ring. If F is a free left R-module of rank κ, then every $max(|R|, \kappa)$-generated submodule of F is isomorphic to a direct sum of at most κ left ideals.* ∎

Corollary 2.1.3. *Let R be a left semihereditary ring and n a natural number. If F is a free left R-module of infinite rank, then every n-generator submodule of F is isomorphic to a direct sum of finitely many n-generator left ideals.* ∎

2.1 Hereditary rings

Corollary 1.2 shows again that every submodule of a free module of rank n over a PID is free of rank at most n (Proposition 1.4.5). Secondly, if R is an integral domain in which every finitely generated left ideal is principal, i.e. R is a left *Bezout domain*, then by Corollary 1.3, every finitely generated submodule of a free left R-module of rank n is again free of rank at most n.

We also note a symmetric form of Corollary 1.3 due to Bergman, which will be used later. To state it, we define a ring to be *weakly semihereditary* if, given maps $\alpha : P_0 \to P_1, \beta : P_1 \to P_2$ between finitely generated projective modules, such that $\alpha\beta = 0$, there is a direct summand Q of P_1 such that im $\alpha \subseteq Q \subseteq \ker \beta$. In terms of matrices, R is weakly semihereditary if for any $r \times n$ matrix A and $n \times s$ matrix B over R such that $AB = 0$, there exists an idempotent matrix E such that $AE = A, EB = 0$.

By applying the duality * (or replacing E by $I - E$ in the matrix condition) we see that this condition is left–right symmetric. In a left semihereditary ring, if α, β are as above, then im β is a projective module, by Corollary 1.3, hence P_1 splits over im β, $P_1 \cong \operatorname{im} \beta \oplus \ker \beta$ and im $\alpha \subseteq \ker \beta$, so the ring is weakly semihereditary; by symmetry the same holds for right semihereditary rings. Using a theorem of Kaplansky, we can now prove

Theorem 2.1.4. *Over a weakly semihereditary ring every projective module is a direct sum of finitely generated modules.*

Proof. Let P be a projective module, say $P \oplus Q = F$ is free, and let A be a finite subset of P. We first show that A is contained in a finitely generated direct summand of P.

The elements of A involve only finitely many coordinates in F, hence A lies in a finitely generated free direct summand F_0 of F. Let $|A| = n$ and take $\alpha : R^n \to F_0$ as the homomorphism mapping a standard basis to A; if the projection $F \to Q$, restricted to F_0 is denoted by β, then $\alpha\beta = 0$, hence $F_0 = P' \oplus P''$, where im $\alpha \subseteq P' \subseteq \ker \beta = P \cap F_0$. Thus P' is a direct summand in F_0 and contained in P, hence it is a direct summand of P and $P' \supseteq \operatorname{im} \alpha \supseteq A$, as we wished to show.

When P is countably generated, by $e_1, e_2, \ldots$ say, we can complete the proof as follows. Suppose we already have a direct decomposition $P = P_n \oplus P'_n$, where P_n contains $e_1, \ldots, e_n$ and is finitely generated. By the first part there is a decomposition $P = P' \oplus P''$, where P' is finitely generated and contains e_{n+1} and a generating set for P_n. Then P_n is contained in P' and is a direct summand in P, hence it is a direct summand in P' (see the Appendix, Lemma A.2), so on setting $P_{n+1} = P'$, we have $P = P_{n+1} \oplus Q'$, where P_{n+1} contains $e_1, \ldots, e_{n+1}$ and is finitely generated. By induction on n, P has a submodule of such a direct sum containing all the e's, but this set generates P, so P is a direct sum of finitely generated modules. Now the conclusion follows by applying Kaplansky's theorem, that every projective module is a direct sum of countably generated modules (Kaplansky [58]). ∎

Exercises 2.1

1. Let R be a right semihereditary ring and let $A \in {}^m R^n$. Show that the right annihilator of A in R_n has the form ER_n, where E is an idempotent $n \times n$ matrix.
2. (Kaplansky [58]) Let M be a countably generated module over a ring R. Assume that each direct summand N of M is such that any $x \in N$ can be embedded in a finitely generated direct summand of N; show that M is a direct sum of finitely generated modules.
3. Show that a ring R is weakly semihereditary if and only if for any finitely generated projective left R-module $P \neq 0$, a finite subset A of P and a finite subset B of P^* such that $AB = 0$, there exists a direct decomposition $P = P' \oplus P''$ such that $A \subseteq P'$ and $P'B = 0$.
4. Show that any right Noetherian weakly semihereditary ring is right hereditary.
5. (Bergman [72a]) Show that if in the definition in Exercise 3 of 'weakly semihereditary ring' we delete the condition that B resp. A be finite, then we obtain a characterization of left resp. right semihereditary rings.
6. If R is weakly semihereditary and $M_1 \subseteq M_2 \subseteq \ldots$ is an ascending chain of finitely generated modules whose union M is projective, show that there is a cofinal chain $P_1 \subseteq P_2 \subseteq \ldots$ such that each P_i is a direct summand of M (the P_i and M_j are *cofinal* in M if each M_j is contained in some $P_{j'}$ and each P_i is contained in some $M_{i'}$).

2.2 Firs and α-firs

Within the class of hereditary rings the projective modules occurring can still be very varied, and to take a simple case we shall assume that all projective modules are free. To exclude pathologies we also assume IBN for our ring.

Thus we are led to define a *free right ideal ring*, or *right fir* for short, as a ring in which all right ideals are free of unique rank. *Left firs* are defined similarly and a *fir* is just a left and right fir. We note that a right (or left) fir necessarily has IBN, since it is either right Ore or contains free right ideals of any rank, by Proposition 0.7.6.

Our first observation is that by Corollary 1.2, in a fir submodules of free modules are free; however, it need no longer be true that the rank of the submodule is bounded by the rank of the free module; this follows from the previous remark about right ideals.

The property of being a fir is preserved by localization, as the next result shows.

2.2 Firs and α-firs

Proposition 2.2.1. *Let R be a right fir and T a right Ore set in R. Then the localization R_T is again a right fir.*

Proof. Let $\mathfrak{a}$ be a right ideal of R_T; then $\mathfrak{a} \cap R$ is a right ideal of R, and hence is free, with a basis $(v_\beta)_{\beta < \alpha}$. It is clear that the v_β generate $\mathfrak{a}$ as right ideal of R_T; for $v_\beta \in \mathfrak{a}$ and if $x \in \mathfrak{a}$, then $xd \in \mathfrak{a} \cap R$ for some $d \in T$, so $xd = \sum v_\beta a_\beta$ and hence $x = \sum v_\beta a_\beta d^{-1}$. We claim that the v_β are right linearly independent over R_T; for if not, let $\sum v_\beta b_\beta = 0 (b_\beta \in R_T)$ be a non-trivial relation, where almost all but not all the b_β are 0. Then we can bring the b's to a common denominator, and multiplying up, we obtain a relation $\sum v_\beta c_\beta = 0 (c_\beta \in R)$, where $c_\beta \neq 0$ if and only if $b_\beta \neq 0$. But this contradicts the linear independence of the v_β over R, hence they form a free generating set of $\mathfrak{a}$. Moreover, if $\mathfrak{a}$ is finitely generated, then it has a finite basis, and since every basis arises from one of $\mathfrak{a} \cap R$ over R, all have the same number of elements, so R_T is indeed a right fir. ∎

We next investigate the presence of chain conditions in firs. It is easy to see that a fir is not Noetherian except in the rather special case of a PID. Nevertheless there is a chain condition that holds in all firs, namely the ACC on n-generator right (or left) ideals. We begin by treating the Ore case:

Proposition 2.2.2. *For any ring R the following conditions are equivalent:*

(a) R is a right fir and a right Ore domain,
(b) R is a principal right ideal domain,
(c) R is a right Noetherian right fir,
(d) R is a right Bezout domain with ACC on principal right ideals.

Proof. (a) ⇒ (b) follows because a right Ore domain cannot have a right ideal that is free of rank > 1. (b) ⇒ (c) is clear; (c) ⇒ (a) follows by Corollary 0.7.7. Now it is clear that (b) ⇒ (d), and (d) ⇒ (c), because (d) implies ACC on finitely generated right ideals, so the ring is right Noetherian. ∎

This result shows in particular that a commutative fir is just a PID. In treating the general case it is useful to cast our net a little wider. Let us define a *right α-fir*, for any cardinal α, as a ring in which all α-generated right ideals are free, of unique rank, and similarly for left α-firs. As in the case of firs, Theorem 1.1 shows that in a right α-fir every α-generated submodule of a free module is free.

To investigate chain conditions, let us say that a ring has right ACC_n, if it satisfies ACC on n-generator right ideals; following Bonang [89], we shall say that a ring with right ACC_n for all n has right pan-ACC. Similarly, a right module with ACC_n is a module with ACC on n-generator submodules and a right module with ACC_n for all n is called a module with pan-ACC.

We note that an α-fir is a β-fir, for any $\beta < \alpha$, so for the strongest results on α-firs we take a hypothesis with α as small as possible. We shall come to the case of finite α in the next section; when α is infinite, the results are then best stated for $\aleph_0$-firs.

Theorem 2.2.3. *Let R be a right $\aleph_0$-fir. Then any finitely related right R-module has pan-ACC, in particular R has right pan-ACC.*

Proof. Suppose first that F is a free right R-module; given any infinite strictly ascending chain

$$N_1 \subset N_2 \subset \ldots, \tag{1}$$

of n-generator submodules of F, their union $F_0 = \cup N_i$ is countably but not finitely generated, and hence is free of countable rank. Take a basis $u_1, u_2, \ldots$ and for a fixed n let P be the submodule generated by $u_1, \ldots, u_{n+1}$. Then $P \subseteq N_i$ for some i and P is a direct summand of F_0, hence of N_i by the Appendix, Lemma A.2. But N_i is free of rank at most n, and so cannot have a direct summand of rank $n+1$, which is a contradiction. This proves the assertion for free right R-modules.

Now let $M = F/L$ be finitely related, where F is free and L is m-generated, say. Then for any ascending chain (1) in M we can write $N_i = F_i/L$. If in (1) each N_i is n-generated, then F_i is $(n+m)$-generated; by the first part of the proof the sequence (F_i) becomes stationary, hence so does (1). ∎

Theorem 2.3 can also be used to factorize matrices over $\aleph_0$-firs if they have no zero-divisors as factors. This will be done in Chapter 3 in a more general context, where this rather cumbersome condition on the factors is expressed in a different form.

Exercises 2.2

1. Give a direct proof that every principal right ideal domain is a right fir.
2. If R is an $\aleph_0$-fir and a right Ore domain, show that R is a principal right ideal domain.
3. A non-zero ring without IBN in which all right ideals are free is called a *right metafir*; if all finitely generated right ideals are free, we have a *metasemifir*. Which results of this section carry over to metafirs or to metasemifirs?
4. In a (two-sided) fir, let $\mathfrak{a}$ be a two-sided ideal and $\mathfrak{b}$ a right ideal. Examine the possible relations between the ranks of $\mathfrak{a} + \mathfrak{b}$ and $\mathfrak{b}$.
5*. (A. H. Schofield) Let R be a right $\aleph_0$-hereditary ring such that for any infinite sequence $P_1, P_2, \ldots$ of non-zero finitely generated projective right R-modules the number of generators of $P_1 \oplus \cdots \oplus P_n$ is unbounded as $n \to \infty$. Show that any projective right R-module has pan-ACC.
6°. Show that a right ideal $\mathfrak{c}$ in a fir is join-irreducible if and only if $\mathfrak{c}$ is principal and the ring is local. Investigate meet-irreducible right ideals in firs.

2.3 Semifirs and *n*-firs

We now turn to α-firs, where α is finite. When $\alpha = n$, we speak of an *n-fir*, while a ring which is an *n*-fir for all natural numbers *n* is called a *semifir*. Here we need not distinguish between left and right, as we shall see in Theorem 3.1. In spite of evident similarities with firs there are significant differences, which we shall discuss here.

In the study of semifirs we shall need to consider relations of the form

$$x.y = x_1 y_1 + \cdots + x_n y_n = 0. \tag{1}$$

Such a relation is said to be *trivial*, if for each $i = 1, \ldots, n$, either $x_i = 0$ or $y_i = 0$. Of course every non-zero ring has non-trivial relations (1), e.g. $x = (a, ab)$, $y = (bc, -c)^T$, for any $a, b, c \in R$, yields a relation which is non-trivial unless $a = 0$, $c = 0$ or $ab = bc = 0$. Let us say that the *n*-term relation (1) is *trivialized* by the invertible matrix P if the relation $xP^{-1}.Py = 0$ is trivial; the passage to this relation is called an *inessential modification* and a relation is called *trivializable* if an invertible matrix trivializing it exists. It turns out that the rings in which all relations are trivializable are just the semifirs.

Theorem 2.3.1. *For any non-zero ring R and any natural number n, the following conditions are equivalent:*

(a) *every m-term relation $\sum_{i=1}^{m} x_i y_i = 0$ where $m \leq n$, can be trivialized by a square matrix,*
(b) *given $x_1, \ldots, x_m \in R (m \leq n)$ which are right linearly dependent, there exist $m \times m$ matrices P, Q over R such that $PQ = I$ and the vector $(x_1, \ldots, x_m)P$ has zero as its first component,*
(c) *any right ideal of R generated by $m \leq n$ right linearly dependent elements has a generating family of fewer than m elements,*
(d) *any right ideal on at most n generators is free, of unique rank, i.e. R is an n-fir,*
(e) *if φ is a map of $^m R$ into a free right R-module ($m \leq n$), then for some $r \leq m$ there is an automorphism μ of $^m R$ which induces an isomorphism $\nu : \mathrm{im}\,\varphi \to {}^r R$ and whose restriction to $\ker \varphi$ provides an isomorphism with $^s R$, where $s = m - r$. Thus we have the following commutative diagram:*

$$\begin{array}{ccccc} \ker \varphi & \longrightarrow & {}^m R & \longrightarrow & \mathrm{im}\,\varphi \\ \downarrow \mu|\ker \varphi & & \downarrow \mu & & \downarrow \nu \\ {}^s R & \longrightarrow & {}^s R \oplus {}^r R & \longrightarrow & {}^r R. \end{array} \tag{2}$$

Here the maps in the bottom line are the natural inclusion and projection. Moreover, these conditions are equivalent to their left–right duals.

The existence of a commutative exact diagram (2) with isomorphisms μ, ν is expressed by saying that the two rows of the diagram are *isomorphic*; in that case $\mu|\ker \varphi$ is clearly also an isomorphism.

Proof. We shall prove (a) $\Rightarrow$ (b) $\Rightarrow$ ... $\Rightarrow$ (e) $\Rightarrow$ (a). Since (a) is left–right symmetric, the conditions must then be equivalent to their left–right duals. The implication (a) $\Rightarrow$ (b) is evident.

(b) $\Rightarrow$ (c). Let $\mathfrak{a}$ be a right ideal, generated by $x_1, \ldots, x_m$ and suppose that the x's are right linearly dependent. By (b) there exist $P, Q \in R_m$ such that $PQ = I$, and if $x' = xP$, where $x = (x_1, \ldots, x_m)$, then $x'_1 = 0$. Clearly $x'_2, \ldots, x'_m \in \mathfrak{a}$ and since $x = xPQ = x'Q$, $\mathfrak{a}$ is generated by $x'_2, \ldots, x'_m$.

(c) $\Rightarrow$ (d). If $\mathfrak{a}$ is an n-generator right ideal, (c) allows us to reduce the number of generators until we get a linearly independent set; so $\mathfrak{a}$ will be free on at most n generators. Let $m \leq n$ be the least integer such that $\mathfrak{a} \cong {}^m R$ and suppose that ${}^m R \cong {}^{m+k} R$ for some $k > 0$. Then $\mathfrak{a}$ has a surjective endomorphism φ with kernel ${}^k R$. If we take a set of m generators of $\mathfrak{a}$, their images under φ will again generate $\mathfrak{a}$ but will be linearly dependent, because $\ker \varphi \neq 0$. Hence by the previous argument, $\mathfrak{a}$ will be free on $m' < m$ generators, a contradiction, which proves that $\mathfrak{a}$ has unique rank.

(d) $\Rightarrow$ (e). Given φ as in (e), an induction shows that $\operatorname{im} \varphi$ is free, say $\operatorname{im} \varphi \cong {}^r R$. Then the exact sequence

$$0 \to \ker \varphi \to {}^m R \to \operatorname{im} \varphi \to 0 \tag{3}$$

splits; so $\ker \varphi$ is a direct summand of ${}^m R$, hence also free, say $\ker \varphi \cong {}^s R$. Thus (3) is isomorphic to the bottom line in (2); since ${}^m R$ has unique rank, it follows that $r + s = m$.

(e) $\Rightarrow$ (a). Given an m-term relation $x.y = 0$, where $y \neq 0$, let $\varphi : {}^m R \to R$ be the mapping $v \mapsto v.y$. By (e) $\operatorname{im} \varphi \cong {}^r R$, where $r > 0$, hence $\ker \varphi \cong {}^s R$, where $s = m - r < m$; thus since $x \in \ker \varphi$, then for some invertible matrix P, xP has a zero component. If the relation $xP.Py^{-1} = 0$ is non-trivial, we can repeat the process; after at most n steps the relation has been trivialized. ∎

Of the above conditions we shall find (a) and (e) the most useful tools, while (b) is the easiest to verify. Conditions (c) and (d) are weaker in appearance than the others, since they do not explicitly assume anything about the modules R^n.

We note a useful consequence of Theorem 3.1. If we have a relation $XY = 0$, where X is an $r \times n$ matrix and Y an $n \times s$ matrix, the relation is called *trivial* if for each $i = 1, \ldots, n$ either the ith column of X or the ith row of Y is 0; now the notion of trivialization can be defined as before. The following result is an easy consequence of Theorem 3.1; in fact the given condition is sufficient as well as necessary.

Corollary 2.3.2. *In a semifir any matrix relation $XY = 0$ can be trivialized; moreover, for any $X \in {}^r R^n$ there exists an invertible $n \times n$ matrix P such that the non-zero columns of XP are right linearly independent; likewise there exists an invertible $r \times r$ matrix Q such that the non-zero rows of QX are left linearly independent. Hence in a semifir any full matrix is regular.* ∎

For non-zero expressions the trivializability condition of Theorem 3.1(a) takes the following form:

Corollary 2.3.3. *Given an expression $c = ab$ in a semifir R, where $a = (a_1, \ldots, a_n), b = (b_1, \ldots, b_n)^T$, there exists $P \in GL_n(R)$ such that $aP = (0, a', a'')$, $P^{-1}b = (b'', b', 0)^T$, where the row a' is right regular and the column b' is left regular.* ∎

Proof. We first transform a to the form $(0, a^*)$, where a^* is right regular; if b takes the form $(b'', b^*)^T$, with a corresponding subdivision, we transform a^* and b^{*T} to the form (a', a'') and $(b', 0)^T$, where b' is left regular, and of course a' is still right regular. ∎

It is clear that an n-fir either has IBN or is of type (h, k), where $h > n$ (see Section 0.1). Hence by Theorem 3.1(e), every n-generator projective module is free; a ring with this property is called *n-projective-free*. Thus we obtain

Corollary 2.3.4. *Any semifir is projective-free and hence an Hermite ring. More specifically, any n-fir is n-projective-free and so is n-Hermite.* ∎

From Theorem 3.1(d) we also obtain the following characterization of semifirs:

Corollary 2.3.5. *A ring R is a semifir if and only if R has invariant basis number and every finitely generated right (or equivalently, left) ideal is free.* ∎

Of the chain of n-fir conditions, the case $n = 1$ is really too general to be of interest, since a 1-fir is just an integral domain. On the other hand, 2-firs form an important class, e.g. in the commutative case they already comprise all semifirs, and so form the class of Bezout domains. By convention a 0-fir is understood to be a non-zero ring. Since every n-fir is also an n'-fir for all $n' \leq n$, we shall generally choose n as small as possible in our hypotheses and as large as possible in our conclusions.

The following rank formula is often useful:

Proposition 2.3.6. *Let R be an n-fir and A, B any submodules of a free R-module such that $A \oplus B$ is n-generated. Then there is a split exact sequence*

$$0 \to A \cap B \to A \oplus B \to A + B \to 0, \qquad (4)$$

all modules occurring are free and

$$rk(A + B) + rk(A \cap B) = rk\, A + rk\, B. \tag{5}$$

Proof. We map $A \oplus B$ to $A + B$ by the rule $(a, b) \mapsto a - b$; clearly the kernel is $A \cap B$ and we obtain the result by applying Theorem 3.1(e). ∎

By Theorem 3.1(c), a 2-fir can be characterized as an integral domain in which any two right linearly dependent elements generate a principal right ideal: the case $m = 1$ of (c) asserts that R is an integral domain; $m = 2$ is the above ideal condition, which is also reformulated in the next result. For any integral domain R and $c \in R^\times$ we shall denote by $L(cR, R)$ the set of principal right ideals containing cR; similarly for $L(Rc, R)$.

Theorem 2.3.7. *For any integral domain the following conditions are equivalent:*

(a) R is a 2-fir,
(b) for any $a, b \in R^\times$ we have $aR \cap bR = mR$ for some $m \in R$, while $aR + bR$ is principal if and only if $m \neq 0$,
(c) in the lattice $\mathrm{Lat}_R(R_R)$ of all right ideals of R, the set $L(cR, R)$ of principal right ideals of R containing a given $c \in R^\times$ forms a sublattice,
(d) any two principal right ideals that intersect non-zero have a principal sum,

(a^o)–(d^o) *the left-right duals of* (a)–(d).

Proof. (a) $\Rightarrow$ (b). Let $a, b \in R^\times$; by Proposition 3.6, $aR + bR$ and $aR \cap bR$ are free and have ranks adding up to $\mathrm{rk}(aR) + \mathrm{rk}(bR) = 2$. Clearly $\mathrm{rk}(aR + bR) \geq 1$, so $\mathrm{rk}(aR \cap bR) \leq 1$, i.e. $aR \cap bR = mR$ for some $m \in R$. Moreover, $m \neq 0$ if and only if $\mathrm{rk}(aR + bR) = 1$, i.e. when $aR + bR$ is principal.

The implications (b) $\Rightarrow$ (c) $\Rightarrow$ (d) are clear. To prove (d) $\Rightarrow$ (a), assume (d): then any two elements that are right linearly dependent generate a principal right ideal, hence R is a 2-fir, by Theorem 3.1(c). Finally the symmetry is clear from (a). ∎

Since any totally ordered set is a lattice, we have, by condition (d):

Corollary 2.3.8. *Any rigid domain is a 2-fir.* ∎

In a commutative 2-fir, i.e. a commutative Bezout domain, $L(cR, R)$ is a sublattice of the lattice of all fractional principal ideals, a lattice-ordered group and hence distributive (see Birkhoff [67], p. 294). In Chapter 4 we shall investigate 2-firs in which $L(cR, R)$ is distributive; of course for any 2-fir R, $L(cR, R)$ is modular, as sublattice of $\mathrm{Lat}_R(R_R)$.

2.3 Semifirs and n-firs

In any integral domain R, two non-zero elements a, b are said to have a *least common right multiple* (LCRM) m if $aR \cap bR = mR \neq 0$; clearly such m, if it exists, is unique up to a unit right factor and it reduces to the usual LCM when R is commutative. A corresponding definition applies to the least common left multiple (LCLM). Suppose now that R is a 2-fir; from Theorem 3.7(b) it is clear that any two right commensurable elements in R have an LCRM. In particular, this holds for right comaximal elements.

Corollary 2.3.9. *In a 2-fir any two non-zero right comaximal elements are right commensurable and hence have an LCRM.*

Proof. Given $a, b \in R^\times$, if $aR + bR = R$, then $aR \cap bR = mR \neq 0$ by Theorem 3.7(b). ∎

These results may be used to derive a normal form for fractions over a 2-fir:

Proposition 2.3.10. *Let a, b be left commensurable elements of a 2-fir R and assume that R is a subring of a ring S in which every factor of b is invertible. Then the element $s = ab^{-1}$ of S can be written in the form $a'b'^{-1}$, where a', b' are left comaximal in R, and if also $s = pq^{-1}$, where $p, q \in R$, then $(p, q)^T \in (a', b')^T R$. In particular, if p, q are left comaximal, then $(p, q)^T$ is right associated to $(a', b')^T$.*

Proof. By Theorem 3.7(d), $Ra + Rb = Rd$ for some $d \in R^\times$, say $a = a'd, b = b'd$. Then $Ra' + Rb' = R$, so a' and b' are left comaximal in R and b', d are invertible in S, hence $s = ab^{-1} = a'dd^{-1}b'^{-1} = a'b'^{-1}$. Further, $Ra' \cap Rb' = Rm \neq 0$, hence $m = b_0 a' = a_0 b'$; it follows that $a_0 = b_0 s$ and clearly $b_0 R \cap a_0 R = mR$. Thus if we have $s = pq^{-1}$, where $p, q \in R$, then $a_0 q = b_0 p = mr = b_0 a'r = a_0 b'r$, so $(p, q)^T = (a', b')^T r$, as claimed. ∎

Occasionally we shall want to impose a stronger condition on our rings. Let G_n be a subgroup of the general linear group $GL_n(R)$ such that G_{n+1} contains all the $n + 1$ natural images of G_n; if the ring R is such that for any $m \leq n$, every m-term relation in R can be trivialized by a member of G_m, R will be called a *strong G_n-ring*; if this holds for all integers n, R is called a *strong G-ring*. For example, a strong GL-ring is just a semifir, by Theorem 3.1(a). These terms will be used in particular for $G_n = E_n(R)$, the group generated by all $n \times n$ elementary matrices, differing from the unit matrix only in one off-diagonal place. We note that a strong E_n-ring is the same as a strong GE_n-ring, where $GE_n(R)$ is the group generated by $E_n(R)$ and the diagonal invertible matrices. By a G_n-ring we shall mean a ring R such that $GL_n(R) = G_n(R)$; if this holds for all n, we speak of a G-ring. We note that a strong GE-ring is a GE-ring, but a strong E-ring need not be an E-ring.

We also note the following homological characterization of semifirs:

Theorem 2.3.11. *A ring R is a semifir if and only if it is weakly semihereditary and projective-free. In this case every projective (left or right) R-module is free; in particular, every projective left (or right) ideal is free. For a semifir R moreover, any finitely generated submodule of a free R-module is free, of unique rank.*

Proof. If R is a semifir, then it is right semihereditary, hence weakly semihereditary; further it is projective-free and every finitely generated projective module, as a submodule of a free module, is itself free.

Conversely, assume that R is weakly semihereditary and projective-free, and consider a relation (1) in R. Using it we can define a map $f : R^n \to R$ by $(a_1, \ldots, a_n) \mapsto \sum a_i y_i$; then $x = (x_1, \ldots, x_n) \in \ker f$ by hypothesis, and since R is weakly semihereditary, there is a decomposition $R^n = P \oplus Q$ such that $x \in P$ and $Py = 0$. Since R is projective-free, P and Q are free, with bases $u_1, \ldots, u_r$ for P and $u_{r+1}, \ldots, u_n$ for Q. The matrix $U = (u_1, \ldots, u_n)^T$ is invertible and since $x \in P$, $x = x'_1 u_1 + \cdots + x'_r u_r = (x'_1, \ldots, x'_r, 0, \ldots, 0)U$, while $Uy = (0, \ldots, 0, u_{r+1}y, \ldots, u_n y)^T$. Thus U trivializes the relation $xy = 0$ and this shows R to be a semifir. By Theorem 1.4 every projective R-module is a direct sum of finitely generated projective modules, which are free because R is projective-free.

To prove the final assertion we note that a semifir is semihereditary and by Corollary 1.2, every finitely generated submodule of a free left R-module is a direct sum of finitely generated left ideals, which are projective and hence free. ∎

For example, using Theorem 3.11, we see that any local ring that is weakly semihereditary is a semifir, by Corollary 0.3.8. Moreover, by Theorem 1.1 we have

Corollary 2.3.12. *A ring is a right (α-)fir if and only if it is right (α-)hereditary and projective-free.* ∎

In any ring R, a family $a_1, \ldots, a_n \in R$ is *right linearly dependent* if there is a relation

$$a_1 b_1 + \cdots + a_n b_n = 0, \quad b_i \in R, \tag{6}$$

which is non-trivial, i.e. where the b_i are not all 0. If one of the a_i is linearly dependent on the rest, this means that there is a relation (6) in which some b_i is a unit. In that case $a_1, \ldots, a_n$ will be called *right unit-linearly dependent*. With these definitions we have

Theorem 2.3.13. *Let R be a ring. Then R is a local ring and a semifir if and only if R is an integral domain and every right linearly dependent family in R is right unit-linearly dependent.*

Proof. Assume that R is a local ring and a semifir. Then R is non-zero, and for any non-trivial relation (6) there is an invertible matrix $P = (p_{ij})$ trivializing (6), by Theorem 3.1; thus $\sum a_i p_{ij} = 0$ for some j. Since P is invertible over the local ring R, each column contains a unit, so for any given j, there exists i such that p_{ij} is a unit, hence the a_i are right unit-linearly dependent.

Conversely, assume that R satisfies the given conditions. Then in any right linearly dependent family one element can be written as a linear combination of the rest and an induction on the number of elements shows R to be a semifir. To verify that R is a local ring, assume the contrary. Then there exist non-units $a, b \in R$ such that $a + b = 1$. So $a, b \neq 0, 1$ and $ab = (1-b)b = b(1-b)$, hence a, b are right linearly dependent, and by hypothesis, unit-linearly, so one must be linearly dependent on the other, say $b = au$. Therefore $a(1 + u) = 1$, but this would mean that a is a unit, a contradiction, which shows R to be a local ring (Proposition 0.3.5). ∎

Since being weakly semihereditary is a categorical property, we can apply Theorem 0.4.8 to obtain the following Morita-invariant description of semifirs:

Theorem 2.3.14. *For any ring R the following conditions are equivalent:*

(a) R is a full matrix ring over a semifir,
(b) R is Morita-equivalent to a semifir,
(c) R is weakly semihereditary and projective-trivial. ∎

The trivialization procedure of Theorem 3.1 may also be applied to matrix products in which there is merely a block of zeros:

Lemma 2.3.15. (Partition lemma) *Let R be an n-fir and let $A \in {}^r R^n$, $B \in {}^n R^s$ such that AB has an $r' \times s''$ block of zeros as shown:*

$$AB = \begin{pmatrix} C' & 0 \\ C & C'' \end{pmatrix} \begin{matrix} r' \\ r'' \end{matrix}$$
$$s' s''$$

where r', r'', s', s'' indicate the numbers of rows and columns, respectively. Then there exists $T \in GL_n(R)$ such that

$$AT^{-1} = \begin{pmatrix} A_1 & 0 \\ A_3 & A_4 \end{pmatrix}, \quad TB = \begin{pmatrix} B_1 & 0 \\ B_3 & B_4 \end{pmatrix}, \tag{7}$$

where A_1 has r' rows and B_4 has s'' columns.

Proof. Partition A, B as $A = (A', A'')^T, B = (B', B'')$, where A', A'' have r', r'' rows respectively and B', B'' have s', s'' columns respectively. Then $A'B'' = 0$, hence by Theorem 3.1 there exists $T \in GL_n(R)$ such that the first n' rows in TB'' are 0 and all columns after the first n' in $A''T^{-1}$ are 0, so AT^{-1}, TB have the form shown in (7). ∎

Let R be a semifir and M a left R-module, with presentation

$$0 \to K \to F \to M \to 0,$$

where F is free and K is a submodule of F. If M is finitely presented, then F, K may be taken to be finitely generated and K is then free, so the characteristic of M is given by

$$\chi(M) = \text{rk } F - \text{rk } K.$$

If M is finitely generated but not finitely related, we put $\chi(M) = -\infty$, while for M not finitely generated we put $\chi(M) = \infty$. This defines the characteristic of M in all cases, although care is needed in adding characteristics, since both ∞ and $-\infty$ can occur.

The following reduction, familiar from field theory, is also of interest.

Proposition 2.3.16. *Let R be an n-fir. Then any nilpotent $n \times n$ matrix A over R has a conjugate that is strictly upper triangular with zeros on the main diagonal, and in particular, $A^n = 0$.*

Proof. We may assume without loss of generality that $A \neq 0$. Let m be the least integer such that $A^{m+1} = 0$ and write $A^m = B$. Then $B \neq 0$ and $AB = 0$; thus the columns of A are right linearly dependent. Hence, by Corollary 3.2 there exists $U \in GL_n(R)$ such that the first column of AU is 0, and so the first column of $U^{-1}AU$ is also 0. Deleting the first row and column from $U^{-1}AU$ we obtain an $(n-1) \times (n-1)$ matrix that is again nilpotent; by induction on n it is conjugate to a matrix that is strictly upper triangular; hence so is A. Now the last part follows easily. ∎

Let us see what becomes of our definitions in the commutative case. More generally, let us take a right Ore domain R: in R any two elements are right linearly dependent, therefore any free right ideal has rank at most 1. Hence if a right Ore domain is a 2-fir, Theorem 3.1(c) shows that any two elements generate a principal right ideal; hence every finitely generated right ideal is principal, i.e. we have a right Bezout domain. For Bezout domains we have the following analogue of Proposition 2.2.

2.3 Semifirs and n-firs

Proposition 2.3.17. *For any ring R the following conditions are equivalent:*

(a) R is a semifir and a right Ore domain,
(b) R is a 2-fir and a right Ore domain,
(c) R is a right Bezout domain,
(d) R is an integral domain in which every 2-generator right ideal is principal.

Proof. (a) $\Rightarrow$ (b) and (c) $\Rightarrow$ (d) are clear and (b) $\Rightarrow$ (c) follows from earlier remarks. To prove (d) $\Rightarrow$ (a), if (d) holds, an easy induction shows that R is a semifir, and if $a, b \neq 0$ but $aR \cap bR = 0$, then $aR + bR$ is free and not principal, which contradicts (d); hence R is also right Ore and (a) follows. ∎

The commutative case yields

Corollary 2.3.18. *Every commutative 2-fir is a Bezout domain.* ∎

Thus for right (or left) Ore domains our chain of conditions from 2-fir to semifir collapses to a single condition. By contrast, in the general case there are n-firs that are not $(n + 1)$-firs, for each n, as examples in Section 2.11 will show.

The remark after Corollary 1.3 showed that for a left Bezout domain R, every finitely generated submodule of R^n is free (of rank at most n). There is no corresponding result for finitely generated left R-modules over a *right* Bezout domain, but we have the following partial analogue:

Proposition 2.3.19. *Let R be a right Bezout domain. Then any finitely generated torsion-free left R-module is free.*

Proof. Let M be torsion-free and generated by $u_1, \ldots, u_n$, where n is minimal. Suppose that $\sum a_i u_i = 0$, where not all the a_i vanish; then $\sum a_i R = dR \neq 0$ for some $d \in R$, say $a_i = da_{1i}$, and $\sum a_{1i} R = R$. Since a right Bezout domain is Hermite, the row $(a_{11}, \ldots, a_{1n})$ can be completed to an invertible $n \times n$ matrix $A = (a_{ij})$. It follows that M is also generated by $v_1, \ldots, v_n$, where $v_i = \sum a_{ij} u_j$. But $dv_1 = \sum da_{1i} u_i = \sum a_i u_i = 0$, and since M is torsion-free, $v_1 = 0$, so M is generated by $v_2, \ldots, v_n$, which contradicts the minimality of n. ∎

When we come to chain conditions, it is clear that there can be no such conditions for a general semifir, since there are none even in the commutative case. However, imposing one chain condition will entail others; thus we note

Proposition 2.3.20. *A semifir with right ACC_n satisfies ACC_n on free right modules.*

Proof. This clearly holds for a right Ore domain R, for R will then be right Bezout, hence right principal (see Exercise 1.3.3). Otherwise an ascending chain of n-generated submodules of a free module is contained in a free module of countable rank, and this is isomorphic to a right ideal of R, by Proposition 0.7.6, and here ACC_n holds by hypothesis. ∎

Over a semifir we have the following description of flat modules:

Proposition 2.3.21. *Let R be a semifir and U a right R-module. Then U is flat if and only if every finitely generated submodule of U is free, i.e. U is semifree.*

Proof. Clearly U is flat whenever all its finitely generated submodules are flat; this is so when the latter are free, so the condition is sufficient. To show its necessity suppose that U is flat and let V be a finitely generated submodule. We have to show that V is free, so let us take a generating set $v_1, \ldots, v_n$ of V, where n is minimal. The conclusion will follow if the v_i are linearly independent, so assume that there is a non-trivial dependence relation $vx = 0$, where $v = (v_1, \ldots, v_n) \in V_n$ and $0 \neq x \in {}^n R$. By Corollary 3.2 there exists $P \in GL_n(R)$ such that the non-zero entries of Px are left linearly independent. Replacing x by Px and v by vP^{-1} and renumbering the components if necessary, we may assume that for some r, $1 \leq r \leq n$, $x_1, \ldots, x_r$ are left linearly independent, while $x_j = 0$ for $j > r$. It follows that the map from R^r to R given by $(z_1, \ldots, z_r) \mapsto \sum_1^r z_i x_i$ is injective. Since U is flat, the induced map $U \otimes R^r \to U \otimes R \cong U$ is again injective; this maps $(u_1, \ldots, u_r)$ to $\sum_1^r u_i x_i$ from U^r to U. But $(v_1, \ldots, v_r)$ is in the kernel, so $v_i = 0$ for $i = 1, \ldots, r$; this contradicts the minimality of n and it shows that V is free on $v_1, \ldots, v_n$, as claimed. ∎

It is clear that a semifir R is coherent (Appendix B.(xi)), hence by Theorem B.10, R^I for any set I is flat and by Proposition 3.19 we obtain a strengthening of Corollary 3.4 (see also the remarks after Theorem 5.1.5 for the corresponding statement for n-firs):

Corollary 2.3.22. *Let R be a semifir and I any set. Then every finitely generated submodule of R^I is free, i.e. R^I is semifree.* ∎

If in the proof of Proposition 3.21, R is a right Bezout domain, then $r = 1$ and we can weaken the hypothesis by assuming U to be torsion-free instead of flat:

Corollary 2.3.23. *Over a right Bezout domain R a left module is semifree and hence flat if and only if it is torsion-free.* ∎

We conclude this section with a result on what may be called 'α-complete' direct limits of α-firs. A partially ordered set will be called α-*directed* if every subset of cardinality at most α has an upper bound. A directed system over an α-directed set is also called an α-*directed* system.

Proposition 2.3.24. *Let α be any cardinal greater than 1. Then the direct limit of any α-directed system of α-firs is again an α-fir.*

Proof. Let $\{R_i\}$ be the given system of α-firs, with maps $f_{ij} : R_i \to R_j$, and put $R = \varinjlim R_i$. When $\alpha = n$ is finite, the assertion is simply that the direct limit of any directed system of n-firs is an n-fir; this follows easily from the characterization of n-firs given in Theorem 3.1(a). Thus we may take α to be infinite. Further, by the finite case, R will be a semifir, so it remains to show that given any set $X \subseteq R$, of cardinality at most α, $RX = \{\sum a_i x_i | a_i \in R, x_i \in X\}$ will be a free left ideal of R.

Since our system is α-directed we can find a ring $R_{i'}$ such that each $x \in X$ has an inverse image x' in $R_{i'}$. Write X' for the set of all these x'. For each finite subset Y of X' and each $i \geq i'$ consider the rank of the left ideal $R_i(Yf_{i'i})$. For fixed Y this rank is non-increasing in i, so it ultimately equals a minimum, which it attains for some i depending on Y. Since there are no more than α finite subsets Y of X', we can find $i'' \geq i'$ such that all the ranks $\mathrm{rk}(R_i Yf_{i'i})$ have their minimum value for $i \geq i''$. Put $X'' = X'f_{i'i''}$ and let B be a basis for the left ideal $R_{i''}X''$ of the α-fir $R_{i''}$. We claim that for all $i \geq i''$ and each finite $C \subseteq B$, $Cf_{i''i}$ is left linearly independent in R_i. It will follow that the image of C in R is left linearly independent, hence the image of B will be left linearly independent, and so will form a basis of RX.

Thus assume that $Cf_{i''i}$ is linearly dependent. Pick a finite subset Y of X'' such that $C \subseteq R_{i''}Y$; then $R_{i''}C$, being a direct summand in $R_{i''}B \supseteq R_{i''}Y$, will be a direct summand in $R_{i''}Y$. Hence C can be extended to a basis C' of $R_{i''}Y$. But $Cf_{i''i}$ is linearly dependent, so $\mathrm{rk}(R_{i''}Yf_{i''i}) < |C'| = \mathrm{rk}(R_{i''}Y)$, and this contradicts the choice of i''. ∎

Exercises 2.3

1. For each $n \geq 1$, determine which of the following are n-firs: $k, k[x], k[x, y]$ (k a commutative field), $\mathbb{Z}, \mathbb{Z}[x]$.
2. For which n is it true that every subring of an n-fir is an n-fir? Give an example of an integral domain which cannot be embedded in a semifir.
3. For any $n \geq 1$, show that a direct limit of n-firs is an n-fir. Does this result extend to semifirs?
4. Is the inverse limit of a system of semifirs always a semifir? (*Hint*: Note that the intersection of any directed system of subrings may be written as an inverse limit.)
5. Give a proof of Proposition 1.4.1 using Theorem 3.1.
6. Let R be an n-fir and S a subring that is also a homomorphic image of R, under a homomorphism fixing S (i.e. S is a *retract* of R). Show that S is again an n-fir.

7*. (G. M. Bergman) Let R be an n-fir and S a set of R-linear automorphisms of R^n. Show that the set of fixed elements is free and is a direct summand of R^n.
8. Let R be an n-fir. Given a relation $x_1 y_1 + \cdots + x_n y_n = 0$ in R, show that $\mathrm{rk}(\sum x_i R) + \mathrm{rk}(\sum R y_i) \leq n$. Does this remain true if $x_i \in {}^r R, y_i \in R^s$?
9. Show that a semifir is a right fir if and only if it is right hereditary. (*Hint*: Use Theorem 1.4.)
10*. Let $R = \mathbb{Z} + x\mathbb{Q}[x]$ be the ring of all polynomials in an indeterminate x with rational coefficients and integral constant term. Show that R is a Bezout domain, but not principal, although all maximal ideals are finitely generated.
11*. Give an example of a semifir R, not a left fir, in which all maximal left ideals are finitely generated. (*Hint*: Try the ring in Exercise 10.)
12. Show that for any n-fir and any R-module P, $P^m \cong R^m$, where $1 \leq m \leq n$, implies $P \cong R$.
13. Let R be a right Bezout domain with left ACC_2. Show that if every finitely generated torsion-free right R-module is free, then R is left and right Bezout.
14. Let R be a semifir and F a free R-module. Show that the intersection of two finitely generated submodules of F is finitely generated. (*Hint*: Remember Proposition 3.6.)
15. Show that every strong GE_n-ring is a strong E_n-ring.
16*. (G. M. Bergman) Let R be a ring for which R^n has unique rank. If every n-term relation in R can be trivialized, show that R is an n-fir. (Without the uniqueness of the rank of R^n it can be shown that R is weakly semihereditary, all finitely generated projective modules are free and if R has type (h, k), then $n = i + jk, 2h - 1 \leq i < h + k, j \geq 1$.)
17. For any local ring R define $G_n(R)$ as the group of all invertible matrices whose entries below the main diagonal lie in the maximal ideal. If R is a commutative discrete valuation ring, show that it is a strong G-ring. If R is a local ring, what can be said about the form of R when R is a strong G-ring?
18. Show that a left (or right) Ore domain is a strong G-ring if and only if it is a strong G_2-ring.
19°. Let R be a semifir, K a subfield of R and α an automorphism of R such that for each $n \neq 0$, the fixed ring of α^n is K. Find conditions for $K[x, x^{-1}; \alpha]$ to be a semifir.
20°. Is every weakly semihereditary local ring a semifir?
21°. Is every non-Ore right fir semiprimitive (i.e. with zero Jacobson radical)? Can a non-Ore fir be simple?
22. Show that if every relation in a ring R can be trivialized, then every finitely generated right ideal of R is free and R has IBN, i.e. R is a semifir (thus the condition in Theorem 3.1(a) that the matrices are square is not needed. *Hint*: Take a minimal generating set of the right ideal).

2.4 The weak algorithm

As we have seen in Section 2.2, firs may in a sense be regarded as a natural generalization of principal ideal domains, and it now remains to find some

2.4 The weak algorithm

examples. Just as a ring can often be recognized as a principal ideal domain by means of the division algorithm (see Section 1.2), so we shall find a characteristic property generalizing the division algorithm and possessed by many firs; this is the weak algorithm, to which we now turn.

Any generalization of the division algorithm will necessarily depend on the form of the value function. We shall not make the most general choice, but take our function to be a *filtration*. This means that we have a mapping from R to $\mathbb{N} \cup \{-\infty\}$ with the properties:

V.1. $v(x) \geq 0$ for all $x \neq 0$, $v(0) = -\infty$,
V.2. $v(x - y) \leq \max\{v(x), v(y)\}$,
V.3. $v(xy) \leq v(x) + v(y)$,
V.4. $v(1) = 0$.

These rules essentially state that $-v(x)$ is a *pseudo-valuation*, though we shall not use that term. If equality holds in V.3, we have a degree-function, as defined in Section 1.1; then $-v(x)$ is a valuation, as usually defined. This will mostly be the case, so we shall also call $v(x)$ the *degree* of x.

Given any filtration on R, let us write $R_{(h)}$ for the set of elements of degree at most h; the $R_{(h)}$ are subgroups of the additive group of R such that

(i) $0 = R_{(-\infty)} \subseteq R_{(0)} \subseteq R_{(1)} \subseteq \ldots$,
(ii) $\cup R_{(h)} = R$,
(iii) $R_{(i)} R_{(j)} \subseteq R_{(i+j)}$,
(iv) $1 \in R_{(0)}$.

Conversely, any series of subgroups $R_{(h)}$ of the additive group of R satisfying (i)–(iv) leads to a filtration v, given by $v(x) = \min\{h | x \in R_{(h)}\}$, as is easily seen. We remark that every ring has the *trivial* filtration

$$v(x) = \begin{cases} 0 & \text{if } x \neq 0, \\ -\infty & \text{if } x = 0. \end{cases}$$

Let R be a filtered ring, with filtration v. Given an element a of R and a family (a_i) ($i \in I$) of elements, a is said to be *right v-dependent* on the family (a_i) if $a = 0$ or if there exist $b_i \in R$, almost all 0, such that

$$v(a - \sum a_i b_i) < v(a) \quad \text{and} \quad v(a_i) + v(b_i) \leq v(a) \text{ for all } i. \tag{1}$$

In the contrary case a is said to be *right v-independent* of the (a_i). We note that dependence on a family is unaffected by adjoining 0 to or removing 0 from the family.

A family (a_i) in R is said to be *right v-dependent* if there exist elements $b_i \in R$, almost all 0, such that

$$v\left(\sum a_i b_i\right) < \max\{v(a_i) + v(b_i)\}, \qquad (2)$$

or if some $a_i = 0$; otherwise the family is *right v-independent*. We note that any right v-independent family will be right linearly independent over R. Of course the converse need not hold; in fact linear dependence is just the special case of v-dependence obtained by taking v to be the trivial filtration.

If an element of a family (a_i) is right v-dependent on the rest, the family is clearly v-dependent. Let us call a finite family $(a_1, \ldots, a_m)$ *strongly right v-dependent* if, given an ordering such that $v(a_1) \leq \ldots \leq v(a_m)$, some a_i is right v-dependent on $a_1, \ldots, a_{i-1}$; a general family is *strongly right v-dependent*, if this is true of some finite subfamily. It follows that a strongly right v-dependent family is right v-dependent, but the converse is not generally true; in fact the converse constitutes the 'weak algorithm', as expressed in the following

Definition A ring R with a filtration v is said to satisfy the *n-term weak algorithm* relative to v (for a positive integer n), if R is non-trivial and any right v-dependent family of at most n members of R is strongly right v-dependent. If R satisfies the n-term weak algorithm for all n, we shall say that R satisfies the *weak algorithm* for v.

We note that if a is right v-dependent on a family (a_i), then by (1), $b_i = 0$ whenever $v(a_i) > v(a)$, hence a is right v-dependent on the a_i of degree at most $v(a)$. If moreover, every element of degree zero is a unit (which will normally be the case), then a family (a_i) is strongly right v-dependent precisely when some a_i is right v-dependent on the rest.

For example, the 1-term weak algorithm for R states that $v(ab) < v(a) + v(b)$ implies that $a = 0$ or $b = 0$, in other words, $v(ab) = v(a) + v(b)$ for all $a, b \neq 0$, i.e. v is a degree-function, and so R is an integral domain. When the 2-term weak algorithm holds in R, v is a degree-function and for any right v-dependent $a, b \in R$ such that $v(a) \geq v(b)$, there exists $c \in R$ such that $v(a - bc) < v(a)$. In other words, the division algorithm holds for any pair of right v-dependent elements. Suppose further, that $v(a) = 0$; the family $(a, 1)$ is right v-dependent, since $a.1 - 1.a = 0$ and $v(a) = v(1)$, hence $v(1 - ab) < 0$, for some $b \in R$, i.e. $ab = 1$. Since R is an integral domain, it follows that a is a unit; so when the 2-term weak algorithm holds, every element of degree zero is a unit.

At first sight it looks as if the weak algorithm refers to the right-hand side, but in fact the notion is left–right symmetric, as we shall now show.

2.4 The weak algorithm

Proposition 2.4.1. *A filtered ring satisfies the n-term weak algorithm if and only if the opposite ring does.*

Proof. Let R be a filtered ring with n-term weak algorithm. Given two families of non-zero elements $a_1, \ldots, a_m, b_1, \ldots, b_m (m \leq n)$ such that

$$v\left(\sum a_i b_i\right) < \max\{v(a_i) + v(b_i)\},$$

we have to show that the b_i are strongly left v-dependent. For $m = 1$ there is nothing to prove, so we may assume that $m > 1$. If $\max\{v(a_i) + v(b_i)\} = k$, we may omit any terms a_i, b_i for which $v(a_i) + v(b_i) < k$, and so we may assume that

$$v(a_i) + v(b_i) = k \quad \text{for} \quad i = 1, \ldots, m. \tag{3}$$

Further, the terms may be renumbered so that

$$v(a_1) \leq \ldots \leq v(a_m), \quad \text{and hence} \quad v(b_1) \geq \ldots \geq v(b_m).$$

We shall use a double induction, on m and k. By the weak algorithm some a_i is right v-dependent on the preceding a's, say for $i = m$, without loss of generality; using j as an index running from 1 to $m - 1$, we have

$$v\left(a_m - \sum a_j c_j\right) < v(a_m), \quad v(a_j) + v(c_j) \leq v(a_m). \tag{4}$$

Write $a'_m = a_m - \sum a_j c_j$; then

$$\sum a_i b_i = \sum a_j b_j + \left(a'_m + \sum a_j c_j\right) b_m = \sum a_j (b_j + c_j b_m) + a'_m b_m. \tag{5}$$

Now by (4), $v(a'_m) < v(a_m)$, and so $v(a'_m) + v(b_m) < k$. Further, $v(c_j b_m) \leq v(c_j) + v(b_m) \leq v(a_m) - v(a_j) + v(b_m) = k - v(a_j) = v(b_j)$, by (4) and (3). Hence $v(b_j + c_j b_m) \leq v(b_j)$ and so

$$\max_j\{v(a_j) + v(b_j + c_j b_m)\} \leq k. \tag{6}$$

If equality holds in (6), we can omit the last term on the right of (5) and use induction on m, while for strict inequality in (6) we can use induction on k. In either case we find that for some j, say $j = 1, b_1 + c_1 b_m$ is left v-dependent on the rest, hence b_1 is left v-dependent on $b_2, \ldots, b_m$ as we wished to show. ∎

We note that for any $n' \leq n$, the n-term weak algorithm entails the n'-term weak algorithm. Thus in any filtered ring R we shall be able to prove more about R, the larger n is. Let us define the *dependence number* of R relative to the filtration v, written $\lambda_v(R)$, as the greatest integer n for which the n-term weak algorithm holds, or ∞ if it holds for all n. Thus $\lambda_v(R) = \infty$ means that the weak algorithm holds for v in R, while $\lambda_v(R) \geq 1$ means that the 1-term weak algorithm holds, i.e. v is a degree-function. In particular, such a ring will be an integral domain. For this reason the notion of dependence as defined here is of no interest for rings other than integral domains. In fact, we shall almost exclusively be concerned with integral domains in this book.

In any filtered ring R, the set $R_{(0)} = \{a \in R | v(a) \leq 0\}$ is clearly a subring. If moreover, $\lambda_v(R) \geq 2$, then as we saw, every element of degree 0 is a unit. It follows that $R_{(0)}$ is a field whenever R satisfies the 2-term weak algorithm for v. For this reason, in considering filtered rings, we shall usually confine our attention to rings where $R_{(0)}$ is a field (not necessarily commutative).

To illustrate the notion of dependence let us consider the commutative case or, a little more generally, the case of Ore domains. In a right Ore domain any set of more than one element is clearly right v-dependent. Hence, if $\lambda_v(R) \geq 2$, the familiar division algorithm holds, in the form A of Section 1.2. Conversely, if the classical division algorithm holds in R, then any element of R is right v-dependent on any non-zero element not of higher degree, and this in turn shows that $\lambda_v(R) = \infty$. These results are summed up in

Proposition 2.4.2. *For any filtered right Ore domain R there are exactly three possibilities:*

(i) $\lambda_v(R) = 0$: v *is not a degree-function,*
(ii) $\lambda_v(R) = 1$: v *is a degree-function, but the division algorithm does not hold,*
(iii) $\lambda_v(R) = \infty$: v *is a degree-function and the division algorithm holds in R.* ■

In contrast to this result, for non-Ore domains λ_v can have any positive integer value, by the results of Section 2.11 or also SF, Section 5.7. All this is of course in strict parallel with the n-fir condition (see Proposition 2.2).

In order to describe the connexion between the weak algorithm and semifirs we shall need a general result on filtered rings:

Lemma 2.4.3. *Let R be a filtered ring. Then any n-term row vector over R can be reduced by a member of $E_n(R)$ acting on the right, to a row whose non-zero components are not strongly right v-dependent.*

Proof. Let $u \in R^n$ and suppose its non-zero components are strongly right v-dependent; thus a non-zero term is right v-dependent on the rest. Then by adding to it an appropriate right linear combination of the remaining terms, we can reduce its degree without affecting the degrees of the other terms. Clearly this operation corresponds to a right action by a member of $E_n(R)$. We repeat the process until no such terms remain; it must terminate since the set of values of v is well-ordered: $-\infty, 0, 1, 2, \ldots$ and so is the sequence of degrees $(v(u_1), \ldots, v(u_n))$ in the lexicographic ordering. ∎

Now let R be a ring satisfying the n-term weak algorithm relative to a filtration v. Then for $m \leq n$, any m-tuple can be reduced by a member of $E_m(R)$ to one in which the non-zero terms are not strongly right v-dependent, hence right v-independent and so linearly independent. This establishes

Theorem 2.4.4. *If R is a filtered ring satisfying the n-term weak algorithm, then R is a strong E_n-ring, and in particular an n-fir. Moreover, any filtered ring with weak algorithm is a semifir and a strong E-ring.* ∎

In fact a filtered ring with weak algorithm is a fir, as we shall see in a moment (Theorem 4.6). As a further consequence of Lemma 4.3 we have

Corollary 2.4.5. *Let R be a filtered ring with weak algorithm. Given an expression*

$$c = \sum_{i=1}^{n} a_i b_i, \qquad (7)$$

where $a_1, \ldots, a_n$ are right linearly independent and $b_1, \ldots, b_n$ are left linearly independent, there exists a matrix $P = (p_{ij}) \in E_n(R)$ such that on writing $a'_j = \sum a_i p_{ij}, b_i = \sum p_{ij} b'_j$, we have $v(a'_i) + v(b'_i) \leq v(c)$.

Proof. By elementary transformations we obtain $a'_j = \sum a_i p_{ij}$ such that $a'_1, \ldots, a'_n$ are right v-independent, but this just means that (2) cannot hold, i.e. $v(a'_i) + v(b'_i) \leq v(c)$, as claimed. ∎

Let us consider more closely the structure of a right ideal $\mathfrak{a}$ in a filtered ring R, where $R_{(0)}$ is a field. A family B of elements of $\mathfrak{a}$ will be called a *weak v-basis* for $\mathfrak{a}$ if (i) all elements of $\mathfrak{a}$ are right v-dependent on B, and (ii) B is not strongly right v-dependent. It is easily seen, using the well-ordering of the range of v, that a weak v-basis of $\mathfrak{a}$ generates $\mathfrak{a}$ as a right ideal; but in a general filtered ring it need be neither v-independent nor a minimal generating set.

When $R_{(0)}$ is a field K, every right ideal $\mathfrak{a}$ of R has a weak v-basis, which may be constructed as follows. For any integer $h \geq 0$, $\mathfrak{a}_{(h)} = \mathfrak{a} \cap R_{(h)}$ is clearly a right K-space; moreover, the set $\mathfrak{a}'_{(h)}$ of all elements of $\mathfrak{a}_{(h)}$ that are right v-dependent on $\mathfrak{a}_{(h-1)}$ is also a right K-space. For evidently $\mathfrak{a}'_{(h)}$ is closed under right multiplication by elements of K; closure under addition is clear if the sum has degree h, while if it does not, the sum lies in $\mathfrak{a}_{(h-1)}$. Now for each $h \geq 0$, choose a minimal set $B_{(h)}$ spanning $\mathfrak{a}_{(h)}$ over $\mathfrak{a}'_{(h)}$, i.e. representatives for a K-basis of $\mathfrak{a}_{(h)}/\mathfrak{a}'_{(h)}$, and put $B = \cup B_{(h)}$. By induction on h it follows that every member of $\mathfrak{a}_{(h)}$ is right v-dependent on B, for all h. However, B itself is not strongly v-dependent, by the minimality condition in our choice of $B_{(h)}$; thus B is a weak v-basis for $\mathfrak{a}$.

Conversely, every weak v-basis of $\mathfrak{a}$ must have the property that its elements of degree h form a right K-basis of $\mathfrak{a}_{(h)}$ (mod $\mathfrak{a}'_{(h)}$). Hence any two weak v-bases of a given right ideal $\mathfrak{a}$ have the same number of elements in each degree h, viz. $\dim_K(\mathfrak{a}_{(h)}/\mathfrak{a}'_{(h)})$. This number will be called *the number of v-generators in degree h* and denoted by $r_h(\mathfrak{a})$. It is clear that when the weak v-basis of $\mathfrak{a}$ is right v-independent, then $\mathfrak{a}$ is free of rank $\sum r_h(\mathfrak{a})$.

If $\mathfrak{a}$ and $\mathfrak{b}$ are two right ideals of R such that $\mathfrak{a} \supset \mathfrak{b}$ and s is the least degree for which $\mathfrak{a}_{(h)} \neq \mathfrak{b}_{(h)}$, then $r_h(\mathfrak{a}) = r_h(\mathfrak{b})$ for $h < s$, while $r_s(\mathfrak{a}) > r_s(\mathfrak{b})$, provided that $r_s(\mathfrak{b})$ is finite.

Although every right ideal in a filtered K-ring R has a weak v-basis, this basis need not be right v-independent. But if $\mathfrak{a}$ happens to have a v-independent generating set B say, then B must be a weak v-basis. For, given $a \in \mathfrak{a}$, where $a = \sum b_i c_i$ ($b_i \in B, c_i \in R$), we have $v(a) = \max\{v(b_i) + v(c_i)\}$ by the v-independence of the b_i; it follows that $v(a - \sum b_i c_i) = -\infty$ is a relation of v-dependence of a on B. Hence all the elements of $\mathfrak{a}$ are v-dependent on B, while B is not strongly v-dependent, by definition.

So far R was any filtered ring in which $R_{(0)}$ is a field. We now strengthen this assumption by imposing the weak algorithm. Then any weak v-basis of a right ideal $\mathfrak{a}$ is right v-independent, so $\mathfrak{a}$ is free, with any weak v-basis as free generating set. We claim that $\mathfrak{a}$ has a unique rank; for the proof it is enough to take the case where $\mathfrak{a}$ is finitely generated (see BA, Proposition 4.6.4). Consider any basis of $\mathfrak{a}$, not necessarily a weak v-basis. By treating in turn the basis elements of degrees $0, 1, 2, \ldots$ we find that this basis can always be transformed to a weak v-basis by a sequence of elementary transformations, and so it has the same number of elements as the latter. Thus $\mathfrak{a}$ has unique rank; by Proposition 4.1 the same holds for all left ideals and so we have proved

Theorem 2.4.6. *Every filtered ring with weak algorithm is a fir, and each left or right ideal has a v-independent basis.* ∎

We note that the IBN for rings with a weak algorithm also follows from Theorem 4.4. Of course, as we saw in Proposition 1.3.1, a ring with a right division algorithm is a right PID; here we do not have a two-sided conclusion, owing to the asymmetry of the division algorithm.

Next we turn to the case of a ring with n-term weak algorithm. As in Theorem 4.4 we find again that this is an n-fir, but in addition we shall see that it also satisfies left and right ACC_n. The result will follow from a general property of weak v-bases in filtered rings:

Lemma 2.4.7. *Let R be a filtered ring such that $R_{(0)}$ is a field. Then R satisfies (left and right) pan-ACC.*

Proof. Let $\mathfrak{a}$ be a right ideal which is generated by a family of at most n elements. Then $\mathfrak{a}$ has a weak v-basis $a_1, \ldots, a_m (m \leq n)$. We associate with $\mathfrak{a}$ the n-tuple $(v(a_1), \ldots, v(a_m), \infty, \ldots, \infty)$ (with $(n-m)\infty$'s) as 'indicator'. Clearly this indicator is independent of the choice of weak v-basis. If $\mathfrak{a} \supset \mathfrak{b}$, the indicator of $\mathfrak{a}$ will be smaller, in the lexicographic ordering, than that of $\mathfrak{b}$. Since the set of these indicators is well-ordered, the ideals satisfy ACC. By symmetry this also holds for left ideals. ∎

For rings with n-term weak algorithm, where $n \geq 2$, $R_{(0)}$ is a field, as we have seen, so this leads to the following result:

Proposition 2.4.8. *Let R be a filtered ring with n-term weak algorithm, where $n \geq 2$. Then R is an n-fir and satisfies left and right pan-ACC. This conclusion still holds for $n = 1$, provided that $R_{(0)}$ is a field.* ∎

Exercises 2.4

1. Verify that a ring R has a filtration v such that $\lambda_v(R) \geq 1$ if and only if R is an integral domain.
2. Show that for any filtered ring with 2-term weak algorithm $R_{(0)}$ is a field. Give an example of a ring with 1-term weak algorithm but not satisfying left or right ACC_1. (*Hint*: Use Lemma 4.7.)
3. Define filtered modules over a filtered ring R and introduce the notion of weak algorithm for filtered modules. Show that every module satisfying the weak algorithm is free; what does the existence of a module ($\neq 0$) with weak algorithm imply about R?
4. Investigate rings satisfying the weak algorithm relative to the trivial filtration.
5. Show that the weak algorithm holds in a filtered ring R if and only if (i) in every right v-dependent family one member is right v-dependent on the rest, and (ii) every element of degree 0 is a unit.
6°. Generalize Hasse's characterization of PIDs (see Exercise 1.3.8) to firs.
7°. Investigate the notion of a weak algorithm relative to a function ϕ more general than a filtration.
8. Extend Corollary 4.5 to the case where the a's and b's are not necessarily linearly independent.

2.5 Monomial K-bases in filtered rings and free algebras

In this section we shall prove an analogue of Theorem 1.2.6, which will describe the rings with a weak algorithm. For any filtered ring R we define the 'formal degree' of an expression $\sum_i a_{i1} \ldots a_{in_i}$ as

$$\max{}_i\{v(a_{i1}) + \cdots + v(a_{in_i})\}.$$

Clearly the actual degree of an element of R never exceeds the formal degree of any expression for it. We also note that the definition of v-independence of a family states that the degree of elements represented by certain expressions should equal the formal degree of these expressions.

Let R be a filtered ring for which $R_{(0)}$ is a field K. A family X of elements of R is called a *monomial right K-basis* if the monomials in X span R as right K-space and are not strongly right v-dependent. A corresponding definition applies to a monomial *left K-basis*. It is clear that any element of a monomial right K-basis has a strictly positive degree. Such a family X may be constructed recursively as follows.

For each $h > 0$ denote by $R'_{(h)}$ the right K-subspace of $R_{(h)}$ spanned by the products ab, where $a, b \in R_{(h-1)}$ and $v(a) + v(b) \leq h$. Now choose a minimal family X_h spanning $R_{(h)}$ (mod $R'_{(h)}$) over K, i.e. a family of representatives for a right K-basis of $R_{(h)}/R'_{(h)}$ and put $X = \cup X_h$. To show that X has the properties stated above, suppose that X is strongly right v-dependent, say

$$x \equiv \sum x_j b_j (\mathrm{mod}\, R_{(h-1)}), \tag{1}$$

where $v(x_j) + v(b_j) \leq v(x) = h$. Any terms $x_j b_j$ with $v(x_j) < h$ lie in $R'_{(h)}$, so (1) takes the form $x \equiv \sum x_j \beta_j$ (mod $R_{(h)}'$), where $\beta_j \in K$ and $v(x_j) = h$ whenever $\beta_j \neq 0$. But this contradicts the construction of X; so no element of X is right v-dependent on the rest, i.e. X is not strongly right v-dependent. Now an easy induction on the degree shows that the monomials in X span R as right K-space; more precisely, the monomials of formal degree at most h span $R_{(h)}$. Thus we see that every filtered ring R for which $K = R_{(0)}$ is a field has a monomial right K-basis. As in the case of weak v-bases of right ideals, we see that the cardinality of a monomial right K-basis (and more precisely, the number of elements of a given degree) is independent of the choice of basis. Correspondingly a monomial left K-basis can be constructed for R, by symmetry.

We now show that the monomial right K-basis is v-independent precisely when the weak algorithm holds. Given any family X of elements in a K-ring R, if R is spanned as right K-space by the monomials in X, we can define a filtration on R by assigning to each $x \in X$ a positive integer as degree and to each element a of R the minimum of the formal degrees of the right K-linear expressions in X representing a. We shall denote by $x_I = x_1 x_2 \ldots x_n$ a monomial in X.

2.5 Monomial K-bases in filtered rings and free algebras

Theorem 2.5.1. *Let R be a filtered ring such that $R_{(0)}$ is a field K, and let X be a monomial right K-basis for R. Then the following conditions are equivalent:*

(a) R satisfies the weak algorithm,
(b) X is right v-independent,
(c) the degree of any expression $\sum x_I a_I (a_I \in K)$ is equal to its formal degree.

When these conditions are satisfied, the monomials in X form a right K-basis for R.

Proof. (a) $\Rightarrow$ (b). If R satisfies the weak algorithm, then any monomial right K-basis X of R is clearly right v-independent and each $x \in X$ has positive degree.

(b) $\Rightarrow$ (c). Since $R_{(0)} = K$, X has no elements of degree 0. If (c) does not hold, then the monomials in X are right K-linearly dependent, say $\sum x_I a_I = 0$. By splitting off the left-hand factor from X in each x_I we can write this as

$$\sum x a_x + \alpha = 0 \quad (x \in X, a_x \in R, \alpha \in K)\alpha.$$

By the v-independence of X, each $a_x = 0$ and so $\alpha = 0$. Now an induction on the formal degree shows that the given relation was trivial. Thus the monomials in X are right K-linearly independent; hence they form a right K-basis for R and (c) holds.

To prove (c) $\Rightarrow$ (a) we show that R satisfies the left-hand analogue of the weak algorithm, which by Proposition 4.1 is equivalent to the weak algorithm itself. Let us consider how monomial *terms*, i.e. scalar multiples of monomials multiply in R. The product $(x_1 \ldots x_i \alpha)(y_1 \ldots y_j \beta)$ can be written $(x_1 \ldots x_i)(\alpha y_1 \ldots y_j \beta)$. If we write the second factor as a right K-linear combination of monomials, little can be said about the terms that will occur, except that we know their degrees. However, in the product all terms will clearly have $x_1 \ldots x_i$ as a left factor.

Let us fix a monomial $x_1 \ldots x_h$ of degree r and define the *right transduction* for this monomial as the right K-linear map $a \mapsto a^*$ of R into itself which sends any monomial of the form $x_1 \ldots x_h b$ to b and all other monomials to 0. Thus a^* is the 'right cofactor' of $x_1 \ldots x_h$ in the canonical expression for a. For any $a \in R$ we have $v(a^*) \leq v(a) - r$, because the degree of a equals its formal degree. Further, if $a, b \in R$, and $v(b) = s$, then

$$(ab)^* \equiv a^* b \pmod{R_{(s-1)}}. \tag{2}$$

This is clear if a is a monomial of degree at least r; in fact we then have equality. If a is a monomial of degree less than r, the right-hand side of (2) is 0 and so (2) holds as a congruence. If $a = w\alpha$, where w is a monomial and $\alpha \in K^\times$, then $(ab)^* = (w\alpha b)^* = (w(\alpha b))^* \equiv w^*(\alpha b) = (w\alpha)^* b$, by the case previously proved, and the congruence is mod $R_{(s-1)}$, where $s = v(b) = v(\alpha b)$. Now (2) follows by linearity.

Assume now that $b_1, \ldots, b_n$ is a left v-dependent family, i.e.

$$v\left(\sum a_i b_i\right) < d = \max\{v(a_i) + v(b_i)\}. \tag{3}$$

We have to show that the b_i are strongly left v-dependent, i.e. taking the b_i ordered so that $v(b_1) \geq \cdots \geq v(b_n)$, we must show that some b_i is left v-dependent on those that follow. By omitting terms if necessary we may assume that $v(a_i) + v(b_i) = d$ for all i, hence $v(a_1) \leq \cdots \leq v(a_n)$.

Let $x_1 \ldots x_h$ be a product of maximal degree $r = v(a_1)$ occurring in a_1 with non-zero coefficient α and denote the right transduction for $x_1 \ldots x_h$ by *. In the expression $\sum a_i{}^* b_i$ the ith term differs from $(a_i b_i)^*$ by a term of degree $< v(b_i) \leq v(b_1)$. Hence the sum will differ by a term of degree less than $v(b_1)$ from $(\sum a_i b_i)^*$, which has degree $\leq v(\sum a_i b_i) - r < d - r = v(b_1)$. Therefore $v(\sum a_i{}^* b_i) < v(b_1)$, and this gives a relation of left v-dependence of b_1 on the remaining b_i, since $a_1{}^* = \alpha \in K^\times$. ∎

More generally, let K be a field and V a K-bimodule and define its tensor powers

$$F_n = V^n = V \otimes \cdots \otimes V \quad (n \text{ terms for } n > 0, F_0 = K). \tag{4}$$

We define the direct sum $F = F_0 \oplus F_1 \oplus \ldots$ as a K-ring with multiplication induced by the natural isomorphism $F_m \otimes F_n \cong F_{m+n}$ ($m, n = 0, 1, \ldots$); the ring F so obtained will be called the tensor K-ring on V and we write $F = K[V]$. If X is a right K-basis for V, then the free monoid on X, denoted by X^*, is a right K-basis for F, with the products of n factors from X as a basis for F_n. Taking X to be indexed: $X = \{x_i\}$, we can write the general element of X^* as

$$x_I = x_{i_1} x_{i_2} \ldots x_{i_n}, \tag{5}$$

where $I = (i_1, \ldots, i_n)$ runs over all finite sequences of subscripts (including the empty sequence, to represent 1). Clearly each element of F is uniquely expressible in the form

$$f = \sum x_I \alpha_I \ (\alpha_I \in K, \text{ almost all } 0). \tag{6}$$

This ring F is called the *tensor K-ring* over X (with base field k) and is denoted by $K_k\langle X \rangle$.

By the *natural filtration* on $K[V]$ we understand the filtration obtained by assigning the degree 1 to the elements of X and using the induced formal degree. By definition it then follows from Theorem 5.1 that $K[V]$ satisfies the weak algorithm. Bearing in mind Theorem 4.6, we obtain

2.5 Monomial K-bases in filtered rings and free algebras

Corollary 2.5.2. *Let K be a field and V a K-bimodule. Then the tensor K-ring on V, $K[V]$, satisfies the weak algorithm relative to the natural filtration; hence it is a fir.* ∎

An important example is obtained by taking V as the K-space with basis $X = \{x_1, \ldots, x_d\}$ and left K-action defined by $cx_i = x_i c (c \in K)$. The ring thus obtained is denoted by $K\langle X \rangle$ or $K\langle x_1, \ldots, x_d \rangle$ and is called the *free K-ring* on X, or in case K is commutative, the *free K-algebra* on X. By Corollary 5.2 the free K-ring satisfies the weak algorithm; in particular, for the free algebra we obtain a characterization in this way:

Theorem 2.5.3. *Let R be an algebra over a commutative field k, with a filtration v such that $R_{(0)} = k$. Then R is the free associative k-algebra on a set X with the formal degree induced from $v : X \to \mathbb{N}$ if and only if the weak algorithm holds in R for v.*

Proof. By hypothesis, k is contained in the centre of R and when the weak algorithm holds, the form of the elements described in Theorem 5.1 shows R to be the free k-algebra on X. The converse has already been noted above. ∎

Let K be a field and R, S any K-rings; then we can define the coproduct $P = R *_K S$ of R and S over K as the pushout of the maps $K \to R$, $K \to S$ (see Appendix B(ii)). In fact it is not necessary for K to be a field, for the coproduct to be defined, but when we do have a field K, then the coproduct is *faithful* and *separating*, i.e. R, S are embedded in P and their intersection is K. This follows by taking any right K-bases X of $R \setminus K$ and Y of $S \setminus K$ and noting that the formal products

$$m = w_1 \ldots w_n \tag{7}$$

with factors from $X \cup Y \cup \{1\}$ form a right K-basis of P. If moreover, R and S are both filtered with $R_{(0)} = S_{(0)} = K$, then P has a natural filtration obtained by taking $P_{(n)}$ to be spanned by products from monomial right K-bases of R and S of total degree n. Now we have the following result whose proof generalizes that of Theorem 5.1:

Theorem 2.5.4. *Let R, S be filtered K-rings with weak algorithm, where $R_{(0)} \cong S_{(0)} \cong K$. Then their coproduct $R *_K S$ with the natural filtration also satisfies the weak algorithm.*

Proof. The coproduct $P = R *_K S$ has the right K-basis (7); so every element of P is a sum of monomial terms $w_1 \ldots w_n \alpha$, where the w_i are taken from monomial right K-bases in R and S and $\alpha \in K$, and any product $(w_1 \ldots w_n \alpha)(w'_1 \ldots w'_m \beta)$ can be written $(w_1 \ldots w_n)(\alpha w'_1 \ldots w'_m \beta)$. Thus for any product $m = w_1 \ldots w_n$ of degree r we can define again the *right transduction* as the right K-linear map $a \mapsto a^*$ of P into itself that sends any monomial $w_1 \ldots w_n b$ to b and all other monomials to 0. Now exactly the same proof as for Theorem 5.1 (c) $\Rightarrow$ (a) shows that P satisfies the weak algorithm. ∎

By taking the free K-ring as one of the factors we obtain the following consequence:

Corollary 2.5.5. *Let R be a ring with weak algorithm, where $R_{(0)} = K$. Then for any set X the free R-ring $R_K\langle X \rangle$ satisfies the weak algorithm and hence is a fir.* ∎

In particular, if D is any field and K a subfield, then the free D-ring $D_K\langle X \rangle$ is a ring satisfying the weak algorithm.

In connexion with Theorem 5.1 we note that, given a field K and a K-ring R with a subset X such that the monomials in X form a right K-basis for R, if we assign arbitrary degrees to members of X and give elements of R their formal degrees when expressed in terms of this basis, this will not necessarily define a filtered ring structure on R. The main reason is that for $\alpha \in K, x \in X$, the element αx, when expressed as a right linear combination of monomials in X, may not have the same formal degree as x. When the weak algorithm holds, as in the above example of free associative algebras, or even in the case of tensor rings on X, this cannot happen. But in general it may not be possible to assign a suitable filtration; e.g. we may have $\alpha x = x^2 + y$ ($\alpha \in K, x, y \in X$). In the proof of Theorem 5.1 essential use was made of the fact that v was given as a filtration on R.

The tensor ring $K[V]$ is an example of an augmented K-ring, that is a K-ring R of the form $R = K \oplus I$, for an ideal I, the *augmentation ideal*. This augmentation ideal I is a K-bimodule and has all the properties of a ring, except that it lacks a unit element, but the standard procedure for adjoining a unit element yields the ring R. Given R, we recover I as follows. The unit element 1 of R is uniquely determined, and now K is the set of all scalar multiples of 1. So we can form the quotient K-bimodule R/K, which is I. This argument can be used to show that V is determined up to isomorphism by $K[V]$:

Theorem 2.5.6. *Let K be a field and V a K-bimodule. Then the tensor ring $K[V]$ determines V up to K-bimodule isomorphism.*

Proof. Suppose that U, V are K-bimodules such that there is a K-ring isomorphism $K[U] \cong K[V]$. We can write $K[U] = K \oplus I$, where $I = U \oplus U^2 \oplus \ldots$; similarly $K[V] = K \oplus J$, where $J = V \oplus V^2 \oplus \ldots$. Then $J \cong K[V]/K \cong K[U]/K \cong I$; hence $U \cong I/I^2 \cong J/J^2 \cong V$. ∎

2.5 Monomial K-bases in filtered rings and free algebras

In the case of a tensor K-ring over X we have $V = (K° \otimes_k K)^{(X)}$, and this shows the cardinality of X to be determined by V, hence we have

Corollary 2.5.7. *For any field K with subfield k and any set X, the cardinality of X is determined by the tensor K-ring $K_k\langle X \rangle$.* ∎

The cardinality of X is called the *rank* of $K_k\langle X \rangle$. Since the tensor ring $K[V]$ is a graded ring, we also have an order-function, where the *order* of an element $f \in K[V]$ is defined as the lowest degree of any terms occurring in f. With the help of this function we can form the completion of $K[V]$, denoted by $K[[V]] = \prod_{n=0}^{\infty} V^n$ and called the *power series K-ring* over V. Its elements are infinite series $f = f_0 + f_1 + \cdots (f_n \in V^n)$. For a family X we shall denote the completion of $K_k\langle X \rangle$ by $K_k\langle\langle X \rangle\rangle$. In the case where X is infinite, it is sometimes advantageous to assign degrees to the elements of X in such a way that only finitely many are of degree $< n$, for any integer n. The ring so obtained will be larger than the completion where all elements of X have degree 1.

The description in Theorem 5.1 of rings with a weak algorithm is not very explicit, for although it enables us to write down many examples, it does not provide a method for constructing *all* rings with a weak algorithm. We shall now describe such a method, but in order to do so we need another concept.

By a *truncated filtered ring* of height h (briefly, h-truncated ring), $R_{(h)}$, we shall mean a finite chain of abelian groups

$$0 = R_{(-\infty)} \subseteq R_{(0)} \subseteq R_{(1)} \subseteq \ldots \subseteq R_{(h)},$$

with a function called *multiplication* defined on $\cup \{R_{(i)} \times R_{(j)} | i + j \leq h\}$ such that:

T.1. For $i + j \leq h$, multiplication restricted to $R_{(i)} \times R_{(j)}$ is a biadditive function with values in $R_{(i+j)}$,
T.2. For $i + j + k \leq h$, multiplication is associative on $R_{(i)} \times R_{(j)} \times R_{(k)}$,
T.3. $R_{(0)}$ contains the neutral element for multiplication, 1.

Here we have used the same symbol, $R_{(h)}$, for the last term of the defining chain and the total structure.

By a morphism $f : R_{(h)} \to S_{(k)}$ of truncated filtered rings we mean a map f respecting addition, multiplication and unit element, such that $R_{(i)}f \subseteq S_{(i)}$; in particular, $h \leq k$.

Let $R_{(h)}$ be a truncated filtered ring and let $a \in R_{(h)}$. We define the *degree*, $v(a)$, of a as the least i such that $a \in R_{(i)}$. From the definition of multiplication in $R_{(h)}$ we see that ab is defined precisely when $v(a) + v(b) \leq h$.

With every filtered ring and every integer h there is associated a truncated filtered ring of height h, obtained by 'forgetting the terms of degree $> h$'.

This 'truncation' functor has a left adjoint, associating with every h-truncated ring $R_{(h)}$ the universal filtered ring $U(R_{(h)})$ generated by $R_{(h)}$. More generally, for every h-truncated ring and each $k \geq h$ there is a universal k-truncated ring.

It is clear how to define 'v-dependence' and 'strong v-dependence' for a family of elements of a truncated filtered ring, exactly as for ordinary filtered rings. Of course the value of $\max_i \{v(a_i) + v(b_i)\}$ in the relations considered must not exceed h. If every v-dependent family is strongly v-dependent, the ring is said to satisfy the weak algorithm. Theorem 5.1 is then easily seen to go through in this context; in particular, for any monomial right K-basis of a truncated filtered K-ring with weak algorithm the degree of any element is equal to its formal degree in X.

Given a truncated filtered ring $R_{(h)}$ satisfying the weak algorithm, let us denote the field $R_{(0)}$ by K and construct X as in the discussion preceding Theorem 5.1. We denote by $R'_{(h+1)}$ the right K-space having as right K-basis the monomials in X of formal degree $\leq h+1$ and for degree-function the formal degree obtained by expressing elements in terms of this basis, using the degrees in $R_{(h)}$ of the elements of X. We claim that $R'_{(h+1)}$ is the universal truncated filtered ring of height $h+1$ for $R_{(h)}$.

Indeed, the space $R'_{(h+1)}$ will clearly have the desired universal property if it can be given a truncated filtered ring structure extending that of $R_{(h)}$. It is clear how we must define addition, right multiplication by elements of K and left multiplication by elements of X to obtain this structure, so it remains to define left multiplication of monomials in X by elements of K, i.e. to give a left K-space structure to $R'_{(h+1)}$. Since we are given this structure on $R_{(h)}$, it suffices to define products αu, where $\alpha \in K$ and u is a monomial of degree $h+1$ in X. Such a monomial can be written as $u = xw$, where $x \in X$ and w is a monomial of lower degree. In $R_{(h)}$ we have $\alpha x = \sum x_I \beta_I$, where $v(x_I) \leq v(x)$ and for each I we have $\beta_I w = \sum x_J \gamma_{JI}$, $v(x_J) \leq v(w)$. We then put $\alpha u = \alpha xw = \sum x_I x_J \gamma_{JI}$. To show that this leads to an $(h+1)$-truncated filtered ring structure on $R'_{(h+1)}$ we need to verify the associative law. By linearity this only needs to be checked for triple products of monomial terms. We consider four cases, where $\alpha, \beta, \gamma \in K$, $u \neq 1$ is a monomial and p, q are arbitrary elements of $R'_{(h+1)}$ of degrees such that products below are defined.

(1) $(up)q = u(pq)$. Here the left multiplication by u is just the formal product of monomials.
(2) $(\alpha u)p = \alpha(up)$. If $u = xu'$, where $x \in X$ is of degree r, and $\alpha x = \sum y_i \delta_i$, then $(\alpha u)p = (\alpha xu')p = (\sum y_i \delta_i u')p$, which equals $\alpha(up)$ by associativity in $R_{(h+1-r)}$.

2.5 Monomial K-bases in filtered rings and free algebras

(3) $(\alpha\beta)u = \alpha(\beta u)$. If $u = xu'$, where $x \in X$ is of degree r, then $(\alpha\beta)x = \alpha(\beta x)$ by associativity in $R_{(r)}$ and now the result follows by associativity in $R_{(h+1-r)}$.

(4) Finally $(\alpha\beta)\gamma = \alpha(\beta\gamma)$ by associativity in K.

Using these four cases we can now verify the associativity of triple products of monomial terms without difficulty.

By Theorem 5.1 (in its extended form, for truncated filtered rings), $R'_{(h+1)}$ now satisfies the weak algorithm and its h-truncation will be $R_{(h)}$. Our aim is to find the most general form for an $(h + 1)$-truncated ring $R_{(h+1)}$ with weak algorithm, having $R_{(h)}$ as its h-truncation. If $R_{(h+1)}$ is such a ring, it is clear from the method of proof of Theorem 5.1 that the family X constructed for $R_{(h)}$ can be enlarged to a corresponding family X' for $R_{(h+1)}$. Since the monomials in X' must be right K-linearly independent, $R_{(h+1)}$ will have $R'_{(h+1)}$ embedded in it, i.e. the map given by the universal property will be injective.

We shall take for $R_{(h+1)}$ any K-bimodule containing $R'_{(h+1)}$ as subbimodule and extend v to it by setting it equal to $h + 1$ outside $R_{(h)}$. In this way $R_{(h+1)}$ becomes a truncated filtered ring of height $h + 1$. For the only multiplications that need to be defined on the elements of degree $h + 1$ are their products with members of $R_{(0)} = K$, and the conditions T.1–T.3 that they must satisfy are just the conditions for a K-bimodule. Further, $R_{(h+1)}$, as a truncated filtered ring, will satisfy the weak algorithm because on enlarging X by adjoining a minimal generating set for the right K-space $R_{(h+1)}$ (mod $R'_{(h+1)}$) we obtain a generating set for $R_{(h+1)}$ satisfying Theorem 5.1.

Since any filtered ring R with a weak algorithm can be written in a unique way as $\cup R_{(h)}$, where each $R_{(h)}$ is an h-truncated ring satisfying the weak algorithm, and equal to the h-truncation of $R_{(h+1)}$, it follows that R may be constructed in this way. Thus we have proved

Theorem 2.5.8. *Any filtered ring with weak algorithm can be constructed by the following steps:*

(0) choose an arbitrary field K to be $R_{(0)}$,

   ...

(h) given $R_{(h-1)}$, form the universal extension $R'_{(h)}$ as above, let $R_{(h)}$ be any K-bimodule containing $R'_{(h)}$ as subbimodule and consider $R_{(h)}$ as truncated filtered ring,

   ...

(∞) define $R = \cup R_{(h)}$ as the required ring. ∎

We note that at step (1) we have $R'_{(1)} = R_{(0)}$, so this step is simply: choose a K-bimodule containing K. Of course the structure of K-bimodules over a field K

is itself a non-trivial topic, for K-bimodules are effectively $(K^\circ \otimes_\mathbb{Z} K)$-modules and a ring of the form $(K^\circ \otimes_\mathbb{Z} K)$ can have a highly complicated structure.

Exercises 2.5

1. Give a direct proof that the centre of a non-Ore filtered integral domain with a weak algorithm is a field (see also Section 6.4).
2. If R is a ring with weak algorithm, show that $R_{(0)}$ is complemented by a right ideal. Give examples to show that $R_{(0)}$ may not have a complement that is a two-sided ideal. (*Hint*: Try skew polynomial rings.)
3. Let R be a K-ring with a subset X whose monomials form a right K-basis for R. Given a filtration on R such that $R_{(0)} \subseteq K$, show that X is right v-independent if and only if every element of positive degree in R is v-dependent on X. Deduce that R satisfies the weak algorithm relative to this filtration.
4. Define a filtration on the free algebra $R = k\langle X\rangle$ for which the weak algorithm does not hold (*Hint*: Regard R as the universal associative envelope of the free Lie algebra on X.)
5. In $\mathbb{Z}\langle X\rangle$ define the *content* $c(f)$ of f by $c(0) = 0$, while for $f \neq 0$ $c(f)$ is the HCF of all the coefficients. Prove Gauss's lemma in the form $c(fg) = c(f)c(g)$ (*Hint*: Imitate the proof in the commutative case.)
6. State and prove an analogue of Theorem 5.1 for truncated filtered rings.
7. For any filtered ring R write $T_h R$ for the h-truncation obtained from it, and for an h-truncated filtered ring $R_{(h)}$ denote by $UR_{(h)}$ the universal filtered ring. Given an h-truncated filtered ring $R_{(h)}$, show that the canonical map $R_{(h)} \to T_h U R_{(h)}$ is surjective but not necessarily injective.
8. Let F be $\mathbb{Q}$ or $\mathbb{F}_p$ (p prime). Show that every F-bimodule is a direct sum of copies of F as F-bimodule. Deduce that every filtered F-ring R with weak algorithm such that $R_{(0)} = F$ is a free algebra.
9. Show that the Jacobson radical of a tensor K-ring $R = K[U]$ is zero. Let I be the augmentation ideal of $K[U]$; show that $\cap I^n = 0$, and when $R = K\langle X\rangle$, find an ideal not contained in I with the same property (see also Section 5.10).
10. Let kF be the group algebra (over a commutative field k) of a free group F. Show how to compute the rank of F in terms of kF.
11. Given a commutative field k and two disjoint sets X and Y, the mixed free k-algebra on X, Y, Y^{-1} is defined as the k-algebra $k\langle X, Y, Y^{-1}\rangle$ generated by X, Y and the inverses of elements of Y, which is universal for Y-inverting maps of $X \cup Y$ into k-algebras. Prove that the numbers of invertible and of non-invertible free generators in a mixed free algebra are independent of the choice of free generators.
12. Let $R = k\langle x_1, \ldots, x_r\rangle$ be the free k-algebra, d_i the derivation mapping x_j to δ_{ij} and $D = \sum x_i d_i$ the derivation that is the identity on the x_i. Show that for any element $a = \sum a_n$ of R, where a_n is homogeneous of degree n, $a^D = \sum n a_n$. (This result generalizes Euler's theorem on homogeneous functions.)
13. (Jategaonkar [69c]; Koshevoi [70]) Let R be an integral domain that is not right Ore. If $xR \cap yR = 0$ ($x, y \neq 0$), show that x and y generate a free algebra (over

$\mathbb{Z}$ or $\mathbb{F}_p$); deduce that an integral domain contains a free algebra on two free generators unless it is a left and right Ore domain.

Given a left but not right Ore domain, obtain an embedding of the free algebra of rank 2 in a field.

14. In $R = k\langle x, y\rangle$ show that the elements $xy^r (r = 0, 1, \ldots)$ form a free generating set of the subalgebra $k + xR$, and deduce that the free algebra of countable rank can be embedded in the free algebra of rank 2.

15. In a free algebra $R = k\langle X\rangle$ show that every non-zero Lie element, i.e. sum of repeated commutators, is an atom. (*Hint*: Let L be the free Lie algebra on X, with a k-basis B which may be taken totally ordered and note that by the Birkhoff–Witt theorem, $k\langle X\rangle$ is isomorphic to the universal associative enveloping algebra U_L of L, with a basis formed by all ascending monomials in B. Now observe that an elements of R is in L if and only if it is linear in B.)

16. Find the centralizer of X in $\mathbb{Q}_\mathbb{Z}\langle X\rangle$. What happens in the general case $K_k\langle X\rangle$? (See Dicks [77], p. 575.)

17. Let $R = k\langle x, y, y^{-1}\rangle$ be the mixed free algebra. Show that the elements $x_i = y^{-i}xy^i (i \in \mathbb{Z})$ form a free set. Denoting by A the subalgebra generated by these elements and by α the automorphism defined on A by $a \mapsto y^{-1}ay$, show that $R \cong A[y, y^{-1}; \alpha]$.

18. Let $R = k\langle x, y\rangle$ and write $[ab^{(1)}] = ab - ba$, $[ab^{(r+1)}] = [[ab^{(r)}]b]$. Show that the subalgebra S generated by the elements $[yx^{(r)}](r = 0, 1, \ldots)$ is freely generated by these elements. If δ is the derivation on S defined by $a \mapsto [ax^{(1)}]$, show that $R \cong S[x; 1, \delta]$.

19*. (McLeod [58]) Let k be a commutative field of characteristic zero. Show that the subalgebra (without 1) of $k\langle x, y\rangle$ generated by all commutators is an ideal. Show also that this fails to hold in finite characteristic or for more than two free generators.

20. (Andrunakievich and Ryabukhin [79]) For any word w in $X = \{x_1, x_2, \ldots\}$ define its *length* $l(w)$ as the number of its factors x_i and the *weight* $p(w)$ as the largest suffix of any x_i occurring. A word w is *light* if $p(w) < l(w)$ and *heavy* if no subword is light. If I is the ideal generated by all light words, show that $k\langle X\rangle/I$ has a basis consisting of all heavy words and this algebra is prime but locally nilpotent.

21. Let D be a field which is a k-algebra, with subalgebras A, B that are isomorphic via a k-linear isomorphism φ. Define D as an A-bimodule with the usual right multiplication and with left multiplication $a.u = (a\varphi)u$ ($u \in D, a \in A$). Show that the tensor D-ring $D[M]$, where $M = D \otimes_A D$, has an element $t \neq 0$ such that $at = t(a\varphi)$ for all $a \in A$. (This is an example of the HNN-construction, see SF, Section 5.6.)

2.6 The Hilbert series of a filtered ring

In Section 2.4 we met the notion of a filtered ring; we shall also need to consider graded rings and we briefly recall the definition. A *graded ring* is a ring H expressible as a direct sum of abelian groups: $H = H_0 \oplus H_1 \oplus H_2 \oplus \ldots$ such that the multiplication maps $H_i \times H_j$ into H_{i+j}. It follows that H_0 is a subring and each H_i is a H_0-bimodule.

If R is a filtered ring, where $R_{(0)}$ is a field K, we consider the associated graded ring $\mathrm{gr}R = \{\mathrm{gr}_n R\}$, where $\mathrm{gr}_n R = R_{(n)}/R_{(n-1)}$. If each $\mathrm{gr}_n R$ is finite-dimensional as right K-space, say $\dim_K(R_{(n)}/R_{(n-1)}) = \alpha_n$, then we can form the formal power series

$$H(R:K) = \sum \alpha_n t^n. \tag{1}$$

It is called the *Hilbert series* of R. Using Theorem 5.1, we can calculate the Hilbert series of any filtered ring with weak algorithm:

Proposition 2.6.1. *Let R be a filtered ring with weak algorithm, where $R_{(0)} = K$. Given a monomial right K-basis $X = \cup X_n$ for R, define $\lambda_n = |X_n| = \dim_K(R_{(n)}/R'_{(n)})$ in the notation of Section 2.5; further, put $H(X) = \sum \lambda_n t^n$. Then*

$$H(R:K) = (1 - H(X))^{-1}. \tag{2}$$

Proof. We saw that a right K-basis of $R_{(n)}/R_{(n-1)}$ is formed by the set of all monomials x_I of degree n. Each sequence $(n_1, \ldots, n_r)$ such that $n_1 + \cdots + n_r = n$ gives rise to $\lambda_{n_1} \ldots \lambda_{n_r}$ monomials of degree n, hence

$$\alpha_n = \sum \lambda_{n_1} \ldots \lambda_{n_r},$$

where the summation is over all ordered partitions of n. It follows that

$$H(R:K) = \sum \lambda_{n_1} t^{n_1} \ldots \lambda_{n_r} t^{n_r},$$

i.e. (2). ∎

In a moment we shall give another proof of (2). Meanwhile we note the special case of free algebras:

Corollary 2.6.2. *Let R be the free k-algebra of rank r, $R = k\langle x_1, \ldots, x_r\rangle$ where all the x_i have degree 1. Then $H(X) = rt$ and therefore $H(R:k) = (1 - rt)^{-1}$.* ∎

We return to a filtered K-ring R and consider the Hilbert series of a right ideal. Given any right ideal $\mathfrak{a}$ of R, let $\beta_n = r_n(\mathfrak{a}) = \dim(\mathfrak{a}_{(n)}/\mathfrak{a}'_{(n)})$ (See Section 2.4) and $\gamma_n = \dim_K(\mathfrak{a}_{(n)}/\mathfrak{a}_{(n-1)})$. We note that β_n and γ_n are bounded by $\dim(R_{(n)}) = \alpha_n$, hence we may define the Hilbert series

$$H(\mathfrak{a}:R) = \sum \beta_n t^n, \quad H(\mathfrak{a}:K) = \sum \gamma_n t^n.$$

If R satisfies the weak algorithm, these series are related by the formula

$$H(\mathfrak{a}:K) = H(\mathfrak{a}:R)H(R:K). \tag{3}$$

2.6 The Hilbert series of a filtered ring

For if (e_λ) is a weak right K-basis of $\mathfrak{a}$, then a right K-basis of $\mathfrak{a}_{(n)}/\mathfrak{a}_{(n-1)}$ is given by the family of all $e_\lambda x_I$ of degree n. Hence we have $\gamma_n = \sum \beta_i \alpha_{n-i}$ and (3) follows.

We can use (3) to give another proof of (2). We note that (2) is essentially a statement about grR, so we may assume R to be a graded ring. Then XR is the augmentation ideal, so that we have $R = K \oplus XR$ as right K-spaces, and $H(XR : R) = H(X), H(R : K) = 1 + H(XR : K)$. Inserting these values in (3) (for $\mathfrak{a} = XR$), we obtain $H(R : K) = 1 + H(X).H(R : K)$, from which (2) follows.

We now turn to modules over filtered rings with a weak algorithm and derive a presentation that in some cases provides information about the characteristic of the module. We recall that for any ring R and any set B, the free right R-module on B is written $R^{(B)}$; more explicitly, if g_b denotes the generator indexed by $b \in B$, its elements have the form of a finite sum $\sum g_b r_b (r_b \in R, b \in B)$.

Theorem 2.6.3. *Let R be a filtered ring with weak algorithm, where $R_{(0)} = K$, and let M be a right R-module. If $\{g_b | b \in B\}$ is the basis of $R^{(B)}$ corresponding to a right K-basis B of M and X is a monomial right K-basis of R, then there is an exact sequence*

$$0 \to R^{(B \times X)} \xrightarrow{\beta} R^{(B)} \xrightarrow{\alpha} M \to 0, \qquad (4)$$

where $\alpha : g_b \mapsto b \, (b \in B)$ and $\beta : (b, x) \mapsto g_b x - \sum g_c \lambda_{c,x}$ if $bx = \sum c \lambda_{c,x}$ in M ($\lambda_{c,x} \in K, b, c \in B$).

Proof. We have $\beta\alpha = 0$ by the definition of α and β. The cokernel of β is the right R-module with generators g_b and defining relations $g_b x = \sum g_c \lambda_{c,x}$, hence the right K-space spanned by the g_b is already an R-module and so is all of coker β. Thus the natural surjection coker $\beta \to M$ is an isomorphism. So far we have only used the facts that B is a right K-basis of M and that the monomials in X span R.

It remains to show that β is injective. The module $R^{(B \times X)}$ has a basis $u_{b,x} (b \in B, x \in X)$ and the general non-zero element has the form

$$s = \sum u_{b,y} x_1 \ldots x_m \lambda_{b,yx_1\ldots x_m}.$$

Put

$$n = \max\{v(yx_1 \ldots x_m) | \lambda_{b, yx_1\ldots x_m} \neq 0 \text{ for some } b\},$$

and consider $s\beta$. By the definition of β we have $u_{b,x}\beta = g_b x +$ terms of lower degree, hence the terms of highest degree (viz. n) in $s\beta$ are

$$\sum g_b y x_1 \ldots x_m \lambda_{b, y, x_1\ldots x_m},$$

summed over all terms with $v(yx_1 \ldots x_m) = n$. This is non-zero, hence $s\beta \neq 0$ and so β is injective. ∎

If X and B are finite, the characteristic of M as R-module is, by (4),

$$\chi_R(M) = (1 - |X|) \dim_K(M). \tag{5}$$

For example, if $R = k\langle x_1, \ldots, x_r \rangle$, $\mathfrak{a}$ is a right ideal of finite rank d as right R-module and $\dim_K(R/\mathfrak{a}) = m$, then taking $M = R/\mathfrak{a}$, we have $\chi_R(M) = 1 - d$ and (5) becomes

$$d - 1 = (r - 1)m. \tag{6}$$

This formula, due to Lewin, and known as the *Schreier–Lewin formula*, is a precise analogue of Schreier's formula for free groups. It tells us that in a free algebra of finite rank, any right ideal $\mathfrak{a}$ of finite codimension is finitely generated. This result holds in fact more generally.

Thus let R be any k-algebra, generated by r elements over k, and write F for the free k-algebra of rank r, so that $R \cong F/\mathfrak{n}$ for some ideal $\mathfrak{n}$ of F. Any right ideal $\mathfrak{a}$ of R corresponds to a right ideal $\mathfrak{A}$ of F containing $\mathfrak{n}$ such that $R/\mathfrak{a} \cong F/\mathfrak{A}$ as right k-spaces. In particular, if $\mathfrak{a}$ is of finite codimension m over k, then so is $\mathfrak{A}$ and by (6), $\mathfrak{A}$ is then free as right F-module, of rank $d = (r - 1)m + 1$. Hence $\mathfrak{a} = \mathfrak{A}/\mathfrak{n}$ can be generated by d elements; moreover, if R is infinite-dimensional over k and c is any regular element of R, then cR is infinite-dimensional and so meets $\mathfrak{a}$ non-trivially. Thus we obtain

Corollary 2.6.4. *Let R be a k-algebra generated by r elements over the commutative field k. Then any right ideal $\mathfrak{a}$ of finite codimension m in R over k can be generated by $(r - 1)m + 1$ elements. Further, if $[R : k] = \infty$, then $cR \cap \mathfrak{a} \neq 0$ for any regular element c in R. In particular, in a free algebra of finite rank every right ideal of finite codimension is right large and finitely generated.* ∎

Exercises 2.6

1. Let R, S be filtered algebras over a commutative field k and $T = R \otimes S$ their tensor product over k. Show that if the Hilbert series of R, S are defined, then so is that of T and $H(T : k) = H(R : k)H(S : k)$.
2. Let R be a filtered ring, where $R_{(0)}$ is a field K. Show that if $H(R : K)$ is defined, then it has an inverse $1 - L(R : K)$, where $L(R : K)$ is a power series in t with integer coefficients and zero constant term. Verify that $L(K[x] : K) = t$. If R, S are filtered rings such that $H(R : K)$ and $H(S : K)$ are both defined, and P is their coproduct over K, show that $H(P : K)$ is also defined and $L(P : K) = L(R : K) + L(S : K)$.
3*. Let $M = C \otimes_R C$ where C is an R-ring and consider the tensor C-ring $T = C[M]$. Show that T is the coproduct of an ordinary polynomial ring over C and a

complex-skew polynomial ring over C. How does this generalize to finite Galois extensions?

4. (Lewin [69]) Show that any two-sided ideal in a free algebra R has the same rank as left and as right R-module.

5. Let $R = k\langle x_1, \ldots, x_r\rangle$ be a free algebra and $\mathfrak{a}$ a non-zero principal right ideal. Then the formula (6) gives

$$\dim_k(R/\mathfrak{a}) = (1-r)^{-1}(1-1) = 0.$$

Explain this paradox.

6. Examine the relation (3) between Hilbert series when $R = K[x; \alpha, \delta]$ is a skew polynomial ring. Do the same for the Hilbert series of the opposite ring when α is not surjective.

7. Define the notion of a filtered module M (over a filtered ring R) satisfying the weak algorithm. Show that any submodule M' with the induced filtration again satisfies the weak algorithm. Under suitable hypotheses define a Hilbert series $H(M:R)$ and show that $H(M:R_{(0)}) = H(M:R)H(R:R_{(0)})$. If $M'' = M/M'$ has the induced filtration, show that

$$H(M'':R_{(0)}) = [H(M:R) - H(M':R)].H(R:R_{(0)}).$$

8*. For any K-ring R the universal derivation bimodule $\Omega_{R/K}$ may be defined as the kernel of the multiplication map $x \otimes y \mapsto xy$ in R so that we have an exact sequence

$$0 \to \Omega_{R/K} \longrightarrow R \otimes_K R \longrightarrow R \to 0,$$

(See FA, Section 2.7 or Exercise 1.1.12), where $\lambda : dx \mapsto x \otimes 1 - 1 \otimes x$. When $R = K\langle X, Y, Y^{-1}\rangle$ is the mixed free algebra, verify that $\Omega_{R/K} = R^{(X \cup Y)}$ and hence obtain the exact sequence

$$0 \to (M \otimes_K R)^{(X \cup Y)} \to M \otimes_K R \to M \to 0.$$

When G is the free group on Y and $X = \emptyset$, then $R = KG$ is the group algebra. Let H be any subgroup of G and put $M = K[H\backslash G]$, where $H\backslash G = \cup gH$ is the coset decomposition. Deduce Schreier's formula in the form $1 - \mathrm{rk}H = (G:H)(1 - \mathrm{rk}G)$.

9*. Let R be a K-ring with a subset X of r elements such that the monomials in X span R as right K-space. Show that for any right R-module M there is a sequence (4), exact except possibly at $R^{(B \times X)}$. If M is s-generated as right R-module, deduce that $M \cong R^s/N$, where N is $(mr+s)$-generated. If $\mathfrak{a}$ is a right ideal of finite codimension m in R over K, show that $\mathfrak{a}$ can be generated by $(mr+1)$ elements.

2.7 Generators and relations for $GE_2(R)$

Our next objective is to treat the analogue of the Euclidean algorithm that exists in rings with 2-term weak algorithm. One of the main consequences is that such a ring is always a strong E_2-ring (see Section 2.3 and Theorem 4.4); moreover, there is a convenient normal form for the elements of $GE_2(R)$, which leads to a presentation of $GL_2(R)$. Since many of the formulae are valid in quite general rings, we shall digress in this section to discuss generators and relations in $GE_2(R)$ for general rings.

For brevity let us write

$$T(x) = \begin{pmatrix} 1 & 0 \\ x & 1 \end{pmatrix}, \quad Q = \begin{pmatrix} 0 & 1 \\ -1 & 0 \end{pmatrix}, \quad [\alpha, \beta] = \begin{pmatrix} \alpha & 0 \\ 0 & \beta \end{pmatrix},$$

$$E(x) = QT(x) = \begin{pmatrix} x & 1 \\ -1 & 0 \end{pmatrix},$$

and put $D(\alpha) = [\alpha, \alpha^{-1}]$. We note the following relations between these matrices, valid over any ring R, for any $x, y \in R$, $\alpha, \beta \in U(R)$:

$$T(x + y) = T(x)T(y), \quad E(x + y) = -E(x)E(0)E(y), \quad E(0) = Q, \tag{1}$$

$$D(\alpha) = -E(\alpha)E(\alpha^{-1})E(\alpha), \tag{2}$$

$$E(x)[\alpha, \beta] = [\beta, \alpha]E(\beta^{-1}x\alpha), \quad [\alpha, \beta][\alpha', \beta'] = [\alpha\alpha', \beta\beta']. \tag{3}$$

We also observe that $Q = T(-1)T(1)^T T(-1) \in E_2(R)$ (where the superscript T indicates the transpose) and conversely, $T(x) = Q^{-1}E(x), T(x)^T = E(-x)Q^{-1}$. Thus $GE_2(R)$ is generated by $E_2(R)$ and all $[\alpha, \beta]$. Further we note the following consequences of (1)–(3):

$$Q^2 = -I, E(1)^3 = -I, E(-1)^3 = I, \tag{4}$$

$$E(x)^{-1} = QE(-x)Q, \tag{5}$$

$$E(x)E(y)^{-1} = E(x - y)Q^{-1} = -E(x - y)Q, \tag{6}$$

$$E(x)E(y)^{-1}E(z) = E(x - y + z), \tag{7}$$

$$E(x)E(\alpha)E(y) = E(x - \alpha^{-1})D(\alpha)E(y - \alpha^{-1}), \tag{8}$$

where $x, y, z \in R$ and $\alpha \in U(R)$. Using (3) and (5), we can bring any element A of $GE_2(R)$ to the form

$$A = [\alpha, \beta]E(a_1)\ldots E(a_n). \tag{9}$$

If $a_i = 0$ for some $i \neq 1, n$, this relation can be shortened by (1), while if $a_i \in U(R)$ for $i \neq 1, n$, it can be shortened using (8) and then (3) to bring $D(\alpha)$ to the left. Thus in any ring R we can express any matrix A in $GE_2(R)$ in the form (9), where $\alpha, \beta \in U(R)$, $a_i \in R$ and such that for $1 < i < n$, a_i is not 0 or a unit. Such an expression for A will be called a *standard form* for A. In the next section we shall see that in any ring R with 2-term weak algorithm there is a unique standard form for each $A \in GE_2(R)$; this will be shown to hold more generally in any ring R with a degree function such that $R_{(0)}$ is a field.

2.7 Generators and relations for $GE_2(R)$

If R is a ring with unique standard form for GE_2, then in any relation in this group the left-hand side can be brought to standard form and by uniqueness this must be I, i.e. any relation can be transformed to the trivial relation I = I using only (1)–(3). This proves

Proposition 2.7.1. *In any ring R with unique standard form for GE_2 the relations (1)–(3) form a complete set of defining relations for $GE_2(R)$.* ∎

The sufficient condition given here is not necessary, since it does not hold for the ring of integers $\mathbb{Z}$, which however satisfies the conclusion (see Exercise 3).

The expression (9) can also be used to describe comaximal relations; for this purpose it is more convenient to replace $E(x)$ by the matrix $P(x)$, given by

$$P(x) = \begin{pmatrix} x & 1 \\ 1 & 0 \end{pmatrix}.$$

This matrix is no longer in $E_2(R)$, since its determinant is -1, but it belongs to $E_2^*(R)$, the *extended elementary group*, defined as the group generated by $E_2(R)$ and $1 \oplus -1$. The reader should have no difficulty in writing out the analogues of (1)–(3) for $P(x)$ instead of $E(x)$, and in this way we obtain for each $A \in GE_2(R)$ the standard form

$$A = [\alpha, \beta] P(a_1) \ldots P(a_n), \quad a_i \in R, \alpha, \beta \in U(R), \quad a_i \neq 0 \text{ for } 1 < i < n. \tag{10}$$

We observe that $A \in E_2^*(R)$ if it is given by (10) with $\beta = \pm \alpha^{-1}$. To obtain explicit formulae for the product in (10) we shall define a sequence of polynomials p_n in non-commuting indeterminates $t_1, t_2, \ldots$ with integer coefficients. The p_n are defined by the recursion formulae:

$$p_{-1} = 0, \quad p_0 = 1, \tag{11}$$

$$p_n(t_1, \ldots, t_n) = p_{n-1}(t_1, \ldots, t_{n-1}) t_n + p_{n-2}(t_1, \ldots, t_{n-2}). \tag{12}$$

For $n \geq 0$, the subscript of p_n indicates the number of arguments, and so may be omitted when the arguments are given explicitly. We shall do so in what follows and write the subscript only when the arguments are omitted. We assert that

$$P(t_1) \ldots P(t_n) = \begin{pmatrix} p(t_1, \ldots, t_n) & p(t_1, \ldots, t_{n-1}) \\ p(t_2, \ldots, t_n) & p(t_2, \ldots, t_{n-1}) \end{pmatrix}. \tag{13}$$

This is clear for $n = 1$; the general case follows by induction, since on writing $p_i = p(t_1, \ldots, t_i)$, $p_i' = p(t_2, \ldots, t_{i+1})$, we have

$$\begin{pmatrix} p_{n-1} & p_{n-2} \\ p_{n-2}' & p_{n-3}' \end{pmatrix} \begin{pmatrix} t_n & 1 \\ 1 & 0 \end{pmatrix} = \begin{pmatrix} p_n & p_{n-1} \\ p_{n-1}' & p_{n-2}' \end{pmatrix}.$$

From the symmetry of (13) it is clear that the p's may also be defined by (11) and

$$p_n(t_1, \ldots, t_n) = t_1 p_{n-1}(t_2, \ldots, t_n) + p_{n-2}(t_3, \ldots, t_n) \, . \tag{14}$$

Either definition shows that p_n may be described as the sum of $t_1 t_2 \ldots t_n$ and all terms obtained by omitting one or more pairs of adjacent factors $t_i t_{i+1}$. This mode of forming the p_n might be called the *leapfrog construction*; the number of terms in p_n is f_n, the nth Fibonacci number. The first few polynomials are

$$p_1 = t_1, \quad p_2 = t_1 t_2 + 1, \quad p_3 = t_1 t_2 t_3 + t_1 + t_3,$$
$$p_4 = t_1 t_2 t_3 t_4 + t_1 t_2 + t_1 t_4 + t_3 t_4 + 1 \, .$$

Equivalently, p_n may be described as the polynomial part of the formal product (when expanded):

$$(t_1 + t_2^{-1})(t_2 + t_3^{-1}) \ldots (t_{n-1} + t_n^{-1}) t_n \, . \tag{15}$$

From (12) it easily follows that

$$p_n(0, t_2, \ldots, t_n) = p_{n-2}(t_3, \ldots, t_n),$$
$$p_n(1, t_2, \ldots, t_n) = p_{n-1}(t_2 + 1, t_3, \ldots, t_n) \, , \tag{16}$$

while (14) yields

$$p_n(t_1, \ldots, t_{n-1}, 0) = p_{n-2}(t_1, \ldots, t_{n-2}),$$
$$p_n(t_1, \ldots, t_{n-1}, 1) = p_{n-1}(t_1, \ldots, t_{n-2}, t_{n-1} + 1) \, . \tag{17}$$

When the t's are allowed to commute, the p's just reduce to the continuant polynomials, used in the study of continued fractions, and we shall use the term *continuant (polynomial)* also to describe the p's in the general case.

It is easily verified that the inverse of $P(x)$ is given by $P(x)^{-1} = P(0)P(-x)P(0)$; hence the inverse of $P(t_1) \ldots P(t_n)$ is given by

$$P(0)P(-t_n) \ldots P(-t_1)P(0)$$
$$= \begin{pmatrix} 0 & 1 \\ 1 & 0 \end{pmatrix} \begin{pmatrix} p(-t_n, \ldots, -t_1) & p(-t_n, \ldots, -t_2) \\ p(-t_{n-1}, \ldots, -t_1) & p(-t_{n-1}, \ldots, -t_2) \end{pmatrix} \begin{pmatrix} 0 & 1 \\ 1 & 0 \end{pmatrix}. \tag{18}$$

It is clear that

$$p(-t_1, \ldots, -t_n) = (-1)^n p(t_1, \ldots, t_n) \, , \tag{19}$$

hence (18) reduces to

$$[P(t_1) \ldots P(t_n)]^{-1} = (-1)^n \begin{pmatrix} p(t_{n-1}, \ldots, t_2) & -p(t_{n-1}, \ldots, t_1) \\ -p(t_n, \ldots, t_2) & p(t_n, \ldots, t_1) \end{pmatrix}. \tag{20}$$

Comparing this formula with (13), we obtain

Lemma 2.7.2. *The continuant polynomials satisfy*

(i) $p(t_1, \ldots, t_n) p(t_{n-1}, \ldots, t_2) - p(t_1, \ldots, t_{n-1}) p(t_n, \ldots, t_2) = (-1)^n$ and
(ii) $p(t_1, \ldots, t_n) p(t_{n-1}, \ldots, t_1) - p(t_1, \ldots, t_{n-1}) p(t_n, \ldots, t_1) = 0$. ∎

Of course this lemma can also be proved directly by induction; it corresponds to the well-known relations between successive convergents to a continued fraction.

We shall now use these formulae to analyse comaximal relations in GE_2-rings. We recall that in any ring R a relation

$$ab' = ba'$$

is called *comaximal* if there exist $c, d, c', d' \in R$ such that $da' - cb' = ad' - bc' = 1$. Suppose now that R is weakly 2-finite. Then by Proposition 0.5.6, there exists $A \in GL_2(R)$ such that

$$A = \begin{pmatrix} a & b \\ * & * \end{pmatrix}, \quad A^{-1} = \begin{pmatrix} * & -b' \\ * & a' \end{pmatrix}, \quad (21)$$

where the asterisks denote unspecified elements. Similarly, if R is a 2-Hermite ring, then every relation of comaximality $ad' - bc' = 1$ arises, by Proposition 0.5.6, from a pair of mutually inverse matrices

$$A = \begin{pmatrix} a & b \\ * & * \end{pmatrix}, \quad A^{-1} = \begin{pmatrix} d' & * \\ -c' & * \end{pmatrix}. \quad (22)$$

This leads to the following explicit formulae for comaximal relations in a GE_2-ring:

Proposition 2.7.3. *Let R be any ring and use $x_1, \ldots, x_n, y, z$ to denote elements of R and α, β units in R.*

(i) *If R is a weakly 2-finite GE_2-ring, then every comaximal relation in R has the form*

$$\alpha p(x_1, \ldots, x_n) p(x_{n-1}, \ldots, x_1) \beta = \alpha p(x_1, \ldots, x_{n-1}) p(x_n, \ldots, x_1) \beta ; \quad (23)$$

(ii) *if R is a 2-Hermite GE_2-ring, then every equation of comaximality can be written as*

$$\alpha p(x_1, \ldots, x_n) p(x_{n-1}, \ldots, x_2) \alpha^{-1} (-1)^n$$
$$- \alpha p(x_1, \ldots, x_{n-1}) p(x_n, \ldots, x_2) \alpha^{-1} (-1)^n = 1 ; \quad (24)$$

(iii) *if R is a strong E_2-ring, then every equation $rs = uv$, where r, s are not both 0 and u, v are not both 0, can be written*

$$y p(x_1, \ldots, x_n) p(x_{n-1}, \ldots, x_1) z = y p(x_1, \ldots, x_{n-1}) p(x_n, \ldots, x_1) z . \quad (25)$$

Proof. (i) In a weakly 2-finite ring every comaximal relation $ab' = ba'$ arises from a pair of mutually inverse matrices (21). Since R is a GE_2-ring, we can write A in the form (10):

$$A = [\alpha, \beta^{-1}]P(x_1)\ldots P(x_n), \tag{26}$$
$$A^{-1} = P(0)P(-x_n)\ldots P(-x_1)P(0)[\alpha^{-1}, \beta], \tag{27}$$

and now (23) follows on combining (26) and (27).

(ii) This follows similarly from (26), (27) and the form (22) for matrices arising from a relation of comaximality.

(iii) Since a strong E_2-ring is a 2-fir, every relation $rs = uv$ is obtained from a comaximal relation $ab' = ba'$, in the form $ya.b'z = yb.a'z$. The result now follows by applying (i) and remembering that 2-firs are weakly 2-finite. ∎

The significance of this proposition becomes clearer if we make the following definitions. We recall from Section 0.5 that two elements a, a' in any ring R are said to be *GL-related* if there exists $A \in GL_2(R)$ such that A has a as (1,1)-entry and A^{-1} has a' as (2,2)-entry. If such A can be found in $GE_2(R)$ we say that a, a' are *GE-related*. Thirdly, if A can be found in $E_2(R)$, then a, a' are said to be *E-related*. This means that A has the form

$$A = D(\alpha)E(a_1)\ldots E(a_n).$$

Clearly E-related elements are GE-related and GE-related elements are GL-related. Moreover, by (13) and (20) a, a' are E-related if and only if there exist $x_1, \ldots, x_n \in R, \alpha \in U(R)$ such that $a = \alpha p(x_1, \ldots, x_n), a' = p(x_n, \ldots, x_1)\alpha$. By Corollary 0.5.5 we see that a, a' are GL-related if and only if they are stably associated. In GE_2-rings we further have

Proposition 2.7.4. *In a GE_2-ring R, for any two elements $a, a' \in R$, the following assertions are equivalent:*

(a) a is GL-related to a',
(b) a is GE-related to a',
(c) a is E-related to an associate of a'.

Proof. The equivalence of (a) and (b) is immediate. When (b) holds, then by (10), $a = \alpha p(x_1, \ldots, x_n), a' = p(x_n, \ldots, x_1)\beta^{-1}$, whence (c) follows, and the converse is clear. ∎

In order to compare E-related elements we make use of the following formulae, which follow from the leapfrog construction of continuants.

2.7 Generators and relations for $GE_2(R)$

Let $x_1, \ldots, x_n$ be any elements of a ring and α a unit. Then we have for odd n,

$$p(x_1\alpha, \alpha^{-1}x_2, x_3\alpha, \ldots, \alpha^{-1}x_{n-1}, x_n\alpha) = p(x_1, \ldots, x_n)\alpha, \qquad (28)$$

$$p(\alpha x_1, x_2\alpha^{-1}, \alpha x_3, \ldots, x_{n-1}\alpha^{-1}, \alpha x_n) = \alpha p(x_1, \ldots, x_n), \qquad (29)$$

while if n is even,

$$p(x_1\alpha, \alpha^{-1}x_2, x_3\alpha, \ldots, x_{n-1}\alpha, \alpha^{-1}x_n) = p(x_1, \ldots, x_n), \qquad (30)$$

$$p(\alpha x_1, x_2\alpha^{-1}, \alpha x_3, \ldots, \alpha x_{n-1}, x_n\alpha^{-1}) = \alpha p(x_1, \ldots, x_n)\alpha^{-1}. \qquad (31)$$

Further, an easy calculation shows that for any n:

$$p_n(x_1, \ldots, x_n) = p_{n+1}(x_1, \ldots, x_{n-1}, x_n - 1, 1). \qquad (32)$$

This formula allows us to change the parity of n in any representation of an element by a continuant, as in the proof of the next result:

Proposition 2.7.5. *In any ring R, if a is E-related to a' and α is a unit, then (i) $a\alpha$ is E-related to $a'\alpha$, (ii) αa is E-related to $\alpha a'$ and (iii) a is E-related to $\alpha^{-1}a'\alpha$; hence αa is E-related to $a'\alpha$.*

Proof. Let $a = p(x_1, \ldots, x_n), a' = p(x_n, \ldots, x_1)$. Then (i) follows by (28) if n is odd; if n is even, we can replace it by $n+1$, using (32) for a and the left-hand analogue of (32) for a', and then applying the preceding argument. Similarly, (ii) follows from (29). To prove (iii) we first ensure that n is even (using (32)) and then apply (30) and (31). Now the last part follows because αa is E-related to $\alpha.\alpha^{-1}a'\alpha = a'\alpha$, by (i) and (ii). ∎

We note that this proposition may be used to give another proof of Proposition 7.4. In the case of a free algebra there is a bound on the length of continuant polynomials for a given pair of E-related elements, which may be stated as follows:

Proposition 2.7.6. *Let a, a' be two E-related elements in the free algebra $k\langle X\rangle$. Then a, a' have the same degree d, say, and these elements can be written in the form*

$$a = \alpha p(a_1, \ldots, a_r), a' = \alpha p(a_r, \ldots, a_1), \quad a_i \in k\langle X\rangle, \alpha \in k,$$

where $r \leq d + 2$.

Proof. The normal form (24) or equivalently, (9) shows that a_i may be taken to be neither 0 nor a unit for $1 < i < r$. It follows that the degree of a (and that of a') is $d \geq r - 2$, whence $r \leq d + 2$. ∎

Exercises 2.7

1. Prove that $p_n(x_1, \ldots, x_n) = p_r(x_1, \ldots, x_r)p_{n-r}(x_{r+1}, \ldots, x_n) + p_{r-1}(x_1, \ldots, x_{r-1})p_{n-r-1}(x_{r+2}, \ldots, x_n)$, for any n and $1 \le r \le n$.
2. Derive the formulae for $P(x)$ corresponding to (4)–(8), with Q replaced by $P = e_{12} + e_{21}$ and $D(\alpha)$ replaced by $C(\alpha) = [\alpha, -\alpha^{-1}]$.
3. Verify that (1)–(3) is a complete set of defining relations for $GE_2(\mathbb{Z})$, but show that the standard form (9) is not unique, by finding an expression for $E(2)E(-2)E(2)$ in terms of $E(2)$ and $E(3)$.
4. (After Brenner [55]; see also Farbman [95]) Show that the matrices $T(\alpha)$ and $T(\beta)^{\mathrm{T}}$ generate a free group if α, β are any real numbers such that $\alpha\beta \ge 4$.
5. By considering characteristic polynomials find all pairs of complex numbers u, v such that $(P(u)P(v))^n = 1$.
6*. Show that in $k\langle x, y, z, t\rangle$ the matrix $\begin{pmatrix} x & y \\ z & t \end{pmatrix}$ cannot be written as a product of elementary and diagonal matrices.
7*. Let $R = A_1(k)$ be the Weyl algebra on x and y over a field k of characteristic not 2. Show that the matrix $\begin{pmatrix} yx & x^2 \\ y^2 & xy \end{pmatrix}$ is invertible but not in $GE_2(R)$. (*Hint*: Compute the inverse in $k(y)[x; 1,']$ and use the filtration by degree for the last part.)
8. (Helm [83]) Show that (1)–(3) is a complete set of defining relations for $GE_2(R)$ provided that for any two non-units a, b in R there exists a unit α such that $a + \alpha$ and $b + \alpha^{-1}$ are units.
9. In any ring R, prove the identity $E(x)E(y)[1, 1 - yx] = [1 - xy, 1] \times E(0)E(y)E(x)E(0)^{-1}$, for any $x, y \in R$ such that at least one of $1 - xy, 1 - yx$ is a unit in R.
10. Show that any local ring R is a GE-ring and that there is a relation of the form $[\alpha, \beta]P(a)P(b)P(c)P(d) = I$, where b, c are neither zero nor units, unless R is a field.
11. Let R be a totally ordered ring (i.e. a ring with a total ordering compatible with addition and multiplication by positive elements). Given $a_1, \ldots, a_n \in R$ such that $a_i > 0$ for $1 \le i \le n$, show that $p(a_1, \ldots, a_r) > 0$. Show that this still holds if $a_1 \ge 0$ and $a_i > 0$ for $2 \le i \le n$, provided that $n \ge 2$.
12*. Let R be a totally ordered ring such that $a > 0$ implies $a \ge 1$. Show that R has a unique standard form (26) for GE_2 (see Cohn, [66b]), subject to $x_1 \ge 0, x_i > 0$ ($1 < i < n$) and when $n = 2, x_1, x_2$ are not both zero.
13*. Let D be a field with a central subfield k and put $R = D_k\langle x \rangle$. Find all elements E-related to x. Under what conditions is an element p of $k[x]$ an atom in R? (*Hint*: Note that $xp = px$ and use the natural homomorphism $D_k\langle x \rangle \to D[x]$ to show that x cannot be comaximal with any non-unit factor of p.)
14*. By using Exercise 13, show that for any $a \in k[x]$, if in $D_k\langle x \rangle, a = bc$, then for some $\alpha \in D^\times$ $b\alpha, \alpha^{-1}c \in k[x]$.
15°. (Bergman [67]) In a free algebra, is every element that is E-related to a square itself a square? (See Exercise 4.3.10.)
16*. In a free algebra R, show that GL-related elements are E_2'-related, where E_2' is the derived group of $E_2(R)$ (see Cohn [66b], Theorem 9.3).

17°. (G. M. Bergman) The continuant polynomials may be regarded as providing a general solution for the equation $ab' = ba'$. Find a general solution to the equation in $r + s$ unknowns $a_1, \ldots, a_r, b_1, \ldots, b_s : p(a_1, \ldots, a_r) = p(b_1, \ldots, b_s)$.
18. (H. Minkowski) Show that any matrix A in $SL_2(\mathbb{Z})$ has the form (10) with $[\alpha, \beta] = I$.
19. The Fibonacci series is defined by $f_0 = f_1 = 1$, $f_{n+1} = f_n + f_{n-1}$. Express the entries of $P(1)^n$ in terms of the f_ν.

2.8 The 2-term weak algorithm

We shall now develop the usual Euclidean algorithm, using the 2-term weak algorithm. In particular, everything that is said will apply to classical Euclidean domains in which the algorithm is defined relative to a degree-function, as well as to free algebras, which possess a weak algorithm by Theorem 5.3.

Let R be any filtered ring with 2-term weak algorithm. Given an equation

$$ab' = ba' \neq 0 \tag{1}$$

in R, if we choose $q_1 \in R$ such that $v(a - bq_1)$ is minimal, we find by an easy induction (as in Section 1.2) that $v(a - bq_1) < v(b)$. Thus

$$a = bq_1 + r_1, \quad v(r_1) < v(b), \tag{2}$$

where q_1, r_1 are unique, by Proposition 1.2.3. Substituting from (2) into (1), we find $r_1 b = (a - bq_1)b' = b(a' - q_1 b')$. If we put $r_1' = a' - q_1 b'$, this may be written

$$r_1 b' = br_1' . \tag{3}$$

By (3) and (2), $v(b) + v(r_1') = v(r_1) + v(b') < v(b) + v(b')$, hence $v(r_1') < v(b')$, so there is complete symmetry (as we know there must be, by Proposition 4.1). It may happen that $r_1 = 0$, but by (3) this is so if and only if $r_1' = 0$. If this is not the case, we can apply the same reasoning to (3) and so obtain the familiar chain of equations of the Euclidean algorithm. More precisely, we obtain two such chains, one for left and one for right division:

$$\begin{aligned} a &= bq_1 + r_1, & a' &= q_1 b' + r_1', & r_1 b' &= br_1' \\ b &= r_1 q_2 + r_2, & b' &= q_2 r_1' + r_2', & r_2 r_1' &= r_1 r_2', \\ r_1 &= r_2 q_3 + r_3, & r_1' &= q_3 r_2' + r_3', & r_3 r_2' &= r_2 r_3', \\ &\cdots & &\cdots & &\cdots \end{aligned} \tag{4}$$

Note that whereas the remainders r_i, r_i' on the two sides are in general distinct, the quotients q_i are the same. The degrees of the remainders decrease strictly:

$$v(b) > v(r_1) > v(r_2) > \ldots, \qquad v(b') > v(r_1') > v(r_2') > \ldots, \qquad (5)$$

so the remainders must vanish eventually. Let n be the least integer such that $r_{n+1} = 0$. Since $r_{n+1}r_n' = r_n r_{n+1}'$, it follows that $r_{n+1}' = 0$; if we had $r_k' = 0$ for some $k \leq n$, then by symmetry $r_k = 0$, which contradicts the definition of n. Hence both chains in (4) end at the same step, i.e. r_{n+1}' is the first vanishing remainder of the right-hand division, and the last two rows of (4) read

$$r_{n-2} = r_{n-1}q_n + r_n \qquad r_{n-2}' = q_n r_{n-1}' + r_n', \qquad r_n r_{n-1}' = r_{n-1}r_n', \\ r_{n-1} = r_n q_{n+1}, \qquad r_{n-1}' = q_{n+1}r_n', \qquad r_{n+1} = r_{n+1}' = 0. \qquad (6)$$

From (4), (6) and the inequalities (5) we see that $v(q_i) > 0$ for $2 \leq i \leq n+1$, while $v(q_1) > 0$ if and only if $v(a) > v(b)$.

Let us again write $P(x) = \begin{pmatrix} x & 1 \\ 1 & 0 \end{pmatrix}$, for any $x \in R$. Then we can express equations (4) and (6) as follows:

$$(a \quad b) = (r_n \quad 0)P(q_{n+1})P(q_n)\ldots P(q_1),$$
$$\begin{pmatrix} a' \\ b' \end{pmatrix} = P(q_1)P(q_2)\ldots P(q_{n+1})\begin{pmatrix} r_n' \\ 0 \end{pmatrix}. \qquad (7)$$

These equations make it evident that r_n is a common left factor of a and b and since the P's are invertible, it is actually a highest common left factor (HCLF). Likewise r_n' is a highest common right factor (HCRF) of a' and b'. In particular it follows that R is a strong GE_2-ring.

In the algorithm (4) the remainders, and hence the quotients are unique, subject to the inequalities (5), and these inequalities will certainly hold if we perform the algorithm in the fewest possible number of steps. In this case, moreover, $q_2, \ldots, q_{n+1}$ have positive degree and when $v(a) > v(b)$, then q_1 also has positive degree. On changing notation and bearing in mind (13) of 2.7, we thus obtain

Proposition 2.8.1. *Let R be a filtered ring with 2-term weak algorithm. If a, b are right commensurable elements of R, then there are expressions*

$$a = up(x_1, \ldots, x_n), b = up(x_1, \ldots, x_{n-1}), \qquad u, x_1, \ldots, x_n \in R, \qquad (8)$$

and if n is chosen minimal, then $x_1, \ldots, x_{n-1}$ are non-zero non-units and the expressions (8) for a, b are unique. Moreover, x_n is non-zero if and only if $v(a) \geq v(b)$, with equality if and only if x_n is a unit.

Proof. We need only observe that n can be reduced if x_i for some i in the range $1 \leq i < n$ is 0 or a unit; the uniqueness now follows by the uniqueness of the remainders. ∎

Let us return to the case of a dependence relation (1) between a and b. If (8) holds, we have of course

$$a' = p(x_n, \ldots, x_1)v, \quad b' = p(x_{n-1}, \ldots, x_1)v \quad \text{for some } v \in R^\times ,$$

so the relation (1) takes the form of (23) of 2.7, with α, β replaced by u, v.

We remark that when $q_2, \ldots, q_{n+1}$ are all of positive degree in the algorithm (4), then the degrees of the remainders must be strictly decreasing. Thus the expression (8) will also be unique if instead of prescribing n to be minimal we require $x_1, \ldots, x_{n-1}$ to be non-zero non-units. Thus in every invertible 2×2 matrix A we can reduce the first row by (7) and so write A uniquely as

$$\begin{pmatrix} \alpha & 0 \\ u & \beta \end{pmatrix} P(x_1) \ldots P(x_n), \text{ where } x_1, \ldots, x_{n-1} \in R^\times \setminus U(R) ,$$

and where $\alpha, \beta \in U(R)$ because A is invertible. Now

$$\begin{pmatrix} \alpha & 0 \\ u & \beta \end{pmatrix} = [\alpha, \beta] P(0) P(\beta^{-1} u) ,$$

hence

$$A = [\alpha, \beta] P(0) P(\beta^{-1} u) P(x_1) \ldots P(x_n),$$

and this form is unique, with the proviso that the first two P's are to be omitted if $u = 0$, or transformed by the relation $P(x) P(\alpha) P(y) = P(x + \alpha^{-1}) C(\alpha) P(y + \alpha^{-1})$ (corresponding to (8) of Section 2.7) if u is a unit. Summing up, we have

Proposition 2.8.2. *Any filtered ring R with 2-term weak algorithm is a strong E_2-ring and the standard form for $GE_2(R)$ is unique.* ∎

Exercises 2.8

1. Let $R = k\langle x_1, x_2, \ldots \rangle$ be the free algebra of finite or countable rank. Show that $GL_2(R) \cong GL_2(k[x])$. (*Hint*: Find a k-linear map that preserves the defining relations, see Cohn [66b].)
2. Given two right comaximal elements a, b in a 2-fir R, show that the equation $ax - by = f$ has a solution (x, y) for any $f \in R$. More precisely, prove the existence of $a', b', c', d' \in R$ such that $ab' = ba', ad' - bc' = 1$ and the general solution (x, y) has the form $x = d'f + b'g, y = c'f + a'g$, where g is arbitrary in R.

3. Let F be a free group. Show that the group algebra kF is a strong E-ring. (*Hint*: If F is free on X, show that every matrix over kF is stably associated to a matrix over $k\langle X\rangle$, see Section 5.8.)

4*. (Bergman [71a]) Let R be a filtered ring with 2-term weak algorithm and S any monoid of (ring-)endomorphisms of R. Show that the set of fixed points under the action of S is a strong E_2-ring (see Proposition 6.8.1).

5. In $R = k\langle x, y\rangle$ show that $(1 - xy)R \cap (1 - yx)R = 0$. Does the same hold for $k\langle x, x^{-1}, y, y^{-1}\rangle$?

6. In the ring $R = k\langle x, y, u, u^{-1}, v\rangle$ find non-zero elements a, b such that $au^{-1}x = xu^{-1}b$. (*Hint*: Put $t_1 = x$, $t_2 = y + v$, $t_3 = u$ and examine the continuant polynomials.)

7. (C. Reutenauer) In $R = \mathbb{Z}\langle X\rangle$ show that $p(a_1, \ldots, a_n)$ and $p(a_n, \ldots, a_1)$ have the same content, as defined in Exercise 5.5. (*Hint*: Use Exercise 5.5 and apply induction on n to Proposition 7.3.)

8*. (G. M. Bergman) Show that in a free algebra $k\langle X\rangle$ the number of elements E-related to a given element is finite. [*Hint*: If $a \in k\langle X\rangle$ has degree d, it is enough to show that there are only finitely many ways of writing a as $p(a_1, \ldots, a_r)$ with $r \leq d + 2$. If not, then the infinitely many distinct sets $(\alpha, a_1, \ldots, a_r)$ such that $\alpha p(a_1, \ldots, a_r) = a$ form an algebraic subset of a finite-dimensional k-space. Since its image under the map to $\alpha p(a_r, \ldots, a_1)$ is infinite, it contains an algebraic curve mapping to a curve under that map. Let L be the function field of this curve; some coefficient of $\alpha p(a_r, \ldots, a_1)$ must be transcendental over k and so have a pole. This means that there is a valuation on L taking a negative value at this coefficient. Now apply Exercise 7.]

2.9 The inverse weak algorithm

The classical division algorithm, as described for the polynomial ring $k[x]$ in Section 1.2, depended essentially on the degree-function $d(a)$ defined in this ring. If instead we use the order-function $o(a)$ defined in Section 1.5, we have an analogous statement, with the opposite inequality:

Given $a, b \in k[x]$ such that $o(b) \leq o(a) < \infty$, there exist q and a_1 such that

$$a = bq + a_1, \quad o(a_1) > o(\alpha). \tag{1}$$

The process can be repeated, but since $\mathbb{N}$ has no maximal element, there is no reason why the process should terminate. However, we can pass to the completion of the ring $k[x]$, namely the formal power series ring $k[[x]]$. Here a repetition of the step (1) leads to a convergent process, and in fact one can make deductions about divisibility in $k[[x]]$ that are quite similar to (and often stronger than) the consequences of the classical division algorithm. Indeed, the ring $k[[x]]$ displays such simple divisibility behaviour that its connexion with the algorithm (1) is usually forgotten. In the non-commutative case we do, however, obtain non-trivial results from the 'inverse algorithm'.

2.9 The inverse weak algorithm

By an *inverse filtration* on a ring R we shall mean a function v such that:

I.1. $v(x) \in \mathbb{N}$ for $x \in R^\times$, $v(0) = \infty$,
I.2. $v(x - y) \geq \min\{v(x), v(y)\}$,
I.3. $v(xy) \geq v(x) + v(y)$.

If equality holds in I.3, we have an order-function as defined in Section 1.5.

Writing $R_{[n]} = \{x \in R | v(x) \geq n\}$, we find that the inverse filtration takes the form

$$R = R_{[0]} \supseteq R_{[1]} \supseteq R_{[2]} \supseteq \ldots, \quad R_{[i]}R_{[j]} \subseteq R_{[i+j]}, \quad \cap R_{[n]} = 0 \; ; \quad (2)$$

we can again form the associated graded ring $\mathrm{gr} R = \{\mathrm{gr}_n R\}$, where $\mathrm{gr}_n R = R_{[n]}/R_{[n+1]}$ $(n = 0, 1, \ldots)$. To give an example, let R be a ring with an ideal $\mathfrak{a}$ such that $\cap \mathfrak{a}^n = 0$; then (2) holds with $R_{[n]} = \mathfrak{a}^n$; this is called the $\mathfrak{a}$–*adic* filtration on R.

In an inversely filtered ring the notions of (strong) v-dependence and (n-term) weak algorithm can be defined just as in an ordinary filtered ring, bearing in mind that all the inequalities have to be reversed; we shall refer to it as the *(n-term) inverse weak algorithm*. As before, the inverse weak algorithm is left–right symmetric.

For an inversely filtered ring R we define the *inverse dependence number* $\mu_v(R)$ as the greatest integer n for which the n-term weak algorithm holds, or ∞ if it holds for all n.

Given $a \in R^\times$, if $v(a) = n$, we write $\bar{a} = a + R_{[n+1]} \in \mathrm{gr}_n R$; $\bar{0}$ is not defined. If R satisfies the 2-term inverse weak algorithm, then $\mathrm{gr}_0 R = R/R_{[1]}$ is a field (and hence $R_{[1]}$ is a maximal ideal). This means that the following general principle applies to such rings; in the case of the ordinary weak algorithm there is a corresponding principle, which we were able to use without stating it formally, because in that case our ring actually *contained* a field:

Lemma 2.9.1. (Exchange principle) *Let R be an inversely filtered ring such that $R/R_{[1]}$ is a field. Given $a, a' \in R$ and $A \subseteq R$, if $v(a) \leq v(a')$ and a is right v-dependent on $A \cup \{a'\}$, then either a is right v-dependent on A or a' is right v-dependent on $A \cup \{a\}$.*

Proof. For $a = 0$ this holds trivially; when $a \neq 0$, there exist by hypothesis, $a_i \in A \cup \{a'\}$, $b_i \in R$ such that

$$\bar{a} = \sum_1^n \bar{a}_i \bar{b}_i \quad \text{in gr } R;$$

thus $v(a_i) + v(b_i) = v(a)$. If no a_i equals a', this shows a to be right v-dependent on A. Otherwise let $a' = a_1$, say; since $v(a) \leq v(a')$, we have equality here and so $v(b_1) = 0$, $\bar{b}_1 \in \mathrm{gr}_0 R$ and by hypothesis $\bar{b}_1$ is a unit, say $\bar{b}_1 \bar{c} = 1$. It follows that

$$\bar{a}_1 = \bar{a}_1 \bar{b}_1 \bar{c} = \bar{a}\bar{c} - \sum_z^n \bar{a}_i \bar{b}_i \bar{c},$$

and so $a' = a_1$ is right v-dependent on $A \cup \{a\}$. ∎

The earlier remarks show that this principle holds whenever $\mu_v(R) \geq 2$. By assuming it explicitly we shall find that some of our results can be extended to arbitrary inversely filtered rings.

Let R be an inversely filtered ring. The chain (2) may be taken as the neighbourhood base at 0 for a topology on R, and we can form the *completion* of R, denoted by $\hat{R}$. Explicitly we have $\hat{R} = C/\mathfrak{N}$, where C is the ring of all Cauchy-sequences in R and $\mathfrak{N}$ the ideal of sequences converging to 0 (see BA, Section 9.2, p. 314). The ring $\hat{R}$ again has an inverse filtration and there is a natural embedding $R \to \hat{R}$ respecting the filtration; we shall usually take this embedding to be an inclusion. If this inclusion is an isomorphism, R is said to be *complete*; for example, $\hat{R}$ is always complete. In any case, the induced mapping of graded rings $\mathrm{gr}\, R \to \mathrm{gr}\, \hat{R}$ is easily seen to be an isomorphism, hence $\mu_v(\hat{R}) = \mu_v(R)$; thus R satisfies the n-term weak algorithm if and only if $\hat{R}$ does.

Frequently $\hat{R}$ is a local ring; we record a sufficient condition for this to happen:

Proposition 2.9.2. *Let R be a complete inversely filtered ring. Then any $x \in R$ is invertible if and only if its image in $R/R_{[1]}$ is invertible, hence R is a local ring if and only if $R/R_{[1]}$ is. In particular, R is a local ring whenever $R/R_{[1]}$ is a field.*

Proof. Clearly any unit in R maps to a unit in $R/R_{[1]}$; conversely, if $\bar{a}$ has a right inverse $\bar{b}$, then $ab = 1 - c$, where $c \in R_{[1]}$, so $c^n \in R_{[n]}$ and $\sum c^n$ is convergent. Now $ab(\sum c^n) = 1$, so every element not in $R_{[1]}$ has a right inverse. Let $au = 1$; then $u \notin R_{[1]}$ and so $ua' = 1$ for some a', and $a' = aua' = a$, which shows that a has the inverse u. If R is a local ring with maximal ideal $\mathfrak{m}$, then $\mathfrak{m} \supseteq R_{[1]}$ and so $R/R_{[1]}$ is a local ring with maximal ideal $\mathfrak{m}/R_{[1]}$. Conversely, if $R/R_{[1]}$ is a local ring, then any element outside the maximal ideal has an inverse $(\mathrm{mod}\, R_{[1]})$ and so by the first part, it has an inverse in R, showing R to be a local ring. Now the last part is clear. ∎

Before we can apply the inverse weak algorithm we still need two general reduction lemmas. To obtain the best results we shall take our rings complete:

Lemma 2.9.3. *Let R be a complete inversely filtered ring and $a, a_1, \ldots, a_n$ any elements of R. Then there exist $b_1, \ldots, b_n \in R$ such that $v(a_i) + v(b_i) \geq v(a)$ and $a - \sum a_i b_i$ is either 0 or right v-independent of $a_1, \ldots, a_n$.*

Proof. Assume that we can find $b_{i,k}$ such that $v(b_{i,k}) \geq v(a_i) - v(a)$ and

$$v\left(a - \sum a_i b_{i,k}\right) \geq v(a) + k .$$

If $a - \sum a_i b_{i,k}$ is right v-independent of the a_i, the result follows; otherwise we can subtract a right linear combination of the a_i of formal order $\geq v(a) + k$ to get an element $a - \sum a_i b_{i,k+1}$ of order $\geq v(a) + k + 1$. If this holds for all k, then $b_{i,k}$ converges to an element b_i by completeness and $a = \sum a_i b_i$. ∎

We can now obtain an analogue of Lemma 4.3, but in a much stronger form. We recall that $P_n(R)$ is the subgroup of $GL_n(R)$ consisting of all signed permutation matrices, and write $Tr_n(R)$ for the subgroup of upper unitriangular matrices, i.e. matrices having 1's on the main diagonal and 0's below it. Clearly $Tr_n(R)$, like $P_n(R)$, is a subgroup of $E_n(R)$.

Lemma 2.9.4. *Let R be a complete inversely filtered ring such that $R/R_{[1]}$ is a field, and let $a_1, \ldots, a_n \in R$. Then there exists $P \in P_n(R)Tr_n(R)$ such that $(a_1, \ldots, a_n)P = (a'_1, \ldots, a'_r, 0, \ldots, 0)$, where the a'_i are not strongly right v-dependent.*

Proof. Let $a'_1 = a_i$ be any element of least value. Applying Lemma 9.3, we can modify the $a_j (j \neq i)$ by right multiples of a'_1 so as to make them 0 or right v-independent of a'_1. This can only increase their values, so $v(a'_1)$ will still be minimal. Let a'_2 be of least value among the resulting elements other than a'_1; by another application of Lemma 9.3 we can make all the elements other than a'_1, a'_2 zero or right v-independent of a'_1, a'_2. Continuing this process, we get a sequence $a'_1, a'_2, \ldots, a'_n$ that will clearly be the image under a unitriangular matrix of a certain ordering of $a_1, \ldots, a_n$. Since

$$v(a'_1) \leq v(a'_2) \leq \ldots \leq v(a'_n) ,$$

by construction, all zeros will occur at the end. Now suppose that some a'_i is right v-dependent on the remaining a'_j. By the exchange principle we conclude that some a'_j is right v-dependent on those preceding it, but this contradicts the construction. Hence no a'_i is right v-dependent on the rest, i.e. they are not strongly right v-dependent. ∎

We now impose the inverse weak algorithm to obtain an analogue of Theorem 4.4.

Theorem 2.9.5. *Let R be a complete inversely filtered ring with n-term inverse weak algorithm, where $n \geq 1$. Then R is a local ring, a strong E_n-ring and an n-fir. In particular, if the inverse weak algorithm holds in R, then R is a local ring, a strong E-ring and a semifir.*

Proof. This is clear when $n = 1$. For $n > 1$ it follows from Lemma 9.4, by using the n-term inverse weak algorithm. ∎

Weak v-bases for right ideals of inversely filtered rings can be defined as before (see Section 2.4) and constructed similarly; the definition, the construction and Lemma 4.7 can all be stated in gr R, using the right ideal of leading terms of members of the right ideal $\mathfrak{a}$ under construction. If R is complete and B is a finite weak v-basis of $\mathfrak{a}$, then B is a generating set for $\mathfrak{a}$; in the general case it is no longer true that B generates $\mathfrak{a}$ but the right ideal generated is merely dense in $\mathfrak{a}$. As in Proposition 4.8 we obtain

Proposition 2.9.6. *Let R be a complete inversely filtered ring with n-term inverse weak algorithm, for some natural number n, and in case $n = 1$ assume also that $R/R_{[1]}$ is a field. Then R has left and right ACC_n.* ∎

In Section 2.4 we saw that a filtered ring with weak algorithm is a fir (Theorem 4.6). This is not to be expected here; for a ring with inverse weak algorithm we find instead that it is a kind of 'topological fir':

Proposition 2.9.7. *Let R be a complete inversely filtered ring with inverse weak algorithm. Then any right ideal $\mathfrak{a}$ of R contains a right v-independent set B such that BR is dense in $\mathfrak{a}$.*

Proof. Let B be a weak v-basis for $\mathfrak{a}$. By the inverse weak algorithm it is right v-independent and any $a \in \mathfrak{a}$ is right v-dependent on B. It follows that for any $b \in B$ and any natural number k there exist elements $c_{b,k} \in R$ such that

$$v\left(a - \sum b c_{b,k}\right) > k,$$

where b runs over B and the sum is finite for any given k. As $k \to \infty$, $c_{b,k}$ converges to c_b, say, and we obtain $a = \sum b c_b$ in the completion of R. The sum here may be infinite, but it is convergent in the sense that only finitely many terms of value $\leq k$ occur for any k; this just means that BR is dense in $\mathfrak{a}$. ∎

The construction of the monomial right K-basis in the remarks preceding Theorem 5.1 was essentially carried out in gr$R$ and so can be repeated here. But instead of finite sums we must now allow infinite convergent series of monomial terms, with coefficients chosen from a set of representatives of $R/R_{[1]}$.

From Proposition 9.7 we have (as in Section 2.5)

2.9 The inverse weak algorithm

Theorem 2.9.8. *Let R be an inversely filtered ring. Then R satisfies the inverse weak algorithm if and only if $R/R_{[1]}$ is a field K and R has a right v-independent weak v-basis for $R_{[1]}$, as right ideal. In this case, if $\bar{K}$ is a set of representatives of K in R (with 0 represented by itself, for simplicity), and $X = \{x_i\}$ is a monomial right K-basis for $R_{[1]}$, then any element in the completion $\hat{R}$ can be uniquely expressed as a convergent series*

$$\sum x_I \alpha_I, \quad \text{where } \alpha_I \in \bar{K}, \tag{3}$$

and where $I = (i_1, \ldots, i_n)$ runs over all finite suffix-sets and $x_I = x_{i_1} \ldots x_{i_n}$. Conversely, all such expressions represent elements of $\hat{R}$. ∎

Here the sum (3) is understood to be convergent if for each integer k, the set $\{I \mid v(x_I) \leq k \text{ and } \alpha_I \neq 0\}$ is finite. We remark that the finite sums (3) form a dense subgroup of $\hat{R}$.

The most important example of a ring with inverse weak algorithm is the power series ring in a number of non-commuting indeterminates over a field. We shall indicate briefly how the inverse weak algorithm can be used to characterize such rings.

Let R be an inversely filtered K-ring, where K is a field. Then $R/R_{[1]}$ is a K-ring in a natural way; if it is equal to K, we shall call R a *connected* inversely filtered K-ring. This just means that $R = K \oplus R_{[1]}$.

For simplicity let us take a commutative field k, form the free k-algebra $R = k\langle X \rangle$ on a set X and denote by v the usual order-function on R, given by the terms of least degree. This defines an inverse filtration on R, for which R is a connected k-algebra. More generally, we may assign different degrees to the various elements of X; thus we take $X = \cup X_i$ itself to be graded, X_i being the set of elements of degree i (before we had the special case $X = X_1$). This is particularly useful when X is infinite and each X_i is chosen to be finite. When X is graded in any way, we have a function $v : X \to \mathbb{N}_{>0}$ and the resulting completion $k\langle\langle X \rangle\rangle$ is the power series ring in the graded set X over k.

The power series ring over a field has the following characterization, analogous to the characterization of free algebras in Section 2.5.

Proposition 2.9.9. *If k is a commutative field and R a complete inversely filtered connected k-algebra, then R is a power series ring in a graded set X over k if and only if R satisfies the inverse weak algorithm.*

Proof. If $R = k\langle\langle X \rangle\rangle$ has an inverse filtration as defined above, then $gr R = k\langle X \rangle$ and this ring satisfies the inverse weak algorithm; hence the same holds for R.

Conversely, if R satisfies the inverse weak algorithm, then by Theorem 9.8, $R_{[1]}$ has a right v-independent weak v-basis X; hence the k-algebra generated by X is free on X and is dense in R. Therefore R is equal to its completion $k\langle\!\langle X\rangle\!\rangle$. ∎

Corollary 2.9.10. *In a power series ring $k\langle\!\langle X\rangle\!\rangle$ in a graded set X over k, any closed subalgebra satisfying the inverse weak algorithm is again a power series ring in a graded set over k.* ∎

Here 'closed' refers of course to the topology: every convergent series of terms in the subalgebra has a sum in the subalgebra.

Corollary 2.9.11. *Let R be a complete inversely filtered connected k-algebra, where k is a commutative field. Then R is a power series ring in a single variable over k if and only if $R \neq k$ and for any $a, b \in R^\times$ such that $v(a) \geq v(b)$, a is right v-dependent on b.*

Proof. Any ring R satisfying the hypotheses has inverse weak algorithm and so, by Proposition 9.9 is of the form $k\langle\!\langle X\rangle\!\rangle$ for some $X \neq \varnothing$. If X contains more than one element, say $x, y \in X$, $x \neq y$, then neither of x, y is right v-dependent on the other. This contradicts the hypothesis, hence $|X| = 1$. Conversely, $k[[x]]$ clearly satisfies the hypothesis. ∎

The rings in Corollary 9.11 turned out to be commutative. If we consider general commutative inversely filtered rings (or even Ore domains), we again find that the weak algorithm already follows from the 2-term weak algorithm. Let us define a right *principal valuation ring*, PVR for short, as an integral domain R with a non-unit p such that every non-zero element of R has the form $p^r u (r \geq 0, u \in U(R))$. When R is commutative, this just reduces to the usual definition of a PVR as a discrete rank 1 valuation ring. In general these rings need not be commutative, so that we may have a right (or left) PVR. These rings arise when the 2-term inverse weak algorithm holds in Ore domains:

Proposition 2.9.12. *Let R be a complete inversely filtered right Ore domain. Then the 2-term inverse weak algorithm holds in R if and only if R is either a field or a right principal valuation ring.*

Proof. Clearly the 2-term inverse weak algorithm holds in any right PVR, and it remains to prove the converse. If $R_{[1]} = 0$, then $R = \operatorname{gr}_0 R$ is a field. Otherwise take $p \in R_{[1]}$ such that $v(p)$ has the least value. Given $a \in R$, a and p are right linearly dependent by hypothesis, hence they are right v-dependent and so by Lemma 9.3, either $a = pc$ or $p = ac$, for some $c \in R$. If a is a non-unit, then $v(a) \geq v(p)$, thus in the second case $v(c) = v(p) - v(a) \leq 0$, hence

c is then a unit, so we have in any case $a \in pR$ whenever a is a non-unit. If we have

$$a = p^r u, \qquad (4)$$

then $v(a) \geq rv(p)$, hence r is bounded and if we choose it maximal in (4), then u must be a unit. Thus R is a right PVR and this clearly satisfies the inverse weak algorithm. ∎

A second important case (which actually includes the case of power series rings in a finite ungraded set) arises from rings with an $\mathfrak{a}$-adic filtration. Given a ring R and an ideal $\mathfrak{a}$ of R, let us write

$$K = \mathrm{gr}_0 R = R/\mathfrak{a}, \quad M_n = \mathrm{gr}_n R = \mathfrak{a}^n/\mathfrak{a}^{n+1}, \qquad (5)$$

so that K is a ring and M_n is a K-bimodule. We first give a general condition for $\mathrm{gr}\, R$ to be a tensor ring:

Theorem 2.9.13. *Let R be a ring and $\mathfrak{a}$ an ideal in R. If in the notation (5), $M = M_1$ is free as right K-module and*

$$\mathfrak{a} \otimes_R \mathfrak{a} \cong \mathfrak{a}^2, \qquad (6)$$

then the graded ring associated with the $\mathfrak{a}$-adic filtration, $\mathrm{gr}R = \oplus M_n$, is a tensor ring:

$$\mathrm{gr}R \cong K[M].$$

Proof. Since $M = M_1 = \mathfrak{a}/\mathfrak{a}^2$ is free as right K-module, we can take a K-basis and lift it back to a subset Y of $\mathfrak{a}$. This gives a map $f : R^{(Y)} \to \mathfrak{a}$, defined by $(a_y) \mapsto \sum y a_y$, such that the induced map $f \otimes_R K : K^{(Y)} \to M_1$ is an isomorphism. For each n we have the induced homomorphism

$$f \otimes_R R/\mathfrak{a}^n : (R/\mathfrak{a}^n)^{(Y)} \to \mathfrak{a}/\mathfrak{a}^{n+1}, \qquad (7)$$

which we shall show by induction to be an isomorphism, for all n. This holds for $n = 1$, so assume that $n \geq 1$ and that (7) is an isomorphism. Applying $\otimes_R \mathfrak{a}$ to (7), we obtain an isomorphism

$$f \otimes \mathfrak{a}/\mathfrak{a}^{n+1} : (\mathfrak{a}/\mathfrak{a}^{n+1})^{(Y)} \to \mathfrak{a} \otimes \mathfrak{a}/\mathfrak{a}^{n+1} \cong \mathfrak{a} \otimes \mathfrak{a} \otimes R/\mathfrak{a}^n$$
$$\cong \mathfrak{a}^2 \otimes R/\mathfrak{a}^n \cong \mathfrak{a}^2/\mathfrak{a}^{n+2},$$

where all tensor products are taken over R and (6) has been used in the last but one step. We thus have an exact commutative diagram

$$\begin{array}{ccccccccc} 0 & \to & (\mathfrak{a}/\mathfrak{a}^{n+1})^{(Y)} & \to & (R/\mathfrak{a}^{n+1})^{(Y)} & \to & (R/\mathfrak{a})^{(Y)} & \to & 0 \\ & & \downarrow & & \downarrow & & \downarrow & & \\ 0 & \to & \mathfrak{a}^2/\mathfrak{a}^{n+2} & \to & \mathfrak{a}/\mathfrak{a}^{n+2} & \to & \mathfrak{a}/\mathfrak{a}^2 & \to & 0 \end{array}$$

We have just seen that the first column is an isomorphism, the third column is an isomorphism by the case $n = 1$ of (7), hence the middle column is an isomorphism. So (7) holds with n replaced by $n + 1$, and hence for all n.

Thus $\mathfrak{a}/\mathfrak{a}^{n+1}$ is free as right $R/\mathfrak{a}^n$-module, and so the sequence

$$0 \to \mathfrak{a}^{n-1}/\mathfrak{a}^n \to R/\mathfrak{a}^n$$

remains exact under the operation $\mathfrak{a}/\mathfrak{a}^{n+1} \otimes_{R/\mathfrak{a}^n} - \cong \mathfrak{a} \otimes_R -$. Hence we have an embedding

$$\mathfrak{a} \otimes_R \mathfrak{a}^{n-1}/\mathfrak{a}^n \to \mathfrak{a}/\mathfrak{a}^{n+1} . \tag{8}$$

Here the image is $M_n = \mathfrak{a}^n/\mathfrak{a}^{n+1}$, so we have an isomorphism $\mathfrak{a} \otimes_R M_{n-1} \cong M_n$. Since $\mathfrak{a} \otimes_R - \cong \mathfrak{a}/\mathfrak{a}^2 \otimes_{R/\mathfrak{a}} - \cong M \otimes_K -$, it follows that $M \otimes_K M_{n-1} \cong M_n$; therefore by induction, $M_n \cong M^{\otimes n}$ and the result follows. ∎

It is easy to derive conditions for the $\mathfrak{a}$-adic filtration to satisfy the inverse weak algorithm:

Corollary 2.9.14. *Let R be a ring with an ideal $\mathfrak{a}$ such that (i) $R/\mathfrak{a}$ is a field, (ii) $\cap \mathfrak{a}^n = 0$ and (iii) $\mathfrak{a} \otimes_R \mathfrak{a} \cong \mathfrak{a}^2$. Then R satisfies the inverse weak algorithm relative to the $\mathfrak{a}$-adic filtration. In particular this holds for a semifir that is a local ring with a maximal ideal $\mathfrak{m}$, finitely generated as right ideal, such that $\cap \mathfrak{m}^n = 0$.*

Proof. By Theorem 9.13, $\mathrm{gr} R \cong K[M]$, where $K = R/\mathfrak{a}$ is a field. Hence $\mathrm{gr} R$ satisfies the inverse weak algorithm, by Theorem 9.8, and clearly the same holds for R itself. When R is a semifir and a local ring, whose maximal ideal $\mathfrak{m}$ is finitely generated as right ideal, then $\mathfrak{m}$ has a right linearly independent generating set $u_1, \ldots, u_n$, say. Now the mapping $\mathfrak{m} \otimes_R \mathfrak{m} \to \mathfrak{m}^2$ can be described as $\sum u_i \otimes a_i \mapsto \sum u_i a_i$, where $a_i \in \mathfrak{m}$, and this is an isomorphism by the linear independence of the u's. Thus (iii) holds as well as (i), (ii) and the result follows. ∎

By Corollary B.8 of the Appendix the relation $\mathfrak{a} \otimes_R \mathfrak{a} \cong \mathfrak{a}^2$ holds, for example, if $\mathfrak{a}$ is flat as right (or left) R-module. In Section 5.10 we shall see that in a fir R any proper ideal $\mathfrak{a}$ satisfies $\cap \mathfrak{a}^n = 0$, and $\mathfrak{a}$, being free, is then flat, so Corollary 9.14 applies to a fir whenever (i) is satisfied.

We note a second case in which the conditions may be applied.

Corollary 2.9.15. *Let R be an augmented k-algebra, with augmentation ideal $\mathfrak{a}$, and let V be the complement of $\mathfrak{a}^2$ in $\mathfrak{a}$ as k-space. If $\mathfrak{a} \otimes_R \mathfrak{a} \cong \mathfrak{a}^2$, then the natural map $f : k[V] \to R$ is injective.*

2.9 The inverse weak algorithm

Proof. Write $S = k[V]$ and denote the augmentation ideal in S by $\mathfrak{b}$. We have $\cap \mathfrak{b}^n = 0$, so if f is not injective, then there exists $c \in \mathfrak{b}^n, c \notin \mathfrak{b}^{n+1}$ for some n and $cf = 0$. Hence cf maps to 0 in $\mathfrak{a}^n/\mathfrak{a}^{n+1}$, but the induced map $\mathfrak{b}^n/\mathfrak{b}^{n+1} \to \mathfrak{a}^n/\mathfrak{a}^{n+1}$ decomposes into a sequence of isomorphisms

$$\mathfrak{b}^n/\mathfrak{b}^{n+1} \cong V^{\otimes n} \cong (\mathfrak{a}/\mathfrak{a}^2)^{\otimes n} \cong \mathfrak{a}^n/\mathfrak{a}^{n+1},$$

which is a contradiction, because $c \neq 0$ in $\mathfrak{b}^n/\mathfrak{b}^{n+1}$. ∎

We now come to an important relation between a free algebra and its completion, but some preparation is necessary. We need to deal with row and column vectors from a ring. The components of such a vector will generally be denoted by a Latin suffix, thus a has the components a_i and a_λ has the components $a_{\lambda i}$. The precise range will be indicated in brackets when it is not clear from the context. We shall also continue to use the notation ab for the product of a row a and a column b, thus $ab = \sum a_i b_i$.

Let $R \subseteq S$ be a pair of rings. Given $a \in S^n, b \in {}^n S$, the product $ab = \sum a_i b_i$ is said to lie *trivially* in R, if for each $i = 1, \ldots, n$, either a_i and b_i lie in R or $a_i = 0$ or $b_i = 0$. Further, the subring R of S is said to be *totally n-inert* in S if for all $m \leq n$ and any families (a_λ) of rows in S^m and of columns (b_μ) in ${}^m S$ such that $a_\lambda b_\mu \in R$ for all λ, μ, there exists $U \in GL_m(S)$ such that on writing $a'_\lambda = a_\lambda U, b'_\mu = U^{-1} b_\mu$, each product $a'_\lambda b'_\mu$ lies trivially in R. If R is totally n-inert in S for all $n \geq 1$, we say that R is *totally inert* in S.

Theorem 2.9.16. (Inertia theorem) *Any free algebra $k\langle X \rangle$ over a commutative field k is totally inert in its power series completion $k\langle\!\langle X \rangle\!\rangle$.*

Proof. Let us put $R = k\langle X \rangle$, $\hat{R} = k\langle\!\langle X \rangle\!\rangle$; these rings are inversely filtered by the order-function v and satisfy the inverse weak algorithm. We now define

$$R_{[1]} = \{f \in R | v(f) > 0\}, \quad \hat{R}_{[1]} = \{f \in \hat{R} | v(f) > 0\},$$

so that $\hat{R}_{[1]}$ is the closure of $R_{[1]}$ in $\hat{R}$. By the inverse weak algorithm we have $R/R_{[1]} \cong \hat{R}/\hat{R}_{[1]} \cong k$. The set X is a weak v-basis of $R_{[1]}$ as right ideal, moreover $R_{[1]}$ is free as right ideal, with X as basis. Clearly X is also a weak v-basis of $\hat{R}_{[1]}$, so each element $f \in \hat{R}_{[1]}$ can be written uniquely as a convergent series

$$f = \sum x f_x, \quad \text{where } f_x \in \hat{R}. \tag{9}$$

We note that $v(f) > \min\{v(f_x) | f_x \neq 0, x \in X\}$ and $f \in R_{[1]}$ if and only if all the f_x lie in R and are almost all zero.

Let $A \subseteq \hat{R}^n, B \subseteq {}^n \hat{R}$ be such that $AB \subseteq R$; we may enlarge A, B to be maximal in the sense that each consists of all the rows (resp. columns) mapped into R by multiplication with the other. Then A is a left R-submodule of $\hat{R}^n$ and B a right R-submodule of $^n\hat{R}$. Further, the image $\bar{A}$ of A in $(\hat{R}/\hat{R}_{[1]})^n = k^n$ is a left K-space of dimension s, where $s \leq n$. By $\hat{R}$-column operations on A (and the corresponding row operations on B) we may assume that A contains $e_1, \ldots, e_s$, part of the standard basis for row vectors, while any component after the first s in any element of A is a non-unit in $\hat{R}$, i.e. has positive order.

Consider the case $s < n$. We claim that for all $a = (a_1, \ldots, a_n) \in A, a_n = 0$. For if not, let us choose $a \in A$ so as to minimize $v(a_n)$. By adding left R-multiples of $e_1, \ldots, e_s$ to a we may suppose that each a_i has positive order. Hence we can write

$$a = \sum xa_x, \quad \text{where } a_x \in \hat{R}^n. \tag{10}$$

We claim that all the a_x lie in A. For, given $b \in B$, we have $\sum xa_x b = ab \in R$, hence $a_x b \in R$ and by the maximality of A we find that $a_x \in A$. By (10) the nth component of a_x must have lower order than a_n. This contradicts the minimality of $v(a_n)$, so $a_n = 0$ for all $a \in A$. We can now omit the final component in A, B and reach the desired conclusion by induction on n.

There remains the case $s = n$. Then $R^n \subseteq A \subseteq \hat{R}^n$, hence $B \subseteq {}^n R$. By symmetry we may also assume that the image $\bar{B}$ of B in $^n(\hat{R}/\hat{R}_{[1]}) = {}^n(R/R_{[1]})$ has dimension n, and so is all of $^n(R/R_{[1]})$. If $b \in B$ has image $\bar{b} = 0$, then in terms of a left weak v-basis Y of R_1 we have $b = \sum b_y y$, where $b_y \in {}^n R$ and the b_y are almost all 0. As before we find that $b_y \in B$, hence

$$^n(R/R_{[1]}) = \bar{B} = B/BR_{[1]} \cong B \otimes_R R/R_{[1]}. \tag{11}$$

Now $R = k\langle X \rangle$ is a fir, so B is a free right R-module, of rank n, by (11). If we take any right R-basis of B, we have n columns forming an $n \times n$ matrix P over R. By (11) the columns of P (mod $R_{[1]}$) form a k-basis of $^n k (\cong {}^n(R/R_{[1]}))$, i.e. P is invertible (mod $R_{[1]}$) and hence invertible over $\hat{R}$. Now $B = P(^n R)$, so $P^{-1}B = {}^n R$ and hence $AP = R^n$, as we wished to show. ∎

We remark that this result holds also for the tensor ring $K_k\langle X \rangle$ and its completion. If R is a local semihereditary ring and its maximal ideal $\mathfrak{m}$ is finitely generated as right ideal, with basis $x_1, \ldots, x_r$, then R satisfies the inverse weak algorithm relative to the $\mathfrak{m}$-adic filtration and $\hat{R}$ is the power series ring on $x_1, \ldots, x_r$. Thus we obtain

Corollary 2.9.17. *Let R be a local semihereditary ring with a maximal ideal $\mathfrak{m}$ and suppose that $\mathfrak{m}$ is finitely generated as right ideal. Then R satisfies the inverse weak algorithm relative to the $\mathfrak{m}$-adic filtration if and only if $\bigcap \mathfrak{m}^n = 0$; moreover its completion $\hat{R}$ is a power series ring in which R is embedded as a totally inert subring.* ∎

2.9 The inverse weak algorithm

There are other subalgebras of $k\langle\langle X\rangle\rangle$ that we may sometimes wish to consider, in particular the algebra of all rational, and that of all algebraic power series.

The least subalgebra of $k\langle\langle X\rangle\rangle$ containing X and closed under taking inverses when they exist in $k\langle\langle X\rangle\rangle$ is denoted by $k\langle\langle X\rangle\rangle_{\text{rat}}$ and is called the algebra of *rational power series*. It is clear that every element with non-zero constant term has an inverse, hence $k\langle\langle X\rangle\rangle_{\text{rat}}$ is a local ring. In Chapter 7 we shall see that $k\langle\langle X\rangle\rangle_{\text{rat}}$ consists of the components of the solutions u of the matrix equation

$$u = Bu + b, \quad B \in \mathfrak{M}_n(k\langle X\rangle), \quad b \in {}^n k\langle X\rangle, \tag{12}$$

where B has zero constant term.

An element f of $k\langle\langle X\rangle\rangle$ is said to be *algebraic* over $k\langle X\rangle$ if it is of the form $f = \alpha + u_1$, where $\alpha \in k$ and u_1 is a component of the solution u of a system of equations

$$u_i = \phi_i(u_1, \ldots, u_n, x_1, \ldots, x_r), \quad i = 1, \ldots, n, \tag{13}$$

where ϕ_i is a (non-commutative) polynomial in the u's and x's without constant terms or terms of degree 1 in the u's. The set of all algebraic elements of $k\langle\langle X\rangle\rangle$ will be denoted by $k\langle\langle X\rangle\rangle_{\text{alg}}$. We remark that for commutative rings this reduces to the usual definition, e.g. $k[[x]]_{\text{alg}}$ is the relative algebraic closure of $k[x]$ in $k[[x]]$ (see Exercise 18). The set $k\langle\langle X\rangle\rangle_{\text{alg}}$ may be described in the following terms.

Theorem 2.9.18. *In the free power series ring $k\langle\langle X\rangle\rangle$, any system (13), where each ϕ_i is a polynomial without constant term or terms of degree 1 in the u's, has a unique solution with components of positive order, and the set $k\langle\langle X\rangle\rangle_{\text{alg}}$ is a subalgebra of $k\langle\langle X\rangle\rangle$; it is a local ring, the non-units being the elements of positive order.*

Proof. Denote by $u_{i\nu}$ the component of u_i of degree ν in the x's. Then by equating homogeneous components in (13) we find $u_{i\nu} = \phi_{i\nu}(u, x)$. Here $\phi_{i\nu}$ is the sum of the terms of degree ν; by hypothesis, for any term $u_{j\mu}$ occurring in $\phi_{i\mu}$ we have $\mu < \nu$, therefore the components $u_{i\nu}$ are uniquely determined in terms of the $u_{j\mu}$ with $\mu < \nu$, while $u_{i0} = 0$, by hypothesis. Thus (13) has a unique solution satisfying $o(u_i) > 0$.

If $u_i = \phi_i(u, x)(i = 1, \ldots, m)$, $v_j = \psi_j(v, x)(j = 1, \ldots, n)$ are two such systems, then to show that $u_1 + v_1, u_1 v_1$ are algebraic, we combine the above systems of equations for u_i, v_j with $w = \phi_1 + \varphi_1$, $w = \phi_1 \varphi_1$, respectively. Thus we have indeed a subalgebra. Further, the system $w = \phi_1 + u_1 w$ shows that $(1 - u_1)^{-1} - 1$ is algebraic, hence so is $(1 - u_i)^{-1}$ and it follows that we have a local ring. For the last assertion take an element f of order 0 in $k\langle\!\langle X\rangle\!\rangle_{\text{alg}}$, say $f = \alpha + u, \alpha \in k^\times$; then $f = \alpha(1 + \alpha^{-1}u)$, hence $f^{-1} = \alpha^{-1}(1 + \alpha^{-1}u)^{-1}$ and this shows f to be a unit in $k\langle\!\langle X\rangle\!\rangle_{\text{alg}}$. ∎

We have the inclusions

$$k\langle X\rangle \subset k\langle\!\langle X\rangle\!\rangle_{\text{rat}} \subset \langle\!\langle X\rangle\!\rangle_{\text{alg}} \subset k\langle\!\langle X\rangle\!\rangle .$$

It is not hard to see that all these inclusions are strict; this is already clear when $|X| = 1$.

In Section 2.5 we defined the notion of transduction for $k\langle X\rangle$ and it is clear how this extends to $k\langle\!\langle X\rangle\!\rangle$. Let us show that both $k\langle\!\langle X\rangle\!\rangle_{\text{rat}}$ and $k\langle\!\langle X\rangle\!\rangle_{\text{alg}}$ admit all transductions. It will be enough to examine the (left or right) cofactor of a generator x since the cofactor of a monomial can then by obtained by iteration.

Let u_1 be a component of a solution of (12), where B has zero constant term, and for a given $x \in X$ let $p \mapsto p'$ be the transduction 'left cofactor of x'. Clearly the constant term α of u equals that of b, as we see by putting $u = \alpha$ in (12). Now if $u = \alpha + v$, where v has zero constant term, then v again satisfies an equation of the form (12), namely $v = Bv + b_1$, where $b_1 = b - \alpha + B\alpha$. Thus we may assume that we have a system (12) in which u has zero constant term. If we now apply $'$ we find

$$u' = [Bu + b]' = Bu' + b' ,$$

hence u' satisfies a system of the form (12) and so is again rational, but of lower degree than u, so the result follows by induction on the degree.

Next consider the algebraic case. Thus we take $u = (u_i)$ to be a solution of (13), where ϕ_i contains no constant terms or terms of degree 1 in the u's. We modify u by subtracting its linear terms, so that it contains no terms of degree less than 2. Further we may assume the alphabet X to be finite. Each u_i now has the form $u_i = \sum u_{ix}x$, where the u_{ix} have no constant terms, and we can write $\phi_i = \sum \phi_{ij}u_j + \sum \phi_{ix}x$. The equations (13) now take the form

$$u_{ix} = \sum \phi_{ij}u_{jx} + \phi_{ix} .$$

If we express ϕ_{ij}, ϕ_{ix} in terms of the x and u_{ix} we obtain a system as before, except that some of the ϕ_{ix} may have constant terms. But this would lead to constant terms in the u_{ix} and hence linear terms in the u_i, contradicting the fact that there are no such terms. So we have indeed a system of the form (13) for the u_{ix}, showing that the u_{ix} are again algebraic. This result may be applied as follows:

2.9 The inverse weak algorithm

Proposition 2.9.19. *Any subring of $k\langle\langle X\rangle\rangle$ containing k and admitting inverses for all elements of zero order and all X-transductions is a semifir. In particular, $k\langle\langle X\rangle\rangle_{rat}$ and $k\langle\langle X\rangle\rangle_{alg}$ are semifirs.*

Proof. Let R be a subring satisfying the conditions and consider a relation in R:

$$\sum a_i b_i = 0, \quad \text{where } a_i, b_i \in R. \tag{14}$$

If some a_i has a non-zero constant term, it is a unit in R and we can trivialize (14). Otherwise let p be a monomial of shortest length occurring in any a_i, say in a_1 and let $f \mapsto f'$ be the transduction 'right cofactor of p'. Then we have $\sum a'_i b_i = 0$, and here a'_1 has a non-zero constant term α, say, where α is the coefficient of p in a_1; moreover, $a'_i \in R$ by hypothesis. So a'_1 is a unit and this allows b_1 to be expressed as a right linear combination of $b_2, \ldots, b_r$, therefore (14) can be trivialized. The rest is clear since, as we have seen, both $k\langle\langle X\rangle\rangle_{rat}$ and $k\langle\langle X\rangle\rangle_{alg}$ admit transductions. ∎

In Chapter 7 we shall see that $k\langle\langle X\rangle\rangle_{rat}$ is actually a fir; by contrast $k\langle\langle X\rangle\rangle_{alg}$ is not a fir when $|X| > 1$ (see Proposition 5.9.9).

Exercises 2.9

1. State and prove an analogue of Proposition 4.2 for the inverse dependence number.
2. Investigate rings satisfying the inverse weak algorithm relative to the trivial inverse filtration.
3. Find an extension of Lemma 9.4 to the case of an infinite family (a_i).
4. Investigate a converse for the exchange principle (Lemma 9.1).
5. Verify the inverse weak algorithm for the following rings and their completions:
   (i) $\mathbb{Z}$ with $v = v_p$, the p-adic valuation for a prime p,
   (ii) $k\langle X\rangle$, where k is a field and v is the degree of the least non-zero terms.
6. Show that every complete inversely filtered connected K-ring with an inverse division algorithm (1) is a power series ring in one indeterminate x, with the commutation rule

$$ax = xa^\alpha + x^2 a^{\delta_1} + x^3 a^{\delta_2} + \ldots,$$

   where α is an endomorphism of K and $(\delta_1, \delta_2, \ldots)$ is a higher α-derivation.
7*. Let A be the group algebra over k of the free group on X and define an order-function with values in $\mathbb{Z} \cup \{\infty\}$ in terms of the total degree:

$$d(x_1^{\varepsilon_1} \ldots x_n^{\varepsilon_n}) = \sum \varepsilon_i.$$

   Show that A has a completion $\hat{A}$ relative to this function and that $k\langle\langle X\rangle\rangle$ can be embedded in $\hat{A}$.

8. Let R be a complete inversely filtered ring with $\mu_v(R) \geq 2$. Show that $R^\times$ is a rigid monoid.

9. Let R be as in Exercise 8. If $a, b, a' \in R^\times$ are such that $v(a) \geq 0$ and $ab = ba'$, find $u, v \in R$ and $r \geq 0$ such that $a = uv, a' = vu, b = a^r u = u a'^r$. Show that for any $b \in R^\times$ there is an order-preserving mapping of the set of left ideals of the eigenring $E(b)$ into the set of left ideals of R of the form Ra', where $ab = ba'$. Deduce that in a power series ring the eigenring of a non-zero element is an Artinian local ring.

10. If X is a finite set, the elements of $k\langle\!\langle X\rangle\!\rangle$ can be described as formal infinite sums $\sum x_I a_I$, where x_I ranges over all monomials in X. Show that for infinite X this is no longer true, but when X is countable, there is a ring whose elements are all formal power series in X, and which is obtained as the completion of $k\langle X\rangle$, when suitable degrees have been assigned to the elements of X. (To extend this result to arbitrary sets X one would need transfinite degrees.)

11. (Jooste [71]) Let R be a complete inversely filtered ring with inverse weak algorithm and let d be any derivation of R with kernel N. Show that any family of elements of N left linearly dependent over R is also left linearly dependent over N. Deduce that for any family (y_i) of elements of N, $N \cap \sum R y_i = \sum N y_i$. Hence show that N is a semifir and that R is flat as an N-module.

12*. (Jooste [71]) Show that in any complete inversely filtered ring R with n-term inverse weak algorithm the kernel N of any derivation of R is an n-fir. (*Hint*: Use Exercise 11.)

13. Show that $k\langle\!\langle X\rangle\!\rangle$ is faithfully flat over $k\langle\!\langle X\rangle\!\rangle_{\text{rat}}$ and $k\langle\!\langle X\rangle\!\rangle_{\text{alg}}$. (Recall that a ring R is *left faithfully flat* over a subring S if and only if R/S is flat as left S-module; see Bourbaki [72].)

14. Show that in a semifir any finitely generated left or right ideal satisfies (6).

15. Let $R = k\langle\!\langle X\rangle\!\rangle$ be the power series ring on a countable set $X = \{x_1, x_2, \ldots\}$. Show that the set $\mathfrak{a}_{[n]}$ of all elements of order $\geq n$ is an ideal in R and that $R = \mathfrak{a}_{[0]} \supset \mathfrak{a}_{[1]} \supset \ldots$. Show also that $\mathfrak{a}_{[i]}\mathfrak{a}_{[j]} \subset \mathfrak{a}_{[i+j]}$ (*Hint*: For example, for $i = j = 1$ show that $\sum x_n^{n+1}$ lies in $\mathfrak{a}_{[2]}$ but not in $\mathfrak{a}_{[1]}^2$.)

16°. In Theorem 9.13, can the condition on M_1 (to be free) be omitted, or replaced by the condition that M_1 be projective?

17°. Investigate inversely filtered k-algebras with inverse weak algorithm which are not connected.

18. For any commutative field k, show that $k[[x]]_{\text{rat}}$ is the rational function field $k(x)$ and $k[[x]]_{\text{alg}}$ is the relative algebraic closure of $k[x]$ in $k[[x]]$.

19*. (G. M. Bergman) In $k\langle\!\langle x, y\rangle\!\rangle$ consider the subalgebra S generated by x, y and $z = \sum x^n y^n$. Show that the augmentation ideal of S has a 2-element weak v-basis, even though it is free of rank 3. (*Hint*: To show that x, y, z are right linearly independent, take a relation and apply a transduction with respect to x.)

20*. (G. M. Bergman) In $k\langle\!\langle x, y\rangle\!\rangle$ define z as in Exercise 19, let T be the subalgebra generated by x, y, xz and $\hat{T}$ be its completion. Show that T is not 1-inert in $\hat{T}$.

21°. (G. M. Bergman) Let R be a free algebra and $\hat{R}$ its power series completion. If an element of R is a square in $\hat{R}$, is it associated (in $\hat{R}$) to the square of an element of R?

22°. (G. M. Bergman) Let $A = \mathbb{R}[t]$ and put $\mathfrak{a} = (t^2 + 1)A$. Show that the $\mathfrak{a}$-adic completion is $\mathbb{C}[[x]]$, where $x = t^2 - 1$. What is the completion of $\mathbb{R}\langle s, t \rangle$, when $\mathfrak{a}$ is the kernel of the homomorphism to the quaternions, mapping s to i and t to j?

2.10 The transfinite weak algorithm

In Section 1.2 we saw that the classical division algorithm can be defined for any ring R with a function from R to $\mathbb{N}$, or more generally, for any ordinal-valued function. However, in defining the weak algorithm we used the additivity of the values $v(x)$ in an essential way, and so were limited to $\mathbb{N}$. Nevertheless, the definition of the weak algorithm can be modified so as to apply to ordinal-valued functions. The resulting notion is not left–right symmetric (in contrast to the ordinary weak algorithm); this makes it suited for studying rings that lack left–right symmetry in some respect. Instead of filtrations we need to study more general functions whose precise form is suggested by Theorem 1.2.2.

Let R be a ring with a function w defined on it, satisfying the following conditions:

T.1. w maps $R^\times$ to a section of the ordinals, $w(1) = 0$, $w(0) = -1$,
T.2. $w(a - b) \leq \max\{w(a), w(b)\}$,
T.3. $w(ab) \geq w(a)$ for any $b \in R^\times$.

From T.3 it follows that R must be an integral domain. Moreover, Proposition 1.2.3 shows that when a division algorithm of the form $a = bq + r$, with $w(r) < w(b)$ exists, then the remainder r is unique.

We shall refer to a function w satisfying T.1–T.3 as a *transfinite right degree function* on R and call $w(a)$ the *degree* of a. Given such a function w on R, we shall say that a family $\{a_i\}$ of elements of R is *right w-dependent* if some a_i is 0 or if there exist elements $b_i \in R$, almost all 0, such that

$$w\left(\sum a_i b_i\right) < \max_i \{w(a_i b_i)\}.$$

Otherwise the family $\{a_i\}$ is called *right w-independent*. An element $a \in R$ is said to be *right w-dependent* on a family $\{a_i\}$ if $a = 0$ or if there exist $b_i \in R$, almost all 0, such that

$$w\left(a - \sum a_i b_i\right) < w(a), \quad w(a_i b_i) \leq w(a) \text{ for all } i \ . \tag{1}$$

Otherwise a is *right w-independent* of the a_i. If in (1), $b_j \neq 0$, then $w(a_j) \leq w(a_j b_j) \leq w(a)$; hence in any dependence (1), a is already right w-dependent on the a_i of degree $\leq w(a)$. We also note that a is right w-dependent on the empty set if and only if $a = 0$.

Now *strong right w-dependence* is defined as for degree-functions. Again we see that any strongly right w-dependent family is right w-dependent and we have the:

Definition. A ring with a transfinite right degree w is said to satisfy the *transfinite right weak algorithm* if any right w-dependent family is strongly right w-dependent.

We emphasize that in contrast to the case considered in Section 2.4, the transfinite weak algorithm does not possess left–right symmetry. In what follows we shall only be concerned with *right* w-dependence and so we often omit the qualifying adjective.

Let R be a ring with a transfinite right degree w satisfying the transfinite weak algorithm; it follows as before that the set $K = \{x \in R \mid w(x) \leq 0\}$ is a field. Given a right ideal $\mathfrak{a}$ of R, let us well-order the elements of $\mathfrak{a}$ in any way so that elements of smaller degree precede those of larger degree and omit any element right w-dependent on earlier ones. The resulting set B is a right w-independent basis of $\mathfrak{a}$, for it is clearly a generating set, and if $w(\sum_1^n a_i b_i) < \max_i\{w(a_i b_i)\}$ where $a_i \in B$ and the a_i occur in the order $a_1, \ldots, a_n$, so that $w(a_1) \leq \ldots \leq w(a_n)$, then by the transfinite weak algorithm, for some i,

$$a_i = \sum_1^{i-1} a_j c_j + a_i', \quad \text{where } w(a_i') < w(a_i) . \tag{2}$$

Hence a_i is right w-dependent on earlier elements, contradicting the fact that $a_i \in B$. Thus B is right w-independent, and *a fortiori* right R-linearly independent, hence a basis of $\mathfrak{a}$. To show that R has IBN, suppose that $\mathfrak{a}$ is generated by $a_1, \ldots, a_n$, where $w(a_1) \leq \ldots \leq w(a_n)$ say, and that these elements are linearly dependent. Then (2) holds for some i and we can replace a_i by a_i'. By induction on the n-tuple $(w(a_1), \ldots, w(a_n))$ we find that $\mathfrak{a}$ can be generated by fewer than n elements, hence by Theorem 3.1, $\mathfrak{a}$ has unique rank and so R has IBN. Remembering Theorem 2.3, we obtain

Theorem 2.10.1. *Any ring with a transfinite right degree function w satisfying the transfinite right weak algorithm is a right fir, and hence has right pan-ACC. Moreover, every right ideal has a right w-independent basis.* ■

To obtain examples of this construction we shall take certain monoid rings and we begin by looking at the monoids that we shall need. Let S be a conical rigid monoid with right ACC_1. On S we can define a partial preordering by left divisibility:

$$u \leq v \text{ if and only if } v = us \text{ for some } s \in S. \tag{3}$$

2.10 The transfinite weak algorithm

Since S has cancellation and is conical, this is in fact a partial ordering and it satisfies the DCC by the right ACC_1 on S. Moreover, for any $s \in S$, the lower segment determined by s, i.e. $\{x \in S \mid x \leq s\}$ is totally ordered, by rigidity, and hence by ACC_1 is well-ordered and so is order-isomorphic to an ordinal number, which we shall denote by $w(s)$ and call the *transfinite degree* (defined by left divisibility). From the definition it is clear that we have

$$w(u) \leq w(v) \text{ implies } w(cu) \leq w(cv), \quad \text{for } u, v, c \in S, \qquad (4)$$

$$w(b) \leq w(c) \text{ implies } w(bu) \leq w(cu), \quad \text{for } b, c, u \in S. \qquad (5)$$

Now consider the monoid ring $R = kS$; we extend w to R by putting

$$w\left(\sum \lambda_s s\right) = \max\{w(s) | \lambda_s \neq 0\}.$$

As usual, by a *leading term* of $a \in kS$ we mean a term of highest degree. Then it is easily seen that (4), (5) still hold for $b, c \in kS$, $u, v \in S$. Moreover, it is clear that T.1 and T.2 hold, and T.3 also follows, for if $a = \sum \lambda_i u_i$, $b = \sum \mu_j v_j$ and $w(ab) < w(a)$, let u_1 be a leading term of a. Then $w(ab) < w(u_1) \leq w(u_1 v_j)$ for all j; thus each term $u_1 v_j$ must be cancelled by a sum of other terms $u_i v_k$. Choose j so that v_j is not a right factor of any v_k; we have $u_1 v_j = u_i v_k$, where by the choice of u_1 (and rigidity) $u_1 = u_i s$, $s \neq 1$, hence $v_k = s v_j$, a contradiction. Thus we have indeed a transfinite degree function. We claim that it satisfies the right transfinite weak algorithm.

Let $a_1, \ldots, a_n$ be a right w-dependent family; thus

$$w\left(\sum a_i b_i\right) < \max_i \{w(a_i b_i)\}. \qquad (6)$$

If one of the a_i is 0, there is nothing to prove, so we may exclude that case. We again suppose the a_i numbered so that

$$w(a_1) \leq \ldots \leq w(a_n). \qquad (7)$$

We shall show that some a_i is right w-dependent on $a_1, \ldots, a_{i-1}$. By omitting superfluous terms we may suppose that

$$w(a_1 b_1) = \ldots = w(a_n b_n) = \alpha.$$

Since cancellation holds in S, the left cofactor of an element u in a product vu is well-defined, and for a given $u \in S$ we can define the left transduction $a \mapsto a^*$ in R, where for any $a \in R$, a^* is the left cofactor of u, i.e. $a = a^*u + \ldots$, where dots denote products not ending in u.

We shall need two properties of transductions.

Lemma 2.10.2. *Let S be a conical rigid monoid with right ACC_1 and let $R = kS$ be the monoid algebra with the transfinite degree function w defined* by left divisibility. Given $u \in S, u \neq 1$, let * denote the left transduction with respect to u. Then

(i) for any $a, b \in R$,

$$(ab)^* = ab^* + \text{ terms of degree } < w(a), \tag{8}$$

(ii) if $a, b, c \in R$, where c includes u among its leading terms, and $w(a) < w(bc)$, then $w(a^*) < w((bc)^*)$.

Proof. (i) By linearity we need only check this claim when $b \in S$. If $b = tu$ for some $t \in S$, then $ab = atu$ and so $(ab)^* = at = ab^*$, hence (8) holds then. Otherwise we have $u = tb$, for some $t \in S, t \neq 1$, by rigidity. Now write $a = a_1 + a_2$, where a_1 is the sum of all terms $\lambda_s s$ occurring in a such that sb has u as a right factor. Then $a_1 b = cu = ctb$ for some $c \in R$. Hence $(ab)^* = (a_1 b)^* + (a_2 b)^* = c$, $b^* = 0$ and $w(c) < w(a_1) \leq w(a)$, so (8) is again satisfied.

(ii) Write $a = a_1 u + a_2$, where no term in a_2 has u as a right factor. Then $a^* = a_1$, $(bc)^* = b+$ lower terms and $w(a_1 u) \leq w(a) < w(bc) = w(bu)$, where we have used (4). Hence by (5) $w(a^*) = w(a_1) < w(b) = w((bc)^*)$, as claimed in (ii). ∎

We shall apply the lemma to (6), where (7) is assumed to hold. Among the leading terms of b_n pick one, u say, which is not a right factor of any other leading term of b_n; it is clear that u cannot be a right factor of any non-leading term of b_n. Denote the coefficient of u in b_n by λ. If * is the left transduction with respect to u, then by (8) we have

$$\sum a_i b_i^* = \left(\sum a_i b_i\right)^* + \text{ terms of degree } < w(a_n). \tag{9}$$

Since $w(\sum a_i b_i) < w(a_n b_n)$ and $(a_n b_n)^* = \lambda a_n +$ terms of lower degree, it follows by (ii) of Lemma 10.2 that $w((\sum a_i b_i)^*) < w(a_n)$, and with (9) this shows that $w(\sum a_i b_i^*) < w(a_n)$; since $b_n^* = \lambda \neq 0$, this shows a_n to be right w-dependent on $a_1, \ldots, a_{n-1}$, and so we obtain

Theorem 2.10.3. *Let S be a conical rigid monoid with right ACC_1. Then the monoid algebra kS satisfies the right transfinite weak algorithm relative to the partial ordering by left divisibility, and hence is a right fir.* ∎

As an example to illustrate Theorem 10.3 consider the monoid S generated by $y, x_i (i \in \mathbb{Z})$ subject to the relations

$$yx_i = x_{i-1} \quad (i \in \mathbb{Z}). \tag{10}$$

It is easily checked that the elements of S can be uniquely expressed in the form

$$x_{i_1} \ldots x_{i_r} y^m, \quad \text{where } i_\rho \in \mathbb{Z}, \ r, m \geq 0 . \tag{11}$$

Clearly S is conical, the form (11) of its elements shows that it is rigid and we have

$$x_{i_1} \ldots x_{i_r} y^m S \supset x_{j_1} \ldots x_{j_s} y^n S$$

if and only if $r \leq s, i_\rho = j_\rho (\rho = 1, \ldots, r), m \leq n$; this shows that right ACC_1 holds in S. Therefore Theorem 10.3 can be applied to show that the monoid algebra $R = kS$ is a right fir, but it is not a left fir, because the left ACC_1 does not hold in R:

$$Rx_0 \subset Rx_1 \subset Rx_2 \subset \ldots .$$

We note that y is right large in R, in fact we have $Ry \subseteq yR$, i.e. y is right invariant in R.

Exercises 2.10

1. Show that in a ring with right transfinite weak algorithm any element of degree 0 is a unit and any element of degree 1 is an atom.
2. Let F be the free group on x and y and let T be the submonoid generated by y and all elements $y^{-n}x (n = 0, 1, 2, \ldots)$. Verify that T is isomorphic to the monoid S defined after Theorem 10.3. Carry out the proof that this monoid has right ACC_1.
3. Let R be a ring with right transfinite weak algorithm. If $p \in R$ is an element of degree 1 and $\mathfrak{a}$ is a proper right ideal containing p, show that $\mathfrak{a}$ has a right w-independent basis including p.
4. Prove an analogue of Lemma 4.7 for the transfinite degree function.
5. Show that the only non-trivial monoid with right cancellation in which (3) is a well-ordering is $\mathbb{N}$. What are the monoids with this property but instead of right cancellation having left cancellation? (See Cohn [61a]).
6°. (Samuel [68]) Let R be a commutative ring with transfinite algorithm. Does R necessarily have an integer-valued algorithm? What is the answer if the residue-class field (mod p), for every atom p, is known to be finite?
7. Let R be the ring defined at the end of Section 2.10 (with the defining relations (10)). Show that the cyclic left R-module R/Rx_0 is Artinian but not Noetherian (see Cohn [97b]).
8*. Let S be the monoid on $x_i, y_i, z (i \in \mathbb{Z})$ with defining relations $x_i = zx_{i-1}, y_i = zy_{i-1}$. Show that the monoid ring $R = kS$ is a right fir, but that the lattice of principal right ideals containing $x_1 R$ is not complete.
9. (Cedó [88]) Let M be the monoid generated by x, y subject to the relation $yxy = x$. Show that M is conical and rigid, but that the monoid algebra kM is not a semifir. (*Hint*: Use the relation $(1 - x)(1 - y) = (1 - y)(1 + xy)$ in kM.) Does left or right ACC_1 hold in kM?

2.11 Estimate of the dependence number

Many situations require counter-examples in the form of an n-fir that is not an $(n+1)$-fir. Usually it is easy to see that the proposed example is not an $(n+1)$-fir, but it is less easy to prove that it is an n-fir and our aim in this section is to derive a result that in certain cases allows the dependence number to be estimated and hence shows the given ring to be an n-fir for appropriate n.

For example, to construct a ring that is weakly n-finite but not weakly $(n+1)$-finite, one would take a k-algebra R on $2(n+1)^2$ generators $a_{ij}, b_{ij}(i, j = 1, \ldots, n+1)$ and form matrices of order $n+1$, $A = (a_{ij})$, $B = (b_{ij})$ with defining relations

$$AB = I. \tag{1}$$

Theorem 11.2 below shows R to be an n-fir and hence weakly n-finite. If it were an $(n+1)$-fir, it would be weakly $(n+1)$-finite, but that is not so, since in R we have

$$BA \neq I. \tag{2}$$

Intuitively it is clear that the relation (1) does not entail the relation $BA = I$, and this can be made into a rigorous proof using a normal form argument. Another method is to interpret A, B as endomorphisms of a free left R-module V of infinite rank. We take a basis $\{u_{vi}\}$ of V indexed by $\mathbb{N} \times \{n+1\}$ and define linear mappings α, β of V into itself:

$$u_{vi}\alpha = \sum a_{ij}u_{v+1,j}, \quad u_{vi}\beta = \begin{cases} \sum_j b_{ij}u_{v-1,j} & \text{if } v > 0, \\ 0 & \text{if } v = 0. \end{cases}$$

Then it is easily verified that $\alpha\beta = 1$, $\beta\alpha \neq 1$; since R can be represented in $\mathrm{End}(V)$ by mapping $A \mapsto \alpha$, $B \mapsto \beta$, it follows that $BA \neq I$.

Let R be a k-algebra (k a commutative field), generated by a set U; thus if $F = k\langle U\rangle$, we have $R \cong F/\mathfrak{a}$, where $\mathfrak{a}$ is the ideal of relations holding in R. On F we have the usual degree-function $d(f)$. Let us write $a \mapsto \underline{a}$ for the natural homomorphism $F \to R$ and define a filtration v on R by

$$v(r) = \inf\{d(a)|\underline{a} = r\}.$$

Given $r, s \in R$, choose $a, b \in F$ such that $\underline{a} = r$, $\underline{b} = s$ and $v(r) = d(a)$, $v(s) = d(b)$. Then

$$v(r+s) \leq d(a+b) \leq \max\{d(a), d(b)\},$$
$$v(rs) \leq d(ab) = d(a) + d(b).$$

2.11 Estimate of the dependence number

Hence we obtain

$$v(r+s) \leq \max\{v(r), v(s)\}, \quad v(rs) \leq v(r) + v(s);$$

clearly $v(1) = 0$, so v is indeed a filtration on R. For any $a \in F$ we define $v(a) = v(\underline{a})$. Then it is easily seen that v is also a filtration on F.

Consider a set of defining relations for R. If one of them is linear, say $\sum \alpha_i u_i + \beta = 0$ ($u_i \in U, \alpha_i, \beta \in k$), then we can use it to eliminate one of the generators u_i, because k is a field. Thus we may assume that there are no linear relations. We shall take the generating set U to be of the form $U = X \cup Y$, where X is indexed by $\mathbb{N} \times I : X = \{x_{vi}\}$; the elements of X have degree 1, while those of Y have degree δ, where δ is a small positive number whose exact size will be fixed later. The defining relations of R are all taken to be of the form

$$\sum_{0}^{n} x_{vi} x_{vj} = b_{ij}, \tag{3}$$

where b_{ij} is an expression in the members of Y. The relation (3) is assigned the *index* (i, j). Now δ is chosen so small that the total degree of b_{ij} is less than 2.

Each element $f \in F = k\langle U \rangle$ is a polynomial in U. It is said to be in reduced form for the suffix 0, briefly 0-*reduced*, if no term in f contains a factor $x_{0i}x_{0j}$, for any of the defining relations (3). Any $f \in F$ can be brought to 0-reduced form by applying the moves

$$x_{0i} x_{0j} \mapsto b_{ij} - \sum_{v \neq 0} x_{vi} x_{vj} \tag{4}$$

arising from (3), whenever possible. It is clear that for any f a 0-reduced form is reached in a finite number of moves. In general there may be more than one such reduced form for a given f; if for each $f \in F$ there is just one 0-reduced form, we shall call it the *normal form* of f (for the suffix 0), and we also say that a normal form for the suffix 0 exists. Similarly, for each suffix $\mu = 1, 2, \ldots, n$ a μ-reduced form and a normal form may be defined, using instead of (4) the moves

$$x_{\mu i} x_{\mu j} \mapsto b_{ij} - \sum_{v \neq \mu} x_{vi} x_{vj} . \tag{4_μ}$$

The normal form of f for μ, when it exists, will be denoted by $[f]_\mu$.

We can now give conditions on a presentation for the n-term weak algorithm to exist:

Theorem 2.11.1. *Let R be a k-algebra generated by a set $U = X \cup Y$, where $X = \{x_{vi}\} (i \in I, v = 0, 1, \ldots, n)$, with a set of defining relations indexed by* some subset of I^2:

$$\sum_{v=0}^{n} x_{vi} x_{vj} = b_{ij},$$

where b_{ij} is an expression in the elements of Y. Further, assume that a normal form exists for each $n = 0, 1, \ldots, n$ satisfying N_v: For any $u, v, w \in U$, if uv is v-reduced, then the terms of highest value in $u[vw]_v$ are in normal form for v.

Then $\lambda_v(R) \geq n$; thus R satisfies the n-term weak algorithm and hence is an n-fir.

Proof. Let us write

$$H^r = \{f \in R | v(f) \leq r\}.$$

We have to prove that the n-term weak algorithm holds; thus if

$$f_1 g_1 + \ldots + f_m g_m \equiv 0 \,(\text{mod}\, H^{r-1}), \quad v(f_\alpha) + v(g_\alpha) = r, \tag{5}$$

where $m \leq n$, we have to show that some g_α is left v-dependent on the g's of value not exceeding $v(g_\alpha)$.

In order to reduce (5) we shall work with the normal form for the suffix 0 and therefore write $[f]$ for $[f]_0$ in what follows. We shall also need to make use of transductions: given $f \in R$ and $u \in U$, we shall denote the right cofactor of u in the expression $[f]$ by $(^u f)$.

We note that under the given hypotheses N_v actually holds for any $u, v, w \in R$. By linearity we need verify this only when u, v, w are products of generators. Now the reducibility of a product $f.g$, where g is a reduced monomial with leftmost factor $x_{\mu i}$ depends only on μ (whether $\mu = v$) and on j (whether there is a relation (i, j) such that f has a rightmost factor $x_{\mu j}$), but (4_μ) shows that the terms of highest value resulting from a reduction have the same 'left-hand data' as the original word.

Now return to (5): if $v(f_\alpha) = 0$ for some α, then f_α is a unit and the result is clear; so we may assume that $v(f_\alpha) > 0$ for all α. We take the terms of highest value in f_α and g_α in normal form; put $v(f_\alpha) = s$ and let

$$f_\alpha \equiv \sum_u u(^u f_\alpha) \,(\text{mod}\, H^{s-1}) \quad (u \in U) \tag{6}$$

be the expression of f_α in normal form (mod terms of lower value than $v(f_\alpha)$). Inserting this expression in (5), we obtain a congruence that may be written as

$$\sum_u u[(^u f_\alpha) g_\alpha] \equiv 0 \,(\text{mod}\, H^{r-1}). \tag{7}$$

2.11 Estimate of the dependence number

If $v(f_\alpha) > 1$ for all α, then the left-hand side of (7) has all its terms of highest value in normal form, for there can be no reduction in $u[(^u f_\alpha)g_\alpha]$ unless this was already possible in f_α, by N_0. From the uniqueness of the normal form it now follows that the coefficient of each u in (7) must vanish, and going back to the original form of this coefficient we find that

$$\sum_\alpha (^u f_\alpha)g_\alpha \equiv 0 \pmod{H^{r-v(u)-1}}. \tag{8}$$

Now the result follows by induction on r.

There remains the case where for some α, say $\alpha = 1$, $v(f_1) = 1$ and $f_1 g_1$ is not 0-reduced. This means that for some $i \in I$, say $i = 1$, f_1 contains a term x_{01} and $x_{01}g_1$ is not 0-reduced. Let us write

$$f_\alpha = x_{01}(^{x_{01}} f_\alpha) + f'_\alpha, \tag{9}$$

and

$$g_\alpha = \sum_j x_{0j}(^{x_{0j}} g_\alpha) + g'_\alpha. \tag{10}$$

Substituting from (9) and (10) into (5), we obtain

$$\sum_\alpha \left(x_{01}(^{x_{01}} f_\alpha) + f'_\alpha \right) \left(\sum_j x_{0j}(^{x_{0j}} g_\alpha) + g'_\alpha \right) \equiv 0 \pmod{H^{r-1}}.$$

Now reduce the terms of the first sum and equate the terms of value r with x_{01} as left-hand factor. We may assume the f_α so numbered that $(^{x_{01}} f_\alpha)$ has value 0 for $\alpha = 1, \ldots, s$ and positive value for $\alpha > s$. Then, since

$$x_{01}x_{0j} = -\sum_\nu x_{\nu 1}x_{\nu j} + b_{1j},$$

we have

$$-\sum_{\beta=1}^s \sum_{\nu=1}^n \sum_j (^{x_{01}} f_\beta)\left(x_{\nu 1}x_{\nu j}(^{x_{0j}} g_\beta) + b_{0j}(^{x_{0j}} g_\beta) \right) + \sum_{\beta=1}^s x_{01}(^{x_{01}} f_\beta)g'_\beta$$

$$+ \sum_{\gamma=s+1}^m x_{01}(^{x_{01}} f_\gamma)g_\gamma + \sum_{\alpha=1}^m f'_\alpha g_\alpha \equiv 0 \pmod{H^{r-1}}.$$

Equating cofactors of x_{01} we find

$$\sum_{\beta=1}^s (^{x_{01}} f_\beta)g'_\beta + \sum_{\gamma=s+1}^m (^{x_{01}} f_\gamma)g_\gamma \equiv 0 \pmod{H^{r-2}}. \tag{11}$$

Now by hypothesis $(^{x_{01}} f_1)$ is a non-zero scalar, λ say; write

$$g_1^* = g_1 + 1/\lambda \left(\sum_{\beta=2}^s (^{x_{01}} f_\beta)g_\beta + \sum_{\gamma=s+1}^m (^{x_{01}} f_\gamma)g_\gamma \right), \tag{12}$$

so by (11), on putting $\lambda^{-1}({}^{x_{01}}f_\beta)({}^{x_{0j}}g_\beta) = h_{j\beta}$, we find

$$g_1^* = \sum_{\beta j} x_{0j} h_{j\beta} + \text{ terms of lower value.} \tag{13}$$

Now (12) represents an elementary transformation of the g's that does not affect g_α for $\alpha \neq 1$. Hence if $f_1 g_1$ is not 0-reduced, we can by an elementary transformation bring g_1 to the form g_1^* given by (13) without disturbing the other g's. Since the x_{0j} are all distinct, $f_1 g_1$ is now μ-reduced for any suffix $\mu \neq 0$. Next take $\mu = 1$; as before there is a term $f_\alpha g_\alpha$ that is not 1-reduced. By what has been said, $\alpha \neq 1$ and after suitable renumbering of the f's and g's we may take $\alpha = 2$. The same argument now shows that there is an elementary transformation bringing g_2 to the form $g_2^* = \sum x_{1j} h_{j\beta} +$ terms of lower value, without disturbing the other g's. By induction on t we may therefore suppose that g_β is replaced by

$$g_\beta^* = \sum x_{\beta-1,j} h_{j\gamma}^\beta + \text{ lower terms } (\beta = 1, \ldots, t-1). \tag{14}$$

Now some $f_\alpha g_\alpha$ is not reduced. This means that $\alpha \neq 1, \ldots, t-1$ and by renumbering the f's and g's we may take $\alpha = t$. As before we can bring g_t to a form g_t^* given by (14) without affecting the $g_\beta (\beta \neq t)$. This applies for $t = 1, 2, \ldots, m$ and since $m \leq n$, we eventually reach a contradiction. So we are reduced to the first case and the result follows. ∎

The hypotheses of Theorem 11.1, although rather cumbersome to state, are quite natural ones holding in many cases, and when they do hold they are usually easy to verify. We shall do so in one important case, that of matrix reduction, which was discussed in Section 0.2. We recall that $\mathfrak{W}_n(R)$ is obtained by regarding the elements of R as $n \times n$ matrices.

Theorem 2.11.2. *Let R be a non-zero k-algebra (k a commutative field). Then $\mathfrak{W}_{n+1}(R)$ has a filtration v for which the n-term weak algorithm holds.*

Proof. It is clear that there is a natural homomorphism $\lambda : R \to \mathfrak{F}_{n+1}(R) = R_k^* \mathfrak{M}_{n+1}(k)$. The elements of R will be denoted by a, b, etc. and their images $a\lambda, b\lambda$ (which are matrices) by $(a_{\mu\nu})$, $(b_{\mu\nu})$, etc. Our task is to establish the n-term weak algorithm on $\mathfrak{W}_{n+1}(R)$ relative to a suitable filtration. The defining relations of R may be taken in the form $ab = c$. As a matrix equation this reads

$$\sum_\nu a_{r\nu} b_{\nu s} = c_{rs}$$

and this is of the form (3). If we regard the moves (4) as a 'straightening process', we see that every expression can be 0-reduced, by an induction on the number of 'bad' factors $a_{r_0} b_{0s}$. To show that we actually get a normal form we have to check that for any expression containing two 'bad' factors the results of the two ways of reducing them can themselves be reduced to a common value, and here it is clearly enough to consider the case where the two factors in question overlap (this is in effect an application of the diamond lemma, see Bergman [78a] or FA, Section 1.4). Thus we have to examine products of the form $a_{r0}b_{00}c_{0s}$. If we add certain (uniquely reducible) terms, we obtain

$$\sum a_{r\mu}b_{\mu\nu}c_{\nu s},$$

and here we can reduce either the first pair of factors and get $\sum_\nu (ab)_{r\nu}c_{\nu s}$, which by a further reduction gives $(abc)_{rs}$; or reducing the second pair first we get $\sum_\mu a_{r\mu}(bc)_{\mu s}$, which again reduces to $(abc)_{rs}$. Thus we get equality (essentially by the associative law) and this establishes the existence of a normal form for the index 0; the same argument holds for $\mu = 1, \ldots, n$.

It remains to verify N_ν; again we need only consider N_0. Thus uv is 0-reduced, so a typical term will be $a_{\lambda\mu}b_{\nu\rho}$, where μ, ν are not both 0 if ab occurs in a defining relation. If vw is not reduced, then $\rho = 0$ and w will have the form $c_{0\sigma}$, so $[b_{\nu 0}c_{0\sigma}] = (bc)_{\nu\sigma} - \sum_{\kappa \neq 0} b_{\nu\kappa}c_{\kappa\sigma}$ and when we multiply on the left by $a_{\lambda\mu}$ it remains 0-reduced because μ, ν are not both 0. Thus all the conditions of Theorem 11.1 are satisfied and so the n-term weak algorithm holds in $\mathfrak{W}_{n+1}(R)$. ∎

Instead of interpreting the elements of R as square matrices of order $n + 1$ we can also take them to be rectangular matrices. Thus if each generator u is interpreted as an $n_u \times m_u$ matrix, where the numbers n, m are such that the defining relations, as matrix expressions, make sense, and if further, $m_u, n_u > n$, then the resulting ring again has n-term weak algorithm. The proof is essentially the same and so is left to the reader.

It is clear from Theorem 11.2 that the ring described at the beginning of this section (with matrices A, B of order $n + 1$ satisfying (1) and (2)) is an n-fir. In the same way we can construct a ring over which every finitely generated module can be generated by n elements, but where n cannot be replaced by $n - 1$. We need only take an $(n + 1) \times n$ matrix A and an $n \times (n + 1)$ matrix B with the relation

$$AB = I. \tag{16}$$

By Theorem 11.2 the resulting ring $U_{n,n+1}$ is an $(n - 1)$-fir, hence weakly $(n - 1)$-finite, but from Section 0.1 we know that on writing $R = U_{n,n+1}$, we have $R^n \cong R^{n+1} \oplus P$ for some R-module P, hence $R^N \cong R^{N+1} \oplus P$ for all $N \geq n$ and it follows that every finitely generated R-module is n-generated. However, R^n cannot be generated by fewer than n elements, because R is weakly

$(n-1)$-finite. Here the remark (on rectangular matrices) following Theorem 11.2 is not needed since the relation (16) can be written $A^i B_j = \delta_{ij}$, where A^i is the ith row of A and B_j is the jth column of B. However, this remark is needed for the next example.

Let A be $m \times n$ and B be $n \times m$ and consider the ring $R = V_{m,n}$ with generators $a_{i\lambda}, b_{\lambda i}$ satisfying

$$AB = I, \quad BA = I.$$

If $m < n$, R is an $(m-1)$-fir but does not have IBN; in fact it may be shown to be of type $(m, n-m)$ (see Exercise 2 below).

Exercises 2.11

1. Prove Theorem 11.2 when the generators are interpreted as rectangular matrices.
2. If $V_{m,n}$ (as defined in the text) is of type (h, k), show that $m \geq h, n \equiv m \pmod{k}$. Use the fact that $V_{m,n}$ is an $(m-1)$-fir to show that $m = h$. (The fact that $k = n - m$ can be shown by a trace argument, see Cohn [66a] or Corner [69].)
3. (Bowtell [67a]). Let A be $n \times m$ and B be $m \times n$, where $n > m > 2$. Consider the ring R with the entries of A, B as generators and defining relations $(AB)_{ij} = 0$ for $i \neq j, i, j = 1, \ldots, n$. Show that R is an $(m-1)$-fir but is not embeddable in a field. (Note that Theorem 11.2 does not apply as it stands.)
4. (Klein [69]) Let A be $n \times n$ and consider the ring R generated by the entries of A with defining relation $A^{n+1} = 0$. Show that $A^n \neq 0$ and that R is an $(n-1)$-fir. Apply Theorem 3.16 to deduce that R cannot be embedded in an n-fir.
5. Let $A_\nu, B_\nu (\nu = 1, 2, \ldots)$ be $n \times n$ matrices and consider the ring generated by their entries with defining relations

$$A_{\nu-1} = A_\nu B_\nu (\nu = 1, 2, \ldots).$$

   Show that R is a semifir and that each A_ν is non-invertible. Moreover, show that $\lambda_\nu(R) \geq n - 1$ and deduce that R satisfies right ACC_{n-1} but not ACC_n.
6. Let A be $n \times n$ and consider the ring R generated by the entries of A with defining relation $A^2 = A$. Show that R is an $(n-1)$-fir but that there is an n-generator projective module that is not free.
7. Construct an $(n-1)$-fir with an n-generator module that is stably free but not free.
8. (Montgomery [83]) Let k be a commutative field and $E = k(\alpha)$ a simple commutative extension of degree $n+1$, with minimal equation $x^{n+1} + a_1 x^n + \ldots + a_{n+1} = 0$ for α ($a_i \in k$). Let R be the k-algebra generated by $x_0, x_1, \ldots, x_n, y_0, \ldots, y_n$ with defining relations obtained by equating the powers of α in $(\sum x_i \alpha^i)(\sum y_j \alpha^j) = 1$. Show that R is weakly n-finite but that $R \otimes_k E$ is not weakly 1-finite.
9. (Malcev [37]) Let A, B be 2×2 matrices and consider the ring R generated by their entries with defining relations $AB = e_{12}$. Show that R is an integral domain but is not embeddable in a field. Show that R is not embeddable in a 2-fir. (*Hint*: Use the partition lemma.)

10. (M. Kireczi) Use Exercises 2 and 0.1.16 to show that there is no homomorphism $V_{1,r} \to V_{m,n}$ when $r > 1$, $1 < m < n$.

Notes and comments on Chapter 2

The generalities on hereditary rings were for the most part well known. Thus Corollaries 1.2 and 1.3 (without reference to κ) occur in Cartan and Eilenberg [56], p. 13, see also Guazzone [62]; the proof of Theorem 1.1 is modelled on these cases. Kaplansky [58] proved that for any ring, every projective module can be written as a direct sum of countably generated modules, and he deduced a commutative form of Lemma 4.7: over a commutative semihereditary ring every projective module is isomorphic to a direct sum of finitely generated ideals. This was proved for left modules over a *left* semihereditary ring by Albrecht [61] and over a *right* semihereditary ring by Bass [64]. Both results are included in Theorem 1.4, which is taken from Bergman [72b].

Firs, i.e. free ideal rings, and semifirs (originally called 'local firs') were introduced in Cohn [64b]; the account in Sections 2.2 and 2.3 is based on that source and on Bergman [67]. The partition lemma 3.15 is taken from Cohn [82a]; it is used mainly in Chapter 7. The notion of n-fir arose out of an idea in Cohn [66a], which leads to a general method of constructing n-firs that are not $(n + 1)$-firs, see Cohn [69c], Bergman [74b] and SF, Section 5.7. The generalization to α-firs is due to Bergman [67], who proved Proposition 3.20, generalizing the fact (proved by Cohn [67]) that any left fir has left pan-ACC. The fact that free modules over firs satisfy pan-ACC can be viewed as a result in universal algebra (see Baumslag and Baumslag [71]). The characterization of local rings that are semifirs (Theorem 3.13) is taken from Cohn [92a]. Theorem 3.14 first occurs in Cohn [66c]. We remark that the hypotheses of Theorem 3.14 are symmetric and show incidentally that for a projective-trivial ring, 'left semihereditary' is equivalent to 'right semihereditary'. For general rings this is no longer so, as an example in Chase [61] shows. Klein [69] has proved that for any integral domain R over which every nilpotent $n \times n$ matrix A satisfies $A^n = 0$, the monoid $R^\times$ can be embedded in a group. Thus Proposition 3.16 can be used to prove that the monoid of non-zero elements in any semifir is embeddable in a group. This holds more generally for any 2-fir (see Gerasimov [82]). For an application of Proposition 3.21 see Jensen [69]. Prest [83] has shown that for any existentially complete prime ring R, every non-zero finitely generated projective module is isomorphic to R; thus such a ring is a 'metasemifir' of type $(1,1)$ (see Exercise 1.9).

2-Firs were defined (under the name 'weak Bezout rings') and studied in Cohn [63a]; the present (weaker) form of their definition is due (independently) to Bergman [67], Bowtell [67a] (see Exercise 3.1.5), Williams [68a] and for right Bezout rings (i.e. rings in which every 2-generator right ideal is principal), Lunik [70]. This observation is essentially Theorem 3.7 (a) $\Rightarrow$ first part of (b). Commutative Bezout domains are studied (under the name 'anneau de Bezout') by Jaffard [60] and Bourbaki [72], Chapter 7; the name is intended to convey that any two coprime elements a, b satisfy the 'Bezout relation' $au - bv = 1$.

The notion of a weak algorithm was introduced by Cohn [61a, 63b] as a simplified and abstract version of what was observed to hold in the free product of fields (Cohn [60]). It was rediscovered by Bergman [67] and later greatly generalized in the coproduct theorems of Bergman [74a] (see SF, Chapter 5). Our presentation is based on all these sources; in particular the original definitions have been modified as suggested by Bergman [67] so as to be closer to the usual notion of dependence; the term 'strong dependence' is new (it was first used in IRT). The n-term weak algorithm was introduced by Cohn [66a, 69b], where Proposition 4.9 was proved. The proof in the text, using weak v-bases, is due to Bergman [67]. For a right ideal in a free algebra, a weak v-basis is always a Gröbner basis (see e.g. Fröberg [97]), though the converse need not hold.

The dependence number as defined in Section 2.4 seems more natural than the notion defined in FR.1 which was larger by one. For any ring R the *dependence number* $\lambda(R)$ may be defined independently of any filtration as the supremum of the $\lambda_v(R)$ as v runs over all filtrations of R. This is a positive integer or ∞, defined for any ring; e.g. $\lambda(R) \geq 1$ if and only if R is an integral domain. Bergman [67] gives examples of rings R such that $\lambda_v(R)$ is finite but unbounded, as v ranges over the filtrations on R; thus $\lambda(R) = \infty$, but R has no filtration for which the weak algorithm holds.

The results of Section 2.5, characterizing free algebras by the weak algorithm, are due to Cohn [61a], though the presentation in Results 5.1–3 largely follows Bergman [67]. This proof uses transductions, whose use in the study of rational and algebraic power series goes back to Nivat [68] and Fliess [70a]. Theorem 5.4, showing that the weak algorithm in filtered K-rings extends to the coproduct, was first proved by Bergman [67], p. 211; see also Williams [69b]. The problem of constructing all rings with a weak algorithm was raised by Cohn [63b] and solved by Bergman [67], whose presentation we follow in Section 2.5. The analogue of Schreier's formula in Section 2.6 was obtained by Lewin [69] as a corollary of the theorem that submodules of free modules are free, which Lewin proves for free algebras by a Schreier-type argument; it was also found by D. R. Lane (unpublished). Its extension to rings with a weak algorithm first appeared in FR.1; the present version owes simplifications to Bergman and Dicks (see also Dicks [74]). Hilbert series (sometimes called Poincaré series, also (following M. Lazard) gocha, in FR.1) have been studied intensively for commutative graded rings in the 1980s, particularly the cases of rationality (see Roos [81]). On Corollary 2.4 see also Rosenmann and Rosset [94].

The Euclidean algorithm in a (non-commutative) integral domain with a division algorithm was treated by Wedderburn [32]; the presentation in Section 2.8, valid for any ring with 2-term weak algorithm, follows Cohn [63b], and the description of $GE_2(R)$ in 2.7 is taken from Cohn [66b]. Proposition 7.1 is improved by Menal [79], who shows that R has unique standard form for GE_2 if and only if R is 'universal' for GE_2 (i.e. the conclusion of Proposition 7.1 holds) and the subring generated by $U(R)$ is a field.

The inverse weak algorithm (Section 2.9) was introduced by Cohn [62a], where Results 9.7–12 are proved; 9.13–14 are taken from Cohn [70c], but the present version of 9.13, with the present simple proof, is due to Dicks [74]. The exchange principle (Lemma 9.1) and Results 9.3–6 are due to Bergman [67], who also proves the inertia theorem (9.16) for free algebras in the case of 1-inertia. The case of total inertia was new in FR.1, though with some gaps in the proof, which were filled in FR.2, following Cohn and Dicks [76]. A special case of Theorem 9.16 was proved by Tarasov [67], who showed that an atom (without constant term) in the free algebra remains one in the power series ring. For an application of the inertia theorem to represent a free radical ring by power series see Cohn [73b]. Corollary 9.17 is taken from Cohn [92b]. Theorem 9.18 answers a question raised by Fliess (in correspondence) and first appeared in FR.2.

2.11 Estimate of the dependence number

The transfinite weak algorithm is taken from Cohn [69b] and was suggested by a method of Skornyakov [65], used there to construct one-sided firs. The construction in the text is based on that by Skornyakov. In FR.1 the right transfinite weak algorithm in the example was checked directly; the proof of Theorem 10.3, applying to a whole class of monoid rings, was new in FR.2. More generally, Kozhukhov [82] has shown that the monoid algebra kS is a right fir if S is a rigid monoid with right ACC_1 such that the subgroup of units in S is a free group and for any unit u and non-unit a, $au \in Sa$ implies that $u = 1$. Kozhukhov shown further that these conditions on S are necessary as well as sufficient for kS to be a right fir.

The HNN-construction (Higman, Neumann and Neumann [49]) starts from a group G with two isomorphic subgroups A, B via an isomorphism $\phi : a \mapsto a\phi$ and constructs a group G_1 containing G and an element t such that $at = t.a\phi (a \in A)$; G_1 is obtained by freely adjoining t to G subject to these relations. Of course it has to be proved that G is embedded in G_1. The analogue for rings mentioned in Exercise 5.21 was introduced and used by Macintyre [79]; for the field analogue see Cohn [71a] and SF, Section 5.5 (see also Exercise 7.5.18).

A 'ringoid', i.e. a small preadditive category, may be regarded as a ring with several objects, and many of the results on general rings, appropriately modified, still hold for ringoids (see Mitchell [72]). By analogy, left free ideal ringoids ('left firoids') may be defined and many of the results of this chapter proved in this context. Thus Faizov [81] obtains analogues of Theorem 3.1 and constructs firoids by a form of the weak algorithm. Wong [78] defines, for any small category $\mathcal{C}$ and any ring R, the *category ring* $R\mathcal{C}$ (analogous to the group ring). By a *bridge category* Wong understands the free category freely generated by an oriented graph with arrow set A and inverses for a certain subset B of A. For the moment let us call a category a *delta* if all its morphisms are isomorphisms and its only endomorphisms are the identities. Now Wong [78] proves that if R is any ring and $\mathcal{C}$ a small category, then $R\mathcal{C}$ is a firoid if and only if either (i) $\mathcal{C}$ is a delta and R is a fir, or (ii) $\mathcal{C}$ is a bridge category and R is a field. Thus for any free monoid $M \neq 1$, the monoid ring RM is a fir if and only if R is a field. In a similar vein, Dicks and Menal [79] have shown that for any non-trivial group G and any ring R, the group ring RG is a semifir if and only if R is a field and G is locally free. More generally, Menal [81] has shown that for any monoid $M \neq 1$, finitely generated over its unit group, and any ring R, the monoid ring RM is a semifir if and only if R is a field and M is the free product of a free monoid and a locally free group. It will follow from Corollary 7.11.8 that the group algebra of a free group is a fir.

Theorem 11.1 was proved in Cohn [69c], generalizing a particular case from Cohn [66a]. Theorem 11.2 is a special case of results in Bergman [74b]. The k-algebra with matrices A, B of order $n + 1$ satisfying $AB = BA = I$ is an n-fir, by Theorem 11.2, in fact it is a fir, by results in Bergman [74b], see also SF, Section 5.7.

3
Factorization in semifirs

For the study of non-commutative unique factorization domains we begin by looking at the lattice of factors and the notion of similarity for matrices in Section 3.1. The resulting concept of non-commutative UFD, in Section 3.2, is mainly of interest for the factorization of full $n \times n$ matrices over $2n$-firs; thus it can be applied to study factorization in free algebras. Another class, the rigid UFDs, forming the subject of Section 3.3, generalizes valuation rings and is exemplified by free power series rings. We also examine various direct decomposition theorems (Sections 3.4 and 3.5), but throughout this chapter we only consider square (full) matrices, corresponding to torsion modules over semifirs. The factorization of rectangular matrices, which is much less well developed, will be taken up in Chapter 5.

3.1 Similarity in semifirs

To study factorizations in non-commutative integral domains it is necessary to consider modules of the form R/aR. We recall from Section 1.3 that two right ideals $\mathfrak{a}, \mathfrak{a}'$ of a ring R are similar if $R/\mathfrak{a} \cong R/\mathfrak{a}'$. In the case of principal ideals the similarity of aR and $a'R$ (for regular elements a and a') just corresponds to the similarity of the elements a and a' as defined in Section 0.5 (see Proposition 0.5.2 and the preceding discussion, as well as Section 1.3).

In a semifir it is possible to simplify this condition still further. Let R be any ring and $A \in {}^mR^n$; the matrix A is said to be *left full* if for any factorization $A = PQ$, where $P \in {}^mR^r$, $Q \in {}^rR^n$, we necessarily have $r \geq m$. Clearly this can only hold when $m \leq n$. Right full matrices are defined similarly and a left and right full matrix is just a *full* matrix, as defined in Section 0.1. Over a semifir we have the following relation between full and regular matrices:

3.1 Similarity in semifirs

Lemma 3.1.1. *Over a semifir every right full matrix is right regular.*

Proof. Let R be a semifir and suppose that $A \in {}^m R^n$ satisfies $AB = 0$ for some $B \neq 0$. Then there is a $T \in GL_n(R)$ such that $AT = (A', 0)$, where A' is $n \times n - 1$. But then $A = A'(I_{n-1}, 0)T^{-1}$, showing that A is not right full. ∎

We note that this argument holds more generally for an $m \times n$ matrix in any n-fir. We also note

Proposition 3.1.2. *An $m \times n$ matrix (over any ring) with an $r \times s$ block of zeros cannot be right full unless $r + s \leq m$. In particular, an $n \times n$ matrix with an $r \times s$ block of zeros, where $r + s > n$, cannot be full.*

Proof. Let A be $m \times n$, with an $r \times s$ block of zeros in the north-east corner, say. Then

$$A = \begin{pmatrix} T & 0 \\ U & V \end{pmatrix} = \begin{pmatrix} T & 0 \\ 0 & I \end{pmatrix} \begin{pmatrix} I & 0 \\ U & V \end{pmatrix},$$

where T is $r \times (n - s)$ and V is $(m - r) \times s$. This expresses A as a product of an $m \times (m + n - r - s)$ by an $(m + n - r - s) \times n$ matrix, so if A is right full, then $m + n - r - s \geq n$, i.e. $r + s \leq m$. ∎

It is clear that the conclusion still holds if the zeros do not occur in a single block but are distributed over r rows and s columns, not necessarily consecutive. An $n \times n$ matrix with an $r \times s$ block of zeros, where $r + s > n$, is called *hollow*.

An $m \times n$ matrix A over any ring R is called *left prime* if in any equation

$$A = PQ, \quad \text{where } P \in R_m, Q \in {}^m R^n, \tag{1}$$

P necessarily has a right inverse; right prime matrices are defined correspondingly. Any right invertible matrix is left prime, for if $A = PQ$ and $AB = I$, then $P.QB = I$, so P is right invertible. For a square matrix the converse holds: a square matrix that is left prime is right invertible, as we see by taking $Q = I$ in (1). Thus, over a weakly finite ring any square matrix that is left prime is a unit. Any left prime matrix over a weakly finite ring is left full, for if not, say $A \in {}^m R^n$ has the form $A = P_1 Q_1$, where $P_1 \in {}^m R^r$, $r < m$, then we can write $A = PQ$, where $P = (P_1, 0) \in R_m$ and since P has a column of zeros, it is not a unit in R_m and so has no right inverse. This shows that A cannot be left prime. So for a matrix over a weakly finite ring we have the implications

$$\text{right invertible} \Rightarrow \text{left prime} \Rightarrow \text{left full}. \tag{2}$$

A matrix relation

$$PQ = 0, \quad \text{where } P \in {}^r R^n, Q \in {}^n R^s \tag{3}$$

is called *right comaximal* if P has a right inverse, *left coprime* if P is left prime and *left full* if P is left full, with similar definitions on the other side. A *full* relation is a relation that is left and right full. By (2) we have, for any matrix relation (3), the implications

$$\text{right comaximal} \Rightarrow \text{left coprime} \Rightarrow \text{left full}.$$

This terminology also applies to relations of the form

$$AB' = BA', \qquad (4)$$

where A is $r \times n'$, B is $r \times n''$, A' is $n'' \times s$ and B' is $n' \times s$, since this may be brought to the form (3) by writing it as

$$(A \ \ B)\begin{pmatrix} -B' \\ A' \end{pmatrix} = 0.$$

If in (3), (4) we have $r + s = n = n' + n''$, the relation is said to be *balanced*. In any relation (4) the indices satisfy $i(A) - i(A') = i(B) - i(B') (= n - r - s)$; so this relation is balanced if and only if either A, A' or equivalently, B, B' have the same index.

Proposition 3.1.3. *Let R be an n-fir and $P \in {}^rR^n$, $Q \in {}^nR^s$ such that (3) holds as a full relation. Then*

$$r + s \leq n, \qquad (5)$$

and when (3) is balanced, so that equality holds in (5), then by cancelling full matrices from the left and right in (3) we get a coprime relation, which is also comaximal. In terms of a matrix relation (4) this states that if (4) is a full relation, then $i(A') \leq i(A), i(B') \leq i(B)$, and (4) is balanced if and only if either inequality becomes an equality.

Proof. By Corollary 2.3.2, (3) may be trivialized; after modifying P and Q we may assume that $P = (P', 0), Q = (Q', Q'')^\mathrm{T}$, where P' is a right regular $r \times t$ matrix, Q' is a $t \times s$ matrix and Q'' an $(n - t) \times s$ matrix. Now (3) becomes $P'Q' = 0$ and so $Q' = 0$, because P' is right regular. By hypothesis, P is left full, hence so is P' and it follows that $r \leq t$. By symmetry Q'' is right full, hence $s \leq n - t$, and (5) follows. When equality holds in (5), then $r = t, s = n - t$, hence P', Q'' are then square and so are full. Cancelling them, we obtain the relation $(I, 0)(0, I)^\mathrm{T} = 0$, which is clearly comaximal. Hence the original relation (3) becomes comaximal after cancelling full factors on the left and right. When our relation has the form (4), the inequality (5) can be restated in terms of the indices. ∎

3.1 Similarity in semifirs

The inequality (5) is a special case of the law of nullity, which we shall meet again in Section 5.5. The condition on the index is necessary, since over $k\langle x, y, z, t\rangle$ we have

$$(x \quad y)\begin{pmatrix} z \\ t \end{pmatrix} = (x \quad y)\begin{pmatrix} z \\ t \end{pmatrix},$$

a coprime relation that is not comaximal and of course (x, y) is not stably associated to $(z, t)^T$.

As we shall see later (Section 5.9), if R satisfies left ACC$_r$ and right ACC$_s$ then we can cancel full matrices on the left and right of (3) so as to obtain a coprime relation. Proposition 1.3 shows that for a balanced relation (3) over an n-fir we do not need ACC. As an easy consequence we have

Corollary 3.1.4. *Over any ring R a comaximal matrix relation (4) is coprime. When R is a semifir, then conversely, any balanced coprime relation is comaximal. If R is an n-fir and (4) is a relation between $n \times n$ matrices that is left coprime and right full, then A, B are stably associated to left factors of A', B' respectively.*

Proof. The first part is clear from Proposition 1.3. To prove the last part, we have a balanced full relation; by Proposition 1.3 we obtain a comaximal relation by cancelling a full matrix C on the right. If the result is $AB_0 = BA_0$, then $A' = A_0C$, $B' = B_0C$ and by Proposition 0.5.6, A, B are stably associated to A_0, B_0 respectively, which gives the desired conclusion. ∎

Sometimes two elements a and a' of a ring R are called 'similar' if the right ideals aR and $a'R$ are similar; as we have seen, for regular elements this condition is left–right symmetric. We then have three names: 'stably associated', 'GL-related' and 'similar', which for regular elements of a ring (or for regular square matrices) mean the same thing. Nevertheless it is convenient to use them all, depending on the context. Thus 'stably associated' is mainly used for matrices, 'similar' for regular elements, while 'GL-related' is used in the context of relations in $GL_2(R)$, as in Section 2.7.

Another consequence of Proposition 1.3 is the fact that the product of full matrices over a semifir is full. We shall prove a slightly more general result:

Proposition 3.1.5. *Over a semifir R any product of left full matrices is left full.*

Proof. Let $A \in {}^rR^m$ and $B \in {}^mR^n$ be left full and suppose that

$$AB = PC, \quad \text{where } P \in {}^rR^s, C \in {}^sR^n,$$

and $s < r$. We write $(B, C)^{\mathrm{T}} = (B', C')^{\mathrm{T}} Q$, where $B' \in {}^m R^t$, $C' \in {}^s R^t$, $Q \in {}^t R^n$; if this factorization is chosen so that t is minimal, then $(B', C')^{\mathrm{T}}$ is right full, hence the relation $AB' = PC'$ is full, because A is left full. Hence by Proposition 1.3, $r + t \leq m + s$, so $m - t \geq r - s > 0$, i.e. $t < m$, but this contradicts the fact that B is left full. ■

A further consequence of Proposition 1.3 is most easily proved by looking at modules.

Proposition 3.1.6. *Let R be a semifir, $A \in {}^m R^n$ a right prime matrix and $A' \in {}^{m'} R^{n'}$ a left full matrix. Then*

$$\{(B, B') | AB' = BA'\} = \{(AN, NA') | N \in {}^n R^{m'}\}. \tag{6}$$

Proof. If M and M' are the left R-modules presented by A and A' respectively, then the homomorphisms $M \to M'$ correspond to the pairs B, B' on the left of (6). By the hypotheses, A is right prime, hence right full and A' is left full. Hence M is a module such that all its quotients have negative characteristic, whereas M' has positive characteristic; the only such homomorphism is the zero map, hence (6) follows. ■

Exercises 3.1

1. Show that Proposition 1.4.3 holds for any 2-fir and any submonoid S consisting of right large elements.
2. (a) Let $\mathfrak{a}$, $\mathfrak{b}$ be two right ideals in a ring R such that $\mathfrak{a} + \mathfrak{b} = R$. Verify that the sequence

$$0 \to \mathfrak{a} \cap \mathfrak{b} \xrightarrow{\lambda} \mathfrak{a} \oplus \mathfrak{b} \xrightarrow{\mu} R \to 0$$

is split exact, where $\lambda(x) = (x, x)$, $\mu(x, y) = x - y$.

(b) Let $\mathfrak{a}$, $\mathfrak{a}'$ be similar right ideals of R and assume that $\mathfrak{a}' = \{x \in R | cx \in \mathfrak{a}\}$, where c is regular (see Proposition 1.3.6). Show that

$$\mathfrak{a} \oplus R \cong \mathfrak{a}' \oplus R.$$

(c) Give an example of a ring containing similar right ideals for which there is no such isomorphism and give examples of similar right ideals that are not isomorphic.

3. Show that in a 2-fir any right ideal similar to a principal right ideal is itself principal. More generally, show that this holds in any 2-Hermite domain. (*Hint*: Use Proposition 0.3.2.)
4. The *permanent* of a square matrix A is obtained by changing all the minus signs in the determinant expansion to $+$. Show that the permanent of A with non-negative

real entries is zero if and only if A is hollow, possibly after permuting its rows and its columns (Frobenius–König theorem, see Minc [78], p. 31).

5. (Bowtell [67a]) If $\begin{pmatrix} a & b \\ c & d \end{pmatrix}$ and $\begin{pmatrix} d' & -b' \\ -c' & a' \end{pmatrix}$ are two mutually inverse matrices, show that $aR \cap bR = ad'bR + bc'aR$. Deduce that if R is an integral domain in which the sum of any two principal right ideals with non-zero intersection is principal (i.e. R is a 2-fir), then this intersection is again principal.

6. Show that any element GL-related to a zero-divisor is itself a zero-divisor, and that any element GL-related to a unit is a unit. Explicitly, if a, a' occur in mutually inverse matrices as in Exercise 5, and a is a unit, then $a'^{-1} = d - ca^{-1}b$.

7. In a weakly 1-finite ring, show that any element GL-related to 0 is 0. Is there a converse?

8. In a 2-fir let $c = ab$. Given c' similar to c, find a' similar to a and b' similar to b such that $c' = a'b'$. Does this hold in more general rings?

9. In any ring, if $AB' = BA'$ is a full relation, show that $i(A) = -i(B') \geq 0 \geq i(A') = -i(B)$.

10. Show that for any finitely generated right ideal $\mathfrak{a}$ in a semifir R and any $c \in R$, $\mathrm{rk}(\mathfrak{a} \cap cR) \leq \mathrm{rk}(\mathfrak{a})$, with equality if and only if $\mathfrak{a} + cR$ is principal.

11. Let R be a ring and R_∞ the ring of infinite matrices over R that differ from a scalar matrix only in a finite square. Show that any two regular matrices in R_∞ that are similar are associated.

12*. (Fitting [36]) Let R be any ring and $a, b \in R$. Show that aR and bR are similar if and only if the matrices $A = \begin{pmatrix} a & 0 & 0 & 0 \\ 0 & 1 & 0 & 0 \end{pmatrix}$ and $B = \begin{pmatrix} b & 0 & 0 & 0 \\ 0 & 1 & 0 & 0 \end{pmatrix}$ are associated.

13. Let k be a commutative field of characteristic 2 and α the endomorphism that sends each element to its square. Show that in the skew polynomial ring $k[x; \alpha]$ there are just two similarity classes of elements linear in x. Generalize to the case of characteristic $p > 2$.

14°. Examine similarity classes of polynomials of higher degree in the ring of Exercise 13; also the case of more general skew polynomial rings.

15. Show that in any skew polynomial ring, similar elements have the same degree.

16. In a free algebra, show that if two homogeneous elements are similar, then they are associated.

17. (G. M. Bergman) In the complex-skew polynomial ring $C[x; -]$, show that $x^4 + 1$ can be written as a product of two atomic factors in infinitely many different ways. By considering the factors, deduce the existence of a similarity class of elements that contains infinitely many elements that are pairwise non-associated (see also Williams [68b]).

18. In a 2-fir R, if $au - bv = c$ and a, c are right comaximal, show that $ua' - b'w = c'$, where a' is similar to a, b' to b and c' to c. (*Hint*: Take a comaximal relation $ac' = ca'$, multiply the given equation by a' on the right and use the fact that $aR \cap bR$ is principal as a right ideal.)

19. Show that over a commutative ring the determinant of any non-full matrix is zero. Does the converse hold? (*Hint*: Try $\mathbb{Z}[x, y]$; see also Section 5.5.)

20. Show that in any ring the intersection of all maximal right ideals similar to a given one is a two-sided ideal. (*Hint*: Recall Corollary 1.3.7.)

3.2 Factorization in matrix rings over semifirs

We saw in Section 1.3 that to study factorization in a principal ideal domain R it is convenient, in considering an element $c \in R^\times$, to look at the module R/cR. This is a module of finite composition length and the Jordan–Hölder theorem immediately yields the fact that R is a UFD, at least in the commutative case, but it continues to hold for non-commutative PIDs with the appropriate definition of UFD, as in Section 1.3. Since every square matrix over a PID is associated to a diagonal matrix, by Theorem 1.4.7, the notion of unique factorization even extends to matrices in that case. However, for firs this is no longer so and we shall need to consider, together with the ring R, the $n \times n$ matrix ring R_n.

Let R be a semifir; an R-module M is said to be a *torsion module* if it is finitely generated, such that $\chi(M) = 0$ and every submodule M' satisfies $\chi(M') \geq 0$, or equivalently, $\chi(M/M') \leq 0$. As explained in Section 0.8, we shall only use the term 'torsion module' in this sense, while the usual sort will be called a 'module of torsion elements'. Any n-generator torsion module M over a semifir R has a presentation

$$0 \to R^n \to R^n \to M \to 0, \tag{1}$$

and we shall also call a torsion module with this presentation an *n-torsion module*.

Given any other presentation of M (by a free module of rank n)

$$0 \to Q \to R^n \to M \to 0,$$

we have, by Schanuel's lemma, $R^n \oplus Q \cong R^{2n}$, and since R is an Hermite ring, $Q \cong R^n$. An alternative description of torsion modules is given in

Proposition 3.2.1. *Let R be a semifir. Then*

(i) *Any finitely presented left R-module M has a presentation*

$$0 \to R^m \to R^n \to M \to 0,$$

and M is a torsion module if and only if $m = n$ and the presenting matrix is full.

(ii) *Given any short exact sequence of R-modules,*

$$0 \to M' \to M \to M'' \to 0, \tag{2}$$

if any two modules are torsion, then so is the third.

(iii) *In any homomorphism f between torsion modules, $\ker f$ and $\coker f$ are again torsion modules.*

Proof. (i) Let M be presented by the matrix A. If M is torsion, then $\chi(M) = 0$, so A is square, say $n \times n$. In any factorization $A = PQ$, P corresponds to a submodule M' of M with quotient defined by Q, and since $\chi(M') \geq 0$, P has non-negative index, therefore A is full. Conversely, a full matrix A ensures that $\chi(M) = 0$ and $\chi(M') \geq 0$, so M is torsion.

(ii) Given an exact sequence (2), we have $\chi(M') - \chi(M) + \chi(M'') = 0$, hence if two of these numbers are zero, so is the third. Suppose now that M' and M'' are torsion modules and let N be a submodule of M. Then

$$\chi(N) = \chi(N \cap M') + \chi(N/(N \cap M')). \tag{3}$$

Since $N \cap M' \subseteq M'$, the first term on the right is non-negative, and so is the second, because $N/N \cap M' \cong (N + M')/M'$ and the right-hand side is a submodule of $M/M' \cong M''$. It follows that M is a torsion module.

Next suppose that M in (2) is a torsion module. Any submodule N of M' is also a submodule of M, hence $\chi(N) \geq 0$ and if M'' is torsion, this shows M' to be torsion. Now any quotient module Q of M'' is also a quotient of M, hence $\chi(Q) \leq 0$, and when M' is torsion, this shows M'' to be torsion.

(iii) Consider a homomorphism between torsion modules, $f : M \to N$. This gives rise to an exact sequence

$$0 \to \ker f \to M \to N \to \operatorname{coker} f \to 0.$$

The alternating sum of the characteristics is 0 and $\chi(M) = \chi(N) = 0$, hence

$$\chi(\ker f) = \chi(\operatorname{coker} f).$$

Here the left is ≥ 0 and the right is ≤ 0, so both sides are 0. Further, any submodule of $\ker f$ is a submodule of M and so has non-negative characteristic, while any quotient of $\operatorname{coker} f$ is a quotient of N and so has non-positive characteristic, therefore $\ker f$ and $\operatorname{coker} f$ are both torsion, as claimed. ∎

We observe that for n-torsion modules this holds more generally over any n-fir.

For any semifir R we shall denote the category of all right torsion modules and all homomorphisms between them by Tor_R; similarly we write ${}_R\operatorname{Tor}$ for the category of left torsion modules. Since every full matrix over R is regular, there is a correspondence between Tor_R and ${}_R\operatorname{Tor}$ in which the right module defined by A corresponds to the left module defined by A. This correspondence is actually a duality (see also Section 5.3).

Theorem 3.2.2. *Let R be a semifir. Then Tor_R and ${}_R\operatorname{Tor}$ are dual categories, i.e. there are contravariant functors $F : \operatorname{Tor}_R \to {}_R\operatorname{Tor}$ and $G : {}_R\operatorname{Tor} \to \operatorname{Tor}_R$*

such that FG and GF are each naturally isomorphic to the identity functors on Tor_R, $_R\text{Tor}$ respectively.

Proof. Let M be a right R-module, presented by a full matrix A. If A is $n \times n$ and $S = R_n$, we have $M = S/AS$ and its dual is $FM = S/SA$. Given an R-homomorphism $f : M \to N$ between right R-modules, we can choose n so that $M = S/AS$, $N = S/BS$ for full matrices A, B over R. Then f is determined by a matrix C such that for some matrix C' we have

$$CA = BC'. \qquad (4)$$

Conversely, any pair (C, C') satisfying (4) defines a homomorphism $M \to N$ by the rule

$$x(\text{mod } AS) \mapsto Cx(\text{mod } BS),$$

where x is a column vector. Now C is determined by f up to an element of BS; thus if $C_1 = C + BZ$, then by (4), $C_1 A = CA + BZA = B(C' + ZA)$. This shows C' to be determined by f up to an element of SA, and so it defines a unique R-homomorphism

$$F(f) = S/SB \to S/SA.$$

Clearly we have $F(fg) = F(g)F(f)$ and $F(1_{S/AS}) = 1_{S/SA}$; thus F is a contravariant functor. In particular, if f is an isomorphism, then so is $F(f)$, hence $F(S/AS)$ depends only on the isomorphism class of S/AS, not on A itself. Now $G : {}_R\text{Tor} \to \text{Tor}_R$ is defined by symmetry and a routine verification shows that $FG \cong 1$, $GF \cong 1$. ∎

We again note that Theorem 2.2 holds more generally for n-torsion modules over n-firs. The following is a more general interpretation of the result: given a factorization of a full matrix C over a semifir R,

$$C = A_1 \ldots A_r,$$

we associate with (5) the series of right ideals of the matrix ring S (containing C) over R

$$S \supseteq A_1 S \supseteq A_1 A_2 S \supseteq \ldots \supseteq A_1 \ldots A_r S = CS,$$

and the corresponding series of torsion quotient modules

$$S/A_1 S, \ A_1 S/A_1 A_2 S \cong S/A_2 S, \ldots, S/A_r S.$$

We also have the series of left ideals

$$S \supseteq SA_r \supseteq SA_{r-1} A_r \supseteq \ldots \supseteq SA_1 \ldots A_r = SC,$$

and torsion quotients

$$S/SA_r, \ldots, S/SA_1.$$

In discussing a factorization (5) we can for most purposes use either of these series and Theorem 2.2 is a general way of expressing this symmetry. It will be described as the *factorial duality*. For the case of elements (1×1 matrices) we see that it holds in any integral domain (= 1-fir).

Theorem 3.2.3. *For any semifir R, the category Tor_R of torsion modules is closed under sums and intersections, in fact it is an abelian category. Further, a torsion submodule of an n-torsion module is again an n-torsion module.*

Proof. Given $f : M \to N$ in Tor_R, $\ker f$ and $\mathrm{coker} f$ are in Tor_R by Proposition 2.1, and $\mathrm{im}\, f = M/\ker f$ is also in Tor_R. Further Tor_R admits direct sums, again by Proposition 2.1. Now if N, N' are torsion submodules of a torsion module M, then $N + N'$ is a submodule of M and a quotient of $N \oplus N'$, hence $N + N'$ is a torsion module; further, by Proposition 2.1, $(N + N')/N' \cong N/(N \cap N')$ is torsion, and so is $N \cap N'$. Hence Tor_R as full subcategory of Mod_R is abelian, by Appendix Proposition B.1. For the last part let M be an n-torsion module with a torsion submodule M'. If M is presented by an $n \times n$ matrix C, then there is a factorization $C = AB$, where A is a matrix presenting M'. If A is $n \times r$, then $r \geq n$, because M is torsion and so C is full; we have $r = n$, since M' is torsion, so M' is also n-torsion. ∎

This result has numerous applications. In the first place we can derive an analogue of Schur's lemma, using the notion of a simple object in Tor_R. A torsion module M is said to be Tor-*simple* if it is a simple object in Tor_R, i.e. if $M \neq 0$ and no submodule of M, apart from 0 and M, lies in Tor_R. For example, if $R = k\langle x, y \rangle$ is a free algebra, then R/xR is Tor-simple, though of course far from simple as R-module. If we write down the factorization corresponding to a chain of submodules in Tor_R, we see that over a semifir R, a torsion module defined by a matrix A is Tor-simple if and only if A is a *matrix atom*, i.e. it is a full matrix that cannot be written as a product of two square non-unit matrices. Schur's lemma now takes the following form:

Proposition 3.2.4. *Let R be a semifir. Then for any matrix atom A over R, the eigenring of A is a field.*

Proof. We have seen that A is a matrix atom precisely if the module M defined by it is Tor-simple. Any non-zero endomorphism f of M must have kernel and cokernel 0, by Proposition 2.1, hence f must be an automorphism. ∎

This result holds more generally for an $n \times n$ matrix atom over a $2n$-fir. In some cases a more precise statement is possible. Thus let R be a semifir that is an algebra over a commutative field k. Then the eigenring of any matrix over R is again an algebra over k. Let A be a matrix atom, so that its eigenring E is a field over k. If λ is any element of E that is transcendental over k, then the elements $(\lambda - \beta)^{-1}$ for $\beta \in k$ all lie in E and are linearly independent over k. This shows that the dimension of E over k is at least $|k|$. If the dimension of R over k is less than $|k|$, this cannot happen (at least when k is infinite), so we obtain

Proposition 3.2.5. *If R is a semifir that is an algebra over a commutative field k, of dimension less than $|k|$, then the eigenring of any matrix atom is algebraic over k.*

Proof. For infinite k this follows from the above remarks; when k is finite, the conclusion follows directly. ∎

As an immediate consequence we have

Corollary 3.2.6. *If R is a semifir that is an algebra of at most countable dimension over an uncountable commutative field k that is algebraically closed, then the eigenring of any matrix atom is k itself.* ∎

Proposition 2.5 and Corollary 2.6 again hold for $n \times n$ matrix atoms over $2n$-firs. Over a semifir with left and right ACC_n every ascending chain of m-generator submodules of R^n, for any fixed $m \leq n$, becomes stationary, by Proposition 2.3.20, and the same is true for descending chains, by duality. Hence every torsion module in Tor_R has finite length, and by the Jordan–Hölder theorem, any two composition series are isomorphic. In terms of matrices a composition series corresponds to a *complete factorization*, i.e. a factorization into atomic factors, so this result may be stated as

Theorem 3.2.7. *Over a semifir with left and right ACC_n every full $n \times n$ matrix admits a complete factorization into matrix atoms, and any two such complete factorizations are isomorphic.* ∎

Here we can again replace the class of semifirs by the wider class of $2n$-firs; moreover, for $\aleph_0$-firs and in particular, for firs, the result holds without assuming chain conditions. A semifir in which every full matrix admits a factorization into matrix atoms will be called *fully atomic*; a $2n$-fir in which every full $m \times m$ matrix for $m \leq n$ admits a factorization into matrix atoms will be called *n-atomic*.

In the applications of this theorem the following entirely elementary result will be of use. We recall that an *elementary matrix* is a matrix differing from the unit matrix in just one off-diagonal entry.

Proposition 3.2.8. *Any upper (or lower) triangular matrix (over any ring) is a product of diagonal matrices and elementary matrices.*

Proof. This is an almost trivial verification, which is best illustrated by the 3×3 case:

$$\begin{pmatrix} a & a' & a'' \\ 0 & b & b' \\ 0 & 0 & c \end{pmatrix} = \begin{pmatrix} 1 & 0 & 0 \\ 0 & 1 & 0 \\ 0 & 0 & c \end{pmatrix} \begin{pmatrix} 1 & 0 & a'' \\ 0 & 1 & b' \\ 0 & 0 & 1 \end{pmatrix} \begin{pmatrix} 1 & 0 & 0 \\ 0 & b & 0 \\ 0 & 0 & 1 \end{pmatrix}$$
$$\begin{pmatrix} 1 & a' & 0 \\ 0 & 1 & 0 \\ 0 & 0 & 1 \end{pmatrix} \begin{pmatrix} a & 0 & 0 \\ 0 & 1 & 0 \\ 0 & 0 & 1 \end{pmatrix}.$$ ■

Given a fully atomic semifir R, let us consider the different factorizations of an element in more detail. Given $c \in R^\times$, if we have two complete factorizations of c, we pass from one to the other by a series of steps in each of which one side in a parallelogram of the factor lattice $L(cR, R)$ is replaced by the opposite side (as in the usual proof of the Jordan–Hölder theorem). In detail, we have a comaximal relation $ab' = ba'$, where a, b, a', b' are atoms, and $aR + bR = R, aR \cap bR = mR$, and in one factorization

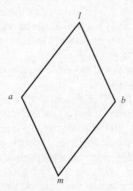

we replace an occurrence of ab' by ba'.

We shall call this passage from ab' to ba' a *comaximal transposition* and say that a, b' (in that order) are *comaximally transposable* if there exist elements a', b such that $ab' = ba'$ is a *balanced* comaximal relation. More generally, these terms will be used when a, b, a', b' are full matrices. The above discussion may be summed up as follows.

Proposition 3.2.9. *Every fir is fully atomic. In a fully atomic semifir every full matrix C has a complete factorization into a product of matrix atoms, and* *given any two complete factorizations of C, we can pass from one to the other by a series of comaximal transpositions.* ∎

Exercises 3.2

1. Show that in any integral domain the following are equivalent: (a) right ACC_1, (b) ACC on principal right ideals generated by left factors of a fixed non-zero element and (c) DCC on principal left ideals generated by right factors of a fixed non-zero element.
2. Show that in a 2-fir with left ACC_1 any family of elements, not all 0, have a highest common left factor.
3°. Let R be a k-algebra with generators (in matrix form) $c_i = (c_{i1}, \ldots, c_{in}) \in R^n$ and $A_i = (a_{rs}^{(i)}) \in R_n$ and defining relations $c_{i-1} = c_i A_i (i = 1, 2, \ldots)$. Show that R is a semifir (by expressing it as a direct union of free algebras). If $\mathfrak{a}_i$ denotes the right ideal generated by the components of c_i, show that $\mathfrak{a}_1 \subset \mathfrak{a}_2 \subset \ldots$ and deduce that R does not satisfy right ACC_n. Is it true that every full $n \times n$ matrix over R is a product of matrix atoms?
4. Let R be an integral domain that is a k-algebra, $a \in R^\times$ and let c be an element of the eigenring of a that is algebraic over k, thus $ca = ac'$, $f(c) = ab$ for some monic polynomial f over k and $b \in R$. Prove that $f(c') = ba$ and $c'b = bc$.
5. Let R be a 2-fir and $M = R/cR$ a 1-torsion module. Show that every finitely generated submodule of M is a torsion module if and only if c is right large. Show also that the qualifier 'finitely generated' can be omitted if R is atomic.
6. Show that any element stably associated to a right large element of a 2-fir is again right large.
7. Over a 2-fir R, show that R/cR is simple as R-module if and only if c is a right large atom.
8. In any ring R, given $a, b \in R$, show that there is an automorphism of R as right R-module that maps aR to bR if and only if a is left associated to b.
9. Show that for a principal ideal domain the notion of torsion module as defined in Section 3.2 agrees with the definition of a module of torsion elements in Section 0.8. Show that over a simple principal ideal domain the class of 1-torsion modules admits direct sums.
10°. Find a class of 2-firs for which the class of 1-torsion modules admits direct sums.
11. Let R be a 2-fir. Show that $\mathrm{Hom}_R(R/Ra, R/Rb) \neq 0$ if and only if there is a non-unit right factor of a similar to a left factor of b.
12*. Let A be the group algebra of the free group on x, y, z over a commutative field k and let R be the subalgebra generated by $z^{-n}x, z^{-n}y, z(n = 1, 2, \ldots)$. Show that R is a right fir but not a left fir. Show also that the pair $\{x, y\}$ has no highest common left factor.
13. Let R be an atomic 2-fir in which every principal right ideal is an intersection of maximal principal right ideals. Show that every 1-torsion module is a direct sum of Tor-simple 1-torsion modules. Show also that in general there are infinitely many pairwise non-associated elements stably associated to a given atom.

14. Show that any atomic right Bezout domain is a principal right ideal domain. Deduce that a ring is a principal ideal domain if and only if it is an atomic left and right Bezout domain.
15. In the Weyl algebra $A_1(k)$ on x and y, where $\mathrm{char}(k) = 0$, verify that $1 + xy$ is an atom and is stably associated to xy. Deduce that $xyx + x$ has two complete factorizations of different lengths.
16. Show that $\mathbb{Z}\langle x, y\rangle$ is not a UFD, by considering factorizations of $xyx + 2x$. What forms can the eigenring of an atom take?
17. Show that any permutation of $1, 2, \ldots, n$ can be written as a product of at most $n(n-1)/2$ transpositions of neighbouring numbers. Deduce that in an atomic 2-fir, given any non-zero non-unit c, we can pass from any one complete factorization of c to any other by a series of at most $n(n-1)/2$ comaximal transpositions.
18°. Verify that in the free algebra $k\langle x, y\rangle$, x and $yx + 1$ are comaximally transposable. Find an element in $k\langle X\rangle$ with two factorizations of length $n \geq 3$, where $n(n-1)/2$ comaximal transpositions are needed to pass from one to the other.
19*. Let R be a semifir with an involution *. Show that by combining it with transposition it can be extended to an involution on the $n \times n$ matrix ring over R. Let A be a symmetric matrix over R (i.e. $A^* = A$). Show that if A is not full, of inner rank r, then it has the form $A = PUV^*P^*$, where P is $n \times r$, U and V are $r \times r$ and $UV^* = VU^*$ is a balanced comaximal relation.

3.3 Rigid factorizations

Let us return to the definition of a UFD and consider more closely in what respects it differs from the definition in the commutative case. As we saw, given two complete factorizations,

$$c = a_1 \ldots a_r \quad \text{and} \quad c = b_1 \ldots b_r \qquad (1)$$

of an element c, necessarily of the same length, there is a permutation $i \mapsto i'$ of $1, \ldots, r$ such that

$$R/a_i R \cong R/b_{i'} R. \qquad (2)$$

However, we note that

(i) a_i and $b_{i'}$ are not necessarily associated and
(ii) in general we do not obtain c by writing the b's in the order $b_{1'} b_{2'} \ldots b_{r'}$.

To this extent, unique factorization in non-commutative domains, as here defined, is a more complicated phenomenon than in the commutative case, but there is a more restrictive concept that is sometimes useful.

Let R be an integral domain. We recall from Section 0.7 that an element $c \in R^\times$ is called *rigid* if $c = ab' = ba'$ implies $aR \subseteq bR$ or $bR \subseteq aR$. In other words, c is rigid if the lattice $L(cR, R)$ is a chain. If every non-zero element of

R is rigid, R is called a *rigid* domain. In that case $R^\times$ is a rigid monoid in the sense of Section 0.7 and by Theorem 0.7.9 we have

Proposition 3.3.1. *Let R be a rigid domain. Then $R^\times$ can be embedded in a group.* ∎

We observe that a rigid domain is necessarily a 2-fir, by Corollary 2.3.8. Now a *rigid* UFD is defined as an atomic rigid domain. For example, in the commutative case a rigid domain is just a valuation ring and a rigid UFD is a principal valuation ring. More generally, a rigid domain is right Ore if and only if all the principal right ideals form a chain; such a ring is called a *right chain ring*. However, a non-commutative rigid domain can be much more general than a chain ring.

Our main source of rigid UFDs stems from the following result.

Theorem 3.3.2. *Any complete inversely filtered ring with 2-term inverse weak algorithm is a rigid UFD.*

Proof. Let R be a ring satisfying the hypothesis. By Theorem 2.9.5, R is a 2-fir; if $ab' = ba' \neq 0$ and $v(a) \leq v(b)$ say, then a, b are right v-dependent and so, by the same theorem, $aR \supseteq bR$, hence R is rigid. By Proposition 2.9.6 it has left and right ACC_1 and so is atomic. Thus R is a rigid UFD. ∎

An obvious example of a rigid UFD (other than a valuation ring) is a formal power series ring $k\langle\langle X\rangle\rangle$ in any number of variables.

We go on to describe rigid domains in more detail; for this we need the following lemma, which is also useful elsewhere.

Lemma 3.3.3. *Let R be an Hermite ring. Two matrices $A \in {}^mR^r$, $B \in {}^rR^n$ are comaximally transposable if and only if there exist matrices $X \in {}^rR^m$, $Y \in {}^nR^r$ such that*

$$XA - BY = I. \tag{3}$$

Proof. Suppose we have a balanced comaximal relation

$$AB = B'A'. \tag{4}$$

By Proposition 0.5.6 there exist mutually inverse matrices $P = \begin{pmatrix} A & B' \\ Y & T \end{pmatrix}$, $P^{-1} = \begin{pmatrix} X & -B \\ Z & A' \end{pmatrix}$, hence $XA - BY = I$, i.e. (3). Conversely, given (3), since R is an Hermite ring, the matrix $(X, -B)$ is completable, so we obtain $Q = \begin{pmatrix} X & -B \\ * & * \end{pmatrix}$ with inverse $Q^{-1} = \begin{pmatrix} A & * \\ * & * \end{pmatrix}$ and equating the

(1, 2)-entries in the equation $Q^{-1}Q = I$ we obtain (4); by construction this relation is comaximal. ∎

As a first consequence we characterize the Jacobson radical J of a semifir, or more generally, of a matrix ring over an Hermite ring.

Theorem 3.3.4. *Let R be a total matrix ring over an Hermite ring. Then $J(R)$ consists of those elements of R that cannot be comaximally transposed with any non-unit.*

Proof. Suppose that $c \in R$ cannot be comaximally transposed with any non-unit. For any $x \in R$ we have the proper comaximal relation

$$c(xc + 1) = (cx + 1)c, \qquad (5)$$

hence $cx + 1$ is a unit for all $x \in R$ and this shows that $c \in J(R)$. Conversely, suppose that $c \in J(R)$ and that b, c can be comaximally transposed. Then by Lemma 3.3, $cx + yb = 1$ for some $x, y \in R$; by the choice of c, $yb = 1 - cx$ is a unit, and since R is weakly finite, b must be a unit. A similar argument holds if c, b can be comaximally transposed. ∎

Corollary 3.3.5. *Let R be an atomic 2-fir. Then a right large element of R lies in $J(R)$ if and only if it has every atom of R as a left factor.*

Proof. If $c \in R$ is right large and has every atom as a left factor, then in any equation (5) $cx + 1$ must be a unit, for otherwise it would have an atomic left factor p; by hypothesis this is also a left factor of c, and hence of 1, which is a contradiction. Thus we see as before that $c \in J(R)$. Conversely, if $c \in J(R)$, let p be an atom that is not a left factor of c. Since c is right large, we have $cR \cap pR \neq 0$; by Theorem 2.3.7 this intersection is principal, so we have a coprime relation $cp' = pc'$. By Corollary 1.4 this is comaximal, so p' is a non-unit and c, p' are comaximally transposable, which contradicts Theorem 3.4. ∎

We remark that in a principal right ideal domain this result applies to every non-zero element.

Secondly we shall obtain sufficient conditions for an element (or matrix) in a semifir to be rigid.

Proposition 3.3.6. *Let R be an $n \times n$ matrix ring over a 2n-fir and let c be a full matrix in R such that any two neighbouring non-unit factors occurring in a factorization of c generate a proper ideal of R. Then c is rigid. If c has an atomic factorization, then it is enough to check a single complete factorization of c; there will only be one complete factorization when c is rigid.*

Proof. If c is not rigid, then $c = ab' = ba'$, where neither of a, b is a left factor of the other. By Proposition 1.3, we obtain a comaximal relation $a_1 b_2 = b_1 a_2$ by cancelling full matrices on the left and right. Thus $a = da_1, b = db_1$, where a_1, b_1 are right comaximal non-units; moreover, $m = a_1 b_2 = b_1 a_2$ is an LCRM of a_1 and b_1. It follows that $a' = a_2 e, b' = b_2 e$ and $a_1 b_2 = b_1 a_2$ is a comaximal transposition occurring in $c = da_1 b_2 e$. By Lemma 3.3, $xa_1 - b_2 y = 1$ for appropriate $x, y \in R$, so the ideal generated by a_1 and b_2 is improper, a contradiction. In the atomic case we can pass from any complete factorization to any other by a series of comaximal transpositions, by Proposition 2.9; since no comaximal transpositions can occur when c is rigid, there will only be one such factorization. ∎

Now rigid domains are described by

Theorem 3.3.7. *An integral domain is rigid if and only if it is a 2-fir and a local ring.*

Proof. We have seen (Corollary 2.3.8) that any rigid domain is a 2-fir, and clearly no two non-units are comaximally transposable, hence by Theorem 3.4, $J(R)$ includes all non-units, i.e. R is a local ring. Conversely, in a 2-fir that is a local ring, any two non-units generate a proper ideal, hence by Proposition 3.6, every non-zero element is rigid. ∎

Adding atomicity, we obtain

Corollary 3.3.8. *A ring is a rigid UFD if and only if it is an atomic 2-fir and a local ring.* ∎

The description of commutative rigid UFDs, namely as principal valuation rings, can be extended to right Ore domains. It is easily verified that any right principal valuation ring is a rigid UFD and conversely, a rigid UFD is a right principal valuation ring if and only if any two atoms are right associated. The next theorem gives conditions for this to happen.

Theorem 3.3.9. *Let R be a rigid UFD. Then R is a right principal valuation ring if and only if it contains a non-unit c such that $cR \cap pR \neq 0$ for every atom p. In particular, this holds when R contains a right large non-unit.*

Proof. Suppose the rigid UFD R contains an element c satisfying the given conditions. For any atom p of R we have $cR \cap pR \neq 0$, hence either $cR \subseteq pR$ or $pR \subseteq cR$. Since c is a non-unit, the second alternative would mean that $cR = pR$, so in any case $cR \subseteq pR$, i.e. c has every atom as left factor. By rigidity this means that all atoms are right associated. Given any atom p, every $a \in R^\times$ can be written as $a = p^k u$ and by choosing k maximal we ensure that

u is a unit. Therefore R is a right PVR. Conversely, in a right PVR the unique atom is a right large non-unit. ∎

From the normal form of the elements it is clear that in a right PVR every right ideal is two-sided. By symmetry we obtain

Corollary 3.3.10. *If R is a rigid UFD with a left and right large non-unit, then R has a unique atom p (up to unit factors) and every left or right ideal of R is two-sided, of the form $p^n R = R p^n$.* ∎

For matrices over local semifirs the factorization is no longer rigid, but we can say a little more than in the general case. We begin with a general lemma. We shall say that a matrix A is *in* an ideal $\mathfrak{a}$ if all its entries lie in $\mathfrak{a}$.

Lemma 3.3.11. *(i) Let R be a ring and $\mathfrak{a}$ a proper ideal of R such that $R/\mathfrak{a}$ is weakly finite. If a matrix A over R is in $\mathfrak{a}$, then A is not stably associated to a matrix of smaller size.*

(ii) If R is a weakly finite ring and

$$AB' = BA' \qquad (6)$$

is a comaximal relation, where $A, A' \in {}^r R^m$, and A or A' is in $J(R)$, then B, B' are invertible and hence A, A' are associated.

Proof. (i) Let $A \in {}^r R^m$, $A' \in {}^s R^n$ and denote the images in $\bar{R} = R/\mathfrak{a}$ by bars. If A, A' are stably associated, then so are $\bar{A}, \bar{A}'$. We shall interpret these matrices as homomorphisms between free left R-modules. By hypothesis, $\bar{A} = 0$, hence coker $\bar{A} \cong \bar{R}^m$, therefore coker $\bar{A}' \cong \bar{R}^m$ and so $\bar{R}^n$ splits over coker $\bar{A}' : \bar{R}^n \cong \bar{R}^m \oplus \operatorname{im} \bar{A}'$. This shows that $m \leq n$ and a dual argument shows that $r \leq s$.

(ii) $\bar{R} = R/J(R)$ is again weakly finite, for if $\bar{X}\bar{Y} = I$, then $XY = I + C$ is invertible, with inverse D, say; then $DXY = I$, hence $YDX = I$, but $\bar{D} = \bar{D}\bar{X}\bar{Y} = I$, therefore $\bar{Y}\bar{X} = I$ and it follows that $\bar{R}$ is weakly finite. Suppose that A is in $J(R)$. Over $\bar{R}$ we have a comaximal relation $\bar{A}\bar{B}' = \bar{B}\bar{A}'$ and $\bar{A} = 0$, so by comaximality $\bar{B}$ is a unit, hence $\bar{A}' = 0$ and $\bar{B}'$ is also a unit. But any matrix invertible mod $J(R)$ is invertible over R, hence B, B' are invertible and so A, A' are associated. ∎

Let R be a local ring; it is clear that R is weakly finite, for if $XY = I$, then $\bar{X}\bar{Y} = I$, hence $\bar{Y}\bar{X} = I$, so YX is invertible and it follows that $YX = I$. Therefore the index of a matrix is preserved by stable association. In particular, any matrix stably associated to a square matrix is itself square. For a square matrix A we define the *level* as the least value of the order of any matrix stably associated to A. For example, A has level 0 if and only if it is a unit.

Proposition 3.3.12. *Let R be a local ring with maximal ideal* $\mathfrak{m}$ *and let* $A \in {}^m R^n$ *have the image* $\bar{A}$ *over* $R/\mathfrak{m}$ *of rank r. Then A is stably associated to an* $(m-r) \times (n-r)$ *matrix* A' *in* $\mathfrak{m}$, *but to no matrix of smaller size than* A'. *In particular, if A is* $n \times n$, *then its level is* $n - r$.

Proof. Since $\bar{A}$ has rank r, it is stably associated to the $(m-r) \times (n-r)$ zero matrix, but to no matrix of smaller size. Thus we have

$$U \begin{pmatrix} A & 0 \\ 0 & I \end{pmatrix} V \equiv \begin{pmatrix} 0 & 0 \\ 0 & I \end{pmatrix} \pmod{\mathfrak{m}},$$

where U, V are invertible (mod $\mathfrak{m}$). It follows that U, V are invertible over R and we have

$$U \begin{pmatrix} A & 0 \\ 0 & I \end{pmatrix} V = \begin{pmatrix} P_1 & P_2 \\ P_3 & P_4 \end{pmatrix},$$

where P_1, P_2, P_3 are in $\mathfrak{m}$ and $P_4 \equiv I \pmod{\mathfrak{m}}$; hence P_4 is invertible over R. By row and column operations we can reduce P_4 to I and P_2, P_3 to 0 and thus find that $A \oplus I$ is associated to $A' \oplus I$, where A' is in $\mathfrak{m}$. Thus A is stably associated to an $(m-r) \times (n-r)$ matrix A' in $\mathfrak{m}$, but clearly to no matrix of smaller size, by Lemma 3.11(i).

In particular, if A is $n \times n$, then the level of A is $n - \mathrm{rk}\, \bar{A}$. ∎

We conclude this section by another example of a one-sided fir; this depends on the following lemma that uses an idea of Chase [62].

Lemma 3.3.13. *Let R be an integral domain with UGN and an element* $p \neq 0$ *such that* $\cap p^n R = 0$. *Then* $R^{\mathbb{N}}$, *as left R-module, is not projective.*

Proof. Let us write $A^{(i)} \cong R$, $A = \prod_{i=1}^{\infty} A^{(i)}$, $A_r = \prod_{r+1}^{\infty} A^{(i)}$, so that $R^{\mathbb{N}} = A \cong A_r$ for all r. If A were projective, there would be a left R-module B such that

$$A \oplus B \cong \oplus_I C_\alpha, \quad \text{where } C_\alpha \cong R \text{ and } \alpha \text{ runs over a set } I. \tag{7}$$

Let us denote by $f_\alpha : A \to C_\alpha$ the projection from A on the summand C_α. We first show that there exists $r \geq 1$ such that

$$A_r f_\alpha = 0 \quad \text{for almost all } \alpha. \tag{8}$$

For suppose this is not so; then for each r, $A_r f_\alpha \neq 0$ for infinitely many α. We shall construct an increasing sequence $\{n_i\}$ of positive integers and sequences $\{\alpha_i\}, \{x_i\}$ such that $\alpha_i \in I$, $x_i \in p^{n_{i-1}} A_i$ and

$$x_k f_{\alpha_i} = 0 \quad \text{for } k < i, \, x_i f_{\alpha_i} \notin p^{n_i} C_{\alpha_i}. \tag{9}$$

Choose $\alpha_1 \in I$ such that $A_1 f_{\alpha_1} \neq 0$ and set $n_0 = 0$; then $A_1 f_{\alpha_1} \nsubseteq P^{n_1} C_{\alpha_1}$ for some $n_1 \geq 0$, hence there exists $x_1 \in p^{n_0} A_1$ such that $x_1 f_{\alpha_1} \notin p^{n_1} C_{\alpha_1}$ and (9) holds for $i = 1$.

Suppose now that x_k, α_k, n_k have been constructed for $k < i$ to satisfy (9). Since $x_1, \ldots, x_{i-1}$ lie in the sum of a finite subfamily of the C_α, there exist $\beta_1, \ldots, \beta_r \in I$ such that $x_k f_\alpha = 0$ for $\alpha \neq \beta_1, \ldots, \beta_r$ and for all $k < i$. So we may choose $\alpha_i \neq \beta_1, \ldots, \beta_r$ such that $A_i f_{\alpha_i} \neq 0$, hence $A_i f_{\alpha_i} \nsubseteq p^m C_{\alpha_i}$ for some m. Let us put $n_i = n_{i-1} + m$; then there exists $x_i \in p^{n_{i-1}} A_i$ such that $x_i f_{\alpha_i} \notin p^{n_i} C_{\alpha_i}$. Now the sequences x_k, α_k, n_k for $k \leq i$ satisfy (9), so the construction is complete.

Let us write $x_k = (x_k^{(i)})$, where $x_k^{(i)} \in A^{(i)}$. Since $x_k \in p^{n_{k-1}} A_k$, it follows that $x_k^{(i)} = 0$ for $k > i$, so $x^{(i)} = \sum_{k=1}^{\infty} x_k^{(i)}$ is a well-defined element of $A^{(i)}$. Further, since $n_0 \leq n_1 \leq \ldots$, there exists $y_r^{(i)} \in A^{(i)}$ such that $x^{(i)} = x_1^{(i)} + \cdots + x_r^{(i)} + p^{n_r} y_r^{(i)}$, so on writing $x = (x^{(i)})$, $y_r = (y_r^{(i)})$, we have

$$x = x_1 + x_2 + \cdots + x_r + p^{n_r} y_r \quad \text{for all } r \geq 1. \tag{10}$$

From (9) it is clear that the α_i are all distinct, hence there exists α_r such that $x f_{\alpha_r} = 0$. Writing x in the form (10) and applying f_{α_r}, we find that

$$x_r f_{\alpha_r} = -p^{n_r}(y_r f_{\alpha_r}) \in p^{n_r} C_{\alpha_r},$$

but this contradicts (9) and so (8) is established.

Suppose now that $A_r f_\alpha = 0$ except for $\alpha = \beta_1, \ldots, \beta_n$; then on writing

$$C' = C_{\beta_1} \oplus \cdots \oplus C_{\beta_n}, \quad C'' = \oplus C_\alpha,$$

we have $A_r \subseteq C'$. Now (7) may be written as

$$A^{(1)} \oplus \cdots \oplus A^{(r)} \oplus A_r \oplus B \cong C' \oplus C''.$$

Since $A_r \subseteq C'$, it follows that A_r is complemented in C', i.e. there exists D such that $A_r \oplus D \cong C'$. Now $C' \cong R^n$ is finitely generated, whereas $A_r \cong R^\mathbb{N}$ is not, because R has UGN. So we have reached a contradiction and this shows that $R^\mathbb{N}$ cannot be projective. ∎

We shall use this result to construct a one-sided fir as follows. Let K be a field with an endomorphism α that is not surjective and consider the skew power series ring $R = K[[x; \alpha]]$. This is a right principal valuation ring with maximal ideal xR. Moreover, R is atomic and so is a rigid UFD; clearly R also has UGN.

To show that R is not a left fir it will be enough to find a left ideal that is not free. In fact we shall find a left ideal isomorphic to $R^\mathbb{N}$; the result then follows by Lemma 3.13 (with $p = x$). To obtain such a left ideal we only need a sequence

(u_n') tending to 0 (in the filtration topology), which is left v-independent. For then $\sum f_n u_n \in R$ for all $(f_n) \in R^{\mathbb{N}}$, and by the v-independence the map

$$(f_n) \mapsto \sum f_n u_n$$

is an isomorphism from $R^{\mathbb{N}}$ to a left ideal of R. For any $c \in K \setminus K^\alpha$ the sequence (xcx^n) has the required properties. Clearly it tends to 0, and if $\sum f_n xcx^n = 0$, then by cancelling a power of x on the right, if possible, we may assume that $f_0 \neq 0$. If we extend α to an endomorphism of R by putting $x^\alpha = x$, then

$$\sum x^{n+1} f_n^{\alpha^{n+1}} c^{\alpha^n} = 0,$$

and so $xf_0^\alpha c \in R^\alpha$, whereas not all the coefficients of $xf_o^\alpha c$ lie in K^α. This contradiction shows that the xcx^n are left v-independent, and so we have proved

Theorem 3.3.14. *Let K be a field with a non-surjective endomorphism α. Then the skew power series ring $K[[x; \alpha]]$ is a right fir (in fact a right principal valuation ring), but not a left fir.* ∎

Exercises 3.3

1. Show that a direct limit of rigid UFDs is again rigid, but not necessarily a UFD.
2*. Let $R = k[x; \alpha]$, where k is a commutative field of characteristic p and α is the pth power map. Find all elements of low degree that are rigid.
3. Let R be a rigid domain with right ACC_1 and a left and right large non-unit. Show that R is a principal valuation ring.
4. Let R be a semifir and consider a matrix relation $PA' = AP'$, where P, A are left coprime and A, A' are square. Show that if A' is full and P is in $J(R)$, then A is invertible. Let $ab = pc$, where $b \neq 0$ and p is an atom of R contained in $J(R)$. Show that if a is a non-unit, then $a \in pR$.
5. Show that in a rigid UFD any relation $ca = ac'$ between non-zero non-units holds if and only if $c = a_0 b, c' = ba_0, a = c^r a_0 = a_0 c'^r$ for some $r \geq 1$.
6. Show that Lemma 3.3 and Theorem 3.4 hold for $2n$-firs and for matrices of the appropriate size.
7. (Koshevoi [66]) An ideal $\mathfrak{p}$ in a ring R is called *strongly prime* if $R/\mathfrak{p}$ is an integral domain. Let R be an atomic 2-fir and $\mathfrak{p}$ a strong prime ideal in R. Show that $\mathfrak{p}$ contains, with any atom a, all atoms similar to a. Let $c \in R$ have the complete factorizations $c = a_1 \ldots a_r = b_1 \ldots b_r$; show that if factors not in $\mathfrak{p}$ are omitted, then the same number of factors remain in each product and corresponding terms are similar.
8. Let R be a right hereditary local ring but not a valuation ring. Show that its centre is a field (Cohn [66d], see also Section 6.4).

9. Show that when $|X| > 1$, then $R = k\langle\langle X\rangle\rangle$ has left ideals isomorphic to $R^\mathbb{N}$ and hence is not a fir.
10. Let R be an integral domain with UGN and let $p \in R^\times$ be such that $\cap p^n R = 0$. Write $P = R^\mathbb{N}$ and S for the submodule of P with almost all components zero. Show that P/S is not projective.
11. Show that every homogeneous Lie element in the free algebra $k\langle X\rangle$ is an atom. (*Hint*: Use a basis of the universal associative envelope of the free Lie algebra consisting of ascending monomials.)
12°. Can the multiplicative monoid of every UFD be embedded in a group?
13°. Investigate firs with finitely many atoms, and those with finitely many matrix atoms.
14*. (G. M. Bergman) Consider the formal power series ring $S = k\langle\langle x_{11}, x_{12}, x_{21}, x_{22}, y_1, y_2\rangle\rangle$ with the homomorphism f defined in matrix form $X \mapsto X, y \mapsto Xy$. Let R be the direct limit of repeated iterations of this homomorphism (i.e. take a countable family $S^{(n)}$ of copies of S with $f : S^{(n)} \to S^{(n+1)}$ and put $R = \varinjlim S^{(-n)}$) and show that R is a local ring and a semifir. Verify that the intersection of the powers of the maximal ideal is non-zero, even though R is atomic and hence a UFD. (*Hint*: Show that Sf is inert in S, and deduce that S is inert in R. See also Sections 1.6 and 5.10.)

3.4 Factorization in semifirs: a closer look

We shall now examine the relation between different factorizations of an element of a 2-fir, or more generally, a full $n \times n$ matrix over a $2n$-fir. For ease of notation we shall take R to be the $n \times n$ matrix ring over a $2n$-fir S and consider an element c of R that is a full matrix over S. A study of the factorizations of c is essentially a study of the factor lattice $L(cR, R)$, but we shall usually express the result directly in terms of factorizations. In speaking of the 'left factors' of an element we shall tacitly understand the equivalence classes under right multiplication by units. In this way the *left* factors of an element c correspond to the principal *right* ideals containing cR:

$$c = ab \text{ (for some } b \in R\text{) if and only if } cR \subseteq aR.$$

Similarly, a chain of principal right ideals from cR to R corresponds to a factorization of c that is determined up to unit factors; we shall call the two factorizations

$$c = a_1 a_2 \ldots a_r \quad \text{and} \quad c = b_1 b_2 \ldots b_r,$$

essentially the same, if $b_i = u_{i-1}^{-1} a_i u_i$, where u_i is a unit and $u_o = u_r = 1$ and *essentially distinct* otherwise.

With these conventions a rigid UFD may be described as an integral domain in which each element has essentially only one atomic factorization. In a general UFD the atomic factorizations of a given element are of course by no means unique, but neither can the factors be interchanged at will. Let us compare the different factorizations of a full $n \times n$ matrix over a $2n$-fir. The isomorphism of two factorizations of a given full matrix is defined as in Section 1.3 for elements. A factorization

$$c = a_1 \ldots a_r \tag{1}$$

is said to be a *refinement* of another,

$$c = b_1 \ldots b_s, \tag{2}$$

if (2) can be obtained from (1) by bracketing some of the a's together; in other words, if (1) arises from (2) by factorizing the b's further. The factors of c correspond to the torsion factor-modules of R/cR; by Theorem 2.3 they admit $+$ and $\cap$ and so form a modular lattice. By the Schreier refinement theorem for modular lattices (Appendix Theorem A.2) we obtain

Theorem 3.4.1. *Over a $2n$-fir, any two factorizations of a full $n \times n$ matrix have isomorphic refinements.* ∎

Looking at the proof of the lattice-theoretic result quoted here, we find that we can pass from the refinement of one chain to that of the other by a series of steps, which each change the chain at a single point, from

$$\ldots \geq x \vee y \geq x \geq x \wedge y \geq \ldots \quad \text{to} \quad \ldots \geq x \vee y \geq y \geq x \wedge y \geq \ldots .$$

This corresponds to a change in the factorizations of the form

$$a_1 \ldots a_i a_{i+1} \ldots a_r \to a_1 \ldots a_{i-1} a'_{i+1} a'_i a_{i+2} \ldots a_r, \tag{3}$$

where $a_i a_{i+1} = a'_{i+1} a'_i$ is a comaximal relation; thus (3) is a comaximal transposition, as defined in Section 3.2. In this way we obtain the following more precise form of Theorem 4.1, which is also a slight generalization of Proposition 2.9.

Theorem 3.4.2. *Over a $2n$-fir, any two factorizations of a full $n \times n$ matrix have refinements that can be obtained from one another by a series of comaximal transpositions of terms.* ∎

As an illustration, let $c = pq = rs$ be two factorizations of an $n \times n$ matrix over a $2n$-fir R and write $pR + rR = uR$, $p = ua$, $r = ub$, so that $aR + bR = R$, and similarly $Rq + Rs = Rv$, $q = b'v$, $s = a'v$. Then our factorizations become $(ua)(b'v) = (ub)(a'v)$, where $ab' = ba'$ is a comaximal transposition.

3.4 Factorization in semifirs: a closer look

In a general ring R a factorization $c = ab$ of a regular element c need not induce a corresponding factorization in a similar element c' – the submodule of R/cR corresponding to aR/cR will again be cyclic, but need not have a cyclic inverse image in R. However, in the $n \times n$ matrix ring over a $2n$-fir, the principal right ideals between cR and R are characterized by the fact that they give rise to n-torsion submodules of R/cR; now an application of the parallelogram law for modular lattices gives

Proposition 3.4.3. *In the $n \times n$ matrix ring R over a $2n$-fir, let c and c' be full matrices that are stably associated. Then the lattices $L(cR, R)$ and $L(c'R, R)$ are isomorphic and the right ideals corresponding to each other under this isomorphism are similar.*

Proof. The first assertion is clear; the second follows because corresponding right ideals are endpoints of perspective intervals. ∎

The isomorphism in Proposition 4.3 can be described explicitly as follows: if $cb' = bc'$ is a comaximal relation for c and c', then to each left factor d of c corresponds the left factor d' of c' given by $dR \cap bR = bd'R$ and to each left factor d' of c' corresponds the left factor d of c given by $dR = cR + bd'R$. These maps are inverse to one another and induce an isomorphism between the lattices of left factors of c and c'.

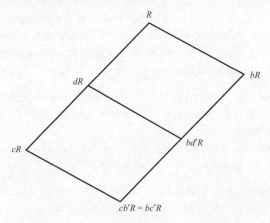

Figure 3.1

However, we note that the actual lattice-isomorphism we get may depend on our choice of comaximal relation – or, equivalently, on our choice of the isomorphism between R/cR and $R/c'R$. For example, take the ring $\mathbb{C}[x; -]$ of complex-skew polynomials; this is a principal ideal domain and hence a fir. The automorphism of the lattice of factors of $x^2 - 1$ induced by the comaximal relation $(x^2 - 1)x = x(x^2 - 1)$ interchanges the factorizations $(x + i)(x + i)$ and $(x - i)(x - i)$ and leaves the factorization $(x + 1)(x - 1)$ fixed. The automorphism induced by the relation

$$(x^2 - 1)1 = 1(x^2 - 1)$$

is of course the identity, while the automorphism induced by

$$(x^2 - 1)i = i(x^2 - 1)$$

interchanges the factorizations $(x + 1)(x - 1)$ and $(x - 1)(x + 1)$.

Over any semifir the factorizations of a given full matrix are closely related to those of its factors; this is best understood by looking first at the situation in lattices. In any lattice L, a *link* or *minimal interval* is an interval $[a, b]$ in L consisting of just two elements, namely its end-points, and no others, thus $a < b$ and no $x \in L$ satisfies $a < x < b$.

In any modular lattice L of finite length there are only finitely many projectivity classes of links, and the homomorphic images of L that are subdirectly irreducible are obtained by collapsing all but one equivalence class of links; these are in fact the simple homomorphic images of L. Here we count two homomorphic images as the same if and only if there is an isomorphism between them, forming a commuting triangle with the homomorphisms from L. If the distinct images are $L_1, \ldots, L_r$, we have a representation of L as a subdirect product of $L_1, \ldots, L_r$. In a distributive lattice of finite length, no two links in any chain are projective, as we shall see in Section 4.4, so the only simple homomorphic image is the two-element lattice $[0, 1]$, also written 2. Hence, by the Birkhoff representation theorem (Appendix Theorem A.8) we obtain

Theorem 3.4.4. (i) *Any modular lattice L of finite length can be expressed as a subdirect product of a finite number of simple modular lattices, viz. the simple homomorphic images of L.*

(ii) *Any distributive lattice L of finite length is a subdirect power of 2; more precisely, if L has length n, it is a sublattice of 2^n.* ∎

There is another representation of modular lattices, to some extent dual to that of Theorem 4.4, that is of use in studying factorizations.

Theorem 3.4.5. *Let L be a modular lattice. Given $a, b \in L$, there is a lattice-embedding $[a \wedge b, a] \times [a, a \vee b] \to L$ given by*

$$(x, y) \mapsto x \vee (b \wedge y) = (x \vee b) \wedge y. \tag{4}$$

Proof. Since $y \mapsto b \wedge y$ is an isomorphism, it follows that $(x, y) \mapsto x \vee (b \wedge y)$ preserves joins and similarly $(x, y) \mapsto (x \vee b) \wedge y$ preserves meets.

3.4 Factorization in semifirs: a closer look

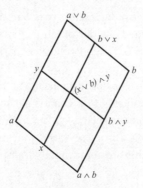

Figure 3.2

Thus (4) is a lattice homomorphism. It is injective because we can recover x and y from the right-hand side: $x \vee (b \wedge y) \vee a = (b \wedge y) \vee a = y \wedge (b \vee a) = y$, $(x \vee b) \wedge y \wedge a = (x \vee b) \wedge a = x \vee (b \wedge a) = x$. ∎

Let L be a lattice and for any $a \in L$ put $(a] = \{x \in L | x \leq a\}$, $[a) = \{y \in L | a \leq y\}$. The proof of Theorem 4.5 suggests that we consider the map $L \to (a] \times [a)$ defined by

$$z \mapsto (z \wedge a, z \vee a). \qquad (5)$$

Of course, in general this is not a lattice-homomorphism, but it clearly is one when L is distributive; further it is then injective, because two elements with the same image are relative complements of a in the same interval, and so must coincide. If moreover, a has a complement, b say, then $(a] = [a \wedge b, a]$, $[a) = [a, a \vee b]$ and if we now invoke Theorem 4.5, we obtain

Proposition 3.4.6. *Let L be a distributive lattice and $a \in L$. Then there is an embedding $L \to (a] \times [a)$ given by (5), and when a has a complement in L, this is an isomorphism.* ∎

The translation of these results into factorizations reads as follows:

Proposition 3.4.7. *Let R be the $n \times n$ matrix ring over a $2n$-fir and let c be a full matrix in R. Then any comaximal relation*

$$c = ab' = ba' \qquad (6)$$

gives rise to an embedding of $L(aR, R) \times L(bR, R)$ into $L(cR, R)$. If $L(cR, R)$ is distributive, this is an isomorphism. ∎

This result gives us a powerful tool relating the factorizations of a and b to those of c. For example, suppose that we have a comaximal relation (6) in which a has a factorization xyz and b has a factorization uv. This gives us chains of lengths 3 and 2 in the lattice of left factors of a, b respectively, and by applying Proposition 4.7 we find that the lattice of left factors of c will have a sublattice of the form shown in Fig. 3.3. Here intervals are marked with the factor of c to which they correspond.

Factorizations are given by paths from the top to the bottom of this diagram; every parallelogram corresponds to a comaximal relation, not only minimal parallelograms, giving relations such as $yu'' = u'y'$, but also larger ones, such as $(y'z')v''' = v'(y''z'')$. Thus, in these various factorizations of c, any factor from a and any factor from b are comaximally transposed.

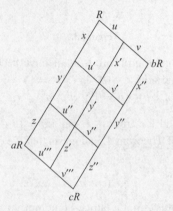

Figure 3.3

Further, if, say x and y are comaximally transposed, $xy = y'x'$, giving the subdiagram of left factors shown in Fig. 3.4, then from Proposition 4.7 we get a corresponding expanded diagram of factors of c, including comaximal parallelograms like $xy = y'x'$.

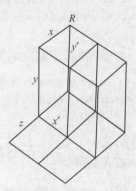

Figure 3.4

However, when $L(cR, R)$ is not distributive, the embedding of Proposition 4.7 need not be an isomorphism. In terms of factorizations this means that some factors of a and b may be comaximally transposed in more ways than the one induced by the comaximal relation (6). For example, in Fig 3.3 some of the parallelograms may be replaced by the diagram of Fig. 3.5. An example of such behaviour occurs in the complex-skew polynomial ring. Here $x^2 - 1 = (x + 1)(x - 1) = (x - 1)(x + 1)$ is a comaximal relation in which each factor is an atom, yet its full diagram of factorizations is of the form of Fig 3.5, because $x^2 - 1 = (x + u)(x - \bar{u})$ for any u on the complex unit circle.

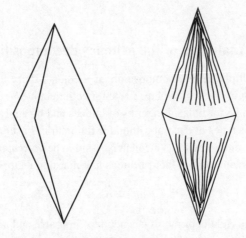

Figure 3.5

Of course this cannot happen when $L(cR, R)$ is distributive; as we have seen, it is then a subdirect power of 2. This case will be studied in more detail in Section 4.4.

Exercises 3.4

1. Show that the group algebra (over k) of the additive group of rational numbers (written multiplicatively) is a non-atomic Bezout domain. (*Hint*: Write the elements as $\sum c_\alpha x^\alpha$, $\alpha \in \mathbb{Q}$, and express the ring as a directed union of polynomial rings.)
2. Show that the ring of power series $\sum c_\alpha x^\alpha$ ($\alpha \in \mathbb{Q}, \alpha \geq 0$) with well-ordered support is an atomless Bezout domain.
3. (A. H. Schofield) Let $X = \{x_1, x_2, \ldots\}$ and consider the free power series ring $P = k\langle\langle X \rangle\rangle$, where each x_i has degree 1. Verify that the endomorphism $\theta : x_i \mapsto x_1 x_{i+1}$ on P maps non-units to non-units and that every non-unit in im θ has x_1 as a proper left factor. Deduce that the direct limit of the system (P, θ^n) is an atomless semifir, which is neither left nor right Ore.
4. In a 2-fir, if an element c can be written as a product of two atoms in at least three different ways, show that all the atomic factors of c are similar. Generalize the result to n factors.
5. Over a 2-fir R consider the equation $p \oplus q = AB$, where p, q are dissimilar atoms and A, B are non-invertible matrices. Show that this factorization is equivalent to

either $(p \oplus 1)(1 \oplus q)$ or $(1 \oplus q)(p \oplus 1)$. What are the possibilities when p is similar to q?

6°. Develop a theory of UFDs that are not necessarily 2-firs (see Brungs [69a]; Cohn, [70a, 73c]).

7. Over a fully atomic semifir, if A is a full matrix that is stably associated to AU, show that U is a unit (without atomicity this need not hold, see Section 1.6).

8°. Investigate non-commutative 2-firs in which any two elements with no common similar factors can be comaximally transposed.

3.5 Analogues of the primary decomposition

Besides the multiplicative decomposition of elements there is the primary decomposition of ideals, which plays a role for commutative Noetherian rings. Much of this can be formulated in terms of lattices and by applying it to Tor_R we obtain various types of decomposition for full matrices over semifirs.

In a weakly 1-finite ring R, an element c is said to be *decomposable* if it has two proper factorizations (i.e. factorizations into non-unit factors)

$$c = ab' = ba', \tag{1}$$

which are left and right coprime; if c is not decomposable and a non-unit, it is said to be *indecomposable*. This term (to be distinguished from 'factorizable into a product of non-unit factors') recalls the fact that for any (in)decomposable element c the module R/cR, equivalently, R/Rc, is (in)decomposable.

If c has two proper factorizations (1) that are left (right) coprime, c is said to be *left (right) decomposable*; if not, and c is a non-unit, it is *left (right) indecomposable*. Clearly any decomposable element is both left and right decomposable; hence any element that is either left or right indecomposable is indecomposable. For example, in an atomic 2-fir a non-unit c is right indecomposable if and only if every complete factorization ends in the same right factor (see Proposition 5.9). It is clear that a left and right decomposable element need not be decomposable (see Exercise 4). We also note that a 2-fir that is a local ring is rigid, by Theorem 3.7, hence in such a ring every element not 0 or a unit is indecomposable.

If R is the $n \times n$ matrix ring over a $2n$-fir, the definitions may be rephrased as follows:

(i) *a full matrix c in R is right decomposable if and only if there exist $a, b \in R$ such that*

$$cR = aR \cap bR, \quad \text{where } cR \neq aR, bR, \tag{2}$$

(ii) *a full matrix c in R is decomposable if and only if there exist $a, b \in R$ such that*

$$cR = aR \cap bR \text{ and } aR + bR = R, \quad \text{where } cR \neq aR, bR. \quad (3)$$

When c is invertible, we have $cR = R$; this case will usually be excluded. We observe that the definitions depend only on the module type of R/cR, as the following result shows.

Proposition 3.5.1. *Let R be the $n \times n$ matrix ring over a semifir and let c be a full non-invertible matrix in R. Then*

(i) *c is right decomposable if and only if R/cR is an irredundant subdirect sum of two torsion modules,*
(ii) *c is left decomposable if and only if R/cR is a sum of two torsion modules,*
(iii) *c is decomposable if and only if R/cR is a direct sum of two torsion modules.*

In particular, if c is (left, right) decomposable, then so is any matrix similar to it.

Proof. (i) Let c be right decomposable, say (2) holds. Then there is a monomorphism

$$R/cR \to R/aR \oplus R/bR. \quad (4)$$

The result of composing this with the projection onto either of the summands is surjective, but neither is injective; hence R/cR is an irredundant subdirect sum of R/aR and R/bR. Conversely, given any subdirect sum representation of R/cR, the kernels of the projection modules are again torsion modules, say $R/aR, R/bR$, which will satisfy (2).

(ii) If c is left decomposable, we have a relation (1), where a, b are left coprime. Hence we have

$$R/cR = (aR + bR)/cR = aR/cR + bR/cR \cong R/b'R + R/a'R; \quad (5)$$

thus R/cR is a sum of two torsion submodules. Conversely, when R/cR is a sum of two torsion modules, say $R/a'R$ and $R/b'R$, then (5) read from right to left, shows that (1) holds with a and b left coprime.

(iii) In particular, when c is decomposable, then the sum in (5) is direct, because $aR \cap bR = cR$. The converse is clear. ∎

We now consider the representations of an element or matrix that correspond to subdirect sum decompositions and direct decompositions of a module, with more than two terms. Let R be the $n \times n$ matrix ring over a $2n$-fir and c a full matrix in R; suppose that we have an irredundant representation

$$cR = a_1 R \cap \ldots \cap a_r R, \tag{6}$$

where each a_i is right indecomposable. Such an equation means that the natural map

$$R/cR \to \oplus R/a_i R \tag{7}$$

obtained by combining the maps $R/cR \to R/a_i R (i = 1, \ldots, r)$ is injective. Thus we have a subdirect sum representation of R/cR. Such a representation certainly exists if R is atomic. The irredundancy of (6) means that no term on the right of (7) can be omitted, while the right indecomposability of the a_i shows that each module $R/a_i R$ is subdirectly irreducible, by Proposition 5.1. If we now apply the Kurosh–Ore theorem for modular lattices (Appendix Theorem A.7), we obtain

Theorem 3.5.2. *Let R be the $n \times n$ matrix ring over a fully atomic semifir. Then for each full non-invertible matrix c in R, cR has an irredundant representation (6), where each a_i is right indecomposable, and if a second such decomposition of cR is given, $cR = b_1 R \cap \ldots \cap b_s R$, then $r = s$ and the $b_j R$ may be exchanged against the $a_i R$, i.e. after suitably renumbering the b's we have for $i = 1, \ldots, r$,*

$$cR = a_1 R \cap \ldots \cap a_i R \cap b_{i+1} R \cap \ldots \cap b_r R. \qquad \blacksquare$$

Similarly, cR has a representation (6), where each a_i is indecomposable and such that

$$a_i R \neq R, \quad a_i R + \left(\bigcap_{i \neq j} a_j R \right) = R, \quad (i = 1, \ldots r). \tag{8}$$

Let us call a representation (6) satisfying these conditions a *complete direct decomposition* of cR. When (8) holds, the map (7) is an isomorphism, so that a complete direct decomposition of cR corresponds to a direct sum representation

$$R/cR \cong R/a_1 R \oplus \cdots \oplus R/a_r R. \tag{9}$$

The first condition in (8) shows that each term on the right is non-zero, while the fact that each a_i is indecomposable means (by Proposition 5.1) that each module $R/a_i R$ is indecomposable in the category Tor_R. We shall express this by saying that each $R/a_i R$ is Tor-*indecomposable*; thus R/cR is Tor-indecomposable (by Proposition 5.1) precisely when c is indecomposable. Since the terms in any direct sum decomposition are always torsion modules (Theorem 2.3), this is

3.5 Analogues of the primary decomposition

equivalent to R/cR being indecomposable as module. If we now apply the Krull–Schmidt theorem for modular lattices (Appendix Theorem A.6) to the lattice of principal right ideals containing cR, we obtain

Theorem 3.5.3. *Let R be the $n \times n$ matrix ring over a fully atomic semifir. Then for each full non-invertible matrix c in R, cR has a complete direct decomposition (6) (i.e. satisfying (8)) and if a second such decomposition $cR = b_1 R \cap \ldots \cap b_s R$ is given, then $r = s$ and the b's may be exchanged against the a's. Moreover, the a's and b's are similar in pairs; thus after suitably renumbering the b's we have a set of coprime relations*

$$ua_i = b_i u_i, \tag{10}$$

where u corresponds to a unit in the eigenring $E(cR)$.

Proof. All except the last part follows from the preceding remarks. Now let $u \in I(cR)$ correspond to the automorphism of R/cR transforming the isomorphism (9) into the corresponding relation with the b's. Then there is a coprime relation $uc = cu'$, say, and if $c = b_i b_i' = a_i a_i'$, then we have $ua_i a_i' = b_i b_i' u'$; since u, c are right comaximal, so are u, b_i for $i = 1, \ldots, r$. If the b's are now renumbered so that $R/a_i R \to R/b_i R$ in the automorphism, then we have the coprime relations (10), for some $u_i \in R$. ∎

We note that Theorems 5.1–5.3 hold more generally for the $n \times n$ matrix ring over an n-atomic $2n$-fir. We also note that since any left or right indecomposable element is indecomposable, it follows that any complete direct decomposition (6) can be refined to an irredundant decomposition. Therefore the decomposition of Theorem 5.3 can have at most as many terms as that of Theorem 5.2.

The Krull–Schmidt theorem for modules can be proved either by lattice theory, as above, or by Fitting's lemma, which states that for an indecomposable R-module M of finite length the endomorphism ring $\text{End}_R(M)$ is *completely primary*, i.e. a local ring in which the maximal ideal is nilpotent (see FA, lemma 4.1.1 or IRT, p. 80). Using Proposition 0.6.1, we can restate Fitting's lemma in the following form:

Proposition 3.5.4. *Let R be the $n \times n$ matrix ring over a fully atomic semifir. Then the eigenring of an indecomposable full matrix is completely primary.* ∎

Thus if c is indecomposable and $a, a' \in R$ are such that $ac = ca'$, then either a, c are right comaximal or there exists $b \in R$ and $r \geq 0$ such that $a^r = cb$. When a is assumed to be an atom, we thus find

Corollary 3.5.5. *Let R be as in Proposition 5.4 and let c be an indecomposable full matrix, with idealizer $I(cR)$. If $I(cR)$ contains an atom p, then either p is right comaximal with c or c is right associated to p^r, for some $r \geq 0$.*

Proof. The remarks made earlier show that if c is not right comaximal with p, then $cb = p^r$ for some $b \in R$ and some $r \geq 0$. Thus p defines a nilpotent endomorphism of $E(cR)$ and

$$R \supset cR + pR \supset \ldots \supset cR + p^{r-1}R \supset cR \supseteq p^r R.$$

Successive quotients $(cR + p^i R)/cR$ form a strictly decreasing sequence of torsion submodules of R/cR whose quotients have no torsion submodules (because p is an atom); thus $cR = p^r R$ and so c is right associated to p^r. ∎

Of course in general the atomic factors of an indecomposable element need not all be similar, e.g. take xy in the free algebra $k\langle x, y\rangle$; any atom in the idealizer of xy, such as $1 + xy$, is comaximal with xy. This example also shows that a factor of a member of $I(cR)$ need not be a member of $I(cR)$, so an idealizer can contain composite elements without containing their atomic factors.

We note the special case of a rigid UFD:

Corollary 3.5.6. *In a rigid UFD, the eigenring of any non-zero non-unit is completely primary.*

Proof. By Corollary 3.8, such a ring R is an atomic 2-fir and a local ring. By the remark at the beginning of this section, any non-unit in R is indecomposable, so the result follows from Proposition 5.4. ∎

For cyclic modules there is a criterion for direct decomposability that actually holds quite generally.

Lemma 3.5.7. *Let R be any ring and $a, b \in R$, where a is right regular. Then*

$$R/bR \cong aR/abR, \tag{11}$$

and this is a direct summand of R/abR if and only if there exist $c, d \in R$ such that

$$da - bc = 1. \tag{12}$$

Moreover, in that case we have

$$R/abR \cong R/bR \oplus R/aR. \tag{13}$$

Proof. Write $M = R/abR$ and suppose that

$$M = N_1 \oplus N_2, \quad \text{where } N_1 \cong aR/abR.$$

Denote by u the image of 1 in M; then $u = v_1 + v_2$, where $v_i \in N_i$, further, $N_1 \cong R/bR$ by (11), which follows because a is right regular. Now $N_i = v_i R$ and we have $v_2 x = 0 \Leftrightarrow ux \in N_1 \Leftrightarrow x \in aR$; hence $N_2 \cong R/aR$ and (13) follows.

To establish (12), we have $v_1 \in uaR$, say $v_1 = uad$; hence $u = uad + v_2$ and so $v_2 = u(1-ad)$. Now $u(1-ad)a = 0$, so $(1-ad)a = -abc$ for some $c \in R$, and hence $a(1-da+bc) = 0$. Since a is right regular, we have $da - bc = 1$, i.e. (12).

Conversely, given (12), we put $r = 1 - ad$, $u = uad + u(1-ad)$, and so $M = uadR + urR$. If $uax = ury$, then $ax - ry = abz$; hence $ax - y + ady = abz$, so $y \in aR$. Now $ra = (1-ad)a = a(1-da) = -abc$. It follows that $ura = 0$, hence $ury = 0$ and so $uaR \cap urR = 0$, as claimed. ∎

To apply the result to $2n$-firs, we note that if S is a $2n$-fir, then an n-generator torsion module over S corresponds under the Morita equivalence (Theorem 0.2.4) to a cyclic S_n-module. We thus have

Proposition 3.5.8. *Let R be the $n \times n$ matrix ring over a $2n$-fir and let M, N be cyclic right R-modules defined by full matrices. Then $M \oplus N$ is cyclic if and only if there is a left comaximal pair of full matrices a, b in R such that $M \cong R/aR$, $N \cong R/bR$.*

Proof. If $M \oplus N$ is cyclic, where $M = R/aR$, $N = R/bR$, then (12) holds; hence a, b are comaximally transposable, say $ab = b'a'$. But then a is similar to a' and a', b are left comaximal.

Conversely, if a, b are left comaximal, let $a_1 b = b_1 a$ be a comaximal relation; then a_1 is similar to a and a_1, b are comaximally transposable; hence by the lemma, $R/a_1 R \oplus R/bR$ is cyclic. ∎

We also note the following condition for one-sided decomposability:

Proposition 3.5.9. *Let R be the $n \times n$ matrix ring over a $2n$-fir and c a full matrix in R. Then the following conditions are equivalent:*

(a) c is right indecomposable with an atomic right factor,
(b) the lattice $L(cR, R)$ has a unique minimal element covering cR,
(c) the lattice $L(Rc, R)$ has a unique maximal element covered by R,
(d) c has an atomic right factor, unique up to left associates.

Proof. This follows easily from what went before; the details may be left to the reader. ∎

Using (c) we obtain

Corollary 3.5.10. *If R and c are as in Proposition 5.9 and c is right indecomposable, then so is any right factor of c.* ∎

Of course this result holds more generally in any integral domain; in fact it is true for any cancellation monoid.

Another type of decomposition, sometimes of interest, exists for certain full matrices over semifirs, namely those whose associated torsion module is completely reducible (or semisimple). A full matrix c in the $n \times n$ matrix ring R over a semifir is said to be *fully reducible* if

$$cR = \cap p_i R, \qquad (14)$$

where the p_i are matrix atoms. By passing to the corresponding torsion module we see that c is fully reducible if and only if R/cR is a subdirect product of Tor-simple modules. Thus being fully reducible is again a property of the module type R/cR. If moreover, R is atomic (e.g. if the ground ring is fully atomic), then it is enough to take finitely many atoms on the right of (14). Taking the number of terms to be minimal, we obtain R/cR as a subdirect sum of the $R/p_i R$, and by minimality the sum is actually direct:

$$R/cR \cong R/p_1 R \oplus \cdots \oplus R/p_r R. \qquad (15)$$

By the factorial duality, being fully reducible in R is a left–right symmetric property. Further, any factor of a fully reducible element is again fully reducible, as the representation (15) shows. More generally, the decomposition (15) holds for an $n \times n$ matrix c over an n-atomic $2n$-fir.

In order to relate full reducibility to factorizations we need another concept, which is most easily stated for cancellation monoids. In any cancellation monoid the notion of full reducibility can be defined in analogy to (14). Now let S be a cancellation monoid. Given a factorization of a non-unit c in S into non-units:

$$c = ab, \qquad (16)$$

if every left factor of c is either a left factor of a or a right multiple of a, then a is called a *left block* of c, while c is said to be *cleft*. Dually, b is then a *right block* of c and (16) is called a *block factorization or cleavage* of c. An element is said to be *uncleft* if it is a non-unit with no cleavage into more than one non-unit.

It is clear that block factorizations are 'rigid' in the sense that, given two block factorizations of c, say $c = ab' = ba'$, we have $a = bu$ or $b = au$ for some $u \in S$. Let $c \in S$ be fully reducible; a given factorization of c:

$$c = a_1 \ldots a_r \qquad (17)$$

3.5 Analogues of the primary decomposition

is called a *cleavage* if, for $k = 2, 3, \ldots, r - 1$, every left factor $a_1 \ldots a_k$ is a left block or equivalently, every right factor $a_k \ldots a_r$ is a right block. In the case of two factors this just reduces to the earlier notion.

If c with a cleavage (17) also has an atomic factorization, this can always be obtained by refining (17) without reordering the terms. It follows that the set of left blocks is finite and by taking a cleavage (17) of c with a maximal number of factors the a_i are uncleft, because otherwise we could increase the number of factors by taking a cleavage of a_i. We note the following uniqueness result:

Lemma 3.5.11. *Let S be a cancellation monoid and $c \in S$ an element of finite length, i.e. with a complete factorization. Then the set of its left blocks is finite, yielding a cleavage (17) that is maximal with respect to refinement. The factors a_i are uncleft, and they are the only maximal uncleft factors of c.*

Proof. Suppose that c has an atomic factorization of length n. Then there are at most n left blocks, ordered by left divisibility, and this leads to a cleavage (17) maximal with respect to refinement. Here the factors a_i are uncleft, since a cleavage of a_i would induce a cleavage of c that is a proper refinement of (17). We claim that $a_1, \ldots, a_r$ are the only maximal uncleft factors of c. For if $c = uvw$, where v is uncleft, then a comparison with (17) shows that v must be a factor of some a_i; for if v had a left factor in a_i and a right factor in a_{i+1}, this would lead to a cleavage of v, against the hypothesis. If v is a proper factor of a_i, then it is not a maximal uncleft factor, so the maximal uncleft factors are, up to associates, the a's themselves, as claimed. ∎

To give an example, let us take $R = \mathbb{Z}$, or more generally, any commutative principal ideal domain. If p is an atom such that $p + 1$ is a non-unit, then the element $p^2(p + 1)$ is uncleft, but it has the cleft factor p^2. Let us call an element *totally uncleft* if all its factors are uncleft; thus in a commutative PID the totally uncleft elements are just the 'squarefree' elements. The connexion with fully reducible elements is given by

Proposition 3.5.12. *Let R be the $n \times n$ matrix ring over a semifir. For any full matrix c in R admitting a complete factorization, the following conditions on c are equivalent:*

(a) c is totally uncleft,
(b) c is fully reducible,
(c) any two neighbouring factors in a complete factorization of c can be comaximally transposed.

Proof. (a) $\Rightarrow$ (b). Let c be totally uncleft; then not every complete factorization of c ends in the same atom, say $c = c_1 p_1 = c_2 p_2$, where $Rp_1 \neq Rp_2$.

Then $cR = c_1R \cap c_2R$; now c_1, c_2 as factors of c are again totally uncleft and by induction on the length of c, each c_i is fully reducible, say $c_iR = \cap_j p_{ij}R$. Hence $cR = \cap_{ij} p_{ij}R$, which shows the truth of (b).

(b) $\Rightarrow$ (c). Assume c fully reducible, take a complete factorization $c = a_1 \ldots a_r$ and fix i in the range $2 \leq i \leq r$. Then $a_{i-1}a_i$ is fully reducible, as factor of c, hence there is a comaximal relation $a_{i-1}a_i = a'_i a''_i$.

(c) $\Rightarrow$ (a). This follows because any cleavage of any factor of c, on being refined to a complete factorization, leads to a pair of neighbouring atomic factors that cannot be comaximally transposed. ∎

In a rigid UFD any factorization is clearly a cleavage. Generally let us say that a factorization of a certain type, e.g. into maximal uncleft factors, is *essentially unique* if it is unique up to inessential modification (i.e. by unit factors). Now Lemma 5.11 shows the truth of

Theorem 3.5.13. *Let R be the $n \times n$ matrix ring over a semifir. Then every full matrix c in R of finite length has a factorization*

$$c = a_1..a_r \tag{18}$$

into maximal uncleft factors, and this is essentially unique. ∎

The factorization (18) is always a cleavage; if the factors are atoms, this means that no two neighbours are comaximally transposable. In terms of lattices this means that $L(cR, R)$ is a chain. Thus we have

Corollary 3.5.14. *A full matrix c of finite length in the $n \times n$ matrix ring R over a semifir has a cleavage into atomic factors if and only if $L(cR, R)$ is a chain, or equivalently, if c is rigid.* ∎

Here (as in Proposition 5.12 and Theorem 5.13) the ring can again be taken to be a $2n$-fir.

Exercises 3.5

1. In the complex-skew polynomial ring $R = \mathbb{C}[x; -]$, find all possible irredundant representations (6) of $(x^2 - 1)R$.
2. Show that in any commutative principal ideal domain that is not a local ring there are elements that have no essentially unique factorization into maximal totally uncleft factors.
3. Show that in a commutative UFD every factor of an indecomposable element is again indecomposable, and find a counter-example in the ring $k[x^2, x^3]$. Show also that in $k\langle x, y, z \rangle$ the element $zx(yx + 1)z$ is indecomposable (even left and right indecomposable), but has a decomposable factor.

3.5 Analogues of the primary decomposition 223

4. Find an element in $k\langle x, y, z\rangle$ that is indecomposable but not left or right indecomposable.
5. Show that in an atomic 2-fir any factor of a fully reducible element is fully reducible.
6. An element c of a ring R is called *primary* if $cR = \cap p_i R$, where the p_i are pairwise similar atoms. Show that in the $n \times n$ matrix ring over an n-atomic $2n$-fir every full matrix that is fully reducible can be written uniquely as LCRM of primary matrices.
7. Let S be a cancellation monoid and let $a = a_1 \ldots a_r$ be a factorization into maximal uncleft factors. If b is a left factor of a, with factorization $b = b_1 \ldots b_s$ into maximal uncleft factors, then for some $i \leq \min(r, s)$, $a_1 \ldots a_i S = b_1 \ldots b_i S$, $b_{i+1} \ldots b_s S \supseteq a_{i+1} S$.
8°. Investigate atomic 2-firs (other than principal valuation rings) in which any two atoms are similar (see Faith [73], p. 362).
9. (Beauregard and Johnson [70]) For $i = 1, \ldots, n$ let p_i be the ith prime in $\mathbb{N}$. Verify that the subring R_i of $\mathbb{Q}[x]$ consisting of all polynomials f such that $f(i)$ has a denominator prime to p_i is a Bezout domain. (*Hint:* Replace x by $x - i$). Show further that $R = \cap R_i$ is a Bezout domain. Show that R contains elements that are p_i-prime but not a product of atoms.

Notes and comments on Chapter 3

The notion of full matrix (Section 3.1) was introduced by Cohn [71a]; for left full, left prime, etc. see Cohn [82a], where results 1.3–1.5 are also proved. Much of Section 3.2 follows Cohn [63a,69a,70a], though the strictly cyclic (= cyclic torsion) modules of FR.1 were replaced by torsion modules in FR.2. This corresponds to taking factorizations of full matrices rather than elements, and most of the subsequent results are stated in this more general form, which is usually no harder to prove. Many of the results are just consequences of the fact (Theorem 2.3) that the torsion modules over a semifir form an abelian category. Factorizations of unbounded length have been studied by Beauregard [69], Beauregard and Johnson [70], Brungs [78] and Paul [73]. The proof of Proposition 2.5, based on the linear independence of the $(\lambda - \beta)^{-1}$, where λ is transcendental, is Amitsur's well-known trick (which he apparently noticed while lecturing on complex function theory).

An interesting generalization of UFD, to include the case of $\mathbb{Z}\langle X \rangle$, has been proposed by Brungs [69a]; for another approach see Cohn [70a] and for a survey, Cohn [73c]. A detailed study of similarity, foreshadowing a form of Schanuel's lemma, was undertaken by Fitting [36]; the parts relevant for us are contained in Sections 0.5 and 3.1 (see also Cohn [82a]). For Section 3.3 see Cohn [62a] and Bowtell [67a]. Proposition 3.6 generalizes a result by Koshevoi [66] for free algebras, and Proposition 3.12 is taken from Cohn [85a]. Lemma 3.13 is the special case for integral domains with UGN of a result of Chase [62]; for the application made here, see Cohn [66d].

Section 3.4 is due to Bergman [67] (see also an unpublished manuscript by Bergman, dating from 1969), while Section 3.5 generalizes earlier results of Ore [33a]; see also Feller [60], Johnson [65] and Cohn [69a, 70a, 73c]. For another approach to primary decomposition in non-commutative rings, see Barbilian [56], Chapter 2. Lemma 5.7 generalizes a result proved in FR.1 and is taken from Dieudonné [73], p. 164.

Cozzens and Faith [75] define a V-ring as a ring in which every simple right module is injective (e.g. the commutative V-rings are just the von Neumann regular rings). It can be shown that a right PID is a V-ring if and only if every non-zero element is fully reducible. Every V-ring is a TC-ring, defined as follows: a *test module* for a ring R is a module T such that, for any R-module M, $\operatorname{Hom}_R(M, T) = 0$ implies $M = 0$; if every test module is a cogenerator (i.e. $\operatorname{Hom}_R(-, T)$ is a faithful functor of T), then R is called a *TC-ring*. Let R be a TC-ring and S a simple R-module with injective hull $E(S)$; then $\operatorname{End}_R(E(S))$ is a local ring and a semifir (see Vamos [76]).

4
Rings with a distributive factor lattice

This chapter examines more closely those 2-firs in which the lattice of factors of any non-zero element is distributive. After some generalities in Section 4.1 and their consequences for factor lattices in Section 4.2 it is shown in Section 4.3 that this holds for free algebras and the consequences are traced out in Sections 4.4 and 4.5 while Section 4.6 describes the form taken by eigenrings in this case.

4.1 Distributive modules

Given any ring R, a full subcategory $\mathcal{A}$ of $_R$Mod, the category of all left R-modules, is said to be *admissible* if any kernel or cokernel (taken in $_R$Mod) of a map of $\mathcal{A}$ is again in $\mathcal{A}$. In general $\mathcal{A}$ may not admit direct sums; if it does, we have an abelian category, by Appendix Proposition B.1. A module in $\mathcal{A}$ will be called an $\mathcal{A}$-module; likewise we shall speak of $\mathcal{A}$-submodules, $\mathcal{A}$-quotients, etc., but the reference to $\mathcal{A}$ will sometimes be omitted when it is clear that we are dealing with $\mathcal{A}$-modules. For example, the category of torsion modules over a semifir is admissible and admits direct sums, as well as sums of torsion submodules.

Since an admissible category $\mathcal{A}$ admits kernels and cokernels, it also admits images and coimages and it follows that the set $\mathrm{Lat}_{\mathcal{A}}(M)$ of $\mathcal{A}$-submodules of any R-module M is a lattice, necessarily modular. If $\mathrm{Lat}_{\mathcal{A}}(M)$ is distributive, the module M is said to be $\mathcal{A}$-*distributive*, or simply *distributive*, if the meaning is clear (see Appendix A). The following are some examples of occurrences of distributive modules:

(i) $\mathcal{A} = {}_R$Mod, where R is a commutative Bezout domain (or more generally, a Prüfer domain). Any cyclic R-module is distributive (Jensen [63]).

(ii) $\mathcal{A} = {}_R\text{Mod}$, where R is semisimple Artinian. An R-module M is distributive if and only if each simple module type occurs at most once in M. If each simple module type occurs exactly once, then M is faithful. For this reason R is called *distributively representable* by Behrens [65].

(iii) In every right Artinian algebra of finite representation type the lattice of all two-sided ideals is distributive, see Pierce ([82], p. 104).

(iv) In Section 4.3 we shall see that any 1-torsion module M over a free algebra $R = k\langle X \rangle$ is Tor-distributive. Here it is essential to consider M as an object of ${}_R\text{Tor}^1$ rather than ${}_R\text{Mod}$.

In the rest of this section we shall examine the structure of a distributive module, in preparation for what follows. If M is a distributive module, then clearly any submodule and any quotient of M is again distributive. We begin with a simple characterization of distributive modules, which is often useful. Let M be a module and suppose that $M = M_1 \oplus M_2$. With any homomorphism $\alpha : M_1 \to M_2$ we associate a submodule of M, the *graph* of α:

$$\Gamma(\alpha) = \{(x, x\alpha) | x \in M_1\}. \qquad (1)$$

It is clear that $\Gamma(\alpha) \cap M_2 = 0$, $\Gamma(\alpha) + M_2 = M$; thus $\Gamma(\alpha)$ is a complement of M_2 in M, and it is easily seen that any complement of M_2 in M defines a homomorphism $M_1 \to M_2$ in this way. Since in a distributive module complements are unique, there can then only be one such map, necessarily the zero map. Thus we obtain

Proposition 4.1.1. *Given an admissible module category $\mathcal{A}$, let M_1, M_2 be any $\mathcal{A}$-modules and put $M = M_1 \oplus M_2$. Then each homomorphism $\alpha : M_1 \to M_2$ determines a graph $\Gamma(\alpha)$ given by (1), which is a complement of M_2 and conversely, each complement of M_2 is the graph of a homomorphism $M_1 \to M_2$. Moreover, when M is distributive, then $\text{Hom}(M_1, M_2) = 0$.* ∎

This result shows in particular that a distributive module cannot be of the form N^2, i.e. the square of a non-zero module. Let us call M *square-free* if it has no factor module isomorphic to such a square. This yields the following criteria for distributivity:

Theorem 4.1.2. *For any module M in an admissible category the following conditions are equivalent:*

(a) *M is distributive,*
(b) *$\text{Hom}(P/(P \cap Q), Q/(P \cap Q)) = 0$ for all submodules P, Q of M,*
(c) *M is square-free.*

Proof. (a) $\Rightarrow$ (b). If M is distributive, then so is $(P + Q)/(P \cap Q) \cong P/(P \cap Q) \oplus Q/(P \cap Q)$, and now (b) follows by Proposition 1.1.

(b) ⇒ (c). If (c) is false, then $M \supseteq A \supset B$, where A/B is a square; this means that there exist $P, Q \supset B$ such that $P + Q = A$, $P \cap Q = B$ and $P/B \cong Q/B$, but this contradicts (b).

(c) ⇒ (a). Suppose M is not distributive. Then Lat(M), being modular, has a five-element sublattice of length 2, by Appendix Proposition A.5, i.e. there exist $P_i (i = 1, 2, 3)$ such that $P_i + P_j = A$, $P_i \cap P_j = B$ for $i \neq j$; hence $A/P_i \cong P_j/B$, and so A/B is a square. ∎

Let M be any module, A, B submodules of M and α a homomorphism of some module into M. Then we clearly always have

$$A\alpha^{-1} + B\alpha^{-1} \subseteq (A + B)\alpha^{-1},$$

but equality need not hold. It is easily seen that we have equality when A or B lies in the image of α. Dually, we have, for a homomorphism β from M into some module,

$$(A \cap B)\beta \subseteq A\beta \cap B\beta,$$

and here we have equality if A or B contains ker β, but not generally. In fact, equality (in either case) characterizes the distributivity of M:

Proposition 4.1.3. *For any module M in an admissible category the following conditions are equivalent:*

(a) M is distributive,
(b) for any module P and homomorphism $\alpha: P \to M$,

$$(A + B)\alpha^{-1} = A\alpha^{-1} + B\alpha^{-1} \text{ for all submodules } A, B \text{ of } M,$$

(c) for any module Q and homomorphism $\beta: M \to Q$,

$$(A \cap B)\beta = A\beta \cap B\beta \text{ for all submodules } A, B \text{ of } M.$$

Proof. (a) ⇔ (b). For any submodule S of P containing ker α the correspondence $S \mapsto S\alpha$ is a bijection between the set of submodules of P containing ker α and the set of all submodules of $P\alpha$, with inverse $A \mapsto A\alpha^{-1}$, where A is a submodule of $P\alpha$. Hence (b) holds if and only if $(A + B)\alpha^{-1}\alpha = A\alpha^{-1}\alpha + B\alpha^{-1}\alpha$. Since $A\alpha^{-1}\alpha = A \cap P\alpha$, this is just $(A + B) \cap P\alpha = A \cap P\alpha + B \cap P\alpha$, which is distributivity.

(a) ⇔ (c). Let $K = \ker \beta$; then $A\beta\beta^{-1} = A + K$, and so $x \in (A\beta \cap B\beta)\beta^{-1}$ if and only if $x \in (A + K) \cap (B + K)$. Thus (c) holds if and only if $(A \cap B) + K = (A + K) \cap (B + K)$, which is (a). ∎

We shall need another technical result on homomorphisms between distributive modules.

Lemma 4.1.4. *Let P, Q be any modules in an admissible category and $\alpha, \beta \in Hom(P, Q)$.*
(i) If Q is distributive and X is a submodule of P, then
   (a) $P = (im\,\alpha)\beta^{-1} + (im\,\beta)\alpha^{-1}$, (b) $X = (X \cap X\alpha\beta^{-1}) + (X \cap X\beta\alpha^{-1})$,
(ii) if P is distributive and Y is a submodule of Q, then
   (a) $0 = (ker\,\alpha)\beta \cap (ker\,\beta)\alpha$, (b) $Y = (Y + Y\alpha^{-1}\beta) \cap (Y + Y\beta^{-1}\alpha)$.

Proof. (i) We have $im(\alpha + \beta) \subseteq im\,\alpha + im\,\beta$, hence by Proposition 1.3,

$$P = (im\,\alpha + im\,\beta)(\alpha + \beta)^{-1} = (im\,\alpha)(\alpha + \beta)^{-1} + (im\,\beta)(\alpha + \beta)^{-1}$$
$$= (im\,\alpha)\beta^{-1} + (im\,\beta)\alpha^{-1}.$$

Now (b) follows by setting ν for the inclusion map of X in P and applying (a) to $\nu\alpha, \nu\beta$, thus: $X = X\alpha(\nu\beta)^{-1} + X\beta(\nu\alpha)^{-1} = (X \cap X\alpha\beta^{-1}) + (X \cap X\beta\alpha^{-1})$.

(ii) follows similarly from Proposition 1.3. ∎

A module M is said to be *meta-Artinian* if every non-zero factor of M has a simple submodule; dually, if every non-zero factor of M has a simple quotient, M is called *meta-Noetherian*. Clearly these properties are inherited by submodules and quotients. When M is taken from an admissible category $\mathcal{A}$, it is understood that only submodules and quotients in $\mathcal{A}$ are understood.

For modules satisfying these hypotheses Lemma 1.4 can be extended as follows:

Proposition 4.1.5. *Let $\mathcal{A}$ be an admissible module category admitting sums of submodules. Given $\mathcal{A}$-modules P, Q and $\alpha, \beta \in Hom(P, Q)$,*

(i) if P is meta-Artinian and Q is distributive, then $ker\,\alpha \subseteq ker\,\beta$ implies $im\,\beta \subseteq im\,\alpha$,
(ii) if Q is meta-Noetherian and P is distributive, then $im\,\beta \subseteq im\,\alpha$ implies $ker\,\alpha \subseteq ker\,\beta$.

Proof. Suppose that $ker\,\alpha \subseteq ker\,\beta$ but $im\,\beta \not\subseteq im\,\alpha$, and define

$$A = \sum \{B \subseteq P | B\beta \subseteq B\alpha\}.$$

Clearly A is the largest submodule of P such that $A\beta \subseteq A\alpha$. In particular it follows that $ker\,\alpha \subseteq ker\,\beta \subseteq A$. Since $im\,\beta \not\subseteq im\,\alpha$, we have $A \neq P$. By hypothesis there exists $X \subseteq P$ such that $A \subset X$ and X/A is simple. But $A \subseteq A\alpha\beta^{-1} \subseteq X\alpha\beta^{-1}$ and $A \subseteq X \cap X\alpha\beta^{-1} \subseteq X$. Since $X\beta \not\subseteq X\alpha$ and X/A is simple, we find that $A = X \cap X\alpha\beta^{-1}$. By Lemma 1.4 (i)(b),

$$X = (X \cap X\alpha\beta^{-1}) + (X \cap X\beta\alpha^{-1}) = A + (X \cap X\beta\alpha^{-1})$$

4.1 Distributive modules

and so $X\alpha = A\alpha + (X \cap X\beta\alpha^{-1})\alpha = A\alpha + (X\alpha \cap X\beta) = A\alpha + (X\alpha\beta^{-1} \cap X)\beta = A\alpha + A\beta = A\alpha$. Since $\ker \alpha \subseteq A$, we find that $X = A$, a contradiction; hence $\operatorname{im} \beta \subseteq \operatorname{im} \alpha$. This proves (i); now (ii) follows similarly, by applying dual arguments. ∎

We recall that a submodule of a module M is called *fully invariant* if it is mapped into itself by all endomorphisms of M. We shall see that under suitable finiteness conditions distributive modules have an even stronger property:

Corollary 4.1.6. *Let $\mathcal{A}$ be an admissible module category admitting sums of submodules. (i) If M is a meta-Artinian distributive $\mathcal{A}$-module and N a submodule of M with a homomorphism $\alpha : N \to M$, then $N\alpha \subseteq N$. In particular, M is fully invariant and any two isomorphic submodules of M are equal.*

(ii) If M is a meta-Noetherian distributive module with a submodule N and a homomorphism $\beta : M \to M/N$, then $\ker \beta \supseteq N$, so that β is induced by an endomorphism of M/N. In particular, distinct submodules of M determine non-isomorphic quotients. Moreover, M is fully invariant.

Proof. (i) is an immediate consequence of Proposition 1.5, putting $\alpha = \beta$ there and taking α in (i) to be the inclusion map. Now the first part of (ii) follows similarly; to show that M is fully invariant, let N be a submodule and $\nu : N \to M/N$ the natural map. By the first part, we have $x\nu = 0 \Rightarrow x\beta\nu = 0$ for any $x \in M$ and it follows that $N\beta \subseteq N$. ∎

In Section 4.6 we shall need a further result. We begin with a lemma.

Lemma 4.1.7. *Let $\mathcal{A}$ be an admissible module category admitting sums of submodules. Given an $\mathcal{A}$-module M with submodules U, V, suppose that neither is contained in the other and that U, V each have a unique maximal submodule U', V' respectively. Then*

$$(U + V)/(U' + V') \cong U/U' \oplus V/V'. \tag{2}$$

Proof. The natural maps $U/U' \to (U + V)/(U' + V')$ and $V/V' \to (U + V)/(U' + V')$ give rise to a map $U/U' \oplus V/V' \to (U + V)/(U' + V')$, which is clearly surjective; we have to show that it is injective. Suppose that $([x], [y])$ is in its kernel, where square brackets denote the residue-classes mod U' and mod V'; this means that $x + y \in U' + V'$, hence $x + y = u + v$, where $u \in U'$, $v \in V'$. It follows that $x - u = v - y$, where $x - u \in U$, $v - y \in V$; hence their common value lies in $U \cap V$. By hypothesis the latter is a proper submodule of U and of V, and so by the maximality of U', V' it must lie in $U' \cap V'$. Therefore $x = (x - u) + u \in U'$ and so $[x] = 0$; similarly $[y] = 0$ and our map is injective, hence an isomorphism. ∎

Proposition 4.1.8. *Let $\mathcal{A}$ be an admissible module category admitting sums of submodules, and let M be an $\mathcal{A}$-module with a unique maximal submodule M'. Then for any two homomorphisms $\alpha, \beta : M \to N$, where N is a distributive module, one of $M\alpha$, $M\beta$ is contained in the other.*

Proof. We may assume that $\alpha, \beta \neq 0$; then $\ker \alpha$ is a proper submodule of M, hence $\ker \alpha \subseteq M'$, $M\alpha \cong M/\ker \alpha$, and it follows that $M\alpha$ has a unique maximal submodule, namely $M'\alpha$. Similarly $M\beta$ has the unique maximal submodule $M'\beta$ and if $M\alpha$, $M\beta$ are incomparable, then by Lemma 1.7,

$$(M\alpha + M\beta)/(M'\alpha + M'\beta) \cong M\alpha/M'\alpha \oplus M\beta/M'\beta \cong (M/M')^2;$$

this contradicts the fact that N is square-free and the conclusion follows. ∎

To end this section let us note a useful result on the endomorphism ring of a distributive module that is a direct sum.

Theorem 4.1.9. *Let $\mathcal{A}$ be an admissible module category admitting sums of submodules, and let M be a distributive $\mathcal{A}$-module such that*

$$M = M_1 \oplus \cdots \oplus M_n. \tag{3}$$

Then

$$\mathrm{End}(M) \cong \prod_{i=1}^{n} \mathrm{End}(M_i). \tag{4}$$

Proof. Any endomorphism α of M is represented by a matrix (α_{ij}), where $\alpha_{ij} \in \mathrm{Hom}(M_i, M_j)$, and by Theorem 1.2, $\alpha_{ij} = 0$ for $i \neq j$. ∎

If M is indecomposable of finite length, then $\mathrm{End}(M)$ is completely primary, by Fitting's lemma, and the Jacobson radical of $\mathrm{End}(M)$ may be described as follows:

Corollary 4.1.10. *Let $\mathcal{A}$ be an admissible module category admitting sums of submodules, and let M be a distributive $\mathcal{A}$-module of finite length, with a direct decomposition (3) into indecomposable submodules M_i. Then each $\mathrm{End}(M_i)$ in (4) is completely primary and the Jacobson radical $J(End(M))$ is nilpotent, given by*

$$J(\mathrm{End}(M)) \cong \prod_{i=1}^{n} J(\mathrm{End}(M_i)). \tag{5}$$

Proof. It is clear from (4) that $J(\mathrm{End}(M))$ has the form (5); since each $J(\mathrm{End}(M_i))$ is nilpotent, the same holds for $J(\mathrm{End}(M))$. ∎

Exercises 4.1

1. Show that a finitely generated $\mathbb{Z}$-module is distributive if and only if it is cyclic.
2. (Jensen [63]) Let M be a distributive R-module. If $a, b \in M$, show that either $Ra \subseteq Rb$ or $Rxb \subseteq Ra$ for some $x \in R$ such that $1 - x$ is a non-unit. Deduce that if R is a local ring, then Lat(M) is totally ordered.
3. Let M be a distributive module. Show that if a perspectivity between two chains of modules exchanges two simple factors P, Q, then P and Q cannot be isomorphic.
4. Let M be a distributive module of finite length. By the Jordan–Hölder theorem, any two composition series of M are projective, in the sense that we can pass from one to the other by a series of perspectivities. Use Exercise 3 to show that any such projectivity preserves the order of the factors of a given isomorphism type.
5*. (G. M. Bergman) Given a module M, if any homomorphism of a submodule N of M into M maps N into itself, is M necessarily distributive? [*Hint*: For a counter-example try a three-dimensional vector space over a field k, regarded as left R-module, where R consists of all matrices $\begin{pmatrix} a & 0 & 0 \\ 0 & a & 0 \\ b & c & a \end{pmatrix}$, where $a, b, c \in k$.]
6. Let M be a distributive module of finite length. If $A_1, \ldots, A_n$ are the simple factors (with their multiplicities) occurring in a composition series of M, show that every endomorphism of M maps each term of a composition series into itself. Deduce that there is a homomorphism End(M) $\to \prod_{i=1}^n$ End(A_i), whose kernel is the radical of End(M) and that this radical is nilpotent.
7*. (Vamos [78]) (i) If $M = Ra + Rb$ is distributive, show that $Ra \cap Rb + R(a+b)$ contains Ra and Rb; deduce that $M = Ra \cap Rb + R(a+b)$. (ii) Show that any finitely generated Artinian distributive module is cyclic. [*Hint*: Let A be minimal finitely generated non-cyclic and B minimal of the form $Ra \cap Rb$, where $A = Ra + Rb$. Use (i) to show that $A = R(a+b) + B_0$, where B_0 is a cyclic submodule of B, and that $B_0 \cap R(a+b) \subseteq B$; deduce that $B_0 = B \subseteq R(a+b)$ and obtain a contradiction.]
8. (Stephenson [74]) Let $A = k[x, y]$, $K = k(x, y)$, denote by R, S the subrings of K obtained from A by localizing at all elements prime to x, y respectively, and let $\alpha : R \to S$ be the k-isomorphism interchanging x and y. Show that $T = \left\{ \begin{pmatrix} a & u \\ 0 & a^\alpha \end{pmatrix} \mid a \in R, u \in K \right\}$ is a ring whose left as well as right ideals are totally ordered, hence T is distributive as left or right T-module, but neither left nor right invariant (T is *left invariant* if $cT \subseteq Tc$ for all $c \in T$).
9. (Camillo [75]) Let k be a commutative field and α an endomorphism such that $1 < [k; k^\alpha] < \infty$. Show that the ring $R = \left\{ \begin{pmatrix} x & y \\ 0 & x^\alpha \end{pmatrix} \mid x, y \in k \right\}$ has only $R, J(R)$ and 0 as left ideals, hence Lat$_R(R)$ is distributive, and R is left but not right invariant.

4.2 Distributive factor lattices

From Theorem 2.3.7 we see that a 2-fir may be defined as an integral domain R such that for any $c \in R^\times$ the set $L(cR, R)$ is a sublattice of the lattice of all right ideals of R. In the commutative case this condition simply states that the principal ideals form a sublattice of the lattice of all ideals. In that case we can go over to the field of fractions and consider the principal fractional ideals; by what has been said they form a modular lattice with respect to the ordering by inclusion. Clearly they also form a group under multiplication, and the group operations respect the ordering. Thus we have a lattice-ordered group; such a group is always distributive, as a lattice (Birkhoff [67], p. 294). This suggests that we single out 2-firs with the corresponding property, and we make the following

Definition. An integral domain R is said to have a *distributive factor lattice*, DFL for short, if for each $c \in R^\times$, the set $L(cR, R)$ is a distributive sublattice of the lattice of all right ideals of R.

From the definition (and Theorem 2.3.7) it is clear that a ring with distributive factor lattice is a 2-fir. Moreover, since $L(cR, R)$ is anti-isomorphic to $L(Rc, R)$, by the factorial duality (Theorem 3.2.2), the notion defined here is left–right symmetric. We shall reformulate this condition below in a number of ways, in terms of 1-torsion modules. We begin with a technical lemma.

Lemma 4.2.1. *Let R be a 2-fir and $a \in R^\times$. Then*

(i) *a is right comaximal with an element similar to a if and only if a is right comaximal with ba, for some $b \in R$,*

(ii) *there exist two elements similar to a and right comaximal if and only if there is an equation $xa_1 y + u a_2 v = 1$, where a_1, a_2 are similar to a.*

Proof. (i) The similarity of a and a' can be expressed by the existence of two mutually inverse matrices

$$A = \begin{pmatrix} a & b \\ c & d \end{pmatrix}, \quad A^{-1} = \begin{pmatrix} d' & -b' \\ -c' & a' \end{pmatrix}.$$

Let us replace these matrices by $TA, A^{-1}T^{-1}$, where for some $t \in R$ to be determined later,

$$T = \begin{pmatrix} 1 & 0 \\ t & 1 \end{pmatrix};$$

then the equation of comaximality obtained by equating the (1, 1)-entry in $TA.A^{-1}T^{-1}$ is

$$a(d' + b't) - b(c' + a't) = 1. \tag{1}$$

By hypothesis a and a' are right comaximal, say

$$au - a'v = 1; \qquad (2)$$

hence $a'vc' + c' = auc'$, and taking $t = vc'$ in (1), we find

$$a(d' + b'vc') - bauc' = 1, \qquad (3)$$

which shows a and ba to be right comaximal. Conversely, if a and ba are right comaximal, say $ad' - bac' = 1$, then by Lemma 3.3.3, taking this relation in the form $a.d' - b.ac' = 1$ we have a comaximal relation $ac'.a = a_1v$, where $a_1 \sim a$. Hence a is right comaximal with a_1 and (i) follows.

(ii) For any $p, q \in R$ the comaximal relation $(1 + pq)p = p(1 + qp)$ shows that $1 + pq \sim 1 + qp$. Now if $xa_1y + ua_2v = 1$, then $1 - xa_1y = ua_2v$, hence $1 - a_1yx \sim ua_2v$, say $1 - a_1yx = u'a_3v'$ (see Exercise 3.1.8), where $a_3 \sim a_2 \sim a$; thus $1 - u'a_3v' = a_1yx$, and repeating the process, we have $1 - a_3v'u' = a_4z$, where $a_4 \sim a$. Thus the elements a_3, a_4 similar to a are right comaximal. Conversely, if a_1, a_2 are similar to a and right comaximal, then $a_1R + a_2R = R$. ∎

We now list some conditions for the distributivity of $L(cR, R)$; it turns out to be more convenient to list the negations, i.e. conditions for non-distributivity:

Proposition 4.2.2. *Let R be a 2-fir and $c \in R^\times$. Then the following conditions are equivalent:*

(a) the lattice $L(cR, R)$ is not distributive,
(b) $c = amb$, where $R/mR \cong M^2 \neq 0$ for some 1-torsion module M,
(c) $c = amb$, where $m = a_1a_2 = a_3a_4$ is a comaximal relation in which $a_1, \ldots, a_4$ are all similar non-units,

$(a^o)-(c^o)$ the left–right analogues of (a)–(c).

Proof. By Proposition 1.1, $L(cR, R)$ fails to be distributive precisely if it is not square-free, i.e. (b), or contains a five-element sublattice of length 2, i.e. (c). Now the symmetry holds by the form of (c). ∎

Next we have the following conditions for global distributivity:

Theorem 4.2.3. *Let R be a 2-fir. Then the condition*

(a) for any similar a, a' and any $b \in R$, $baR \cap a'R \neq 0 \Rightarrow ba \in a'R$, implies
(b) for any similar $a, a' \in R$, $aR \cap a'R \neq 0 \Rightarrow aR = a'R$,
 and this implies the following conditions, which are equivalent among themselves and to their left–right analogues:
(c) for each $c \in R^\times$, the lattice $L(cR, R)$ is distributive,

(d) there is no comaximal relation $a_1 a_2 = a_3 a_4$, where $a_1, \ldots, a_4$ are all similar non-units,
(e) if a is a non-unit, $(R/aR)^2$ is not cyclic,
(f) if a, a' are similar non-units, there is no equation $xay + ua'v = 1$,
(g) for any non-unit a, there is no equation $ax + uav = 1$.

Moreover, when R satisfies right ACC_1, then all the conditions (a)–(g) are equivalent.

We see that DFL is the condition (c); a 2-fir satisfying (a) is sometimes said to possess the *right strong* DFL property. *Left strong DFL* is defined similarly and *strong DFL* means 'left and right strong *DFL*'.

Proof. (a) $\Rightarrow$ (b) $\Rightarrow$ (c). Taking $b = 1$ in (a) we find that when $aR \cap a'R \neq 0$, then $aR \subseteq a'R$ and by symmetry, $aR = a'R$, i.e. (b). Now (b) asserts that isomorphic 1-torsion quotient modules of a 1-torsion module have the same kernel; if $L(cR, R)$ is not distributive, it has a five-element sublattice of length 2, hence we can find isomorphic quotient modules with distinct kernels, i.e. non-(c) $\Rightarrow$ non-(b) and so (b) $\Rightarrow$ (c).

To prove (c) $\Rightarrow$ (a) when right ACC_1 holds for R, consider the assertion (a), say $bac = a'u$ for $c, u \in R^\times$. We observe that $R/a'R$ is isomorphic to R/aR, which is a quotient of R/acR; since $bac \in a'R$, left multiplication by b defines a homomorphism $R/acR \to R/a'R$, while the conclusion of (a) states that left multiplication by b defines a homomorphism $R/aR \to R/a'R$. If we combine left multiplication by b with the isomorphism $R/a'R \cong R/aR$, we see that (a) asserts that any homomorphism $R/baR \to R/aR$ is induced by an endomorphism of R/aR. So when (c) holds, then (a) follows by Corollary 1.6 (ii). Now (c) $\Leftrightarrow$ (d) $\Leftrightarrow$ (e) by Proposition 2.2 and (d) $\Leftrightarrow$ (f) $\Leftrightarrow$ (g) by Lemma 2.1. ∎

As an application this yields another criterion for distributivity:

Corollary 4.2.4. *If R is a 2-fir such that the polynomial ring $R[t]$ is 2-Hermite, then R has a distributive factor lattice.*

Proof. To prove the result we shall verify (g) of Theorem 2.3. Thus let $ax - yaz = 1$ in R. Then we have in $R[t]$,

$$a(tz + x) - (t + y)az = 1. \tag{4}$$

Since $R[t]$ is 2-Hermite, the row $(a, t + y)$ and the column $(tz + x, -az)^T$ can be completed to mutually inverse matrices:

$$A = \begin{pmatrix} a & t+y \\ p & g \end{pmatrix}, \quad A^{-1} = \begin{pmatrix} tz+x & -f \\ -az & q \end{pmatrix}, \quad f, g, p, q \in R[t].$$

4.2 Distributive factor lattices

By equating entries in $AA^{-1} = A^{-1}A = I$ we find that $aza = qp$, hence p, q are of degree 0 in t. Further we have

$$af = (t+y)q, \quad p(tz+x) = gaz. \tag{5}$$

By comparing degrees we find that f, g are linear in t; thus we may put $f = f_1 t + f_0$, $g = g_1 t + g_0$, where $f_i, g_i \in R$. By equating coefficients of t in (5) we find $af_1 = q$, $p = g_1 a$, so

$$\begin{pmatrix} a & t+y \\ p & g \end{pmatrix} = \begin{pmatrix} 1 & t+y \\ g_1 & g \end{pmatrix} \begin{pmatrix} a & 0 \\ 0 & 1 \end{pmatrix}.$$

This shows a to be a unit and (g) follows. ∎

We observe that every commutative 2-fir, and in particular, every commutative principal ideal domain, satisfies (a) of Theorem 2.3 and hence also (b)–(g). The implication (c) ⇒ (a) does not hold without chain conditions (see Exercise 15). Some consequences of (a) are listed in the exercises; here we shall consider 2-firs for which (b) fails to hold and show that whenever an element has two similar atomic right factors that are not left associated, then it has (generally) infinitely many. More precisely, we give a lower bound to the number of similarity classes.

Let R be a 2-fir, $c \in R^\times$ and let $c = ab$. Then $M = R/Rc$ has the submodule $N = Rb/Rab \cong R/Ra$. If $c = a'b'$ is another factorization of c, in which $a' \sim a$, then $N' = Rb'/Ra'b' \cong R/Ra'$ is a submodule of M that is isomorphic to N, and $N = N'$ if and only if $Rb = Rb'$, or equivalently, $aR = a'R$, i.e. a and a' are right associated. Suppose now that $N \neq N'$, say $N' \not\subseteq N$. Since N, N' are isomorphic cyclic modules, they have generators u, u' respectively, which correspond under this isomorphism. Given any $\alpha \in \text{End}(N)$, $u\alpha + u'$ generates a submodule N_α of M that is a homomorphic image of N, for the map $f_\alpha : xu \mapsto x(u\alpha + u')$ clearly defines a homomorphism. We note that $f_\alpha \neq 0$, for if $f_\alpha = 0$, then $u\alpha = -u'$, hence $N' = N\alpha \subseteq N$, which is not the case.

Assume now that $N \cap N' = 0$; this means that $Rb \cap Rb' = Rc$ or equivalently, $aR + a'R = R$, i.e. a and a' are right comaximal. In that case the submodules N_α defined by the different endomorphisms of N are distinct, for if $N_\alpha = N_\beta$, then $u\alpha + u' = x(u\beta + u')$, i.e. $(1-x)u' \in N \cap N' = 0$, so $u' = xu'$, hence $u = xu$ and $u\alpha = xu\beta = u\beta$, therefore $\alpha = \beta$. So there are at least as many different submodules N_α as there are elements in $\text{End}(N)$, and each corresponds to a left factor of c similar to a right factor of a. The result may be stated as

Theorem 4.2.5. *Let R be a 2-fir and suppose that $c \in R^\times$ has two left factors a, a' that are similar and right comaximal. Then the number of non-right-*

associated left factors of c that are similar to a right factor of a is at least $|E(Ra)|$. In particular, if a is an atom in R, then the number of non-right-associated left factors of c similar to a is $0, 1$ or at least $|E(Ra)|$. ∎

If R is an algebra over an infinite commutative field k, any eigenring is an algebra over k and hence infinite. Thus we obtain

Corollary 4.2.6. *Let R be a 2-fir that is also a k-algebra, where k is an infinite commutative field. Then the number of non-right-associated left factors of any $c \in R^{\times}$ similar to a given atom is $0, 1$ or infinite.* ∎

Finally we specialize to the case of principal ideal domains. In this case there is a simple criterion for the distributivity of factor lattices. We first describe similar right invariant elements (recall that $c \in R$ is *right invariant* if c is regular and $Rc \subseteq cR$).

Lemma 4.2.7. *In any ring, two right invariant elements that are similar are right associated.*

Proof. Let c be right invariant in R. Then $Rc \subseteq cR$, so cR annihilates the module R/cR. In fact, cR is the precise annihilator, for if a annihilates cR, then $Ra \subseteq cR$ and so $a \in cR$.

Now let $c \sim c'$ and assume that both c, c' are right invariant. Then $R/cR \cong R/c'R$ and equating annihilators, we find that $cR = c'R$, hence $c = c'u, c' = cv$, so $c = c'u = cvu$. Since c is right regular, $vu = 1$; similarly $uv = 1$, therefore c' is right associated to c. ∎

Theorem 4.2.8. *A principal right ideal domain has a distributive factor lattice if and only if every non-zero element is right invariant.*

Proof. Let R be a right PID whose non-zero elements are all right invariant. Then any two similar elements are right associated, by Lemma 2.7, and so by Theorem 2.3, R has a distributive factor lattice.

Conversely, if R has DFL, then its lattice of right ideals is distributive and any homomorphism $\alpha : R \to R/cR$ is such that $\ker \alpha \supseteq cR$, by Corollary 1.6; taking α to be the map $x \mapsto ax + cR$, we therefore have $ac \in cR$. This holds for all $a \in R$, hence $Rc \subseteq cR$ and so c is right invariant. ∎

Exercises 4.2

1. Let R be an atomic 2-fir in which no two similar non-units are right comaximal. Show that R has DFL.
2. Show that a right invariant ring always satisfies condition (a) of Theorem 2.3.

3. Show that an atomic 2-fir in which any two atoms are either right associated or comaximally transposable is right invariant and hence has DFL.
4. If c is a right invariant element of finite length in a 2-fir, show that any element similar to c is right associated to c and hence is again right invariant. Can the condition on the length of c be omitted?
5. Show that a skew polynomial ring over a field has DFL if and only if it is commutative.
6. Let R be a 2-fir with DFL. If S is a subring that is a 2-fir containing all the units of R, show that S has DFL.
7. Let R be an atomic 2-fir with DFL. Show that if all atoms in R are similar, then R is a local ring. Show that the 'atomic' hypothesis can be omitted if we assume instead that 'any two non-units have a similar non-unit factor' (in the terminology of Section 6.4, no two non-units are totally coprime).
8. Let R be a 2-fir with right strong DFL. Show that for any $a, b \in R^\times$, $I(abR) \subseteq I(aR)$.
9*. Show that a 2-fir with right strong DFL and with a right large non-unit has a right invariant element and hence is a right Ore domain. Show that the conclusion does not hold for every 2-fir with DFL.
10°. In a 2-fir, if $xay + ua'v = 1$, where $a \sim a'$, is a necessarily right comaximal with an element similar to a?
11°. When the conditions of Proposition 2.2 hold, does it follow that $xc - cy = 1$ for some $x, y \in R$?
12*. Show that every 2-fir with right strong DFL has elements that are not fully reducible.
13°. Determine the structure of non-commutative invariant principal ideal domains.
14. (Bergman [67]) Let K/k be a Galois extension with group $G = \text{Gal}(K/k)$ and let M be any K-bimodule satisfying $\lambda x = x\lambda (x \in M, \lambda \in k)$. Show that $M = \oplus M_\sigma (\sigma \in G)$, where $M_\sigma = \{x \in M | x\alpha = \alpha^\sigma x\}$. Let R be a k-algebra containing K, but not in its centre. Define R_σ as above and show that for any $x \in R_\sigma$ $x(x - \alpha) = (x - \alpha^\sigma)x$ is a comaximal relation. Deduce that if R is a 2-fir with right strong DFL, then for any $x \in R_\sigma$ ($\sigma \neq 1$), $1 - x$ is a unit. (*Hint*: For the last part replace x by $x\alpha$.)
15. (Brungs and Törner [81]) Let k be a commutative field and let σ be the k-automorphism of the rational function field $k(x, y)$ interchanging x and y. Denoting by (x) the effect of localizing at the set of all elements prime to x, show that the ring $R = k[x, y]_{(x)} + tk(x, y)[[t; \sigma]]$ is a principal ideal domain and a chain ring, and hence has DFL. Verify that $y \in U(R)$ and $ytR \subset tR$, so R satisfies (c), but neither (a) nor (b) of Theorem 2.3.

4.3 Conditions for a distributive factor lattice

In order to find general conditions for an atomic 2-fir R to have a distributive factor lattice we recall from Theorem 2.3 (b) that this is equivalent to requiring that if an element c has two factorizations

$$c = ab = a_1 b_1, \tag{1}$$

with similar right factors b, b_1, then these factors must be left associated. Moreover, if an equation (1) holds in a 2-fir R with b, b_1 similar atoms that are not left associated, then in case R is an algebra over an infinite field, there must be infinitely many right factors similar to b, but pairwise not left associated, by Corollary 2.6. Thus in a sense we have a one-parameter family of factorizations of c; this idea may be formalized by adjoining an indeterminate t to k and showing that (1) leads to a factorization of c in $R \otimes k(t)$ that does not arise from a factorization in R (so that c is not inert in $R \otimes k(t)$). Suppose however that R is 1-inert in $R \otimes k(t)$; this holds, for example, if R is a free algebra. Then this situation cannot occur and we conclude that similar right factors are necessarily left associated. We shall see how this property can be used to provide us with many examples of rings with a distributive factor lattice. Throughout this section, k is a commutative field, and all tensor products are understood to be over k.

Definition   Let R be a k-algebra; any property X of R is said to be *absolute* (over k) if it holds for $R \otimes E$, where E is any algebraic field extension of k; X is said to be *persistent* (over k) if it holds for $R \otimes k(t)$, where t is a central indeterminate.

For example, the free k-algebra $k\langle X \rangle$ is a persistent and absolute fir over k; if E is a commutative field extension of k, then the tensor ring $E_k\langle X \rangle$ (defined in Section 2.4) for $X \neq \emptyset$ is an absolute fir if and only if E is a regular field extension of k (recall that E/k is a regular extension if $E \otimes_k F$ is an integral domain for all commutative field extensions F/k), and a persistent fir if and only if E is algebraic over k.

Proposition 4.3.1.   *Let R be a k-algebra, which is an absolute integral domain over k. Then R is 1-inert in $R \otimes k(t)$. If further, R is a persistent 2-Hermite ring over k, then $R[t]$ is 2-Hermite.*

Proof.   Consider an equation

$$c = ab, \qquad (2)$$

where $c \in R, a, b \in R[t]$. Since R is an integral domain, it follows that $a, b \in R$, so that R is 1-inert in $R[t]$. Now the first assertion will follow if we show that $R[t]$ is 1-inert in $R \otimes k(t)$, so suppose that $c \in R[t]$ has a factorization (2), where now $a, b \in R \otimes k(t)$. If we multiply (2) by the denominators of a and b, we obtain an equation of the form

$$fc = a'b', \quad f \in k[t]^\times, a', b' \in R[t]. \qquad (3)$$

4.3 Conditions for a distributive factor lattice

Here we may assume that a', b' have no non-unit factors in $k[t]$, since any such factor could be cancelled against f. If f has positive degree, let α be a zero of f in k_a, the algebraic closure of k and form $R \otimes k_a$. By hypothesis this is an integral domain and we have $a'(\alpha)b'(\alpha) = 0$, hence either a' or b' is divisible by $t - \alpha$, which is a contradiction. Therefore f has degree zero; so it lies in k and may be absorbed in a'.

Suppose now that R is also a persistent 2-Hermite ring. Then $R \otimes k(t)$ is 2-Hermite, hence weakly 2-finite, and so is $R[t]$, as a subring. It remains to show that every 1×2 matrix over $R[t]$ with a right inverse is completable. Let $a \in R[t]^2$ be right invertible over $R[t]$, say $ab = 1$ for some $b \in {}^2R[t]$. Then we can complete a, b to a pair of mutually inverse matrices over $R \otimes k(t)$, and clearing denominators, we obtain a relation

$$\begin{pmatrix} a \\ a' \end{pmatrix} (b \ b') = \begin{pmatrix} 1 & 0 \\ 0 & f \end{pmatrix}, \quad f \in k[t]^\times, a' \in R[t]^2, b' \in {}^2R[t].$$

If f has degree 0, this shows a to be completable over $R[t]$; otherwise we have by weak 2-finiteness,

$$(b \ b'f^{-1}) \begin{pmatrix} a \\ a' \end{pmatrix} = I,$$

and so $fba + b'a' = f.I$. Therefore, if α is any zero of f in k_a, we have $b'(\alpha)a'(\alpha) = 0$. Here the column b' is multiplied by the row a', so we have a set of four equations; since $R \otimes k_a$ is an integral domain, either $a'(\alpha) = 0$ or $b'(\alpha) = 0$. If $b'(\alpha) = 0$, then b' is divisible by an irreducible factor of f and by cancelling it we can decrease the degree of f. If $a'(\alpha) = 0$, then a' is divisible by an irreducible factor f_1 say, of f, and we can replace a', b' by $a' f_1^{-1}, b' f_1$. Now we can again reduce the degree of f and so complete the proof by induction on the degree of f. ∎

Before we come to the main result, we need another lemma. For any element $a \in R[t]$ we shall indicate the value of a obtained by specializing t to 0 by a subscript: $a_0 = a(0)$.

Lemma 4.3.2. *Let R be a 2-fir that is a k-algebra and let*

$$ab = cd \tag{4}$$

be an equation holding in $R[t]$ such that b, d are left comaximal in $R \otimes k(t), b_0, d_0$ are not both 0 and a_0, c_0 are right comaximal in R. Then b_0, d_0 are left comaximal in R.

Proof. Since a_0, c_0 are right comaximal and $a_0 b_0 = c_0 d_0$, this product is the least common left multiple of b_0 and d_0 in R (it was only to get this conclusion that we had to assume R to be a 2-fir).

Now b, d are left comaximal in $R \otimes k(t)$; therefore we have an equation

$$pb - qd = f, \quad p, q \in R[t], f \in k[t]^\times. \tag{5}$$

If $f_0 = 0$, then $p_0 b_0 = q_0 d_0$ and this is a left multiple of $a_0 b_0 = c_0 d_0$. Hence by subtracting a suitable left multiple of (4) from (5) we can modify p, q so that both become divisible by t. We can then divide p, q, f all by t and obtain an equation of the same form as (5), but with f of lower degree. In this way we reduce the degree of f, and continuing this process we eventually reach a case where $f_0 \neq 0$. Taking the constant term of f as 1, we then find that $p_0 b_0 - q_0 d_0 = 1$, which shows b_0, d_0 to be left comaximal. ∎

We now come to the main result of this section, giving conditions for DFL.

Theorem 4.3.3. *Let R be a k-algebra that is an absolute integral domain and a persistent 2-fir. Then R has the strong DFL property.*

Proof. We have to verify condition (a) of Theorem 2.3. Suppose that R satisfies the hypotheses and that $ba'R \cap aR \neq 0$, where $a \sim a'$; we have a relation

$$ba'c = ad \neq 0,$$

which is right coprime and so c, d are left comaximal. Further, let $au' = ua'$ be a comaximal relation between a and a'. Then in $R[t]$ we have

$$a(dt + u'c) = (bt + u)a'c. \tag{6}$$

Now any common right factor of $dt + u'c$ and $a'c$ in $R \otimes k(t)$ can by 1-inertia be taken in R (Proposition 3.1). Hence it must right-divide $d, u'c$ and $a'c$, i.e. generate a left ideal containing $Rd + Ru'c + Ra'c = Rd + Rc = R$. Thus (6) is in fact left comaximal in $R \otimes k(t)$; further, the constant terms of the right factors are not both 0 and those of the left factors are right comaximal in R. Hence by Lemma 3.2, $u'c$ and $a'c$ are left comaximal, i.e. c is a unit and $ba' \in aR$, so condition (a) of Theorem 2.3 holds, as well as its left–right dual, by symmetry, ensuring strong DFL. ∎

The free k-algebra $k\langle X \rangle$ is clearly an absolute and persistent fir over k; hence we obtain

Corollary 4.3.4. *The free k-algebra $k\langle X \rangle$ has the strong DFL property.* ∎

For non-commutative fields this no longer holds, for even when $|X| = 1$, we have in the principal ideal domain $K[x]$ over the skew field K, for $ab \neq ba$ an equation

$$(x - a)bc - b(x - a)c = 1, \quad \text{where } c = -(ab - ba)^{-1},$$

so by Theorem 2.3 (g), $K[x]$ does not even have DFL. By combining the above result with Theorem 2.8, we obtain

Corollary 4.3.5. *Let R be a k-algebra that is a principal ideal domain. If R is an absolute integral domain and a persistent 2-fir over k, then it is an invariant ring.* ∎

Of course the converse does not hold, since a commutative PID need not be absolute or persistent, as we shall see in Section 4.6. But there is a partial converse to Theorem 3.3.

Proposition 4.3.6. *Let R be a 2-fir with DFL that is an algebra over an infinite field k. Then R is 1-inert in $R \otimes k(t)$.*

Proof. Let $c \in R$ have a factorization over $R \otimes k(t)$; by clearing denominators we can write this in the form

$$fc = ab, \quad a, b \in R[t], f \in k[t]. \tag{7}$$

We shall denote the degrees of a, b, f by p, q, r respectively; further we may assume without loss of generality that a has no non-unit left factor in R and b has no non-unit right factor in R.

Let $\mathfrak{a}$ be the right ideal generated by the coefficients of $a = a(t)$, considered as a polynomial in t. Then for any $p + 1$ distinct elements $\alpha_0, \ldots, \alpha_p \in k$ we can express all the coefficients of $a(t)$ as k-linear combinations of $a(\alpha_0), \ldots, a(\alpha_p)$; hence these $p + 1$ elements generate $\mathfrak{a}$ as a right ideal. If we choose $p + 1$ such values so as to avoid the zeros of f in k (at most r in number), then the $a(\alpha_i)$ are left factors of c; hence $\sum a(\alpha_i)R = \mathfrak{a}$ is principal and by the assumption on a it must be R itself. Similarly, the intersection of $q + 1$ of the right ideals $a(\alpha)R$ is a right ideal containing cR, say eR; we assert that $eR = R$.

If $eR \neq R$, then the principal right ideals between eR and R form a distributive lattice $\neq 0$, and hence (Appendix A) there is a homomorphism of this lattice onto the two-element lattice $2 = [0, 1]$ such that $eR \mapsto 0, R \mapsto 1$. Suppose that $a(\alpha)R$ maps to 0 for more than $p + r$ values of α. Then we can find $p + 1$ values of α, avoiding the r zeros of f, for which $a(\alpha)R$ maps to 0, but as we have seen, their sum maps to 1. This contradiction shows that $a(\alpha)R$ maps to 0 for at most $p + r$ values of α. Similarly, if $a(\alpha)R$ maps to 1 for more than $q + r$ values of α, then by choosing $q + 1$ values avoiding the zeros of f, we find that their intersection eR maps to 0, again a contradiction. But every $a(\alpha)R$ maps to 0 or 1, so there cannot be more than $p + q + 2r$; this contradicts the fact that k is infinite, and it follows that $eR = R$. Thus e is a unit and it follows that $a(\alpha)$ is a unit for any α not a zero of f. Now if f is divisible by t, then either a or b must be divisible by t and we can cancel a power t; so we may assume that $f(0) = 1$. It follows that $a(0)$ is a unit; a similar argument shows that $b(0)$ is a unit, and so $c = a(0)b(0)$ is a unit. Hence a and b are units in $R \otimes k(t)$, as is f, and $c = af^{-1}.b = af^{-1}b.b^{-1}b = c.1$. ∎

We note, however, that in the situation of this proposition R need not be a persistent 2-fir, e.g. take $R = E[x]$, where E is a commutative field extension of k that is not algebraic over k.

Exercises 4.3

1. Let R be a k-algebra that is an absolute integral domain, a persistent 2-fir and a right Ore domain. Show that any two similar elements are right associated.
2. Let R be a k-algebra that is an absolute integral domain and a persistent 2-fir. Show that if two elements of R are similar in $R \otimes k(t)$, then they are similar in R.
3. Let R be a k-algebra that is an integral domain. Show that an atom in R remains an atom in $R \otimes k(t)$.
4*. Let R be an n-fir over an algebraically closed field k. Show that every $n \times n$ matrix over $R[t]$ is inert in $R \otimes k(t)$. [*Hint*: The transforming matrix can be taken to lie in the subgroup generated by $GL_n(R)$ and the diagonal matrices over $k(t)$.] Deduce that R is totally n-inert in $R[t]$.
5*. Let R be a k-algebra that is an absolute n-fir and a persistent $2n$-fir. Given a full matrix c in R_n such that any factor of c in $R[t]_n$ can be reduced to one in R_n on multiplying by an element of $GL_n(R[t])$, show that the corresponding module R_n/cR_n is distributive. Taking $n = 2$ and p, q dissimilar atoms in R, show that $c = p \oplus q$ satisfies the above hypotheses, but not $p \oplus p$. Verify that when $c = p \oplus p$, R_2/cR_2 is not distributive.
6. In the complex-skew polynomial ring $R = \mathbb{C}[x; -]$ show that $x^2 - 1$ has the factorizations $x^2 - 1 = (x - u)(x + \bar{u})$, where u ranges over the unit circle. Obtain the corresponding factorizations over $\mathbb{C}(t)[x; -]$ with $u = (t + i)(t - i)^{-1}$ and show that these cannot be pulled back to R.
7. Let R be a k-algebra, M an R-module, P a submodule of M with inclusion map $i: P \to M$ and $f: P \to M$ a homomorphism. Assume that the image of the homomorphism $i + tf$ of $P \otimes k(t)$ into $M \otimes k(t)$ is of the form $N \otimes k(t)$ for some $N \subseteq M$. Show that $Pf \subseteq P$ and hence obtain another proof of Theorem 3.3.
8°. Investigate Proposition 3.6 when the field k is finite.
9. (Beauregard [80]) Let E be a commutative field with an automorphism α and F a subfield mapped into a proper subfield of itself by α. Write $R =$

$E[[x; \alpha]]$, $P = \{f \in R | f(0) \in F\}$. Verify that R is a local ring and a principal ideal domain; hence show that P is an Ore domain that is right invariant but not left invariant.
10. In a free algebra, if $ab = bc^2$, show that a is a square (see Exercise 2.7.15).
11. Show that in a 2-fir with DFL, for any atom c and any $n \geq 1$, c^n is indecomposable.
12°. Investigate atomic 2-firs in which the length of indecomposable elements is bounded.
13. Let R be an atomic 2-fir with DFL. Show that if R is a matrix local ring, then its capacity must be 1, i.e. it is a scalar local ring.
14. Show that if R is a persistent 2-Hermite ring over k, then any equation $ax + uav = 1$ implies that a is a unit (see Theorem 2.3 (g)).
15*. In a ring R with DFL, show that if $a, b, c \in R$ are such that any two of aR, bR, cR have a non-zero intersection, then $aR \cap bR \cap cR \neq 0$. Show also that $aR \cap bR \neq 0$, $aR \cap cR \neq 0$ is not enough. (*Hint*: Take $b = au + 1$, $c = av + 1$ for suitable u, v.) Find a generalization to n terms.

4.4 Finite distributive lattices

In an atomic 2-fir with distributive factor lattice, the left factors of a given non-zero element form a distributive lattice of finite length. For a closer study of this lattice we shall in this section describe its structure in terms of partially ordered sets.

Let us denote by $\mathcal{P}os$ the category of finite partially ordered sets, with isotone (i.e. order-preserving) maps as morphisms. By $\mathcal{DL}_f(0, 1)$, or $\mathcal{DL}$ for short, we shall denote the category of all distributive lattices of finite length, with lattice homomorphisms, i.e. maps preserving meet, join, 0, 1, as morphisms. In each of these categories $2 = \{0, 1\}$ denotes the chain of length 1. We begin by defining two functors between these categories.

Take $P \in \mathcal{P}os$ and consider $P^* = \mathrm{Hom}_{\mathcal{P}os}(P, 2)$; this set P^* may be regarded as a finite distributive lattice, namely a sublattice of 2^P. The elements of P^* – isotone maps from P to 2 – may also be described by the subsets of P mapped to 1. They are precisely the *upper segments* of P, i.e. subsets X with the property

$$a \in X, b \geq a \quad \text{implies} \quad b \in X.$$

An upper segment of the form $\mu_a = \{x \in P \mid x \geq a\}$ is said to be *principal*. We observe that the partially ordered set of all principal upper segments of P is isomorphic to P, as member of $\mathcal{P}os$.

Clearly each $\alpha \in P^*$ is completely determined by the upper segment mapped to 1, and every upper segment defines such a map. Hence P^* may be identified with the set of all upper segments of P.

Next take $L \in \mathcal{DL}$ and write $L^* = \mathrm{Hom}_{\mathcal{DL}}(L, 2)$. Here we can regard L^* as a partially ordered set, writing $f \leq g$ if and only if $xf \leq xg$ for all $x \in L$. The elements of L^* may be characterized by the subsets they map to 1. Given $f \in L^*$, let $a \in L$ be the meet of all x satisfying $xf = 1$. Then $af = 1$ and a is the unique minimal element with this property. Clearly $a > 0$ and if $a = x \vee y$, then $1 = xf \vee yf$, hence $xf = 1$ or $yf = 1$, i.e. $x \geq a$ or $y \geq a$; thus a is *join-irreducible*:

$$a \neq 0 \text{ and } a = x \vee y \quad \text{implies} \quad x = a \text{ or } y = a.$$

Conversely, any join-irreducible element a gives rise to an $f \in L^*$, defined by the rule: $xf = 1$ if and only if $x \geq a$. We may thus identify L^* with the partially ordered set of the join-irreducible elements of L.

Our object will be to show that * establishes a duality between $\mathcal{DL}$ and $\mathcal{Pos}$. Before coming to the main result we need a lemma on objects in $\mathcal{DL}$, but this is just as easily proved in a more general setting:

Lemma 4.4.1. *In any lattice with minimum condition, each element is the join of the join-irreducible elements below it.*

Proof. Let L be a lattice with minimum condition and suppose it does not satisfy the conclusion. Then we can find $a \in L$ such that a is not the join of join-irreducible elements. If we take a to be minimal with this property, then a cannot be join-irreducible and $a \neq 0$, because 0 is the join of the empty family. Hence $a = b \vee c$, where $b < a, c < a$. By the minimality of a, both b and c are joins of join-irreducible elements, hence so is $a = b \vee c$. ∎

Theorem 4.4.2. *The categories $\mathcal{Pos}$ and $\mathcal{DL}$ are dual to each other, via the contravariant functors*

$P \mapsto P^* = $ *lattice of upper segments of P,*
$L \mapsto L^* = $ *set of join-irreducible elements of L.*

Moreover, if P and L correspond, then $\cap, \cup$ on upper segments of P correspond to $\vee, \wedge$ in L and the length of L equals $|P| + 1$.

Proof. It is clear that two contravariant functors are defined between these categories by means of $\mathrm{Hom}(-, 2)$; it only remains to show that $P^{**} \cong P$, $L^{**} \cong L$.

Let $P \in \mathcal{Pos}$; then P^* consists of all upper segments of P. If $\alpha \in P^*$ and $a_1, \ldots, a_r$ are the different minimal elements of the upper segment α, then $x \in \alpha$ if and only if $x \geq a_1$ or ... or $x \geq a_r$. Hence $\alpha = \mu_{a_1} \vee \ldots \vee \mu_{a_r}$, where μ_c is the principal upper segment defined by c. This shows α to be join-irreducible if and only if it is principal, and so P^{**}, the set of join-irreducible upper segments of

4.4 Finite distributive lattices

P, is just the set of principal upper segments of P, which we saw is isomorphic to P.

Next, given $L \in \mathcal{DL}$, consider L^*, the partially ordered set of its join-irreducible elements. This set determines L, by Lemma 4.1: each $a \in L$ can be represented by the set of all join-irreducible elements $\geq a$, and the set of join-irreducible elements occurring are just the upper segments, thus $L^{**} \cong L$.

Let L and P correspond under this duality and suppose that P has n elements. Then we can form a chain in L by picking a maximal element $a_1 \in P$, next a maximal element a_2 in $P\setminus\{a_1\}$, etc. It follows that every chain in L has $n+1$ elements. ∎

It is clear that every P^* is finite, as subset of 2^P, hence we obtain

Corollary 4.4.3. *Any distributive lattice of finite length n is finite, with at most 2^{n-1} elements.* ∎

The interest in the duality described in Theorem 4.2 resides in the fact that for any $L \in \mathcal{DL}$ and $P \in \mathcal{Pos}$ that correspond under the duality, P is usually much simpler than L. For example, a Boolean algebra corresponds to a totally unordered set (see Exercise 2), the lattice on the left corresponds to the set on the right,

and the free distributive lattice on three generators, a lattice of length 6 with 18 elements (see Grätzer [78], p. 38) corresponds to the three-peak crown:

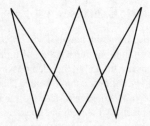

There is another way of describing the correspondence of Theorem 4.2 that is of importance for us in what follows.

In a distributive lattice of finite length, every link is projective to exactly one link with join-irreducible upper end-point. For let $L = P^*$; any link in P^* has the form $[S, S \cup \{x\}]$, for an upper segment S and an element x such that all elements $> x$ lie in S. Any link perspective to it has the form $[S', S' \cup \{x\}]$ with the same x and the set of all these links clearly has a lowest member, namely $[T, T \cup \{x\}]$, where T is the set of all elements $> x$; clearly $T \cup \{x\}$ is join-irreducible.

It follows that the join-irreducible elements correspond to the projectivity classes of links. If our lattice has length n, then there are just n join-irreducible elements and hence n projectivity classes of links. Since each class has a representative in each chain (by the Jordan–Hölder theorem), there is exactly one representative from each class in each chain. Thus we have proved most of

Proposition 4.4.4. *Let L be a distributive lattice of length n. Then the links of L fall into exactly n projectivity classes and each chain in L contains exactly one link from each class. Moreover, each projectivity class contains a unique lowest link, and its upper end-point is join-irreducible. If we partially order the set of projectivity classes of links using the partial ordering of the corresponding join-irreducible elements, the resulting partially ordered set is order-isomorphic to L^*. Given projectivity classes of links α and β, we have $\alpha < \beta$ if and only if the link from α occurs below the link from β in every chain.*

Proof. Only the last assertion remains to be proved. Let $a, b \in P$ and let α, β be the projectivity classes of links corresponding to a, b respectively. If a, b are incomparable in P, then we can form chains in P^* in which the representative of α lies lower than that of β, and chains in which it lies higher, depending on whether we choose a before b or b before a in forming the chain. But if $a < b$ in P, then we must choose a before b and hence in every chain the representative of α lies lower than that of β. ∎

Exercises 4.4

1. Show that a modular lattice has finite length if and only if every chain in it is finite. Give examples of (i) an infinite modular lattice of finite length and (ii) a general lattice, all of whose chains are finite, but their lengths are unbounded.
2. Show that a finite distributive lattice is complemented if and only if the corresponding partially ordered set is totally unordered. (*Hint*: In a Boolean algebra, the join-irreducible elements are precisely the minimal non-zero elements.)
3. Show that a finite distributive lattice is indecomposable (as a direct product) if and only if the corresponding partially ordered set is connected (i.e. any two elements can be joined by a chain of comparable elements).
4. Examine how the correspondence of Theorem 4.2 is affected if we take (i) lower instead of upper segments, (ii) meet- instead of join-irreducible elements and (iii) make both these changes.

5. Let L be a modular lattice of finite length in which any two projective intervals are perspective. Show that L must be distributive. Determine all such lattices, using Theorem 4.2.
6. Show (using Proposition 4.4) that in a 2-fir with DFL an element of length n has at most $n!$ essentially distinct factorizations (see Section 3.4).

4.5 More on the factor lattice

Let R be an atomic 2-fir with a distributive factor lattice. For each $c \in R^\times$, $L_c = L(cR, R)$ is a distributive lattice of finite length. We shall write P_c for the corresponding partially ordered set L_c^*. Each complete factorization of c:

$$c = p_1 p_2 \ldots p_n \tag{1}$$

corresponds to a chain in L_c; if $c = q_1 \ldots q_n$ is another atomic factorization of c, then p_i is said to be *equivalent* to q_j if we can pass from the link $[p_1 \ldots p_i R, p_1 \ldots p_{i-1} R]$ to the link $[q_1 \ldots q_j R, q_1 \ldots q_{j-1} R]$ by a series of comaximal transpositions. Here p_i refers not to an element of R but to its occurrence in the factorization (1) of c; thus in xyx (in a free algebra) the two factors x are inequivalent. Since comaximal transpositions correspond to perspectivities in L_c, the equivalence classes of (occurrences of) atomic factors correspond to projectivity classes of links in L_c, and thus to elements of P_c. We shall refer to an equivalence class of atomic factors of c as an abstract *atomic factor* of c; thus P_c may be thought of as the set of abstract atomic factors of c. By Proposition 4.4, each abstract atomic factor has just one representation in each complete factorization of c, and of two abstract atomic factors, p and q say, p precedes q, $p < q$, if p occurs on the left of q in every complete factorization of c. On the other hand, when p, q are incomparable, then they may be comaximally transposed whenever they occur next to each other in a complete factorization. Every complete factorization is completely determined by the order in which the abstract factors occur; in particular, an element with n factors cannot have more than $n!$ complete factorizations.

Any expression of c as a product $c = ab$ corresponds to a decomposition of P_c into a lower and a complementary upper segment, which may be identified with P_a, P_b respectively. Given two factorizations

$$c = ab' = ba', \tag{2}$$

we see that the highest common left factor and the least common right multiple of a, b will correspond to the intersection and union respectively, of P_a, P_b. In particular, a comaximal relation (2) for c corresponds to an expression of P_c as a union of two disjoint lower segments, which means a partition of its diagram into two disconnected components. We note also that in this case $L_c \cong L_a \times L_b$, in agreement with Proposition 3.4.6.

Recalling that projective links in L_c correspond to similar factors, we see that with every element of $P_c = L_c^*$ we can associate a similarity class of atoms in R. Abstract factors corresponding to the same similarity class must be comparable in P_c because similar atoms cannot be comaximally transposed in R (by Theorem 2.3), hence every similarity class forms a chain within P_c. It follows that the only automorphism of P_c preserving similarity classes is the identity; hence the same holds for L_c. Thus for any similar elements c and c' the isomorphism between L_c and $L_{c'}$ (and between P_c and $P_{c'}$) is unique. We state this conclusion as

Proposition 4.5.1. *Let R be an atomic 2-fir with a distributive factor lattice. Then for any two similar elements c, c' of R there is a unique isomorphism $L_c \to L_{c'}$ between the factor lattices preserving the similarity classes associated with the links in $L_c, L_{c'}$.* ∎

If $f : R \to R'$ is a homomorphism of atomic 2-firs with DFL, then for any $c \in R$ such that $c \notin \ker f$, we get a lattice homomorphism from L_c to L_{cf}: the obvious map preserves HCLFs because it preserves comaximality and it preserves LCRMs by the factorial duality. By Theorem 4.2, a homomorphism in the opposite direction is induced from P_{cf} to P_c.

In a commutative principal ideal domain, or indeed in any commutative UFD, two atoms are coprimely transposable if and only if they are non-associated. It follows that the only possible structures for the sets P_c in this case are disjoint unions of finite chains. For example, in $\mathbb{Z}$, $720 = 2^4.3^2.5$, hence P_{720} consists of three chains, of lengths 4, 2 and 1. By contrast, in the non-commutative case all possible structures for P_c can be realized:

Theorem 4.5.2. *Let $A_n = k < x_1, \ldots, x_n >$ be the free k-algebra of rank n. Given any partially ordered set P of n elements, there exists $c \in A_n$ with $P_c \cong P$.*

Proof. The case $n = 0$ is clear, so assume that $n > 0$ and let α be any element of P. By induction on the number of atomic factors in α we may assume that we have found $c' \in A_{n-1}$ such that $P_{c'} \cong P' = P \setminus \{\alpha\}$.

Write $P' = U \cup V \cup W$, where U is the set of elements $< \alpha$ in P, V the set of elements incomparable with α and W the set of elements $> \alpha$. Clearly $U, U \cup V$ and $U \cup V \cup W$ are lower segments of P'; they correspond to left factors a, ab and $abd = c'$ of c'. We put $c = a(bx_n + 1)bd$ and claim that $P_c \cong P$.

In the first place $bx_n + 1$ is an atom, since it is linear in x_n and in any factorization the term independent of x_n must divide 1. We now identify P_c with P by letting the factors of a, b and d correspond as in the identification of $P_{c'}$ and

4.5 More on the factor lattice

P', and letting the abstract atomic factor to which $bx_n + 1$ belongs correspond to α. It remains to check that the partial ordering of P_c agrees with that of P.

Since ab is a left factor of c and bd is a right factor, the orderings on the corresponding subsets of P_c will agree with those on P_{ab} and P_{bd}, as required. The new abstract factor is incomparable with the factors of b, because of the comaximal relation $(bx_n + 1)b = b(x_n b + 1)$. Now the partial ordering will be completely determined if we show that the factor corresponding to $bx_n + 1$ lies above all factors of a and below all factors of d. By symmetry it suffices to prove the first statement.

Suppose the contrary; then for some non-unit right factor e of a, we would have a comaximal relation $e(bx_n + 1) = fe'$. Now we obtain a ring homomorphism $A_n \to A_{n-1}$ by putting $x_n = 0$; this will preserve comaximal relations and hence it maps f to an element similar to $b.0 + 1 = 1$, i.e. a unit. However f itself is similar to the non-unit $bx_n + 1$ in A_n and so must involve x_n. But then the product $e(bx_n + 1) = fe'$ will involve monomial terms in which x_n occurs, but is not the last factor (since e' is a non-unit). This is a contradiction, and it shows that every factor of a lies below $bx_n + 1$. ■

In fact all these partially ordered sets may already be realized in A_2. We shall prove this by showing that A_n (for any $n \geq 1$) can be embedded as a 1-inert subring in A_2:

Theorem 4.5.3. *The free algebra of countable rank can be embedded 1-inertly in the free algebra of rank 2.*

Proof. Let $F = k\langle Z \rangle$, where $Z = \{z_0, z_1, \ldots\}$. Since F is free on Z, the mapping $\delta : z_i \mapsto z_{i+1} (i = 0, 1, \ldots)$ extends to a unique derivation of F. We form the skew polynomial ring $H = F[x; 1, \delta]$; from the commutation rule

$$ax = xa + a^\delta \quad (a \in F) \tag{3}$$

and the definition of δ we find that

$$z_{i+1} = z_i{}^\delta = z_i x - x z_i = [z_i, x]. \tag{4}$$

We claim that H is the free k-algebra on x, z_0. For it is clearly generated by x and z_0 over k; to show that x, z_0 are free generators, we establish a homomorphism $\beta : H \to G = k\langle x, y \rangle$ such that $x \mapsto x, z_0 \mapsto y$. We begin by defining $\beta : Z \to G$ by

$$\beta : z_n \mapsto [\ldots [y, x], \ldots, x] \text{ with } n \text{ factors } x.$$

Since F is free on Z, this map extends to a homomorphism $\beta' : F \to G$. Moreover, we have $z_n^\delta \beta' = z_{n+1} \beta' = [\ldots [y, x], \ldots, x] = [z_n \beta', x]$ (where there are $n+1$ factors x). Hence if δ_x is the inner derivation defined by x in G, we have $\delta\beta' = \beta'\delta_x$. Now the defining relations of H in terms of F are just the equations (3), which may be written $\delta = \delta_x$. Hence on H we have $\delta_x \beta' = \beta'\delta_x$; thus the defining relations of H are preserved by β' and so β' may be extended to a homomorphism β of H into G. Since G is free on x, y, this shows H to be free on x, z_0 as claimed. Moreover, we see that β is surjective, hence it is an isomorphism between H and G.

It remains to show that the inclusion $F \to H$ is 1-inert. Given $c \in F$, suppose that in H we have $c = ab, a, b \in H$. We can write $a = x^r a_0 + \ldots, b = x^s b_0 + \ldots$, where $a_0, b_0 \in F$ and dots denote terms of lower degree in x. Then $c = ab = x^{r+s} a_0 b_0 + \ldots$; by uniqueness, $r + s = 0$, hence $r = s = 0$ and $a, b \in F$ and it follows that c is inert in H. ∎

In Chapter 7, when we come to construct a universal field of fractions for every free algebra, we shall find that the above embedding of F in H extends to an embedding of their universal fields of fractions (Theorem 7.5.19).

Exercises 4.5

1. Let R be an atomic 2-fir with DFL. Show that any factorization $c = a_1 \ldots a_n$ corresponds to an isotone map of P_c into the ordered set of $n + 1$ elements.
2. A subset X of a partially ordered set is called *convex* if $x, y \in X, x < a < y$ implies $a \in X$. If R is an atomic 2-fir with DFL and for $c \in R$, P_c denotes the set of similarity classes of atomic factors as before, show that a subset X of P_c is convex if and only if c has a factorization $c = aub$, where $P_u = X$; if $c = a'u'b'$ is another factorization with $P_{u'} = X$, show that u' is obtainable from u by a series of comaximal transpositions.
3. Let R be an atomic 2-fir with DFL. Given two factorizations $c = ab = a'b'$ of an element c of R, if each similarity class contributes at least as many terms to a factorization of a as it does to a factorization of a', show that $a \in a'R$.
4. Find elements in $k\langle x, y\rangle$ with factor lattices corresponding to the following partially ordered sets:

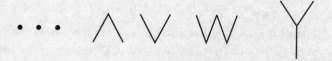

5*. Find an element in the free algebra $k\langle x, y, z\rangle$ whose factor lattice is the free distributive lattice (with 0 and 1) on three free generators.
6*. (M. L. Roberts) Let * be the anti-automorphism of $R = k\langle x, y\rangle$ interchanging x and y, and for matrices it is combined with transposition. Show that if $P = \begin{pmatrix} y & xy^2 \\ x & yx^2 \end{pmatrix}$, then P^*P is an atom of C, where C is the subalgebra of R generated by $u + u^*(u \in$

R). Deduce that P^*P is not inert in R. Show also that the embedding of C in R does not preserve full matrices (i.e. is not honest). $\Big[$*Hint*: Use Q^*Q, where
$$Q = \begin{pmatrix} y & xy^2 & xyxy^2 \\ x & yx^2 & yxyx^2 \end{pmatrix}.\Big]$$

4.6 Eigenrings

We have seen in Section 0.6 that the eigenring of a matrix is just the endomorphism ring of the module defined by the matrix. In the case of a fir we shall find that the eigenrings are as a rule small, only exceptionally are they large. So one can expect two kinds of results on eigenrings: (a) eigenrings are nearly always small and (b) under suitable conditions eigenrings are large. Thirdly, in the case of a ring with DFL we have (c) the consequences of distributivity for the eigenring. Our results will mainly be under headings (a) and (c). Thus when A is a regular matrix over R, we shall show that (i) when R is a persistent semifir over k, then the eigenring of A is algebraic over k (Theorem 6.9) and (ii) when R is a free k-algebra, then the eigenring of A is finite-dimensional over k. This will be proved in a special case (column matrices) in Proposition 6.12, the general case being reserved for Chapter 5.

As before, all our rings will be algebras over a commutative field k; hence the eigenring of an element or a matrix will also be a k-algebra. If the eigenring of a matrix A is k itself, we shall say that A has a *scalar* eigenring.

From Corollary 1.10 we obtain the following result on the structure of eigenrings:

Proposition 4.6.1. *Let R be an atomic 2-fir with distributive factor lattice. Then the eigenring of any $c \in R^\times$ is a direct product of a finite number of completely primary rings.* ∎

Beyond this rather general fact it seems difficult to apply the results on distributive modules to the study of arbitrary rings with DFL. In what follows we shall therefore put further restrictions on the ring; most of these will be satisfied by free algebras.

Let R be a k-algebra; we recall that R is said to be *algebraic* over k, if every element of R satisfies a polynomial equation over k. If the matrix ring R_n is algebraic over k, for all $n \geq 1$, then R will be called *matrix algebraic* over k. Of course for commutative R this is the same as 'algebraic'. When R is a skew field, this condition can be expressed in terms of the rational function field $R(t)$, as we shall see in Proposition 6.7, but no examples are known of algebraic fields that are not matrix algebraic. However, for eigenrings we shall be able to reduce one condition to the other by means of the following lemma:

Lemma 4.6.2. *Let R be a ring and $A \in {}^mR^n$ any matrix with left eigenring E; then the diagonal sum $C = A \oplus \cdots \oplus A$ (r terms) has left eigenring E_r.*

Proof. E is the endomorphism ring of the left R-module $M = R^n/(R^m)A$, while $C = A \oplus \cdots \oplus A$ defines the left R-module $R^{rn}/(R^{rm})C \cong M^r$; this clearly has endomorphism ring E_r. ∎

A matrix A over a k-algebra R is called *algebraic* over k if it satisfies a polynomial equation over k; A is *transcendental* over k if for every non-zero polynomial f over k, the matrix $f(A)$ is regular; this term will mainly be used when R itself is a field. In general a matrix is neither transcendental nor algebraic, but we always have a decomposition; to derive it we need a result known as the see-saw lemma:

Lemma 4.6.3. *Let R, S be k-algebras and M an (R, S)-bimodule. Given $a \in R, b \in S$, assume that there is a polynomial f over k such that $f(a)$ is a unit, while $f(b) = 0$. Then for any $m \in M$, the equation*

$$ax - xb = m \tag{1}$$

has a unique solution $x \in M$.

Proof. In $End_k(M)$ write $\lambda_a : x \mapsto ax$, $\rho_b : x \mapsto xb$; then (1) may be written

$$x(\lambda_a - \rho_b) = m. \tag{2}$$

By hypothesis $f(\lambda_a)$ is a unit, $f(\rho_b) = 0$ and λ_a, ρ_b commute. Hence if we define $\phi(s, t)$ in commuting variables s, t by

$$\phi(s, t) = \frac{f(s) - f(t)}{s - t},$$

then $\phi(\lambda_a, \rho_b)(\lambda_a - \rho_b) = (\lambda_a - \rho_b)\phi(\lambda_a, \rho_b) = f(\lambda_a) - f(\rho_b) = f(\lambda_a)$, and this is a unit, hence (2) and with it (1) has a unique solution in M. ∎

The result may be restated in matrix form. Consider the matrix ring $\begin{pmatrix} R & M \\ 0 & S \end{pmatrix}$; given a matrix in this ring:

$$\begin{pmatrix} a & u \\ 0 & b \end{pmatrix}, \tag{3}$$

where $f(a)$ is a unit and $f(b) = 0$, we can find a conjugate of (3) in diagonal form. For if we transform (3) by $I + xe_{12}$, we obtain

$$\begin{pmatrix} a & u + ax - xb \\ 0 & b \end{pmatrix}$$

and by the lemma, the north-east block is 0 for a suitable choice of x.

Proposition 4.6.4. *Let K be a skew field that is a k-algebra. Then every square matrix over K is conjugate to a diagonal sum of an algebraic and a transcendental matrix.*

Proof. Let $A \in K_n$ and consider $V = {}^nK$ as a $(K, k[t])$-bimodule, in which the action of t for a given K-basis $u_1, \ldots, u_n$ is given by

$$u_i t = \Sigma a_{ij} u_j, \quad A = (a_{ij}). \tag{4}$$

Since $K \otimes k[t] = K[t]$ is a principal ideal domain, V has a unique submodule V_0 of torsion elements with torsion-free quotient, which, being finitely generated, is free. Let V_1 be a complement of V_0, so that

$$V = V_0 \oplus V_1. \tag{5}$$

Using a basis adapted to the decomposition (5), we find that A takes the form

$$\begin{pmatrix} A_0 & A' \\ 0 & A_1 \end{pmatrix},$$

where A_0 is algebraic and A_1 transcendental. By Lemma 6.3 and the remark following it we can reduce A' to 0 and so obtain the desired conclusion. ∎

Let us consider the following special case of (4):

$$u_i t = u_{i+1} \quad (i = 1, \ldots, n-1), \quad u_n = u_1 a_1 + \cdots + u_n a_n, \quad \text{where } a_i \in K.$$

The corresponding matrix has the form

$$A = \begin{pmatrix} 0 & 1 & 0 & 0 & \ldots & 0 \\ 0 & 0 & 1 & 0 & \ldots & 0 \\ & \ldots & & & \ldots & \\ 0 & 0 & \ldots & \ldots & 0 & 1 \\ a_1 & a_2 & \ldots & \ldots & a_{n-1} & a_n \end{pmatrix}. \tag{6}$$

This matrix is called the *companion matrix* of the polynomial

$$f = t^n - a_1 - ta_2 - \ldots - t^{n-1}a_n. \tag{7}$$

As is easily verified, $tI - A$ is stably associated to f, and it follows that $f(A) = 0$ and A has the invariant factors $1, 1, \ldots, 1, f$.

To find a criterion for algebraicity we shall use the normal form obtained in Theorem 1.4.7. As we have seen there, if $A \in K_n$, then $tI - A$ is associated to $\text{diag}(\lambda_1, \ldots, \lambda_n)$, where $\lambda_1, \ldots, \lambda_n$ are the invariant factors of A and $\lambda_{i-1} || \lambda_i$. When k is the precise centre of K, this leads to a criterion for A to be algebraic or transcendental. An element c of a PID is said to be *bounded*, if it divides an invariant element. If c has no bounded factor apart from units, it is said to be *totally unbounded*. Given a monic polynomial $f = t^r + a_1 t^{r-1} \ldots + a_r$ in

$K[t]$, suppose that f is invariant; then $bf = fb'$ for any $b' \in K$, a comparison of degrees shows that $b' \in K$ and by comparing coefficients of t^r we see that $b' = b$. Since b was arbitrary in K, it follows that the coefficients of f lie in the centre of K. Since every non-zero polynomial is associated to a monic polynomial, it follows that every invariant polynomial is associated to a monic polynomial over the centre of K.

Theorem 4.6.5. *Let K be a field with centre k and let $A \in K_n$ have invariant factors $\lambda_1, \ldots \lambda_n$. Then*

(i) A is algebraic over k if and only if λ_n is bounded, equivalently, λ_n divides a polynomial with coefficients in k,

(ii) A is transcendental over k if and only if λ_n is totally unbounded, and then $\lambda_1 = \ldots = \lambda_{n-1} = 1$.

Proof. (i) Since $K[t]$ is a principal ideal domain, we can apply Theorem 1.4.7 to obtain the relation

$$P(tI - A)Q^{-1} = \operatorname{diag}(\lambda_1, \ldots, \lambda_n) \quad (\lambda_{i-1} || \lambda_i), \tag{8}$$

where P, Q are invertible matrices. Since $tI - A$ is regular, the diagonal elements on the right of (8) are all non-zero. Suppose that λ_n is bounded, say $\lambda_n \mid f$, where f is an invariant polynomial. By the above remarks f may be taken to be a monic polynomial with coefficients in k. Since each λ_i divides λ_n and hence divides f, there is a diagonal matrix D such that $DP(tI - A)Q^{-1} = fI$, hence $DP(tI - A) = fQ$. Dividing fI by $tI - A$, we find

$$fI = H(tI - A) + L, \tag{9}$$

where H, L are polynomials in A with coefficients in $k[t]$, k respectively. Here we can put $t = A$ and so obtain $L = f(A)$. Thus we have $Q^{-1}DP(tI - A) = fI = H(tI - A) + f(A)$, whence

$$f(A) = (Q^{-1}DP - H)(tI - A).$$

If $Q^{-1}DP \neq H$, the right-hand side will contain terms in t, whereas the left-hand side does not; hence $Q^{-1}DP - H = 0$ and we conclude that $f(A) = 0$. Conversely, if f is a polynomial over k satisfied by A, then $fI = H(tI - A)$ for some polynomial H and it follows that f is a bound for λ_n.

(ii) Suppose that λ_n is not totally unbounded, say it has a bounded factor p, with bound p^*. Then the module V defined by A has a non-zero element annihilated by p and so also by p^*. Now p^* is invariant, hence with coefficients in k and $p^*(A)$ is singular, so A cannot be transcendental. Conversely, if A is not transcendental, then V has a non-zero element annihilated by an invariant polynomial, so some invariant factor λ_i has a non-unit factor that is bounded, hence λ_n then has a bounded factor. Thus A is transcendental if and only if λ_n is totally unbounded, and then no other λ_i can be a non-unit, because this would give rise to a non-unit invariant element dividing λ_n. ∎

Corollary 4.6.6. *A skew field K that is a k-algebra is matrix algebraic over k if and only if its centre is algebraic over k and every non-zero polynomial over K is bounded.*

Proof. Let C be the centre of K. If K is matrix algebraic over k, then K is algebraic over k, and so is C. Further, any polynomial $f \neq 0$ is the sole invariant factor $\neq 1$ of its companion matrix B_f. By hypothesis B_f is algebraic, hence by Theorem 6.5, f is bounded. Conversely, when these conditions hold, take any square matrix over K; all its invariant factors are bounded, so A is algebraic over C and hence also over k. ∎

We note another condition for a field K to be matrix algebraic, in terms of the rational function field $K(t)$, which is sometimes useful.

Proposition 4.6.7. *Let K be a skew field that is a k-algebra. Then K is matrix algebraic over k if and only if $K(t) = K \otimes_k k(t)$.*

Proof. Clearly we have the inclusion

$$K \otimes_k k(t) \subseteq K(t); \qquad (10)$$

we have to find when equality holds. Suppose first that k is the exact centre of K. By Corollary 6.6, K is matrix algebraic if and only if every non-zero polynomial is bounded. But this just means that every element of $K(t)$ can be written as a fraction with denominator in $k[t]^\times$, which is precisely the condition for equality in (10).

Now let C be the centre of K. Then $C \supseteq k$ and by what we have shown, K is matrix algebraic over C if and only if $K \otimes_C C(t) = K(t)$. Assume that K is matrix algebraic over k. Then K is matrix algebraic over C and C is algebraic over k; hence every polynomial over K divides a polynomial over C, which in turn divides a polynomial over k, and so equality holds in (10). Conversely, when equality holds in (10), then $K \otimes C(t) = K(t)$. Hence K is matrix algebraic over C, and (10) also shows that every polynomial over C divides a polynomial over k; applying this result to $t - \alpha (\alpha \in C)$, we see that C is algebraic over k, and it follows that K is matrix algebraic over k, as we had to show. ∎

We can now return to the study of eigenrings. For our first main result we need a form of the inertia lemma:

Lemma 4.6.8. *Let S be a ring containing a central regular element t such that $\cap t^n S = 0$, and such that $R = S/tS$ is a semifir. If the induced map $GL_n(S) \to GL_n(R)$ is surjective for all $n \geq 1$, then every matrix over S is inert in $S[t^{-1}]$.*

Here $S[t^{-1}]$ denotes of course the ring obtained from S by adjoining an inverse of t; since t is central and regular, S is embedded in $S[t^{-1}]$.

Proof. Write $x \mapsto x_0$ for the natural homomorphism $S \to S/tS = R$; this amounts to putting $t = 0$. We take $A \in {}^m S^n$ and suppose that over $S[t^{-1}]$:

$$A = PQ, \quad \text{where } P \text{ is } m \times r \text{ and } Q \text{ is } r \times n. \tag{11}$$

If P or Q is 0, there is nothing to prove, so we may assume $P, Q \neq 0$. Now every non-zero matrix B over $S[t^{-1}]$ can be written in the form $t^v B'$, where $v \in \mathbb{Z}$, B' has entries in S and $B'_0 \neq 0$. Hence on changing notation, we can rewrite (11) as

$$A = t^{-v} PQ, \quad \text{where } P \in {}^m S^r, Q \in {}^r S^n, P_0, Q_0 \neq 0. \tag{12}$$

We have to show that v can be taken to be ≤ 0, so suppose that $v > 0$; we shall show how to replace v by $v - 1$ in (12). If $v > 0$, then $P_0 Q_0 = 0$; since R is a semifir, we can find a matrix $U_1 \in GL_r(R)$ trivializing this relation, and by hypothesis we can lift U_1 to $U \in GL_r(S)$. Hence on replacing P, Q by $PU, U^{-1}Q$ we find that for some $s \geq 1$ all the columns in P_0 after the first s are 0, while the first s rows of Q_0 are 0. We now multiply P on the right by $V = tI_s \oplus I_{r-s}$ and Q on the left by V^{-1}; then P becomes divisible by t, while Q still has entries in S. In this way we can, by cancelling t, replace v by $v - 1$ in (12) and after v steps we obtain the same equation with $v = 0$; this shows A to be inert in $S[t^{-1}]$. ∎

We note that the condition '$GL_n(S) \to GL_n(R)$ is surjective' is satisfied under any of the following assumptions:

(i) $R = S/tS$ is a retract of S, i.e. there is a homomorphism $R \to S$ such that the composition with the natural homomorphism $R \to S \to S/tS$ is the identity,
(ii) R is a GE_n-ring: $GL_n(R)$ is generated by elementary and diagonal matrices, or
(iii) t lies in the Jacobson radical of S.

The verification is straightforward and may be left to the reader.

To illustrate the lemma, let R be any semifir, $R[[t]]$ the ring of formal power series in a central indeterminate t and $R((t))$ the ring of formal Laurent series. Then by the lemma, every matrix over $R[[t]]$ is inert in $R((t))$. Secondly, let R

be a semifir that is a k-algebra, and denote by $R[t]_{(t)}$ the localization of $R[t]$ at the set of all polynomials in t over k with non-zero constant term. Then every matrix over $R[t]_{(t)}$ is inert in $R \otimes k(t)$.

We can now prove our first main result on algebraic eigenrings:

Theorem 4.6.9. *Let R be a k-algebra that is a persistent semifir over k. Then the eigenring of any regular matrix over R is matrix algebraic over k.*

Proof. If $A \in {}^m R^n$ is regular, then $B = A \oplus \cdots \oplus A$ (r terms) is again regular, and to show that A has a matrix algebraic eigenring we must show, by Lemma 6.2, that the eigenring of B is algebraic, for all $r \geq 1$. So it is enough to show that the eigenring of A is algebraic and then apply the result to B.

Take $P \in R_n, P' \in R_m$ such that $AP = P'A$. Then in $R \otimes k(t)$ we have

$$A(I - tP) = (I - tP')A. \tag{13}$$

Let us show that A and $I - tP'$ are left coprime. If Q is a square common left factor, we have

$$(A, I - tP') = Q(S, T) \quad \text{over } R \otimes k(t). \tag{14}$$

By Lemma 6.8 and the remark following it we obtain such a factorization over $R[t]_{(t)}$, and by moving any denominators from (S, T) to Q we may assume that S has entries in $R[t]$. If we now put $t = 0$ in (14), we obtain $Q_0 T_0 = I$. Since R is weakly finite, Q_0 is invertible over R, hence Q is invertible over $R[[t]]$. Over this ring we can therefore rewrite the equation $A = QS$ as $S = Q^{-1}A$. But S has entries in $R[t]$ and A is regular over R; it follows that Q^{-1} involves only finitely many powers of t, and so has entries in $R[t] \subseteq R \otimes k(t)$. This shows that Q is invertible over $R \otimes k(t)$ and so A and $I - tP'$ are left coprime.

By symmetry A and $I - tP$ are right coprime; thus (13) is a coprime relation and hence comaximal (Corollary 3.1.4). Replacing t by $u = t^{-1}$, we obtain a relation

$$A(uI - P) = (uI - P')A,$$

still comaximal in $R \otimes k(t) = R \otimes k(u)$. Writing down a relation of left comaximality and clearing denominators in u, we obtain

$$CA + D(uI - P) = fI, \quad C \in {}^n R[u]^m, D \in R[u]_n, f \in k[u]^\times. \tag{15}$$

We now write all powers of u on the right of the coefficients and substitute P for u. This is permissible since the substitution $u \mapsto P$ respects right multiplication by matrix polynomials whose coefficients are matrices commuting with P. If $C = \Sigma C_i u^i$, then the first term in (15) is $\Sigma C_i A P^i = \Sigma C_i P'^i A$, while the second term vanishes. Thus (15) reduces to $GA = f(P)$ (where $G = \Sigma C_i P'^i$), which means that P satisfies an equation mod $R_m A$, and this holds for all P, P' satisfying (13), hence the eigenring $E(A)$ is algebraic over k. By the initial remark, $E(A)$ is also matrix algebraic, as we had to show. ∎

Corollary 4.6.10. *Let R be a persistent semifir over an algebraically closed field k. Then every matrix atom of R has a scalar eigenring.*

Proof. We know that the eigenring of a matrix atom is a field, by Schur's lemma (Proposition 3.2.4), and the only algebraic skew field extension of k is k itself. ∎

For a two-sided ideal $\mathfrak{a}$ the eigenring is just the residue-class ring modulo $\mathfrak{a}$; thus we obtain

Corollary 4.6.11. *Let R be a persistent semifir over k and $\mathfrak{a}$ a two-sided ideal of R, non-zero and finitely generated as left ideal. Then $R/\mathfrak{a}$ is matrix algebraic over k.*

Proof. Let $u_1, \ldots, u_r$ be a basis of $\mathfrak{a}$ as free left R-module; then the column $u = (u_1, \ldots, u_r)^T$ is regular, and by Theorem 6.9 its eigenring is matrix algebraic over k. ∎

Theorem 6.9 can be applied to free algebras but, as already mentioned, there is a stronger result in this case.

Proposition 4.6.12. *Let $R = k\langle X \rangle$ be the free k-algebra on a set X. If $\mathfrak{a}$ is a non-zero left ideal and $\mathfrak{b}$ is a finitely generated left ideal, then $\mathrm{Hom}_R(R/\mathfrak{a}, R/\mathfrak{b})$ is finite-dimensional over k.*

Proof. Let $H = \mathrm{Hom}_R(R/\mathfrak{a}, R/\mathfrak{b})$; as we have seen in Section 0.6, $H = I/\mathfrak{b}$, where $I = \{x \in R | \mathfrak{a}x \subseteq \mathfrak{b}\}$. We shall enlarge I by choosing a non-zero element c in $\mathfrak{a}$ and defining $I' = \{x \in R | cx \in \mathfrak{b}\}$ clearly it will be enough to show that $I'/\mathfrak{b}$ is finite-dimensional over k. Let $u_1, \ldots, u_r$ be a basis of $\mathfrak{b}$; then $y \in I'$ precisely if

$$cy = \sum y_i u_i \quad \text{for some } y_i \in R . \tag{16}$$

Clearly the u_i involve only finitely many of the free generators; we write $X = X' \cup X''$, where X' is the finite subset of generators occurring in the u's and X'' is its complement in X. We assign the degree 1 to each member of X' and let d be the maximum of the degrees of $u_1, \ldots, u_r$ in X'; further we assign the degree $d + 1$ to each member of X''. Then it is clear that the space F of elements of R of degree at most d is finite-dimensional. By (16) and the weak algorithm we have, for each $y \in I'$,

$$y = \Sigma f_i u_i + y', \quad \deg y' < d \, . \tag{17}$$

This shows that $I' \subseteq \mathfrak{b} + F$, and it follows that

$$H = \operatorname{Hom}_R(R/\mathfrak{a}, R/\mathfrak{b}) = I/\mathfrak{b} \subseteq I'/\mathfrak{b} \subseteq (\mathfrak{b} + F)/\mathfrak{b} \cong F/(F \cap \mathfrak{b}) \, .$$

Hence H is finite-dimensional, because this is true of F. ∎

In particular, taking $\mathfrak{b} = \mathfrak{a}$, we obtain

Corollary 4.6.13. *If R is a free k-algebra and $\mathfrak{a}$ is a finitely generated non-zero left ideal, then $\operatorname{End}_R(R/\mathfrak{a})$ is finite-dimensional over k.* ∎

This result shows in particular that the eigenring of any non-zero element in a free algebra is finite-dimensional; this will be proved for any regular matrices in Section 5.8. The same reasoning yields a converse to Corollary 2.6.4:

Corollary 4.6.14. *Let R be a free k-algebra. Then every finitely generated left ideal that is left large has finite codimension over k.*

Proof. Let $\mathfrak{a}$ be a left ideal in R satisfying the hypothesis, and let $u_1, \ldots, u_r$ be a basis of $\mathfrak{a}$. Given any $y \in R^\times$, we have $Ry \cap \mathfrak{a} \neq 0$, hence we again have an equation (16) for some $c \neq 0$. Now it follows as before that $R/\mathfrak{a}$ is finite-dimensional, as we had to show. ∎

In general the endomorphism ring of a distributive module need not be commutative; it may not even be invariant, as we saw in Exercise 1.8. However this is true for free algebras; in fact it holds under slightly wider hypotheses.

Proposition 4.6.15. *Let R be a k-algebra that is an atomic 2-fir with a distributive factor lattice. If each atom of R has a scalar eigenring, then the eigenring of every non-zero element of R is commutative.*

Proof. Let $M = R/Rc$, where $c \in R^\times$. We have to show that $E(Rc) \cong \operatorname{End}_R(M)$ is commutative; this holds by hypothesis when c is an atom, so we may use induction on the length of c. Every $\alpha \in E(Rc)$ maps each 1-torsion submodule of M into itself, so if M is the sum of its proper submodules, we can embed $E(Rc)$ into the direct product of the corresponding endomorphism rings, hence $E(Rc)$ is then commutative. The alternative is that the sum of all proper submodules of M is a unique maximal submodule M', say. By hypothesis, every endomorphism of M/M' is induced by multiplication by an element of k, hence every endomorphism of M is of the form $\lambda + \alpha$, where $\lambda \in k$ and $M\alpha \subseteq M'$. It is therefore enough to show that any two non-surjective endomorphisms of M commute. Let us take such endomorphisms α, β of M. By Proposition 1.8, one of $M\alpha, M\beta$ is contained in the other, say $M\alpha \subseteq M\beta \subseteq M'$.

Now α maps M into $M\beta \cong M/\ker\beta$, hence by Corollary 1.6 (ii), α is induced by an endomorphism α' of $M\beta$, i.e. $\alpha = \beta\alpha'$. Let β_1 be the restriction of β to $M\beta$; then we have

$$\beta\alpha = \beta\beta\alpha' = \beta\beta_1\alpha',$$

but on $M\beta$ all endomorphisms commute, by the induction hypothesis. Hence $\beta\beta_1\alpha' = \beta\alpha'\beta_1 = \alpha\beta_1 = \alpha\beta$, and so $\alpha\beta = \beta\alpha$, as claimed. ∎

Corollary 4.6.16. *Let R be a k-algebra that is an atomic 2-fir and remains one under arbitrary field extension. Then the eigenring of any non-zero element is commutative.*

Proof. Let k' be an algebraically closed field extension of cardinality greater than $\dim_k R$. Then all atoms in $R \otimes k'$ have scalar eigenrings (Corollary 3.2.6) and $R \otimes k'$ has DFL, by Theorem 3.3, hence we can apply the result just proved (and the change-of-rings formula, Proposition 0.6.2) to reach the conclusion. ∎

This corollary shows that in a free k-algebra all eigenrings of non-zero elements are commutative. Hence the result also holds for matrices that are stably associated to elements, but it does not extend to general matrices (see Exercise 8).

For examples of non-commutative eigenrings let us take the complex-skew polynomial ring $\mathbb{C}[x; -]$ and consider the elements $x^2 + 1$ and $x^2 - 1$. Both are invariant (even central), hence their eigenrings are quotients of the whole ring by the ideals they generate. The element $x^2 + 1$ is an atom, so the quotient is a field (Proposition 3.2.4); this is easily seen to be the field of quaternions. The element $x^2 - 1$ is a product of two atoms, neither of them invariant, hence the eigenring is a 2×2 matrix ring over $\mathbb{C}$. For the algebra $\mathbb{Z}\langle x, y\rangle$ the eigenrings can be very different, as Exercise 3.2.16 shows.

Exercises 4.6

1. In an atomic 2-fir characterize the elements whose eigenring has zero radical.
2. (Cohn [69a]) Show that in the free algebra $\mathbb{R}\langle x, y\rangle$ the element $a = xy^2x + xy + yx + x^2 + 1$ is an atom, but does not remain one under extension to $\mathbb{C}$. Deduce that the eigenring of a is $\mathbb{C}$. Find an element in the idealizer mapping to i.
3. (Roberts [82]) Let $k = \mathbb{F}_3(t)$, and in $k\langle x, y\rangle$ examine $c = x^2yxyxyx^2 + x^2yxyx + x^2y^2x^2 + xyxyx^2 + tx^3 + x^2y + xyx + yx^2 + 1$. Verify that c is an

atom but splits on adjoining a cube root of t. In $\mathbb{Q}\langle x, y\rangle$ consider the element obtained from c by replacing t by 2; show that it is an atom but splits on adjoining a cube root of 2.

4°. (Ikeda [69]) In a free algebra of finite rank, is every ideal that is maximal as left ideal finitely generated as left ideal? (By Corollaries 6.14 and 2.6.4 this is equivalent to the question: is every skew field that is finitely generated as k-algebra necessarily finite-dimensional over k?)

5*. (H. Bass) Let R be a commutative principal ideal domain containing a field k. Show that R is a persistent PID if and only if every prime ideal in $R[t]$ that is not minimal among the non-zero primes (i.e. of height >1) meets $k[t]^\times$. Deduce that the condition that $R/\mathfrak{p}$ be algebraic over k for any non-zero prime ideal $\mathfrak{p}$ is sufficient as well as necessary for $R \otimes k(t)$ to be a principal ideal domain.

6. Apply Exercise 5 to test whether R is a persistent PID in the following cases: (i) $R = k(x)[y]$, (ii) R is a PID whose residue-class fields are algebraic over k and (iii) $R = k[[t]]$.

7. (G. M. Bergman) In the ring of integral quaternions show that the eigenring of each atom is commutative, but that this need no longer hold for general (non-zero) elements.

8. Let p be an atom in a free algebra. Show that the eigenring of the matrix diag (p, p) is not commutative.

9. Let R be a persistent semifir over k and $\mathfrak{a} \neq 0$ an ideal containing an invariant element. Show that $R/\mathfrak{a}$ is matrix algebraic over k.

10°. Consider fields that are k-algebras, where k is a commutative field. Find a field that is algebraic but not matrix algebraic over k.

Notes and comments on Chapter 4

Most of the results in this chapter are due to Bergman and the author, and were first published in FR.1. In particular, Theorem 2.3 was obtained by the author in 1964 (unpublished) and he conjectured that it applied to free algebras. This conjecture was proved by Bergman in 1966. Much of the chapter is contained in Bergman [67], especially the later version, and Sections 4.4 and 4.5 follow this source (and other unpublished work of Bergman in the 1960s) rather closely. In 1966 Bergman proved that the tensor ring $E_k\langle X\rangle$ has a distributive factor lattice whenever E/k is a purely inseparable commutative field extension, where the inseparability cannot be omitted. This result was never published, but in 1981 the author, using results from Bergman [74a], found a shorter proof (see Cohn [89a]). The results of Section 4.4, of course, go back further, e.g. Birkhoff [67], though our presentation follows Bergman [67]. The latter also contains Theorem 5.2, while Theorem 5.3 is taken from another unpublished manuscript of Bergman (ca. 1968), with a new proof, taken from Cohn [90].

The material of Section 4.1 went through several versions and was improved as a result of discussions with Bergman and Stephenson; FR.2 followed Camillo [75] and Stephenson [74] in presenting properties of distributive modules. In this work the order has been changed, putting the criterion for distributivity (based on results of Roos [67]) at the beginning (Results 5.1 and 5.2). Further, 'semi-Artinian' has been replaced by the stronger notion 'meta-Artinian' (likewise for Noetherian) to correct an error and give a smoother presentation. Corollary 2.4 is taken from Cohn [82c] and Theorem 2.5 generalizes (and simplifies) a result of Noether and Schmeidler (1920).

A 2-fir such that two similar right commensurable elements are right associated was called 'uniform' in FR.1, but this term is used in other senses now and has therefore been discarded. A 2-fir R was called 'conservative' if R and $R \otimes k(t)$ are 2-firs and R is 1-inert in $R \otimes k(t)$; the place of this term has now been taken by the terms 'absolute' and 'persistent' as in the text. Thus Theorem 3.3, the main result of Section 4.3, originally stated that a conservative 2-fir is uniform.

From the results of Section 4.4 it follows that every distributive lattice (not necessarily finite) is isomorphic to a *ring of sets*, i.e. a lattice of subsets of a set under the operations $\cap, \cup$. For since a finite distributive lattice is finite, any distributive lattice can be represented as the inverse limit of the sets representing its finite subsets (the Birkhoff–Stone representation theorem, see Grätzer [78], p. 64).

In Theorem 5.3 the construction has been changed from FR.2, which leads to a shorter proof (and an application in Section 7.5, see Cohn [90]). The see-saw lemma and its consequences (Results 6.3–6.6) previously formed part of Section 8.5. Theorem 6.9 was first proved for elements in commutative PIDs by H. Bass (in a letter to the author in 1964). FR.1 contained a version for elements of a persistent 2-fir; the present version is taken from Cohn [85b]. Theorem 6.12 has a new shorter proof communicated by Bergman. Proposition 6.15 is also due to Bergman, dating back to ca. 1968 (unpublished), while Corollary 6.14 was proved by Rosenmann and Rosset [91].

5
Modules over firs and semifirs

Just as firs form a natural generalization of principal ideal domains, so there is a class of modules over firs that generalizes the finitely generated modules over principal ideal domains. They are the positive modules studied in Section 5.3; they admit a decomposition into indecomposables, with a Krull–Schmidt theorem (in fact this holds quite generally for finitely presented modules over firs), but it is no longer true that the indecomposables are cyclic. On the other hand, there is a dual class, the negative modules, and we shall see how the general finitely presented module is built up from free modules, positive and negative modules. A basic notion is that of a bound module; this and the duality, essentially the transpose, also used in the representation theory of algebras, are developed in Sections 5.1 and 5.2 in the more general context of hereditary rings. In the special case of free algebras, the endomorphism rings of finitely presented bound modules are shown to be finite-dimensional over the ground field. This result, first proved by J. Lewin, is obtained here by means of a normal form for matrices over a free algebra, due to M. L. Roberts, and his work is described in Section 5.8.

A second topic is the rank of matrices. Several notions of rank are defined, of which the most important, the inner rank, is studied more closely in Section 5.4. Over a semifir the inner rank obeys Sylvester's law of nullity. This leads to a natural generalization of semifirs: the Sylvester domains, first defined by W. Dicks and E. Sontag. They and some variants form the subject of Section 5.5 and 5.6.

In Section 5.7 we compare the different factorizations of a rectangular matrix over a semifir. Here the results are less complete, although in some ways parallel to the square case. There is an analysis of factorizations, which throws some light on the limitations to be expected.

The remainder deals with various chain conditions in Section 5.9 and the intersection theorem for ideals in firs in Section 5.10.

5.1 Bound and unbound modules

Let R be any ring and $\mathcal{T}, \mathcal{F}$ two classes of left R-modules such that

(i) $X \in \mathcal{T}$ if and only if $\mathrm{Hom}_R(X, Y) = 0$ for all $Y \in \mathcal{F}$,
(ii) $Y \in \mathcal{F}$ if and only if $\mathrm{Hom}_R(X, Y) = 0$ for all $X \in \mathcal{T}$.

If we view Hom as a bifunctor (i.e. a functor in two arguments) on the category of modules, (i) and (ii) express the fact that $\mathcal{T}$ and $\mathcal{F}$ are annihilators of each other, and we shall sometimes write $\mathcal{T} = {}^\perp\mathcal{F}, \mathcal{F} = \mathcal{T}^\perp$. There is a certain parallel here with the concept of orthogonality in a metric linear space, but by contrast to that case, Hom is *not* symmetric in its two arguments. Any $\mathcal{T}$ and $\mathcal{F}$ satisfying (i) and (ii) are called a *torsion class* and its associated *torsion-free* class respectively, and the pair $(\mathcal{T}, \mathcal{F})$ is called a *torsion theory*. Given any class $\mathcal{C}$ of R-modules, we obtain a torsion theory by setting $\mathcal{F} = \mathcal{C}^\perp, \mathcal{T} = {}^\perp\mathcal{F}$; this is called the torsion theory *generated* by $\mathcal{C}$; thus $\mathcal{T}$ is the smallest torsion class containing $\mathcal{C}$. Analogously the torsion theory *cogenerated* by $\mathcal{C}$ is formed by setting $\mathcal{T} = {}^\perp\mathcal{C}, \mathcal{F} = \mathcal{T}^\perp$; here $\mathcal{F}$ is the smallest torsion-free class containing $\mathcal{C}$. We shall be particularly interested in the torsion theory cogenerated by R.

Thus we define an R-module M to be *bound* if

$$M^* = \mathrm{Hom}_R(M, R) = 0.$$

This means that there are no linear functionals on M apart from 0. The modules in the corresponding torsion-free class are said to be *unbound*. An unbound module can also be defined as a module with no non-zero bound submodule. For if N satisfies this condition and M is any bound module, then so is any homomorphic image of M, hence $\mathrm{Hom}_R(M, N) = 0$, and so N is unbound. Conversely, if N has a bound submodule $N' \neq 0$, then $\mathrm{Hom}_R(N', N) \neq 0$.

In every torsion theory the classes $\mathcal{T}, \mathcal{F}$ admit certain operations; for the bound and unbound modules this is easily verified directly:

Proposition 5.1.1. *Over any ring R, the class of bound modules is closed under the formation of homomorphic images, module extensions, direct limits and hence direct sums.* ∎

Proposition 5.1.2. *Over any ring R, the class of unbound modules contains all free modules and is closed under the formation of submodules and arbitrary direct products (hence under inverse limits and direct sums), and under module extensions.* ∎

Let M be any R-module; by Proposition. 1.1, M has a unique maximal bound submodule M_b, viz. the sum of all bound submodules of M, and M/M_b has no non-zero bound submodules. Dually, M_b may also be characterized as the least submodule of M with unbound quotient.

To give an example, over $\mathbb{Z}$ (or more generally, any principal ideal domain), M_b is just the submodule of all torsion elements of M. Moreover, for a finitely generated $\mathbb{Z}$-module M we have a decomposition

$$M = M_b \oplus F, \quad \text{where } F \text{ is free.} \tag{1}$$

Such a decomposition exists in fact over any semifir:

Theorem 5.1.3. *Let R be a semifir and M a finitely generated R-module. Then M has a decomposition (1), where M_b is the maximal bound submodule and F is free; here M_b is unique and F is unique up to isomorphism. Moreover, M^* is free and $\operatorname{rk} M^* = \operatorname{rk} F$.*

Proof. Let us write $M = M_0 \oplus F$, where F is a free summand of maximal rank. This is possible because the rank of F is bounded by the number of generators of M. If M_0 is not bound, there is a non-zero homomorphism $f : M_0 \to R$, and since R is a semifir, $\operatorname{im} f$ is free. Thus we can split off a free module from M_0 but this contradicts the maximality of $\operatorname{rk} F$. Hence M_0 is bound and $M/M_0 \cong F$ is unbound, therefore $M_0 = M_b$ is the maximal bound submodule, and (1) is established. The uniqueness is clear, and dualizing (1) we find that $M^* \cong F^*$, hence M^* is free and $\operatorname{rk} M^* = \operatorname{rk} F^* = \operatorname{rk} F$. ∎

The unique submodule M_b in (1) is called the *bound component* of M.

It is clear that a corresponding result holds for n-generator modules over n-firs. In particular, this leads to a condition for a module to be bound:

Corollary 5.1.4. *A module over a fir (or a finitely generated module over a semifir, or an n-generator module over an n-fir) R is bound if and only if it does not contain R as a direct summand.* ∎

In the same situation a finitely generated module is unbound if and only if it is free. More generally, a module over a fir is unbound if and only if it is a direct limit of free modules; thus the class of unbound modules over a fir R may be described as the closure of $\{R\}$, i.e. the class of modules obtained from R by taking submodules, extensions and direct products.

For the moment let us write $\mathcal{B}, \mathcal{U}$ for the classes of bound and unbound modules, respectively (over any ring R). Sometimes we wish to consider a wider class than $\mathcal{U}$ (and a corresponding narrower class than $\mathcal{B}$). Given $n \geq 1$, if every bound n-generator submodule of a module N is zero, N is said to be *n-unbound*, and the class of all such N is written $\mathcal{U}_n$. It is clear that

$$\mathcal{U}_1 \supseteq \mathcal{U}_2 \supseteq \ldots \supseteq \mathcal{U}.$$

The class $\mathcal{B}_n = {}^\perp \mathcal{U}_n$ consists of all modules all of whose non-zero quotients have a non-zero bound n-generator submodule, and it is easily verified that $\mathcal{B}_n^\perp = \mathcal{U}_n$. Thus $(\mathcal{B}_n, \mathcal{U}_n)$ is a torsion theory and

$$\mathcal{B}_1 \subseteq \mathcal{B}_2 \subseteq \ldots \subseteq \mathcal{B}.$$

It follows that Proposition 1.1 also holds for $\mathcal{B}_n$ and Proposition 1.2 for $\mathcal{U}_n$.

Over an integral domain the 1-unbound modules are just the modules without non-zero torsion elements (i.e. torsion-free in the classical sense). This follows from the next result, which describes the n-unbound modules over n-firs:

Theorem 5.1.5. *Let R be an n-fir. Then an R-module M is n-unbound if and only if every n-generator submodule of M is free.*

Proof. Let M be n-unbound, $n > 0$ and assume the result for integers less than n. Any submodule N of M generated by n elements is a homomorphic image of R^n, say $N = (R^n)f$. Let $g : N \to R$ be a non-zero homomorphism; then $fg : R^n \to R$ is non-zero and by Theorem 2.2.1 (e), im(fg) is free of some rank $r > 0$; thus im$(fg) = R^r$ and applying an appropriate automorphism to R^n we find maps

$$R^r \oplus R^{n-r} \to N \to R^r,$$

such that the first map is still surjective, while the composition is the projection onto R^r. Thus we obtain the decomposition $N = R^r \oplus (R^{n-r})f$. If $r = n$, f is an isomorphism; otherwise by induction $(R^{n-r})f$ is free and hence so is N. The converse is clear from the definition of n-unbound. ■

Of particular interest for our purpose is the fact that by Proposition 1.2 for $\mathcal{U}_n$, all direct powers R^I are n-unbound, so that, by Theorem 1.5, when R is an n-fir, any n-generator submodule of R^I is free. This generalizes Corollary 2.3.22.

Bound modules satisfy chain conditions under fairly mild restrictions. To state them we recall that a left (or right) hereditary ring is certainly weakly semihereditary, so we can apply Theorem 2.1.4 to conclude that any projective module over a left hereditary ring is a direct sum of finitely generated modules.

Theorem 5.1.6. *Let R be a left hereditary ring, and let M be a finitely related left R-module. Then each submodule of M is a direct sum of a finitely presented module and a projective module.*

Proof. By hypothesis, $M = F/N$, where F is free and N is finitely generated. Every submodule of M has the form P/N, where $N \subseteq P \subseteq F$ and here P is projective, because R is left hereditary. By Theorem 2.1.4, P is a direct sum of finitely generated modules and so it contains a finitely generated direct summand P' containing N (because N was finitely generated). Writing $P = P' \oplus P''$, we have the exact sequence

$$0 \to P'/N \to P/N \to P/P' \cong P'' \to 0.$$

Since P'' is projective, this sequence splits and P', N are both finitely generated, hence P'/N is finitely presented. ∎

Clearly a bound module contains no non-zero projective submodule as a direct summand; so we find

Corollary 5.1.7. *Over a left hereditary ring any bound submodule of a finitely related left module M is finitely presented, hence M satisfies ACC on bound submodules.* ∎

Taking the module itself to be bound, we obtain

Corollary 5.1.8. *Over a left hereditary ring every finitely related bound left module is finitely presented.* ∎

Exercises 5.1

1. Show that for any ring R and any left ideal $\mathfrak{a}$ of R, $R/\mathfrak{a}$ is bound if and only if the right annihilator of $\mathfrak{a}$ is zero. Deduce that for a matrix defining a left R-module M the matrix must be left regular; M is bound if and only if the matrix is right regular.
2. Show that a finitely related bound module over any ring is finitely presented. Deduce that a projective module P is finitely generated provided that it has a finitely generated submodule N such that P/N is bound.
3. Show that the class of bound left R-modules admits submodules if and only if the injective hull of R (as left module) is unbound. (*Hint*: Recall that the injective hull is the maximal essential extension.)
4. Let R be a left hereditary ring and M a finitely related left R-module. Show that the projective submodules of M form an inductive system (i.e. admitting unions of ascending chains).
5*. Let R be an integral domain and M an R-module. Show that a homomorphism $M \to R$ annihilates all torsion elements; deduce that any module consisting entirely of torsion elements is bound. Show that every finitely generated bound R-module consists of torsion elements if and only if R is both left and right Ore.

6°. (G. M. Bergman) Characterize integral domains over which every bound left module is generated by torsion elements.

7. Let K be a field and R a subring containing two right R-linearly independent elements a, b. Verify that the submodule $Ra^{-1} + Rb^{-1}$ of K is bound, as left R-module. Deduce that K is not semifree as left R-module (see also Exercise 0.8.3).

8. Show that over a left Bezout domain R, a finitely generated left R-module without non-zero torsion elements (i.e. torsion-free) need not be free. (*Hint*: Use Exercise 7.)

9. Find an example of a module that is n-unbound for all $n \geq 1$, yet not unbound. (*Hint*: Over a semifir this requires a module M whose finitely generated submodules are free, while $M^* = 0$.)

10. Let R be an n-fir and $\mathcal{C}_n$ the class of all n-unbound left R-modules with ACC$_n$. Show that R must have left ACC$_n$ for $\mathcal{C}_n$ to be non-trivial. When R has left ACC$_n$ show that $\mathcal{C}_n$ contains all free modules and is closed under (i) the formation of submodules, extensions and unions of ascending sequences with quotients of successive terms in $\mathcal{C}_n$ (and hence direct sums) and (ii) inverse limits (and hence direct products).

11. (Bergman [67]) (i) Let k be a field with an endomorphism α that is not surjective and define $R' = k[y; \alpha]$. Show that yR' is a two-sided ideal that is free as left ideal; further, show that if (u_i) is a left basis for k over k^α, then (yu_i) is a left basis for yR' over R'.

    (ii) Let $R = k^\alpha + yR'$ be the subring of polynomials with constant term in k^α. Verify that R is a right Ore domain with the same field K of fractions as R', but R is not left Ore (note that even R' is not left Ore).

    (iii) Take $x \in k \setminus k^\alpha$ and let $f : R'^2 \to R'$ be the linear functional $(a, b) \mapsto axy - by$. Then f also defines a linear functional f_R (by restriction) on R^2. Show that ker $f_R = yR'(1, x) = \Sigma R'yu_i(1, x)$ and deduce that a submodule of an R-module of rank 1 can have infinite rank (for a suitable choice of k and α). Take a particular subscript $i = 1$ say, and put $M = R^2/R'yu_1(1, x)$. Verify that f_R induces a map $f_M : M \to R$ and by showing that ker f_M is embeddable in R^2, verify that M is unbound. Show also that $K \otimes M = K^2/K(1, x)$ and hence that M, although unbound, is not embeddable in $K \otimes M$. Conclude that M is also not embeddable in R^I for any set I.

12. Let R be a right Ore domain, M and N left R-modules and $f : M \to N$ a homomorphism. If N is finitely generated and coker f is bound, verify that rk $N \leq$ rk M. Does this remain true without the condition that N be finitely generated? (*Hint*: Try $N = K$.)

13. Show that for any finitely generated modules M, N over a semifir R, if $M \oplus R \cong N \oplus R$, then $M \cong N$.

14*. (G. M. Bergman) Let A be the subgroup of $\mathbb{Q}^N$ consisting of all sequences (x_n) such that $nx_n \in \mathbb{Z}$ for all n and $x_n \in \mathbb{Z}$ for almost all n, while the sum over all the non-integral x_n is integral. Verify that A is unbound, but cannot be embedded in any direct power of $\mathbb{Z}$.

15°. (G. M. Bergman) Characterize the rings R such that every finitely generated unbound R-module can be embedded in a direct power of R.

16°. Can the hypothesis on R (to be left hereditary) in Theorem 1.6 be omitted? (See also Proposition 0.5.1).

5.2 Duality

We shall now establish a duality for bound modules over hereditary rings, from which it will follow that any finitely presented bound module satisfies DCC as well as ACC on bound submodules. We shall present the actual results so as to apply to arbitrary rings and only specialize at the end.

Let us call a module M over any ring R *special* if it is finitely presented and of projective dimension at most 1. Thus a special R-module M is a module with a resolution

$$0 \to Q \to P \to M \to 0,$$

where P, Q are finitely generated projective R-modules. Over a semifir the special modules are just the finitely presented modules. If we dualize the above exact sequence, taking M to be a left R-module, we get the exact sequence of right R-modules

$$0 \to M^* \to P^* \to Q^* \to \text{Ext}(M, R) \to 0,$$

bearing in mind that $\text{Ext}(P, R) = 0$, because P is projective. Let us assume that M is bound, so that $M^* = 0$. Dualizing once more and defining the *transpose* as $\text{Tr}(M) = \text{Ext}(M, R)$, we obtain the following commutative diagram with exact rows:

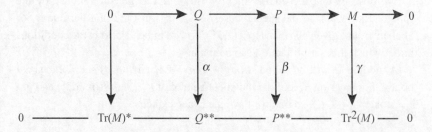

Here α and β are isomorphisms, because P and Q are finitely generated projective. This allows us to define γ and to show that it is an isomorphism too (by the 5-lemma); likewise we conclude that $\text{Tr}(M)^* = 0$. These remarks suggest the truth of

Proposition 5.2.1. *For any ring R there is a duality Tr between the categories of special bound left R-modules and special bound right R-modules, such that if $M = \text{coker}(Q \to P)$, where P, Q are finitely generated projective and $Q \subseteq P$, then $Tr(M) = \text{coker}(P^* \to Q^*)$.*

Proof. Consider a mapping $f : M \to M'$ between two special bound left R-modules, given by presentations of the above form

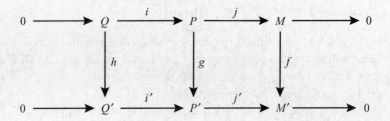

By the projectivity of P, jf lifts to a map $g : P \to P'$ and $h = g|Q$ maps into Q'. The map f is completely determined by the pair (g, h), for any such pair of maps induces a map of the cokernels. Moreover, two such pairs (g, h) and (g_1, h_1) give the same map f if and only if there exists $e : P \to Q'$ such that $g_1 - g = ei'$, $h_1 - h = ie$ (thus (g, h) and (g_1, h_1) are 'homotopic'). Hence the category of special bound left R-modules is equivalent to the category of maps $Q \to P$, which are injective with injective duals under homotopy-equivalence and whose morphisms are homotopy-classes of commuting squares. *This* category is clearly dual to the corresponding category of maps between projective right R-modules, and hence to the category of special bound right R-modules. ■

Here, as in every duality, monomorphisms correspond to epimorphisms; since we are dealing with module categories, this means that injective maps correspond to surjective maps under this duality and vice versa.

We observe that the duality of Proposition 2.1 is given explicitly by the functor $\mathrm{Ext}(-, R)$. In the special case $R = \mathbb{Z}$ it is just the familiar duality of abelian groups given by $A \mapsto \mathrm{Hom}\,(A, \mathbb{Q}/\mathbb{Z})$. We indicate briefly the conditions under which this simplification can be made.

A module M will be called *strongly bound* if M and all its submodules are bound. For any ring R, let E be the injective hull of R, as left R-module and put $T = \mathrm{coker}\,(R \to E)$, so that we have the exact sequence

$$0 \to R \to E \to T \to 0. \tag{1}$$

It is easily verified that M is strongly bound if and only if $\mathrm{Hom}\,(M, E) = 0$ (using the fact that R is an essential submodule of E). By (1) we have, since E is injective, the exact sequence

$$0 \to \mathrm{Hom}(M, R) \to \mathrm{Hom}(M, E) \to \mathrm{Hom}(M, T) \to \mathrm{Ext}(M, R) \to 0. \tag{2}$$

However, when M is strongly bound, then $\mathrm{Hom}\,(M, E) = 0$ and so we have

Proposition 5.2.2. *Let R be any ring and define E, T as above. Then for any strongly bound module M,*

$$\mathrm{Ext}(M, R) \cong \mathrm{Hom}(M, T). \qquad\qquad ■ \tag{3}$$

Over a principal ideal domain (commutative or not) every finitely generated bound module is strongly bound and we can therefore use Proposition 2.2 to express the duality in terms of Hom. Moreover, in this case E is just the field of fractions of R. This is an R-bimodule, hence T is then also an R-bimodule and it follows that (for a left R-module M) (3) is actually a right R-module isomorphism. When R is a left fir but not left principal, there will always be finitely generated modules that are bound but not strongly bound. Even then we can use the exact sequence (2) to describe $\mathrm{Ext}(M, R)$ as the cokernel of the mapping $\mathrm{Hom}(M, E) \to \mathrm{Hom}(M, T)$. However, we cannot in general expect to write $\mathrm{Ext}(M, R)$ in the form $\mathrm{Hom}(M, I)$ for some I, i.e. the functor $\mathrm{Ext}(-, R)$ is not representable, because $\mathrm{Ext}(-, R)$ is not left exact, unless we restrict the class of bound modules further.

Returning to Proposition 2.1, let us apply the result to hereditary rings. In the first place, every finitely presented module is now special. Moreover, by Corollary 1.7, every finitely related bound module satisfies ACC on bound submodules (necessarily finitely presented by Corollary 1.8), and applying the above duality, we find that the module satisfies DCC for bound submodules. Thus we obtain

Theorem 5.2.3. *Let R be a left hereditary ring. Then any finitely related bound left R-module satisfies ACC and any finitely presented bound right R-module satisfies DCC on bound submodules. In particular any finitely related bound module over a left and right hereditary ring satisfies both chain conditions for bound submodules.* ∎

Exercises 5.2

1. What becomes of the duality of Proposition 2.1 in the case where R is left self-injective, i.e. injective as left R-module?
2. For this exercise only, let us call a module M *extra-special* if it has a presentation $0 \to P \to P \to M \to 0$, where P is finitely generated projective. Show that in the duality of Proposition 2.1, extra-special bound modules correspond to extra-special bound modules.
3. Let R be a left fir. If every bound left R-module is strongly bound, show that R is left principal. Is it enough to assume the condition for finitely generated bound modules?
4. Let R be a left Ore domain and K its field of fractions. Show that K is the injective hull of R. Does this remain true for more general rings that are embeddable in fields?
5*. Let R be a two-sided fir and E its injective hull as left R-module. Describe the R-bimodule structure of E and compare it with the injective hull of R as right

R-module. (*Hint*: Observe that right multiplication by $a \in R$ is an endomorphism that extends to an endomorphism of E by injectivity.)

6. Let R be a left hereditary ring and M a finitely related left R-module. Show that if P is a maximal projective submodule of M, then M/P is strongly bound. Deduce that for every finitely related module M there is an exact sequence

$$0 \to P \to M \to Q \to 0,$$

where P is projective and Q is finitely generated and strongly bound.

7. Give an example of a strongly bound module over $k\langle x, y \rangle$ that is not finite-dimensional over k.

8. Let R be an integral domain and $\mathfrak{a}$ a left ideal. Show that $R/\mathfrak{a}$ is strongly bound if and only if R is an essential extension of $\mathfrak{a}$.

9. Let R be an integral domain and E its injective hull as left R-module. Show that for any left R-module M there exists a subset X of M such that every map $X \to E$ extends to a unique homomorphism $M \to E$; thus the natural embedding $^X R \to E$ extends to a homomorphism $M \to E$.

10*. Does Theorem 2.3 hold for fully atomic semifirs? (*Hint*: See Proposition 9.6 below.)

11. (A. H. Schofield) Show that for bound left R-modules M, N over a hereditary ring R, $\mathrm{Hom}_R(M, N) \cong \mathrm{Tor}_1^R(\mathrm{Tr}\, M, N)$. (*Hint*: Resolve M and note that $\mathrm{Hom}_R(P, N) \cong P^* \otimes N$ for finitely generated projective P.)

5.3 Positive and negative modules over semifirs

We have already met torsion modules in Section 3.2, where we saw that over a principal ideal domain they reduce to finitely generated modules of torsion elements, while many of the properties of the latter carry over to torsion modules over semifirs. In this section we apply the results of Sections 5.1 and 5.2 to study two classes of finitely presented modules over semifirs: the positive modules, which over principal ideal domains correspond to finitely generated modules, and their duals, the negative modules, which have no analogue in the classical case.

Throughout this section all modules occurring will be finitely presented modules over semifirs; in that case the characteristic of a module, as defined in Section 0.6, is an integer, positive, negative or zero, and by Proposition 0.5.2, the characteristics are additive on short exact sequences. Moreover, for any semifir R, the category of all finitely presented left R-modules is an abelian subcategory of $_R\mathrm{Mod}$, by Appendix Theorem B.12.

Definition. Let R be a semifir and M an R-module.

(i) If M is finitely presented and $\chi(M') \geq 0$ for all submodules M' of M, then M is said to be *positive*.

5.3 Positive and negative modules over semifirs

(ii) If M is finitely presented and $\chi(M'') \leq 0$ for all quotients M'' of M, then M is said to be *negative*.

(iii) If M is positive and $\chi(M') > 0$ for all non-zero submodules M', then M is called *strictly positive;* if M is negative and $\chi(M'') < 0$ for all non-zero quotients M'', then M is said to be *strictly negative*.

Any submodule of characteristic 0 of a positive module is a torsion module and the same holds for any quotient of characteristic 0 of a negative module. Therefore a strictly positive module may also be defined as a positive module without non-zero torsion submodules and a strictly negative module is a negative module with no non-zero torsion quotients. We note further that a torsion module is just a module that is both positive and negative. Writing $_R\text{Pos}$, $_R\text{Neg}$, $_R\text{Tor}$ for the categories of positive, negative and torsion left R-modules, we thus have

$$_R\text{Tor} = {_R}\text{Pos} \cap {_R}\text{Neg}.$$

Over a principal ideal domain any finitely generated module is positive and there are no negative modules apart from torsion modules, because the characteristic does not assume negative values in this case (see Section 2.3). By contrast, a semifir that is not an Ore domain will always have modules of arbitrary negative characteristic, by Proposition 0.7.6.

We have seen that torsion modules (over semifirs) are presented by full matrices, and we now examine the presenting matrices of positive and negative modules. Let M be a left module presented by the $m \times n$ matrix C, which can be taken to be left regular; it is clear that M is bound if and only if C is also right regular. Let M' be a submodule with quotient $M'' = M/M'$; then M', M'' may be presented by matrices A,B such that $C = AB$ (Proposition 0.5.2). Recalling the definitions of left (right) full and prime matrices, from Section 3.1, we obtain

Proposition 5.3.1. *Let R be a semifir and M a finitely presented left R-module, with presenting matrix C. Then C is left regular, and it is also right regular if and only if M is bound. Further, M is positive if and only if C is left full, and negative if and only if C is right full. Finally, M is strictly positive (resp. negative) if and only if C is left (resp. right) prime.* ∎

These results suggest that there should be a duality between positive and negative modules, and this is in fact the case, provided that we restrict ourselves to bound modules. We note that in the decomposition of a module,

$$M = M_b \oplus R^n, \tag{1}$$

if M is positive, then so is M_b, the bound component. Of course, a negative module is always bound, by definition, since R has no non-zero submodules of characteristic ≤ 0.

Theorem 5.3.2. *Let R be a semifir. Then $Tr = Ext(-, R)$ provides a duality between the category of bound positive left R-modules and the category of negative right R-modules.*

Proof. The result follows by a straightforward application of Proposition 2.1. ∎

In order to establish chain conditions we shall assume that R is a fir. Then we can apply Theorem 2.3 to obtain

Theorem 5.3.3. *Let R be a left fir. Then any positive bound left R-module satisfies ACC on bound submodules and a negative right R-module satisfies DCC on bound submodules. In particular, over a two-sided fir any positive bound or negative module satisfies both chain conditions on bound submodules.* ∎

Later, in Section 5.9, we shall meet other chain conditions valid over certain semifirs.

Proposition 5.3.4. *Let R be a semifir. Then Pos, Neg and the class of bound modules all admit extensions and hence finite direct sums. Further, Pos admits submodules and Neg admits quotients within the category of all finitely presented R-modules.*

Proof. In the short exact sequence

$$0 \to M' \to M \to M'' \to 0, \qquad (2)$$

assume that M', M'' are positive and $N \subseteq M$. Then $\chi(N \cap M') \geq 0$ and

$$N/(N \cap M') \cong (N + M')/M' \subseteq M/M' \cong M'',$$

hence $\chi(N/N \cap M') \geq 0$, therefore $\chi(N) = \chi(N/N \cap M') + \chi(N \cap M') \geq 0$. This shows M to be positive. Now assume M', M'' to be bound and let $f : M \to R$ be a homomorphism. Then $f \mid M' = 0$, hence f is induced by a homomorphism $\bar{f} : M'' \to R$, which must be 0, so $f = 0$ and M is bound. Thus Pos and the class of bound modules both admit extensions; by duality the same holds for Neg. The remaining assertions are clear from the definitions. ∎

Let us consider a finitely presented bound module M over a fir R; by Theorem 2.3, any chain of bound submodules in M is finite. Moreover, any bound submodule of M is finitely presented, by Corollary 1.8 applied to (2), and so has finite characteristic. Let M_1 be a submodule of least characteristic, $\chi(M_1) = h$,

5.3 Positive and negative modules over semifirs

say. Then M_1 is negative, for any submodule N of M_1 satisfies $\chi(N) \geq h$, hence $\chi(M_1/N) \leq 0$. Since the set of negative submodules of M satisfies both chain conditions and admits sums, there is a unique maximal negative submodule M^- of M. Any submodule N of M satisfies $\chi(N) \geq \chi(M^-)$, so if $N \supseteq M^-$, then $\chi(N/M^-) \geq 0$, with strict inequality unless $N = M^-$; this shows M/M^- to be strictly positive.

Dually we can find a least submodule M^+ with bound positive quotient M/M^+, and M^+ is strictly negative; therefore $M^+ \subseteq M^-$. Of course M^-/M^+ is both positive and negative and so is a torsion module. Thus we have

Theorem 5.3.5. *Let M be a finitely presented bound module over a fir R. Then there is a chain*

$$0 \subseteq M^+ \subseteq M^- \subseteq M, \qquad (3)$$

where M^- is the greatest negative submodule of M and M^+ the least submodule with positive quotient M/M^+. Moreover, M/M^- is strictly positive, M^+ is strictly negative and M^-/M^+ is a torsion module. ∎

This result has an interpretation in terms of matrices, which we shall meet in Proposition 4.7.

If we now impose the left Ore condition, we have a principal left ideal domain, and here the positive modules admit quotients as well as submodules. However, negative modules are absent; more precisely, they reduce to modules of torsion elements, as do bound positive modules. Thus in the Ore case we obtain

Proposition 5.3.6. *The left torsion modules over a left Bezout domain are precisely the finitely presented modules generated by torsion elements, and all their elements are torsion. Moreover, in an exact sequence (2), if M is a torsion module and M' is finitely generated, then M', M'' are torsion modules.*

Proof. Over a left Bezout domain the characteristic is non-negative, by the remarks after Corollary 2.1.3. Now a torsion module M is certainly finitely presented and $\chi(M) = 0$; if $x \in M$ is torsion-free, then $\chi(Rx) = 1$ and so $\chi(M/Rx) = -1$, which is a contradiction, so all elements of M are torsion. Conversely, if M is finitely presented and consists of torsion elements, its rank is 0 and so is its characteristic. This remark also shows that $\chi(M') = \chi(M'') = 0$ in (2), and when M' is finitely generated, then both M' and M'' are finitely presented. ∎

Corollary 5.3.7. *Over a left or right Bezout domain, every n-generator torsion module has a chain of torsion submodules of length n, whose quotients are cyclic torsion modules.*

Proof. Clearly an n-generator module over any ring has a chain of submodules of length n with cyclic quotients. When R is left Bezout and M is torsion, these quotients are torsion by the proposition; for right Bezout domains the result follows by duality. ∎

Over a principal ideal domain a more precise decomposition can be obtained, as we saw in Section 1.4.

So far we have confined our attention to finitely generated modules, as that is the most interesting case (for us). However, it is also possible to extend the notions defined here. Let us briefly mention the result for torsion modules; the extension to positive and negative modules is entirely similar.

For any semifir R we define the category $_R\text{Tor}^\uparrow$ of *general torsion modules* as consisting of those left R-modules M in which every finite subset is contained in a finitely generated torsion submodule. Then $_R\text{Tor}^\uparrow$ (as a full subcategory of $_R\text{Mod}$) is again an abelian category; moreover it has exact direct limits and a generator, i.e. it is a *Grothendieck category* (see Cohn [70b] for proofs) and may be obtained as the completion of $_R\text{Tor}$. Dually one defines the category of *protorsion modules* $_R\text{Tor}^\downarrow$ to consist of all inverse limits of finitely generated torsion modules and all continuous homomorphisms (relative to the natural topology on the inverse limit). Now the functor $\text{Tr} = \text{Ext}\,(-, R)$ establishes a duality between the categories $_R\text{Tor}^\uparrow$ and $\text{Tor}_R^\downarrow$ (see Cohn [70b]).

In Section 3.5 we saw that the Krull–Schmidt theorem applies to torsion modules over semifirs. This amounts to considering the factorization of (square) full matrices. When we come to consider rectangular matrices, we find that a similar result holds; we shall state it as a Krull–Schmidt theorem for finitely presented modules (Theorem 3.9). Our first task is to prove a form of Fitting's lemma.

Lemma 5.3.8. *Let M be a finitely presented bound indecomposable module over a fir R. Then $End_R(M)$ is a completely primary ring.*

Proof. We have to show that every endomorphism of M is either nilpotent or invertible. So let α be an endomorphism of M that is not an automorphism; then $M\alpha$ is a bound submodule of M and we have the descending chain

$$M \supseteq M\alpha \supseteq M\alpha^2 \supseteq \ldots.$$

By Theorem 3.3 this chain becomes stationary, say $M\alpha^n = M\alpha^{n+1} = \ldots$. On $M\alpha^n$, α^r is a surjective endomorphism for any $r \geq 0$, and we have the exact sequence

$$0 \longrightarrow \ker \alpha^r \cap M\alpha^n \longrightarrow M\alpha^n \xrightarrow{\alpha^r} M\alpha^n \longrightarrow 0. \qquad (4)$$

Put $N_r = \ker \alpha^r \cap M\alpha^n$; then by (4), $\chi(N_r) = 0$ and clearly,

$$N_1 \subseteq N_2 \subseteq \ldots . \qquad (5)$$

If we take the bound components only, we get an ascending chain, by Theorem 1.3, which again becomes stationary, by Theorem 3.3, say $(N_m)_b = (N_{m+1})_b = \ldots$. Since $\chi(N_m) = 0$, we have

$$N_m \cong (N_m)_b \oplus R^k, \quad \text{where } k = -\chi((N_m)_b),$$

and in $M/(N_m)_b$ we have the ascending chain of modules $N_\mu/(N_m)_b$, each isomorphic to R^k. But $M/(N_m)_b$ satisfies ACC_k, by Theorem 2.2.2, so this chain also becomes stationary, say $N_p = N_{p+1} = \ldots$. Now the rest of the proof follows along the usual lines:

Let $r = \max(n, p)$; then for any $x \in M$, $x\alpha^r \in M\alpha^{2r}$, say $x\alpha^r = y\alpha^{2r}$, hence $x = y\alpha^r + z$, where $z \in \ker \alpha^r$, so we have

$$M = M\alpha^r + \ker \alpha^r. \qquad (6)$$

If $x \in M\alpha^r \cap \ker \alpha^r$, then $x = y\alpha^r$ and $y\alpha^{2r} = x\alpha^r = 0$, so $y \in \ker \alpha^{2r} = \ker \alpha^r$, hence $x = y\alpha^r = 0$. This shows the sum (6) to be direct, and by the indecomposability of M, either $M\alpha^r = 0$ and α is nilpotent, or $\ker \alpha^r = 0$, $M\alpha^r = M$ and α is an automorphism. ∎

We recall that any R-module M is indecomposable if and only if $\mathrm{End}_R(M)$ contains no idempotents $\neq 0, 1$. In particular, when R is an integral domain, then $\mathrm{End}_R(R) \cong R$ contains no idempotents $\neq 0, 1$ and so R is indecomposable as left or right R-module. The Krull–Schmidt theorem for firs now takes the following form:

Theorem 5.3.9. *Let M be a finitely presented module over a fir R. Then there exists a decomposition*

$$M \cong M_1 \oplus \cdots \oplus M_r \oplus R^h, \qquad (7)$$

where each M_i is bound indecomposable, and R is indecomposable. Given a second such decomposition of M:

$$M \cong M_1' \oplus \cdots \oplus M_s' \oplus R^k,$$

we have $h = k, r = s$ and there is a permutation $i \mapsto i\sigma$ of $1, \ldots, r$ such that $M'_i \cong M_{i\sigma}$.

Proof. We have $M = M_b \oplus R^k$, and here k is uniquely determined as the rank of M/M_b (or also of M^*) because R has IBN. Thus we need only decompose M_b; by Theorem 3.3 there exists a complete decomposition, and since each component has a local endomorphism ring, by Lemma 3.8, the conclusion follows by Azumaya's form of the Krull–Schmidt theorem (see e.g. FA, Theorem 4.1.6 or IRT, Theorem 2.31). ∎

We also note an analogue of Schur's lemma. Let us call an R-module M *minimal bound* if M is non-zero bound, but no proper non-zero submodule is bound; over a semifir this means by Theorem 1.3 that every finitely generated proper submodule is free. Now we have the following form of Schur's lemma:

Proposition 5.3.10. *Let R be a semifir and M a finitely presented minimal bound R-module. Then $End_R(M)$ is a field.*

Proof. Consider an endomorphism $\alpha : M \to M$; its image is again bound, hence it is 0 or M. Suppose that $\alpha \neq 0$; then $\operatorname{im} \alpha = M$ and we have an exact sequence

$$0 \longrightarrow \ker \alpha \longrightarrow M \xrightarrow{\alpha} M \longrightarrow 0. \tag{8}$$

Now $\ker \alpha \neq M$; hence it is free and by comparing characteristics in (8), it has rank 0; therefore $\ker \alpha = 0$ and thus α is an automorphism. ∎

We conclude this section with an application of the above results (due to Bergman [2002]), namely the embedding of any fir in a field. This question will be taken up again later, in Chapter 7, in a more general context, but the proof given here is more direct.

We have seen in Section 3.3 that the endomorphism ring of a simple torsion module is a field. For strictly positive or negative modules this need not hold; for example, if M is any strictly positive module, then so is $M \oplus M$, but its endomorphism ring has nilpotent elements. However, we have the following result:

Proposition 5.3.11. *Let R be a semifir and M, N any strictly positive R-modules, such that $\chi(M) = 1$. Then any non-zero homomorphism $f : M \to N$ is injective.*

Proof. We have an exact sequence

$$0 \longrightarrow \ker f \longrightarrow M \xrightarrow{f} N \longrightarrow \operatorname{coker} f \longrightarrow 0.$$

If $\ker f \neq 0$, then $\chi(\ker f) > 0$, hence $\chi(\mathrm{im}\, f) = 1 - \chi(\ker f) \leq 0$, and so $\mathrm{im}\, f = 0$. ∎

In particular, taking $N = M$, we obtain

Corollary 5.3.12. *Let R be a semifir and M a strictly positive R-module of characteristic 1. Then $\mathrm{End}_R(M)$ is an integral domain.* ∎

Here $\mathrm{End}_R(M)$ need not be a field, since, for example, R itself is strictly positive of characteristic 1, but the endomorphism ring is a field for bound left R-modules over a semifir with right pan-ACC, under the conditions of Corollary 3.12 (see Exercise 10 below).

Now let R be a fir; by a *pointed R-module* we shall understand a pair (M, c) consisting of an R-module M and a non-zero element $c \in M$. We consider the category $\mathcal{L}$ of all pointed strictly positive left R-modules of characteristic 1. For example, $(R, 1)$ is a pointed module in $\mathcal{L}$. The morphisms $(M, c) \to (M', c')$ of $\mathcal{L}$ are the homomorphisms $f : M \to M'$ such that $cf = c'$. We observe that between any two modules M, M' of $\mathcal{L}$ there is at most one morphism; for if f' is another, then $f - f' : M \to M'$ is a homomorphism vanishing on c and hence zero, by Proposition 3.11. Thus $\mathcal{L}$ is a preordering, with $(R, 1)$ as least element.

Lemma 5.3.13. *The category $\mathcal{L}$ is a directed preordering.*

Proof. Let $(T, a) \to (M, b), (T, a) \to (N, c)$ in $\mathcal{L}$ be given and consider $M \oplus N$. This is again strictly positive and it contains the submodule $T_1 = \{(x, -x) \mid x \in T\}$, which is isomorphic to T, hence of characteristic 1. Let S be a maximal submodule of characteristic 1 containing T_1 and write $P = (M \oplus N)/S$. Then we see (as in the proof of Theorem 3.5) that P has characteristic 1 and is strictly positive, with distinguished element $(b, 0) \equiv (0, c) \pmod{S}$; moreover it is the unique largest quotient of $M \oplus N$ with this property, by the maximality of S; therefore it is the pushout (in the category $\mathcal{L}$). This shows $\mathcal{L}$ to be directed. ∎

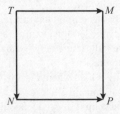

We now form the direct limit U of this directed system $\mathcal{L}$ of pointed modules, identifying the distinguished elements. We observe that U is a left R-module containing R as submodule, for given any member (M, c) of $\mathcal{L}$, there is a unique embedding $(R, 1) \to (M, c)$. Now fix any non-zero element u of U; there is a strictly positive submodule of characteristic 1 containing u and the submodule generated by u is free: $Ru \cong R$. Now U can also be obtained as the direct limit of all the pointed strictly positive modules of characteristic 1, with $(R, 1)$ replaced by (R, u). This means that U has an automorphism mapping 1 to u. Since u was any non-zero element of U, we see that $E = \mathrm{End}_R(M)$ is transitive on the non-zero elements of U and it contains R as a subring. Moreover, if $f, g \in E$ and $1f = 1g$, then $f - g$ is not injective and hence vanishes, by Proposition 3.11. Thus each $f \in E$ is determined by its effect on $1 \in R$. Given $f \in E^\times$, there exists $g \in E$ with $1.fg = 1$, hence $fg = 1$ and this shows E to be a field. Thus we obtain

Theorem 5.3.14. *Every fir can be embedded in a field.* ∎

Here the finiteness conditions holding in a fir were needed in the proof. In Chapter 7 we shall see that every semifir R can be embedded in a field; moreover, there is a universal field of fractions for R, having any other field containing R as a specialization.

Exercises 5.3

1. Let R be a semifir and $n \geq 1$. Taking the minimal projective over R_n to have rank $1/n$, show that any finitely presented module over R_n has a characteristic of the form $r/n, r \in \mathbb{Z}$. How are the characteristics of modules over R and over R_n related, which correspond under the category-equivalence?
2. Show that over a semifir, every torsion submodule of an n-torsion module can be generated by n elements. Deduce that a torsion module over a fully atomic semifir satisfies ACC for torsion submodules.
3. Let M be an n-generator module over a semifir and M' any submodule. Show that M' can be generated by $n - \chi(M/M')$ elements. (*Hint*: Use the diagram of Proposition 0.5.2).
4. Let R be a principal ideal domain and $c \in R^\times$. A non-unit left factor b of c will be called *inessential* if c has a non-unit left factor left coprime to b, otherwise it is *essential*. Show that the cyclic left torsion modules defined by two elements c, c' have isomorphic injective hulls if and only if c and c' have an essential left factor (up to similarity) in common.
5°. Does the result of Exercise 4 hold when R is a fir?
6. Let R be a principal ideal domain and $c \in R^\times$. Show that the injective hull of R/Rc can be expressed as a limit of cyclic modules. Give an example in $k\langle x, y \rangle$ where this fails.
7. Let $R = k\langle x, y, z \rangle$ and consider the torsion modules $M = R/Rxz, N = R/Ryz$. Verify that both have R/Rz as quotient, and denote the pullback by T. Show that

T is not a torsion module, but that it has a unique largest torsion submodule. Show that the lattice of torsion submodules of T is finite and isomorphic to that of $R/R(xy+1)xz$.

8. Let R be a semifir, $a, b \in R^\times$ and $A = R/Ra, B = R/Rb$. Show that $\mathrm{Ext}^1(B, A) \cong \mathrm{Tr}\,(B) \otimes A \cong R/(Ra + bR)$. Deduce that every extension of A by B splits if and only if $Ra + bR = R$.

   If $R = k\langle x, y \rangle$, take $a = x, b = y$ and find the non-split extensions apart from the cyclic one.

9. Let R be a semifir, $a, c \in R^\times$, $A = R/Ra$, $C = R/cR$. Show that $\mathrm{Tor}_1(C, A) \cong (cR \cap Ra)/cRa$.

10. Let R be a semifir with right ACC_n. Show that if M is a bound strictly positive left R-module generated by n elements, such that $\chi(M) = 1$, then $\mathrm{End}_R(M)$ is a field.

11°. (Bergman [67]). A module is said to be *polycyclic*, if it has a finite chain of submodules with cyclic factors. Does there exist a semifir, not left or right Bezout, over which every torsion module is polycyclic?

12°. (Bergman [67]). Determine the class of semifirs over which every polycyclic torsion module can be written as a direct sum of cyclic torsion modules (observe that this includes all principal ideal domains, by Theorem 1.4.10).

13. Let M, N be strictly negative modules over a semifir R such that $\chi(N) = -1$. Show that any non-zero homomorphism from M to N is injective.

14*. Let M, N be non-zero bound modules over a semifir R. If for all non-zero bound submodules $M' \subseteq M, N' \subseteq N$ we have $\chi(M') + \chi(N') > \chi(M)$, show that the kernel of any non-zero homomorphism from M to N is free. Deduce that if P is a finitely presented non-zero R-module such that for every finitely generated non-zero proper submodule P' of P, $2\chi(P') > \chi(P)$, then $\mathrm{End}_R(P)$ is an integral domain. Hence obtain another proof of Corollary 3.12.

15. Let R be a semifir and ${}_R\mathcal{F}$ the full subcategory of ${}_R\mathrm{Mod}$ whose objects are the finitely presented modules. Use the characteristic to give a direct proof that ${}_R\mathcal{F}$ is an abelian subcategory of ${}_R\mathrm{Mod}$.

16. Let R be a semifir that is right Bezout. Show that every finitely generated bound left R-module is negative.

17. Let R be a semifir. Show that if every left torsion R-module has finite length (in the lattice of all submodules), then R is a principal left ideal domain.

18°. Investigate (i) firs for which there are only finitely many simple torsion modules (up to isomorphism) and (ii) firs over which the indecomposable torsion modules have finite length (within the category Tor).

5.4 The ranks of matrices

The rank of a matrix is a numerical invariant that is certainly familiar to the reader, at least for matrices over a commutative ring. We shall need an analogue for matrices over more general rings; there is then more than one invariant that can lay claim to generalize the usual rank. This is not surprising, since even in the commutative case one defines row and column rank separately and *then* proves them equal. We shall introduce three different notions of rank, the row rank, column rank and inner rank. Of these, the last, already encountered in Section 0.1, will mainly be used, but the others are sometimes of interest, and our first task is to describe their relationship to each other. For a smoother development we begin with some remarks on the decomposition of free modules over n-firs.

Let M be a module over any ring R and N a submodule of M. Then we define the *closure* $\bar{N}$ in M as the intersection of the kernels of all the linear functionals on M that vanish on N. If $\bar{N} = M$, we say that N is *dense* in M; this is the case precisely when M/N is bound, in the sense of Section 5.1. If $\bar{N} = N$, we say that N is *closed*; this is so if and only if N is the kernel of a mapping into a direct power of R. Hence if N is closed in M, then M/N is unbound.

If R is an n-fir, then the submodules of R^n that are annihilators of finite families of linear functionals are direct summands of R^n, by Theorem 2.3.1 (e). Moreover, any given direct sum decomposition of R^n has the form $R^n \cong R^r \oplus R^s$, for a unique pair of integers r, s satisfying $r + s = n$. It follows that any chain of direct summands of R^n is finite, of length at most n; but any closed submodule of R^n is the intersection of a descending chain of direct summands, and so is itself a direct summand. This proves

Proposition 5.4.1. *Let n be a positive integer. For any n-fir R, the closed submodules of R^n are precisely the direct summands. They form a lattice in which every maximal chain has length n, and the height of any member of this lattice is its rank as a free module. Moreover, if M is any closed submodule of R^n, then the linear functionals vanishing on M form a free module of rank $n - \mathrm{rk}\, M$.* ∎

We now give two examples to show that the relation between the ranks of a module and its closure is not very close in general; other examples will appear in the exercises.

Let R be any integral domain; if $x_1, \ldots, x_m$ are *left* linearly independent elements of R, then the elements $(x_1, 0, \ldots, 0), \ldots, (x_m, 0, \ldots, 0)$ of R^n generate a free submodule of rank m, but its closure $(R, 0, \ldots, 0)$ has rank 1. Secondly, let $y_1, \ldots, y_n$ be *right* linearly independent; then it is clear that no non-zero linear functional of R^n vanishes on $(y_1, \ldots, y_n)$, so the submodule generated by this element is dense in R^n. Here we have a submodule of rank 1 whose closure has rank n.

Let us now turn to matrices. An $m \times n$ matrix $A = (a_{ij})$ over a ring R may be interpreted in different ways, as

(a) a right R-module homomorphism of columns ${}^n R \to {}^m R$,
(b) a left R-module homomorphism of rows $R^m \to R^n$,

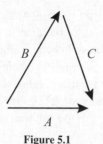
Figure 5.1

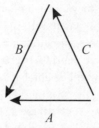
Figure 5.2

(c) an element of the (R_m, R_n)-bimodule ${}^m R \otimes R^n$,
(d) an (R, R)-bilinear mapping $R^m \times^n R \to R$.

Each interpretation leads to a notion of rank, of which some coincide when the ring is suitably specialized.

Proposition 5.4.2. *Let R be an n-fir and $A \in {}^m R^n$. Then the following numbers are equal and do not exceed n:*

(i) *the rank of the submodule of ${}^m R$ spanned by the n columns of A, i.e. rk im A under interpretation (a),*
(ii) *$n - $ rk ker A under interpretation (a),*
(iii) *the rank of the closed submodule generated by the rows of A under interpretation (b).*

Proof. The equality of (i) and (ii) is clear, as well as the fact that the common value cannot exceed n. To prove the equality of (i) and (iii), suppose that the closure of the image of the map $A : R^m \to R^n$ has rank r, the number described in (iii). Then we can factor A as in Fig. 5.1 above. Here B has dense image (its target is the closure of its image); this means that as a matrix it is right regular, while C is the inclusion in a direct summand. Dualizing, we obtain maps represented by the same matrices and still factoring A, as shown in Fig. 5.2, where C is the projection onto a summand and B, being right regular, defines an injective mapping. Hence the image of A in ${}^m R$ is isomorphic to ${}^r R$, where r is the number defined in (i). ∎

The common value of these numbers is called the *column rank* of A. This definition makes it clear that the column rank of the matrix representing a given linear transformation is independent of the choice of bases. In particular, we see that the column rank is unchanged by elementary operations. However, we have had to restrict the class of rings (to n-firs) to ensure that the image is a free module. Sometimes one defines the *column nullity* of A as rk ker A; unlike the column rank this is invariant under stable association.

Over an m-fir we can similarly define the *row rank* of A. Now case (i) of Proposition 4.2 (and its left–right dual) relates these numbers to the usual row and column ranks of matrices, over a commutative integral domain, say. In the case of an $m \times n$ matrix A over a max(m, n)-fir we have both kinds of ranks and in general they will be unrelated. This is clear from the examples given earlier, if we note that the row rank of A equals the rank of $R^m.A$, while the column rank equals the closure thereof. Denoting the row and column ranks by ρ_r, ρ_c, respectively, we have for a product of matrices:

$$\rho_r(AB) \le \rho_r(A), \quad \rho_c(AB) \le \rho_c(B).$$

It is also possible to formulate characterizations of these two types of rank in terms of interpretations (c) and (d) of the matrix A and the reader may wish to do this, but they are less simple. Various versions of the definitions of row and column rank may also be formulated for matrices over rings other than n-firs, but they are not in general equivalent. However, the inner rank, defined in Section 0.1, is naturally defined over any ring. Moreover, it is left–right symmetric and will be of importance later. Our next result gives some equivalent ways of describing it:

Proposition 5.4.3. *Let R be any ring and $A \in {}^m R^n$. Then the following four numbers are equal and do not exceed min(m,n):*

(i) *the least r such that the map A (under interpretation (a) or (b)) can be factored through R^r,*
(ii) *the least r such that A can be written $\sum_1^r b_i \otimes c_i$, under interpretation (c),*
(iii) *the least r such that the image of A in R^n is contained in a submodule generated by r elements (interpretation (b)) and*
(iv) *the least r such that the image of A in ${}^m R$ is contained in a submodule generated by r elements (interpretation (a)).*

Further, this number does not exceed $\rho_r(A)$ in an m-fir and does not exceed $\rho_c(A)$ in an n-fir.

Proof. This is a straightforward consequence of the definitions, details of which may be left to the reader. ∎

From (ii) we see that the number described in this proposition is just the inner rank $\rho(A)$. It is most conveniently determined as the least r such that

$$A = BC, \text{ where } B \in {}^m R^r, C \in {}^r R^n. \tag{1}$$

Any factorization of A as in (1) with $r = \rho(A)$ is a rank factorization of A, as defined earlier. Using rank factorizations, we can verify the following inequalities without difficulty:

$$\rho(A \oplus B) \leq \rho A + \rho B, \qquad (2)$$

$$\rho(AB) \leq \min\{\rho A, \rho B\}, \qquad (3)$$

$$\rho((A'\ A'')) \geq \max\{\rho A', \rho A''\}. \qquad (4)$$

We also note that for any $a \in R$, $\rho a = 0$ if and only if $a = 0$, for then (and only then) can a be written as a product of a 1×0 by a 0×1 matrix. For any $a \neq 0$ in R we have of course $\rho a = 1$. From (4) and its transpose we see that the inner rank of any matrix is at least the rank of any submatrix. Like other ranks, the inner rank is unchanged when the matrix is multiplied on the left or right by an invertible matrix. Over any field it reduces to the usual rank; more generally, we have

Proposition 5.4.4. *Over a left Bezout domain the inner rank of any matrix equals its row rank.*

Proof. We know that in any case $\rho A \leq \rho_r A$; we have to prove equality here. Let $\rho A = r$; using interpretation (b), we can factor A by R^r. The image in R^r is a finitely generated submodule of R^r, hence by Proposition 0.8.3 and the sentence that precedes it, of rank at most r, therefore the image in R^n also has rank at most r, i.e. $\rho_r A \leq r$. ∎

In a two-sided Bezout domain, by symmetry, the row rank, column rank and inner rank of any matrix are all equal. Recalling the definition of a left full matrix from Section 3.1, we conclude from Proposition 4.4,

Corollary 5.4.5. *A matrix over a left Bezout domain is left regular if and only if it is left full.* ∎

This result will be generalized in Section 5.5, where the precise class of rings satisfying the conclusion of this corollary will be determined, at least in the commutative case.

We note that over any ring, the full matrices are just the $n \times n$ matrices of inner rank n, for all $n = 1, 2, \ldots$. Over a ring with UGN, every invertible matrix is full. For weakly finite rings we have the following more general result:

Proposition 5.4.6. *Let R be a weakly finite ring and consider a partitioned matrix over R:*

$$M = \begin{pmatrix} A & B \\ C & D \end{pmatrix}, \quad \text{where } A \in R_r.$$

If A is invertible, then $\rho M \geq r$, with equality if and only if $CA^{-1}B = D$.

Proof. The inequality is an immediate consequence of (4) and the remark following it. Now by passing to associates, which leaves the inner rank unchanged, we have

$$M \to \begin{pmatrix} I & A^{-1}B \\ C & D \end{pmatrix} \to \begin{pmatrix} I & A^{-1}B \\ 0 & D - CA^{-1}B \end{pmatrix} \to \begin{pmatrix} I & 0 \\ 0 & D - CA^{-1}B \end{pmatrix}.$$

If $\rho M = s$, so that $s \geq r$, this matrix can be written

$$PQ = \begin{pmatrix} P' \\ P'' \end{pmatrix} (Q' \quad Q''),$$

where P' is $r \times s$ and Q' is $s \times r$. Thus $I = P'Q'$; if $r = s$, then $Q'P' = I$, by weak finiteness, but $P'Q'' = 0$, so $Q'' = 0$; similarly $P'' = 0$ and so $D - CA^{-1}B = P''Q'' = 0$. The converse is clear. ■

We also note the following restatement of Theorem 3.5 in terms of matrices:

Proposition 5.4.7. *Let A be a regular matrix of inner rank r over a fir R. Then there is a factorization*

$$A = PA_0Q,$$

where $A_0 \in R_r$ is full, P is right prime and Q is left prime. Moreover, if $A = P'Q'$ is any rank factorization of A, then $P' = PU$, $Q' = VQ$, $A_0 = UV$, for some matrices $U, V \in R_r$.

Proof. This is a straightforward translation, which may be left to the reader. ■

Sometimes it is possible to describe the inner rank in terms of full matrices alone. This will be useful for us later; we shall need the following lemma:

Lemma 5.4.8. *Let R be any ring.*

(i) Given a matrix A over R, suppose that A is left full but does not remain left full when the first column is omitted. Then there is a factorization

$$A = B \begin{pmatrix} 1 & 0 \\ 0 & C \end{pmatrix},$$

where B is (square) full and C is left full, with one less row and column than A.

(ii) If every full matrix over R is regular, then every left full matrix is left regular and every right full matrix is right regular

Proof. (i) Let A be $m \times n$ and write $A = (a, A')$, where a is $m \times 1$, A' is $m \times (n-1)$ and A' is not left full. Then we can write $A' = DC$, where D is $m \times (m-1)$ and C is $(m-1) \times (n-1)$; now

$$A = (a \quad A') = (a \quad D)\begin{pmatrix} 1 & 0 \\ 0 & C \end{pmatrix}$$

is the desired factorization. Clearly if $B = (a, D)$ is not full or C is not left full, then A cannot be left full.

(ii) Let $A \in {}^m R^n$ be left full; we shall use induction on n. If A remains left full when the first column is omitted, the result follows by induction. Otherwise, by (i), there is a factorization $A = B(1 \oplus C)$, where B is full and C is left full. Suppose that $xA = 0$ for some row x. Then $xB(1 \oplus C) = 0$, hence $xB = 0$, again by induction, and so $x = 0$, as claimed. Similarly every right full matrix is right regular. ∎

Theorem 5.4.9. *Let R be a ring such that the set of all full matrices over R is closed under products (where defined) and diagonal sums. Then the inner rank of a matrix over R is the maximum of the orders of its full submatrices.*

Proof. We may take R to be non-zero, since otherwise there is nothing to prove. Let A be an $m \times n$ matrix of inner rank r, with a rank factorization $A = BC$. We shall show by induction on $m + n$ that A has a full $r \times r$ submatrix (clearly the largest possible). Since the product of two full matrices, where defined, is again full, by assumption, we need only show that B and C each have a full $r \times r$ submatrix, and by symmetry it suffices to consider C. Now C is left full; if it remains left full when the first column is omitted, then by induction the truncated matrix, hence C itself, has a full $r \times r$ submatrix. So we may assume that C does not remain full when the first column is omitted. By Lemma 4.8 (i) there is a factorization $C = D(1 \oplus E)$, where D is full and E is left full. By the induction hypothesis E has a full $(r-1) \times (r-1)$ submatrix, and since the full matrices admit diagonal sums, $1 \oplus E$ has a full $r \times r$ submatrix, and its product with D is the desired full $r \times r$ submatrix of C. ∎

A ring homomorphism is said to be *honest* if it keeps full matrices full. With this definition we have

Corollary 5.4.10. *Every honest homomorphism of a ring into a field preserves the inner rank.*

Proof. The hypothesis ensures that the class of all full matrices over R is closed under products and diagonal sums; moreover, a non-full matrix necessarily maps to a non-full matrix, so the result follows by applying Theorem 4.9. ∎

There is a criterion for projective-freeness in terms of full matrices that is often useful:

Proposition 5.4.11. *Let R be a non-zero ring such that every full matrix over R is regular. Then R is projective-free.*

Proof. The conclusion will follow by Proposition 0.4.7 if we can show that R has IBN and that all idempotent matrices split. Let E be an idempotent matrix over R and take a rank factorization $E = AB$. Since this is a rank factorization, A is right full and B is left full. By hypothesis (and Lemma 4.8) it follows that A is right regular and B is left regular. Now E is idempotent, so $ABAB = AB$; cancelling A on the left and B on the right, we obtain $BA = I$, which shows that E splits.

It remains to show that R has IBN. If this is not so, suppose that $R^n \cong R^{n+k}$ and that n is the least value for which this is true for some $k > 0$. Then there is an $n \times (n+k)$ matrix P and an $(n+k) \times n$ matrix Q that are mutually inverse. Let us write

$$P = (U \quad V), \quad Q = \begin{pmatrix} X \\ Y \end{pmatrix},$$

where U, X are $n \times n$, V is $n \times k$ and Y is $k \times n$. Thus $XU = I, YV = I, XV = 0, YU = 0$; therefore U is not full. We take a rank factorization $U = AB$, where A is $n \times r$, B is $r \times n$ and $r < n$. We have $XAB = XU = I_n$, $BXAB = B$, and cancelling B on the right, we find $BXA = I_r$. It follows that $R^n \cong R^r$, which contradicts the hypothesis on n; hence R has IBN and so is projective-free. ∎

The converse of Proposition 4.11 is false, as an example in Section 5.5 will show. However, the hypothesis is satisfied for all semifirs, as we saw in Section 3.1.

Exercises 5.4

1. Let R be any ring with a field of fractions K (i.e. a field containing R as a subring and generated by R as a field). Define the *K-rank* $\mathrm{rk}_K M$ of any right R-module K as the dimension over K of $M \otimes K$. If R is a left Ore domain with field of fractions K and M is a finitely generated right K-module, show that $\mathrm{rk}_K M = \mathrm{rk}_K M^*$, and that $\mathrm{rk}_K M$ is the maximum length of chains of closed submodules in M.
2. (Klein [72a]) Let $n \geq 1$ and let R be a ring such that any chain of closed submodules of R^n has length at most n; show that any nilpotent $n \times n$ matrix A over R satisfies $A^n = 0$. If R is an integral domain, prove the converse (see also Proposition 2.3.16).

5.4 The ranks of matrices

3. Let R be a ring such that $R = \sum_1^n x_i R$; show that $(x_1, \ldots, x_n)$ generates a direct summand of R^n and hence a closed submodule that is free of rank 1.
4. Let R be a non-zero ring. Show that there is a universal bound for the inner rank of matrices over R if and only if R does not satisfy UGN.
5. (Bergman [67]) Let R be an integral domain and x, y two right linearly independent elements of R. Show that each of $(x, y, 1)$, $(0, 0, 1)$ generates a closed submodule of R^3, but their sum is dense in R^3. Show that if R is any non-Ore 3-fir, then the lattice of closed submodules of R^3 is not modular.
6. (Bergman [67]) Given rings $R \subseteq S$ and an R-module M, a submodule M' is said to be S-closed if it is the zero-set of a family of maps $M \to S$; thus 'R-closed' is the same as 'closed'. Show that every closed submodule of M is S-closed. If R is a right Ore domain and K its field of fractions, show that every K-closed submodule of a finitely generated left R-module is closed; further, if M is a finitely generated right R-module, show that there is a natural bijection between the set of K-closed submodules of M and the K-submodules of $M \otimes K$. Deduce that if R is a left and right Ore domain with field of fractions K and M is a finitely generated right R-module, then the lattice of closed submodules of M is isomorphic to $\mathrm{Lat}_K(M \otimes K)$, and hence is modular.
7. Let R be a semifir and M a left R-module with generators $e_1, \ldots, e_n$ and defining relations $\sum a_{vi} e_i = 0$, where $A = (a_{vi})$ is a given matrix. Show that M is bound if and only if $\rho_c(A) = n$. When $R = k\langle x, y \rangle$, give an example of a bound module with positive characteristic.
8*. Show that a ring R is projective-free if and only if, for any $n \times n$ idempotent matrix E of inner rank r, $\rho(I - E) \leq n - r$. Show also that a non-zero ring R is Hermite if and only if, for any $A \in {}^rR^n$, $B \in {}^nR^r$ such that AB is invertible, and any $C \in {}^nR^m$, $AC = 0$ implies $\rho C \leq n - r$.
9*. Give a proof of Proposition 4.11 by proving directly that every finitely generated projective module is free, of unique rank. (*Hint*: Use Lemma 0.3.3).
10°. Give an example of a projective-free ring that is not an integral domain. (This would provide a counter-example to the converse of Proposition 4.11. For another counter-example, see Section 5.5.)
11. Find all rings over which every full matrix is invertible.
12. Show that over a semifir any matrix stably associated to a full matrix is again full.
13. Let R be an n-fir. If $\rho \begin{pmatrix} A & C \\ 0 & B \end{pmatrix} \leq n$, show that $\rho \begin{pmatrix} A & C \\ 0 & B \end{pmatrix} \geq \rho A + \rho B$. Show also that if (A, B, C) is any matrix over R such that $\rho(A, B) + \rho(A, C) \leq n$, then $\rho(A, B, C) \leq \rho(A, B) + \rho(A, C) - \rho(A)$.
14. Let R be a left Ore domain. Show that any $r \times s$ matrix over R, where $r > s$, annihilates a non-zero row. Defining the row rank, resp. column rank, of a matrix as the maximum number of linearly independent rows, resp. columns, show that over a two-sided Ore domain, the row and column ranks of an $r \times s$ matrix are equal (their common value may be called the *outer rank*).
15. Show that a commutative integral domain for which the inner and outer ranks coincide has the property: $a | b_1 b_2 \Rightarrow a = a_1 a_2$, where $a_i | b_i$. (*Hint*: Consider 2×2 matrices; rings with the stated property are called *primal*.)

16. (McAdam and Rush [78]) Let R be a commutative integral domain with field of fractions K. Show that the mapping $R \to K$ preserves the inner rank of 2×2 matrices if and only if the inverse of every finitely generated fractional ideal of R is semicyclic (i.e. every finitely generated submodule is contained in a cyclic module), or equivalently, R is primal.
17. Let $f : R \to S$ be a ring homomorphism and $f_n : R_n \to S_n$ the induced homomorphism of matrix rings. Show that if f preserves the inner rank, then so does f_n.
18*. Let R be a fir that is not right principal and let $\mathfrak{a}$ be a two-sided ideal whose rank as right ideal is finite, equal to $r > 1$. If $u = (u_1 \ldots u_r)$ is a basis of $\mathfrak{a}$, verify that the map $\rho : R \to R_r$ defined by $au = u.\rho(a)$ is a homomorphism. Show that this map is never honest.

5.5 Sylvester domains

In Section 5.4 we saw that the various ranks defined there coincide for Bezout domains and this raises the problem of determining the precise class of rings for which this holds. Even in the commutative case this includes more than just Bezout domains, since the ranks agree, for example, for the polynomial ring $k[x, y]$, as will follow from Corollary 5.5. However, it fails for $k[x, y, z]$, as is shown by the matrix

$$\begin{pmatrix} 0 & z & -y \\ -z & 0 & x \\ y & -x & 0 \end{pmatrix}, \tag{1}$$

which is full, yet a (left and right) zero-divisor (see the remarks after (8) below and Exercise 4).

In this section we shall introduce a class of rings, following Dicks and Sontag [78], which in the commutative case gives a complete answer to these questions. We begin by recalling part of Proposition 3.1.3, which we shall formulate as follows:

Proposition 5.5.1. (Law of nullity) *Let R be an n-fir and $P \in {}^r R^n$, $Q \in {}^n R^s$. If*

$$PQ = 0, \tag{2}$$

then

$$\rho P + \rho Q \leq n. \tag{3}$$

Proof. Let $P = P'A$ and $Q = BQ'$ be rank factorizations for P, Q; then P' is right full and hence right regular, by Lemma 3.1.1. Similarly, Q' is left regular,

5.5 Sylvester domains

and so (2) implies the full relation AB = 0. Hence by applying Proposition 3.1.3, we obtain (3). ∎

This result suggests the following:

Definition. A non-zero ring R is called a *Sylvester domain* if for any $P \in {}^rR^n$, $Q \in {}^nR^s$ such that $PQ = 0$, it follows that $\rho P + \rho Q \le n$.

As an immediate consequence of this definition we obtain Sylvester's law of nullity for the inner rank:

Corollary 5.5.2. *Let R be a Sylvester domain and $A \in {}^rR^n$, $B \in {}^nR^s$. Then*

$$\rho A + \rho B \le n + \rho(AB). \tag{4}$$

Hence the product of two full matrices (of the same order) is full.

Proof. If $AB = PQ$ is a rank factorization, so that Q has $\rho(AB)$ rows, then

$$(A \quad P)\begin{pmatrix} B \\ -Q \end{pmatrix} = 0,$$

hence by Proposition 5.1, since $(A \quad P)$ has $n + \rho(AB)$ columns,

$$n + \rho(AB) \ge \rho(A \quad P) + \rho\begin{pmatrix} B \\ -Q \end{pmatrix} \ge \rho A + \rho B.$$

This proves (4) and now the last part clearly follows. ∎

Taking $n = r = s = 1$, we see that there are no zero-divisors, so a Sylvester domain is indeed an integral domain. By Proposition 5.1, every semifir is a Sylvester domain, though again the converse does not hold. We begin by deriving some consequences of the law of nullity:

Lemma 5.5.3. *Let R be a Sylvester domain. Then R is projective-free and for any matrices A, B over R,*

$$\rho(A \oplus B) = \rho A + \rho B. \tag{5}$$

Further, if A, B, C are any matrices over R with the same number of rows and if $\rho(A, B) = \rho(A, C) = \rho A$, then

$$\rho(A \quad B \quad C) = \rho A. \tag{6}$$

Proof. The definition shows that every full matrix over R is regular, hence R is projective-free, by Proposition 4.11. To prove (5) we have, by suitably partitioning a rank factorization of $A \oplus B$,

$$\begin{pmatrix} A & 0 \\ 0 & B \end{pmatrix} = \begin{pmatrix} P \\ P' \end{pmatrix}(Q \quad Q'),$$

where P has the same number of rows as A and its number of columns is $\rho(A \oplus B)$. Since $PQ' = 0$, we have by (3),

$$\rho(A \oplus B) \geq \rho P + \rho Q' \geq \rho(PQ) + \rho(P'Q') = \rho A + \rho B \geq \rho(A \oplus B),$$

where the last step follows by (2) of Section 5.4, and this proves (5). To establish (6) we partition rank factorizations of $(A, B), (A, C)$ as $(A, B) = D(E, E'), (A, C) = F(G, G')$; by hypothesis, $A = DE = FG$ are also rank factorizations of A. Thus the number of columns of (D, F) is $2\rho A$ and since

$$(D \quad F)\begin{pmatrix} -E \\ G \end{pmatrix} = 0,$$

we have, by (3),

$$2\rho A \geq \rho(D \quad F) + \rho\begin{pmatrix} -E \\ G \end{pmatrix} \geq \rho D + \rho G \geq \rho(DE) + \rho(FG)$$
$$= \rho A + \rho A.$$

Thus equality holds throughout; in particular, $\rho A = \rho(D, F)$. Clearly, $\rho A \leq \rho(A, B, C)$, and since

$$(A \quad B \quad C) = (D \quad F)\begin{pmatrix} E & E' & 0 \\ 0 & 0 & G' \end{pmatrix},$$

we have $\rho(A, B, C) \leq \rho(D, F) = \rho A$. ■

We now come to a result providing a source of Sylvester domains that are not semifirs:

Theorem 5.5.4. *Let A be a commutative principal ideal domain and X any set. Then $A\langle X \rangle$ is a Sylvester domain.*

Proof. Writing K for the field of fractions of A, we have a homomorphism $A\langle X \rangle \to K\langle X \rangle$ and it will be enough to show that this map is inner rank preserving, for then the law of nullity will hold in $A\langle X \rangle$.

To prove that the inner rank is preserved it suffices to show that every matrix over $A\langle X \rangle$ is inert in $K\langle X \rangle$, and since $K = \cup A[c^{-1}]$, where c ranges over $A^\times$, we need only verify inertia for $A[c^{-1}]\langle X \rangle$. Writing c as a product of atoms and using induction, we need to show that every matrix over $A\langle X \rangle$ is inert in $A[p^{-1}]\langle X \rangle$, where p is an atom in A. We put $S = A\langle X \rangle$; p is a central atom in S and $S/pS = k\langle X \rangle$, where $k = A/pA$ is a field. Hence $R = S/pS$ is a fir, and it is a GE-ring, even an E-ring, by Theorem 2.4.4, so we can apply the inertia lemma (Lemma 4.6.8) to conclude that every matrix of S is inert in $S[p^{-1}]$. ■

If in this theorem we take $X = \{x\}$ and $A = k[y]$, where k is a commutative field, we find that $A\langle X\rangle = k[x, y]$, and we obtain

Corollary 5.5.5. *The polynomial ring in two variables over any commutative field is a Sylvester domain.* ∎

For three variables this is no longer true, as is shown by the fact that the matrix in (1) is full, yet a zero-divisor (see the remarks after (8)). A source of Sylvester domains is the result of Dicks and Sontag [78], which states that the coproduct of any family of Sylvester domains over a skew field is again a Sylvester domain; this is also a consequence of lemma 5.7.5 of SF. By applying this result to Corollary 5.5 above we see that the k-algebra generated by $x_1, \ldots, x_4$ with defining relations $x_1 x_2 = x_2 x_1$, $x_3 x_4 = x_4 x_3$ is a Sylvester domain. As another result of Dicks and Sontag [78], we have

Theorem 5.5.A. *For any positive integers m, r, n let $R(m, r, n)$ be the k-algebra on $r(m + n)$ generators written as an $m \times r$ matrix $X = (x_{ij})$ and an $r \times n$ matrix $Y = (y_{jk})$ with defining relations in matrix form $XY = 0$. Then*

(i) every full matrix over $R(m, r, n)$ is left regular if and only if $r > n$,
(ii) every full matrix over $R(m, r, n)$ is regular if and only if $r > \max(m, n)$,
(iii) $R(m, r, n)$ is a Sylvester domain if and only if $r \geq m + n$. ∎

We omit the proof, which uses Bergman's coproduct theorems, and merely observe that the ring $R(m, r, n)$ is an $(r - 1)$-fir, but not an r-fir, by a slight modification of Theorem 2.11.2, or also SF, theorem 5.7.6.

In order to study Sylvester domains, we shall for a moment consider a somewhat wider class: the rings whose full matrices are left regular. This is a one-sided class; if we impose the corresponding right-hand condition as well, by requiring full matrices to be regular, we obtain projective-free rings, by Proposition 4.11, but this class is still larger than the class of Sylvester domains. Some further definitions will be needed. For any ring R, an R-module will be called *spatial* if for every $n \geq 1$, any set of n R-linearly dependent elements lies in a submodule generated by $n - 1$ elements. An easy induction shows that in a spatial module every finite subset is contained in a free submodule, i.e. every spatial module is semifree; in particular, a spatial module is always flat. We shall define the *local rank* of any R-module M as the least integer n such that every finite subset of M lies in an n-generator submodule; of course M may very well not have a local rank. To give an example, the rational field $\mathbb{Q}$ as $\mathbb{Z}$-module has local rank 1, while the real field $\mathbb{R}$ has no local rank, and for a finitely generated $\mathbb{Z}$-module the local rank is just the minimal generating number.

Theorem 5.5.6. *In any ring R the following conditions are equivalent:*

(a) every full matrix is left regular,
(b) every left full matrix is left regular,
(c) every flat left R-module is spatial,
(d) every free left R-module is spatial,
(e) the right annihilator of any non-zero $m \times n$ matrix has local rank $< n$,
(f) the right annihilator of any non-zero vector in R^n has local rank $< n$.

When these conditions hold, then the kernel of any homomorphism between spatial right R-modules is again spatial.

Proof. We shall prove (a) $\Rightarrow$ (b) $\Rightarrow$ (e) $\Rightarrow$ (f) $\Rightarrow$ (a) and (b) $\Rightarrow$ (c) $\Rightarrow$ (d) $\Rightarrow$ (b). (a) $\Rightarrow$ (b) follows as in the proof of Lemma 4.8 (ii). (b) $\Rightarrow$ (e): Let $A \neq 0$ be an $m \times n$ matrix. Any finite set of columns in the right annihilator of A can be written as a matrix B such that $AB = 0$. Now for any rank factorization $B = CD$, D is left full and so is left regular by (b), hence $AC = 0$. Again by (b), C is not left full, but it is right full, by definition, therefore it has at most $n - 1$ columns. Now the columns of $B = CD$ lie in the submodule of the right annihilator of A generated by the columns of C, which shows that the right annihilator of A has local rank at most $n - 1$.

(e) $\Rightarrow$ (f) is clear, and (f) $\Rightarrow$ (a): Let A be an $n \times n$ matrix and suppose that $XA = 0$ for some non-zero $X \in R^n$. Then the columns of A lie in the right annihilator of X, which by (f) has local rank less than n. Thus the columns of A lie in an $(n - 1)$-generator submodule of ${}^n R$, so A is not full.

(b) $\Rightarrow$ (c). Let M be a flat left R-module and take any finite linearly dependent subset of M, arranged as a column $X \in {}^n M$, say. By hypothesis there is a non-zero $A \in R^n$ such that $AX = 0$. Since M is flat, this comes from an R-relation, say $X = BY$, where $B \in {}^n R^m$, $Y \in {}^m M$ and $AB = 0$. Thus B is not left regular and by (b) cannot be left full, say $B = CD$, where $C \in {}^n R^{n-1}$, $D \in {}^{n-1} R^m$. Then $X = BY = C(DY)$, so the components of X lie in the $(n - 1)$-generator submodule of M generated by the entries of DY, and this shows M to be spatial.

(c) $\Rightarrow$ (d) is obvious, and (d) $\Rightarrow$ (b) by the following chain of statements, each of which implies the next. (1) Every free left R-module is spatial. (2) For all $m, n \geq 1$, any set of m left R-linearly dependent elements of R^n lies in an $(m - 1)$-generator submodule. (3) For all $m, n \geq 1$, any $m \times n$ matrix that is not left regular is the product of an $m \times (m - 1)$ by an $(m - 1) \times n$ matrix. (4) Every matrix that is not left regular is not left full. (5) Every left full matrix is left regular.

5.5 Sylvester domains

Assume now that (a)–(f) hold and that $\alpha : M \to N$ is a homomorphism between spatial right R-modules. Any finite linearly dependent subset of $\ker \alpha$ can be written as a row, say $X \in (\ker \alpha)^m$. We write $X = YA$, where $Y \in M^r$, $A \in {}^rR^m$ and A, Y are chosen so that r is minimal. Then A has inner rank r, i.e. it is left full and so left regular. Since M is spatial, we have $r < m$ and the elements of Y are linearly independent. Now $\alpha(Y) \in N^r$ and there is again a factorization $\alpha(Y) = ZB$, where the elements of Z are linearly independent, thus Z is right regular. We have

$$0 = \alpha(X) = \alpha(YA) = \alpha(Y)A = ZBA;$$

since Z is right regular and A is left regular, it follows that $B = 0$, so $\alpha(Y) = 0$. Hence $Y \in (\ker \alpha)^r$ and this shows $\ker \alpha$ to be spatial. ∎

Thus for the rings of this theorem, spatial, flat and semifree left modules are the same. We note that the class of rings described here is wider than the class of Sylvester domains, since (a) is satisfied in any Sylvester domain. Of course the conclusions of Theorem 5.6 are not left–right symmetric, i.e. there exists a ring R such that R satisfies the conditions but R° does not, and Theorem 5.A provides examples of rings satisfying these conditions but not their left–right duals. The relation with Sylvester domains is more closely described by

Corollary 5.5.7. *Every Sylvester domain has weak global dimension at most 2 and is projective-free, and every flat module is spatial, hence semifree.*

Proof. Over a Sylvester domain R, every full matrix is regular, so the conditions of Theorem 5.6 and their left–right duals hold. It follows that every flat R-module is spatial, hence semifree, and the kernel of any homomorphism between flat modules is flat; hence w.gl.dim.$R \leq 2$. Further, R is projective-free by Proposition 4.11. ∎

It is not known whether conversely, the conditions of Theorem 5.6 (and its left–right dual) will ensure that R is a Sylvester domain, but the ring will be projective-free, by Proposition 4.11. If R is projective-free of weak global dimension at most 1, and moreover right coherent, then every finitely generated right ideal is finitely related and flat, hence projective, and so is free, of unique rank. So we have

Corollary 5.5.8. *Every projective-free ring that is right coherent of weak global dimension at most 1 is a semifir.* ∎

For Ore domains there is a converse to Theorem 5.6 and Corollary 5.7. In that case the class of rings described in Theorem 5.6 coincides with the class of Sylvester domains.

Theorem 5.5.9. *For any two-sided Ore domain R with field of fractions K, the following conditions are equivalent:*

(a) R is a Sylvester domain,
(b) $w.gl.dim.R \leq 2$ and every flat R-module is semifree,
(c) the right annihilator of any matrix is semifree,
(d) the right annihilator of any row vector is semifree,
(e) every full matrix is left regular,
(f) the embedding $R \to K$ preserves the inner rank.

$(a°)–(f°)$ *The left–right duals (a)–(f).*

Proof. For any ring R we clearly have (a)$\Rightarrow$(b) $\Rightarrow$ (c) $\Rightarrow$ (d).

(d) $\Rightarrow$ (e) for any right Ore domain R with field of fractions K: Let $X \in R^n$, $X \neq 0$; if the right annihilator N of X contains n left R-linearly independent elements, then they are left K-linearly independent, but this is impossible because $X \neq 0$. By (d), N is semifree, so it has local rank at most $n - 1$ and now (e) follows by Theorem 5.6.

(e) $\Rightarrow$ (f) for any left Ore domain R with field of fractions K: Any full matrix over R is left regular over R and remains so over K, hence it has a left inverse over K. By Corollary 4.10 the embedding $R \to K$ preserves the inner rank, so (f) holds.

(f) $\Rightarrow$ (a). Sylvester's law of nullity holds in K, therefore it holds in R, so R is a Sylvester domain. Now (a°)–(f°) follow by the evident symmetry of (a). ∎

Let R be an Ore Sylvester domain. Then for any Ore subset S the localization R_S is again Sylvester, by the same reasoning as for (e) $\Rightarrow$ (f) above. Hence we obtain

Corollary 5.5.10. *The localization of any Ore Sylvester domain at an Ore set is again a Sylvester domain.* ∎

To illustrate the result of Theorem 5.9, let R be any ring and M a left R-module, with a finite free resolution

$$\cdots \longrightarrow F_2 \xrightarrow{\alpha_1} F_1 \xrightarrow{\alpha_0} F_0 \longrightarrow M \to 0,$$

where F_i is free of rank n_i. If the matrix representing α_1 can be written as

$$A = BC, \tag{7}$$

where B is right regular and C is left regular, then the rows of C generate the left annihilator of α_0 and are left R-linearly independent, hence ker α_0 is then free and so pd$(M) \leq 2$. Thus for any finitely generated module (over a Noetherian ring, say) of homological dimension at least 3 we obtain a matrix A that cannot be expressed as in (7). For example, if $R = k[x, y, z]$, where k is a commutative field, then there is a minimal resolution for k (Koszul resolution)

$$0 \to R \xrightarrow{A_2} R^3 \xrightarrow{A_1} R^3 \xrightarrow{A_0} R \to k \to 0, \tag{8}$$

with presenting matrices

$$A_2 = (x \quad y \quad z), \quad A_1 = \begin{pmatrix} 0 & z & -y \\ -z & 0 & x \\ y & -x & 0 \end{pmatrix}, \quad A_0 = \begin{pmatrix} x \\ y \\ z \end{pmatrix}.$$

If A_1 were not full, a minimal factorization as in (7) would allow us to replace (8) by a shorter sequence, in contradiction to the fact that $\text{pd}_R k = 3$. Thus A_1 is full, but clearly it is neither left nor right regular. Since $k[x, y, z]$ is projective-free (the Quillen–Suslin proof of Serre's conjecture, that polynomial rings over a field are projective-free see e.g. Lam [78]), this shows the converse of Proposition 4.11 to be false.

Sometimes we shall want to consider the nullity condition used to define Sylvester domains for infinite matrices; here the inner rank of a matrix with infinitely many rows or columns (or both) is again defined in terms of rank factorizations. Thus we consider the following condition. For all natural numbers r, n and all sets I,

$$\text{if } A \in {}^r R^n, B \in {}^n R^I \text{ and } AB = 0, \text{ then } \rho A + \rho B \leq n. \tag{9}$$

Taking I finite, we see that any non-zero ring satisfying (9) is a Sylvester domain. Moreover, if $\rho B = t$, say, we have a rank factorization $B = CD$, where C is $n \times t$ and D is $t \times I$. Applying (9) to the relation $AC.D = 0$ and remembering that $\rho D = t$, we find that $AC = 0$. This shows R to be right coherent. Conversely, if R is a right coherent Sylvester domain and $AB = 0$ as in (9), then $B = CD$, where C is $n \times t$, for some t, and $AC = 0$. Hence we have $\rho A + \rho B \leq \rho A + \rho C \leq n$, which shows that (9) holds in R. We thus obtain

Proposition 5.5.11. *The class of non-zero rings satisfying (9) consists of all right coherent Sylvester domains.* ∎

In order to describe coherent Sylvester domains we shall need the following characterization of rings of weak global dimension at most 2:

Lemma 5.5.12. *A ring R has weak global dimension at most 2 if and only if the dual of every finitely presented left R-module is flat.*

Proof. Let M be a finitely presented left R-module, say

$$F_1 \xrightarrow{\alpha} F_0 \to M \to 0, \tag{10}$$

where F_0, F_1 are free of finite rank. Then by dualizing we obtain the exact sequence

$$0 \longrightarrow M^* \longrightarrow F_0^* \xrightarrow{\alpha^*} F_1^*, \qquad (11)$$

for the dual M^*. If w.gl.dim.$R \leq 2$, then M^* is flat, as kernel of α^*. Conversely, if the condition of the lemma holds, then the right annihilator of every matrix is flat. To reach the conclusion it suffices to show that any homomorphism $F_1 \to F_0$ of free right R-modules has a flat kernel K. Since flatness is a local condition, it is enough to verify that any finite subset X of K lies in a flat submodule of K. Now X lies in a finitely generated free submodule F_1' of F_1 and the image of F_1' lies in a finitely generated free submodule F_0' of F_0. Thus the kernel of the map $F_1' \to F_0'$ is the right annihilator of a matrix, hence it is flat, and as a submodule of K containing X it has the requisite properties. ∎

Since any Sylvester domain has weak global dimension at most 2 (by Corollary 5.7), the conditions of the next result hold in any right coherent Sylvester domain.

Proposition 5.5.13. *For any ring R the following conditions are equivalent:*

(a) R is right coherent and has weak global dimension at most 2,
(b) the dual of every finitely presented left R-module is finitely generated projective,
(c) the dual of every right R-module is flat.

Proof. (a) ⇔ (b). R is right coherent if and only if the dual of every finitely presented left R-module is finitely presented (see Appendix Theorem B.9). Using Lemma 5.12, we see that (a) is equivalent to the dual of every finitely presented left R-module being finitely presented and flat, hence projective, i.e. (b).

(a) ⇒ (c). Here we shall use the characterization of right coherent rings as rings for which the dual of every free right module is flat. For any right R-module M there is a presentation (10), where F_0, F_1 are free right R-modules. Dualizing gives M^* as the kernel of a homomorphism between flat modules (as in (11)), hence M^* is flat, because the weak global dimension is at most 2. Now (c) ⇒ (a) follows because coherence is clear from the characterization just used, while the condition on the weak global dimension follows from the left–right dual of Lemma 5.12. ∎

By adding right coherence we obtain a consequence of Theorem 5.9.

Theorem 5.5.14. *For any two-sided Ore domain R the following conditions are equivalent:*

(a) R is a right coherent Sylvester domain,
(b) the dual of every right R-module is semifree,
(c) R is projective-free, $w.gl.dim.R \leq 2$ and R is right coherent,
(d) the dual of every finitely presented left R-module is free,
(e) the right annihilator of every matrix is free,
(f) the right annihilator of every row vector is free.

Proof. (a) $\Rightarrow$ (b). When (a) holds, R has weak global dimension at most 2, so by Proposition 5.13 the dual of every right R-module is flat, and by Theorem 5.9, semifree.

(b) $\Rightarrow$ (c) $\Rightarrow$ (d) is clear from Proposition 5.13, and (d) $\Rightarrow$ (e) $\Rightarrow$ (f) follows directly.

(f) $\Rightarrow$ (a). If the right annihilator of every row vector is free, then R is a Sylvester domain by Theorem 5.9, and it is right coherent, because over a right Ore domain a free submodule of a free right module of finite rank again has finite rank. ■

As for Corollary 5.10 we see that any localization of a right coherent Ore Sylvester domain is again a right coherent Ore Sylvester domain.

Exercises 5.5

1. Show that, for any free group F and any commutative PID A, the group ring AF is a Sylvester domain.
2. Let R be a ring. Show that any non-zero matrix $A \in {}^mR^n$ has a right annihilator that is free of rank $< n$ and a left annihilator free of rank $< m$ if and only if R is left and right coherent and every full matrix is regular.
3. (Dicks and Sontag [78]) Show that every Sylvester domain of weak global dimension at most one is a semifir.
4. Give a direct proof that the matrix A_1 in (8) is full. (*Hint*: Examine the 2×2 submatrices of the factors in a rank factorization of A_1.) Show also that $A_1 \oplus x$ is not full. Further, show that the result of multiplying the last column of A_1 by z is not full.
5. In any Sylvester domain, prove Frobenius' inequality: if AB and BC are defined, then $\rho(AB) + \rho(BC) \leq \rho(ABC) + \rho B$. (*Hint*: Take rank factorizations $B = FG$ and $ABC = PQ$ and consider the relation $AF.GC - P.Q = 0$.)
6. Let R be a right coherent Sylvester domain and $\mathfrak{a}$ a non-zero finitely generated right ideal. Show that $\chi(\mathfrak{a}) > 0$; deduce that any negative right R-module is bound.
7°. Let R be a right coherent Sylvester domain. Which finitely generated right R-modules of positive characteristic and homological dimension 1 can be embedded as right ideals in R? Which can be embedded as submodules in free R-modules?
8. Show that the property of being a coherent Ore Sylvester domain is preserved by localization.

9°. Let R be a Sylvester domain whose centre C is not a field and let k be the field of fractions of C. What can be said about $R \otimes_C k$? Under what conditions is it a semifir?

10. Show that if R is a persistent semifir, then $R[x]$ is a Sylvester domain (see Cohn [82c]). Show that any coherent Sylvester domain is a strictly positive left module, i.e. $\chi(\mathfrak{a}) > 0$ for any finitely presented left ideal $\mathfrak{a}$ of R (see Cohn [82d]).

11*. Show that if R, S are semifirs, both k-algebras, their tensor product $R \otimes_k S$ need not be a Sylvester domain (see Cohn [97a]).

12. Fill in the details of the following proof that every Sylvester domain has UGN. Let $I_n = PQ$ be a rank factorization, where Q has $n-1$ rows. If P_0 denotes the first $n-1$ rows of P and Q_1 the last $n-1$ columns of Q, show that P_0, Q_1 are full and $\rho(P_0 Q_1) = n - 2$; hence obtain a contradiction.

5.6 Pseudo-Sylvester domains

Sylvester domains share at least some of the good properties of semifirs, the main one being that they have a universal field of fractions, to be described in Chapter 7. It turns out that this property holds for an even wider class of rings, the pseudo-Sylvester domains, whose definition resembles that of Sylvester domains, but with the inner rank replaced by the stable rank, defined in Section 0.1. Moreover, these rings arise naturally in the study of localizations; we shall therefore briefly discuss them here.

We recall that in any ring R the stable rank $\rho^* A$ of a matrix A, defined as the limit of $\rho(A \oplus I_n) - n$, is finite precisely when R has UGN and it then satisfies the inequality

$$0 \leq \rho^* A \leq \rho A. \tag{1}$$

Further, when the stable rank exists in R, then

$$\rho^*(A \oplus I_r) = \rho^* A + r, \quad \text{for all } r \geq 0, \tag{2}$$

while

$$\rho^*(A \oplus I_r) = \rho(A \oplus I_r) \quad \text{for all sufficiently large } r. \tag{3}$$

Thus by taking the diagonal sum of A with a unit matrix of sufficiently high order we can *stabilize* the rank, i.e. make the stable rank equal to the inner rank. We shall call A *stabilized* if $\rho^* A = \rho A$.

Definition. (i) An *S-ring* is a ring for which the stable rank is defined and satisfies the law of nullity. (ii) A *pseudo-Sylvester domain* is a weakly finite S-ring.

We recall from Proposition 0.1.3 that every non-zero matrix over a ring R has positive stable rank if and only if R is weakly finite. Thus in a Sylvester domain $\rho^* a = 1$ for any $a \neq 0$, and so the law of nullity shows that a pseudo-Sylvester domain is indeed an integral domain. It also follows that every Sylvester domain is a pseudo-Sylvester domain, as the terminology suggests. More precisely we have

Proposition 5.6.1. *A ring is a Sylvester domain if and only if it is an S-ring and an Hermite ring.*

Proof. Any Sylvester domain is clearly an S-ring, and being projective-free (Corollary 5.7), it is Hermite. Conversely, if R is an S-ring and Hermite, then by Proposition 0.4.4 the stable rank exists and agrees with the inner rank, so we have a Sylvester domain. ∎

Since an Hermite ring is weakly finite, a Sylvester domain is certainly a pseudo-Sylvester domain, but examples will soon show that the latter form a wider class. We shall find that much of the theory of Sylvester domains has a parallel for pseudo-Sylvester domains. We first note

Proposition 5.6.2. *Over a pseudo-Sylvester domain every finitely generated projective module is stably free.*

Proof. Let P be a finitely generated projective left R-module, defined as cokernel of the idempotent matrix E; then $P \oplus R^m \oplus 0^n$ is given by the cokernel of the idempotent matrix $F = E \oplus 0_m \oplus I_n$ and we have $I - F = (I - E) \oplus I_m \oplus 0_n$. By taking m, n large enough we can ensure that F and $I - F$ are stabilized. If F is $N \times N$, $\rho F = r$, $\rho(I - F) = s$, then $r + s \leq N$. Let $F = A_1 B_1$, $I - F = A_2 B_2$ be rank factorizations; then

$$I = (A_1 \quad A_2) \begin{pmatrix} B_1 \\ B_2 \end{pmatrix},$$

and weak finiteness shows that $N \leq r + s$, and $A = (A_1, A_2)$, $B = (B_1, B_2)^T$ are mutually inverse; therefore $B_1 A_1 = I$ and $P \oplus R^m$ is free. ∎

In order to describe pseudo-Sylvester domains more closely we shall prove an analogue of Theorem 4.9, which requires the next lemma. Here a matrix A is called *stably left full* if $A \oplus I_r$ is left full for all $r \geq 1$; similarly for stably right full. An $n \times n$ matrix A is called *stably full* if $A \oplus I_r$ is full for all $r \geq 0$ or equivalently, $\rho^* A = n$.

Lemma 5.6.3. *Let R be a ring with UGN in which the product of stably full matrices (where defined) is stably full. Suppose that C is a stably left full $r \times n$*

matrix over R such that the matrix consisting of the last s columns, where $s \leq r \leq n$, has stable rank s. Then C has a stably full $r \times r$ submatrix that includes the last s columns.

Proof. We shall use induction on $n - s$. When $s = n$, C is stably full and there is nothing to prove, so we may assume that $n > s$. Let $C = (C_1, C')$, where C_1 is the first column of C. If C' is stably left full, then by the induction hypothesis it has a submatrix with the desired property. Otherwise we can find $t \geq 1$ such that $C' \oplus I_t$ is not left full, but of course, $C \oplus I_t$ is left full. Hence, by Lemma 4.8, we have $C \oplus I_t = D(1 \oplus E)$, where D is stably full and E is stably left full. Clearly the submatrix of E consisting of any subset of its columns has stable rank at least equal to that of the corresponding submatrix of $C \oplus I_t$; in particular the submatrix consisting of the last $s + t$ columns is stably full. Now E has $(n - 1) + t$ columns and $((n - 1) + t) - (s + t) = n - s - 1 < n - s$, so we can apply induction to find a stably full submatrix M of E of order $(n - 1) + t$ that contains the last $s + t$ columns of E. Then $1 \oplus M$ is a stably full submatrix of $1 \oplus E$ containing the last $s + t$ columns, so $D(1 \oplus M)$ is a submatrix of columns of $C \oplus I_t$ containing the last $s + t$ columns, and it is stably full, by our hypothesis on products. If we omit the last t columns and the corresponding rows (which after the removal of those columns consist of zeros), we get a stably full submatrix of columns of C including the last s columns, as required. ∎

We can now prove the analogue of Theorem 4.9.

Proposition 5.6.4. *Let R be a ring with UGN in which the product of stably full matrices over R (where defined) is stably full. Then the stable rank of a matrix over R is the maximum of the orders of its stably full submatrices.*

Proof. Let A be a matrix over R of stable rank r and suppose the rank is stabilized by I_s; thus we can write $A \oplus I_s = BC$, where B is stably right full and C is stably left full, both of rank $r + s$. Now the matrices consisting of the last s rows of B and the last s columns of C have stable rank s, because their product is I_s. By applying Lemma 6.3 to C and its left–right dual to B, we get stably full submatrices consisting of $r + s$ columns of C including the last s columns and $r + s$ rows of B including the last s rows. Their product is a submatrix of $A \oplus I_s$ of order $r + s$ including I_s; by the hypothesis on products its stable rank is $r + s$, so it is of the form $N \oplus I_s$, where N is an $r \times r$ submatrix of A, which is of stable rank r, as required. ∎

We shall also need an analogue of Lemma 5.3.

Lemma 5.6.5. *Let R be an S-ring. Then*

(i) for any matrices A, B over R, $\rho^*(A \oplus B) = \rho^* A + \rho^* B$,
(ii) If A, B, C have the same number of rows and $\rho^*(A, B) = \rho^*(A, C) = \rho^* A$, then

$$\rho^*(ABC) = \rho^* A.$$

Proof. (i) Write $A' = A \oplus I_m$, $B' = B \oplus I_n$, where m, n are chosen so large that $A', B', A' \oplus B'$ are all stabilized. Now the proof of Lemma 5.3 shows that $\rho(A' \oplus B') = \rho A' + \rho B'$, hence $\rho^*(A \oplus B) = \rho^* A + \rho^* B$, as claimed. Similarly, in (ii) we replace A by $A' = A \oplus I_n$ with n chosen so large that $A', (A', B'), (A', C')$ are all stabilized, where $B' = (B, 0)^T, C' = (C, 0)^T$. ∎

By Proposition 6.2, over a pseudo-Sylvester domain every finitely generated projective module is stably free; further, it can be shown that any pseudo-Sylvester domain R satisfies w.gl.dim.$R \leq 2$, and for Ore domains this condition is sufficient as well as necessary for R to be a pseudo-Sylvester domain (see Cohn and Schofield [82]). Thus for any field D, even skew, the polynomial ring $D[x, y]$ in two central indeterminates is a pseudo-Sylvester domain, likewise the first Weyl algebra $A_1(k)$ over any commutative field k. Neither of these rings is a Sylvester domain, unless D is commutative.

Exercises 5.6

1. Prove Sylvester's law of nullity for the stable rank in a pseudo-Sylvester domain, in the form of Corollary 5.2.
2. Show that if for any matrix A over a ring R, $\rho^* A$, defined as $\lim[\rho(A \oplus I_n) - n]$, is not identically $-\infty$, then it is non-negative and this is so precisely when R has UGN.
3. Verify that the Weyl algebra $A_1(k)$ is a pseudo-Sylvester domain but not a Sylvester domain.
4. Use Lemma 6.5 to show that $I = \{a \in R | \rho^* a \leq 0\}$ is an ideal and use Proposition 0.1.3 to deduce that for any ring R, R/I is the maximal weakly finite homomorphic image. Hence obtain another proof of Proposition 0.1.2.
5. Let R be the k-algebra generated by a, b with the defining relation $ba = 1$. Show that R has UGN but is not weakly finite. Verify that $\rho^*(1 - ab) = 0$.
6°. Does the law of nullity hold for the stable rank in the ring R of Exercise 5? (An affirmative answer would provide an example of an S-ring that is not a pseudo-Sylvester domain).
7. (G. M. Bergman) Suppose that R is a weakly semihereditary ring with UGN. Show that R is (a) a Sylvester domain if and only if every finitely generated projective R-module is free (hence if and only if R is a semifir); (b) a pseudo-Sylvester domain

if and only if every finitely generated projective module is stably free; (c) an S-ring if and only if for every finitely generated projective R-module P, there exists a finitely generated R-module Q such that $Q \oplus R^n \cong R^n$ for some $n \geq 1$, for which $P \oplus Q$ is stably free.

8°. (G. M. Bergman) Do the results of Exercise 7 (without the remark on semifirs in (a)) hold under weaker conditions than being weakly semihereditary?

5.7 The factorization of matrices over semifirs

In Section 3.2 we described the factorization of full matrices over semifirs and we saw there that with the appropriate finiteness assumptions matrix rings over semifirs behave very much like UFDs. By contrast the factorization of non-full or rectangular matrices is less well explored and no complete analogue of Theorem 3.2.7 is known. We begin our study with a closer analysis of comaximal relations.

Proposition 5.7.1. *Let R be a ring with UGN. Given a comaximal relation*

$$PQ = 0, \quad P \in {}^rR^n, Q \in {}^nR^s, \tag{1}$$

we have

$$r + s \leq n \tag{2}$$

and there exist an $n \times r$ matrix P' and an $s \times n$ matrix Q' such that

$$\begin{pmatrix} P \\ Q' \end{pmatrix} (P' \quad Q) = \begin{pmatrix} I & 0 \\ 0 & I \end{pmatrix}, \tag{3}$$

Proof. Since (1) is comaximal, there exist matrices P_1, Q_1 such that $PP_1 = I$, $Q_1 Q = I$; putting $P' = P_1$, $Q' = Q_1 - Q_1 P_1 P$, we obtain (3), and now (2) follows by UGN of R. ∎

If the relation (1) is balanced, we have equality in (2), so that the matrices on the left of (3) are square. If moreover, R is weakly finite, the matrices on the left of (3) are mutually inverse. In this case we have

$$P'P + QQ' = I. \tag{4}$$

If we take our relation in the form $AB' - BA' = 0$, this leads to the following result:

Proposition 5.7.2. *Let R be a weakly finite ring. In any balanced comaximal relation*

$$C = AB' = BA', \tag{5}$$

C is a least common right multiple of A, B and a least common left multiple of A', B'.

Proof. Writing $P = (A, B), Q = (-B', A')^{\mathrm{T}}$, we have a balanced comaximal relation of the form (1), hence by Proposition 7.1 there exist P', Q' such that (3) holds. Since the relation is balanced, we have $r + s = n$, and by weak finiteness (4) now follows. Suppose now that $PS = 0$; then $S = (P'P + QQ')S = Q(Q'S)$. Hence if T is a common right multiple of A and B, say $AS_1 = BS_2 = T$, then $PS = 0$, where $S = (-S_1, S_2)$, so S is indeed a right multiple of Q. The rest follows by symmetry. ∎

Suppose now that (1) is not balanced and assume further that R is an Hermite ring. Then the matrices on the left of (3) can be completed to be mutually inverse.

Proposition 5.7.3. *Let R be an Hermite ring. Given a comaximal relation*

$$C = AB' = BA', \quad A \in {}^r R^m, B \in {}^r R^n, A' \in {}^n R^s, B' \in {}^m R^s, \tag{6}$$

if this is not balanced, then we can add columns to A' and B' to obtain a comaximal relation, i.e. there exist $A'' \in {}^n R^t, B'' \in {}^m R^t, C'' \in {}^r R^t$, where $r + s + t = m + n$, such that

$$(C \quad C'') = A(B' \quad B'') = B(A' \quad A''),$$

and the right-hand equation is a balanced comaximal relation. Dually, we can add rows to A and B to obtain an $(r + t) \times s$ matrix with two factorizations forming a balanced comaximal relation, or more generally, add t' rows to A, B and t'' columns $(t' + t'' = t)$ to A', B' to obtain an $(r + t') \times (s + t'')$ matrix with two factorizations forming a comaximal relation. ∎

In a semifir any relation may be expressed in terms of a comaximal relation. In order to do this we need a normal form for matrices in terms of the column rank ρ_c. We observe that the result gives nothing new when the given matrix is right full, or more generally, right regular. We recall that an *inessential modification* of a product AB is the replacement of A, B by $AU, U^{-1}B$, where U is an invertible matrix; if in a factorization only inessential modifications are possible, it is *essentially unique*.

Proposition 5.7.4. *Let R be a semifir and $A \in {}^r R^m$ a matrix of column rank t. Then $t \leq m$ and A may be written as*

$$A = PA', \quad \text{where } P \in {}^r R^t, A' \in {}^t R^m, \tag{7}$$

P is right regular and A' is right invertible, and subject to these conditions this factorization is essentially unique. Moreover, there is a balanced comaximal relation

$$A'B'' = 0,$$

where B'' is left invertible, such that for any matrix B with m rows, $AB = 0$ if and only if

$$B = B''C \text{ for some } C. \tag{8}$$

Proof. Given A as stated, we can find $T \in GL_m(R)$ such that AT has its first t columns right linearly independent and the rest 0. By suitably partitioning T, we have

$$A(T_1 \quad T_2) = (P \quad 0),$$

where $P \in {}^rR^t$ is right regular, and if $T^{-1} = (U, V)^T$, then $A = (P, 0)T^{-1} = PU$. Thus we obtain (7), and U is right invertible because $UT_1 = I$. If we also have $A = QW$, where Q is right regular and W is right invertible, denote by U', W' right inverses for U, W; then $Q = PUW' = PE$, say; similarly $P = QWU' = QF$, hence $P = QF = PEF$, so $EF = I$ and similarly $FE = I$; hence (7) is essentially unique.

Since A' is right invertible, there exist A'', B', B'' such that $(A', A'')^T$ and (B', B'') are square and mutually inverse. Hence $A'B'' = 0$ is a balanced comaximal relation. Moreover, if $AB = 0$, then $PA'B = 0$, hence $A'B = 0$, therefore $B = (B', B'')(A', A'')^T B = B''A''B$, so that (8) holds with $C = A''B$. ∎

We can now prove the basic lemma for the analysis of relations in a semifir:

Lemma 5.7.5. *Let R be a semifir. Given a relation*

$$AB' = BA', \quad A \in {}^rR^m, B \in {}^rR^n, A' \in {}^nR^s, B' \in {}^mR^s, \tag{9}$$

suppose that (A, B) has column rank t and $(B', A')^T$ has row rank u. Then

$$t + u \leq m + n, \tag{10}$$

and there exist $P \in {}^rR^t, A_1 \in {}^tR^m, B_1 \in {}^tR^n$ such that P is right regular and

$$(A \quad B) = P(A_1 \quad B_1). \tag{11}$$

Further, there is a balanced comaximal relation $A_1B'_1 = B_1A'_1$, with A'_1, B'_1 depending only on A, B and not on A', B', such that $B' = B'_1Q, A' = A'_1Q$, for some Q.

Proof. This is an immediate consequence of Proposition 7.4, by writing (9) as $(A, B)(-B', A')^T = 0$, while (10) follows by the law of nullity. ∎

We note that the assertion of Lemma 7.5 is asymmetric; the formulation of the left-hand analogue is left to the reader.

In factorizing a rectangular matrix the role of an atom is taken over by an unfactorable matrix; to define this concept we shall take a closer look at factorizations.

Let R be a semifir as before. A factorization of a matrix C over R:

$$C = AB \tag{12}$$

is said to be *proper* if A has no right inverse and B has no left inverse. Clearly any matrix having a proper factorization cannot be invertible.

A factorization (12) is called *regular* if A is right and B is left regular. Such a factorization is proper if and only if neither of A, B is a unit. For if A or B is a unit, (12) is not proper. Conversely, if (12) is not proper, say A has a right inverse: $AA' = I$, then $A(A'A - I) = 0$; since A is right regular, $A'A = I$, so A has the inverse A'; similarly if B has a left inverse.

Given any factorization (12) of C over a semifir, we can always ensure by an inessential modification that A becomes $(A_1, 0)$, where A_1 is right regular. If B becomes $(B_1, B_2)^{\mathrm{T}}$, then $C = AB = A_1 B_1$, where now A_1 has fewer columns than A unless A itself was right regular. If we now repeat this operation on B we obtain a regular factorization (12) of C. We note that the operations carried out on A, namely multiplying on the right by an invertible matrix and omitting some columns, do not affect the property of having a right inverse, while if B has no left inverse, this remains true, so a proper factorization remains proper. This shows the truth of

Proposition 5.7.6. *Let R be a semifir and C any matrix over R. Then any factorization (12) can, by an inessential modification and omitting some zero columns of A and the corresponding rows of B, followed by an inessential modification and omitting some zero rows of B and the corresponding columns of A, be brought to the form of a regular factorization*

$$C = A_1 B_1. \tag{13}$$

Moreover, if (12) was proper, then so is (13). ∎

Definition. A matrix C over a ring R is said to be *unfactorable* if it is a regular non-unit that cannot be written as a product of two regular non-units.

It is clear that this property is preserved by stable association. We also note that since a one-sided unit is always a zero-divisor, an unfactorable matrix, being regular, has neither a left nor a right inverse. By Proposition 7.6, over a semifir R a matrix C is unfactorable if it is a regular non-unit and in any factorization $C = AB$, either A has a right inverse or B has a left inverse, i.e. C has no proper factorization. Moreover, given any matrix C over a semifir R, we can always find a left invertible matrix P, a right invertible matrix Q and a regular matrix C_0 such that

$$C = PC_0Q,$$

by applying Proposition 7.4 and its left–right dual. It is easily seen that a finitely presented module over a semifir is minimal bound if and only if its matrix is unfactorable.

We can now prove a refinement of the analysis of relations in Lemma 7.5, for use with unfactorable matrices.

Lemma 5.7.7. *Let R be a semifir and C a regular matrix over R. Given two factorizations of C:*

$$C = AB' = BA', \tag{14}$$

where A is unfactorable and BA' is a proper factorization, either there is a balanced comaximal relation

$$AB_1 = BA_1, \quad \text{where } A' = A_1Q, B' = B_1Q, \tag{15}$$

or after inessential modifications and omitting the zero rows of A' and the corresponding columns of B, there is a matrix U such that $B = AU, B' = UA'$.

Proof. By Lemma 7.5 we have $(A, B) = P(A_0, B_0), (B', A')^T = (B_1, A_1)^TQ$ and $A_0B_1 = B_0A_1$ is balanced comaximal. Now because A is unfactorable, either P is right invertible; then so is (A, B), because (A_0, B_0) is right invertible, and we have a balanced comaximal relation $AB_1 = BA_1$, or A_0 is left invertible. In the latter case, by an inessential modification on the left of (14) we can take $A = P(I, 0)^T$, hence $P = (A, P')$ for some P', and by an inessential modification on the right of (14), and omitting zero rows of A' and the corresponding columns of B, we ensure that A' is left regular. Writing $B_0 = (U, V)^T$, with a partition corresponding to that of P, we have $B = PB_0 = AU + P'V$, hence $AB' = BA' = (AU + P'V)A'$, i.e. $A(B' - UA') - P'VA' = 0$. By Lemma 7.5, $P = (A, P')$ is right regular, hence $B' = UA', VA' = 0$. Since A' is left regular, $V = 0$ and so $B = PB_0 = AU$ as claimed. ∎

Corollary 5.7.8. *If in (14) of Lemma 7.7 both factorizations are proper and A, B are both unfactorable, then either (15) holds or we may take U to be a unit, so that A, B are right associated. Thus either we have a balanced comaximal relation $AB_1 = BA_1$, where $A' = A_1Q, B' = B_1Q$ for some matrix Q or AB'*

5.7 The factorization of matrices over semifirs

and BA' can be reduced to the same form by inessential modification and omitting zero rows of the right-hand factor and the corresponding columns from the left-hand factor.

Proof. In this case, if there is no balanced comaximal relation (15), we have $B = AU$, $A = BT$, hence $A = BT = AUT$ and A is right regular, hence $UT = I$; similarly $TU = I$ and so U is invertible. ∎

We saw in Section 3.1 that a comaximal relation (1) is balanced if and only if A and A' have the same index, and by Proposition 0.5.6, in any weakly finite ring two matrices A, A' are stably associated if and only if they occur in a balanced comaximal relation (1). If the product AB' occurs in a factorization and we replace it by BA', related to AB' by a balanced comaximal relation (1), we shall call this change a *comaximal transposition*. It is clear that for full (square) matrices this reduces to the definition given in Section 3.2.

Let C be a regular matrix over a semifir R and consider a factorization

$$C = A_1 A_2 \ldots A_r. \tag{16}$$

By inessential modification and omitting zero rows or columns and the corresponding columns or rows from the other factor we may assume that $A_1, \ldots, A_r$ are regular. If any A_i is a unit, we combine it with either of its neighbours; thus we may assume that all the A_i are regular non-units. Any such factorization of C corresponds to a chain of submodules with bound quotients; when R is a fir, both chain conditions hold for such chains, by Theorem 2.3; hence there is always a maximal refinement of such a chain. In terms of the factorization (16) this means that the A_i are unfactorable, since a matrix is unfactorable precisely if the module defined by it is minimal bound. Thus we have

Theorem 5.7.9. *Over a fir, any regular matrix has a factorization into unfactorable matrices. More generally, any factorization into regular non-units can be refined to a factorization into unfactorable matrices.* ∎

This proves the existence of complete factorizations. In order to compare two factorizations of a given matrix we apply Lemma 7.5; here a numerical condition on the size of the matrices is needed, which unfortunately does not always hold:

Lemma 5.7.10. *Let R be a semifir and C a regular matrix over R which has two factorizations:*

$$C = AB' = BA', \tag{17}$$

such that one of the following four equivalent conditions is fulfilled:

$$i(A) + i(B) \leq i(C), \tag{18}$$

$$i(C) \leq i(A') + i(B'), \tag{19}$$

$$i(A) \leq i(A'), \tag{20}$$

$$i(B) \leq i(B'). \tag{21}$$

Then either A, B have a common left factor that is not right invertible, or A', B' have a common right factor that is not left invertible, or (17) is a comaximal transposition, in which case equality holds in (18)–(21).

Proof. The equivalence of (18)–(21) is easily verified, using the fact that $i(A) + i(B') = i(C) = i(B) + i(A')$. Moreover, when (17) is comaximal, then $i(A) + i(B) \geq i(C)$, so in that case (18) is just the condition for (17) to be balanced.

Suppose now that no common left or right factor exists in (17). Let C be $m \times n$ and let (A, B) have column rank t. By Lemma 7.5 there exists $P \in {}^m R^t$, such that P is right regular and a common left factor of A and B. By hypothesis, $PQ = I$ for some Q; thus P is right regular right invertible, hence a unit, and so may be absorbed in A, B. By Lemma 7.5 there is now a balanced comaximal relation $AB_1 = BA_1$ and $A' = A_1 Q, B' = B_1 Q$ for some Q. Further, $i(Q) = i(AB') - i(AB_1) = i(C) - i(A) - i(B) \geq 0$ by (18). By hypothesis, Q has a left inverse, so by weak finiteness it is square and hence a unit. It follows that (17) is indeed a comaximal transposition. ∎

This result suggests that we can compare two factorizations of C provided that the left-hand factors are not too 'skew', i.e. they do not both have a large index. As a typical example where a comparison is impossible (owing to the failure of (18)), consider the free k-algebra on eight generators a, b, c, d, p, q, r, s and write $X = \begin{pmatrix} a & b \\ c & d \end{pmatrix}, Y = \begin{pmatrix} p & q \\ r & s \end{pmatrix}$. Then by taking the first row of the comaximal relation $X.(I + YX) = (XY + I).X$, we obtain the unbalanced relation

$$(a \quad b) \begin{pmatrix} 1 + pa + qc & pb + qd \\ ra + sc & 1 + rb + sd \end{pmatrix} = (1 + ap + br \quad aq + bs) \begin{pmatrix} a & b \\ c & d \end{pmatrix}.$$

Exercises 5.7

1. Let $AB' = BA'$ be a comaximal relation over a ring with UGN. Show that $i(A') \leq i(A), i(B') \leq i(B), i(A') + i(B') \leq i(AB') \leq (A) + i(B)$, with equality in one place (and hence in all) if and only if the relation is balanced.

2. Let R be the $n \times n$ matrix ring over a $2n$-fir and let a, b be elements of R with a full common right multiple. Show that $aR \cap bR$ and $aR + bR$ are both principal.
3. Show that an unfactorable $n \times n$ matrix over any ring R is an atom in R_n. Does the converse hold? Does it hold when R is commutative?
4. Prove the assertion made before Lemma 7.7, that a matrix over a semifir is unfactorable if and only if the module defined by it is minimal bound. Deduce that every minimal bound module over a semifir is strictly positive.
5. Let R be a semifir and $A \in {}^m R^n$ an unfactorable matrix; show that the submodule of R^n generated by the rows of A is a direct summand of every finitely generated proper submodule of R^n containing it. Does the converse hold?
6. Let A be an invertible matrix over a semifir and consider a block decomposition $A = (A', A'', A''')^T$ with the corresponding decomposition $A^{-1} = (B', B'', B''')$. If $A'B''' = 0$, then $A''B'' = A''B''$ is a comaximal relation, but not balanced, unless A', B''' were vacuous.
7°. Develop a theory of unique factorization rings that allows for the factorization of zero-divisors, taking e.g. the factorization of matrices over a principal ideal domain as a model (see Section 1.3).

5.8 A normal form for matrices over a free algebra

In the polynomial ring $k[x]$ over a commutative field k it is easy to write down a normal form for polynomials under association: each non-zero polynomial is associated to precisely one monic polynomial. In the free k-algebra of rank greater than 1 the polynomials have a more complicated form, but now it is more natural to permit matrices as well and ask for a normal form under stable association. In particular, this allows us to take the matrix to be linear in the variables. This is the process of 'linearization by enlargement', also called *Higman's trick* (see Higman [40]). As a typical case let us take the (m, n)-entry of an $m \times n$ matrix. If this has the form $f + ab$, then by taking the diagonal sum with 1 and applying elementary transformations, we obtain successively

$$\begin{pmatrix} f+ab & 0 \\ 0 & 1 \end{pmatrix} \to \begin{pmatrix} f+ab & a \\ 0 & 1 \end{pmatrix} \to \begin{pmatrix} f & a \\ -b & 1 \end{pmatrix},$$

where only the last two entries of the last two rows are shown. By repeated application this allows us to reduce any matrix over $k\langle X \rangle$ to one in which the elements of X occur to at most the first degree. Such a matrix is called *linear*; taking $X = \{x_1, \ldots, x_d\}$ for simplicity, we see that every linear $m \times n$ matrix has the form

$$A = A_0 + \sum_1^d A_i x_i, \quad \text{where } A_0, A_i \in {}^m k^n. \tag{1}$$

A matrix is called *right monic* if it is linear of the form (1), where $A_1, \ldots, A_d$ are right comaximal (i.e. $(A_1, \ldots, A_d)$ has a right inverse); a *left monic* matrix is defined similarly and a *monic* matrix is one that is left and right monic. Thus any (left or right) monic matrix is necessarily linear. Monic matrices have the following good property:

Lemma 5.8.1. *Let A be a right monic matrix over $k\langle X \rangle$. Then the homogeneous component of degree 1 in A is left regular; in particular, A is left regular but not a unit.*

Proof. Suppose that $A = A_0 + \sum A_i x_i$ and $\sum C A_i x_i = 0$; then $\sum C^\lambda A_i x_i = 0$, where C^λ is the sum of the terms of highest degree in C. Equating left cofactors of x_i we obtain $C^\lambda A_i = 0$, so $C^\lambda (A_1, \ldots, A_d) = 0$ and therefore $C^\lambda = 0$, since the A_i are right comaximal. This shows $A^\lambda = \sum A_i x_i$ to be left regular; it follows that A itself is left regular but not a unit, for if $CA = 0$ or I, then $C^\lambda A^\lambda = 0$, hence $C^\lambda = 0$ and so $C = 0$. ∎

On the other hand, it is not enough to assume $A_0, A_1, \ldots, A_d$ to be right comaximal, as the example $(1, x_1)^T$ shows, which is left annihilated by $(x_1, -1)$.

The next result represents the weak algorithm for matrices:

Lemma 5.8.2. *Let A be left monic and B right monic over $R = k\langle X \rangle$. If P, Q are any matrices over R such that*

$$PB = AQ, \tag{2}$$

then there exists a matrix C over R such that

$$P = AC + P_0, \quad Q = CB + Q_0, \tag{3}$$

where P_0, Q_0 are matrices over k.

Proof. Let us write $A = A_0 + \sum A_i x_i$, $B = B_0 + \sum B_i x_i$ and again denote the matrix of highest terms in any matrix T by T^λ. On equating the highest terms in (2), we obtain

$$\sum P^\lambda B_i x_i = \sum A_i x_i Q^\lambda. \tag{4}$$

By Lemma 8.1, B^λ is left regular and A^λ is right regular; hence P and Q have the same degree. If $\deg Q = 0$, then $\deg P = 0$, so then there is nothing to prove. We may therefore assume that $\deg Q \geq 1$ and write $Q^\lambda = \sum Q_i x_i$, where Q_i is over R. Equating left cofactors of x_i in (4), we find $P^\lambda B_i = A^\lambda Q_i$, hence

$$P^\lambda (B_1, \ldots, B_d) = A^\lambda (Q_1, \ldots, Q_d).$$

5.8 A normal form for matrices over a free algebra

Let D be a right inverse of $(B_1, \ldots, B_d)$; then D is over k. We put $C = (Q_1, \ldots, Q_d)D$ and obtain

$$P^\lambda = P^\lambda(B_1, \ldots, B_d)D = A^\lambda C. \tag{5}$$

Now by (2) we have

$$(P - AC)B = A(Q - CB).$$

This is an equation of the same form as (2), but $P - AC$ has lower degree than P, by (5) and $\deg(Q - CB) < \deg(Q)$, so the result follows by induction on the degree of Q. ∎

We can now establish the existence of a normal form for matrices over $k\langle X\rangle$.

Theorem 5.8.3. *Let $R = k\langle X\rangle$ be a free k-algebra. Any matrix over R is stably associated to a matrix $A \oplus 0$, where A is monic (and 0 need not be square). In particular, any (left) regular matrix is stably associated to a (right) monic matrix. Moreover, if $A \oplus 0$, $A' \oplus 0$ are two matrices that are stably associated and such that A, A' are monic, then the number of rows of A, A' agree, as do those of $A \oplus 0$, $A' \oplus 0$; likewise for the columns and there exist invertible matrices P, Q over k such that $PA' = AQ$.*

Proof. By the linearization process described earlier we reach a linear matrix, which can be chosen so as to have the form $A \oplus 0$, where A has the linear form (1) and is $m \times n$, with m, n minimal. We claim that A is monic. For if $(A_1, \ldots, A_d)$ has no right inverse say, then this $m \times nd$ matrix over k has rank less than m and so by elementary row operations the last row can be reduced to 0. If now the last row of A_0 is also 0, we can reduce m by 1, contradicting the minimality (this amounts to writing A as $B \oplus N$, where N is the 1×0 matrix). So the last row of A_0 is not 0, and by column operations over k and further row operations over R we find that A is associated to $A' \oplus I$, where A' is again linear. So A is stably associated to A', but this again contradicts the minimality of $m + n$. Hence A is right monic; by symmetry it is also left monic. Thus the existence of the form (1) is established. If the original matrix was left regular, then so is $A \oplus 0$ and it follows that 0 is a matrix of 0 rows, so $A \oplus 0$ is right monic. We remark that if A is invertible, it is stably associated to the (0×0) null matrix.

Suppose now that A and A' are both monic and $A \oplus 0$, $A' \oplus 0$ are stably associated; then $F \oplus 0$ and $F' \oplus 0$ are associated, where $F = A \oplus I_r$, $F' = A' \oplus I_s$, for suitable r and s. Thus there exist invertible matrices U, V over R such that for appropriate partitioning

$$\begin{pmatrix} U_1 & U_2 \\ U_3 & U_4 \end{pmatrix} \begin{pmatrix} F & 0 \\ 0 & 0 \end{pmatrix} = \begin{pmatrix} F' & 0 \\ 0 & 0 \end{pmatrix} \begin{pmatrix} V_1 & V_2 \\ V_3 & V_4 \end{pmatrix}. \tag{6}$$

It follows that $U_3F = 0 = F'V_2$. Now A, A' are regular, hence so are F, F' and therefore $U_3 = 0 = V_2$. Let F be $m \times n$, $F'm' \times n'$ and $F \oplus 0$, $F' \oplus 0$ both $t \times u$. Then U is $t \times t$, U_3 is $(t - m') \times m$, so by Proposition 3.1.2, $t - m' + m \le t$, i.e. $m \le m'$. Similarly V is $u \times u$ and V_2 is $n' \times (u - n)$, so $n' + u - n \le u$, and hence $n' \le n$. By symmetry, i.e. multiplying both sides of (6) by U^{-1} on the left and V^{-1} on the right, we have $m' \le m, n \le n'$ and so F, F' are both $m \times n$. It follows that U_1, U_4, V_1, V_4 are all square and so are invertible, since this is true of U, V. Moreover $U_1F = F'V_1$, so F, F' are associated and A, A' are stably associated. We thus have a comaximal relation $AB' = BA'$, by Proposition 0.5.7. Hence there exist matrices C, D, C', D' over R such that

$$\begin{pmatrix} A & B \\ C & D \end{pmatrix} \quad \text{and} \quad \begin{pmatrix} D' & -B' \\ -C' & A' \end{pmatrix} \tag{7}$$

are mutually inverse. Further, by Lemma 8.2 there exists P such that

$$B = AP + B_0, \quad B' = PA' + B'_0,$$

where B_0, B_0' are over k. Hence on multiplying the matrices (7) by $\begin{pmatrix} I & -P \\ 0 & I \end{pmatrix}$ on the right and $\begin{pmatrix} I & P \\ 0 & I \end{pmatrix}$ on the left, respectively, we obtain a pair of mutually inverse matrices

$$\begin{pmatrix} A & B_0 \\ C & D_1 \end{pmatrix} \quad \text{and} \quad \begin{pmatrix} D'_1 & -B'_0 \\ -C' & A' \end{pmatrix}.$$

We also have $A'C = C'A$, so by another application of Lemma 8.2 we obtain a pair of inverse matrices

$$\begin{pmatrix} A & B_0 \\ C_0 & D_2 \end{pmatrix} \quad \text{and} \quad \begin{pmatrix} D'_2 & -B'_0 \\ -C'_0 & A' \end{pmatrix},$$

where C_0, C'_0 are over k. Thus we have the equation

$$AD'_2 - B_0 C'_0 = I.$$

Equating highest terms we find $\sum A_i x_i (D_2')^\lambda = 0$. By Lemma 8.2, A^λ is right regular; we conclude that $(D_2')^\lambda = 0$, hence $D'_2 = 0$ and so $-B_0 C'_0 = I$. Thus B_0 has a right inverse over k and by symmetry B'_0 has a left inverse. Hence $i(B) \ge 0 \ge i(B')$; since A, A' are stably associated, they have the same index, hence so do B, B' and therefore $i(B_0) = i(B'_0) = 0$. It follows that B_0, B'_0 are invertible over k and we have $AB'_0 = B_0 A'$, as claimed. ∎

The uniqueness result proved here should be compared with the assertion of Proposition 0.6.5. The proof of Theorem 8.3 shows that once we have reached a linear form for our matrix, if the latter is left regular, we can achieve a right monic form by elementary column operations over k and row operations over R, and similarly on the other side. Thus we have

Corollary 5.8.4. *If A is a left regular linear matrix over $R = k\langle X\rangle$, then there exist invertible matrices P over R and U over k such that $PAU = B \oplus I$, where B is right monic. Here the coefficients of P, U can be chosen to lie in the subfield of k generated by the coefficients of A.* ∎

We can now give the extension of Proposition 4.6.12 promised in Section 4.6 by proving that the eigenring of a regular matrix over a free k-algebra is finite-dimensional over k. Equivalently, we shall show that the endomorphism ring of a finitely presented bound module is finite-dimensional; this will appear as consequence of a more general result:

Theorem 5.8.5. *Let $R = k\langle X\rangle$ be a free k-algebra and let M, N be finitely presented R-modules of which M is bound. Then $\mathrm{Hom}_R(M, N)$ is finite-dimensional over k.*

Proof. As in Proposition 4.6.12 we may take R to be of finite rank. Let M, N be left R-modules with defining matrices A, B, where A may be taken to be monic, by Theorem 8.3, and A is regular (because M is bound), while B is left regular right monic. A homomorphism $f : M \to N$ corresponds to a pair of matrices P, Q such that

$$AQ = PB, \tag{8}$$

and conversely, such a pair of matrices defines a homomorphism, while P, Q and P', Q' define the same homomorphism if and only if $P - P' = AC, Q - Q' = CB$ (see Section 0.6). By Lemma 8.2, if (8) holds, we can write $Q = CB + Q_0$, $P = AC + P_0$, where P_0, Q_0 are over k. Thus the given homomorphism may be represented by P_0, Q_0. But the space of these matrices is a finite-dimensional k-space, hence $\mathrm{Hom}_R(M, N)$ is finite-dimensional, as claimed. ∎

To obtain an estimate for the dimension, let A be $m \times n$ and B $r \times s$. As we saw in Section 0.6, we have

$$\mathrm{Hom}_R(M, N) \cong \mathrm{I}(A, B)/\mathfrak{b}, \tag{9}$$

where $\mathrm{I}(A, B) = \{Q \in {}^nR^s | AQ = PB \text{ for some } P \in {}^mR^r\}$ and $\mathfrak{b}$ is the left R-module spanned by the rows of B. Now Theorem 8.5 shows that

$$\mathfrak{b} \subseteq I(A, B) \subseteq \mathfrak{b} + {}^nk^s ;$$

hence

$$I(A, B)/\mathfrak{b} \subseteq (\mathfrak{b} + {}^n k^s)/\mathfrak{b} \cong {}^n k^s/({}^n k^s \cap \mathfrak{b}),$$

and so

$$\dim \operatorname{Hom}_R(M, N) \leq \dim {}^n k^s = ns. \qquad (10)$$

We also note that the condition that M be bound cannot be omitted, since for example, $\operatorname{Hom}_R(R, N) \cong N$, but a finitely presented module, even bound, need not be finite-dimensional over k. For example, if $R = k\langle x, y\rangle$, then R/Rx is bound but has the k-basis $1, y, xy, x^2 y, \ldots$.

Taking $N = M$ in Theorem 8.5, we obtain the desired generalization of Proposition 4.6.12:

Corollary 5.8.6. *Let M be a finitely presented bound module over $R = k\langle X\rangle$. Then $\operatorname{End}_R (M)$ is finite-dimensional over k.* ■

The monic normal form can also be used to describe factorizations of full matrices over $k\langle X\rangle$. First we need a bound on the degree of the factors, where the *degree* of a matrix $P = (p_{ij})$ is defined as $d(P) = \max\{d(p_{ij})\}$.

Lemma 5.8.7. *Let $R = k\langle X\rangle$ be the free k-algebra on $X = \{x_1, \ldots, x_d\}$ and let C be an $m \times n$ matrix over R with a rank factorization*

$$C = AB. \qquad (11)$$

Then there is an invertible matrix P such that $d(AP) \leq d(C)$, $d(P^{-1}B) \leq d(C)$.

Proof. Consider the free k-algebra S on $x_1, \ldots, x_d, y_1, \ldots, y_m, z_1, \ldots, z_n$; there is an embedding of R in S, defined by $x_i \mapsto x_i$, which is honest, since R is a retract of S. We extend the degree on R to S by putting $d(y_i) = d(z_i) = 1$.

In S we have, on writing $y = (y_1, \ldots, y_m), z = (z_1, \ldots, z_n)^{\mathrm{T}}, yCz = (yA)(Bz)$. Since (11) is a rank factorization, A is right regular and B is left regular, therefore, by Corollary 2.4.5 there is an invertible matrix P such that on writing $AP = (a'_{ij}), P^{-1}B = (b'_{ij})$, we have

$$d\left(\sum y_i a'_{ij}\right) + d\left(\sum b'_{jk} z_k\right) \leq d\left(y_i c_{ij} z_j\right), \qquad (12)$$

where $C = (c_{ij})$. If we denote by α_j the degree of the jth column of A' and by β_j the degree of the jth row of B', then (12) becomes $\alpha_j + 1 + \beta_j + 1 \leq d(C) + 2$, hence $\alpha_j \leq d(C)$, and taking the maximum over j, we find that $d(A') \leq d(C)$; similarly $d(B') \leq d(C)$. ■

5.8 A normal form for matrices over a free algebra

We remark that if in this lemma C is full, then any factorization into square matrices is a rank factorization. We can now describe the possible factorizations of a full linear matrix over $k\langle X\rangle$.

Theorem 5.8.8. *Let $R = k\langle X\rangle$ and let $C \in R_n$ be a monic matrix that is full. Then C is not an atom if and only if either $n = 0$ or $n > 1$ and there exist $P, Q \in GL_n(k)$ and $r, s > 0$ such that $r + s = n$ and*

$$PCQ = \begin{pmatrix} A & 0 \\ D & B \end{pmatrix}, \quad A \in R_r, B \in R_s, D \in {}^sR^r. \tag{13}$$

Proof. If (13) holds, then

$$PCQ = \begin{pmatrix} A & 0 \\ 0 & I \end{pmatrix}\begin{pmatrix} I & 0 \\ D & I \end{pmatrix}\begin{pmatrix} I & 0 \\ 0 & B \end{pmatrix},$$

and A, B are again monic, hence (by Lemma 8.1) not invertible. We therefore have a non-trivial factorization of PCQ and hence of C.

Suppose conversely that $C = FG$ is a factorization of C. Any square factor of C is again full, and if it is a non-unit, its degree is positive; moreover, by Lemma 8.7 we may take F to be of degree 1. Being full, F is regular, so by Corollary 8.4 there exist $P \in GL_n(k), U \in GL_n(R)$ such that $PFU = A \oplus I$, where A is monic $r \times r$; since F is a non-unit, we have $r > 0$ and since G is a non-unit, $r < n$. Hence

$$PC = \begin{pmatrix} A & 0 \\ 0 & I \end{pmatrix}\begin{pmatrix} G' \\ G'' \end{pmatrix}. \tag{14}$$

Now PC is linear, hence G'' is linear and G' has degree 0, again by the regularity of A^λ (Lemma 8.1). Further, G' has r rows and since each factor in (14) is full, G' has rank r, so there exists $Q \in GL_n(k)$ such that $G'Q = (I_r, 0)$. It follows that

$$PCQ = \begin{pmatrix} A & 0 \\ 0 & I \end{pmatrix}\begin{pmatrix} I & 0 \\ D & B \end{pmatrix} = \begin{pmatrix} A & 0 \\ D & B \end{pmatrix},$$

and this is of the required form. ∎

This result also provides a bound on the length of factorizations in terms of n. We remark that if in a linear matrix A the cofactor of some x_i is the unit matrix, then A is monic, by definition; it is also full, for if I is the cofactor of x_1, say and we specialize $x_2, \ldots, x_d$ to 0, A becomes $A_0 + Ix_1$. If we now specialize x_1 to an element of $k(t)$, an infinite field containing k, which is not an eigenvalue of $-A_0$, we obtain a non-singular matrix, hence A was full. For such a matrix the criterion of Theorem 8.8 takes the following form:

Corollary 5.8.9. *Let $R = k\langle X \rangle$ and let $A = A_0 + \sum A_i x_i$ be a linear $n \times n$ matrix over R such that $A_i = I$ for some $i \geq 0$. Then A is an atom if and only if $A_0, \ldots, A_d$ act irreducibly on k^n.*

Proof. Clearly A is monic and full, so by Theorem 8.8 it is not an atom if and only if there exist $P, Q \in GL_n(k)$ such that

$$PA_i Q = \begin{pmatrix} B_i & 0 \\ D_i & C_i \end{pmatrix}, \quad i = 0, 1, \ldots, d. \tag{15}$$

By hypothesis $A_i = I$ for some $i \geq 0$, so in particular PQ has the block rectangular form (15), and hence so does $PA_i P^{-1} = PA_i Q(PQ)^{-1}$. Now the result follows by (15). ∎

This result makes it easy to construct matrix atoms; we record one important case:

Corollary 5.8.10. *Let $R = k\langle X \rangle$ be a free algebra of rank at least $N = n^2$ and let $A_1, \ldots, A_N$ be a k-basis of $\mathfrak{M}_n(k)$. Then $A = \sum A_i x_i$, where the x_i are distinct elements of X, is an absolute matrix atom; in fact it remains an atom under all commutative field extensions of k.*

Proof. Since the A_i form a basis, we have $I_n = \sum \alpha_i A_i$ for suitable $\alpha_i \in k$, where $\alpha_1 \neq 0$, say. If we make a linear change of generators in R by writing $y_1 = \sum \alpha_i x_i$, $y_j = x_j (j \neq 1)$, then A satisfies the hypothesis of Corollary 8.9 relative to the y's and hence is an atom; clearly it remains one under any extension of the ground field. ∎

Let us return to eigenrings for a moment. We have seen that eigenrings of regular matrices (over free algebras) are finite-dimensional over k (Theorem 8.5), a matrix atom has as eigenring a field (Proposition 3.2.4) and for a non-zero element the eigenring is commutative (Proposition 4.6.15). The latter no longer holds for matrices, for we shall see that any finite-dimensional k-algebra can occur as eigenring of a regular matrix, and any skew field finite-dimensional over k can occur as eigenring of a matrix atom.

Theorem 5.8.11. *Let k be a commutative field and F a finite-dimensional k-algebra. Then there exists a torsion module M over a free k-algebra R of finite rank, such that $\text{End}_R(M) \cong F$. Moreover, if F is a field, then M can be taken to be Tor-simple.*

Proof. Let $[F : k] = n$ and embed F in $E = \text{End}_k(F) \cong k_n$ by letting F act on itself by left multiplications. We denote the image of F in E by F' and its centralizer in E by G. Since F acts bicentrally on itself, the centralizer of G

5.8 A normal form for matrices over a free algebra

is F'. Now G is finitely generated as k-algebra, by $A_0, A_1, \ldots, A_m \in k_n$, say, where we may take $A_m = I$ without loss of generality. Let $R = k\langle x_1, \ldots, x_m\rangle$ and put $A = A_0 + \sum A_i x_i \in R_n$; it is clear that A is full and monic. By (9) and (10), if M is the module presented by A, we have

$$\operatorname{End}_R(M) \cong I(A) \cap k_n = \{P \in k_n | PA_i = A_i Q, i = 0, 1, \ldots, m \text{ and some } Q\}. \tag{16}$$

Since $A_m = I$, we have $Q = P$ on the right of (16); therefore $\operatorname{End}_R(M)$ is the centralizer of the A_i, hence of G, and so $\operatorname{End}_R(M) \cong F$.

Suppose now that F is a field, finite-dimensional over k. If k is a finite field, F is a commutative field extension of k (by Wedderburn's theorem), and we can write $F = k(\alpha)$ for some $\alpha \in F$. If the minimal polynomial of α over k is p, then F is the endomorphism ring of the simple torsion module $R/p(x)R$, where $R = k[x]$.

There remains the case when k is infinite. Let $[F:k] = n$; take a k-basis $f_1, \ldots, f_n$ of F and with n distinct elements $\lambda_1, \ldots, \lambda_n$ of k define matrices $A_0, A_1 \in F_n$ by

$$A_0 = \operatorname{diag}(\lambda_1, \ldots, \lambda_n), \quad A_1 = \sum f_i e_{ii} + \sum_{i>1} e_{i1} + \sum_{j>1} e_{1j}.$$

We claim that A_0, A_1 generate F_n as k-algebra. Since the λ_i are distinct, the subalgebra generated by A_0 contains all diagonal matrices over k, in particular it contains each e_{ii}; hence the subalgebra generated by A_0, A_1 also contains $e_{i1} = e_{ii} A_1 e_{11}$ and $e_{1j} = e_{11} A_1 e_{jj}$, and so also $e_{ij} = e_{i1} e_{1j}$. Thus it contains k_n; it also contains $f_i e_{ii} = e_{ii} A_1 e_{ii}$ and so contains all of F_n.

Now F_n has just one simple left F_n-module, S say, up to isomorphism, and $[S:k] = n^2$. Consider the embedding $F_n \to \operatorname{End}_k(S) \cong \mathfrak{M}_{n^2}(k)$; since F_n acts irreducibly on S, A_0 and A_1 act irreducibly on $\mathfrak{M}_{n^2}(k)$ and if their images in $\mathfrak{M}_{n^2}(k)$ are A'_0, A'_1 then $P = A'_0 + x_1 A'_1 + x_2 I$ is a matrix atom, by Corollary 8.9. Moreover, the centralizer of A'_0, A'_1 is the centralizer of F_n acting in $\operatorname{End}_k(S)$ and so is isomorphic to F; hence if M is the module over $R = k\langle x_1, x_2\rangle$ defined by P, then M is Tor-simple and

$$\operatorname{End}_R(M) \cong I(P) \cap k_n.$$

As before this is the centralizer of A'_0, A'_1, i.e. F, as we had to show. ∎

Exercises 5.8

1. Verify that the proof of Theorem 8.3 shows every linear matrix over $k\langle X\rangle$ to be associated to a matrix of the form $A \oplus 0 \oplus I$, where A is monic.

2°. Extend Lemma 8.2 to matrices over $K_k\langle X\rangle$ (*Hint*: Try the form $A = A_0 + \sum A_{ij}x_i u_j$, where the $\{u_j\}$ form a left k-basis of K.)
3. (M. L. Roberts) Let $R = K_k\langle X\rangle$ be a tensor K-ring. Show that every full matrix over R is stably associated to a matrix $C + \sum A_i x_i B_i$, where the A_i are right comaximal, the B_i are left comaximal and in each term $A_i x_i B_i$ the columns of A_i are linearly independent over k, and likewise the rows of B_i.
4. Find a monic matrix over $k\langle X\rangle$ that is not full. (*Hint*: Try a hollow matrix.)
5. (G. M. Bergman) Find a full linear matrix of the form (1) such that no k-linear combination of $A_1, \ldots, A_d$ is a regular element of k_n (*Hint*: Linearize an element of $k\langle X\rangle$ that maps to 0 under every homomorphism $k\langle X\rangle \to k$.)
6. (G. M. Bergman) Let $R = K\langle X\rangle$, where K is an infinite-dimensional persistent division algebra (e.g. a commutative infinite algebraic field extension). Show that $\mathrm{End}_R(R/xR)$ is infinite-dimensional over k, even though R/xR is a finitely presented bound module.
7. Use Theorem 8.11 to find a matrix over $R\langle X\rangle$ with $\mathbb{C}$ as eigenring, and a matrix with the quaternions as eigenring.
8. (Cohn [76b]) Let K be a field that is a k-algebra. Show that two square matrices A, B over K are conjugate over k if and only if $xI - A$ is stably associated to $xI - B$ over $K_k\langle X\rangle$. Deduce that if a matrix P is stably associated to $xI - A$, then A is determined by P up to conjugation by a matrix over k.

5.9 Ascending chain conditions

In Section 3.2 we saw that any $2n$-fir with left and right ACC_n is n-atomic. To prove a corresponding result in Section 5.7 we needed to assume that R is a fir; but merely assuming left ACC_n will enable us to split off a 'maximal' left factor from any matrix of inner rank n:

Proposition 5.9.1. *Let R be a $2m$-fir and A an $m \times n$ matrix with a rank factorization*

$$A = PA', \quad \text{where } P \in {}^m R^r, A' \in {}^r R^n, \tag{1}$$

and where A' is left prime. Then for any other rank factorization $A = P'A''$ there is an $r \times r$ matrix V such that $A'' = VA'$, $P = P'V$. An expression (1) with A' left prime exists whenever R has left ACC_m or is r-atomic.

Proof. Let $A = PA' = P'A''$, where P, P' have r columns and A' is left prime. Then

$$(P \quad -P')\begin{pmatrix} A' \\ A'' \end{pmatrix} = 0, \quad \text{hence } \rho(P \quad -P') + \rho\begin{pmatrix} A' \\ A'' \end{pmatrix} \leq 2r,$$

by the law of nullity, but each summand is at least r, so $\rho(A', A'')^{\mathrm{T}} = r$ and we

have
$$\begin{pmatrix} A' \\ A'' \end{pmatrix} = \begin{pmatrix} U \\ V \end{pmatrix} C, \quad \text{where } U, V \in R_r, C \in {}^r R^n.$$

Since A' is left prime, U is a unit and so may be absorbed in C. Thus $C = A'$, $A'' = VA'$, $A = P'A'' = P'VA' = PA'$. Since A' is left full, it is left regular and so $P = P'V$, as claimed.

Now assume left ACC_m and write $A = P_0 A_0$, where $P_0 \in {}^m R^r$, $A_0 \in {}^r R^n$ and A_0 is left full. Since $r \le m$, we can choose a maximal r-generator submodule of R^n containing $R^r A_0$; this is of the form $R^r A'$, where the rows of A' are the generators. The equation (1) follows, and A' is left prime by the maximality of $R^r A'$.

When R is r-atomic, we write $A = P_1 A_1$, where P_1 is $m \times r$ and A_1 is $r \times n$; then A_1 is left full. Let A_1' be a full $r \times r$ submatrix of A_1; then in any factorization $A_1 = P_2 A_2$, where $P_2 \in R_r$, $A_2 \in {}^r R^n$, P_2 is a left factor of A_1' and so the number of terms in a complete factorization of P_2 is bounded by the corresponding number for A_1'. Thus we can ensure that A_2 is left prime by taking P_2 with a maximal number of factors. ∎

By restating the result in terms of modules we see that for the rings considered every finitely presented module has a largest positive quotient, and dually, a largest negative submodule; this was proved for the special case of firs in Theorem 3.5.

Theorem 5.9.2. *Let R be a $2m$-fir that has left ACC_m or is m-atomic, and let M be an m-generator submodule of a free left R-module F. Denote by r the least integer for which there is an r-generator submodule between M and F, thus $r \le m$. Then there is a greatest r-generator submodule N between M and F.* ∎

Here 'greatest' is understood in the sense that N contains every r-generator submodule of F containing M. Thus N/M is the greatest negative submodule of F/M. A dual argument shows that every finitely presented right R-module has a largest positive quotient. Thus we have

Corollary 5.9.3. *Let R be a $2m$-fir that has left and right ACC_m or is m-atomic, and let F be a free R-module of rank at most m with an m-generator submodule M. Then F/M has a largest positive quotient and a largest negative submodule.* ∎

We now consider another chain condition that entails pan-ACC and holds in all firs. A module M (over any ring R) is said to satisfy $\text{ACC}_{\text{dense}}$ if every ascending chain of finitely generated submodules of M with bound quotients (or equivalently: dense inclusions) must break off. If a ring R satisfies $\mathrm{ACC}_{\mathrm{dense}}$ as left (or right) module over itself, we shall say that R satisfies *left* (or *right*) $\mathrm{ACC}_{\mathrm{dense}}$.

To clarify the relation between $\mathrm{ACC}_{\mathrm{dense}}$ and ACC_n we shall need a result on chains with bound quotients. In a partially ordered set, two subsets of elements will be called *cofinal* if each element of either is majorized by some element of the other set.

Lemma 5.9.4. *In an n-fir R, let M be a module with a sequence*

$$M_1 \subseteq M_2 \subseteq \ldots \tag{2}$$

of submodules, each free of rank at most n, and assume that (2) is not cofinal with any sequence of submodules all free of rank less than n. Then M_i/M_{i-1} is bound for all sufficiently large i.

Proof. Any M_i/M_{i-1} that is not bound will have R as a direct summand, by Corollary 1.4, and its complement corresponds to a submodule M_i' of smaller rank than M_i, so of rank less than n, such that $M_{i-1} \subset M_i' \subset M_i$. If this happens for infinitely many i, then the sequence M_i' is cofinal, but of rank less than n. Hence the conclusion follows. ∎

Suppose now that R is an n-fir with left $\mathrm{ACC}_{\mathrm{dense}}$. Given an ascending chain of n-generator left ideals of R, let $m \leq n$ be the least integer for which our chain is cofinal with a chain of left ideals that are free of rank at most m. Applying Lemma 9.4 to this chain, we see that ultimately it has bound quotients and so becomes constant. This proves

Corollary 5.9.5. *For any n-fir, left $\mathrm{ACC}_{\mathrm{dense}}$ implies left ACC_n. In particular, a semifir with left $\mathrm{ACC}_{\mathrm{dense}}$ satisfies left pan-ACC.* ∎

The next result elucidates the role of $\mathrm{ACC}_{\mathrm{dense}}$ in modules over semifirs:

Proposition 5.9.6. *Let R be any ring and M an R-module. If every countably generated submodule of M is free, then M satisfies $\mathrm{ACC}_{\mathrm{dense}}$. Conversely, if R is a semifir and M an R-module satisfying $\mathrm{ACC}_{\mathrm{dense}}$ and such that all finitely generated submodules of M are free, then every countably generated submodule of M is free.*

Proof. Let M be a left R-module all of whose countably generated submodules are free and consider a chain

$$M_1 \subseteq M_2 \subseteq \ldots$$

of dense inclusions of finitely generated submodules in M. Their union M' is countably generated and hence free. Since M_1 is a finitely generated submodule of M', it involves only finitely many members of a basis of M', and so is contained in a free direct summand of finite rank, N, of M', say $M' = N \oplus N'$, where N' is also free. Let $p : M' \to N'$ be the projection onto N'. For any linear functional α on N', $p\alpha$ is a linear functional on M', zero on M_1, hence by density zero on each M_i, and so zero on M'. It follows that $N' = 0$, so $M' = N$ is finitely generated and our chain must terminate.

Conversely, assume that R is a semifir, M satisfies $\text{ACC}_{\text{dense}}$, and all its finitely generated submodules are free. Let N be a submodule generated by countably many elements $u_1, u_2, \ldots$. Put $N_0 = 0$ and for each $i \geq 0$ let us recursively construct N_{i+1} as a maximal finitely generated submodule of N in which $N_i + Ru_{i+1}$ is dense; this is possible by $\text{ACC}_{\text{dense}}$. We claim that each N_i is a direct summand in N_{i+1}. Indeed, let N_i' be a direct summand of N_{i+1} containing N_i that is free of least possible rank. Any linear functional α on N_i' that is zero on N_i will have for kernel a direct summand of N_i' (and hence of N_{i+1}), which contains N_i and is free of smaller rank than N_i' (by Theorem 2.2.1 (e)), unless $\alpha = 0$. But N_i' was of minimal rank, hence $\alpha = 0$ and N_i is dense in N_i'. By construction of N_i we therefore have $N_i' = N_i$.

We thus have $N_{i+1} = N_i \oplus P_{i+1}$, say, where P_{i+1}, being finitely generated, is free. Hence $N = \cup N_i = P_1 \oplus P_2 \oplus P_2 \ldots$ is also free. ∎

Bearing in mind Theorem 1.5, we can state the result as

Corollary 5.9.7. *Let R be a semifir. Then a countably generated R-module is free if it has ACC_{dense} and is n-unbound for all n. Moreover, a semifir is a right $\aleph_0$-fir if and only if it satisfies ACC_{dense}.* ∎

The first part also provides a sharpening of Theorem 2.2.3.

Corollary 5.9.8. *If R is a left $\aleph_0$-fir, then any free left R-module satisfies ACC_{dense} and hence pan-ACC.* ∎

As a consequence we can show that $k\langle\langle X \rangle\rangle$ is not an $\aleph_0$-fir; we saw in Exercise 3.3.9 that it is not a fir.

Proposition 5.9.9. *If X is a set with more than one element, then $k\langle\langle X \rangle\rangle$ is not an $\aleph_0$-fir; likewise the subring $k\langle\langle X \rangle\rangle_{alg}$ of algebraic power series is not an $\aleph_0$-fir.*

This is in contrast to $k\langle\langle X \rangle\rangle_{\text{rat}}$, which will be shown to be a fir in Theorem 7.11.7.

Proof. Let x, y be distinct elements of X and consider the element $v = \sum_0^\infty x^i yxy^i$. It is clear that v satisfies the equation

$$v = xvy + yx, \qquad (3)$$

and moreover, (3) determines v uniquely as an algebraic power series. Let $\mathfrak{a}_n$ be the left ideal of $k\langle\langle X\rangle\rangle$ generated by $x, xy, xy^2, \ldots, xy^n, vy^{n+1}$. These elements are left linearly independent, for if $\sum a_i xy^i + bvy^{n+1} = 0$, then a_i must vanish as left cofactor of $xy^i (i \leq n)$, so also $b = 0$. Hence $\mathfrak{a}_n$ is free on these elements as basis. Now $\mathfrak{a}_n/\mathfrak{a}_{n-1}$ is bound, for if λ is any linear functional that vanishes on $xy^i (i < n)$ and vy^n, then

$$0 = (vy^n)\lambda = (xvy^{n+1} + yxy^n)\lambda = x(vy^{n+1}\lambda) + y(xy^n)\lambda,$$

therefore $vy^{n+1}\lambda = xy^n\lambda = 0$, and so $\lambda = 0$. On the other hand, $xy^n \notin \mathfrak{a}_{n-1}$, so $\mathfrak{a}_{n-1} \subset \mathfrak{a}_n$. This shows that $k\langle\langle X\rangle\rangle$ does not satisfy ACC$_\text{dense}$ and so by Corollary 9.8 it is not a left $\aleph_0$-fir.

Now consider $R_1 = k\langle\langle X\rangle\rangle_\text{alg}$; here the argument is the same: we have a semifir, and again obtain a sequence of left ideals violating left ACC$_\text{dense}$. ∎

Thus, the power series ring is an example of a semifir satisfying pan-ACC but not ACC$_\text{dense}$. By contrast, in Bezout domains ACC$_\text{dense}$ can be replaced by pan-ACC:

Proposition 5.9.10. *Over a right Bezout domain R, any torsion-free (= 1-unbound) left R-module with pan-ACC has ACC$_\text{dense}$. Hence a right Bezout domain with left pan-ACC is a left $\aleph_0$-fir.*

Proof. Let M be a torsion-free left R-module and take a chain

$$M_1 \subseteq M_2 \subseteq \ldots \qquad (4)$$

of finitely generated submodules with dense inclusions. By Proposition 2.3.19, any finitely generated torsion-free left R-module is free. If for some $i > 1$, rk M_{i-1} < rk M_i, then the induced map $K \otimes M_{i-1} \to K \otimes M_i$ is not surjective, where K is the field of fractions of R; hence there is a non-zero K-linear functional on $K \otimes M_i$ that vanishes on $K \otimes M_{i-1}$. By right multiplication with an appropriate element of R we may assume that the induced map takes M_i into R. Thus we have a linear functional on M_i that is zero on M_{i-1} without vanishing, and this contradicts the fact that M_i/M_{i-1} is bound. Thus all the ranks in (4) must be equal, to n say, and so the sequence terminates by ACC$_n$. ∎

Proposition 9.10 cannot be simplified by taking ACC_n for a fixed n only, since even over $\mathbb{Z}$ the conditions $\text{ACC}_n (n = 1, 2, \ldots)$ are independent (see Exercise 9). If we combine the first part of Proposition 9.10 with Corollary 9.7, we obtain a somewhat surprising conclusion:

Corollary 5.9.11. *Let R be a right Bezout domain with left pan-ACC. Then any countably generated left R-module embedded in a direct power R^I is free.* ∎

This shows, for example, that every countably generated subgroup of $\mathbb{Z}^I$ is free, although of course $\mathbb{Z}^I$ is not free, unless I is finite (by Lemma 3.3.13).

Exercises 5.9

1. (Continuation of Exercise 1.10) If R is a semifir and $\mathcal{C}$ the class of R-modules that are n-unbound for all n and have $\text{ACC}_{\text{dense}}$, show that $\mathcal{C}$ is closed under the operations listed in (i) but not, in general, (ii). Show that R must have left $\text{ACC}_{\text{dense}}$ for $\mathcal{C}$ to be non-trivial. If R is right Bezout, show that $\mathcal{C} = \cap \mathcal{C}_n$ and hence in this case $\mathcal{C}$ admits (ii) too.
2. Let R be a right Ore domain. Show that if a torsion-free left R-module satisfies ACC on submodules of rank at most n, for all n, then it also satisfies $\text{ACC}_{\text{dense}}$. (*Hint*: Imitate the proof of Proposition 9.10.)
3*. (Bergman [67]) Let R be an integral domain that is not a right Ore domain. Show that $R^\mathbb{N}$ as left R-module does not have $\text{ACC}_{\text{dense}}$. (*Hint*: Let $a, b \in R$ be right linearly independent and define $e_i, f_i \in R^\mathbb{N}$ by $e_i \pi_j = \delta_{ij}$, $f_i \pi_j = a^{j-i} b$, where $a^r = 0$ for $r < 0$ and π_j is the projection on the jth factor. Verify that $R f_i$ is dense in $R e_i + R f_{i+1}$ and deduce that the $M_i = R e_1 + \cdots + R e_i + R f_{i+1}$ form a strictly ascending chain of dense inclusions.)
4. Verify that the first part of Proposition 9.1 holds for Sylvester domains.
5. Let R be a semifir and $\mathfrak{a} = \cup \mathfrak{a}_i$, where $\mathfrak{a}_i$ is a finitely generated left ideal properly containing $\mathfrak{a}_{i-1}$ as a dense submodule. Show that $\mathfrak{a}$ is countably generated but not free.
6. Let kF be the group algebra over k of the free group F on x, y, z and let R be the subalgebra generated by z and all $z^{-n} x, z^{-n} y (n = 1, 2, \ldots)$. Show that R is a right but not left fir (see Theorem 2.10.3). Show also that (x, y) cannot be written as $u(x', y')$, where (x', y') is left prime (see Cohn [82a]).
7. Using Exercise 1.2, show that any left hereditary integral domain has left ACC_1.
8. Prove the following converse of Proposition 9.10: If R is a semifir over which each torsion-free left R-module with pan-ACC satisfies $\text{ACC}_{\text{dense}}$, then R is right Bezout.
9*. (Bergman [67]) (i) For any prime number p denote by $\mathbb{Z}_p$ the ring of p-adic integers and by $\mathbb{Q}_p = \mathbb{Z}_p [p^{-1}]$ its field of fractions (the p-adic numbers), so that $\mathbb{Z}[p^{-1}] \cap \mathbb{Z}_p = \mathbb{Z}$. Let $1, x_1, \ldots, x_n$ be any $\mathbb{Z}$-linearly independent elements of

$\mathbb{Z}_p$ and define a subgroup G of $\mathbb{Q}_p^{n+1}$ by the equation

$$G = (\mathbb{Z}_p^{n+1} + (1, x_1, \ldots, x_n)\mathbb{Q}_p) \cap \mathbb{Z}[p^{-1}]^{n+1}.$$

Verify that any finitely generated subgroup of G can be generated by $n + 1$ elements. Show further that for any $h > 0$ there exists $a = (1, a_1, \ldots, a_n) \in \mathbb{Z}^{n+1}$ such that $p^{-h}a \in G$ and deduce that G is not finitely generated. Hence show that G does not satisfy ACC_{n+1}.

(ii) Let G be defined as in (i) and suppose that C is a union of n-generator subgroups in G. Show that C is annihilated by a non-zero $\mathbb{Q}$-linear functional λ with coefficients in $\mathbb{Z}$. Show further that C is contained in $p^{-h}\mathbb{Z}^{n+1}$, where p^h is the highest power of p dividing $\lambda((1, x_1, \ldots, x_n))$. Deduce that C is finitely generated and that G has ACC_n. This shows that the conditions $ACC_n (n = 1, 2, \ldots)$ form a strictly increasing sequence, even over $\mathbb{Z}$.

10°. Find an ACC (of the type ACC_{dense}) such that every semifir satisfying this condition on the left is a left fir, but such that not every semifir satisfying the condition is a left PID.

11°. Find an example of a left fir that is elementarily equivalent to a fir, but is not two-sided.

12°. For a left Ore domain, does left ACC_n for some n, or left ACC_{dense} imply the corresponding condition for free modules?

13. In Proposition 9.4, show that the M_i/M_{i-1} are torsion modules for all large i.

14. An *involution* of a ring R is an anti-automorphism whose square is the identity. If R is any ring with an involution $*$, verify that the map $A \mapsto A^H$, where the entries of A^H are $a_{ij}^H = a_{ji}^* (A = (a_{ij}))$, is an involution of R_n. Let R be a semifir with involution $*$ and let $A \in R_n$ satisfy $A^H = A$. Show that if $\rho A = r$, then $A = PA_1P^H$, where $P \in {}^nR^r$ and $A_1^H = A_1$. If moreover R is a fir, show that P may be taken to be right prime.

15°. Does the conclusion of Corollary 9.11 hold for left $\aleph_0$-firs?

5.10 The intersection theorem for firs

In this section we shall apply the ACC for bound submodules of finitely related modules over a fir to show that the intersection of the powers of a proper ideal in a fir is zero. In fact we shall prove a slightly more general result that will be needed in Chapter 6.

We begin by considering an arbitrary ring R. If $\mathfrak{a}_1, \mathfrak{a}_2, \ldots$ is any sequence of (left, right or two-sided) ideals of R, we define

$$\prod_{i \geq 1} \mathfrak{a}_i = \bigcap_{n \geq 1} \mathfrak{a}_1 \mathfrak{a}_2 \ldots \mathfrak{a}_n.$$

Clearly the left-hand side is a (left, right, two-sided) ideal whenever all the $\mathfrak{a}_i$ are.

5.10 The intersection theorem for firs

Lemma 5.10.1. *Let $\mathfrak{a}_1$ be a right ideal and let $\mathfrak{a}_2, \mathfrak{a}_3, \ldots$ be a sequence of two-sided ideals of a ring R, and assume that $\mathfrak{a}_1$ is free as right R-module, with basis (e_λ). Then any $a \in \Pi \mathfrak{a}_i$ can be written as*

$$a = \sum e_\lambda a_\lambda, \quad \text{where } a_\lambda \in \prod_{i \geq 2} \mathfrak{a}_i. \tag{1}$$

Proof. Given $n > 0$, we have $a \in \mathfrak{a}_1 \mathfrak{a}_2 \ldots \mathfrak{a}_n$, by hypothesis, so there is an expression $a = \sum u_{i_1} u_{i_2} \ldots u_{i_n}$, where $u_{i_r} \in \mathfrak{a}_r$; in particular there is an expression $a = \sum e_\lambda a_\lambda$ with

$$a_\lambda \in \mathfrak{a}_2 \ldots \mathfrak{a}_n. \tag{2}$$

Since the e_λ are right linearly independent, the a_λ are independent of n, and (2) holds for all n, which is the assertion. ∎

We next give a condition for a homomorphic image of a non-zero bound module to be non-zero.

Lemma 5.10.2. *Let R be a ring and $E \in {}^m R^r$, $B \in {}^r R^n$ such that EB is left regular and $R^r/R^m E$ is a non-zero bound module. Then $R^r B/R^m EB$ is also a non-zero bound module.*

Proof. $R^r B/R^m EB$ is a homomorphic image of $R^r/R^m E$ and so is also bound. If it were 0, we would have $B = CEB$ for some $C \in {}^r R^m$. Hence $EB = ECEB$ and since EB is left regular, it follows that $EC = I$, therefore $R^m E$ is a direct summand in R^r, so $R^r/R^m E$ is projective, contradicting the assumption that it is non-zero and bound. ∎

We shall want to apply this result in the following form:

Corollary 5.10.3. *Let R be a ring and $\mathfrak{a}$ a proper non-zero free right ideal of R, with basis (e_λ). Given $a_1, \ldots, a_r \in \mathfrak{a}$, left and right linearly independent over R, write*

$$a_i = \sum e_\lambda b_{\lambda i}. \tag{3}$$

Then $\sum Rb_{\lambda i} / \sum Ra_j$ is a non-zero bound module.

Proof. In Lemma 10.2 we shall take for E the row vector whose entries are some of the e_λ including all those occurring with a non-zero coefficient in (3), and let B be the $r \times n$ matrix formed by the b's in (3). The left linear independence of the a's means that EB is left regular; hence E is also left regular. Since the a_i are right linearly independent, E is also right regular; hence $R^r/R^m E$ is bound. Moreover, since $\mathfrak{a}$ is a proper right ideal, E has no right inverse; by regularity it has no left inverse and so $R^r/R^m E \neq 0$. It follows by Lemma 10.2 that $R^r B / R^m E B \neq 0$, and it is bound, as homomorphic image of $R^r / R^m E$. ∎

We can now prove a general form of the intersection theorem:

Theorem 5.10.4. *Let R be a left fir and $\mathfrak{a}_1, \mathfrak{a}_2 \ldots$ a sequence of proper two-sided ideals that are free as right ideals. Then $\prod_{i \geq 1} \mathfrak{a}_i = 0$.*

Proof. Suppose that $\prod \mathfrak{a}_i \neq 0$ and let $\mathfrak{c}_1$ be a finitely generated non-zero left ideal contained in $\prod \mathfrak{a}_i$. Let $a_1, \ldots, a_r$ be a basis of $\mathfrak{c}_1$ and (e_λ) a basis of $\mathfrak{a}_1$ as right ideal, and write $a_i = \sum e_\lambda b_{\lambda i}$. Then $b_{\lambda i} \in \prod_{j \geq 2} \mathfrak{a}_j$ by Lemma 10.1, hence $\mathfrak{c}_2 = \sum R b_{\lambda i} \subseteq \prod_{j \geq 2} \mathfrak{a}_j$ and by Lemma 10.3, $\mathfrak{c}_2 / \mathfrak{c}_1$ is bound and non-zero. By induction we obtain a sequence of finitely generated left ideals $\mathfrak{c}_n$ such that $\mathfrak{c}_n \subseteq \prod_{i \geq n} \mathfrak{a}_i$ and we have the strictly ascending chain

$$\mathfrak{c}_1 \subset \mathfrak{c}_2 \subset \ldots$$

with bound quotients. Thus we have an infinite ascending sequence of bound submodules of $R / \mathfrak{c}_1$, which contradicts Corollary 1.7. ∎

In particular, taking R to be a two-sided fir, we find that the conclusion holds for any sequence of proper two-sided ideals. Taking all ideals equal, we obtain

Corollary 5.10.5. *In a two-sided fir the intersection of the powers of any proper two-sided ideal is zero.* ∎

Exercises 5.10

1. Give a direct proof of Corollary 10.5 for principal ideal domains.
2. Let R be an integral domain with left ACC_1. Show that any proper two-sided ideal $\mathfrak{a}$ that is principal as right ideal satisfies $\cap \mathfrak{a}^n = 0$.
3. Let $\mathfrak{a}$ be a finitely generated ideal in a commutative ring R. If $\mathfrak{a}^2 = \mathfrak{a}$, show that $\mathfrak{a} = eR$ for an idempotent e.
4°. Find a generalization of Exercise 3 to the non-commutative case.
5. Show that in a left fir R, if $\mathfrak{a}\mathfrak{b} \subseteq \mathfrak{a}'\mathfrak{b}$ for two-sided ideals $\mathfrak{a}, \mathfrak{a}'$ and a left ideal $\mathfrak{b}$, then $\mathfrak{a} \subseteq \mathfrak{a}'$ (see also Section 6.6).
6. Show that the conclusion of Corollary 10.5 does not hold for the one-sided fir constructed in Section 2.10.
7. Let F be a free group and kF the group algebra of F over k. If $\mathfrak{a}$ is the augmentation ideal (induced by the kernel of the homomorphism $F \to 1$), show that $\cap \mathfrak{a}^n = 0$. Define the lower central series of F recursively by $\gamma_1 F = F$, $\gamma_{n+1} F = [\gamma_n F, F]$ (commutator subgroup), and show that for any $u \in \gamma_n F$, $u \equiv 1 (\text{mod} \mathfrak{a}^n)$; deduce that $\cap \gamma_n F = 1$ (Magnus' theorem).

8. (G. M. Bergman) Show that in the situation of Lemma 10.3, for the free algebra the modules $R^r/R^m E$ and $R^r B/R^m EB$ need not be isomorphic (*Hint*: If $R = k\langle x, y\rangle$, take $E = (x, y)$, $B = (1, 0)^T$.)

Notes and comments on Chapter 5

Sections 5.1 and 5.2, based on Cohn [73d, 77a], describe the background on modules in the more general setting of hereditary rings. The notion of torsion class used in Section 5.1 is basic in the study of torsion theories (see e.g. Stenström [75]), but we need only the most elementary properties; in any case the usual treatment deals mainly with hereditary torsion theories (in which the torsion class admits subobjects), and so does not apply here. The transpose $\mathrm{Tr}(M) = \mathrm{Ext}^1_R(M, R)$ has also been used by M. Auslander in the study of Artin algebras.

Torsion modules over firs were first described in Cohn [67] and in FR.1 formed the basis of the factorization of full matrices. The positive and negative modules (corresponding to left and right full matrices) are studied in Cohn [77a, 82a]. The Krull–Schmidt theorem for finitely presented modules over firs (Theorem 3.9) was new in FR.2; it has also been obtained independently by Schofield (unpublished). The application in Theorem 3.14, giving a conceptual proof of the embedding of a fir in a field, is due to Bergman [2002].

The treatment in 5.4 essentially follows Bergman [67], but Theorem 4.9 (which is used in Section 5.5) is taken from Cohn [74b], and Proposition 4.11 is new (Proposition 4.11 of FR.2 has become Lemma 0.3.3). Sylvester domains were introduced by Dicks and Sontag [78], and Section 5.5 is based on this source, but the presentation has been modified here so as to be independent of the results of Chapter 7. The law of nullity first occurs (for the case of fields) in Sylvester [1884]; in the case of semifirs it first appeared in FR.1. In studying localization (Chapter 7) we shall need to consider pseudo-Sylvester domains; their properties, described in Section 5.6 are taken from Cohn and Schofield [82]. Lemma 6.3 is due to Bergman, and is used here in the proof of Proposition 6.4. Some of the results from Section 5.6 of FR.2 are now to be found in Sections 0.1 and 0.4. The analysis of matrix relations and factorization in Section 5.7 was mostly new in FR.2; see also Cohn [82d].

The normal form that is the subject of Section 5.8 is due to Roberts [84] and all the results of this section are taken from this source, except for Lemma 8.1, which is implicit in Roberts [82], where a more general form of Exercise 8.3 also occurs. The uniqueness of Theorem 8.2 was proved in a special case in Cohn [76b]. Lemma 8.7 on the factorization of matrices over a free algebra, and Corollary 8.6 on which it depends, are due to Schofield, who is also responsible for the elegant proof of Theorem 8.11 (replacing an earlier proof by Roberts that required a larger rank in (ii)).

The study of ascending chain conditions in Section 5.9 is taken from Bergman [67], with results 9.1–9.3 added from Cohn [82a]. Proposition 9.6 was previously stated for ideals in $\aleph_0$-firs, while Proposition 9.9 was an exercise in FR.1. Corollary 9.7 generalizes Pontryagin's theorem: A countably generated torsion-free abelian group with pan-ACC is free (Pontryagin [39], p. 168). In the case of $\mathbb{Z}$, related results have been obtained by Specker [50], who shows, for example, that in the subgroup B of $\mathbb{Z}^I$ consisting of all bounded sequences, every subgroup of cardinal at most $\aleph_1$ is free. Dubois [66] and independently Nöbeling [68] have shown that B itself is free, and more recently Bergman [72a] has given a very brief proof of this fact. The intersection theorem in Section 5.10 was first proved in the case where all the $\mathfrak{a}_i$ are equal, by Cohn [70c], and this appeared in FR.1. It was greatly generalized by Bergman [72b,73]; a special case of his results yields Theorem 10.4 as well as a slight simplification of the proofs. Lemma 10.2 was suggested by G. M. Bergman, as a way of proving Corollary 10.3. Proposition 5.10.12 of FR.2 on α-directed systems of α-firs, has now become Proposition 2.3.24.

The specialization lemma that occupied Section 5.9 in FR.2 is now part of Section 7.8.

6
Centralizers and subalgebras

The first topic of this chapter is commutativity in firs. We shall find that any maximal commutative subring of a 2-fir with strong DFL is integrally closed (Corollary 1.2), and the same method allows us to describe the centres of 2-firs as integrally closed rings and make a study of invariant elements in 2-firs and their factors in Sections 6.1 and 6.2. The well-known result that a simple proper homomorphic image of a principal ideal domain is a matrix ring over a skew field is generalized here to atomic 2-firs (Theorem 2.4). In Section 6.3 the centres of principal ideal domains are characterized as Krull domains. Further, the centre of a non-principal fir is shown to be a field in Section 6.4.

Secondly we look at subalgebras and ideals of free algebras in Section 6.6; by way of preparation submonoids of free monoids are treated in Section 6.5. A brief excursion into coding theory shows how the Kraft–McMillan inequality can be used to find free subalgebras, and the fir property of free algebras is again derived (Theorem 6.7). Section 6.7 is devoted to a fundamental theorem on free algebras: Bergman's centralizer theorem (Theorem 7.7).

Section 6.8 deals with invariants under automorphisms of free algebras, and Section 6.9 treats the Galois correspondence between automorphism groups and free subalgebras, as described by Kharchenko. The final section, 6.10, brings a result on the structure of $\mathrm{Aut}(k\langle x, y\rangle)$, showing all these automorphisms to be tame (Czerniakiewicz–Makar-Limanov theorem), by exhibiting this group as a free product with amalgamation.

6.1 Commutative subrings and central elements in 2-firs

Just as commutative 2-firs have a rather special form, so it is possible to say more about commutative subrings of 2-firs. Assuming strong DFL, we shall show that maximal commutative subrings are integrally closed. We recall that if $A \subseteq A'$ are commutative integral domains, then an element $y \in A'$ is *integral over A* if there is a monic equation for y with coefficients in A: $y^n + a_1 y^{n-1} + \cdots + a_n = 0$ ($a_i \in A$). Equivalently, the A-module generated by the powers of y is finitely generated over A. The set of all elements of A' integral over A forms a subring $\bar{A}$ of A', the *integral closure* of A in A', and A is *integrally closed* in A' if $\bar{A} = A$. By a *finite integral* extension of A in A' we understand a ring B between A and A' that is finitely generated as A-module. Clearly it then follows that all elements of B are integral over A. Suppose that B is a finite integral extension of A in its field of fractions k. Then we can write $B = \sum_1^n A u_i$, where $u_i \in k$, or equivalently, $B = \sum u_i A$. The u_i may be brought to a common denominator, say $u_i = a_i d^{-1}$ ($a_i, d \in A$); then $B = \Sigma A a_i d^{-1}$, and so $Bd \subseteq A$. This means that the *conductor* of A in B, defined as

$$\mathfrak{f} = \{a \in A | Ba \subseteq A\},$$

is different from 0. We note that $\mathfrak{f}$ may also be described as the largest ideal in A that is also an ideal in B.

We begin with a result extending embeddings of commutative rings to integral extensions.

Proposition 6.1.1. *Let R be a 2-fir, A a commutative subring of R and B a finite integral extension of A in its field of fractions. Then there exists an injection $f : B \to R$ and $e \in R^\times$, such that*

$$ex = xf.e \quad \text{for all } x \in A. \tag{1}$$

If moreover, R has right strong DFL ($baR \cap a'R \neq 0$, where a, a' are similar, implies $ba \in a'R$), then e can be taken to be 1.

Proof. Since B is finite integral over A, there exists $c \in A^\times$ such that $Bc \subseteq A$. Put $\mathfrak{c} = Bc$; this is a finitely generated non-zero ideal of A (and of B). Any two non-zero elements of $\mathfrak{c}$ have a non-zero common multiple in A, hence in R; since R is a 2-fir, it follows that $R\mathfrak{c}$ is principal, say $R\mathfrak{c} = Re$ for some $e \in R^\times$. We have $eA \subseteq R\mathfrak{c}A = R\mathfrak{c} = Re$; hence there is a homomorphism $f_0 : A \to R$ such that

$$ea = af_0.e \quad \text{for all } a \in A.$$

Clearly f_0 is injective; moreover, if $a \in Bb$ ($a, b \in A$), then

$$af_0.e = ea \in Rea = R\mathfrak{c}a \subseteq R\mathfrak{c}Bb \subseteq R\mathfrak{c}b = Reb = Rbf_0.e,$$

hence $af_0 \in R.bf_0$. By Proposition 0.7.5 there exists an injective homomorphism $f : B \to R$ extending f_0, and this proves (1).

Suppose now that R has right strong DFL. Let $q \in B$, say $q = ad^{-1}, a, d \in A$. Then $a = qd$, so $af = qf.df$, hence

$$qf.ed = qf.df.e = af.e = ea \in eR.$$

By right strong DFL, $qf.e \in eR$, say $qf.e = e.r$. So we have

$$erd = qf.ed = ea,$$

therefore $rd = a$. Thus $a = qd$ implies $a = rd$ for some $r \in R$, and so by Proposition 0.7.5 (with f = identity) we find that $q \mapsto r$ is a well-defined homomorphism $f : B \to R$ such that $af = a$ for all $a \in A$, as claimed. ∎

Corollary 6.1.2. *In a 2-fir with right strong DFL every maximal commutative subring is integrally closed in its field of fractions.*

Proof. Let R be a 2-fir with right strong DFL, A a maximal commutative subring and B a finite integral extension of A in its field of fractions. By Proposition 1.1, there is an injection $f : B \to R$ that reduces to the identity on A. But B is commutative, so we have $B = A$, by the maximality of A. ∎

Corollary 6.1.3. *The centre of a 2-fir is integrally closed in its field of fractions.*

Proof. Let R be a 2-fir, denote its centre by C and suppose that B is a finite integral extension of C in its field of fractions. By Proposition 1.1 there exists an embedding $f : B \to R$ such that $af = a$ for all $a \in C$ (because C is the centre), so we may assume that $C \subseteq B \subseteq R$. Let $b \in B$, say $a = cb (a, c \in C)$; then for any $r \in R$,

$$cbr = ar = ra = rcb = crb,$$

hence $br = rb$, so $b \in C$. This shows that $C = B$, which establishes the result. ∎

To obtain further information on the centres of atomic 2-firs we consider the set $\mathrm{Inv}(R)$ of invariant elements of R, i.e. regular elements c such that $cR = Rc$. We recall from Proposition 1.4.6 that in any ring R a regular element c is invariant if and only if the left and right ideals of R generated by c are both two-sided; moreover, if R is an integral domain and $aR = Ra'$, then the proof of Proposition 1.4.6 shows that a' is associated to a. Such an ideal will be called an *invariant principal ideal*. In a 2-fir we have

Proposition 6.1.4. *In any 2-fir the invariant principal ideals form a sublattice of the lattice of all ideals.*

Proof. Let $aR = Ra, bR = Rb$ be two invariant principal ideals in a 2-fir R. Then $0 \neq ab \in aR \cap bR$, hence $aR + bR = dR = Ra + Rb = Rd'$, and $aR \cap bR = mR = Ra \cap Rb = Rm'$. By Proposition 1.4.6, $dR = Rd, mR = Rm$ and so d and m are invariant, as claimed. ∎

This result shows that in any 2-fir R the monoid $\mathrm{Inv}(R)$ of invariant elements is lattice-ordered by divisibility. Thus any two invariant elements have an HCF and this is the same whether calculated in $\mathrm{Inv}(R)$ or in R; similarly for the LCM.

We now add atomicity to our assumptions; we recall that a *prime* element is an invariant non-unit p such that

$$p|ab \text{ implies } p|a \text{ or } p|b;$$

this has a meaning since left and right divisibility by p coincide, by invariance. Further we define an Inv-*atom* as an invariant element that is an atom within $\mathrm{Inv}(R)$. Thus an invariant atom will always be an Inv-atom, but not conversely. The ring R is said to have *unique factorization of invariant elements* if $\mathrm{Inv}(R)$ is a UF-monoid, as defined in Section 0.9. By applying Theorem 0.9.4 we obtain the following description of UF-monoids:

Theorem 6.1.5. *For any ring R the following conditions are equivalent:*

(a) R is a ring with unique factorization of invariant elements,
(b) R satisfies ACC on invariant principal ideals and any two invariant elements have an HCF in $\mathrm{Inv}(R)$,
(c) R satisfies ACC on invariant principal ideals and any two invariant elements have an LCM in $\mathrm{Inv}(R)$,
(d) $\mathrm{Inv}(R)$ is atomic and every Inv-atom is prime. ∎

We note that e.g. the second part of (b) certainly holds if any two invariant elements have an HCF in R and this HCF is invariant. If we merely know that $a, b \in \mathrm{Inv}(R)$ have a (left or right) HCF d in R, we cannot assert that d is invariant, though there is an important case in which this holds, namely when R is a 2-fir, as Proposition 1.4 shows. If we apply Theorem 1.5 in this case, we obtain a factorization theorem for invariant elements:

Theorem 6.1.6. *Every atomic 2-fir has unique factorization of invariant elements: thus every non-unit invariant element c can be written as a product of Inv-atoms*

$$c = a_1 \ldots a_r \tag{2}$$

and if

$$c = b_1 \ldots b_s \tag{3}$$

6.1 Commutative subrings and central elements in 2-firs

is any other factorization of c into Inv-atoms, then $s = r$ and there is a permutation $i \mapsto i'$ of $1, \ldots, r$ such that b'_i is associated to a_i. Moreover, any order of the factors in (2) can be realized. ∎

In terms of ideals Theorem 1.6 states that in an atomic 2-fir the ideals with invariant generator form a free commutative monoid under ideal multiplication. Of course this is merely a reflection of the fact that a lattice-ordered group with descending chain condition (on the positive elements) is necessarily free abelian (Birkhoff [67], p. 298).

In certain cases the last result can be extended to matrix rings. Let us call an $n \times n$ matrix C over a ring R *right invariant* if it is right invariant as element of R_n, thus C is regular and for each $X \in R_n$ there exists $X' \in R_n$ such that $XC = CX'$. We remark that over any Sylvester domain a right invariant matrix must be full. For if $C = PQ$ is a rank factorization, where P is $n \times r$ and Q is $r \times n$, suppose that $r < n$. Then by Theorem 5.4.9, P contains a full $r \times r$ submatrix, say in the first r rows, so there exists $P' \in {}^nR^{n-r}$ such that $P_1 = (P, P')$ is full and so right regular. Put $Q_1 = (Q, 0)^T$; then $C = P_1 Q_1$ and since C is right invariant, there exists $P_2 \in R_n$ such that $P_1 C = CP_2 = P_1 Q_1 P_2$, hence $C = Q_1 P_2$. But Q_1 is not left regular, so neither is C, a contradiction. This shows that C must be full.

To describe the form of right invariant matrices, we shall use the following two lemmas:

Lemma 6.1.7. *Let R be a ring, $\gamma \in R$ and $U \in GL_n(R)$. If γ is right invariant in R, then γU is right invariant in R_n.*

Proof. Suppose that γ is right invariant. For any $x \in R$ we have $x\gamma = \gamma x'$, for a unique $x' \in R$; hence there is for any $X \in R_n$, a unique $X' \in R_n$ satisfying $X\gamma = \gamma X'$ and it follows that $X\gamma U = \gamma X'U = \gamma U.U^{-1}X'U$. ∎

Lemma 6.1.8. *For any ring R and any $n > 1$ the following conditions are equivalent:*

(a) *there exists $C \in R_n$ that is right invariant but cannot be written in the form $C = \gamma U$, where $\gamma \in R$ is right invariant and $U \in GL_n(R)$,*
(b) *there exists a projective left R-module P such that*
   (b.i) *P is not free of rank 1,*
   (b.ii) *$P^n \cong R^n$, and*
   (b.iii) *P contains a regular element p such that for every $r \in R$ there is a unique endomorphism of P mapping p to rp,*
(c) *R has a two-sided ideal I with zero left annihilator, such that $I^n \cong R^n$, as right R-modules, but I is not free of rank 1.*

Proof. (a) $\Rightarrow$ (b). Let $C \in R_n$ be right invariant; then $XC = CX'$, where $X \mapsto X'$ is an endomorphism of R_n, injective since C is regular. Applied to the matrix units e_{ij} this yields a set of matrices e'_{ij} that again satisfy the relations for matrix units. Writing $P = R_n e'_{11}$, we obtain the isomorphism $R^n \cong P^n$. Thus P satisfies (b.ii).

If we now apply the relation $e_{11}C = Ce'_{11}$ to the row vector $e_1 = (1, 0, \ldots, 0)$ and write $p = e_1 C$, we obtain $p = pe'_{11}$, hence $p \in P$. We next apply the relation $re_{11}C = C(re_{11})'$ (for any $r \in R$) to e_1 and find that $rp = p(re_{11})'$, which is (b.iii), and it remains to prove (b.i). Suppose then that P is free of rank 1, say $P = Rq_1$. Then the above p can be written as γq_1 for some $\gamma \in R$. Writing $q_i = q_1 e_{1i}' (i = 2, \ldots, n)$, we see that $Rq_i \cong P$ and $q_1, \ldots, q_n$ are left linearly independent and hence form a basis of $P^n \cong R^n$. If the matrix mapping e_i to q_i is U, this is invertible and we find that $C = \gamma U$, where the right invariance of C in R_n implies the right invariance of γ in R. This contradicts (a) and it shows that (b.i) holds.

(b) $\Leftrightarrow$ (c) by the duality of projective modules. For the left module map $\psi : R \to P$ given by $r \mapsto rp$ in (b) dualizes to a map of right modules $\psi^* : P^* \to R$; the uniqueness in (b.iii) means that ψ has a dense image and it follows that ψ^* is injective, so that we may regard P^* as a right ideal I of R. The existence part in (b.iii) corresponds to the condition that I is mapped into itself by left multiplication by all $r \in R$, i.e. that I is a two-sided ideal and the regularity of p in (b.iii) amounts to saying that ψ is injective, i.e. I has zero left annihilator. Of course (b.i) and (b.ii) translate to the corresponding conditions in (c) on the right R-module structures of I and I^n.

(b) $\Rightarrow$ (a). Assume (b) and write

$$R^n \cong P_1 \oplus \cdots \oplus P_n, \tag{4}$$

where the P_i are pairwise isomorphic modules satisfying (b.i)–(b.iii). Let $p_1 \in P_1$ have the properties of p in (b.iii), let $p_i \in P_i (i = 2, \ldots, n)$ be the element of P_i corresponding to p_1 under the given isomorphism and denote the matrix mapping each e_i to p_i by C. Since each p_i has zero annihilator, it follows that C is left regular. By (b.iii) (and the isomorphism $P_i \cong P_j$) it follows that for each $X \in R_n$ there exists a unique $X' \in R_n$ such that $XC = CX'$, and the uniqueness of X' shows that C is also right regular. Thus C has been shown to be right invariant. If $e_{11}C = Ce_{11}'$, then e_{11}' is the projection of (4) on P_1. If we had $C = \gamma U$, where γ is right invariant, then $Y = U^{-1} e_{11} U$ would satisfy the equation $e_{11}C = CY$, by which we just characterized the projection onto P_1. Hence we would have $P_1 = Re_1 U$, a free module on $e_1 U$, which contradicts (b.i). Therefore C cannot have the form γU and (a) follows. ∎

Suppose that R is a projective-free ring; then (c) holds neither in R nor its opposite R°; hence we obtain

Corollary 6.1.9. *Let R be a projective-free ring. Then any right invariant matrix C over R has the form*

$$C = \gamma U, \qquad (5)$$

where γ is a right invariant element of R and $U \in GL_n(R)$. If moreover, C is invariant, then it has the form (5), where γ is invariant. ■

Combining this result with Theorem 1.6, we obtain

Theorem 6.1.10. *Over a fir R every invariant matrix can be written as a product of Inv-atoms in R and an invertible matrix over R, and the factorization is unique up to associates and the order of the factors.* ■

Let us return to the case of elements (Theorem 1.6). In order to apply this result to study the centre of a 2-fir we need to recall some facts on valuations. Let A be a commutative integral domain and k its field of fractions; any homomorphism $v : A^\times \to \Gamma_{\geq 0}$ into the positive cone of a totally ordered additive group Γ (with the convention $v(0) = \infty$), such that

$$v(a - b) \geq \min\{v(a), v(b)\},$$

is called a *general valuation*. Such a valuation can always be extended in just one way to a valuation of k, again written v, and the set $k_v = \{x \in k \mid v(x) \geq 0\}$ is a local Bezout domain containing A; k_v is called the *valuation ring* of v. Now the ring $A^c = \cap k_v$, where v ranges over all general valuations on A, consists precisely of all elements of k that are integral over A; thus A^c is just the *integral closure* of A in k, in fact A is integrally closed if and only if $A^c = A$ (see e.g. BA, Section 9.4).

There is a related construction that we shall need here. This arises if instead of general valuations we limit ourselves to $\mathbb{Z}$-valued valuations. Let A be a commutative integral domain and k its field of fractions, as before. If there is a family V of $\mathbb{Z}$-valued valuations on k such that (i) for any $x \in A^\times$, $v(x) \geq 0$ for all $v \in V$, with equality for almost all v and (ii) $A = \cap k_v$, then A is said to be a *Krull domain*. More generally, suppose that there is a family V of $\mathbb{Z}$-valued valuations on A satisfying (i); then $A^* = \cap k_v$ is clearly a Krull domain containing A.

From our earlier remarks it is clear that every Krull domain is integrally closed. Every commutative UFD is a Krull domain: we take V to be the family of valuations associated with the atoms of A. Likewise every Noetherian integrally closed domain is a Krull domain; here V is the class of all valuations associated with the minimal prime ideals of A (see Bourbaki [72], Chapter 7, §1, No. 3).

We now have the following sharpening of a special case of Corollary 1.3.

Theorem 6.1.11. *The centre of an atomic 2-fir is a Krull domain.*

Proof. Let R be an atomic 2-fir, C its centre and k the field of fractions of C. By Theorem 1.6, each $a \in C^\times$ has a decomposition into Inv-atoms

$$a = u \prod p^{\alpha_p} \quad (u \in U(R), \alpha_p \geq 0).$$

Fix the Inv-atom p and consider the function v_p defined on C by

$$v_p(a) = \alpha_p.$$

Clearly this is $\mathbb{Z}$-valued, in fact it is non-negative on C, and $v_p(a) = 0$ for almost all p. Hence $C^* = \cap k_v$ is a Krull domain containing C; we claim that $C^* = C$. Let $d \in C^*$, say $d = ab^{-1}$, where $a = u \prod p^{\alpha_p}, b = v \prod p^{\beta_p}, u, v \in U(R)$ and $\alpha_p \geq \beta_p$. Since $a = db$, we have $u \prod p^{\alpha_p} = dv \prod p^{\beta_p}$; if $\beta_p > 0$, we can cancel p, replacing α_p by $\alpha_p - 1$ and β_p by $\beta_p - 1$. After finitely many steps we find $u_1 \prod p^{\alpha_p - \beta_p} = dv_1$, where u_1, v_1 are units, possibly different from u, v because the p's are merely invariant and not central. It follows that $d \in R$; now we find as in the proof of Corollary 1.3 that $d \in C$, i.e. $C^* = C$. ∎

As we shall see in Section 6.3, this theorem is best possible, in the sense that any Krull domain can occur as the centre of an atomic 2-fir. It follows in particular that any fir has a Krull domain as centre, but in Section 6.4 we shall see that the centres of non-Ore firs are much more restricted.

Exercises 6.1

1. Given integral domains $A \subseteq B$, an element $y \in B$ is called *left integral* over A if the left A-module generated by the powers of y is finitely generated over A. Show that this is so if and only if y satisfies a monic polynomial equation with left coefficients in A.
2°. If every element of a ring B is left integral over a subring A, B is called *left integral* over A. Is the notion of left integral extension transitive?
3. Let A be a right Ore domain with field of fractions K, and let B be an A-subring of K, with conductor $\mathfrak{f}$ of B in A. Show that $\mathfrak{f} \neq 0$ if and only if B is finite right integral over A, and that $\mathfrak{f}$ is the largest ideal in A that is also a left ideal in B.
4. Let $B = k\langle x, y \rangle$ and denote by A the subalgebra generated by x^2 and y^2. Find the set A^c of elements of B left integral over A. Is this set closed under addition or multiplication? Is ΣAa^i a subring for every $a \in A^c$?

5. Let R be an atomic 2-fir with DFL. Show that if A is a (left and right) Ore subring of R and B a finite left integral extension of A in its field of fractions, then there is an embedding of B in R whose restriction to A is the identity map. Deduce that a maximal Ore subring of R is integrally closed.
6. Let I_r be the set of all right invariant elements in an integral domain R and S the set of all left factors (in R) of elements of I_r. Show that S is a right Ore set in R.
7. Let α be an automorphism of a field K such that no positive power of α is inner. Show that the monic invariant elements of the skew polynomial ring $K[x; \alpha]$ are the powers of x. If $\alpha^r (r > 0)$ is inner, but no lower power, find all monic invariant elements.
8. Show that a principal ideal domain is simple if and only if it has no non-unit invariant element.
9. Let K be a field of characteristic 0 and D an outer derivation of K. Show that $K[x; 1, D]$ is simple and hence has no non-unit invariant element.
10. Let R be a 2-fir and $c \in R$ an invariant element. Find the condition on c for R/cR to be (i) simple and (ii) semisimple.
11. Let R be an atomic 2-fir and $c \in R$ a non-unit invariant element. Show that the ring $A = R/cR$ is Artinian and is such that every left (or right) ideal is the annihilator of its annihilator in A (i.e. A is a *quasi-Frobenius ring*).
12°. Is every Artinian principal ideal ring a quasi-Frobenius ring?
13. Let K be a field and R a subring that is a 2-fir; show that the elements $ab^{-1}(a, b \in R, b \neq 0)$ satisfying $axb = bxa$ for all $x \in R$ form an integrally closed subring.
14*. Find a ring in which the multiplication of invariant principal ideals is non-commutative. (*Hint*: Use a non-commutative analogue of Exercise 2.3.10).
15. Show that any right invariant element of a right principal Bezout domain is invariant and has a complete factorization.
16. Show that in a 2-fir with right ACC_1 every left invariant element has a complete factorization.
17. (G. M. Bergman) Let R be the $\mathbb{R}$-algebra of functions on the real line generated by the functions $\sin x$ and $\cos x$, and let S be the subalgebra generated by $\sin 2x$ and $\cos 2x$. Show that the matrix

$$\sin x \begin{pmatrix} \cos x & \sin x \\ -\sin x & \cos x \end{pmatrix}$$

in R_2 lies in fact in S_2 and is an invariant element of that ring, but cannot be written as γU for $\gamma \in S$ and $U \in GL_2(S)$.
18. (G. M. Bergman) Given a Dedekind domain D whose ideal class group has an element of order $n > 1$, show that D_n has an invariant element that is not of the form γU, for $\gamma \in D$ and $U \in GL_n(D)$.
19*. (G. M. Bergman) Let k be a commutative field of characteristic zero, $K = k(y)$ the rational function field in an indeterminate y over k, δ the derivation over k such that $y\delta = 1$ (differentiation with respect to y) and $R = K[x; 1, \delta]$ the ring of differential operators. Writing $p = xy^2 + y$, show that the centralizer of p in R is $k[p]$. Show that $k[p^2, p^3]$ is contained in $_xR = \{u \in R | xu \in Rx\}$, even though $p \notin {_xR}$.

6.2 Bounded elements in 2-firs

We have seen that in an atomic 2-fir the decompositions into Inv-atoms play an important role. In general there is no reason to suppose that Inv-atoms will be atoms, but at least we can factorize them into atoms, and this suggests that we look more closely at the factors of invariant elements. Such factors and the cyclic modules they define are called *bounded* (to be distinguished from 'bound' modules, see Exercise 1). It will be convenient to begin with a general definition and specialize later.

Definition. A right module M over a ring R is said to be *bounded* if there is a regular element $c \in R$ such that $Mc = 0$.

An equivalent definition is to require the annihilator $\mathrm{Ann}(M)$ of M, an ideal in R, to contain a regular element. We note that the direct sum of bounded modules is again bounded; for if $Mc = Nd = 0$, where c, d are regular, then cd is regular and $(M \oplus N)cd = Ncd \subseteq Nd = 0$. An element $a \in R$ is said to be *right bounded* if R/aR is bounded. Thus an element a in an integral domain R is right bounded if and only if aR contains a non-zero two-sided ideal. We remark that any right bounded element is right large, for if a is right bounded, say $Rd \subseteq aR$, and $b \in R^\times$ is given, then $bd \in aR$, so $bR \cap aR \neq 0$.

Suppose now that R is an integral domain. To describe the right bound of $a \in R^\times$ more closely, let $a = bc$; then for any $d \in \mathrm{Ann}(R/aR)$ we have $Rd \subseteq aR = bcR$; in particular, $bd \in bcR$, so $d \in cR$. The same is true if instead of a we take any element similar to a. Thus

$$\mathrm{Ann}(R/aR) \subseteq \cap \{cR \mid c \text{ is a right factor of an element similar to } a\}. \qquad (1)$$

This inclusion can also be rewritten as

$$\mathrm{Ann}(R/aR) \subseteq \cap \{cR \mid c \text{ is similar to a right factor of } a\}. \qquad (2)$$

When R is a 2-fir, we can prove that equality holds in (2):

Theorem 6.2.1. *Let R be a 2-fir and for any $a \in R$, define the right ideal*

$$I = \cap \{cR \mid c \text{ is similar to a right factor of } a\}. \qquad (3)$$

Then

(i) *I is the annihilator of the set of all torsion elements of R/aR; hence*
(ii) *if a is right large (so that all elements of R/aR are torsion), then $I = \mathrm{Ann}(R/aR)$.*
(iii) *The following conditions are equivalent (and so imply the conclusion of (ii)):*

(a) a is right bounded,
(b) I is a non-zero two-sided ideal,
(c) I contains a non-zero two-sided ideal.
(iv) I is closed under left multiplication by all units of R. Hence if R is generated by its units (e.g. if R is a group algebra over a field), then (b) (and hence (a) and (c)), holds whenever $I \neq 0$.

Proof. We have seen that (2) holds for any right bounded element a; if a is not right bounded, then the left-hand side of (2) is zero, so (2) holds for all $a \in R$. Now let x be a torsion element in R/aR and choose u in the right-hand side of (2). We have $xR \cap aR \neq 0$, hence there is a right coprime relation $xa' = ax'$, and here a' is similar to a right factor of a, therefore $u = a'u_1$ for some $u_1 \in R$. Hence $xu = xa'u_1 = ax'u_1 \in aR$, so u annihilates x(mod aR), and since x was any torsion element of R/aR, (i) follows. When a is right large, (ii) follows. Turning to (iii), we see that (a) $\Rightarrow$ (b) by (ii); clearly (b) $\Rightarrow$ (c) and when (c) holds, then since $I \subseteq aR$, (a) follows. Finally (iv) follows because similarity classes are closed under taking associates. ∎

Suppose now that R is an atomic 2-fir and that a is right bounded. Then all the cR on the right of (3) contain a fixed non-zero element, d say, where $d \in \text{Ann}(R/aR)$, and since the lattice $L(dR, R)$ has finite length, it is complete and the intersection on the right of (3) is principal, say $\text{Ann}(R/aR) = bR$. The element b, unique up to right associates, is called the *right bound* of a.

Corollary 6.2.2. *In any integral domain, any left factor of a right invariant element is right bounded. In an atomic 2-fir, conversely, every bounded element a is a left factor of a right invariant element.*

Proof. If $d = ab$ for a right invariant element d, then $Rd \subseteq dR \subseteq aR$, hence $d \in \text{Ann}(R/aR)$, i.e. a is right bounded, with bound $d'R \supseteq dR$. Conversely, for a right bounded element a of an atomic 2-fir R, $\text{Ann}(R/aR) = dR$. Here $Rd \subseteq dR$, so d is right invariant, and $d \in aR$. ∎

An element a in an integral domain R is said to be *bounded* if it is a factor of an invariant element: $c = dab$; it is then also a right factor, for we have $cb' = b'c = b'dab$, hence $c = b'da$. By symmetry it is also a left factor of c. Any bounded element is clearly left and right bounded, and for a bounded element a in an atomic 2-fir R, we have by Proposition 1.4,

$$\text{Ann}(R/aR) = \text{Ann}(R/Ra). \tag{4}$$

Conversely, if in an atomic 2-fir, a is left and right bounded and (4) holds, then $\text{Ann}(R/aR) = dR = Rc$, hence by Proposition 1.4.6, $\text{Ann}(R/aR)$ has an invariant generator and a is then bounded.

Let R be an atomic 2-fir and let $a \in R$ be bounded; the invariant generator of $\text{Ann}(R/aR)$ is then unique up to associates; it will be called the *bound* of a and denoted by a^*. Clearly a^* depends only on the similarity class of a. By Theorem 2.1, a^* can also be defined by

$$a^*R = \cap \, \{cR \mid c \text{ is similar to a right factor of } a\}. \tag{5}$$

The right-hand side can be taken to be a finite intersection, by the DCC in R/a^*R, and if we take it to be irredundant, we obtain a subdirect sum representation of R/a^*R, qua right R-module, by the modules R/cR. This shows that every atomic factor of a^* is similar to a factor of a. By Theorem 2.1 this characterizes the atomic factors of a^* as the atoms similar to factors of a.

Let $a \in R$ be bounded; if $a = bcd$, then c is again bounded, with bound dividing a^*. This follows by observing that R/cR is a quotient of a submodule of R/aR. Likewise the product of any bounded elements is bounded. These facts may be expressed by saying that the modules R/aR, where a is bounded, form the objects of a *dense* subcategory of the category Tor_R^1 of cyclic torsion modules.

The quotient of a 2-fir by an ideal with invariant generator has a rather special form, which is described in

Theorem 6.2.3. *Let R be a 2-fir and c a non-unit right invariant element. Then cR is an ideal in R and R/cR is a ring (not necessarily a domain) in which every finitely generated right ideal is principal; it is a field precisely when c is an atom in R.*

Further, R/cR is Artinian if and only if c is a product of atoms. When these conditions hold, then $R/cR \cong \mathfrak{M}_n(K)$ for a field K and some $n \geq 1$ if and only if cR is maximal among ideals with a right invariant generator.

Proof. Since c is right large and R is a 2-fir, every finitely generated right ideal containing cR is principal, hence of the form dR, where d is a left factor of c. It follows that R/cR is right Bezout and it is a field precisely if c is an atom. This also shows that R/cR is Artinian precisely when c is a product of atoms.

Now let cR be (proper) maximal among ideals with right invariant generator c; write $S = R/cR$ and let $\mathfrak{a}$ be a minimal right ideal of S. Then $S\mathfrak{a}$ is two-sided, hence of the form dR/cR, where dR is two-sided, so d is right invariant and by the maximality of cR we have $dR = R$, i.e. $S\mathfrak{a} = S$. Thus S is a sum of copies of the simple right S-module $\mathfrak{a}$, hence a direct sum, say $S \cong A^n$, where $A \cong \mathfrak{a}$

as right S-module, and so

$$S \cong End_S(A^n) \cong \mathfrak{M}_n(K),$$

where $K = End_S(A)$ is a field, by Schur's lemma (Proposition 3.2.4). ∎

We now examine the case of invariant elements in 2-firs; here we have a more precise statement; some of the next results can be deduced from Theorem 2.3, but in view of its importance we give a separate proof.

Theorem 6.2.4. *Let R be an atomic 2-fir.*

(i) *If c is an invariant element of R, then the quotient ring R/cR is simple if and only if c is an Inv-atom.*
(ii) *Every Inv-atom is a product of similar bounded atoms.*
(iii) *If p is a bounded atom, then its bound p^* is an Inv-atom whose atomic factors are precisely all the atoms similar to p. Moreover, the eigenring K of pR is a field and*

$$R/p^*R \cong \mathfrak{M}_n(K), \quad \text{where } n = l(p^*). \tag{6}$$

Proof. We shall prove the parts in reverse order, beginning with (iii). Let p have bound $p^* = a^*b^*$, where a^*, b^* are non-unit invariant elements. Then p divides either a^* or b^* and so has a smaller bound, a contradiction; thus p^* is an Inv-atom. By Theorem 2.1, $p^*R = \cap p'R$, where p' runs over all elements similar to p, and here we can take a finite intersection. Thus R/p^*R is a submodule of $(R/pR)^N$ for some $N \geq 1$ (in fact p^* is fully reducible by Proposition 3.5.12), and since R/pR is Tor-simple, every torsion submodule has the form $(R/pR)^n$ for some $n \leq N$. Hence

$$R/p^*R \cong (R/pR)^n, \tag{7}$$

as right R-modules. By comparing the lengths of composition series within Tor_R we see that $n = l(p^*)$, and comparing endomorphism rings in (7), we obtain the isomorphism (6), where K, the eigenring of pR, is a field by Schur's lemma.

To prove (ii) let p^* be an Inv-atom and let p be an atom dividing p^*. Then p^* is the bound of p and as (5) shows, an atom divides p^* precisely if it is similar to p.

Finally, to prove (i), we see from (6) that for an Inv-atom c, R/cR is simple. Conversely, if c is invariant but not an Inv-atom, then R/cR has non-trivial quotients and so cannot be simple. ∎

We next look at the direct decompositions of R/aR, where a is bounded. Our first task is to separate out the bounded components in such a decomposition;

for this we need a definition and a lemma. Let us call two elements a, b of a 2-fir R *totally coprime* if no non-unit factor of a is similar to a factor of b, i.e. if R/aR and R/bR have no isomorphic factor modules apart from 0.

Lemma 6.2.5. *A bounded element in a 2-fir R can be comaximally transposed with any element totally coprime to it.*

Proof. Let $a, b \in R$ be totally coprime and suppose that a is bounded, with bound $a^* = a_1 a$, say. Then a^* and b are left coprime and hence right comaximal: $a^* u - bv = 1$. It follows that $u'a^* - bv = 1$ for some $u' \in R$, thus $u'a_1 a - bv = 1$, and so by Lemma 3.3.3, a, b are comaximally transposable. By the symmetry of the situation, b, a are also comaximally transposable. ∎

Let a be any non-zero element in an atomic 2-fir R and take a complete factorization

$$a = p_1 p_2 \ldots p_r. \tag{8}$$

Suppose that a bounded atom occurs in this factorization, say $p_{i_1}, \ldots, p_{i_k}$ are all bounded similar atoms, while the remaining atoms in (8) are not similar to p_{i_1}. By repeated application of Lemma 2.5 we can write a as

$$a = p'_{i_1}, \ldots, p'_{i_k} p'_{j_1} \ldots p'_{j_h},$$

where the p'_i but not the p'_j are similar to p_{i_1}. Bracketing the first k factors together, and the last h, we have $a = bc$, where b, c are totally coprime and b, like p_{i_1}, is bounded. Applying Lemma 2.5 again, we can write this as $a = c'b'$, where b' is similar to b and c' similar to c. Therefore $a = bc = c'b'$ leads to a direct decomposition of a (see Section 3.5):

$$aR = bR \cap c'R, \quad R = bR + c'R.$$

We now repeat this process with a replaced by c' and eventually reach a direct decomposition of a into products of pairwise similar bounded atoms and a term containing no bounded non-unit factors. We shall call an element *totally unbounded* if it has no bounded factor apart from units. So our result may be stated as

Theorem 6.2.6. *Any element $a \neq 0$ of an atomic 2-fir R has a direct decomposition*

$$aR = q_1 R \cap \ldots \cap q_k R \cap uR, \tag{9}$$

where each q_i is a product of similar bounded atoms, while atoms in different q's are dissimilar and u is totally unbounded. Moreover, the q_i and u are unique up to similarity, while uR is absent if and only if a is right bounded.

Proof. The existence of such a decomposition follows from what has just been shown; only the uniqueness still remains to be proved. Now (9) gives rise to a direct decomposition

$$R/aR \cong R/q_1R \oplus \cdots \oplus R/q_kR \oplus R/uR.$$

Here the R/q_iR are uniquely determined as the homogeneous components corresponding to a given Tor-simple bounded isomorphism type, while R/uR contains all Tor-simple submodules of unbounded isomorphism type. Moreover, q_iR is unique as the intersection of all $pR \supseteq aR$, where p runs over all atoms bounded by q_i. Clearly R/aR is bounded precisely when the last term R/uR is absent. ∎

We note that neither u nor the q_i are in general indecomposable in the sense defined in Section 3.5. In the case of principal ideal domains this result, applied to the terms of the decomposition in Theorem 1.4.10, leads to a strengthening of that result:

Proposition 6.2.7. *Let R be a principal ideal domain and M a finitely generated right R-module consisting of torsion elements. Then*

$$M \cong R/q_1R \oplus \cdots \oplus R/q_kR \oplus R/uR,$$

where each q_i is a product of pairwise similar bounded atoms, while u is totally unbounded. The last term R/uR may be absent; this is so if and only if M is bounded. ∎

Let R again be an atomic 2-fir and consider an element $c \in \mathrm{Inv}(R)$. We shall be interested in the decompositions of R/cR. Let us call $c \in \mathrm{Inv}(R)$ *Inv-decomposable* if it has a factorization

$$c = ab, \qquad (10)$$

into non-unit invariant elements a, b that are left (hence also right) coprime; otherwise c is *Inv-indecomposable*. Clearly c is Inv-indecomposable if and only if $cR = Rc$ is meet-irreducible, and when this is so, c is a product of similar atoms, by Theorem 2.6. For a bound element we have the following relationship between its bound and that of its atomic factors.

Lemma 6.2.8. *Let R be an atomic 2-fir and let $q \in R$ be bound with a complete factorization $q = p_1 \cdots p_n$, where all the p_i are atoms similar to p. Then q^* is associated to p^{*r}, where $r \leq n$, with equality if and only if q is rigid.*

Proof. The module R/qR has a submodule lattice that is modular of length n. The action by p^* reduces the length by at least 1, and by exactly 1 precisely when there is only one maximal submodule. By induction the action with p^{*n} reduces it to zero, hence $q|p^{*n}$, and so $q^*|p^{*n}$; thus q^* is associated to p^{*r} where $r \leq n$, and equality holds precisely when the submodules of R/qR form a chain, i.e. when q is rigid. ∎

There is a simple relationship between Inv-indecomposable elements and bounded elements that are indecomposable as defined in Section 3.5.

Proposition 6.2.9. *Let R be an atomic 2-fir. Then*

(i) *an invariant element is Inv-indecomposable if and only if it is associated to a power of an Inv-atom;*

(ii) *if $q \in R$ is bounded and indecomposable, then its bound is Inv-indecomposable, hence q is then a product of similar atoms.*

Proof. (i) follows by Theorem 1.6. To prove (ii) let q be bounded. By Theorem 2.6, q is a product of similar atoms, say $q = p_1 \ldots p_n$, where the p_i are all similar; they all have the same bound p^*, say. By Lemma 2.8, $q^* = p^{*r}$, where $r \leq n$; this shows q^* to be Inv-indecomposable. ∎

Of course an *un*bounded indecomposable element need not be a product of similar atoms, as the example xy in the free algebra $k\langle x, y\rangle$ shows. Further, the converse of Proposition 2.9 (ii) is false, i.e. the bound of a decomposable bounded element need not be Inv-decomposable; thus in Proposition 2.9 (ii), q^* itself may well be decomposable (see Exercise 4).

Let q be a bounded indecomposable element in an atomic 2-fir R; by Proposition 2.9, its bound q^* is Inv-indecomposable, say $q^* = p^{*e}, l(p^*) = h$. By Theorem 2.4, $R/p^*R \cong K_h$, where the field K is the eigenring of an atomic factor of p^*. Now $R/p^*R \cong Q/J(Q)$, where $Q = R/q^*R$ and $J(Q)$ is the Jacobson radical of Q. Since Q is Artinian (Theorem 2.3), we can lift the matrix basis from R/p^*R to R/q^*R (see e.g. FA, Section 4.3), whence $Q \cong L_h$, where $L/J(L) \cong K$, i.e. Q is an Artinian matrix local ring over the scalar local ring L. Note that L, like Q, is Artinian, hence it is completely primary (i.e. all its non-units are nilpotent).

Now take a complete direct decomposition of R/q^*R as right R-module. The summands are necessarily cyclic; thus

$$R/q^*R \cong R/q_1R \oplus \cdots \oplus R/q_kR. \tag{11}$$

Since $R/q^*R \cong L_h$ has a complete direct decomposition into h isomorphic right ideals, we see that $k = h$ and all the R/q_iR are isomorphic to R/qR. Thus $R/q^*R \cong (R/qR)^h$, as right R-modules. Since $l(q^*) = eh = hl(q)$, we see that $l(q) = e$. The result may be summed up as

Theorem 6.2.10. *In an atomic 2-fir R, let $q \in R$ be bounded indecomposable. Express its bound q^* as a power of an Inv-atom, say $q^* = p^{*e}$. Then $l(q) = e$, and if $l(p^*) = h$, then*

$$R/q^*R \cong (R/qR)^h, \qquad (12)$$

as right R-modules, while as a ring, R/q^*R is a full matrix ring over a completely primary ring:

$$R/q^*R \cong \mathfrak{M}_h(L), \quad \text{where } L = End_R(R/qR). \qquad ■ \quad (13)$$

If q, q^* are as in Theorem 2.10, then q^*R is determined by the similarity class of q as the annihilator of R/qR, while R/qR is determined by q^*R as an indecomposable part of R/q^*R. Hence we have

Corollary 6.2.11. *In an atomic 2-fir, two bounded indecomposable elements have the same bound if and only if they are similar.* ■

Next we turn to the question of deciding when a given product of similar bounded atoms is indecomposable. Let p^* be an Inv-atom of length h; for any integer $e \geq 0$ we have by Theorem 2.10, on decomposing p^{*e},

$$R/p^{*e}R \cong (R/q_eR)^h,$$

for some indecomposable element q_e of length e. Thus q_e is a product of e atomic factors that are all similar. Conversely, if p is a bounded atom and p_i is similar to p for $i = 1, \ldots, e$, then $q = p_1 \ldots p_e$ is bounded by p^{*e}, and if its exact bound is p^{*e}, then q is indecomposable. For if q could be decomposed, it would have a smaller bound, as we see by acting on R/qR with p^*. This proves

Proposition 6.2.12. *In an atomic 2-fir R, each bounded indecomposable element is a product of similar atoms. If $p \in R$ is a bounded atom, then a product $q = p_1 \ldots p_e$ of atoms similar to p is indecomposable if and only if p^{*e} is the exact bound of q. Moreover, for any integer $e \geq 1$, a bounded indecomposable element q_e of length e exists such that*

$$R/p^{*e}R \cong (R/q_eR)^h, \quad \text{where } h = l(p^*). \qquad ■ \quad (14)$$

The relation (14) yields the following result for factors of powers of an Inv-atom:

Corollary 6.2.13. *In an atomic 2-fir, if a has the bound p^{*e}, where p^* is an Inv-atom, then a direct decomposition of a has at most $l(p^*)$ terms.* ■

If an element a has the bound p^{*e} and b has the bound p^{*f}, then ab is bounded by p^{*e+f}. Applying this remark to a product of similar bounded atoms, we obtain

Corollary 6.2.14. *In an atomic 2-fir, any factor of a bounded indecomposable element is again bounded indecomposable.* ∎

Now a product of atoms $p_1 p_2$ is indecomposable if and only if they cannot be comaximally transposed. Hence any product $q = p_1 \ldots p_n$ of bounded atoms has no decomposable factors if and only if no pair of adjacent factors can be comaximally transposed, i.e. if q is rigid. By Proposition 2.12, two similar atoms p, p' with a common bound p^* are comaximally transposable if and only if pp' does not divide p^*. Thus we obtain

Corollary 6.2.15. *Let R be an atomic 2-fir and $q = p_1 \ldots p_e$ a bounded product of atoms. Then q is rigid if and only if all the p_i have a common bound p^* and $p_{i-1} p_i$ does not divide p^* for $i = 2, \ldots, e$. In particular, since p^r is rigid, p^2 does not divide p^*.* ∎

There remains the problem of finding which elements are bounded. We shall confine ourselves to the case of an atomic 2-fir R. If p is an atom in R, then $K = \text{End}_R(R/pR)$ is a field and each element a of R defines a K-endomorphism of R/pR by right multiplication. If the natural mapping $R \to R/pR$ is written $x \mapsto \bar{x}$, then the K-endomorphism defined by a is $\rho_a : \bar{x} \mapsto \bar{x}a$. We want to find an upper bound for the K-dimension of ker ρ_a. Clearly, if $a = a_1 \ldots a_r$, then $\rho_a = \rho_{a_1} \ldots \rho_{a_r}$ and by Sylvester's law of nullity for fields,

$$\dim \ker \rho_a \leq \dim \ker \rho_{a_1} + \ldots + \dim \ker \rho_{a_r}. \tag{15}$$

We claim that ρ_a is injective when a has no factor similar to p. Thus assume that $\text{Ker}\rho_a \neq 0$; then there exists $x \notin pR$ such that $xa \in pR$, say

$$xa = py \quad (x \notin pR). \tag{16}$$

Since p is an atom, (16) is left coprime, hence $a = p'a'$, where p' is similar to p. This shows that ρ_a is injective when a has no factor similar to p.

Suppose now that c is similar to p; then any element $x' \in R$ satisfying $x'c = py'$ for some $y' \in R$, while $x' \notin pR$, defines an isomorphism $R/cR \to R/pR$, and any two such isomorphisms differ by an endomorphism of R/pR, i.e. an element of K; hence ker ρ_c is one-dimensional in this case. Going back to (15), we see that ker ρ_a has a dimension at most equal to the number of factors of a that are similar to p. Thus we obtain

Theorem 6.2.16. *Let R be an atomic 2-fir, p an atom in R and $K = End_R(R/pR)$. Then for any $a \in R$, the mapping $\rho_a : \bar{x} \mapsto \bar{x}a$ is a K-endomorphism of R/pR and*

$$\dim_K \ker\rho_a \leq m, \tag{17}$$

where m is the number of factors in a complete factorization of a that are similar to p. ■

Suppose now that p is a right large atom of R. Then for any $x \notin pR$ there is a comaximal relation

$$xp' = px';$$

hence there exists p' similar to p and annihilating $\bar{x}$. Conversely, as we have seen, any p' similar to p annihilates some $\bar{x}$, hence when p is right large and q is any atom, then ρ_q is injective if and only if q is not similar to p. We derive the following consequence.

Let p be a right bounded atom of R, with right bound p^* of length m. Then p is certainly right large, and by (17),

$$\dim_K(R/pR) \leq m. \tag{18}$$

If this inequality were strict, we could find a product of fewer than m factors similar to p which annihilates R/pR, and hence by Theorem 2.1,

$$aR \subseteq \cap\{p'R | p' \text{ is similar to p}\} = p^*R.$$

Thus $a \in p^*R$, but this contradicts the fact that $l(a) < m = l(p^*)$. Hence equality holds in (18). This shows that for a right bounded atom p, R/pR is finite-dimensional (and, of course, p is right large). Conversely, if p is right large and R/pR finite-dimensional over K, then $\rho_{p'}$ is not injective for p' similar to p and by induction we can find $c \in R^\times$ annihilating R/pR, hence p is then right bounded. We thus obtain

Corollary 6.2.17. *Let R be an atomic 2-fir. Then an atom p in R with eigenring K is right bounded if and only if p is right large and $dim_K(R/pR)$ is finite. Moreover, in this case we have*

$$\dim_K(R/pR) = l(p^*).$$ ■

Exercises 6.2

1°. Let R be an integral domain. Show that every bounded R-module is bound, but in general the converse is false. For which class of domains does the converse hold (for finitely generated modules)?

2. Let R be a principal right ideal domain and $c \in R^\times$. If $\cap dR \neq 0$, where d runs over all elements similar to right factors of c, show that c is bounded. Show that this no longer holds for firs, by taking $R = k\langle x, y\rangle$, $c = x$.
3. (G. M. Bergman) If $R = k\langle x, y\rangle$, show that by taking $a = x$ in Theorem 2.1, we have $I = xR$, while for $a = xy + 1$, $I = 0$. (This example shows that the hypothesis on a, to be right large, cannot be omitted.)
4. (Jacobson [43]) Let K be a field of finite dimension over its centre and α an automorphism of K such that α^r is inner for some $r \geq 1$. Show that every nonzero element of the skew polynomial ring $K[x; \alpha]$ is bounded; illustrate this fact in the complex-skew polynomial ring $\mathbb{C}[x; -]$. Give an example of a product of similar bounded atoms that is decomposable in the sense of Section 3.5.
5. Show that for a bounded element in an atomic 2-fir R, 'left indecomposable' = 'indecomposable' (see Section 3.5). If q is such an element, show that q is rigid.
6°. Find an integral domain with an element a that is left and right bounded but not bounded.
7. (Beauregard [74]) Let k be a field with an endomorphism α that is not surjective; denote by $F = k[[x; \alpha]]$ the skew power series ring with coefficients on the right and $cx = xc^\alpha$ (as usual) and $G = F[[y; \alpha]]$ the power series ring with coefficients on the left and $yc = c^\alpha y$. Verify that xy is in the centre of G, and show that any two elements in G have a LCRM and LCLM, but G is not a 2-fir. Show that x and y are bounded and find their right bounds. Show also that (4) does not hold for $a = x$.
8. Let R be any ring and c an invariant element of R. Show that for any $a \in R$ the following are equivalent: (i) a is left bounded and a left factor of c, (ii) a is right bounded and a right factor of c, (iii) a is a left and a right factor of c.
9. Let R be an atomic semifir and $m, n \geq 1$. If p is a bounded atom in R_n, show that its bound p^* is an Inv-atom of R whose atomic factors in R_m are precisely the atoms of R_m stably associated to p. Show that the eigenring of p is a field K, say, and $R_n/p^*R_n \cong K_{rn}$, where $r = l(p^*)$. Deduce that every bounded atom of R_n is stably associated to an element of R.
10. Let R be a fir with infinite centre and put $S = R_n$. Show that any atom p in S with eigenring K is right bounded if and only if S/pS is simple as right S-module and $\dim_K(S/pS)$ is finite.
11. Let R be an atomic 2-fir and c an invariant element of R. Show that R/cR is semisimple if and only if c is not divisible by the square of any Inv-atom.
12. Let R be a ring and for any R-module M define the *tertiary radical* (of 0 in M) as the set of all $x \in R$ that annihilate a large submodule of M. If R is an atomic 2-fir and c is bounded in R, find conditions for R/cR to have a tertiary radical of the form $bR \supseteq cR$. If c is indecomposable, show that $b = p^*$, where p^* is the Inv-atom corresponding to c.
13. Let R be a non-simple principal ideal domain. Show that the quotient by any maximal ideal is simple Artinian.
14. (Jategaonkar [69a]) For any commutative field k let $F = k(t_{m,n})$, where $m \leq n$, $m, n = 1, 2, \ldots$ with the automorphism $\alpha : t_{m,n} \mapsto t_{m+1,n}$ if $m < n$, and $t_{n,n} \mapsto t_{1,n}$. Show that the skew polynomial ring $F[x; \alpha]$ has Inv-atoms of all positive degrees and hence has homomorphic images of the form $\mathfrak{M}_n(C)$ for all $n \geq 1$.

15*. (G. M. Bergman) Let R be an n-fir and M a finitely related bounded R-module on fewer than n generators. Show that M is a torsion module. (*Hint*: Use Theorem 2.3).

6.3 2-Firs with prescribed centre

This section is devoted to proving the converses of Corollary 1.3 and Theorem 1.11, in the following strong form: every integrally closed domain (resp. Krull domain) occurs as the centre of some Bezout domain (resp. principal ideal domain). Since every Bezout domain is a 2-fir, and every principal ideal domain is an atomic 2-fir, this, with Theorem 1.11, completely characterizes the centres of (atomic) 2-firs.

The proof proceeds in two stages.

(i) Given a commutative integrally closed domain C, we construct a commutative Bezout domain A with an automorphism of infinite order whose fixed ring is C; further, when C is a Krull domain, A can actually be chosen to be a principal ideal domain.

(ii) Given a commutative Bezout domain A with an automorphism α of infinite order, we construct a Bezout domain containing A, whose centre is precisely the fixed ring of α. Moreover, when A is a principal ideal domain, the ring containing it can be chosen to be principal.

It is convenient to begin with (ii). The two cases considered require rather different treatment, and we therefore take them separately.

Proposition 6.3.1. *Let A be a commutative Bezout domain with field of fractions K and an automorphism α of infinite order and denote by C the fixed ring of α acting on A. Then the ring $R = A + xK[x; \alpha]$ is a Bezout domain containing A, with centre C.*

Proof. The automorphism α of A extends in a unique way to an automorphism of K, again denoted by α (see Theorem 0.7.8). Now form the skew polynomial ring $S = K[x; \alpha]$ and let R be the subring of all polynomials with constant term in A, thus $R = A + xS$. That R is a subring is clear; we claim that it has the desired properties. In the first place, if $f = \sum x^i a_i$ lies in the centre, then for any $b \in A$, $fb = bf$, and on equating coefficients we find that $a_i b = a_i b^{\alpha^i}$. Now for each $i > 0$ there exists $b \in A$ such that $b^{\alpha^i} \neq b$, hence $a_i = 0$ for $i > 0$ and $f = a_0$. Further, the equation $fx = xf$ shows that $a_0^\alpha = a_0$, so $f = a_0 \in C$, as claimed.

It remains to show that R is Bezout. Let $f, g \in R$; we must show that $fR + gR$ is principal, and here we may assume that f and g have no common left factor of positive degree in x, for if $f = df_1, g = dg_1$ in S, then on multiplying f_1, g_1 by a common left denominator of their constant terms, say $e \in A$ and right multiplying d by e^{-1}, we have reduced the situation to the case where $f_1, g_1 \in R$. Now if $f_1 R + g_1 R = hR$ has been established, then $fR + gR = dhR$. By looking at the highest common left factor of f, g in $K[x;\alpha]$ we find polynomials u, v in the latter ring such that $fu - gv = 1$. On multiplying up by a suitable element of A we obtain an equation

$$fu - gv = \gamma, \quad \text{where } u, v \in R, \gamma \in A^\times. \tag{1}$$

If the constant terms of f, g are λ, μ respectively, say $f = \lambda + f_0, g = \mu + g_0$, where f_0, g_0 have zero constant term, then $\lambda = f - \gamma(\gamma^{-1} f_0) \in fR + gR$ by (1), and similarly $\mu \in fR + gR$. Since A is Bezout, $\lambda A + \mu A = \delta A$ for some $\delta \in A$. Now f_0 and g_0, having zero constant term, are left divisible by δ, hence $f, g \in \delta R$. On the other hand, $\delta \in \lambda A + \mu A \subseteq fR + gR$; hence $fR + gR = \delta R$. Thus the right ideal of R generated by any two of its elements is principal; by symmetry so is the left ideal generated by them, and it follows that R is a Bezout domain. ∎

By combining this result with Corollary 1.3 we get

Corollary 6.3.2. *The fixed ring of an automorphism of infinite order acting on a commutative Bezout domain is integrally closed.* ∎

Of course this result can also be proved directly (see Exercise 1).

Proposition 6.3.3. *Let A be a commutative principal ideal domain with an automorphism α of infinite order, and denote by C the fixed ring of α acting on A. Then the ring of skew formal Laurent series $R = A((x;\alpha))$ is a principal ideal domain containing A, with centre C.*

Proof. We form the ring $R = A((x;\alpha))$ of formal Laurent series $f = \sum x^i a_i$, where $a_i \in A$ and $a_i = 0$ for i less than some k depending on f, with the commutation rule $ax = xa^\alpha$. The verification that the centre of R is C is as before: the equations $fb = bf$ show that $f = a_0$ and now $fx = xf$ shows that $a_0 \in C$. It remains to show that R is a principal ideal domain.

Given a right ideal $\mathfrak{a}$ of R, the leading coefficients of elements of $\mathfrak{a}$ will form a right ideal $\mathfrak{a}^*$ of A that must be principal, so we can choose $f \in \mathfrak{a}$ such that the leading coefficient of f generates $\mathfrak{a}^*$. Let K be the field of fractions of A; then $K((x;\alpha))$ is a field in which $f^{-1}\mathfrak{a}$ is a right R-submodule that contains 1, hence

$$f^{-1}\mathfrak{a} \supseteq R, \tag{2}$$

6.3 2-Firs with prescribed centre

and the leading coefficients of all members of $f^{-1}\mathfrak{a}$ lie in A. We claim that equality holds in (2); for if not, then $f^{-1}\mathfrak{a}$ would contain an element with a coefficient not in A. By subtracting an appropriate member of R, we would obtain a member of $f^{-1}\mathfrak{a}$ with leading coefficient not in A, which is a contradiction. Thus equality holds in (2) and so $\mathfrak{a} = fR$. ∎

We remark that we have to assume that A is principal even to show that R is Bezout. In fact, if we perform the construction with a Bezout domain A, we do not generally get a Bezout domain. Using Theorem 1.11, we again have a corollary (which, as before, can be proved directly).

Corollary 6.3.4. *The fixed ring of an automorphism of infinite order acting on a commutative principal ideal domain is a Krull domain.* ∎

We now come to step (i) of our programme. This is in effect the converse of Corollaries 3.2 and 3.4.

Proposition 6.3.5. *Every integrally closed commutative integral domain occurs as the fixed ring of an automorphism of infinite order acting on a commutative Bezout domain.*

Proof. We first give the basic construction that will in characteristic 0 produce the required ring, and then show how to modify it to get the full result.

Let C be the given domain and K its field of fractions. By hypothesis, $C = \cap K_v$, where v ranges over the family V of all general valuations defined on C. We form the polynomial ring $K[t]$ in an indeterminate t. Each $v \in V$ can be extended to $K[t]$ by putting

$$v\left(\sum t^i a_i\right) = \min_i \{v(a_i)\}.$$

We assert that this is again a valuation. The rule $v(a - b) \geq \min\{v(a), v(b)\}$ is clear; to show that v is multiplicative, let $a = \sum t^i a_i, b = \sum t^j b_j$ and let a_r, b_s be the first coefficient attaining the minimum $v(a), v(b)$ respectively. The product $c = ab$ has coefficients $c_k = \sum a_i b_{k-i}$ and $v(c_k) \geq v(a_r) + v(b_s)$, with equality holding for $k = r + s$, as is easily verified. Thus v is a valuation on $K[t]$ and in fact can be extended to a valuation, again denoted by v, of the field of fractions $K(t)$. We now define

$$A = \cap K(t)_v, \quad \text{where } v \text{ ranges over } V.$$

Thus A consists of all fractions f/g, $f, g \in C[t]$, such that $v(f) \geq v(g) (v \in V)$. A is sometimes called the *Kronecker function ring*.

We claim that A is a Bezout domain: given two elements of A, on multiplying by a common denominator, we may take them to be $f, g \in C[t]$. Now take any n greater than the degree of f in t and form $h = f + t^n g$. Then clearly $v(h) \leq v(f), v(g)$, hence $f/h, g/h \in A$ and so $fA + gA = hA$.

Consider the map α of $K[t]$ defined by $f(t) \mapsto f(t+1)$. This is an automorphism, clearly $v(f(t+1)) \geq v(f(t))$ for any $v \in V$, and by writing down the same inequality with -1 in place of 1, we see that $v(f^\alpha) = v(f)$, hence α extends to an automorphism of A. We observe that α is of infinite order precisely when K has characteristic 0. In that case the fixed field of α acting on $K(t)$ is K, hence the fixed ring of α acting on A is $K \cap A = C$, as required.

For a proof that works in all cases, we modify our construction by starting, not with $K[t]$, but with $K[\ldots, t_{-1}, t_0, t_1, \ldots]$ with countably many indeterminates. As before, we get a Bezout domain, and for our automorphism α we use instead of the translation $t \mapsto t+1$ the substitution $t_n \mapsto t_{n+1}$. Clearly this is of infinite order and the fixed ring in the action on $K(\ldots, t_{-1}, t_0, t_1, \ldots)$ is K, hence the fixed ring of A is C. Thus A is a Bezout domain with an automorphism of infinite order whose fixed ring is C. ∎

Suppose now that C is a Krull domain. Let K be its field of fractions and V the family of valuations defining C. We form A as in the proof of Proposition 3.5, using the family V instead of the family of all valuations. Then it follows as before that A is a Bezout domain with fixed ring C. We claim that now A is in fact a principal ideal domain. Given $a \in A^\times$, it is clear that $v(a) \neq 0$ for only finitely many v, say $v_1, \ldots, v_r$. Now for any factor b of a, $0 \leq v(b) \leq v(a)$ for all $v \in V$, and if $b' \in A$ is such that $v(b) = v(b')$ for all $v \in V$, then b and b' are associated. Hence there are only finitely many classes of factors of a, therefore A is atomic and so is a principal ideal domain. This proves

Proposition 6.3.6. *Every Krull domain occurs as the fixed ring of an automorphism of infinite order acting on a commutative principal ideal domain.* ∎

Putting all the results of this section together, we obtain

Theorem 6.3.7. *Every integrally closed integral domain occurs as the centre of a Bezout domain; every Krull domain occurs as the centre of a principal ideal domain.* ∎

Exercises 6.3

1. Give a direct proof (e.g. by valuation theory) that the fixed ring of an automorphism acting on a commutative Bezout domain R is integrally closed, and is a Krull domain when R is a principal ideal domain.
2*. Does Proposition 3.1 still hold when the automorphism has finite order?

3. Examine why the proof of Proposition 3.1 fails for principal ideal domains and that of Proposition 3.3 fails for Bezout domains.
4. In the proof of Proposition 3.5, if K has finite characteristic, complete the argument by taking in place of $K[t]$ the ring $K[t, t^{1/2}, t^{1/4}, \ldots]$ with automorphism $t \mapsto t^2$. Alternatively, use the ring $K[s, t, s^{-1}, t^{-1}]$ with automorphism $s \mapsto t, t \mapsto st$.
5. Verify that for any commutative field k, the polynomial ring in countably many indeterminates $k[x_1, x_2, \ldots]$ is a Krull domain, but not Noetherian. Deduce the existence of a Noetherian domain with a non-Noetherian centre.
6. Show that the centre of a 2-fir with right ACC_1 is a Krull domain.
7°. Which commutative rings occur as the centres of Sylvester domains? (In view of Theorem 3.7 this asks whether the centre of a Sylvester domain is necessarily integrally closed.)

6.4 The centre of a fir

The results of Section 6.3 give a complete description of the possible centres of principal ideal domains or Bezout domains, as well as some information on the centres of 2-firs, but they leave open the question whether, for example, any Krull domain can occur as the centre of a genuine, i.e. non-Ore, fir. As we shall see, once we assume that our rings are non-Ore, the centre is very much more restricted. Thus the centre of a non-Ore fir is necessarily a field. More generally, this conclusion will hold for any non-Ore 2-fir with right ACC_2.

For the proof of the main result we shall need a technical lemma.

Lemma 6.4.1. *Let S be a simple Artinian ring. Then*

(i) every right ideal of S is principal, with an idempotent generator,
(ii) if $aS = bS (a, b \in S)$, then a and b are right associated,
(iii) if $a, b \in S$ are right comaximal, then $a + by$ is a unit for some $y \in S$,
(iv) given $a, b \in S$, there exist $d, a_1, b_1 \in S$ such that $a = da_1, b = db_1$ and a_1, b_1 are right comaximal.

Proof. (i) and (ii) are an easy consequence of the fact that S is semisimple. It follows that every element is right associated to an idempotent. Thus to prove (iii) we may assume that $b^2 = b$. Suppose that $as + bt = 1$; we have $a = (1 - b)a + ba$, hence on substituting, we find that $(1 - b)as + bas + bt = 1$, i.e.

$$b(as + t) + (1 - b)as = 1 = b + (1 - b).$$

Since $bS \cap (1 - b)S = 0$, it follows that $(1 - b)as = 1 - b$; hence

$$(1 - b)S = (1 - b)asS \subseteq (1 - b)aS \subseteq (1 - b)S,$$

and so $(1-b)S = (1-b)aS$. By (ii), $(1-b)a = (1-b)u$ for a unit u; therefore $u = (1-b)u + bu = (1-b)a + bu = a + b(u-a)$, which is the required form.

(iv) By (i), $aS + bS = eS$, where $e^2 = e$. Now $aS + bS + (1-e)S = eS + (1-e)S = S$, so by (iii), $a + bx + (1-e)y$ is a unit for some $x, y \in S$. Put $a_1 = a + (1-e)y$, $b_1 = b$; then $a_1 + b_1 x$ is a unit and $a = ea_1, b = eb_1$. ∎

We can now prove the main result of this section.

Theorem 6.4.2. *Let R be an atomic 2-fir with right ACC_2 that is not right Ore. Then every invariant element of R is a unit; in particular, the centre of R is a field.*

Proof. Suppose that R satisfies the hypotheses and contains a non-unit invariant element c. If we choose c such that cR is maximal (by right ACC_2), then $S = R/cR$ is simple Artinian, by Theorem 2.3. Further, R contains a maximal ideal of rank 2, say $xR \oplus yR$. If the elements of S corresponding to x, y are written $\bar{x}, \bar{y}$, then by Proposition 4.1 (iv) there exist $e, a, b \in S$ such that a, b are right comaximal and $\bar{x} = ea, \bar{y} = eb$. By (iii) $a + bz = u$ is a unit for some $z \in S$, hence

$$(ea \quad eb)\begin{pmatrix} 1 & 0 \\ z & 1 \end{pmatrix} = (eu \quad eb), \quad (eu \quad eb)\begin{pmatrix} 1 & -u^{-1}b \\ 0 & 1 \end{pmatrix} = (eu \quad 0). \quad (1)$$

The elementary matrices written lift to elementary matrices over R, so there exists $P \in E_2(R)$ such that $(x, y)P = (v, w)$, where $\bar{w} = 0$, so that $w \in cR = Rc$, say $w = sc$. Now

$$xR \oplus yR = vR \oplus wR = vR \oplus scR \subseteq vR + sR;$$

here $vR + sR$ has rank 2, for otherwise we would have $vp = sq \neq 0$ and so $v(pc) = sqc = scq' = wq' \neq 0$, which contradicts the fact that $vR \cap wR = 0$. So $vR + sR$ has rank 2, and $vR + sR \supseteq xR + yR$ by (1), so $vR + sR = xR + yR$, by the maximality of the latter. Therefore $s \in vR + scR$, say $s = vm + scn$; it follows that

$$s(1 - cn) = zm \in sR \cap vR = 0,$$

hence $1 - cn = 0$, and this shows that c is a unit. Now the rest follows easily. ∎

This result applies in particular to firs that are not right Ore, or equivalently, not right principal, so we obtain

Corollary 6.4.3. *Let R be a fir that is not a principal right ideal domain. Then every invariant element of R is a unit, in particular the centre of R is a field.* ∎

Finally we record, for later use, the form taken by the centre of the field of fractions of a principal ideal domain.

Proposition 6.4.4. *Let R be a principal ideal domain, K its field of fractions and C the centre of K. Then C consists of all elements of the form ab^{-1}, a, $b \in R$, such that $b \neq 0$ and*

$$axb = bxa \quad \text{for all } x \in R. \tag{2}$$

Proof. If $ab^{-1} \in C$, then $ab^{-1}y = yab^{-1}$ for all $y \in K$, hence on writing $y = bx$ we obtain (2). Conversely, when (2) holds, then $ab = ba$, hence in K, $ab^{-1} = b^{-1}a$ centralizes R and so also K. ∎

Exercises 6.4

1. Let R be any ring, $\mathfrak{a}$ a right ideal maximal among non-finitely generated right ideals of R and let c be an invariant element not in $\mathfrak{a}$. Show that $\mathfrak{a} \cap cR = \mathfrak{a}c$.
2. (Goursaud and Valette, [75]). Show that if R is right hereditary and c an invariant element of R, then R/cR is right Noetherian. (*Hint*: Use a dual basis.)
3. Let R be a right hereditary integral domain, but not right Ore. Show that R has no non-unit invariant elements.
4. Show that the subring of $\mathbb{Q}\langle x, y\rangle$ consisting of all elements in which the constant term has odd denominator is a non-Ore semifir with non-unit central elements.
5°. Determine the possible centres of 2-firs with right ACC_1.
6. Show that the subalgebra of $k\langle x, y, y^{-1}\rangle$ generated by $x, y, y^{-n}x$ is a non-Ore right fir with the right invariant non-unit y (see section 2.10).
7. Let K be a field with a non-surjective endomorphism α. Show that $K[x; \alpha]$ is a fir that is not left Ore and has the right invariant element x.
8. Show that the subalgebra of $k\langle x, y, y^{-1}\rangle$ generated by y and $y^{-m}xy^{-n} (m, n \geq 0)$ is a semifir, neither left nor right Ore, and with the non-unit invariant element y.
9°. Does there exist a left fir that is not right Ore, with a non-unit right invariant element? Consider the same question for a fir that is not left or right Ore.

6.5 Free monoids

Before discussing subalgebras of free algebras it is helpful to look at free monoids, where we shall meet the same problems, but in a simplified form, since there is only one operation. It is particularly instructive to see what becomes of the weak algorithm in monoids; as we saw in Section 2.5, the weak algorithm may be used to characterize free algebras, and below we obtain a corresponding result for free monoids.

The free monoid on a set X is denoted by X^*. Each element of X^* may be written as a finite sequence of elements of X – with the empty sequence representing 1 – and multiplication consists of juxtaposition. Clearly this representation of elements is a normal form; moreover, 1 is the only unit in X^*, and since it has cancellation, it is conical (see Section 0.9). We shall see in a moment that X^* is rigid, i.e. it has cancellation and $ab' = ba'$ implies that $a \in bX^*$ or $b \in aX^*$. Further, X is the precise set of atoms in X^*; thus the generating set X is uniquely determined in terms of the monoid structure on X^*. It is this fact that accounts for the simplicity of the theory.

Theorem 6.5.1. *A monoid S is free on the set X of its atoms if and only if it is generated by X and is conical and rigid.*

Proof. Let S be free on X; clearly S is generated by X, it is conical and has cancellation. If $ab' = ba'$, express a, b, a', b' in terms of X, say $a = x_1 \ldots x_r, b = y_1 \ldots y_s, a' = y_{s+1} \ldots y_n, b' = x_{r+1} \ldots x_m$; then $x_1 \ldots x_m = y_1 \ldots y_n$ and since S is free on X, we have $m = n$ and $y_i = x_i$, so if $r \leq s$ say, then $b = ac$, where $c = x_{r+1} \ldots x_s$; thus S is rigid.

Conversely, assume that the conditions hold. It will be enough to show that any element of S can be written in just one way as a product of elements of X. For any $a \in S$ there is at least one way of so expressing a; suppose that

$$a = x_1 \ldots x_m = y_1 \ldots y_n, \quad x_i, y_j \in X. \tag{1}$$

By rigidity, $x_1 = y_1 b$ or $y_1 = x_1 b$ for some $b \in S$, say the former holds. Since x_1, y_1 are atoms, b must be a unit and so $b = 1$, i.e. $x_1 = y_1$. Cancelling the factor x_1 and applying induction on $\max(m, n)$, we find that the two expressions for a in (1) agree, so S is indeed free on X. ∎

Here the *length* of a, i.e. the number of factors in a complete factorization of a, is usually denoted by $|a|$. We see that when $|a| \geq |b|$, then $ab' = ba'$ implies that $a = bc$ for some $c \in S$.

Theorem 5.1 may be used to give criteria for a submonoid of a free monoid to be free. Consider, for example, the free monoid on one free generator x (the free cyclic monoid). The submonoid generated by x^2 and x^3 is commutative; if it were free, it would be cyclic, which is clearly not the case. Below we shall obtain a result that makes it easy to find other examples of non-free submonoids of free monoids. In any monoid we define an *ideal* as a subset T such that $aT \cup Ta \subseteq T$ for all a (as for rings) and an *anti-ideal* (also called a *stable* subset) as a subset T such that $a \in T$ whenever $ab, ca \in T$ for some $b, c \in T$, thus

$$TT^{-1} \cap T^{-1}T \subseteq T. \tag{2}$$

We first record some conditions for a submonoid to be an anti-ideal:

Proposition 6.5.2. *For any monoid S and submonoid T the following conditions are equivalent:*

(a) T is an anti-ideal in S,
(b) for all $a \in S\setminus T$, $aT \cap Ta \cap T = \emptyset$.
(c) for all $a \in S, b \in T$, if $ab, ba \in T$, then $a \in T$.

Proof. (a) $\Rightarrow$ (b). If T is an anti-ideal and $a \in S\setminus T$, then either $aT \cap T = \emptyset$ or $Ta \cap T = \emptyset$, hence (b) holds. Next assume (b) and suppose that $b, ab, ba \in T$. Then $ab^2a \in aT \cap Ta \cap T$; hence this set is non-empty and by (b), $a \in T$, which proves (c). Now (c) $\Rightarrow$ (a) is clear. ∎

We also note the following property of anti-ideals:

Corollary 6.5.3. *Let S be a monoid and T a proper anti-ideal in S. Then T contains no ideal of S.*

Proof. Let $\mathfrak{a}$ be an ideal of S. If $\mathfrak{a} \subseteq T$, take $a \in \mathfrak{a}$ and $b \in S\setminus T$; then $ab, ba \in T$, hence $b \in T$, which is a contradiction. ∎

Now the following result provides a supply of non-free submonoids in free monoids:

Theorem 6.5.4. *Let S be a conical rigid monoid. Then for any submonoid T of S the following three conditions are equivalent:*

(a) T is rigid;
(b) T is an anti-ideal in S;
(c) given $a, b, b' \in T$, if $ab' = ba$, then $a = bc$ or $b = ac$ for some $c \in T$.
   Moreover, if S is generated by its atoms, it is free and these conditions are equivalent to
(d) T is free.

Proof. (a) $\Leftrightarrow$ (b). Let $as = b, sa' = b'$, where $a, a', b, b' \in T$; then $asa' = ba' = ab'$, hence by (a), $a = bc$ or $b = ac$ for some $c \in T$. If $a = bc$, then $a = asc$ and so $sc = 1$. If $b = ac$, then $as = ac$ and so $s = c$. In either case $s \in T$, and (b) follows. Now assume (b) and let $ab' = ba'$. By rigidity of S, one of a, b is a right multiple of the other, say $a = bc, a' = cb'$, where $c \in S$. Hence $c \in TT^{-1} \cap T^{-1}T$, and so $c \in T$ by (b), showing T to be rigid, i.e. (a) holds.

(a) $\Leftrightarrow$ (c). Clearly (a) $\Rightarrow$ (c); conversely, let $ab' = ba'$; then $b'a.b'b = b'b.a'b$, so by (c), either $b'a = b'bc$ or $b'b = b'ac$ for some $c \in T$, and by cancelling b' we obtain (a).

(d) ⇔ (a) Clearly (d) ⇒ (a). If S is generated by its atoms, it is free by Theorem 5.1; now T is also generated by its atoms and so if (a) holds, then it is free, i.e. (d) holds. ∎

We note that whereas (b) refers to S, (a) and (c) are intrinsic in T. In a free monoid, let us associate with every element $u \neq 1$ the shortest element of which it is a power (the 'least repeating segment') and call this the *root* of u. For example, $xyxyxy$ has the root xy. Every element $u \neq 1$ has a unique root, which may also be characterized as an element of the shortest positive length commuting with u. From the criteria of Theorem 5.4 we easily obtain the following result, whereby a *free subset* of a monoid we understand a free generating set of the submonoid generated. A corresponding definition applies to free subsets of an algebra.

Corollary 6.5.5. *In a free monoid S the following conditions on a pair of distinct elements a, b different from 1 are equivalent:*

(a) $\{a, b\}$ is a free subset of S,
(b) $ab \neq ba$,
(c) a, b have distinct roots.

Proof. If (c) fails, we have $a = u^r, b = u^s$, hence $ab = ba$ and (b) fails. If (b) fails, so does (a) and finally, if a, b are not free, then there is a non-trivial relation that after cancelling common left factors reduces to

$$av = bw, \qquad (3)$$

where v, w are words in a, b. Suppose that $|b| \leq |a|$; then by rigidity, $a = bc$ for some $c \in S$. Since $a \neq b$, we have $c \neq 1$ and (3) can be replaced by $cv = w$ and v, w are words in b, c. This relation is still non-trivial; by induction on $|a| + |b|$ it follows that b, c have the same root, hence so do a, b. ∎

This result shows that commutativity is an equivalence relation on the set of elements different from 1 in a free monoid, and each equivalence class, with 1 adjoined, is a free cyclic monoid.

In Section 2.4 we saw that the free algebra $k\langle X \rangle$ may also be defined as the monoid algebra of the free monoid on X, kX^*. The weak algorithm holding in the free algebra may be regarded as the analogue of rigidity and it is clear that any free submonoid T generates a free subalgebra kT, though of course, not every free subalgebra of $k\langle X \rangle$ is of this special form. We shall briefly consider free generating sets of free submonoids; this is part of the theory of codes, an interesting recent application of semigroup theory, but an extended treatment would go beyond the framework of this book (see FA, Chapter 10; also Lothaire [97], Berstel and Perrin [85], Lallement [79]).

6.5 Free monoids

Let X be a finite set; as before, we denote the free monoid on X by X^* and we put $X^+ = X^* \setminus \{1\}$. A subset Y of X^+ is called a *code* if it is a free generating set of the submonoid it generates. For example, if $X = \{x, y\}$, then $\{x, xy, xy^2\}$ is a code, as is $\{y, yx, y^2x\}$, but $\{x, xy, yx\}$ is not, because $x.yx = xy.x$. It is also clear why 1 has to be excluded from a code. Suppose that Y is a subset of X^+ that is *not* a code; then we have an equation between two distinct words in the elements of Y and by cancellation we may take this to be of the form $yu = y'v$, where $y, y' \in Y$, $y \neq y'$. By the rigidity of X^* we have $y = y'z$ or $y' = yz$ for some $z \in X^*$, say the former holds. Then we say that y' is a *prefix* of y. A subset Y of X^* is called a *prefix set* if no element of Y is prefix of another, and what has been said shows the truth of

Proposition 6.5.6. *Every prefix set $\neq \{1\}$ in a free monoid is a code.* ∎

For this reason a prefix set $\neq \{1\}$ in a free monoid is also called a *prefix code*. *Suffix sets* are defined by left–right symmetry; apart from $\{1\}$ they are again codes, e.g. $\{x, xy, xy^2\}$ is a suffix set, but not a prefix set. Prefix codes are of particular interest in coding theory since any 'message' in a prefix code can be deciphered reading letter-by-letter from left to right (this is also known as a 'zero-delay code'). By contrast, if a code is not prefix, one may have to read arbitrarily far to decipher a message, e.g. $\{xy, yx, z, zx\}$ is a suffix set, but to decipher a message of the form $zxyxyxy\ldots$ one has to read to the end, to see if it ends in xy or yx.

Let X be a set of r elements; then X^* contains r^n words of length n. We shall again write $|u|$ for the length of u, and for any subset Y of X^+ define its *weight* (possibly ∞) as

$$\mu(Y) = \sum_{y \in Y} r^{-|y|}. \qquad (4)$$

Writing X^n for the set of all words of length n in X^*, we see that $\mu(Y \cap X^n)$ may be regarded as the 'weight' of $Y \cap X^n$ as a fraction of $\mu(X^n) = 1$; in particular, $\mu(Y \cap X^n) \leq 1$.

If Y is finite and $1 \notin Y$, then

$$\mu(Y) \leq \max\{|y| | y \in Y\}, \qquad (5)$$

for if the right-hand side has the value N, then $Y \subseteq X \cup X^2 \cup \ldots \cup X^N$.

For any two subsets Y, Z of X^+ let $YZ = \{yz | y \in Y, z \in Z\}$; then

$$\mu(YZ) \leq \mu(Y)\mu(Z), \qquad (6)$$

and when the right-hand side is finite, equality holds if and only if each member of YZ factorizes uniquely as yz, where $y \in Y, z \in Z$. In particular, for a code Y

we have

$$\mu(Y^n) = \mu(Y)^n, \quad n = 1, 2, \ldots \tag{7}$$

We can now state a necessary condition for a subset of X^+ to be a code:

Theorem 6.5.7. *Let X be a finite set. If Y is a code in the free monoid X^*, then*

$$\mu(Y) \leq 1. \tag{8}$$

Proof. Suppose first that Y is finite, say $\max\{|y| \,|\, y \in Y\} = N$. Then by (7) and (5),

$$\mu(Y)^n = \mu(Y^n) \leq nN, \quad \text{for all } n \geq 1.$$

Hence $\mu(Y) \leq (nN)^{1/n}$ and letting $n \to \infty$, we obtain (8). Since any subset of a code is again a code, the result follows for all finite subsets of a code, and by passing to the limit we see that it holds generally. ∎

Remarks. 1. If $Y = \{y_1, y_2, \ldots\}$, where $|y_i| = n_i$, then (8) takes the form (Kraft–McMillan inequality):

$$\sum r^{-n_j} \leq 1. \tag{9}$$

This necessary condition is not sufficient for Y to be a code, but when it holds, we can always find a prefix code with elements of lengths $n_1, n_2, \ldots$ (see Exercise 16).

2. A *maximal* code is a code that is not a proper subset of any code. By Theorem 5.7 we see that any code Y such that $\mu(Y) = 1$ is necessarily maximal, e.g. if $X = \{x, y\}$, then $\{x, xy, xy^2, y^3\}$ or $\{x, xy, xy^2, \ldots\}$ are maximal codes.

Theorem 5.7 has an analogue for free algebras, which we shall now derive. For any subset Y of $k\langle X\rangle \backslash k$ let us define the *weight* as

$$\mu(Y) = \sum_{y \in Y} r^{-\deg y}.$$

To get a measure for the elements of degree n, we need to take a basis (rather than the set of all elements); so to obtain an analogue of (7) we shall assume that $Y \cup \{1\}$ is k-linearly independent. Then (7) also holds for Y; further, we have

$$\mu(Y) \leq \max\{\deg y \,|\, y \in Y\}. \tag{10}$$

For suppose the right-hand side of (10) is N and take the set of all monomials to be lexicographically ordered. If $y, y' \in Y$ have the same monomial as leading term, we can replace y' by $y' - \alpha y$, with $\alpha \in k$ chosen so that $y' - \alpha y$ has smaller leading term than y'. By repeating this process if necessary, we can bring Y to a form where no two elements have the same monomial as leading term, and the process cannot decrease $\mu(Y)$; moreover, the linear independence of $Y \cup \{1\}$ is preserved. Thus

$$\mu(Y) = \sum r^{-\deg y} \leq \sum_r^N r^i r^{-i} = N,$$

and this proves (10). With these preparations we can state

Theorem 6.5.8. *If Y is a free subset of $k\langle X \rangle$, then $\mu(Y) \leq 1$.*

Proof. As in Theorem 5.7 it is enough to consider the case where Y is finite, say $\max\{\deg y | y \in Y\} = N$. Since Y satisfies (7), we have $\mu(Y^n) = \mu(Y)^n$ and by (10), $\mu(Y^n) \leq nN$, so as before we find $\mu(Y) \leq 1$. ∎

We conclude this section by considering another monoid associated with rings. In any ring R the two-sided ideals form a monoid under the usual multiplication

$$\mathfrak{a}\mathfrak{b} = \left\{ \sum a_i b_i | a_i \in \mathfrak{a}, b_i \in \mathfrak{b} \right\}.$$

It is easy to verify that for principal ideal domains this monoid is free abelian (Exercise 17). We shall show that for a fir it is a free monoid in most other cases.

Theorem 6.5.9. *Let R be a fir that has no non-unit right invariant elements or no non-unit left invariant elements. Then the ideals of R form a free monoid.*

If, as seems plausible, every fir with a non-unit right invariant element is right principal, the conclusion of Theorem 5.9 holds for all firs that are not left and right principal, but it is not known whether this is the case (see Exercise 4.9).

Proof. We have to verify the conditions of Theorem 5.1, and we shall proceed in a number of steps. Some of the properties actually hold under weaker assumptions.

(i) *If R is a right fir, $\mathfrak{a}$ a non-zero right ideal and $\mathfrak{b}, \mathfrak{b}'$ any left ideals, then*

$$\mathfrak{a}\mathfrak{b} = \mathfrak{a}\mathfrak{b}' \Rightarrow \mathfrak{b} = \mathfrak{b}'.$$

Proof. If we write $\mathfrak{a} = \oplus a_i R$, then the equation $\mathfrak{a}\mathfrak{b} = \mathfrak{a}\mathfrak{b}'$ becomes $\oplus a_i \mathfrak{b} = \oplus a_i \mathfrak{b}'$; equating cofactors of a_1 we find $\mathfrak{b} = \mathfrak{b}'$. ∎

(ii) *Let R be a right fir and $\mathfrak{a}, \mathfrak{b}$ ideals of R satisfying $\mathfrak{a} + \mathfrak{b} = R$, $\mathfrak{a}\mathfrak{b} = \mathfrak{b}\mathfrak{a}$. Then either $\mathfrak{a}$ or $\mathfrak{b}$ is principal as right ideal of R. Hence if R has no non-unit* *invariant elements, then*

$$\mathfrak{a} + \mathfrak{b} = R, \mathfrak{a}\mathfrak{b} = \mathfrak{b}\mathfrak{a} \Rightarrow \mathfrak{a} = R \text{ or } \mathfrak{b} = R. \tag{11}$$

Proof. If $\mathfrak{a} = R$ or $\mathfrak{b} = R$, there is nothing to prove, so assume the contrary. We have $\mathfrak{a} \cap \mathfrak{b} = (\mathfrak{a} \cap \mathfrak{b})(\mathfrak{a} + \mathfrak{b}) \subseteq \mathfrak{a}\mathfrak{b} + \mathfrak{b}\mathfrak{a} \subseteq \mathfrak{a} \cap \mathfrak{b}$, hence $\mathfrak{a} \cap \mathfrak{b} = \mathfrak{a}\mathfrak{b}$, and by the second isomorphism theorem,

$$R/\mathfrak{b} \cong (\mathfrak{a} + \mathfrak{b})/\mathfrak{b} \cong \mathfrak{a}/\mathfrak{a}\mathfrak{b}. \tag{12}$$

Now $\mathfrak{a}$ is free as right R-module, say $\mathfrak{a} = \oplus_I a_i R$; hence $\mathfrak{a}/\mathfrak{a}\mathfrak{b} \cong \oplus_I R/\mathfrak{b}$. Since the left-hand side of (12) is finitely generated, the right-hand side must be so too, hence I is finite, i.e. $\mathfrak{a}$ is finitely generated. By symmetry the same holds for $\mathfrak{b}$. Let $\mathfrak{a}, \mathfrak{b}$ have ranks p, q respectively; then by comparing characteristics in (12) we find $1 - q = p(1 - q)$, hence $(p - 1)(q - 1) = 0$ and so $p = 1$ or $q = 1$, i.e. either $\mathfrak{a}$ or $\mathfrak{b}$ is principal.

Now the second assertion follows because any ideal that is principal as right ideal has a right invariant generator. ∎

(iii) *Let R be a fir that has no non-unit right invariant elements or no non-unit left invariant elements. Then the monoid of non-zero ideals of R is rigid.*

Proof. Assume that R has no non-unit right invariant elements, say. Let $\mathfrak{a}, \mathfrak{b}, \mathfrak{a}', \mathfrak{b}'$ be non-zero ideals such that

$$\mathfrak{a}\mathfrak{b}' = \mathfrak{b}\mathfrak{a}'. \tag{13}$$

We have to find an ideal $\mathfrak{c}$ such that $\mathfrak{a} = \mathfrak{b}\mathfrak{c}$ or $\mathfrak{b} = \mathfrak{a}\mathfrak{c}$. Suppose first that $\mathfrak{a} + \mathfrak{b} = R$ and $\mathfrak{a}' + \mathfrak{b}' = R$. Then

$$\mathfrak{b}' = \mathfrak{a}\mathfrak{b}' + \mathfrak{b}\mathfrak{b}' = \mathfrak{b}\mathfrak{a}' + \mathfrak{b}\mathfrak{b}' = \mathfrak{b},$$

and similarly $\mathfrak{a}' = \mathfrak{a}$. Hence $\mathfrak{a}\mathfrak{b} = \mathfrak{b}\mathfrak{a}$ and since R has no non-unit right invariant elements, it follows by (ii) that $\mathfrak{a} = R$ or $\mathfrak{b} = R$. Note that in (13) we only had to assume that $\mathfrak{a}, \mathfrak{b}$ are right ideals and $\mathfrak{a}', \mathfrak{b}'$ left ideals, since this together with the equalities $\mathfrak{a} = \mathfrak{a}', \mathfrak{b} = \mathfrak{b}'$ is enough to make them two-sided.

Next assume merely that $\mathfrak{a} + \mathfrak{b} = R$ and let $\mathfrak{a}' + \mathfrak{b}' = \oplus_I Re_i$. Pick $i' \in I$ and let $\pi : \oplus_I Re_i \to R$ be the left linear functional 'left cofactor of $e_{i'}$.' Then $\mathfrak{a}'\pi + \mathfrak{b}'\pi = R$ and $\mathfrak{a}.\mathfrak{b}'\pi = \mathfrak{b}.\mathfrak{a}'\pi$, where $\mathfrak{a}'\pi, \mathfrak{b}'\pi$ are not both 0, hence neither is 0. By the previous case we can find an ideal $\mathfrak{c}$ such that $\mathfrak{a} = \mathfrak{b}\mathfrak{c}$ or $\mathfrak{b} = \mathfrak{a}\mathfrak{c}$.

Finally the general case can be reduced to the case just considered by putting $\mathfrak{a} + \mathfrak{b} = \oplus_I f_j R$. ∎

(iv) *Let us write $\mathfrak{a} \prec \mathfrak{b}$ to mean $\mathfrak{a} = \mathfrak{c}\mathfrak{b}$, where $\mathfrak{c} \neq R$. Every left fir satisfies ACC for two-sided ideals, with respect to '$\prec$'.*

Proof. Suppose that we have a strictly ascending chain

$$0 \neq \mathfrak{b}_0 \prec \mathfrak{b}_1 \prec \mathfrak{b}_2 \prec \ldots, \tag{14}$$

say $\mathfrak{b}_{i-1} = \mathfrak{a}_i \mathfrak{b}_i$. Then $\mathfrak{b}_0 = \mathfrak{a}_1 \mathfrak{b}_1 = \mathfrak{a}_1 \mathfrak{a}_2 \mathfrak{b}_2 = \ldots$, hence

$$\mathfrak{b}_0 \subseteq \cap \mathfrak{a}_1 \ldots \mathfrak{a}_n,$$

but the intersection on the right is 0, by Theorem 5.10.4, whereas $\mathfrak{b}_0 \neq 0$. This contradiction shows that (14) must break off. ∎

(v) *In any fir, every non-zero ideal is a product of atoms (in the monoid of non-zero ideals).*

Proof. By the left–right analogue of (iv) any proper ideal $\mathfrak{a}$ has a maximal left factor $\mathfrak{p}_1 : \mathfrak{a} = \mathfrak{p}_1 \mathfrak{a}_1$ say. If $\mathfrak{a}_1$ is proper, it has a maximal left factor $\mathfrak{p}_2$, giving $\mathfrak{a} = \mathfrak{p}_1 \mathfrak{p}_2 \mathfrak{a}_2$, and by (iv) this process must terminate; when it does, we have the desired expression of $\mathfrak{a}$ as a product of atoms. ∎

This completes the proof of (v) and with it, of Theorem 5.9. ∎

Exercises 6.5

1. Show that any retract of a free monoid is free.
2. Use Theorem 5.4 to find a procedure for reducing a finite subset Y of a free monoid to a set Y' such that (i) Y and Y' generate the same submonoid T and (ii) if T is free, then Y' is a code.
3. Let $S = \{0, 1, x\}$ with multiplication $x^2 = 0$. Show that S is generated by $\{x\}$ and is conical and rigid (save for cancellation) but not free.
4. Let F be the free group on a finite set X and let X^* be the submonoid generated by X. Classify the submonoids between X^* and F. Does this set of submonoids satisfy ACC? (*Hint*: Consider $S_n = \{x, x^{-n}y\}^*, n = 0, 1, \ldots$.)
5. Show that every submonoid of a free monoid has a unique minimal generating set, consisting of its atoms.
6. Show that in a free monoid, the set of palindromes (i.e. fixed elements under the order-reversing anti-automorphism) of even length generates a free submonoid, with a prefix code as generating set. What about the set of all palindromes of length nk ($n = 1, 2, \ldots$) for a fixed k?
7. Show that the intersection of any family of free submonoids of a free monoid is free. Show more generally that in any conical cancellation monoid, any non-empty intersection of rigid submonoids is rigid, but that neither of the conditions 'conical' or 'cancellation' can be omitted.
8. Prove the following generalization of Theorem 5.1: Let G be a group, F a free monoid and $S = G * F$ their coproduct (= pushout). Show that (i) given units $u, v \in S$, if $uc = cv$ for a non-unit c, then $u = v = 1$; (ii) for any $c \in S$ the number of non-unit factors in a factorization of c is bounded; (iii) S is rigid.

Conversely, show that a monoid satisfying (i)–(iii) is of the form $G*F$, where the group G is unique as the group of units while the monoid F is free and unique up to isomorphism.

9*. Let A^* be the free monoid on a set A, and for any subset X define the *free hull* of X as the intersection of all free monoids containing X. Given any finite subset X of A^+ and its free hull Y^*, the submonoid freely generated by its unique free generating set Y, consider the mapping $\alpha : X \to Y$ defined as follows: for $x \in X$, x^α is the unique word $y \in Y$ such that $x \in yY^*$. Show that α is well-defined (using the fact that Y is a code) and surjective (take $z \in Y \setminus X^\alpha$ and show that $Z = (Y \setminus z)z^*$ is a smaller code than Y). Show that if α is injective, then X is a code, and deduce the Defect Theorem: the free hull Y^* with free generating set Y of a finite subset X of A^*, not a code, satisfies $\text{card}(Y) < \text{card}(X)$ (see Lothaire [97], p. 6).

10. Let S be a monoid in which the relation '$a \leq b$ if and only if $ax = b$ for some $x \in S$' is a well-ordering. If S also has right cancellation ($xz = yz \Rightarrow x = y$), show that S either consists of one element or is isomorphic to $\mathbb{N}$.

11*. Let S be a monoid that is well-ordered under the relation in Exercise 10 and has left cancellation. Show that S has a presentation with generators $x_\alpha (\alpha < \tau)$ for some cardinal τ and defining relations $x_\alpha x_\beta = x_\beta$ if $\alpha < \beta$ (Cohn [61a]).

12. Let X be a set and take the monomials in X to be totally ordered in some way, so as to respect left and right multiplication. Show that in $k\langle X \rangle$ the homogeneous elements with leading coefficient 1 form a free monoid.

13. Show that a subset of $k\langle X \rangle$ is right d-independent whenever its leading terms form a prefix code. Does the converse hold?

14. For any set X and any $n \geq 1$, show that X^n is a maximal code.

15. Give an example of a subset of $X^* \setminus \{1\}$ which fails to be a code, yet satisfies the Kraft–McMillan inequality. (*Hint*: Consider the effect of enlarging X.)

16. Let $X = \{0, 1, \ldots, r-1\}$ and let $1 \leq n_1 \leq n_2 \leq \ldots$ be a sequence of integers satisfying $\sum_i r^{-n_i} \leq 1$. Define s_i as the partial sum: $s_1 = 0$, $s_{i+1} = s_i + r^{-n_i}$ and put $p_i = r^{s_i} s_i$. Verify that p_i is an integer satisfying $0 \leq p_i < r^{n_i}$. Let $y_i \in X^*$ be p_i in the r-adic scale (i.e. to base r). Verify that y_i has at most n_i digits (so by prefixing 0s it can be taken to have exactly n_i digits), and that $\{y_1, y_2, \ldots\}$ is a prefix code.

17. Let R be a principal ideal domain. Show that each non-zero ideal of R is generated by an invariant element. Deduce that $\mathfrak{a}\mathfrak{b} = \mathfrak{b}\mathfrak{a}$ for all ideals $\mathfrak{a}, \mathfrak{b}$. (*Hint*: Recall Theorem 0.9.4 and Proposition 1.4.6). Then conclude that the monoid of non-zero ideals of R is free commutative.

18°. Let R be a principal right ideal domain. Is the monoid of its non-zero ideals necessarily commutative?

19°. Let R be a fir. If R has a non-unit right invariant element, is it necessarily right principal?

20*. (C. Reutenauer) Show that the intersection of free power series subrings of $k\langle\langle X \rangle\rangle$ is a free power series ring. Deduce an analogue of the defect theorem: any finite subset Y of $k\langle\langle X \rangle\rangle$ is either free or is contained in a free power series ring of rank $< \text{card}(Y)$.

6.6 Subalgebras and ideals of free algebras

A subalgebra of a free associative algebra is not necessarily free. This is already clear from the commutative case, by considering the subalgebra of the polynomial ring $k[x]$ generated by x^2 and x^3. In the case of the polynomial ring there is a simple criterion for a subalgebra to be free. The proof uses some ideas from valuation theory (see e.g. BA, chapter 9). In particular we note that all valuation subrings of $k(y)$ except one contain y, and the intersection of these subrings is just $k[y]$.

Proposition 6.6.1. *A subalgebra R of a polynomial ring $k[x]$ is free if and only if it is integrally closed (in its field of fractions).*

Proof. Assume that R is a free subalgebra; then either $R = k$ or $R = k[y]$ for some y transcendental over k. In either case R is integrally closed.

Conversely, let R be integrally closed and denote its field of fractions by K. Since $k \subseteq R$, we have $k \subseteq K \subseteq k(x)$. If $K = k$, then $R = k$ and the result follows. Otherwise, by Lüroth's theorem (see e.g. BA, theorem 11.3.4), K is a simple purely transcendental extension of k. Since R is integrally closed, it is an intersection of valuation rings of K (see Section 6.1). Now any valuation ring of K is of the form $\mathfrak{o}_p \cap K$, where $\mathfrak{o}_p$ is a valuation ring of $k(x)$ (see e.g. BA, Section 9.4). If x is finite at p, then $\mathfrak{o}_p \supseteq k[x] \supseteq R$, whence $R \subseteq \mathfrak{o}_p \cap K$. Thus R is contained in all valuation rings of K over k, except at most one, namely the one obtained from the pole of x, and for this place the residue-class field is k. Since $R \neq k$, R is not contained in all the valuation rings; if $\mathfrak{o}_q$ is the exceptional one, we can choose a generator y of K such that q is a pole of y, for if not, say if y maps to $a \in k$, then we can replace y by $(y - a)^{-1}$. Now R is the intersection of all the other valuation rings of $K = k(y)$ over k, so it follows that $R = k[y]$. ∎

For free algebras on more than one generator no such convenient criterion is known. We can of course use the characterization in terms of the weak algorithm given in Section 2.5. Applied to subalgebras, this yields

Proposition 6.6.2. *Let $F = k\langle X \rangle$ be a free k-algebra. Then a subalgebra R of F is free if and only if there is a filtration on R over k for which R satisfies the weak algorithm.* ∎

The difficulty in applying this criterion lies in the fact that the different degree-functions in a free algebra are not related in any obvious way. If F is a free algebra and R a subalgebra, then any degree-function on F will define a degree on R, and if R satisfies the weak algorithm for this function, then it must be free. However, this sufficient condition is not necessary; thus if R is a free subalgebra of F, then there is always a degree-function on R for which the weak algorithm holds (Proposition 6.2), but this function need not be defined on all of F. Let us call a subalgebra R of F *regularly embedded* if there is a degree-function on F for which both F and R satisfy the weak algorithm. Examples of irregularly embedded free subalgebras are given in Exercises 2 and 3.

There is one case where more information is available, namely when $F = k[x]$ is free of rank 1. We already know a simple test for a subalgebra to be free (Proposition 6.1) and, as we shall see, there is also a simple criterion in terms of the algorithm. For in this case (and only here) the only automorphisms are the affine transformations $x \mapsto ax + b (a, b \in k, a \neq 0)$, therefore the degree-function on F is unique up to a scalar factor. We thus obtain

Proposition 6.6.3. *A subalgebra of $k[x]$ is free if and only if it has the division algorithm relative to the x-degree.* ∎

Sometimes it is possible to obtain conditions for a homogeneous subalgebra to be free. We recall that a homogeneous subalgebra is a subalgebra generated by homogeneous elements (not necessarily all of the same degree). Thus, using Corollary 2.9.15, we have

Theorem 6.6.4. *Let $F = k\langle X \rangle$ be a free algebra and R a homogeneous subalgebra, with augmentation ideal $\mathfrak{a}$. Then the following conditions are equivalent:*

(a) *For every homogeneous subspace V of R such that $\mathfrak{a} = V \oplus \mathfrak{a}^2$, the natural homomorphism $\psi : k\langle V \rangle \to R$ is an isomorphism,*
(b) *R is free on a homogeneous set of free generators,*
(c) *R is free k-algebra,*
(d) *R is right hereditary,*
(e) *$\mathfrak{a}$ is flat as right R-module,*
(f) *the natural map $\mathfrak{a} \otimes_R \mathfrak{a} \to \mathfrak{a}^2$ is an isomorphism.*

Proof. (a) $\Rightarrow$ (b) $\Rightarrow$ (c) $\Rightarrow$ (d) $\Rightarrow$ (e) $\Rightarrow$ (f) are clear ((e) $\Rightarrow$ (f) follows by Appendix Proposition B.8). To prove (f) $\Rightarrow$ (a) we need only show, by Corollary 2.9.15, that ψ is surjective. By hypothesis, $\mathfrak{a} = V + \mathfrak{a}^2$, hence $\mathfrak{a}^n = (V + \mathfrak{a}^2)^n \subseteq V^n + \mathfrak{a}^{n+1}$, and so we have

$$\mathfrak{a}^n = V^n + \mathfrak{a}^{n+1} \quad \text{for all } n \geq 1. \tag{1}$$

It follows, on setting $R' = \operatorname{im} \psi$, that

$$R = R' + \mathfrak{a}^n \quad \text{for all } n \geq 1. \tag{2}$$

For this holds when $n = 1$, and since $V \subseteq R'$, it follows generally by induction, using (1). Now let u be a homogeneous element of R, take $n > \deg u$ and write $u = r' + a$, where $r' \in R', a \in \mathfrak{a}^n$. Since V is homogeneous, so is R' and hence $u = r' \in R', a = 0$. Thus $R' = R$ and ψ is surjective, as claimed. ∎

There is another characterization of homogeneous subalgebras that is often useful:

Theorem 6.6.5. *Let $F = k\langle X\rangle$ be a free k-algebra and R a homogeneous subalgebra. Then the following conditions are equivalent:*

(a) R is free, with a homogeneous free generating set Y which is right F-linearly independent ,
(b) F is free as left R-module,
(c) F is flat as left R-module.

Proof. (a) $\Rightarrow$ (b). Assume (a) and let W be the k-space spanned by Y, so that $W \otimes_k F = YF, R = k\langle W\rangle$. Choose a homogeneous k-space V such that $F = V \oplus YF$; then

$$F \cong V \oplus (W \otimes F) \cong V \oplus (W \otimes V) \oplus (W \otimes W \otimes F)$$
$$\cong (k \oplus W \oplus (W \otimes W) \oplus \ldots) \otimes V$$
$$\cong R \otimes V,$$

and this shows V to be spanned over k by a left R-basis of F.

Thus F is free over R, i.e. (b). (b) $\Rightarrow$ (c) is clear, and to prove (c) $\Rightarrow$ (a), assume that F is flat as left R-module. We take a resolution of k:

$$0 \to \mathfrak{f} \to F \to k \to 0,$$

where $\mathfrak{f}$ is the augmentation ideal of F. Similarly, if $\mathfrak{a}$ is the augmentation ideal of R, we have an exact sequence

$$0 \to \mathfrak{a} \to R \to k \to 0.$$

Since R is free as R-module, we obtain an exact sequence (by the extended Schanuel lemma, Appendix Lemma B.5):

$$0 \to \mathfrak{a} \to \mathfrak{f} \oplus R \to F \to 0.$$

Here $F, R, \mathfrak{f}$ are left R-flat, hence so is $\mathfrak{a}$. Now Theorem 6.4 (e) shows that R is free, with homogeneous generating set Y. Hence Y is a right R-basis of $\mathfrak{a}$. But the embedding

$$0 \to R^{(Y)} = \mathfrak{a} \to R$$

remains exact under the operation $\otimes_R F$ (because F is left flat), therefore Y is right F-linearly independent and (a) follows. ∎

We now turn to examine right ideals in a free algebra. If $F = k\langle X\rangle$ is a free k-algebra and $\mathfrak{r}$ a right ideal in F, then $F/\mathfrak{r}$ is a right F-module; hence it is a k-space and we shall find that it has a basis derived from a prefix code in X. This will lead to another proof that $\mathfrak{r}$ is free as right F-module, but some preparations are necessary.

In every partially ordered set S with DCC there is a natural bijection between lower segments (i.e. complements of upper segments) and antichains (i.e. sets of pairwise incomparable elements): with every lower segment L we associate the set L^O of minimal elements of the complement $S\setminus L$, clearly an antichain, and with every antichain A we associate A^O, the complement of the upper segment generated by A. It is easily checked that these mappings are mutually inverse.

Let X be a finite set and X^* the free monoid on X, partially ordered by left divisibility:

$$a \leq b \text{ if and only if } b = ac \text{ for some } c \in X^*. \tag{3}$$

It is clear that X^* satisfies DCC for this partial ordering. A lower segment in this partial ordering is called a *Schreier set*; such a set is characterized by the fact that with any word it contains all its left factors (prefixes). An antichain in X^* is just a prefix set; thus we have a natural bijection between the prefix sets and the Schreier sets. Starting from a prefix set P, the corresponding Schreier set P^O consists of all words with no prefix in P, while for a Schreier set C the corresponding antichain C^O consists of all words not in C but with every proper prefix in C. We note that the prefix set $\{1\}$ corresponds to the empty Schreier set; thus the prefix codes correspond to the non-empty Schreier sets. A prefix code and its Schreier set lead to a useful factorization of the free monoid:

Proposition 6.6.6. *Let X be any non-empty set. Then the construction described above for the free monoid X^* yields a natural bijection between the prefix codes and the non-empty Schreier sets on X^*, and if P, C correspond in this way, then every element of X^* can be written uniquely as the product of a string of members of P and a member of C:*

$$X^* = P^*C, \tag{4}$$

while P and C are given by

$$C = X^*\setminus PX^*, \tag{5}$$
$$P = CX\setminus C. \tag{6}$$

Moreover for any $c \in C$ and any $w \in X^*$ of length h, the product cw is either in C or of the form pw', where $p \in P$ and $w' \in X^*$ is of length $< h$.

6.6 Subalgebras and ideals of free algebras

Proof. It is clear that (5) and (6) lead to the bijection that has already been described. To establish (4), let $w \in X^*$ and write $w = ab$, where a is the longest left factor of w in P^*; then clearly $b \in C$ and (4) follows. The final assertion is easily verified. ∎

We can now give another proof of the right fir property of rings with a weak algorithm in an explicit form.

Theorem 6.6.7. *Let R be a filtered ring with a weak algorithm, $R_0 = K$ a field. Then any right ideal $\mathfrak{r}$ of R is free. More precisely, there is a Schreier set C that is a left K-basis of R (mod $\mathfrak{r}$); if $P = C^\circ$ is the corresponding prefix set, then for each $p \in P$ there exist unique elements $\alpha_{p,c} \in K (c \in C)$, almost all zero, such that*

$$p - \sum \alpha_{p,c} c \in \mathfrak{r} \, (c \in C),$$

and the elements $y_p = p - \sum \alpha_{p,c} \, c$ form a basis of $\mathfrak{r}$ as free right R-module.

Proof. Let X be a monomial left K-basis of R, constructed as in Section 2.5. Any element of X^* with a prefix in $\mathfrak{r}$ is itself in $\mathfrak{r}$, so let us take the subset Z of X^* consisting of all elements with no prefix in $\mathfrak{r}$. We now build up a Schreier set by induction on the length as follows: at the first stage we choose 1; next we take a subset C_1 of Z consisting of 1 and of elements of length 1 that together with 1 are left K-linearly independent (mod $\mathfrak{r}$). Thus C_1 is a Schreier set. When C_h has been chosen as a Schreier set, to contain C_{h-1} and left linearly independent (mod $\mathfrak{r}$), we add elements of length $h+1$ to preserve these properties to form C_{h+1}, as long as this is possible. This process can only stop when we have a left K-basis of R (mod $\mathfrak{r}$), and this is the desired Schreier set C. Let P be the corresponding prefix set; for each $p \in P$, $C \cup \{p\}$ is still a Schreier set, but is left linearly dependent (mod $\mathfrak{r}$), by the maximality of C; thus we have

$$p - \sum \alpha_{p,c} c \in \mathfrak{r} \quad (c \in C, \alpha_{p,c} \in K). \tag{7}$$

Writing $y_p = p - \sum \alpha_{p,c} c$, we claim that any element r of R can be written in the form

$$r = \sum y_p g_p + \sum \beta_c c, \quad \text{where } g_p \in R, \, \beta_c \in K. \tag{8}$$

By linearity it is enough to prove this when r is a monomial w, and we may further assume that $w \notin C$. By Proposition 6.6 we can then write $w = pu$ for some $p \in P$. Using (7) and the definition of y_p, we find

$$w = y_p u + \sum \alpha_{p,c} cu.$$

Now for any $c \in C$, either $cu \in C$ or $cu = p_1 u_1$, where $|c| < |p_1|, |u_1| < |u|$. In the first case we have achieved the form (8), in the second case we can apply induction on $|u|$ to express $p_1 u_1$ in the same form, and (8) follows.

Now take $r \in \mathfrak{r}$ and apply the natural homomorphism $R \to R/\mathfrak{r}$. Writing the image of $a \in R$ as $\bar{a}$, we have, by (8),

$$0 = \bar{r} = \sum \beta_c \bar{c},$$

but by hypothesis the $\bar{c}$ are left linearly K-independent, hence $\beta_c = 0$ in this case and so $r = \sum y_p g_p$. This shows that the y_p generate $\mathfrak{r}$. To prove their linear independence over R, assume that $\sum y_p g_p = 0$, where the g_p are not all 0. Then by (7)

$$\sum p g_p = \sum \alpha_{p,c} c g_p. \tag{9}$$

Let w be a word of maximum length occurring in the g_p, say in $g_{p'}$. Since P is a prefix code, $p'w$ occurs with a non-zero coefficient, λ say, on the left of (9). Hence we have

$$\lambda = \sum \alpha_{p,c} \mu_{p,c},$$

where $\mu_{p,c}$ is the coefficient of $p'w$ in cg_p. Now $p'w = cu$ can only hold when c is a proper left factor of p', hence $|c| < |p'|, |u| > |w|$, and this contradicts the definition of w. Hence the y_p form a free generating set of $\mathfrak{r}$, as claimed. ∎

In particular, this result shows again that each right ideal in R (and, by symmetry, each left ideal) is free (Corollary 2.5.3). The uniqueness of the rank is clear since R has a homomorphism to K and so has IBN.

We can also use these results to obtain another derivation of the Schreier–Lewin formula (see Section 2.6). To do this we observe that for free algebras, (4) of Proposition 6.6 translates to the relation

$$\left(1 - \sum X\right)^{-1} = \left(1 - \sum P\right)^{-1} C$$

in $F = \mathbb{Z}\langle\!\langle X \rangle\!\rangle$. On multiplying out, we find

$$\left(1 - \sum P\right) = C\left(1 - \sum X\right). \tag{10}$$

Thus if $F/\mathfrak{r}$ has finite dimension r over k and X has finite cardinality d, then the right-hand side of (10) will lie in $\mathbb{Z}\langle X \rangle$, hence so will the left-hand side, which shows P to be finite too. By Theorem 6.7 this tells us that $\mathfrak{r}$ has finite rank, n say, as right F-module. If we map $\mathbb{Z}\langle X \rangle$ to $\mathbb{Z}$ by mapping each $x \in X$ to 1, (10) takes the form of the Schreier–Lewin formula:

$$1 - n = r(1 - d). \tag{11}$$

6.6 Subalgebras and ideals of free algebras

Free subalgebras of a free algebra F have the property that they contain no non-zero ideal of F; this follows because they are anti-ideals. We shall need a result providing us with a supply of anti-ideals. A subring S of a ring R is said to be *unit-closed* in R if every unit of R that lies in S is also a unit in S.

Lemma 6.6.8. *Let R be an integral domain and S a subring. If S is a 2-fir, unit-closed in R, then S is an anti-ideal.*

Proof. Let $a \in R$ and suppose that $b, c \in S^\times$ are such that $ab, ca \in S$. Then in S we have $c.ab = ca.b$, hence $cS + caS$ is principal, say dS, where $d \in S$. Now $dR = cR + caR = cR$, so $c = du$, where u is a unit in R. We have $du = c \in dS$ and so $u \in S$; since S is unit-closed, u is a unit in S. Now $ca \in dS = duS = cS$, therefore $a \in S$, showing S to be an anti-ideal. ∎

In a free k-algebra the units all lie in k and so are contained in every subalgebra; hence we obtain

Theorem 6.6.9. *In a free algebra, every subalgebra that is a 2-fir (in particular every free subalgebra) is an anti-ideal.* ∎

Exercises 6.6

1. Let $F = k\langle x, y\rangle$; show that the subalgebra R generated by $x + y^2$ and y^3 is free on these generators but does not satisfy the weak algorithm relative to the (x, y)-degree. By a suitable change of degree-function show that R is regularly embedded in F.
2. (D. R. Lane) Let $F = k\langle x, y\rangle$; show that the subalgebra generated by $u = (xy)^4 + y(xy)^2$ and $v = (xy)^3$ is free, but not regularly embedded. (*Hint:* Show that $k\langle u, v\rangle$ is regularly embedded in $k\langle y, yx\rangle$, which is regularly embedded in F. Alternatively verify that any degree-function d satisfies $d([u, v]) < d(uv) = d(vu)$ and use Corollary 7.4 below.)
3*. (W. Dicks) Let $F = k\langle x, y\rangle$; show that the subalgebra $u = xyx - y$, $v = uyx$ and $w = uxy$ is free on these generators but is not regularly embedded in F.
4. (R. E. Williams) In a free algebra F, let R be the subalgebra generated by $b_1, \ldots, b_n$ such that $\deg(b_1) = \ldots = \deg(b_n)$, and this is the minimum degree of non-zero elements in R. Show that if the elements $b_1 \ldots b_n$ are linearly independent, they form a free generating set.
5. Let $F_n = k\langle x_1, \ldots, x_n\rangle$ be the free k-algebra of rank n. Show that if there is a surjective homomorphism $\varphi : F_n \to F_m$, then $m \leq n$, with equality if and only if φ is an isomorphism (*Hint:* Look at the terms of low degree.)
6°. (V. Drensky) Assume that $k\langle X\rangle \subset k\langle Y\rangle \subset k\langle X \cup Z\rangle$, as augmented k-algebras. Is $k\langle Y\rangle$ free on a set $X \cup T$ for some T?
7°. (W. E. Clark) Is every retract of a free algebra of rank n free? (For $n = 1, 2$ see Section 6.7).

8. (G. M. Bergman) In $\{x, y\}^*$ consider the submonoid generated by $u = xy, v = yx, z = x^2$ and y. Verify that it satisfies the relations $yz^n u = vz^n y (n = 1, 2, \ldots)$, but that none of these equations is implied by earlier ones. Deduce the existence of a finitely generated subalgebra of $k\langle x, y\rangle$ that is not finitely presented.

9. (Kolotov [78]) Show that in any integral domain, the family of anti-ideals is closed under intersections, unions of chains, and if C is an anti-ideal in B and B in A, then C is an anti-ideal in A. Show also that an anti-ideal of R contains with any right ideal of R its idealizer in R.

10. (G. M. Bergman) Let A be a commutative integral domain and F its field of fractions. Show that a subring of A is an anti-ideal if and only if it is the intersection of A with a subfield of F.

11. Show that in Theorem 6.7, for a right ideal of finite codimension the prefix set P is a maximal code.

12. (W. Dicks) Let $F = k\langle x, y, z\rangle$ and let R be the subalgebra generated by $a = xy, b = xyz + x + z, c = zyx + x + z, d = yx$. Verify that R is an anti-ideal; moreover, for any $s \in F\setminus R$, $sR \cap R = 0$, but that R is not a 2-fir. (*Hint*: Verify that $a + 1$ and b are right comaximal in F and examine the homomorphism of the subalgebra generated by a, d, x, z mapping a, d to -1 and x, z to 0.)

13*. (Bergman [a]) Let $F = k\langle x, y_1, y_2, y_3, z\rangle$, G_1 the subalgebra generated by $x, y_1, y_3, y_2 z, z$ and G_2 the subalgebra generated by $x, xy_1, y_1 y_2 - y_3, y_2, z$. Verify that G_1 and G_2 are each free on the given generating sets, but that $G_1 \cap G_2$ is not a 3-fir, hence not free. (*Hint*: Verify that the relation $x.(y_1 y_2 - y_3)z + xy_1.(-y_2 z) + xy_3.z = 0$ can be trivialized in G_1 and in G_2 but not in $G_1 \cap G_2$.)

14. (Bergman [a]) Show that in a free k-algebra the intersection of any family of free k-subalgebras is a 2-fir, but need not be free.

15. A prefix code on X^* is called *complete* if the right ideal generated by it meets every non-empty right ideal of X^*. Show that in the correspondence between prefix codes and Schreier sets, the finite non-empty Schreier sets correspond to the complete finite prefix codes.

16°. Prove an analogue of Theorem 6.7 for group algebras of free groups and more generally for mixed algebras. Is there an extension for firs that are augmented k-algebras or for rings with weak algorithm?

6.7 Centralizers in power series rings and in free algebras

In a free algebra, few elements commute with each other, and one would expect the centralizer of a non-scalar element to be small. This expectation is borne out, as we shall see in Theorem 7.7 below. A similar question can be raised for the free power series ring, and as this is rather easier to answer, we begin with it.

Theorem 6.7.1. *Let R be an inversely filtered ring that satisfies a 2-term inverse weak algorithm and is complete. Then the centralizer C of any $a \in R$, not zero or a unit, is a complete principal valuation ring.*

6.7 Centralizers in power series rings and in free algebras

If moreover, R is a connected k-algebra, then the centralizer of any non-scalar element of R is a formal power series ring in one variable over k and so is commutative.

Proof. Let $x, y \in R$ be right linearly dependent and $v(x) \geq v(y)$, say; then $x - yz$ and y are right v-dependent for any $z \in R$, hence by Lemma 2.9.3, $x = yz$ for some $z \in R$, so $xR \subseteq yR$.

If a is as stated in the theorem, then $v(a) > 0$; given $x, y \in C$, we have $xa^n = a^n x$ for any $n \geq 0$, hence for sufficiently large n, $a^n R \subseteq xR$ and $a^n R \subseteq yR$. Thus $xR \cap yR \neq 0$, and so, if $v(x) \geq v(y)$, then $x = yz$. Clearly $z \in C$, so $xC \subseteq yC$, i.e. C is a right chain ring. Since R is atomic, by Proposition 2.9.6, and any non-unit of C remains a non-unit in R, it follows that C is atomic and therefore a rigid UFD, by Theorem 3.3.2. It contains a large non-unit, namely a, hence it is a right (and by symmetry left) principal valuation ring, by Theorem 3.3.9, and it is complete because R is.

Now assume that R is connected and let x be an element of least positive degree in C, so that every ideal in C has the form $Cx^n = x^n C$. Then every element of C has the form $x^n v$, where $n \geq 0$ and v is a unit. The additional hypothesis allows us to write $v = \alpha + x^r v'$ ($\alpha \in k, r \geq 0, v'$ a unit), and an induction argument, using the completeness, shows that $C = k[[x]]$. ∎

In particular, the free power series ring satisfies all the hypotheses and we have

Corollary 6.7.2. *The centralizer of a non-scalar element in the free power series ring $k\langle\langle X \rangle\rangle$ is of the form $k[[c]]$ for some element c of positive order.* ∎

We know from Corollary 2.9.11 that any non-scalar element in a free power series ring generates a free power series ring in one variable. Two elements may not generate a free power series algebra, e.g. the complete algebra generated by x^2 and x^3 is not free; here and in the proof below we shall use the term 'free subset' in the abstract sense, i.e. ignoring the filtration. We have the following analogue for part of Corollary 5.5:

Proposition 6.7.3. *Let R be a complete inversely filtered ring with 2-term inverse weak algorithm, which is a connected k-algebra. If x, y are two non-commuting elements of R, say $x = \alpha + x_1, y = \beta + y_1$, where $\alpha, \beta \in k, x_1, y_1 \in R_{[1]}$, then the complete algebra generated by x, y is the free power series ring $k\langle\langle x_1, y_1 \rangle\rangle$.*

Proof. By hypothesis $xy - yx \neq 0$, hence $v(xy - yx)$ is a positive integer n; we shall use induction on n. Since $xy - yx = x_1 y_1 - y_1 x_1$, we may replace x, y

by x_1, y_1 respectively; for simplicity of notation we shall drop the subscript, thus we may assume that $v(x), v(y) > 0$.

If x, y are not free, we have a relation

$$xf + yg + \lambda = 0, \qquad (1)$$

where $f, g \in k\langle\!\langle x, y\rangle\!\rangle$ and $\lambda \in k$. Equating constant terms in (1) we find that $\lambda = 0$, so x, y are right linearly dependent. If $v(x) \geq v(y)$, then

$$x = yz, \qquad (2)$$

for some $z \in R$. Let $z = \gamma + z_1 (\gamma \in k, z_1 \in R_{[1]})$; if we can show that y, z_1 are free, then so are x, y by (2), for the elements y, yz are clearly free in $k\langle\!\langle y, z_1\rangle\!\rangle$. Now

$$0 \neq xy - yx = yzy - yyz = y(zy - yz),$$

hence $yz - zy \neq 0$ and $v(yz - zy) < v(xy - yx)$. So the result follows by induction. ∎

The result may be expressed more strikingly by saying: x and y are free if and only if they do not commute. In particular, since any free algebra can be embedded in a power series ring, we have

Corollary 6.7.4. *Any two non-commuting elements of a free k-algebra form a free set.* ∎

We now go on to consider centralizers in a free algebra $F = k\langle X\rangle$. Let $a \in F\backslash k$ and denote by C the centralizer of a in F. The embedding of F in $k\langle\!\langle X\rangle\!\rangle$ shows that C is commutative, by Corollary 7.2; moreover, C is finitely generated, as module over $k[a]$ or as algebra. For if $d(a) = n$ say, we choose for each integer $\nu = 0, 1, \ldots, n-1$ such that an element of degree $\equiv \nu \pmod{n}$ occurs in C, an element of least degree $\equiv \nu \pmod{n}$ in C. Calling these elements $c_0, \ldots, c_r$ ($r \leq n-1$), we see that every element in C has the same leading term as some $c_i a^j$ ($i = 0, \ldots, r, j = 0, 1, \ldots$), by Corollary 7.2; hence the $c_i a^j$ span C over $k[a]$, and together with a they generate C as k-algebra.

Our aim is to show that C is a polynomial ring over k. In order to establish this fact we shall study homomorphisms of C into polynomial rings. If we look at the leading terms of elements of C we note that all are powers of a given one (essentially by an application of Corollary 5.5), and it turns out that in order to achieve a homomorphism of C into a polynomial ring we need a preordering of the free monoid that lists each word together with all its powers. This is done by introducing 'infinite' words.

6.7 Centralizers in power series rings and in free algebras

Let X be a totally ordered set and W the set of all right infinite words in X, i.e. infinite sequences of letters from X. Given $u \in X^+$, we denote by u^∞ the word obtained by repeating u indefinitely: $u^\infty = uuu\ldots \in W$. Thus we have a mapping $u \mapsto u^\infty$ from X^+ to W that identifies two words if and only if they have the same root* We shall take W to be ordered lexicographically.

Lemma 6.7.5. *Given $u, v \in X^+$, if $u^\infty > v^\infty$, then*

$$u^\infty > (uv)^\infty > (vu)^\infty > v^\infty, \tag{3}$$

and similarly with $>$ replaced by $<$ or $=$ throughout.

Proof. Suppose that $(uv)^\infty > (vu)^\infty$; then

$$(vu)^\infty = v(uv)^\infty > v(vu)^\infty = v^2(uv)^\infty > v^2(vu)^\infty = \ldots \to v^\infty,$$

since the lexicographic order is 'continuous'. Similarly we find that $(uv)^\infty < u^\infty$; therefore (3) follows whenever $(uv)^\infty > (vu)^\infty$. Likewise $(uv)^\infty < (vu)^\infty$ implies

$$u^\infty < (uv)^\infty < (vu)^\infty < v^\infty, \tag{4}$$

while $(uv)^\infty = (vu)^\infty$ implies

$$u^\infty = (uv)^\infty = (vu)^\infty = v^\infty. \tag{5}$$

Now for any u, v exactly one of (3), (4), (5) holds and the assertion follows. ∎

The monoid algebra kX^* is just the free algebra $F = k\langle X \rangle$. Given any periodic word z in W, i.e. an infinite power of a word in X, let us define A_z as the k-subspace of F spanned by the words u satisfying $u = 1$ or $u^\infty \leq z$, and let I_z be the k-subspace spanned by the words u such that $u \neq 1$ and $u^\infty < z$. By Lemma 7.5, A_z is a subalgebra of F in which I_z is a two-sided ideal.

The set of words u in X such that $u^\infty = z$, together with 1, form the set of non-negative powers of an element v that we shall call again the *root* of z. It follows that $A_z/I_z \cong k[v]$.

Proposition 6.7.6. *Let C be a finitely generated subalgebra of a free k-algebra F. If $C \neq k$, then there is a homomorphism f of C into the polynomial ring in one variable over k such that $Cf \neq k$.*

Proof. Let $F = k\langle X \rangle$, where X is totally ordered. Take a finite generating set Y for C and let z be the maximum of u^∞ as u ranges over all monomials

---

* For example, can and cancan.

$\neq 1$ occurring with non-zero coefficient in members of Y. Then $Y \subseteq A_z$, hence $C \subseteq A_z$ and the quotient map $f : A_z \to A_z/I_z \cong k[v]$ is non-trivial on C. ∎

When C is not finitely generated, the result need no longer hold (see Exercise 5).

Consider now the free algebra $F = k\langle X \rangle$. Given a non-scalar element a, its centralizer C is a finitely generated commutative k-algebra, as we have seen. Therefore it can be mapped non-trivially into a polynomial algebra; now C as finite extension of $k[a]$ has transcendence degree 1 over k and so it must be embedded in the polynomial algebra. Being integrally closed (by Corollary 1.2), it must be free (Proposition 6.1), so we have proved Bergman's centralizer theorem:

Theorem 6.7.7. *Let $F = k\langle X \rangle$ be the free k-algebra on X. Then the centralizer of any non-scalar element of F is a polynomial ring in one variable over k.* ∎

Exercises 6.7

1. (Bergman [67]) Let $F = k\langle x, y \rangle$ and let C be the centralizer of an element $a \in F \setminus k$. Using the remarks following Corollary 7.4, show that the valuation on $k[a]$ given by the degree in a is totally ramified on C (i.e. it extends to a unique valuation on C with the same residue-class field).
2. (Schur [1904]) Let R be the ring $k[x; 1, \,']$ of differential operators, where $k = F(t)$ is a rational function field (over a field F of characteristic 0) and $'$ denotes differentiation with respect to t. Show that the centralizer of any element outside F is commutative. (*Hint*: Apply Theorem 7.1 to the completion of R by negative powers of x.)
3. (Bergman [67]) Let X^* be the free monoid on $X = \{x_1, \ldots, x_{n-1}\}$ and W the set of infinite words in X. With each $u = a_1 a_2 \ldots$ in W associate the 'decimal expansion' $\lambda(u) = \sum a_i n^{-i}$ and obtain a formula for $\lambda(u^\infty)$ in terms of $\lambda(u)$ and the length of u. Hence express $\lambda(uv)$ as a convex linear combination of $\lambda(u^\infty)$ and $\lambda(v^\infty)$.
4. With every ring R we can associate another ring R^{ab}, the ring R made abelian, which is obtained by dividing R by the ideal generated by all the commutators. Thus the natural mapping $R \to R^{ab}$ is universal for homomorphisms of R into commutative rings. Given any ring R, denote by S the subring generated by the kernel of the natural mapping $R \to R^{ab}$ (i.e. the commutator ideal). For any $a, b \in R$ write $c = [a, b] = ab - ba$ and establish the identity

$$c[ac, bc] = ca[c, bc] + cb[ac, c] + c^4.$$

Deduce that if $R = k\langle X \rangle (|X| > 1)$, then S^{ab} has nilpotent elements other than 0 (and so cannot be a free algebra or a polynomial ring).

5. Show that S^{ab} in Exercise 4 admits no homomorphism onto a non-scalar subalgebra of a polynomial ring. Use this fact to show that the hypothesis that C be finitely generated in Proposition 7.6 cannot be omitted.

6. In a free algebra over a field of characteristic 0, show that if $ab + ba = 0$, then either a or b is 0 (this follows from Theorem 7.7, but give a direct proof).

7. Show that the commutation relation on $F \setminus k$ is transitive, and hence an equivalence. (*Hint*: Use Theorem 7.7 and Corollary 7.4.)

8. (G. M. Bergman; Kolotov [78]) Let $F = k\langle X \rangle$ be a free algebra. Show that any subalgebra generated by two elements y, z of F is either free on y, z or is contained in a 1-generator subalgebra. If $F = k\langle x, y, z\rangle$, then the elements $u = xyxz + xy$, $u' = zxyx + yx$, $v = xyx$ satisfy the relation $uv = vu'$, but there is no 2-generator subalgebra containing u, u', v. (*Hint*: If there is such a subalgebra B, say, it must be a 2-fir, hence an anti-ideal. Use the given relation to show that $xy, yx \in B$ and deduce in turn that $x, y, z \in B$.)

9°. (Bergman [67]) Let F be a free k-algebra and $\hat{F}$ its completion by power series. Given $a \in F$, denote by C, C' its centralizers in $F, \hat{F}$ respectively. Is C' the closure of C in $\hat{F}$?

10. (G. M. Bergman) Given $a, b \in k\langle X\rangle$, different from zero, show that for any monomial u in the support of a, some right multiple of u is in the support of ab. (*Hint*: Take a longest right multiple of u in supp(a) and multiply it by a longest monomial in supp(b).)

11. (Koshevoi [71]) Let R be a k-algebra; a subalgebra is called *pure* if it contains $u \in R$, whenever it contains a non-scalar polynomial in u. Show that the subalgebra of $k\langle x, y, z\rangle$ generated by x, xy, z, yz is a pure subalgebra, but is not a 2-fir. (*Hint*: Use Exercise 10.)

12*. (Dicks [74]) Let b be any element in a free k-algebra F. Show that the idealizer $I(bF)$ is pure, and hence has the centralizer property of Theorem 7.7.

13°. (G. M. Bergman) Given $X \neq \emptyset$, which submonoids S of X^* have the property that kS is pure in $k\langle X\rangle$?

14°. (G. M. Bergman) Which monoids S (not necessarily embeddable in a free monoid) have the property that the centralizers in the monoid ring kS of non-scalar elements all have the form $k[c]$?

15. Let G be the free group on $s_i (i \in I)$ and consider the map $f : kG \to k\langle\!\langle x_i | i \in I \rangle\!\rangle$ given by $s_i \mapsto 1 + x_i$. Show that any multiplicative commutator q maps to a series $1 + \bar{q} + \ldots$, where $\bar{q}$ is the corresponding additive commutator and dots indicate terms of higher order. Deduce that f is injective, and hence show that any two elements of kG either commute or generate a free subalgebra.

6.8 Invariants in free algebras

Let R be a k-algebra and G a group of k-algebra automorphisms of R. An element r of R is said to be an *invariant* of G if $r^g = r$ for all $g \in G$; if $r^g = \lambda_g r (\lambda_g \in k)$ for all $g \in G$, r is called a *relative invariant*. Thus r is a relative invariant precisely when G stabilizes kr.

The set of all invariants of G is a k-subalgebra of R, denoted by R^G and called the *fixed ring*, or *algebra of invariants* of G. The set of relative invariants is not an algebra, but we shall sometimes refer to the subalgebra generated by this set as the algebra of relative invariants. It is conjectured that if R is a free algebra and G a group of k-algebra automorphisms of R, then both these subalgebras are free. We shall deal with some special cases of these conjectures here. We recall that in any filtered ring the filtration will generally be denoted by v or d. To begin with we shall show that the algebra of invariants is a 2-fir; this holds under rather wider conditions:

Proposition 6.8.1. *Let R be a filtered ring with 2-term weak algorithm and let G be a set of endomorphisms of R. Then the fixed ring R^G is a strong E_2-ring, and hence a 2-fir.*

Proof. Let $a, b \in R^G$ be right commensurable and $v(a) \geq v(b)$, say. By Proposition 2.8.1 there exists a unique sequence $x_0, x_1, \ldots, x_n \in R$ (for some $n \geq 1$) such that

$$a = x_0 p(x_1, \ldots, x_n), \quad b = x_0 p(x_1, \ldots, x_{n-1}), \tag{1}$$

where the p's are the continuant polynomials defined in Section 2.7. For any $\alpha \in G$ we have

$$a = a^\alpha = x_0^\alpha p(x_1^\alpha, \ldots, x_n^\alpha), \quad b = b^\alpha = x_0^\alpha p(x_1^\alpha, \ldots, x_{n-1}^\alpha);$$

by uniqueness, $x_0 = x_0^\alpha, \ldots, x_n = x_n^\alpha$, hence $x_i \in R^G (i = 0, 1, \ldots)$. Moreover, by Section 2.7 there exists $U \in E_2(R^G)$ such that

$$(a, b)U = (x_0, 0), \tag{2}$$

hence R^G is a strong E_2-ring, and (2) shows that any 2-generator right ideal is free, of rank 2, 1 or 0, i.e. R is a 2-fir. ∎

This result does not hold for all 2-firs, or even for all firs, as the following example shows. The ring $\mathbb{C}[e^{it}]$ is a principal ideal domain, hence so is its localization $\mathbb{C}[e^{it}, e^{-it}]$; under complex conjugation its fixed ring is $\mathbb{C}[\cos t, \sin t]$, a Dedekind domain that is not principal (see also Exercise 12). On the other hand, the result shows that for any free algebra $R = k\langle X \rangle$ and any group G of automorphisms of R, the fixed algebra R^G is a strong E_2-ring. One would like to assert that R^G satisfies the weak algorithm, but to establish this claim we shall need to assume that the automorphisms respect the grading.

Consider a graded ring $A = \oplus \mathrm{gr}_n A$; any element $a \in \mathrm{gr}_i A$ is said to have the *degree* $d(a) = i$. Since the $\mathrm{gr}_i A$ only meet in 0, the degree is uniquely defined, except for 0, which has all degrees. The graded ring A is said to have the *n-term weak algorithm* if any right linearly dependent set of at most n elements is strongly right linearly dependent, i.e.

6.8 Invariants in free algebras

Given any right linearly dependent sequence $a_1, \ldots, a_m (m \leq n)$, numbered so that $d(a_1) \leq \ldots \leq d(a_m)$, some a_i is a right linear combination of $a_1, \ldots, a_{i-1}$.

If this condition holds for all n, A is said to have the weak algorithm. With any filtered ring $R = \cup R_{(h)}$ we can associate the graded ring A with components $\mathrm{gr}_i A = R_{(i)}/R_{(i-1)}$, and it is easily verified that R has the (n-term) weak algorithm if and only if the associated graded ring does. For graded rings the expected result is easily obtained:

Proposition 6.8.2. *Let $A = \oplus \mathrm{gr}_i A$ be a graded ring with n-term weak algorithm ($n \geq 1$), and G a group of homogeneous automorphisms of A, i.e. $(\mathrm{gr}_i A)G = \mathrm{gr}_i A$ for all i. Then the fixed ring $A^G = \oplus (\mathrm{gr}_i A)^G$ is a graded ring with n-term weak algorithm.*

Proof. Clearly $A^G = \oplus (\mathrm{gr}_i A)^G$ is a graded ring; now consider a homogeneous relation in A^G:

$$a_1 b_1 + \ldots + a_m b_m = 0,$$

where $m \leq n$, arranged so that $v(a_1) \leq \ldots \leq v(a_m)$, where v is the degree-function, and of course, $v(a_j) + v(b_j)$ is independent of j. By the n-term weak algorithm in A there exists j, $1 \leq j \leq m$, such that a_j is right linearly dependent on $a_1, \ldots, a_{j-1}$:

$$a_j = a_1 c_1 + \ldots + a_{j-1} c_{j-1} \quad (c_h \in A).$$

If we choose the least such j, then $a_1, \ldots, a_{j-1}$ are right A-linearly independent. For any $g \in G$ we have $a_v^g = a_v (v = 1, \ldots, n)$, hence

$$a_1(c_1 - c_1^g) + \ldots + a_{j-1}(c_{j-1} - c_{j-1}^g) = 0.$$

By the independence of $a_1, \ldots, a_{j-1}$ we find that $c_v^g = c_v$, so $c_v \in A^G (v = 1, \ldots, j-1)$, and this establishes the n-term weak algorithm for A^G. ∎

In particular, we may take $A = k\langle X \rangle$, where X is graded in any way, and apply Proposition 2.5.3, to obtain a result of Kharchenko [78] and (independently) Lane [76]:

Theorem 6.8.3. *Let $R = k\langle X \rangle$ be a free k-algebra and G a group of k-automorphisms of R. If G is homogeneous with respect to the grading on R induced by some function $d : X \to \mathbb{N}_{>0}$, then the algebra of invariants of G is free, on a set that is homogeneous with respect to d.* ∎

As a special case we have the standard grading: $d(x) = 1$ for all $x \in X$. The same argument will show for any group G of ring automorphisms of $R = k\langle X \rangle$, R^G is a free k^G-algebra.

Let us next look at an example of relative invariants, and calculate the algebra of relative invariants of the group of all linear automorphisms of $R = k\langle X \rangle$, where $X = \{x_1, \ldots, x_r\}$ and char $k = 0$. Regarding R as a graded algebra, we have $R = \oplus \operatorname{gr}_n R$ where $\operatorname{gr}_n R$ is the homogeneous component of degree n, and $G = GL_r(k)$, viewed as the group of all linear automorphisms of R. Denote by R' the algebra generated by the relative invariants. For each $n \geq 1$, the symmetric group Sym_n and G both act on $\operatorname{gr}_n R$, where for $\sigma \in \operatorname{Sym}_n$, $y_1, \ldots, y_n \in X$ we define

$$(y_1 y_2 \ldots y_n)\sigma = y_{1\sigma} y_{2\sigma} \ldots y_{n\sigma},$$

and extend this action by linearity; we shall also say that Sym_n acts by *place-permutations* on $\operatorname{gr}_n R$. It is easy to see that the actions of Sym_n and G commute, so $\operatorname{gr}_n R$ is a $G \times \operatorname{Sym}_n$-module, while R' is a Sym_n-module. Now consider the standard alternating polynomial of degree r (see e.g. FA, Section 7.5):

$$\delta = S(x_1, \ldots, x_r) = \sum_{\sigma \in \operatorname{Sym}_r} \operatorname{sgn} \sigma x_{1\sigma} \ldots x_{r\sigma} \in A_r. \tag{3}$$

For each $g \in G$, $\delta^g = \det g \cdot \delta$, so δ is a relative invariant of G. Therefore R' contains the set

$$\cup_n \{\delta^n \sigma | \sigma \in \operatorname{Sym}_n\}. \tag{4}$$

It is easy to see that this set is closed under multiplication; hence the space spanned by it is a subalgebra of R'. By the representation theory of the symmetric group (see e.g. James and Kerber [81], Chapter 3) this is all of R', at least when $k = \mathbb{Q}$. To find a k-basis, we restrict the σs in (4) to range over the standard Young tableaux with r rows, all of depth n. By the Frame–Robinson–Thrall formula the number of these tableaux is

$$\frac{(nr)!}{\prod_{i=1}^{r} \prod_{j=1}^{n} (i+j-1)}.$$

In fact R' is a free algebra, by Theorem 8.3, as the fixed ring of $SL_r(k)$. To obtain a free generating set we restrict σ in (4) further by deleting the standard Young tableaux constructed by juxtaposing Young tableaux with fewer columns, e.g. where σ fixes jr, for some $j < n$. For example, when $r = 2$, then $\delta = c = [x_1, x_2]$ and the free generating set has the Hilbert series

$$\sum_{n \geq 0} \frac{1}{n+1} \binom{2n}{n} t^{2n+2} = \tfrac{1}{2}(1 - \sqrt{1-4t^2}) = t^2 + t^4 + 2t^6 + \ldots.$$

So there is one generator of degree 2, viz. c, one of degree 4, $x_1 c x_2 - x_2 c x_1$, etc. (for $r > 2$ the series is probably not algebraic). By applying Proposition 2.6.1 we thus find the Hilbert series $2/(1 + \sqrt{1 - 4t^2})$ for the algebra.

The results obtained so far do not indicate when R^G is finitely generated, but for finite groups of linear automorphisms (i.e. automorphisms induced by a linear transformation of the free generators), we have a precise result:

Theorem 6.8.4. *Let G be a finite group of linear automorphisms of a free algebra of finite rank, $R = k\langle x_1, \ldots, x_r \rangle$. Then R^G is finitely generated as a k-algebra if and only if G consists of scalar transformations. In this case G is cyclic, generated by an automorphism*

$$x_i \mapsto \zeta x_i, \quad (i = 1, \ldots, r), \tag{5}$$

for some primitive mth root of 1, ζ, and R^G is generated by the r^m monomials of length m.

Proof. Assume that G is scalar; it is isomorphic to a finite subgroup of $k^\times$, hence cyclic of order m, say, with generator given by (5). It is clear that R^G is freely generated by the r^m monomials of length m.

Conversely, assume that R^G is finitely generated and let $g \in G$; we have to show that g is scalar. For $r = 1$ there is nothing to prove, so we may take $r \geq 2$. Let k^{alg} be an algebraic closure of k and define $R^{\text{alg}} = R \otimes_k k^{\text{alg}}$; then $G \subseteq GL_r(k^{\text{alg}})$ and since k^{alg} has a k-basis, it follows that $(R^{\text{alg}})^G = R^G \otimes k^{\text{alg}}$ and this is finitely generated as k-algebra. We may therefore take k to be algebraically closed. Further we note that $R_{(n)}$ admits G, since the latter acts linearly, so that $R^G_{(n)}$ has a meaning.

On conjugating G by a suitable $U \in GL_r(k)$ we may take g to be in Jordan normal form, say

$$(\lambda_1 x_1 + \varepsilon_1 x_2, \lambda_2 x_2 + \varepsilon_2 x_3, \ldots, \lambda_{r-1} x_{r-1} + \varepsilon_{r-1} x_r, \lambda_r x_r), \tag{6}$$

where $\lambda_i \in k^\times$ and for $1 \leq i \leq r - 1$, either $\varepsilon_i = 0$, or $\varepsilon_i = 1$ and $\lambda_i = \lambda_{i+1}$. Let $R = \cup R_{(n)}$ be filtered by degree, so that $R^G = \cup R^G_{(n)}$. Since R^G is finitely generated, it must be generated by $R^G_{(N)}$ for some N, and on increasing N if necessary, we may assume that $|G|$ divides N.

Denote by A the set of all monomials $\neq 1$ that occur in the supports of elements of $R^G_{(N)}$. Then $R^G_{(N)}$ lies in the k-space spanned by $A \cup \{1\}$, hence R^G lies in the k-algebra generated by A. Taking $\delta = S(x_1, \ldots, x_r)$ again to be the standard polynomial as in (3) and $\sigma \in \text{Sym}_{Nr}$ to act by place-permutations, we find that $\delta^N \sigma$ is fixed by G, because for any $h \in GL_r(k)$,

$$(\delta^N \sigma)h = (\det h)^N \delta^N \sigma;$$

thus when $h \in G$, $(\det h)^N = 1$, by the choice of N, so that $\delta^N \sigma \in R^G$. This shows that every monomial occurring in the support of $\delta^N \sigma$ is a product of elements of A, and these elements are of length at most N. Since σ was arbitrary in Sym_{N_r}, it follows that every place-permutation of $(x_1 x_2 \ldots x_r)^N$ is a product of elements of A, of length at most N, therefore every monomial of length N has a left factor in A. By considering x_1^N we see that

$$x_1^m \in A \quad \text{for some } m \geq 1. \tag{7}$$

Moreover,

$$\text{there is a monomial } w \text{ such that } wx_1, \ldots, wx_r \text{ all lie in } A; \tag{8}$$

for if this were not so, then we could for every $n \geq 1$ construct a monomial of length n with no left factor in A, which is a contradiction.

By (7) there exists an element u of R^G whose support contains x_1^m, and we can take u to be homogeneous and the coefficient of x_1^m to be 1. Thus for some $\mu \in k$ we have

$$u - x_1^m - \mu x_1^{m-1} x_2 \in V,$$

where V is the k-space spanned by all monomials of length m, other than $x_1^m, x_1^{m-1} x_2$. From the form (6) of g it follows that $Vg \subseteq V$, and so

$$u = u^g \in (\lambda_1 x_1 + \varepsilon_1 x_2)^m + \mu(\lambda_1 x_1 + \varepsilon_1 x_2)^{m-1}(\lambda_2 x_2 + \varepsilon_2 x_3) + V,$$
$$= \lambda_1^m x_1^m + \lambda_1^{m-1} \varepsilon_1 x_1^{m-1} x_2 + \mu \lambda_1^{m-1} \lambda_2 x_1^{m-1} x_2 + V.$$

It follows that $\lambda_1^m = 1$, so $\mu = \lambda_1^{m-1} \varepsilon_1 + \mu \lambda_1^{m-1} \lambda_2$, hence $\lambda_1 \mu = \varepsilon_1 + \mu \lambda_2$, $\varepsilon_1 = \mu(\lambda_1 - \lambda_2)$. But we saw that $\varepsilon_1 = 0$ unless $\lambda_1 = \lambda_2$, so $\varepsilon_1 = 0$ in all cases. Since the order of the Jordan block was arbitrary, g must be diagonal, say $g = (\lambda_1 x_1, \ldots, \lambda_r x_r)$. Hence R^G is spanned by A.

If $u \in R^G$, then u is fixed by g. Consider a monomial w as in (8) and suppose that $wg = \lambda w (\lambda \in k^\times)$; then $(wx_i)g = wx_i$ because $wx_i \in A$. But we have $(wx_i)g = \lambda \lambda_i wx_i$, hence $\lambda_i = \lambda^{-1}$ for $i = 1, \ldots, r$, and so g is a scalar. ∎

We can also describe the Hilbert series of the fixed algebra R^G:

Proposition 6.8.5. *Let G be a finite group of linear automorphisms of $R = k \langle x_1, \ldots, x_r \rangle$ over a commutative field k of characteristic 0 (or at least prime to $|G|$). Then the Hilbert series of the fixed ring R^G is given by*

$$H(R^G : k) = \frac{1}{|G|} \sum_{g \in G} (1 - \mathrm{Tr}(g)t)^{-1},$$

where $Tr : GL_r(k) \to k$ is the trace map.

6.8 Invariants in free algebras

Proof. Let $R = \oplus \mathrm{gr}_n R$ be graded by total degree in the xs. Each $\mathrm{gr}_n R$ is a kG-module and the Hilbert series of R^G is

$$H(R^G : k) = \sum \dim_k (\mathrm{gr}_n R)^G \cdot t^n.$$

The element $e = |G|^{-1} \Sigma g \in kG$ is an idempotent such that $R^G = Re$; therefore $\dim_k (\mathrm{gr}_n R)^G = \mathrm{Tr}(e : \mathrm{gr}_n R \to \mathrm{gr}_n R)$. Now $\mathrm{gr}_n R = (\mathrm{gr}_1 R)^{\otimes n}$, where $\otimes n$ denotes the nth tensor power and g acts on $\mathrm{gr}_n R$ as $g^{\otimes n}$, while $\mathrm{Tr}(g^{\otimes n}) = (\mathrm{Tr}\, g)^n$. Therefore

$$\mathrm{Tr}(e : \mathrm{gr}_n R \to \mathrm{gr}_n R) = \tfrac{1}{|G|} \sum_{g \in G} \mathrm{Tr}(g : \mathrm{gr}_n R \to \mathrm{gr}_n R)$$

$$= \tfrac{1}{|G|} \sum_{g \in G} (\mathrm{Tr}\, g)^n.$$

Hence

$$H(R^G : k) = \sum_n \dim_k (\mathrm{gr}_n R)^G \cdot t^n$$

$$= \sum_n \tfrac{1}{|G|} \sum_{g \in G} (\mathrm{Tr}\, g)^n \cdot t^n$$

$$= \tfrac{1}{|G|} \sum_{g \in G} (1 - \mathrm{Tr}(g) t)^{-1}.$$

∎

Corollary 6.8.6. *If a subgroup G of Sym_r acts on $R = k\langle x_1, \ldots, x_r \rangle$ by permutations of the variables, then*

$$H(R^G : k) = \tfrac{1}{|G|} \sum_m \tfrac{\beta_m}{1 - mt},$$

where β_m is the number of elements of G fixing exactly m of the x_i.

Proof. G acts by permuting the monomials of R, so the dimension of $(\mathrm{gr}_n R)^G$ does not depend on the characteristic of k. In fact a basis of $(\mathrm{gr}_n R)^G$ is given by the elements $\Sigma \{wg | g \in G\}$, where w runs over all monomials of length n. We may therefore take char k to be 0. By Proposition 8.5,

$$H(R^G : k) = \tfrac{1}{|G|} \sum (1 - \mathrm{Tr}(g) t)^{-1},$$

and here $\mathrm{Tr}(g)$ is just the number of xs fixed by g, so there are β_m summands $(1 - mt)^{-1}$, for $m = 0, 1, \ldots$, hence the result. ∎

We remark that by Theorem 8.3 the fixed algebra is free, and a free generating set may be written down explicitly.

Exercises 6.8

1. Find all the invariants and relative invariants of $k[x]$ under the group of all automorphisms.
2. Let A be a graded ring with weak algorithm and G a group of homogeneous automorphisms of R. Find all the relative invariants of R.
3. Let $R = k\langle X \rangle$ be a free algebra of finite rank and G a finite group acting on X. Show that R has a basis of elements $\sum \{wg | g \in G\}$, where w ranges over all monomials that (relative to the lexicographical ordering) are maximal in their orbits. Calling these monomials G-*maximal*, show that the set of all G-maximal monomials is a free monoid, and the indecomposable ones form a prefix code, which is a free generating set of R^G.
4*. (Bergman and Cohn [69]) Let $R = k\langle X \rangle$ and let G be a group acting on X with finite orbits and without fixed points. Show that R^G is a free algebra. If $|X| = r$ and G is precisely doubly transitive on X, so that $|G| = r(r-1)$ and each element of G other than 1 fixes at most one point of X, show that the number of orbits of length n in X^* is $(r-1)^{n-2}(n \geq 2)$, while there is one orbit of length 1. Show also that if γ_n denotes the number of generators of length n, then $\gamma_n = (r-2)(\gamma_{n-1} + \gamma_{n-2})$.
5*. (Dicks and Formanek [82]) For R and G as in Exercise 4 find $H(R^G : k)$ when $|G| = p =$ char k.
6*. (P. Webb) For R and G as in Exercise 4 find $H(R^G : k)$ when $|G| = p^n$, $p =$ char k.
7°. Is $H(R^G : k)$ always rational, for $G \subseteq GL_d(k)$? (From the text and Exercise 6 this is so when char $k = 0$ or when char $k = p \neq 0$ and G is a p-group.)
8. (Almkvist, Dicks and Formanek [85]) Show that $H(R^G : k)$ is rational whenever G has a cyclic Sylow p-subgroup, where $p =$ char k.
9. (Almkvist, Dicks and Formanek [85]) Let G be a compact subgroup of $GL_d(\mathbb{C})$. Show that if $R = \mathbb{C}\langle x_1, \ldots, x_d \rangle$, then

$$H(R^G : \mathbb{C}) = \int_G (1 - t.Tr)^{-1} \quad \text{for } |t| < d^{-1},$$

where $\int_G$ is the normalized Haar measure.
10. (D. R. Lane) Let $R = k\langle x_1, x_2 \rangle$ and let G be the cyclic group generated by an automorphism of the form $(x_1 + p, x_2)$, where $p \in k[x_2]^\times$. Show that $c = pax_1 - x_1 ap \in R^G$ for all $a \in R^G$ and that R^G is the smallest k-algebra with this property containing $k\langle c, x_2 \rangle$. Find a free generating set for R^G.
11. (G. M. Bergman) Let R be a ring, S a subring of R and X a subset of S such that the right ideal XR is R-projective, with dual R-basis $\{\alpha_x : XR \to R | x \in X\}$. Show that if there is an S-bimodule map $\pi : R \to S$ fixing S, then $\{\alpha_x \pi : XS \to S | x \in X\}$ is a dual S-basis for the right ideal XS. Deduce that if S is a subring of a ring R and is a direct summand of R as S-bimodule, then if R is right (semi-)hereditary, so is S.
12. (G. M. Bergman) Let R be a ring and G a group of automorphisms of R such that the orbit rG of any $r \in R$ is finite, of order invertible in R. Show that the averaging map $\pi : R \to R^G$ given by $r\pi = |rG|^{-1} \sum rg$ is an R^G-bimodule map fixing R^G. Deduce that if R is right (semi-)hereditary, then so is R^G.

13. (A. H. Schofield) Let $R = k\langle x_1, x_2\rangle$, where char $k = 0$. If $u = (x_1 + p, x_2 + \alpha)$, where $p \in k[x_2]^\times$, $\alpha \in k^\times$, show that there exists $q \in k[x_2]$ such that $q(x) - q(x + \alpha) = p(x)$ and deduce that u is conjugate to $(x_1, x_2 + \alpha)$.
14°. Show that for any group of automorphisms of the free algebra $k\langle X\rangle$, the ring of all invariants is free.
15°. Does the analogue of Proposition 8.1 hold with '2-term weak algorithm' replaced by 'n-term weak algorithm' and '2-fir' replaced by 'n-fir', e.g. for $n = 3$?
16°. Find an analogue of Proposition 8.5 when $|G|$ is divisible by char k.

6.9 Galois theory of free algebras

Our aim in this section is to establish a Galois correspondence for free algebras. However, it will be instructive to prove the result in a more general setting, to bring out more clearly what special properties of $k\langle X\rangle$ are needed.

Throughout this section R will be an integral domain and End(R) the ring of all endomorphisms of R, as abelian group, and these endomorphisms will be thought of as maps of R, written on the right, with composition $x.y$. We recall that with operators on the right, in any operator equation an injective operator is left regular and a surjective operator is right regular. For each $r \in R$ we identify r with the element of End(R) given by right multiplication by r and thus view R as a subring of End(R). Similarly we regard the opposite ring R^o as a subring of End(R) by identifying the element r^o corresponding to $r \in R$ with the map consisting of left multiplication by r. Thus for $r, s \in R$ we have

$$r.s = rs, \quad r^o.s^o = (sr)^o, \quad r.s^o = s^o.r, \quad rs^o = sr. \tag{1}$$

Moreover, $r = r^o$ if and only if r lies in the centre of R. The group Aut(R) of ring automorphisms is a subgroup of $U(\text{End}(R))$, and it will now be convenient to denote the effect of $g \in \text{End}(R)$ on R by rg rather than r^g and the expression rsg will mean $r(sg)$, not $(rs)g$. We shall be concerned with elements of the form

$$f = \sum g_i.a_i^o.b_i, \quad \text{where } g_i \in \text{Aut}(R), a_i, b_i \in R.$$

Such an element will be called an *R-trace form*; its effect on R is

$$rf = \sum a_i(rg_i)b_i.$$

We also note the rules

$$r.g = g.rg, \quad r^o.g = g.(rg)^o \quad (r \in R, g \in \text{Aut}(R)),$$

which follow from the formulae $(sr)g = (sg)(rg)$, $(rs)g = (rg)(sg)$.

An element g of Aut(R) is said to *X-inner* on a subring S of R, after Kharchenko (= Харченко) if there exist $a, b \in R^\times$ such that $a.g.b^o - a^o.b$

vanishes on S, i.e.
$$b(sg)(ag) = asb \quad \text{for all } s \in S. \tag{2}$$

For $s = 1$ this reduces to $b.ag = ab$; this yields ag and now (2) shows sg to be determined entirely by a and b. Hence if T is an extension ring of R containing $u = a^{-1}b$ and $u^{-1} = b^{-1}a$, we have $sg = u^{-1}su$ in T, so that g is actually induced by an inner automorphism of a certain extension ring T (in fact the Martindale ring of quotients, see Montgomery and Passman [84]); however, the definition does not presuppose the existence of T. If g is X-inner on all of R, we call it X-*inner*. From $b.ag = ab$ we obtain $b^\circ.a^\circ = (ag)^\circ.b^\circ$, and by (1) and the subsequent observations, we have

$$(ag)^\circ.b.b^\circ = b^\circ.a^\circ.b = b^\circ.a.g.b^\circ = b^\circ.g.ag.b^\circ.$$

Since R is a domain, we can cancel b° on the right, and writing $a = a'g^{-1}$, we obtain

$$b^\circ.g.a' = a'^\circ.b, \quad \text{where } a', b \in R^\times.$$

This shows that the property of being X-inner is left–right symmetric. Moreover, since R is closed under conjugation by ab^{-1}, R contains $a' = (ab^{-1})^{-1}a(ab^{-1}) = bab^{-1}$, whence $ba = a'b$, and further, $ab^{-1} = b^{-1}a'$. With the help of this relation it is easily seen that the set of X-inner automorphisms is closed under composition. An automorphism that is not X-inner is called X-*outer*, and a subgroup G of $\text{Aut}(R)$ is called X-*outer* if the only X-inner element of G is 1.

We shall recall the results of Kharchenko's Galois theory, essentially following Montgomery and Passman [84]. We remark that most of these results were actually proved for semiprime rings by Kharchenko [77] and for prime rings by Montgomery and Passman [84]; the restriction to integral domains (which is all we shall need) simplifies many of the arguments.

We leave to the reader the verification that the X-inner automorphisms form a subgroup of $\text{Aut}(R)$. Our first object is an analogue of Dedekind's lemma on the independence of automorphisms; this is prefaced by two lemmas.

Lemma 6.9.1. *Let R be an integral domain and S a subring; further, let $\alpha = \sum g_i.a_i^\circ.b_i$ be an R-trace form and β an element of $R.\alpha.S$, say $\beta = \sum_{j=1}^m r_j.\alpha.s_j$ $(r_j \in R, s_j \in S)$ Then we have $R\beta \subseteq (R\alpha)S$ and*

$$\beta = \sum g_i.a_i^\circ.c_i \quad \text{where } c_i = \sum_{j=1}^m (r_jg_i)b_is_j.$$

Proof. We have $R\beta \subseteq R(R.\alpha.S) \subseteq (R\alpha)S$; now the rest is straightforward. ∎

6.9 Galois theory of free algebras

Lemma 6.9.2. *Let R be an integral domain and S a subring such that $RaS \cap S \neq 0$ for all $a \in R^\times$. Suppose that there is an R-trace form $\alpha = \sum g_i.a_i^o.b_i$ in which not all the b_i are zero. Then there exists a trace form*

$$\beta = \sum g_i.a_i^o.c_i \in R.\alpha.S,$$

such that not all the c_i are zero, and for all i, j such that $c_i, c_j \neq 0$, the automorphism $g_j^{-1}g_i$ is X-inner on S.

Proof. Using Lemma 9.1 and the fact that $Rb_1 S \cap S \neq 0$, we may replace α by an element of $R.\alpha.S$, chosen so as to ensure that $b_1 \in S^\times$. Fix $s \in S^\times$; then we have

$$\alpha.sb_1 - (b_1s)g_1^{-1}.\alpha = \sum g_i.a_i^o.c_i,$$

where $c_i = b_i s b_1 - ((b_1s)g_1^{-1}g_i)b_i$, by Lemma 9.1 with a minimal number of terms. We have $c_1 = 0$, and so $c_i = 0$ for all i, i.e. $b_i^o.b_1 - b_1^o.g_1^{-1}g_i.b_i$ vanishes on S; hence $g_1^{-1}g_i$ is X-inner on S, and so is $g_j^{-1}g_i = (g_1^{-1}g_j)^{-1}g_1^{-1}g_i$. ∎

We shall apply these results as follows.

Proposition 6.9.3. *Let R be an integral domain, G an X-outer subgroup of $\mathrm{Aut}(R)$ and let $\alpha = \sum_1^n g_i.a_i^o.b_i (a_i, b_i \in R, g_i \in G)$ be an R-trace form. Then*

(i) *$R.\alpha.R$ contains an element of the form $g.(\Sigma a_i^o.c_i)$, where the c_i are not all 0 and the sum is over those i for which $g_i = g$,*

(ii) *if $\alpha = 0$ and $g_1.a_1^o, \ldots, g_n.a_n^o$ is a minimal right R-dependent family, then all the g_i are equal, and*

(iii) *if the g_i are distinct and the a_i, b_i are non-zero, then $a.\alpha \neq 0$ for all $a \in R^\times$.*

Proof. We may assume that the sum for α is non-empty and all the b_i are non-zero. By Lemma 9.1 (with $S = R$), any $\beta \in R.\alpha.R$ is of the form $\beta = \sum g_i.a_i^o.c_i$. If we choose a β with the least positive number of non-zero c_i and apply Lemma 9.2, we find that β has the form required in (i), because G is X-outer.

If $\alpha = 0$, then $\beta = 0$ and (ii) follows by minimality. Finally, suppose that $g_1, \ldots, g_n$ are distinct and the a_i are non-zero. By (ii) a minimal right R-dependent subfamily of $g_1.a_1^o, \ldots, g_n.a_n^o$ must consist of a single element, but $g_i.a_i^o.b_i \neq 0$ if $b_i \neq 0$, so any one-element subfamily is right R-independent. It follows that $g_1.a_1^o, \ldots, g_n.a_n^o$ are right R-independent, hence for any $a, b_1, \ldots, b_n \in R^\times$,

$$a.\alpha = \Sigma g_i.a_i^o.(ag_ib_i) \neq 0.$$

∎

For a better understanding of the relation between X-inner automorphisms and the ring elements that determine them, we make the following observation.

Lemma 6.9.4. *Let R be an integral domain and $g \in Aut(R)$.*

(i) *For $a, b, c, d \in R^\times$, consider the following four conditions:*
(a) $a^\circ.b = g.c^\circ.d$, *i.e.*
$$arb = crgd \quad for\ all\ r \in R, \tag{3}$$

(b) $a^\circ.c = a.g.c^\circ$, *i.e.*
$$arc = crg\,ag \quad for\ all\ r \in R, \tag{4}$$

(c) $b.d^\circ = b^\circ.g.d$, *i.e.*
$$drb = bg\,rgd \quad for\ all\ r \in R, \tag{5}$$

(d)
$$ab = cd. \tag{6}$$

Then (a) $\Leftrightarrow$ ((b)&(d)) $\Leftrightarrow$ ((c)&(d)). In particular (a) implies that g is X-inner.

(ii) *For any $p, q, w \in R^\times$, the following conditions are equivalent:*
(e) *The elements $a = p, c = q$ satisfy (b), i.e.*
$$prq = q\,rg\,pg \quad for\ all\ r \in R, \tag{7}$$

(f) *The elements $a = wp, c = wq$ satisfy (b), i.e.*
$$wprwq = wqrg(wp)g \quad for\ all\ r \in R, \tag{8}$$

(g) *The elements $a = pw, c = qwg$ satisfy (b), i.e.*
$$pwrqwg = q\,wg\,rg(pw)g \quad for\ all\ r \in R. \tag{9}$$

Proof. (i) Assume that (a) holds. Then (d) follows by taking $r = 1$. If we now apply (a) to the product $arab$, first replacing r by ra, and then replacing ab by cd, we obtain $c(ra)gd = arcd$. Cancelling d, we have $arc = c(rg)(ag)$, i.e. (b). For the converse, we assume (b) and note that (a) is trivial for $r = 0$, while for $r \neq 0$, we have the equivalences

$$arb = crgd$$
$$\Leftrightarrow arbc = crg\,dc \qquad (\text{right multiply by } c)$$
$$\Leftrightarrow crg\,bg\,ag = crg\,dc \qquad (\text{by}(b))$$
$$\Leftrightarrow bg\,ag = dc. \qquad (\text{left cancel } crg)$$

For $r = 1$ the first line reduces to (d), so assuming (b), (d), the last line, which is independent of r, holds. Hence the first line holds for all r, and (a) follows.

The equivalence (a) ⇔ (c) & (d) follows by symmetry.

(ii) Assume (e), i.e. $prq = qrg\,pg$. Hence $wprq = wq\,rg\,dg$. This is of the form (a) and applying (a) ⇒ (b) we get (f). Conversely, assuming (f): $wp\,rwq = wq\,rg(wp)g$, we cancel w on the left to obtain $pr\,wq = q\,rg(wp)g$; applying (a) ⇒ (b), we find that $prq = q\,rg\,pg$, which is (e). Now (e) ⇒ (b) is proved similarly, using right instead of left multiplication and cancellation. ∎

Proposition 6.9.5. *Let R be an integral domain, G a finite X-outer subgroup of Aut(R) and $a \in R^\times$. Then $R^G aR$ and RaR^G contain non-zero ideals of R and every non-zero ideal of R meets R^G non-trivially.*

Proof. Put $\alpha = \sum g$; then $R\alpha \subseteq R^G$, because $\alpha h = \alpha$ for all $h \in G$. By Proposition 9.3 (i), $R.(\alpha.a).R$ contains an element of the form $\beta = g.c$ for some $c \in R^\times$; therefore

$$Rc = (Rg)c = R\beta \subseteq (R\alpha)aR \subseteq R^G aR. \qquad (10)$$

It follows that $RcR \subseteq R^G aR$ is a non-zero ideal of R contained in $R^G aR$. For any non-zero ideal I choose

$$0 \neq b \in \bigcap_{g \in G} Ig,$$

e.g. $b = (cg_1)\ldots(cg_n)$, where $G = \{g_1, \ldots, g_n\}$; then for all $g \in G$, $bg \in I \subseteq R^G aR$. By Proposition 9.3 (iii), $b.\alpha \neq 0$, hence

$$0 \neq Rb.\alpha \subseteq R^G \cap \sum Rbg \subseteq R^G \cap I.$$

Thus I meets R^G; by symmetry the same result holds for RaR^G. ∎

We shall want to know under what circumstances an X-inner map reduces to the identity; it is convenient to state the conditions more generally for inclusion maps.

Lemma 6.9.6. *Let R be an integral domain, G a finite X-outer subgroup of Aut(R), S an R^G-subring of R and $\sigma : S \to R$ an R^G-ring homomorphism. If there exist $a, b, c, d \in R^\times$ such that*

$$asb = c(s\sigma)d \quad \text{for all } s \in S, \qquad (11)$$

then σ is the inclusion map. In particular, the fixed ring of an X-inner automorphism $\neq 1$ of R cannot contain the fixed ring of a finite group of X-outer automorphisms.

Proof. Given $r \in R$, we have $s = \sum rg \in R^G \subseteq S$, so we have $s\sigma = s$, because σ fixes R^G, and hence $asb = csd$. Thus for all $r \in R$, $\sum a(rg)b =$

$\sum c(rg)d$, i.e.
$$\sum (g.a^\circ.b - g.c^\circ.d) = 0.$$

By Proposition 9.3 (ii), $g.a^\circ$ and $g.c^\circ$ are right R-dependent, for some $g \in G$, say
$$g.a^\circ.e = g.c^\circ.f, \quad \text{where } e, f \in R^\times.$$

This is of the form (3) of Lemma 9.5, with $g = 1$ and b, d replaced by e, f. Hence it implies Equation (4) of Lemma 9.4 (with $g = 1$), i.e. $arc = cra$. Moreover, on putting $s = 1$ in (11), we obtain $ab = cd$, i.e. (6), and so, by Lemma 9.4, we have (3) (with $g = 1$), hence for $s \in S$, $asb = csd$; therefore $s\sigma = s$, which shows σ to be induced by the inclusion map. Since g is a unit in End(R), we have $a^\circ.e = c^\circ.f$, and so
$$\begin{aligned} b.a^\circ.e &= b.c^\circ.f \\ &= c^\circ.bf = c^\circ.ae \\ &= c^\circ.a.e. \end{aligned}$$

Cancelling e, we obtain $b.a^\circ = c^\circ.a$. Hence, for all $s \in S$, we have $c(s\sigma)d = asb = csa$. Putting $s = 1$ we find that $a = d$ and on cancelling a, $\sigma = 1$, so σ is reduced to the inclusion map. To prove the final assertion we take $S = R$, let σ be an X-inner automorphism and for the given equation take an equation of the form (4) defining σ. ∎

These results may be used to derive analogues of the fundamental theorem of Galois theory for fields; we shall limit ourselves here to what is actually needed later.

Proposition 6.9.7. *Let R be an integral domain, G a finite X-outer subgroup of Aut(R) and S an R^G-subring of R.*

(i) If H is the subgroup of G fixing S elementwise, so that
$$R^G \subseteq S \subseteq R^H,$$
then S contains a non-zero ideal of R^H.

(ii) Any injective R^G-homomorphism $\sigma : S \to R$ is the restriction to S of an element of G (unique up to left multiplication by an element of H).

Proof. Let $g_1, \ldots, g_n$ be a left transversal of H in G, so that G is the disjoint union of the cosets $g_i H$. Put $\alpha = \sum g_i$, $\beta = \sum_{h \in H} h$; then $\alpha.\beta = \sum g$. For any $i \neq j$, $g_i^{-1}g_j \notin H$, so $g_i^{-1}g_j$ does not restrict to the inclusion map on S, by the choice of H, and by Lemma 9.6, $g_i^{-1}g_j$ is not X-inner on S. We now apply

6.9 Galois theory of free algebras

Lemma 9.2 (with the help of Proposition 9.5) to conclude that $R.\alpha.S$ contains an element

$$\sum r_j.\alpha.s_j = g.c, \quad \text{where } c \in R^\times \text{ and } g = g_i \text{ for some } i.$$

Since $RcS \cap S \neq 0$, we may assume that $c \in S^\times$, and on replacing α by $g^{-1}\alpha$, we may take $g = 1$. By Proposition 9.3(iii), $c.\beta \neq 0$ and β centralizes R^H and $(R^H)^\circ$, i.e. it is a homomorphism of R^H-bimodules, so

$$0 \neq c.\beta = \sum r_j.\alpha.s_j.\beta$$
$$= \sum r_j.\alpha\beta.s_j, \qquad (12)$$

therefore

$$0 \neq R(c.\beta) \subseteq (R(\alpha\beta))S \subseteq R^G S \subseteq R^H.$$

Since $c.\beta$ is a left R^H-module homomorphism, $R(c.\beta)$ is a non-zero left ideal of R^H contained in S. By symmetry S also contains a non-zero right ideal ϕR and so $S \supseteq R(\beta.c)\phi R$, i.e. S contains a non-zero ideal of R^H and (i) is proved.

Let $\sigma : S \to R$ be an R^G-ring monomorphism. Since the R-trace form described in (12) takes values in S, we may multiply (12) on the right by $s.\sigma$, for any $s \in S$; this and the fact that $c.\beta = \beta.c$, yields

$$\beta.c.s.\sigma = c.\beta.s.\sigma$$
$$= \sum r_j.\alpha\beta.s_j.s\sigma$$
$$= \sum r_j.\alpha\beta\sigma.s_j.s\sigma$$
$$= \sum r_j.\alpha\beta.(s_j\sigma).(s\sigma),$$

because $R(\alpha\beta) \subseteq R^G$ and σ fixes R^G elementwise. Since $\alpha\beta = \sum g$, we have

$$\beta.c.s.\sigma = \sum g.b_g.s\sigma, \qquad (13)$$

where $b_g = \sum (r_j g)(s_j \sigma)$ is independent of s. Since $\beta \neq 0$ and σ is injective, the expression (13) is non-zero, hence $b_{g'} \neq 0$ for some $g' \in G$. Replacing s by sc, we see that

$$\sum g.b_g.(sc)\sigma = \beta.c.sc.\sigma = cs.\beta.c.\sigma$$
$$= cs.\sum g.b_g = \sum g.(cs)g.b_g.$$

By Proposition 9.3(iii), applied to the difference between the initial and final summations above, we obtain $b_g(sc)\sigma = (cs)gb_g$ for all $g \in G$. Applying g'^{-1}, we find

$$(b_{g'}g'^{-1})(s\sigma g'^{-1})(c\sigma g'^{-1}) = cs(b_{g'}g'^{-1}).$$

By Lemma 9.6, $\sigma g'^{-1}$ is the inclusion map on S, and this means that σ extends to g'. ∎

We can now describe Kharchenko's form of the Galois correspondence between groups and rings.

Theorem 6.9.8. *Let R be an integral domain and G a finite X-outer subgroup of Aut(R). Then the following conditions on an R^G-subring S of R are equivalent:*

(a) $S = R^H$, where H is the subgroup of G fixing S elementwise,
(b) S is an anti-ideal of R,
(c) for all $a \in S^\times, r \in R$, if $ar \in S$, then $r \in S$,
(d) for any subring T of R, such that $S \subset T \subseteq R$, S contains no non-zero ideal of T.

Moreover, each R^G-automorphism of S extends to an element of G, and there is a natural bijection between the subgroups H of G and the R^G-subrings of R that are anti-ideals in R, given by

$$H \leftrightarrow R^H. \tag{14}$$

Proof. We begin by proving the last part. Let H be a subgroup of G. Given $r \in R$, if $ar \in R^H$ for some non-zero $a \in R^H$, then for all $h \in H$, $ar = (ar)h = a(rh)$, so $rh = r$ and $r \in R^H$. This shows R^H to be an anti-ideal in R (in fact, it satisfies a stronger, one-sided condition). If H, K are subgroups of G and $R^H = R^K$, then each $h \in H$ fixes R^K, so by Proposition 9.7(ii) (with $S = R, G = K$) h extends to an element of K, i.e. $h \in K$, and thus $H \subseteq K$; by symmetry, $H = K$, so the correspondence (14) is injective.

Now consider any R^G-subring S of R that is an anti-ideal in R. Let H be the subgroup of G fixing S elementwise, so that $S \subseteq R^H$; we shall prove that equality holds. By Proposition 9.7(i), S contains a non-zero ideal I of R^H, but no proper subring does, by Corollary 5.3. Hence $R^H = S$, as claimed. This establishes the bijection and the rest follows by Proposition 9.7(ii). Now the equivalence of (a)-(d) follows from what has been said and the properties of anti-ideals. ∎

To apply this result we need to determine the X-outer automorphisms and anti-ideals in 2-firs. We first show how to get a supply of X-outer automorphisms.

Lemma 6.9.9. *If R is a 2-fir all of whose invariant elements are central, then Aut(R) is X-outer.*

Proof. Suppose that $g \in$ Aut(R) is X-inner, say

$$a^\circ.b = a.g.b^\circ = g.ag.b^\circ. \tag{15}$$

We must show that $g = 1$. Applying this relation to 1, we get $0 \neq ab = b(ag) \in aR \cap bR$, hence
$$aR + bR = dR, \quad a = da_1, b = db_1.$$
Now (15) becomes $a_1^o.d^o.d.b_1 = g.(da_1)g.b_1^o.d^o$, i.e.
$$da_1.r.db_1 = db_1.rg.(da_1)g; \tag{16}$$
by Lemma 9.4 (f) $\Rightarrow$ (e) it follows that we may take $d = 1$, hence $a_1^o.b_1 = a_1.g.b_1^o$. We now have $aR + bR = R$ and hence
$$Rb = aRb + bRb = b(Rg)(ag) + bRb = bR(ag) + bRb$$
$$= b(R(ag) + Rb) = bR.$$
by Lemma 9.4, (g) $\Rightarrow$ (e) we can get rid of any common right factors and so find that $1 \in R(ag) + Rb$. Thus b is invariant, and so by hypothesis, $b^o = b$; by symmetry, $a^o = a$. Cancelling b from (15), we have $a = a^o = a.g$, hence $g = 1$, as claimed. ∎

We recall from Section 6.6 that if S is a 2-fir, unit-closed in R (i.e. $U(R) \cap S = U(S)$), then S is an anti-ideal (Lemma 6.8). As a consequence we obtain

Theorem 6.9.10. *Let R be a filtered K-ring with 2-term weak algorithm, where $K = R_{(o)}$, and whose invariant elements are central. If G is a finite group of K-automorphisms of R, then there is a natural bijection between subgroups of G and R^G-subrings of R that are 2-firs.*

Proof. By Lemma 9.9, G is X-outer; the above remark shows that each K-subring of R that is a 2-fir is an anti-ideal and by Proposition 8.1 each fixed ring is a 2-fir. ∎

For free algebras we deduce the following result:

Theorem 6.9.11. *Let R be a free k-algebra and G a finite group of k-algebra automorphisms of R that are homogeneous with respect to some grading of R for which the weak algorithm holds (e.g. linear automorphisms). Then there is a natural bijection between the subgroups of G and the free subalgebras of R containing R^G.*

Proof. This is an immediate consequence of Theorems 9.10 and 8.3. ∎

Exercises 6.9

1. Prove the results of this section for prime rings.
2. Show that for an integral domain R the X-inner automorphisms form a normal subgroup of $\text{Aut}(R)$.

3. (G. M. Bergman) Let R be an integral domain, $a, b, c, d \in R^\times$, and g a permutation of R fixing 1. Show that if g satisfies the condition of Lemma 9.4 (a), then it is an automorphism of the additive group of R such that $(rs)g = rg.sg$ for all $r \in Ra, s \in R$. Are there similar statements involving b, c, d? (*Hint*: Put $r = ta$, transform *atasb* in two ways and cancel d on the right.)
4. (G. M. Bergman) Continuing Exercise 3, deduce (still assuming Lemma 9.4 (a)), that if Ra and Rc are comparable, as well as bR and dR, then g is a ring automorphism of R.
5°. (G. M. Bergman) With R, a, b, c, d, g as in Exercise 3, how much of Lemma 9.4 remains true under these more general conditions?
6*. (G. M. Bergman) Find an example of an integral domain R, $a, c \in R^\times$ and a permutation g of R satisfying $1g = 1$ and the identities $arc = crg$ $ag = c(ra)g$, such that g is not a ring automorphism of R. (*Hint*: Let S be the mixed free monoid on x, y, z, z^{-1} and M a submonoid of S containing no non-zero power of z. On M define a binary operation $*$ by putting $p * q = pq$ unless $p = axz^i, q = z^j xb$, in which case $p * q = p\bar{q}$, where $\bar{q} = z^{-j}xb$. Verify that this operation is associative and has the 'unique product property', i.e. for any finite subsets U, V of M there exists an element that can be written as $u * v$ for just one $u \in U$ and one $v \in V$. Deduce that the monoid algebra kM is an integral domain and take g to be conjugation by z (an automorphism of M but not of $(M, *)$). Verify the desired identities with $a = yz, c = y$, but show that $(x * x)g \neq xg * xg$.)

6.10 Automorphisms of free algebras

We now turn to examine the structure of the automorphism group of the free algebra. So far the only satisfactory results have been found in rank 2; thus we shall be studying the automorphisms of $k\langle x, y \rangle$. The case of rank 1 is of course well known (and rather trivial): the only automorphisms of $k[x]$ are the affine transformations $x \mapsto ax + b$, where $a, b \in k, a \neq 0$.

More generally, we consider the algebra $R = k\langle X \rangle$, where k is a field and X is any finite set, and define some types of automorphisms that are frequently encountered.

1. The mapping

$$\tau_a : x \mapsto x + a_x, \quad \text{where } x \in X, a_x \in k, \tag{1}$$

defines an automorphism of R, called a *translation*. The group of translations will be denoted by T.

2. Any automorphism of the k-space kX with basis X uniquely defines an automorphism of R:

$$\alpha : x \mapsto \Sigma \alpha_{xy} y, \quad \text{where } \alpha_{xy} \in k, (\alpha_{xy}) \text{ invertible}. \tag{2}$$

These are just the automorphisms leaving kX invariant and are called *linear*.

6.10 Automorphisms of free algebras

3. An automorphism leaving $k + kX$ invariant (and leaving k fixed) is called *affine*; such an automorphism has the form

$$x \mapsto \sum \alpha_{xy} y + \beta_x, \quad \text{where } \alpha_{xy}, \beta_x \in k, (\alpha_{xy}) \text{ invertible}. \tag{3}$$

The group of all affine automorphisms of R will be denoted by A in this section.

4. Let x_0 be a specified element of X and write $X_0 = X \setminus \{x_0\}$. For any $f \in k\langle X_0 \rangle$ there is an automorphism of R sending x_0 to $x_0 + f$ and fixing X_0; it is called an x_0-*based shear*. More generally, an automorphism of R leaving $k\langle X_0 \rangle$ fixed is called a *triangular automorphism;* in such an automorphism $x_0 \mapsto \lambda x_0 + f$, where $\lambda \in k^\times$, $f \in k\langle X_0 \rangle$. The group of these automorphisms will be denoted by Δ.

We remark that all of these types can equally well be defined for the polynomial ring $k[X]$; here the triangular automorphisms are known as *de Jonquières automorphisms*.

Let

$$\alpha : x \mapsto f_x \quad (x \in X) \tag{4}$$

be any endomorphism of R (i.e. a k-linear ring endomorphism). If $f_x(0) = 0$ for all $x \in X$, then α is called *centred* or *augmentation preserving*; in fact it preserves the augmentation ideal XR. If α is given by (4) and τ_a is the translation defined by (1), then

$$\tau_a \alpha : x \mapsto f_x + a_x,$$

and for a suitable choice of $a \in k^X$, viz. $a_x = -f_x(0)$, we can ensure that $\tau_a \alpha$ is centred. This remark is sometimes used to effect a reduction to centred automorphisms, as in the proof of our first result:

Proposition 6.10.1. *Any surjective endomorphism of $k\langle X \rangle$, where X is finite, is an automorphism.*

Proof. Let ϕ be a surjective endomorphism; by composing it with a translation we may take it to be centred. Suppose that ϕ is not injective and let $0 \neq w \in \ker \phi$. Put $d = d(w)$ and denote by $\mathfrak{n}$ the set of all elements of $R = k\langle X \rangle$ of order greater than d. Clearly $\mathfrak{n}$ is an ideal and $\mathfrak{n}\phi \subseteq \mathfrak{n}$, because ϕ is centred. Therefore ϕ induces an endomorphism $\phi_1 : R/\mathfrak{n} \to R/\mathfrak{n}$, which like ϕ is surjective. Since X is finite, $R/\mathfrak{n}$ is finite-dimensional over k, and so ϕ_1 is an automorphism, but $w\phi_1 = w\phi = 0$, a contradiction; hence ϕ is injective and so is an automorphism. ∎

A corresponding result holds for $k[X]$. We remark that since abelianization is a functor, the natural mapping

$$\operatorname{Aut} k\langle X \rangle \to \operatorname{Aut} k[X]$$

is a homomorphism. In fact it can be shown that for $|X| \leq 2$ it is an isomorphism.

The following definition is basic for much of what follows. An automorphism of $k\langle X\rangle$ or $k[X]$ is called *tame* if it can be obtained by composing affine automorphisms and shears; all other automorphisms are called *wild*. The following are some examples of automorphisms not known to be tame. The automorphism $x \mapsto f_x$ will also be written $\{f_x\}$.

1. (D. J. Anick) Let $X = \{x, y, z\}$: for any $p \in k\langle X\rangle$ the endomorphism $\{x + yp, y, z + py\}$ fixes $xy - yz$. So for $p = xy - yz$ we get an automorphism with inverse $\{x - yp, y, z - py\}$.

2. The automorphism of 1. has tame image in $k[X]$, but the following example (due to Nagata) is not known to be tame. Let $X = \{x, y, z\}$; for any $p \in k[X]$, the endomorphism $(x + zp, y + 2xp + zp^2, z)$ fixes $x^2 - yz$, so for $p = x^2 - yz$ we get an automorphism with inverse $(x - zp, y - 2xp + zp^2, z)$.

3. (M. Nagata, D. J. Anick) $X = \{w, x, y, z\}$, $R = k\langle X\rangle$ or $k[X]$. For any $p \in R$, the endomorphism $(w, x + pz, y + wp, z)$ fixes $wx - yz$, so it is an automorphism for $p = wx - yz$. If we replace k by $k[w, z]$, then the same formula gives wild automorphisms of $k[w, z]\langle x, y\rangle$ and $k[w, x, y, z]$.

Our main objective will be to show that every automorphism of $k\langle x, y\rangle$ is tame; we shall do this by presenting the automorphism group of $k\langle x, y\rangle$ as the free product of A and Δ, amalgamating their intersection. As a preparation we recall the definition of a free product with amalgamation. Let G_1 and G_2 be two groups with subgroups F_i of $G_i (i = 1, 2)$ that are isomorphic, say F is a group with isomorphisms $\varphi_i : F \to F_i$. The group generated by the elements of G_1 and G_2 with all the defining relations in G_1 and G_2 as well as the relations $x\varphi_1 = x\varphi_2 (x \in F)$, is called the *free product of G_1 and G_2 amalgamating F_1 with F_2*, and is denoted by $G_1 *_F G_2$. We remark that F may be 1; this is simply called the *free product*. At the other extreme, if $F_1 = G_1$ then the free product reduces to G_2; this case (and the case $F_2 = G_2$) is usually excluded. For simplicity we shall identify F with the subgroups F_1 and F_2. Then the elements of the free product can be expressed in the form

$$au_1 \ldots u_n, \qquad (5)$$

where $a \in F$ and the u_i are alternately from $G_1 \setminus F$ and $G_2 \setminus F$.

Let us put P for this free product; its structure may be described as follows, using the letters E and E^*. For each element (5) we form a set XY, where $X = E$ if $u_1 \in G_1$ and $X = E^*$ if $u_1 \in G_2$, and similarly $Y = E$ if $u_n \in G_1$ and $Y = E^*$ if $u_n \in G_2$. Thus we have a partition of P into five sets: EE, EE^*, E^*E, E^*E^* as well as F (in case the u_i are absent). The following properties are easily verified, where $X, Y, \ldots$ denote E or E^* and $X^{**} = X, \ldots$

6.10 Automorphisms of free algebras

B.1 F is a subgroup of P,
B.2 If $f \in F, g \in XY$, then $fg \in XY$,
B.3 If $g \in XY$, then $g^{-1} \in YX$,
B.4 If $g \in XY, h \in Y^*Z$, then $gh \in XZ$,
B.5 For each $g \in P$ there is an integer $N(g)$ such that for any representation $g = g_1 \ldots g_n (g_i \in X^*X)$ we have $n \le N(g)$,
B.6 $EE^* \ne \emptyset$.

Given any group G, a partition of G into five disjoint sets $F, EE, EE^*, E^*E, E^*E^*$ satisfying B.1–B.6 is called a *bipolar structure* and F is its *core*. Thus any free product has a bipolar structure; conversely, a group with a bipolar structure is either a free product with amalgamation or an HNN-extension. The latter does not concern us here, but see Exercise 6.

In any group G with a bipolar structure an element $g \in G$ is called *reducible* if $g \notin F$ and $g = hk$, where $h \in XZ, k \in Z^*Y$. Otherwise g is *irreducible*, thus an irreducible element is in $F \cup XY$ and not of the form $g = hk$, where $h \in XZ, k \in Z^*Y$. We observe from B.5 that G is generated by its irreducible elements. In a free product with amalgamation the irreducible elements are the products of the form au, where $a \in F$ and $u \in G_i \setminus F$, $i = 1$ or 2; then EE^* contains no irreducible elements, for an element of EE^* has the form $au_1 \ldots u_n$ where u_1 and u_n belong to different factors. Conversely, we shall now show that any group with a bipolar structure such that EE^* contains no irreducible elements is a free product.

Theorem 6.10.2. *Let G be a group with a bipolar structure such that EE^* contains no irreducible elements. Then G is a free product with the core as amalgamated subgroup.*

Proof. Let $g, h \in G$ and suppose that $g \in XZ, h \in ZY$. If g is irreducible, then $gh \in F \cup WY$ for some W, for if $gh \in WY^*$, then $h^{-1} \in YZ$ by B.3 and so $g = gh.h^{-1} \in WZ$, contradicting the fact that g is irreducible. Similarly if h is irreducible then $gh \in F \cup XW$ for some W.

Next we claim that if $g \in XZ, h \in ZY$ are both irreducible, then gh is an irreducible element of $F \cup XY$. For $gh \in F \cup XY$, by what has been shown. If gh is reducible, then $gh \in XY$ and $gh = uv$, where $u \in XV$ and $v \in V^*Y$. Now $h^{-1} \in YZ$ and is irreducible, so we have $vh^{-1} \in F \cup V^*W$, but since $g = u(vh^{-1})$ is irreducible, it follows that $vh^{-1} \in F$; so by B.2, $g = u(vh^{-1}) \in XV$. Further, $h^{-1} = v^{-1}(vh^{-1})$, so $h^{-1} \in YV^*$ by B.3 and B.2, hence $h \in V^*Y$, but this contradicts the fact that $g \in XZ$ and $h \in ZY$. This shows gh to be irreducible.

Suppose now that $f \in F$, $g \in XY$ and g is irreducible. Then $fg \in XY$ by B.2; we claim that fg is irreducible. For if not, then $fg = uv$, where $u \in XV$ and $v \in V^*Y$ for some V. Then $g = (f^{-1}u)v$, where $f^{-1}u \in XV$ and $v \in V^*Y$, contradicting the fact that g is irreducible. Hence fg is irreducible; by a similar argument and B.3, $f^{-1}g^{-1}$ is an irreducible element of YX, so it follows that gf is an irreducible element of XY.

Now define

$$G_1 = F \cup \{x | x \text{ is an irreducible element of } EE\},$$
$$G_2 = F \cup \{x | x \text{ is an irreducible element of } E^*E^*\}.$$

We claim that G_1 and G_2 are subgroups of G. By B.1, B.2, G_1 admits inverses; consider a product hk. If h, k are both in F, then $hk \in F$. If h, k are both irreducible elements of EE, then $hk \in F$ or hk is an irreducible element of EE, by what has been proved. If just one of h, k is in F, then hk is an irreducible element of EE. This shows G_1 to be a subgroup of G and the same argument applies to G_2. Clearly $G_1 \cap G_2 = F$ and it remains to show that G is a free product amalgamating F.

By hypothesis EE^* contains no irreducible elements, hence E^*E which contains their inverses, also has no irreducible elements. Thus $G_1 \cup G_2$ includes all irreducible elements and so forms a generating set for G, so we can write every element of G in the form $g = g_1 \ldots g_n$, where the $g_i \in G_1 \cap G_2$. If $g_i \in F$, where $i < n$, then $g_i g_{i+1}$ is irreducible; similarly if $i = n$, we can use $g_{n-1}g_n$. If g_1, g_2 both lie in G_1 then $g_1 g_2$ is again irreducible and is in G_1, so we can replace $g_1 g_2$ by a single element. In this way we obtain representation

$$g = g_1 \cdots g_n, \tag{6}$$

where the g_i are alternately in G_1 and G_2, and either $n = 1$ and $g_1 \in F$ or each g lies in $G_1 \setminus F$ or $G_2 \setminus F$. Since $EE^* \neq \varnothing$ by B.6, F is a proper subgroup of G_1 and G_2 and the element (6) is not 1 unless $n = 1$ and g_1 is the unit element of F. Hence G is the free product of G_1 and G_2 amalgamating F. ∎

To apply this result we shall need a normal form for the elements of $GL_2(k\langle x, y\rangle)$. It will be no harder to do this for a general ring, so let R be any ring, denote its group of units by U and write $U_0 = U \cup \{0\}$. It will be convenient to have a compact notation for matrices, so let us put (as in Section 2.7), for any $a \in R$, $\alpha, \beta \in U$,

$$E(a) = \begin{pmatrix} a & 1 \\ -1 & 0 \end{pmatrix}, \quad [\alpha, \beta] = \begin{pmatrix} \alpha & 0 \\ 0 & \beta \end{pmatrix}, \quad D(\alpha) = [\alpha, \alpha^{-1}]. \tag{7}$$

These matrices are all invertible and as we saw in Section 2.7, they generate the group $GE_2(R)$. In Section 2.7 it was shown that every $A \in GE_2(R)$ can be

6.10 Automorphisms of free algebras

expressed in the form

$$A = [\alpha, \beta]E(a_1)\ldots E(a_n), \quad a_i \in R, \alpha, \beta \in U, \tag{8}$$

where $a_i \notin U_0$ for $1 < i < n$ and when $n = 2$, a_1, a_2 are not both 0. This expression for A will be called the *standard form*. In the special case of a free algebra such a form exists for all matrices in $GL_2(R)$ and is unique, by Proposition 2.8.2. For the case of one indeterminate this is of course well known, and by a theorem of Nagao [59] one has the free product representation

$$GL_2(k[x]) = T_2(k[x]) *_T GL_2(k), \tag{9}$$

where $T_2(k[x])$ is the group of upper triangular invertible 2×2 matrices over $k[x]$ and $T = T_2(k) = T_2(k[x]) \cap GL_2(k)$. This can be proved by defining a bipolar structure on $GL_2(k[x])$, though a direct proof is quicker. In the same way we have, as an easy consequence of the uniqueness of the expression (8),

Theorem 6.10.3. *For any free algebra $k\langle X \rangle$ over a field k,*

$$GL_2(k\langle X\rangle) = T_2(k\langle X\rangle) *_T GL_2(k). \tag{10}$$

Proof. From (8) we know that the left-hand side is generated by $T_2(k\langle X\rangle)$ and $GL_2(k)$, since

$$\begin{pmatrix} a & 1 \\ -1 & 0 \end{pmatrix} = \begin{pmatrix} 1 & -a \\ 0 & 1 \end{pmatrix}\begin{pmatrix} 0 & 1 \\ -1 & 0 \end{pmatrix},$$

and the uniqueness of (8) yields the free product representation (10); the details may be left to the reader. ∎

We shall also need the following reduction.

Lemma 6.10.4. *Let $P = G *_F H$ be a free product with amalgamated subgroup and consider an extension E of P by a group T. Then $E = G' *_{F'} H'$, where G', H', F' are extensions of G, H, F respectively by T.*

Proof. Consider the following diagram, where the bottom line is the given extension and the last vertical arrow is an isomorphism:

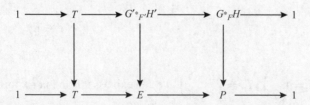

Here $G'*_{F'}H'$ is the pullback of $E \to P$ and $G*_F H \to P$; this follows by forming the pullback with $G*_F H$ replaced by G, H, F in turn. As in an additive category we find that the map $G'^{*}_{F}H' \to G^{*}_{F}H$ has the kernel T and now we can complete the diagram as shown. By the short five-lemma the middle vertical arrow is an isomorphism and so the result follows. ∎

Our main objective is a proof of the theorem of Czerniakiewicz and Makar-Limanov, that every automorphism of $k\langle x, y\rangle$ is tame; we shall prove this by showing that $\mathrm{Aut}(k\langle x, y\rangle)$ is a free product.

Theorem 6.10.5. *The automorphism group of $k\langle x, y\rangle$ is the free product of the group A of affine automorphisms and the group Δ of triangular automorphisms amalgamating their intersection.*

Proof. Let us write $R = k\langle x, y\rangle$, denote the group of all centred automorphisms of R by C and the group of all translations by T. It is clear that T is normal in $\mathrm{Aut}(R)$ with quotient C; thus $\mathrm{Aut}(R)$ is a semidirect product that is described by the exact sequence

$$1 \to T \to \mathrm{Aut}(R) \to C \to 1. \tag{11}$$

Suppose now that we have a free product representation of C. Then by applying Lemma 10.4 to the exact sequence (11) we obtain a free product representation of $\mathrm{Aut}(R)$, so to complete the proof we need only show that

$$C = \Delta_0 *_S L, \tag{12}$$

where $L = A \cap C$ is the group of linear automorphisms, $\Delta_0 = \Delta \cap L$ is the group of centred triangular automorphisms and $S = \Delta_0 \cap L$ is the group of linear triangular automorphisms (generalized shears). For the proof of (12) we shall describe a bipolar structure on C, to which Theorem 10.2 can be applied.

Any element p of $R = k\langle x, y\rangle$ can be written in the form $p = p_1 x + p_2 y + \lambda$, where p_1, p_2 are uniquely determined elements of R and $\lambda \in k$ is likewise unique. Hence every centred automorphism g of R can be expressed in the form

$$\begin{aligned} x^g &= ax + by, \\ & \qquad\qquad\qquad (a, b, c, d \in R). \\ y^g &= cx + dy, \end{aligned} \tag{13}$$

Writing $u = \begin{pmatrix} x \\ y \end{pmatrix}$, $T_g = \begin{pmatrix} a & b \\ c & d \end{pmatrix}$, we can express this in matrix form as

$$u^g = T_g u.$$

6.10 Automorphisms of free algebras

It follows that
$$T_{gh}u = u^{gh} = (T_g u)^h = T_g^h T_h u,$$
where T_g^h is the matrix obtained by letting h act on T_g. Therefore we have
$$T_{gh} = T_g^h T_h. \tag{14}$$
Further, $T_1 = I$, and so each T_g is invertible, with inverse
$$T_g^{-1} = T_{g^{-1}}^g. \tag{15}$$
It follows that $T \in GL_2(R)$, so that we have a mapping $C \to GL_2(R)$, which however is not a homomorphism, but a crossed homomorphism, by (14).

By (8) we have the unique form
$$T_g = [\alpha, \beta] E(a_1) \ldots E(a_n), \tag{16}$$
where $\alpha\beta \neq 0$, $a_i \notin k$ for $1 < i < n$ and when $n = 2$, a_1, a_2 are not both zero. In particular, the matrices of Δ_0 have the form
$$T_g = [\alpha, \beta] E(a) E(0), \tag{17}$$
where $g \in S$ if and only if $a \in k$, and the matrices of L have the form
$$T_g = [\alpha, \beta] E(\lambda) \text{ or } [\alpha, \beta] E(\lambda) E(\mu), \quad \text{where } \alpha, \beta, \lambda, \mu \in k,$$
$$\alpha\beta \neq 0. \tag{18}$$
Here $g \in S$ if and only if the second form applies and $\mu = 0$.

Let us now construct a bipolar form for C. We put $F = S$; given $g \in C$, assume that $g \notin S$ and that T has the form (16). We take $g \in XY$, where $X = E$ if $a_1 \in k$ and $X = E^*$ if $a_1 \notin k$, while $Y = E$ if $a_n \neq 0$ and $Y = E^*$ if $a_n = 0$. It only remains to verify B.1–B.6.

B.1 is clear. To prove B.2, let $f \in F$, $g \in XY$, say $T_f = [\alpha', \beta'] E(\lambda) E(0)$ and T_g is given by (16). Then $T_f^g = T_f$ because α', β', λ are fixed under g. Using (2) and (1) of 2.7, we find
$$T_{fg} = T_f T_g = [\alpha', \beta'][\alpha, \beta] E(\alpha^{-1}\lambda\beta) E(0) E(a_1) \ldots E(a_n)$$
$$= [-\alpha'\alpha, -\beta'\beta] E(a_1 + \alpha^{-1}\lambda\beta) E(a_2) \ldots E(a_n).$$
Now $a_1 + \alpha^{-1}\lambda\beta \in k$ if and only if $a_1 \in k$, so it follows that $fg \in XY$.

B.3. If T is given by (16), then
$$T_g^{-1} = [\alpha', \beta'] E(0) E(a'_n) \ldots E(a'_1) E(0),$$
where α', β' are α, β in some order and a'_i is associated to a_i. Moreover, $T_{g^{-1}} =$

$(T_g^{-1})^{g^{-1}}$ and $a_i'' = (a_i')^{g^{-1}}$ lies in k if and only if a_i does. Thus

$$T_g^{-1} = [\alpha', \beta']E(0)E(a_n'')\ldots E(a_1'')E(0). \tag{19}$$

We have $g \in XE^* \Leftrightarrow a_n = 0 \Leftrightarrow a_n'' = 0$, and this is so if and only if the coefficient of the first factor $E(.)$ in the reduced form of (19) is not in k. Thus $g \in XE^* \Leftrightarrow g^{-1} \in E^*X$; taking complements we find that $g \in XE \Leftrightarrow g^{-1} \in EX$; combining these cases with those obtained by interchanging g and g^{-1} we find that $g \in XY$ if and only if $g^{-1} \in YX$.

B.4. We take g in the standard form (16) and

$$T_h = [\gamma, \delta]E(b_1)\ldots E(b_p).$$

Then

$$T_{gh} = T_g^h T_h = [\lambda, \mu]E(a_1'')\ldots E(a_n'')E(b_1)\ldots E(b_p), \tag{20}$$

where $\lambda, \mu \in k$, $\lambda\mu \neq$, $a_i'' = (a_i')^h$ and a_i' is an associate of a_i. Here (20) may not be in standard form, but we can reach a standard form by applying the formulae (1), (8) and (3) of Section 2.7 a finite number of times. Any terms b_j that remain are unaffected by these changes, while the a_i'' that remain are only changed at most by a unit factor. Thus if a_1'' and b_p are still present after this reduction, then for $g \in XU$, $h \in VZ$ we have $gh \in XZ$. In particular, this will be the case if in the expression (20) for T_{gh} not all the factors $E(.)$ stemming from g, nor all those stemming from h cancel.

We shall now show that if $g \in XY$, $h \in Y^*Z$, then there is no cancellation at all. For suppose that some cancellation takes place in (20) and write $gh = k$. Then $k \neq 1$, because g, h cannot be mutually inverse, by B.3. Hence not all of g and h is cancelled, say not all of g. Write $g = kh^{-1}$; then in the expression for $T_{kh^{-1}}$ not all the factors $E(.)$ from k nor all those from h^{-1} are cancelled. By B.3, $h^{-1} \in ZY^*$, so if $k \in UV$, then by what has been shown, $g \in UY^*$, which contradicts the fact that $g \in XY$. This proves that no cancellation takes place and B.4 follows.

To verify B.5, take T_g in the form (16) and put $g = g_1 \ldots g_r$, where $g_i \in X_{i-1}^* X_i$; then when T_g is expressed as a product corresponding to the g_i, no cancellation takes place. Therefore $r \leq n$ and so B.5 holds with $N(g) = n$.

Finally B.6 is clear, since $E(\lambda)E(a)E(0) \in EE^*$ when $\lambda \in k$, $a \notin k$. Thus B.1–B.6 hold.

Further, any element of EE^* has the form g, where

$$T_g = [\alpha, \beta]E(\lambda)E(a_1)\ldots E(a_n)E(0),$$

and $n \geq 1$ because $g \notin S$. We put $g = hk$, where

$$T_h = [\alpha, \beta]E(\lambda), \quad T_k = E(a_1)\ldots E(a_n)E(0).$$

Here $h \in EE$, $k \in E^*E^*$; this shows that EE^* contains no irreducible elements, and so the conclusion follows by Theorem 10.2. ∎

In the proof of this result the commutativity of k has not been used, so the conclusion holds even if k is non-commutative, but of course it is necessary for the variables to centralize k; this fact was used in deriving (13).

As we remarked earlier, there is a natural homomorphism

$$\varphi \; : \; \operatorname{Aut} k\langle X \rangle \to \operatorname{Aut} k[X].$$

By a theorem of H. W. E. Jung [42], every automorphism of $k[x, y]$ is also tame, so it follows that φ is surjective when $|X| \leq 2$. In fact Czerniakiewicz [71,72] and Makar-Limanov [70a] show that φ is an isomorphism. As a consequence the group $\operatorname{Aut}(k[x, y])$ is the free product of its affine and triangular subgroups, amalgamating their intersection.

For every endomorphism $\alpha = (f_1, f_2)$ of the polynomial ring $k[x_1, x_2]$ we can define the *Jacobian* matrix $J(\alpha) = (\partial f / \partial x)$, and it is easily seen that $J(1) = I$ and $J(\alpha\beta) = J(\alpha)^\beta J(\beta)$. It follows that for an automorphism α, $J(\alpha)$ is invertible, hence $\det(J(\alpha))$ is then an element of $k^\times$. This suggests the following question:

Jacobian problem. *Let α be an endomorphism of $k[x_1, x_2]$, where k is a field of characteristic 0, such that $\det(J(\alpha)) \neq 0$; is α necessarily an automorphism?*

This was solved only recently, by S. Pinchuk [94], who found a counter-example of degree 25 (in his paper the degree was given as 35, due to an oversight). Some partial results were found by Bass, Connell and Wright [82]. Of course in finite characteristic p there is a simple negative answer, as we see by considering the mapping $(x_1, x_2) \mapsto (x_1 + x_1^p, x_2)$.

As a corollary of the above results we can obtain a conjugacy result for finite subgroups.

Theorem 6.10.6. *Every finite subgroup of $\operatorname{Aut}(k\langle x, y \rangle)$ of order invertible in k has a conjugate in L, the subgroup of linear automorphisms.*

Proof. Let G be a finite subgroup of $\operatorname{Aut}(R)$; from the representation in Theorem 10.5 it follows that G has a conjugate in Δ or A. For let $g \in G$ have the normal form

$$g = ag_1 \ldots g_n,$$

and call n the *length* of g. Since g has finite order, g_1 and g_n lie in the same factor and we can decrease the length by passing to a conjugate; in fact g will have a conjugate of length 1 and so lie in one of Δ, A, say in A. If g was chosen of maximal length in G, then it follows that for any $h \in G$ the same conjugate also lies in A, because gh is again of finite order. Thus G has a conjugate in Δ or A. We shall treat these cases in turn.

(i) $G \subseteq \Delta$ Here the action of G leaves the subspaces $kx + k[y]$ and $k[y]$ invariant, so by Maschke's theorem we can choose a G-invariant complement kx' of $k[y]$ in $kx + k[y]$. This provides an automorphism $\alpha : (x, y) \mapsto (x', y)$ such that $\alpha^{-1} G \alpha \subseteq T$.

Explicitly we can write for each $g \in G$,

$$g : (x, y) \mapsto (\lambda_g x + f_g(y), \mu_g y + \nu_g).$$

Put $f(y) = n^{-1} \sum_g \lambda_g^{-1} f_g(y)$, where $n = |G|$ and define $\alpha : (x, y) \mapsto (x - f(y), y)$. By comparing the expressions for x^{gh} we have $\lambda_{gh} = \lambda_g \lambda_h$, $f_{gh} = \lambda_g f_h(y) + f_g(\mu_h y + \nu_h)$, so for any $h \in G$,

$$nf(y) = \sum_g \lambda_{gh}^{-1} f_{gh}(y) = n\lambda_h^{-1} f_h(y) + n\lambda_h^{-1} f(\mu y + \nu).$$

It follows that $\alpha^{-1} h \alpha : (x, y) \mapsto (\lambda x, \mu x + \nu)$, so $\alpha^{-1} G \alpha \subseteq T$, and this reduces the problem to the case where $G \subseteq A$.

(ii) $G \subseteq A$. Now G acts on the space $kx + ky + k$ with invariant subspace k. Again we can find a complement $kx' + ky'$ of k in $kx + ky + k$; now $\alpha : (x, y) \mapsto (x', y')$ is an automorphism such that $\alpha^{-1} G \alpha \subseteq L$. Explicitly we write for $g \in G$,

$$g : (x, y) \mapsto (\lambda_g x + \mu_g y + \nu_g, \lambda'_g x + \mu'_g y + \nu'_g)$$

and define $\alpha : (x, y) \mapsto (x + \nu, y + \nu')$, where $\nu = n^{-1} \sum \nu_g$, $\nu' = n^{-1} \sum \nu'_g$. Then $n\nu = \sum_g \nu_{hg} = \sum_g (\lambda_h \nu_g + \mu_h \nu'_g + \nu_h) = n(\lambda_h \nu + \mu_h \nu' + \nu_h)$ and similarly for ν', therefore $\alpha^{-1} h \alpha : (x, y) \mapsto (\lambda_h x + \mu_h y, \lambda'_h x + \mu'_h y)$ and this shows that $\alpha^{-1} G \alpha \subseteq L$, as required. ∎

By invoking Theorem 6.8.3, we obtain

Corollary 6.10.7. *For any finite group G of automorphisms of $k\langle x, y \rangle$ of order invertible in k, the fixed algebra $k\langle x, y \rangle^G$ of G is free.* ∎

In these results the hypothesis that $|G|$ is prime to char k cannot be omitted. For example, if k is finite, then A is a finite subgroup of $\text{Aut}(R)$, but no conjugate of A lies in L, a proper subgroup of A.

Exercises 6.10

1. Let X be a set, $x_0 \in X$ and $X_0 = X \setminus \{x_0\}$. Show that every automorphism of $k\langle X \rangle$ fixing $k\langle X_0 \rangle$ (or even just leaving it invariant) maps x_0 to $\lambda x_0 + f$, where $\lambda \in k^\times$, $f \in k\langle X_0 \rangle$.

2. Show that for any automorphism α of $k\langle x, y\rangle$, $[x^\alpha, y^\alpha] = \lambda_\alpha [x, y]$, where $\lambda_\alpha \in k^\times$ and $[x, y] = xy - yx$. Verify that $\alpha \mapsto \lambda_\alpha$ is a homomorphism from $\mathrm{Aut}(k\langle x, y\rangle)$ to k. Its kernel will be denoted by $\mathrm{SAut}(k\langle x, y\rangle)$, whose elements are called *special* automorphisms.
3. Show that $\mathrm{SAut}(k\langle x, y\rangle) \cong S\Delta * SA$, where $S\Delta = \Delta \cap S\,\mathrm{Aut}$, $SA = A \cap S\,\mathrm{Aut}$ and the amalgamation is over their intersection.
4. (Russell [80]) Show that for any finite set X, any surjective k-algebra endomorphism of $k[X]$ is an automorphism.
5. In $k\langle x, y\rangle$ put $p = [c, x]y - [c, y]x$, where $c \in k\langle x, y\rangle$. Show that the stabilizer of kp is the group of all linear automorphisms. Find the stabilizer of kq, where $q = [c, y]$.
6. Let G be a group with two subgroups F_1, F_2 that are isomorphic. Define G^* as the group presented by the elements and relations of G, as well as a new element t and the relations $f_1 t = t f_2$, where f_i ranges over F_i and $f_1 \mapsto f_2$ in the isomorphism. Verify that G is embedded in G^* and that f_1 and f_2 are conjugate (G^* is called an *HNN-extension*, see Higman Neumann and Neumann [49]). Show that the elements of G^* have as normal form a product of elements of G and t, t^{-1}. Define a bipolar structure on G^*.

Notes and comments on Chapter 6

Bergman's centralizer theorem (Theorem 7.7, see Bergman [67, 69]) was one of the main results of his thesis. It had been conjectured by the author in the early 1960s (see Cohn [62a] for special cases of Theorem 7.1 and Cohn [63b] for the fact that the centralizer is commutative). The proof of the centralizer theorem depends on the results in Section 6.1 on the integral closure of commutative subrings. These results are proved by Bergman for Ore subrings; this form (also given in FR.1) is outlined in the exercises.

The notions of bounded and invariant element are treated for principal ideal domains by Jacobson [43]; our account in Section 6.1 and 6.2 follows Bowtell and Cohn [71]; for Theorem 2.16 see Carcanague [71]. Corollary 1.9 and Theorem 1.10 were new in FR.2; Theorem 1.11 as well as Corollary 3.2 and the developments in Section 6.3, are taken from Bergman and Cohn [71], with simplified proofs. Ischebeck [71] has extended Theorem 1.11 by showing that the centre of a directly indecomposable hereditary ring is a Krull ring. The conclusion of Theorem 4.2 (every invariant element is a unit) has been established for non-Ore 2-firs with right ACC_2 by M. L. Roberts [89], who also proves Lemma 4.1, but there are examples showing it to be false for 2-firs with right ACC_1 (Bergman and Cohn [71]); see also Exercise 4.3. Theorem 4.2 also ceases to hold when we require c to be merely right invariant (Exercise 4.6) or left invariant (Exercise 4.7). The main problem that remains is whether a non-Ore fir can have non-unit right invariant elements (Exercise 4.9°); its solution could simplify the statement of Theorem 5.9.

Codes, i.e. free subsets of free monoids (also called 'comma-free codes') form a flourishing subject in their own right, see Lothaire [97], Berstel and Perrin [85]. Theorem 5.1 occurs essentially in Levi [44], see also Dubreil-Jacotin [47]; part (b) of Theorem 5.4 was obtained by Schützenberger [59], the rest was proved by Cohn [62b].

The presentation of Section 6.5 and 6.6 follows Cohn [62b], but Results 5.5 and 5.6 were added in FR.2. The Kraft–McMillan inequality was shown by Kraft [49] to be necessary and sufficient for a prefix code to exist; McMillan [56] proved that it holds for every code. The analogue for free algebras, Theorem 5.8, is due to Kolotov [78]. Theorem 5.9 is due to Bergman and Lewin [75]. Proposition 6.1 first appeared in Cohn [64a]; Theorem 6.4 (and Corollary 2.9.15, on which it is based) is taken from Knus [68]; the proof given here is due to Dicks. The part (c) ⇒ (a) of Theorem 6.5 is in Moran [73], part (a) ⇒ (b) was proved by Dicks (unpublished). Theorem 6.7 goes back to Lewin [69]; the present proof, with Proposition 6.6, is due to Berstel and Reutenauer [88]. The notion of anti-ideal was introduced by Kolotov [78], who also proved that free subalgebras of free algebras are anti-ideals (Theorem 6.9), generalizing a result in FR.1, that the subalgebra generated by a non-trivial ideal is not free. The more general form in Lemma 6.8 is due to Bergman (unpublished). Proposition 7.3 appeared originally in FR.1 (it was proved by the author in response to a question by G. Baumslag).

Sections 6.8 and 6.9 were new in FR.2 (as Section 6.10 and 6.11 there) and follow Dicks' exposition in the Bedford College study group 1982–84. Proposition 8.1 is due to Bergman [71a], Theorem 8.3 was proved by Kharchenko [78] [91], chapter 6; it was proved independently by Lane [76]. Results 8.4, 8.5 and 8.6 are taken from Dicks and Formanek [82], but Theorem 8.4 was also found by Kharchenko [84], on whose proof the one in the text is based (with simplifications by Dicks). This type of argument is used in the theory of reductive algebraic groups, see Springer [77]. S. Donkin (unpublished) has shown that the results of the example after Theorem 8.3 still hold for any infinite field. For a finite field k, $GL_r(k)$ is a finite group and there are many relative invariants, e.g. $\sum v^n$ may not lie in the subalgebra generated by δ and its transforms by place-permutations (take $k = \mathbb{F}_2, n = 3, r = 2$).

It may be of interest to compare the algebra of invariants of a finite group G of linear automorphisms of $k\langle x_1, \ldots, x_r\rangle$ with that of $k[x_1, \ldots, x_r]$. We list the results without proof and refer for details (and references) to Stanley [79]:

1. T. E. Molien [1897] showed that in characteristic 0, the Hilbert series is given by

$$H(k[x_1, \ldots, x_r]^G : k) = |G|^{-1} \sum (\det(I - gt))^{-1}.$$

(Note that $\det(I - gt) = 1 - t.\mathrm{Tr}\, g + \cdots \pm t^n$.)

2. In 1916, E. Noether proved that $k[x_1, \ldots, x_r]^G$ is finitely generated; this holds even if G is not linear (but still finite).

3. Coxeter, Shephard and Todd (1954), Chevalley (1955) and Serre (1958) showed that if $k[x_1, \ldots, x_r]^G$ is a polynomial ring, then G can be generated by pseudo-reflexions, i.e. transformations g such that $g - 1$ has rank 1. The converse holds if char k does not divide the order of G, but not in general; see also Nakajima [83]. For example, in $G = \mathrm{Sym}_d \subseteq GL_d(k)$ the transpositions are pseudo-reflexions, they generate G and $k[x_1, \ldots, x_d]^G$ is a polynomial ring, in any characteristic (by the fundamental theorem of symmetric functions).

Galois theory underwent several generalizations (see Jacobson [64]). Kharchenko [84,91] presented a version for prime rings; we follow the trace-form presentation of this theory by Montgomery and Passman [84], adapted here to integral domains.

6.10 Automorphisms of free algebras

Kharchenko's key result is Theorem 9.8; the application to free algebras, Theorem 9.11, appears in Kharchenko [78]. Lemma 9.9 was proved for free algebras by Kharchenko [78] and independently by Martindale and Montgomery [83]; the present form is taken from Bergman and Lewin [75]. Some of the proofs have been simplified here by Lemma 9.4, which was suggested by Bergman.

The fact that all automorphisms of $k\langle x, y\rangle$ are tame was proved by Makar-Limanov [70a] and independently by Czerniakiewicz [71]. The free product representation in Theorem 10.3 is due to Cohn [2002] and is based on the work of Stallings [68], following the exposition in Lyndon and Schupp [77].

7
Skew fields of fractions

This chapter studies ways of embedding rings in fields and more generally, the homomorphisms of rings into fields. For a commutative ring such homomorphisms can be described completely in terms of prime ideals, and we shall see that a similar, but less obvious, description applies to quite general rings.

After some generalities on the rings of fractions obtained by inverting matrices (Section 7.1) and on R-fields and their specializations (Section 7.2), we introduce in Section 7.3 the notion of a matrix ideal. This corresponds to the concept of an ideal in a commutative ring, but has no direct interpretation. The analogue of a prime ideal, the prime matrix ideal, has properties corresponding closely to those of prime ideals, and in Section 7.4 we shall see that the prime matrix ideals can be used to describe homomorphisms of general rings into fields, just as prime ideals are used in the commutative case. This follows from Theorem 4.3, which characterizes prime matrix ideals as 'singular kernels', i.e. the sets of matrices that become singular under a homomorphism into some field.

This characterization is applied in Section 7.5 to derive criteria for a general ring to be embeddable in a field, or to have a universal field of fractions. These results are used to show that every Sylvester domain (in particular every semifir) has a universal field of fractions.

In the rest of this chapter these ideas are used to describe free fields and give another existence proof using the specialization lemma (Section 7.8), obtain localization theorems (Section 7.11), a description of centralizers (Section 7.9) and of the multiplicative group of the universal field of fractions of a semifir (Section 7.10). This requires a comparison of the different representations of a given element of this field (Section 7.6) and it involves a numerical invariant, the depth (Section 7.7), which has no counterpart in the commutative case. Finally, in Section 7.12 we examine a special class of rings, the fully reversible rings, for which the embedding theorems take on a particularly simple form.

7.1 The rational closure of a homomorphism

Let R be a ring; our basic problem will be to study the possible ways of embedding R in a field. Of course there may be no such embedding, and it is more natural to treat the wider problem of finding homomorphisms of R into a field. Even this problem may have no solution, e.g. if $R = A_n$ is a full matrix ring over a non-zero ring A, where $n > 1$, then any image of R is again an $n \times n$ matrix ring and so cannot be a subring of a field.

As a step towards the solution of the basic problem we may take a subset M of R and consider homomorphisms mapping all elements of M to invertible elements, i.e. *M-inverting* homomorphisms. In the commutative case, once we have an $R^\times$-inverting homomorphism to a non-zero ring, we have achieved the embedding in a field, but in general this need not be the case, since elements such as $ab^{-1} + cd^{-1}$ may not be invertible in an $R^\times$-inverting homomorphism. Thus for a general non-commutative ring the M-inverting homomorphisms are not very good approximations to homomorphisms into a field. We shall remedy this defect by inverting, instead of a set of elements, a set of square matrices. For a commutative ring this gives nothing new, since we can invert any square matrix A simply by adjoining an inverse of $\det A$. In the general case this is no longer possible, for even the Dieudonné determinant turns out, on closer examination, to be a *rational* function of the matrix entries (see Section 7.10 below).

Let R be a ring and Σ a set of matrices over R. A homomorphism $f : R \to S$ to another ring S is said to be Σ-*inverting* if every matrix in Σ is mapped by f to an invertible matrix over S. The matrices in Σ need not be square; however, since we are mainly concerned with homomorphisms into fields, we shall usually restrict the matrices to be square. The set of all square matrices over R will be denoted by $\mathfrak{M}(R)$. Given any set $\Sigma \subseteq \mathfrak{M}(R)$ and any Σ-inverting homomorphism $f : R \to S$, we define the Σ-*rational closure* of R in S as the set $R_\Sigma(S)$ of all entries of inverses of matrices in Σ^f, the image of Σ under f; the elements of $R_\Sigma(S)$ are also said to be Σ-*rational* over R. When Σ is the set of all matrices whose images under f have an inverse in S, we also write $R^f(S)$ instead of $R_\Sigma(S)$ and speak of the *f-rational* or simply the *rational* closure.

As we shall see, the *f*-rational closure of a ring R under a homomorphism f is always a subring containing $\mathrm{im}\, f$. For general sets Σ the Σ- rational closure need not be a subring, as we know from the commutative case. If M is a multiplicative subset of a commutative ring R, then as we have seen in Section 0.7, the localization R_M is a ring. Let us call a set Σ of matrices *upper multiplicative* if $1 \in \Sigma$, and whenever $A, B \in \Sigma$, then $\begin{pmatrix} A & C \\ 0 & B \end{pmatrix} \in \Sigma$ for any matrix C of appropriate size; *lower multiplicative sets* are defined similarly (with C in the lower corner). If Σ is upper multiplicative and any matrix in Σ still lies in Σ after any permutation of rows and the same permutation of columns, then Σ is said to be *multiplicative*; clearly such a set is also lower multiplicative.

We first check that the set of all matrices inverted in a homomorphism is multiplicative:

Proposition 7.1.1. *Given any homomorphism of rings, $f : R \to S$, the set of all matrices over R whose image under f is invertible over S is multiplicative.*

Proof. Clearly 1^f is invertible, and if A, B are invertible matrices over S, then for any matrix C of suitable size,

$$\begin{pmatrix} A & C \\ 0 & B \end{pmatrix} \text{ has the inverse } \begin{pmatrix} A^{-1} & -A^{-1}CB^{-1} \\ 0 & B^{-1} \end{pmatrix},$$

and invertibility is unaffected by permuting rows and columns. ∎

The Σ-rational closure can be characterized in various ways. As before, we shall use the notation e_i for the column vector (of length determined by the context) with 1 in the ith place and 0s elsewhere, and e_i^T for the corresponding row vector.

Theorem 7.1.2. *Let R, S be rings and Σ an upper multiplicative set of matrices over R. Given any Σ-inverting homomorphism $f : R \to S$, the Σ-rational closure $R_\Sigma(S)$ is a subring of S containing $\mathrm{im} f$, and for any $x \in S$ the following conditions are equivalent:*

(a) $x \in R_\Sigma(S)$,
(b) x is a component of the solution u of a matrix equation

$$Au - e_j = 0, \quad \text{where } A \in \Sigma^f, \tag{1}$$

(c) x is a component of the solution u of a matrix equation

$$Au - a = 0, \quad \text{where } A \in \Sigma^f, \tag{2}$$

and a is a column with entries in $\mathrm{im} f$,
(d) $x = bA^{-1}c$, where $A \in \Sigma^f$, b is a row and c is a column with entries in $\mathrm{im} f$.

Proof. We first prove the equivalence of the four conditions. (a) ⇒ (b). By definition $R_\Sigma(S)$ consists of the entries of the inverses of matrices in Σ^f. If x occurs as (i, j)-entry of A^{-1}, then it is the ith component of the solution of (1), so (b) holds. (b) ⇒ (c) is clear and (c) ⇒ (d) because when (2) holds, then

7.1 The rational closure of a homomorphism

$u_i = e_i^T A^{-1} a$. To show (d) $\Rightarrow$ (a), let $x = bA^{-1}c$; then we have

$$\begin{pmatrix} 1 & b & 0 \\ 0 & A & c \\ 0 & 0 & 1 \end{pmatrix}^{-1} = \begin{pmatrix} 1 & -bA^{-1} & bA^{-1}c \\ 0 & A^{-1} & -A^{-1}c \\ 0 & 0 & 1 \end{pmatrix},$$

where the matrix whose inverse is taken is again in Σ^f.

To prove that the Σ-rational closure $R_\Sigma(S)$ is a ring containing $\mathrm{im}\, f$ we shall use property (c). Let $a \in \mathrm{im}\, f$; then a satisfies the equation $1.u - a = 0$, which is of the form (2), hence $R_\Sigma(S) \supseteq \mathrm{im}\, f$. Now if u_i is the ith component of the solution of (2) and v_j the jth component of the solution of $Bv - b = 0$, then $u_i - v_j$ is the ith component of the solution of

$$\begin{pmatrix} A & C \\ 0 & B \end{pmatrix} w - \begin{pmatrix} a \\ b \end{pmatrix} = 0,$$

where C has for its jth column the ith column of A and the rest 0. Next, if $v_j = 0$, then $u_i v_j = 0$; otherwise $u_i v_j$ is the ith component of the solution of

$$\begin{pmatrix} A & C \\ 0 & B \end{pmatrix} w - \begin{pmatrix} 0 \\ b \end{pmatrix} = 0,$$

where C has as its jth column $-a$ and the rest 0. This shows that $R_\Sigma(S)$ is closed under subtraction and multiplication, and we have already seen that it contains 1, therefore it is a subring, as claimed. ∎

Let R be a ring, Σ a multiplicative set of matrices over R and $f : R \to S$ a Σ-inverting homomorphism. Then for any $p \in R_\Sigma(S)$ we define the *left depth* $d_l(p)$ as the least n for which there is an $n \times n$ matrix A in Σ^f and a column $c \in {}^n(Rf)$ such that p occurs among the entries of the column $A^{-1}c$; the *right depth* $d_r(p)$ is defined similarly, using rows of bA^{-1}. It is also possible to define an *upper depth* $\bar{d}(p)$, using matrices A^{-1} and a *lower depth* $\underline{d}(p)$ using elements $bA^{-1}c$, but they will not be needed in what follows. It is easily seen that

$$\underline{d}(p) \leq d_l(p) \leq \bar{d}(p), \quad \underline{d}(p) \leq d_r(p) \leq \bar{d}(p), \tag{3}$$

and d_l, d_r cannot differ by more than 1 from each other and the other depths. This follows from the proof of Theorem 1.2 (or also from that of Proposition 1.4 below) and may be left to the reader to verify. An element of right depth 1 has the form ab^{-1}; this expression will be called a *right fraction*, *reduced* if a and b are right coprime. Similarly, an expression $b^{-1}a$ is called a *left fraction*, again *reduced* if a, b are left coprime. In particular, if an element p in the universal field of fractions of a semifir can be written as a right fraction ab^{-1} and as a left fraction $b'^{-1}a'$, then $a'b = b'a$ and this expression may be taken to be comaximal, by Proposition 3.1.3; hence p can then be expressed as a reduced right fraction and as a reduced left fraction.

Often it is convenient to use a different notation for the system (2), by taking as our basic matrix the augmented matrix (a, A). Thus, omitting the reference to f, for simplicity, which amounts to letting R act on S via f, we shall write our system as

$$Au = 0, \quad A \in {}^m R^{m+1}, \tag{4}$$

where u is a vector with first component 1. The columns of A will be indicated by subscripts, thus

$$A = (A_0, A_1, \ldots, A_m) = (A_0, A_\bullet, A_\infty),$$

where $A_\infty = A_m$ is the last column and $A_\bullet = (A_1, \ldots, A_{m-1})$ represents the remaining columns. We shall call (4) an *admissible system* in S and A an *admissible matrix* over S of order $o(A) = m$ for the element p if (4) has a unique solution $u \in {}^{m+1}S$, normalized by the condition $u_0 = 1$, where $u_m = p$. Thus a sufficient condition for an $m \times m + 1$ matrix A over R to be admissible is that the image of the matrix formed by the last m columns is invertible over S; when S is a field, this sufficient condition is also necessary. The last m columns, $(A_\bullet, A_\infty)$ form the *denominator*, the first m columns, $(A_0, A_\bullet)$ form the *numerator*, $A_\bullet$ is called the *core* of p in the representation (4) and we write $u = (1, u_\bullet, p)^T$. Of course these matrices depend not merely on p, but on the choice of A in (4).

We note that matrices over the Σ-rational closure can be obtained as solutions of matrix equations in exactly the same way:

Proposition 7.1.3. *Let R be a ring, Σ a lower multiplicative set of matrices over R and $f : R \to S$ a Σ-inverting homomorphism. Then for any $m \times n$ matrix P over $R_\Sigma(S)$ there exists $r \geq 0$ and $A \in {}^{r+m}(\operatorname{im} f)^{n+r+m}$, $u = (I, U, P)^T \in {}^{n+r+m} S^n$ such that*

$$Au = 0, \quad A = (A_0, A_\bullet, A_\infty), \quad (A_\bullet, A_\infty) \in \Sigma^f, \tag{5}$$

where A_0 is $(r + m) \times n$, $A_\bullet$ is $(r + m) \times r$ and A_∞ is $(r + m) \times m$.

Proof. We have to show that every matrix P is determined by an equation (5). Suppose that P', P'' are determined by matrices A', A'' respectively; then $P = P' + P''$ is determined by the system

$$\begin{pmatrix} A_0' & A_\bullet' & A_\infty' & 0 & 0 \\ A_0'' & 0 & -A_\infty'' & A_\bullet'' & A_\infty'' \end{pmatrix} \begin{pmatrix} I \\ U' \\ P' \\ U'' \\ P \end{pmatrix} = 0.$$

Hence it is enough (by induction on the order of A) to consider a matrix P with a single non-zero entry, say $P = \begin{pmatrix} p & 0 \\ 0 & 0 \end{pmatrix}$. If $Cu = 0$ is an admissible system for p, then we have

$$\begin{pmatrix} C_0 & 0 & C_\bullet & C_\infty & 0 \\ 0 & 0 & 0 & 0 & I_{m-1} \end{pmatrix} \begin{pmatrix} 1 & 0 \\ 0 & I_{n-1} \\ u_\bullet & 0 \\ p & 0 \\ 0 & 0 \end{pmatrix} = 0.$$

∎

The admissible system (4) is more accurately described as a *left* admissible system; it follows by symmetry that the elements of $R_\Sigma(S)$ can equally well be determined in terms of a *right* admissible system $vB = 0$, where B is a matrix of index -1. The next result describes the relation between these two types.

Proposition 7.1.4. *In the situation of Proposition 1.3, let P be an $m \times n$ matrix with a left admissible $(r + m) \times (n + r + m)$ matrix A. Then the $(m + (m + r) + n) \times (n + r + m)$ matrix*

$$\begin{pmatrix} 0 & 0 & -I \\ A_0 & A_\bullet & A_\infty \\ I & 0 & 0 \end{pmatrix}$$

is right admissible for P.

Proof. By hypothesis we have an equation

$$(A_0 \ \ A_\bullet \ \ A_\infty) \begin{pmatrix} I \\ U \\ P \end{pmatrix} = 0. \tag{6}$$

Hence we have

$$\begin{pmatrix} A_0 & A_\bullet & A_\infty \\ I_n & 0 & 0 \end{pmatrix} \begin{pmatrix} I_n \\ U \\ P \end{pmatrix} = \begin{pmatrix} 0 \\ I \end{pmatrix}.$$

The matrix on the left is square, and it has an inverse over S; this means that

$$\begin{pmatrix} A_0 & A_\bullet & A_\infty \\ I & 0 & 0 \end{pmatrix}^{-1} = \begin{pmatrix} 0 & I \\ Y & U \\ Z & P \end{pmatrix}, \text{ say,}$$

where the NW-block on the right is clearly 0. Therefore

$$(I_m \ Z \ P) \begin{pmatrix} 0 & 0 & -I_m \\ A_0 & A_\bullet & A_\infty \\ I & 0 & 0 \end{pmatrix} = 0,$$

and this is an equation of the required form. ∎

The form (d) of Theorem 1.2 is also used sometimes; more generally, we may consider an element u of the form

$$u = d - bA^{-1}c, \tag{7}$$

where $d \in \text{im } f$ and A, b, c are as in Theorem 1.2(d). We shall say that u is *represented* by $r = d - (b, A, c)$, or also by the *display block*, or simply, the *display*:

$$\begin{pmatrix} A & c \\ b & d \end{pmatrix}. \tag{8}$$

When A is invertible, then the matrix (8) is stably associated to $d - bA^{-1}c$, as the following reduction shows:

$$\begin{pmatrix} A & c \\ b & d \end{pmatrix} \to \begin{pmatrix} 1 & c \\ bA^{-1} & d \end{pmatrix} \to \begin{pmatrix} 1 & 0 \\ bA^{-1} & d - bA^{-1}c \end{pmatrix} \to \begin{pmatrix} 1 & 0 \\ 0 & d - bA^{-1}c \end{pmatrix}.$$

The matrix A will be called the *pivot matrix* of the display, its order the *dimension* and d its *scalar term*. The display is called *pure* if its scalar term is 0.

We note the following relation between an admissible matrix A and the element p determined by it:

Proposition 7.1.5. (Cramer's Rule) *Let $f : R \to S$ be a homomorphism of rings. Given $p \in S$, if A is a matrix over im f admissible over S for p, then*

$$(A_\bullet \ -A_0) = (A_\bullet \ A_\infty) \begin{pmatrix} I & u_\bullet \\ 0 & p \end{pmatrix}; \tag{9}$$

thus over S, p is stably associated to the numerator of the system (4). More generally, this holds when p is a matrix over S. If S is weakly finite, then $p = 0$, or in case of a matrix, p is not full whenever its numerator is not full over S.

Proof. The first part is immediate, since $(A_\bullet, A_\infty)$ is invertible over S and the left-hand side of (4) is associated to the numerator, or using (6) when p is a matrix. The second part follows because a non-zero element has positive stable rank over S, by Proposition 0.1.3. ∎

7.1 The rational closure of a homomorphism

For reference we note that if A, B are admissible matrices for p, q respectively, then $(A_\infty, A_\bullet, A_0)$ is an admissible matrix for p^{-1}, $(A_0 - A_\infty, A_\bullet, A_\infty)$ is an admissible matrix for $p + 1$, and

$$\begin{pmatrix} B_0 & B_\bullet & B_\infty & 0 & 0 \\ A_0 & 0 & A_\infty & A_\bullet & A_\infty \end{pmatrix} \text{ and } \begin{pmatrix} B_0 & B_\bullet & B_\infty & 0 & 0 \\ 0 & 0 & A_0 & A_\bullet & A_\infty \end{pmatrix} \quad (10)$$

are admissible matrices for $p - q$ and pq, respectively. As we see by looking at the denominators in (10), with the present conventions we need to take our matrices to be lower multiplicative. We also note that any two elements can be brought to a common denominator; more generally we have

Proposition 7.1.6. *Let R be any ring and Σ a subset of $\mathfrak{M}(R)$ closed under diagonal sums. Then any finite set of elements of the localization R_Σ can be brought to a common denominator.*

Proof. By induction it is enough to prove the result for two elements. Let $p, q \in R_\Sigma$, with admissible matrices $A = (A_0, A_\bullet, A_\infty)$, $B = (B_0, B_\bullet, B_\infty)$ and consider the matrix

$$\begin{pmatrix} A_0 & A_\bullet & A_\infty & 0 & 0 \\ B_0 & 0 & 0 & B_\bullet & B_\infty \end{pmatrix}. \quad (11)$$

From (11) we obtain admissible matrices for p and q by putting B_0, resp. $A_0 = 0$, hence each has an admissible matrix whose denominator is the diagonal sum of the denominators in A and B. ∎

Exercises 7.1

1. Let R be a commutative ring and M a subset of R. Find conditions on M for the set $\{a^f(s^f)^{-1} | a \in R, s \in M\}$, under any M-inverting homomorphism f, to be a subring.
2. Show that the transpose of every invertible matrix is invertible.
3. For any ring homomorphism $f : R \to S$ define the *unit-closure* of R in S as the least subring of S containing im f and closed under forming inverses of elements, when they exist in S. Show that the unit-closure is contained in the rational closure and give examples to show that in general these two closures are distinct.
4. Prove the inequalities (3) for the depths of an element.
5. Let R be a ring, Σ a multiplicative set of matrices over R and $f : R \to S$ a Σ-inverting homomorphism. Show that for any $p, q \in R_\Sigma(S)$ and $a \in \text{im} f$, $a \neq 0$, $d(p - q) \leq d(p) + d(q)$, $d(pq) \leq d(p) + d(q)$, $d(a) = 1$, $d(pa) \leq d(p)$, $d(ap) \leq d(p) + 1$, where d is the left depth. What are the corresponding inequalities for the right depth?

6. Show that for any Σ-localization $R \to R_\Sigma$ and any subset I of R the set of solutions of admissible equations with matrices $A = (A_0, A_\bullet, A_\infty)$, where the entries of A_0 are in I, forms a left ideal of R_Σ.
7. Given a homomorphism $f : R \to S$, if every square matrix from R maps either to a left and right zero-divisor (or 0) or to an invertible matrix over S, show that $R^f(S)$ is such that every non-zero element is either a left and right zero-divisor or invertible in S. If, further, S is an integral domain, deduce that the rational closure of R in S is a field. Show that under a homomorphism of R into a field, the rational closure is a subfield.
8. Given a homomorphism $f : R \to S$, where S is weakly finite, let Σ be the set of all matrices inverted under f. Show that for any square matrices A, B over R, if $\begin{pmatrix} A & C \\ 0 & B \end{pmatrix} \in \Sigma$ for some C, then $A, B \in \Sigma$. Why is weak finiteness necessary?
9. Let R and S be algebras over an infinite field. Given a homomorphism $f : R \to S$, show that for any finite set of elements in the rational closure there exist a matrix A and columns $c_1, \ldots, c_r$ such that the given elements are the last components of the solutions of equations $Ax - c_i = 0, i = 1, \ldots, r$.
10. Show that in a local ring S, the unit-closure of any subring R is equal to its rational closure. (*Hint*: Verify that a matrix over S is invertible if and only if its image in the residue-class field is invertible.) In particular, the unit-closure and the rational closure of $k\langle X \rangle$ in $k\langle\!\langle X \rangle\!\rangle$ are the same.

7.2 The category of R-fields and specializations

Given a ring R, we recall that an R-*ring* is a ring L with a homomorphism $R \to L$. For fixed R, the R-rings form a category in which the morphisms are the ring homomorphisms $L \to L'$ such that the triangle shown is commutative. In particular a field that is an R-ring will be called an R-*field*.

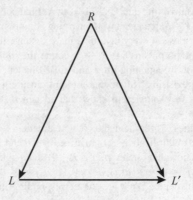

7.2 The category of R-fields and specializations

We shall be concerned with R-fields that as fields are generated by the image of R. Such fields are epimorphic in the sense that the map from R is an epimorphism; we digress briefly to explain the connexion.

A ring homomorphism $f : R \to S$ is called an *epimorphism* (in the category of rings) if for any homomorphisms g, g' from S to some ring T, $fg = fg' \Rightarrow g = g'$. Some equivalent descriptions are given in

Proposition 7.2.1. *For any ring homomorphism $f : R \to S$ the following conditions are equivalent:*

(a) f is an epimorphism,
(b) in the S-bimodule $S \otimes_R S$ we have $x \otimes 1 = 1 \otimes x$ for all $x \in S$,
(c) the multiplication map $S \otimes_R S \to S, x \otimes y \mapsto xy$, is an isomorphism.

Proof. (a) $\Rightarrow$ (b). Consider the split null extension $M = S \oplus (S \otimes_R S)$ with the multiplication $(x, u)(y, v) = (xy, xv + uy)$. The two maps $S \to M$ sending x to $(x, 1 \otimes x)$ and $(x, x \otimes 1)$ are easily seen to be ring homomorphisms that agree on R, so by (a) they are equal, i.e. $x \otimes 1 = 1 \otimes x$ for all $x \in S$. To prove (b) $\Rightarrow$ (c), we note that the multiplication homomorphism maps $\sum x_i \otimes y_i$ to $\sum x_i y_i$; when (b) holds, then $\sum x_i \otimes y_i = \sum x_i y_i \otimes 1 = 1 \otimes \sum x_i y_i$ and so (c) follows. The converse is clear.

(b) $\Rightarrow$ (a). When (b) holds and $g, g' : S \to T$ are two homomorphisms such that $fg = fg'$, then the map $S \otimes S \to T$ given by $x \otimes y \mapsto xg \cdot yg'$ is well-defined, because $fg = fg'$ and maps $x \otimes 1$ to xg and $1 \otimes x$ to xg', but $1 \otimes x = x \otimes 1$, hence $g = g'$ as we had to show. ∎

As a consequence we have

Corollary 7.2.2. *A homomorphism $f : R \to K$ from a ring R to a field K is an epimorphism if and only if K is the field generated by im f.*

Proof. Suppose that the field generated by im f is a proper subfield H of K and that $\{u_i\}$ is a right H-basis of K, where $u_1 = 1$. Then $K \otimes_H K = \sum u_i \otimes K = K \oplus \sum_{i \neq 1} u_i \otimes K$ and this is not isomorphic to K, so f is not an epimorphism. Conversely, if K is the field generated by im f, then by Theorem 1.2, every element of K is the entry of a matrix B that is the inverse of A^f for some matrix A over R. Thus $A^f.B = B.A^f = I$, hence for any homomorphisms g, g' such that $fg = fg'$, A^{fg} has the inverses B^g and $B^{g'}$, which must coincide, and it follows that $g = g'$. ∎

An R-field K for which the map $R \to K$ is an epimorphism will be called an *epic R*-field. An epic R-field K for which the given map $R \to K$ is injective is called a *field of fractions* of R. Of course the elements of K will not generally be left or right fractions, but have a more general form, as we saw in Section 7.1.

The only R-ring homomorphism possible between epic R-fields is an isomorphism. For any homomorphism between fields must be injective, and in this case the image will be a field containing the image of R, hence we have a surjection, and so an isomorphism. This shows the need to consider more general maps. Let us define a *subhomomorphism* between R-fields K, L as an R-ring homomorphism $f : K_f \to L$ from an R-subring K_f of K to L such that any element of K_f not in the kernel of f has an inverse in K_f. This definition shows that K_f is a local ring with maximal ideal $\ker f$, hence $K_f/\ker f$ is a field, isomorphic to a subfield of L, namely $\mathrm{im}\, f$. The latter is a subfield of L containing the image of R in L, hence if L is an epic R-field, then $\mathrm{im}\, f = L$. Thus we obtain

Lemma 7.2.3. *For any ring R, any subhomomorphism to an epic R-field is surjective.* ∎

Two subhomomorphisms from an R-field K to another one, L, are considered equivalent if they agree on an R-subring K_0 of K and the common restriction to K_0 is again a subhomomorphism. It is clear that this is indeed an equivalence relation and this suggests the following

Definition. Let K and L be two R-fields, where R is any ring. An equivalence class of subhomomorphisms from K to L is called a *specialization* from K to L.

We note that the set of all subhomomorphisms defining a given specialization ϕ has an intersection that is again a subhomomorphism defining ϕ; this will be called a *minimal* subhomomorphism and we shall usually represent a specialization by its minimal subhomomorphism, whose domain will be called the *minimal domain* or simply the *domain* of the specialization.

The R-fields and specializations form a category denoted here by $\mathcal{F}_R$. Here it is only necessary to check that the composition of maps is defined and is associative. Given specializations $f : K \to L, g : L \to M$, let K_0, L_0 be the domains of f and g respectively, and put $K_1 = \{x \in K_0 | xf \in L_0\}$, $f_1 = f|K_1$. We assert that $f_1 g : K_1 \to M$ is a subhomomorphism and so defines a specialization. Let us denote the canonical mapping $R \to K$ by μ_K; then we have $\mu_K f = \mu_L$, hence $R\mu_K \subseteq K_1$, so that K_1 is an R-ring. Moreover, if $x \in K_1$ and $xf_1 g \neq 0$, then $xf = xf_1 \neq 0$, so $x^{-1} \in K_0$ and $(x^{-1})f = (xf)^{-1} \in L_0$, hence $x^{-1} \in K_1$. This shows that $f_1 g$ defines in fact a specialization. To prove associativity, consider subhomomorphisms $f : K_1 \to K_2, g : K_2 \to K_3, h : K_3 \to K_4$. The composites $(fg)h$ and $f(gh)$ have the same domain, namely the subset of K_1 mapped by f into the domain of g and by g into the domain of h; clearly $(fg)h$ and $f(gh)$ are both the composite of f, g and h and so they agree on this common domain.

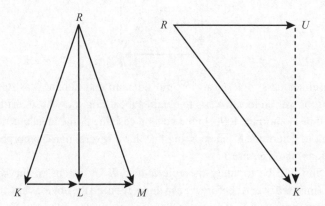

Let $\mathcal{E}_R$ be the full subcategory of $\mathcal{F}_R$ whose objects are the epic R-fields. An initial object in $\mathcal{E}_R$ is called a *universal R-field*. Explicitly a universal R-field is an epic R-field U such that for any epic R-field K there is a unique specialization $U \to K$. Clearly a universal R-field, if it exists at all, is unique up to isomorphism.

In general a ring R need not have a universal R-field, even when it has R-fields; e.g. a commutative ring R has a universal R-field if and only if its nilradical is prime, and we shall obtain an analogous condition for general rings in Theorem 5.2. Suppose that R has a universal R-field U; then R has a field of fractions if and only if U is a field of fractions of R, as a glance at the above triangle shows. In that case we call U the *universal field of fractions* of R.

Let us illustrate these definitions by taking R to be a commutative ring. Then the epic R-fields correspond precisely to the prime ideals of R. Thus, given any epic R-field K, the kernel of the canonical map $\mu_K : R \to K$ is a prime ideal and, conversely, if $\mathfrak{p}$ is a prime ideal of R, then the mapping $R \to \mathcal{F}(R/\mathfrak{p})$, where $\mathcal{F}(A)$ is the field of fractions of the domain A, gives us an epic R-field. The category $\mathcal{E}_R$ in this case is equivalent to the set of all prime ideals of R, with inclusion maps as morphisms. There is a universal R-field if and only if there is a least prime ideal, i.e. the nilradical is prime, and when this is 0 (i.e. when R is an integral domain), we have a universal field of fractions. A similar correspondence exists in the general case, and will be described in Section 7.5 below, once we have identified the objects to be used in place of prime ideals.

The situation can be described by a commutative diagram

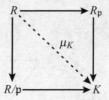

We can either pass to $R/\mathfrak{p}$, an integral domain, and obtain K as its field of fractions, or we can localize at $\mathfrak{p}$ to obtain a local ring $R_\mathfrak{p}$, whose residue-class field will be isomorphic to K. In the general case the prime ideal $\mathfrak{p}$ is no longer sufficient to determine K, since a ring may have several non-isomorphic fields of fractions (see Exercise 13).

Our aim will be to study the epic R-field K by means of the set of all matrices inverted over K; before we can do so, we need the obvious but important remark that for any set Σ of matrices there always exists a *universal Σ-inverting homomorphism:* by this term we understand a homomorphism $\lambda : R \to R_\Sigma$ that is Σ-inverting and such that any Σ-inverting homomorphism f can be factored uniquely by λ, i.e. given $f : R \to S$ such that Σ^f consists of invertible matrices, there is a unique homomorphism $\bar{f} : R_\Sigma \to S$ such that $f = \lambda \bar{f}$. The ring R_Σ is clearly determined up to isomorphism by these conditions; it is called the *universal Σ-inverting ring* or also a *universal localization* of R. Such a ring always exists, for any choice of R and Σ, and it may be constructed as follows. For each $m \times n$ matrix $A = (a_{ij})$ in Σ we take a set of mn symbols, arranged as an $n \times m$ matrix $A' = (a'_{ji})$ and let R_Σ be the ring generated by the elements of R as well as all the a'_{ji} and as defining relations all the relations holding in R, together with the relations, in matrix form,

$$AA' = I, \quad A'A = I \quad \text{for each } A \in \Sigma. \tag{1}$$

The mapping taking each element of R to the corresponding element of R_Σ is clearly a homomorphism $\lambda : R \to R_\Sigma$, which is Σ-inverting, by construction. If $f : R \to S$ is any Σ-inverting homomorphism, we define a homomorphism $\bar{f} : R_\Sigma \to S$ by putting $x\bar{f} = xf$ for all $x \in R$ and for any matrix $A \in \Sigma$ defining $\bar{f}$ on A' by putting $A'\bar{f} = (Af)^{-1}$. This gives a well-defined homomorphism $\bar{f}$, because any relation in R_Σ is a consequence of the defining relations in R and the relations (1), and all these relations also hold in S. Since we are mainly concerned with embeddings in fields, our sets Σ will usually consist of square matrices, but this is not essential in the general situation. We remark that a localization R_Σ in the context of Section 7.1 would be $R_\Sigma(R_\Sigma)$, but unlike that case, R_Σ here is an absolute construction.

7.2 The category of R-fields and specializations

Of course the canonical homomorphism $\lambda : R \to R_\Sigma$ need not be injective and may, in fact, be zero, i.e. R_Σ may be 0, e.g. if Σ contains a zero matrix. However, from the relation $f = \lambda \bar{f}$ we already see that if there is a Σ-inverting homomorphism f that is injective, then λ must be injective. We sum up these results in

Theorem 7.2.4. *Let R be a ring and Σ any set of matrices over R. Then there is a ring R_Σ, unique up to isomorphism, with a universal Σ-inverting homomorphism*

$$\lambda : R \to R_\Sigma. \tag{2}$$

Moreover, λ is injective if and only if R can be embedded in a ring over which all the matrices of Σ have inverses. ∎

The ring R_Σ will be called the *universal localization* of R with respect to Σ.

Let us now consider, for an epic R-field K, in place of $\mathfrak{p} = \ker \mu_K$ the set $\mathcal{P}$ of all square matrices over R that become singular over K. This set $\mathcal{P}$ is called the *singular kernel* of μ_K and is written $\operatorname{Ker} \mu_K$. There is no obvious way of forming an analogue of $R/\mathfrak{p}$, viz. 'the ring obtained by making the matrices in $\mathcal{P}$ singular', but we can form $R_\mathcal{P}$ as the universal Σ-inverting ring R_Σ, where Σ is the complement of $\mathcal{P}$ in the set $\mathfrak{M}(R)$ of all square matrices over R. By abuse of notation we sometimes write this as $R_\mathcal{P}$, in analogy to the commutative case. Our next result describes any epic R-field in terms of its singular kernel, or rather, its complement in $\mathfrak{M}(R)$.

Theorem 7.2.5. *Let R be any ring. Then*

(i) *if Σ is a set of matrices over R such that the universal localization R_Σ is a local ring, then the residue-class field of R_Σ is an epic R-field, and*

(ii) *if K is an epic R-field and Σ the set of all matrices over R whose images in K are invertible, then Σ is multiplicative and R_Σ is a local ring with residue-class field isomorphic to K.*

Proof. Let Σ be a set of matrices over R such that R_Σ is a local ring, and denote its residue-class field by K. By composing the natural mappings we get a homomorphism $R \to R_\Sigma \to K$, and K is generated by the entries of the inverses of images of matrices in Σ, hence it is an epic R-field.

Conversely, let K be any epic R-field and Σ the set of all matrices over R whose images in K are invertible. Then Σ is multiplicative, by Proposition 1.1. Further, by the definition of Σ and by Theorem 2.4, we have an R-ring homomorphism $\alpha : R_\Sigma \to K$, and it will be enough to prove that any element of R_Σ not in $\ker \alpha$ is invertible.

Let $p \in R_\Sigma$ have the admissible system over R_Σ, $Au = 0$, where $(A_\bullet, A_\infty) \in \Sigma$, and so, by Cramer's rule (Proposition 1.5), p is stably associated to $(A_0, A_\bullet)$.

If $p\alpha \neq 0$, then $p\alpha$ is invertible; hence so is $(A_0, A_\bullet)^\alpha$. Therefore $(A_0, A_\bullet) \in \Sigma$ and applying Cramer's rule once more we find that p is invertible in R_Σ; thus every element of R_Σ not in ker α has an inverse. It follows that R_Σ is a local ring with maximal ideal ker α, and its residue-class field is therefore isomorphic to K, as claimed. ∎

The sets Σ for which R_Σ is a local ring may be described as follows, using 'minor' to mean the submatrix corresponding to a subdeterminant.

Proposition 7.2.6. *Let R be a ring and Σ a multiplicative set of matrices. Then R_Σ is a local ring if and only if $R_\Sigma \neq 0$ and for $A \in \Sigma$, if the $(1,1)$-minor of A is not invertible over R_Σ, then $A - e_{11}$ is invertible over R_Σ.*

Proof. Suppose that R_Σ is a local ring and denote its residue-class field by K. Any matrix over R_Σ is invertible if and only if its image is invertible over K, so we need only show that if an invertible matrix A has a non-invertible $(1,1)$-minor, then $A - e_{11}$ is invertible over K. For such an A some non-trivial left linear combination of the rows of its $(1,1)$-minor is zero. If we take the corresponding left linear combination of the last $n-1$ rows of A, we obtain $(c, 0, \ldots, 0)$, where $c \neq 0$, because A is non-singular. We now subtract from the first row of A, c^{-1} times this combination of the other rows and obtain the matrix $A - e_{11}$, which is therefore invertible. Conversely, assume that this condition holds and let u_1 be the first component of the solution of

$$Au - e_1 = 0. \tag{3}$$

If u_1 does not have a left inverse in R_Σ, then by Cramer's rule the numerator of u_1 in (3) cannot then be invertible over R_Σ. This numerator, up to stable association, is just the $(1,1)$-minor of A, hence $A - e_{11}$ is then invertible over R_Σ. We now apply Lemma 0.5.8 with $M = R^n, N = R, s = u = A^{-1}e_1, t = e_1^T$. By hypothesis, $I_n - A^{-1}e_1 e_1^T = A^{-1}(A - e_{11})$ is invertible, hence so is $1 - e_1^T A^{-1} e_1 = 1 - u_1$. Thus for any $x \in R_\Sigma$, either x has a left inverse or $1 - x$ has an inverse, hence R_Σ is a local ring, by Proposition 0.3.5. ∎

In Theorem 2.5 we saw that any epic R-field may be described entirely in terms of matrices over R and their inverses; we now show how to express specializations in terms of the sets of matrices inverted.

Theorem 7.2.7. *Let R be any ring, K_1, K_2 any epic R-fields, Σ_i the set of all matrices over R inverted in K_i and R_i the universal localization R_{Σ_i} with maximal ideal $\mathfrak{m}_i (i = 1, 2)$. Then the following conditions are equivalent:*

(a) *there is a specialization $\alpha : K_1 \to K_2$,*
(b) $\Sigma_1 \supseteq \Sigma_2$,

7.2 The category of R-fields and specializations

(c) *every rational relation over R satisfied in K_1 is satisfied in K_2,*
(d) *there is an R-ring homomorphism $R_2 \to R_1$.*

If there is a specialization from K_1 to K_2 and one from K_2 to K_1, then $K_1 \cong K_2$.

We note the reversal of direction in (d) compared with (a).

Proof. (a) $\Rightarrow$ (b). Let $\mu_i : R \to K_i$ be the canonical homomorphism. Take $A \in \Sigma_2$ and denote its image under μ_i by A_i. Then A_2 has an inverse that is the image of a matrix B over K_1 (by Lemma 2.3): $A_2 B^\alpha = I$; hence $A_1 B = I + C$, where $C^\alpha = 0$. Since the domain of a subhomomorphism defining α is a local ring, $I + C$ has an inverse over R, therefore so does A_1, i.e. $A \in \Sigma_1$.

(b) $\Rightarrow$ (c) is clear and so is (c) $\Rightarrow$ (d), for when (c) holds, then $\lambda_1 : R \to R_1$ is Σ_2-inverting and so may be factored by λ_2.

(d) $\Rightarrow$ (a). Let R_0 be the image of R_2 in R_1. Then the natural homomorphism $R_1 \to K_1$ maps R_0 to $R_0' = R_0/(R_0 \cap \mathfrak{m}_1)$. Now R_0 is a local ring (as homomorphic image of R_2) and $R_0 \cap \mathfrak{m}_1$ is a proper ideal, therefore the natural homomorphism $R_2 \to K_2$ can be taken via R_0', giving a homomorphism from a local subring (namely R_0') of K_1 onto K_2; this is the required specialization. Now the last point follows using (b). ∎

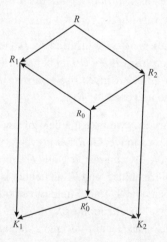

Corollary 7.2.8. *Let R be any ring. Given a minimal subhomomorphism of R-fields, $\phi : K_1 \to K_2$, if $p \in K_1$ is an element with an admissible matrix, this remains admissible over K_2 if and only if p is in the domain of ϕ.*

Proof. Let dom $\phi \subseteq K_1$ be the domain of ϕ, and let A be an admissible matrix for $p \in K_1$ that remains admissible over K_2. Since the denominator of A is invertible over the residue-class field of the local ring dom ϕ, it is invertible over dom ϕ itself, hence $p \in$ dom ϕ. Now consider the set of all elements of K_1 that can be represented by admissible matrices that remain admissible over K_2; this is a subring of dom ϕ that contains the image of R and admits inverses of elements whose image under ϕ is non-zero, hence it coincides with dom ϕ, by the minimality of the latter. ∎

In Theorem 2.7 suppose that K_2 is itself a universal localization; by Lemma 2.3 this means that $\mathfrak{m}_2 = 0$ and $R_2 = K_2$. Then the homomorphism $R_2 \to K_1$ is an isomorphism. Let us call a specialization of R-fields *proper* if it is not an isomorphism. Then we can express the result as

Corollary 7.2.9. *Let R be any ring. If an epic R-field K is itself a universal localization, then K cannot be obtained by a proper specialization from another R-field.* ∎

In particular, this shows that if R has a universal R-field U, then U is the only epic R-field that can be a universal localization. However, U may also be a universal localization of other rings, as the next result shows:

Proposition 7.2.10. *Let R be a ring with a universal R-field U and let Σ be any set of matrices over R such that R_Σ admits a homomorphism to a field. Then U is a universal R_Σ-field.*

Proof. Since the members of Σ are inverted over some field, they are inverted over U, so the homomorphism $R \to U$ can be taken via R_Σ; now any homomorphism from R_Σ to a field K gives rise to a map $R \to K$ and since U is universal for R, it is also universal for R_Σ. ∎

To illustrate Corollary 2.9, we take the field of fractions of a commutative integral domain R; this is a universal R-field (even a field of fractions) and a universal localization of R. By contrast, if E and F are commutative fields, then $R = E \times F$ has two epic R-fields E and F, but neither is a universal R-field. The ring $S = E[x]/(x^2)$ has a unique epic S-field, viz. $S/(x)$, so this is a universal S-field, but not a universal localization of S (because S is itself a local ring). The ring $T = S \times F$ also has E and F as epic T-fields, but only one of them, namely F, is a universal localization and neither is a universal T-field. Other examples are given in the exercises.

Exercises 7.2

1. Show that an epimorphic image of a commutative ring is again commutative. Let R be a commutative ring and Σ a set of square matrices over R. Show that

7.2 The category of R-fields and specializations

$R_\Sigma \cong R_S$, where S is the set of all determinants of members of Σ. Deduce that R_Σ is commutative.

2. Let $K = k(x, y)$ be a rational function field over a commutative field k and let L be any commutative k-field. Given any $a, b \in L$, verify that there is a subhomomorphism from K to L that maps any rational function $f(x, y)/g(x, y)$ such that $g(a, b) \neq 0$ to $f(a, b)/g(a, b)$. If $a, b \in k$, show that this subhomomorphism can be extended to one whose domain includes $u = (x - a)/(y - b)$, which is mapped to an element c of L. Deduce the existence of a specialization defined for $f(u, y)/g(u, y)$ whenever $g(c, b) \neq 0$ and taking this element to $f(c, b)/g(c, b)$.

3. Show that if in a ring homomorphism $f : R \to S$, S is the rational closure of f, then f is epic (in the categorical sense). Show that the inclusion of $k[x, y]$ in $k[x, y, xy^{-1}, xy^{-2}, \ldots]$ is an epimorphism, but the former ring is rationally closed in the latter.

4. (G. M. Bergman) Show that in $\mathfrak{M}_2(k\langle x, y\rangle)$ the inclusion of the subring $\begin{pmatrix} k & V \\ 0 & k \end{pmatrix}$, where V is the subspace spanned by $1, x, y$, is an epimorphism, even though the subring satisfies certain polynomial identities not holding in the whole ring.

5. Show that a commutative ring R has a universal R-field if and only if R has a unique least prime ideal (i.e. the nilradical of R is prime). What is the corresponding condition for left Noetherian rings?

6. For which values of n does the ring $\mathbb{Z}/n\mathbb{Z}$ of integers mod n have (i) a field of fractions, (ii) a universal localization that is a field and (iii) a universal $\mathbb{Z}/n\mathbb{Z}$-field?

7. In the diagram for the proof of Theorem 2.7 let $R = \mathbb{Z}, K_1 = \mathbb{Q}, K_2 = \mathbb{Z}/p\mathbb{Z}$, where p is a prime. Describe the rings and maps in the diagram.

8. Show that the $n \times n$ upper triangular matrix ring R over a field K has n epic R-fields isomorphic to K as fields, but distinct as R-fields. Deduce that for $n > 1$, R has no universal R-field.

9. Show that if R is a left Bezout domain, then for any set Σ of square matrices over R, $R_\Sigma \cong R_S$ for a suitable subset S of R. (*Hint*: Use Exercise 0.7.11.)

10*. Let R be a ring, Φ the set of all full matrices over R and suppose that there exists a Φ-inverting homomorphism into a non-zero weakly finite ring. Show that the universal localization R_Φ is a local ring.

11. Show that a homomorphism $R \to K$ of a ring into a field is an epimorphism (in the category of rings) if and only if K is an epic R-field.

12. A set Σ of square matrices over a ring R is said to be *saturated* if any matrix (not necessarily square) over R that becomes invertible in R_Σ lies in Σ. Show that any saturated set is multiplicative. Given any set Σ of matrices over R, let Σ' be the set of all matrices inverted in R_Σ. Show that Σ' is saturated and that $R_\Sigma \cong R_{\Sigma'}$ (Σ' is called the *saturation* of Σ).

13*. (Fisher [71]) Let k be a commutative field and $K = k(t)$ the rational function field in t over k with the endomorphism α_n induced by $t \mapsto t^n$. Show that if $n > 1$, the k-subalgebra of $K[x; \alpha_n]$ generated by x and $y = xt$ is free on x and y. Hence obtain an embedding of the free algebra $R = k\langle x, y\rangle$ in a field, viz. $K(x; \alpha_n)$. Show that the fields so obtained are 'minimal' objects in the category of fields of fractions of R and are non-isomorphic, as R-fields, for different n.

14*. (G. M. Bergman) Let R be a ring with two non-isomorphic fields of fractions K, L; say L is not a specialization of K. Denote by $\bar{R} = R + tK[t]$ the subring of $K[t]$ consisting of all polynomials with constant term in R. Show that $\bar{R}$ is an Ore domain and so has a field of fractions, viz. $K(t)$, but no universal R-field.

15*. (A. H. Schofield) Show that if R is a ring with a universal R-field, then the polynomial ring $R[t]$ has a universal $R[t]$-field.

16°. Let R be any ring. Given a local R-ring L, if L is the ring generated by R and the inverses of elements not in the maximal ideal, show that it is an epic R-ring. Find an example of an epic R-ring that is local but is not generated in this way.

17°. For which specializations of R-fields is the domain a valuation ring?

18°. Find conditions for the universal R-field to be a universal localization (see Section 7.6 below for a special case).

7.3 Matrix ideals

We have seen in Section 7.2 that just as the epic R-fields for a commutative ring R can be described in terms of their kernels, so in general they are determined by their singular kernels (Theorem 2.5 (ii)). Now the kernels in the commutative case are well known as prime ideals, and it remains to elucidate the singular kernels. This will be done in this section, where we shall describe them by a set of axioms reminiscent of the axioms for prime ideals. The resulting concept of a 'prime matrix ideal' is best studied as a special case of a matrix ideal; this is a set of matrices with properties similar to those of an ideal in a ring. In order to define it we need to describe the appropriate operations on matrices; instead of the sum and product we have the determinantal sum and the diagonal sum. The latter we have already met; we recall that for any two matrices A, B the diagonal sum $A \oplus B$ is always defined, and if A, B are square of orders r, s, respectively, then $A \oplus B$ is square of order $r + s$. We shall use notations such as $A_1 \oplus A_2 \oplus \cdots A_m$ for repeated diagonal sums; if in the last sum all the A_i are equal to A, we also write $\oplus^m A$.

We recall that a square matrix A is *non-full* if it can be written in the form PQ, where P is $n \times r$, Q is $r \times n$ and $r < n$. It follows that a non-full matrix maps to a singular matrix under any homomorphism into a field. If A is non-full, then its diagonal sum with any square matrix is again non-full. For if $A = PQ$ as above, then $A \oplus B = (P \oplus B)(Q \oplus I)$. On the other hand, if A is full, then its diagonal sum with another full matrix need not be full. For example, as we saw in Section 0.1, a ring fails to have UGN precisely if the unit matrix I_n of some order is not full, even though 1 is full in any non-zero ring. We also note that if A is a matrix (full or not) such that $\oplus^r A$ is not full, for some r, then $\oplus^r A$

and hence A itself maps to a singular matrix under any homomorphism into a field.

The second operation to be described is defined only for certain pairs of matrices. Let A and B be two $n \times n$ matrices that differ at most in the first column, say $A = (A_1, A_2, \ldots, A_n)$, $B = (B_1, A_2, \ldots, A_n)$; then we define the *determinantal sum* of A and B with respect to the first column as the matrix $C = (A_1 + B_1, A_2, \ldots, A_n)$. The determinantal sum with respect to another column (or a row) is defined similarly, when it exists. We write

$$C = A \nabla B, \qquad (1)$$

indicating, if necessary, in words the row or column whose elements are being added.

When R is commutative, so that determinants are defined, we have $\det C = \det A + \det B$, whenever A, B, C are related as in 1. On the other hand, over a field, even skew, if two of A, B, C are singular, then so is the third, as is easily seen. Further, any ring homomorphism preserves determinantal sums, therefore if two of A, B, C map to singular matrices under a homomorphism to a field, then so does the third.

Repeated determinantal sums need to be used with care, since the operation is not everywhere defined and *a fortiori* not associative. Thus to say that C is a determinantal sum of matrices $A_1, \ldots, A_m$ means that we can replace a pair of neighbouring matrices among $A_1, \ldots, A_m$ by their determinantal sum (with respect to some row or column) and repeat this process on another neighbouring pair in the resulting set, and so on until we are left with one matrix, namely C. We shall indicate this by writing

$$C = A_1 \nabla \ldots \nabla A_m, \qquad (2)$$

where it is understood that the sum has to be bracketed appropriately to be evaluated. We also note the form of the distributive law:

$$(A \nabla B) \oplus P = (A \oplus P) \nabla (B \oplus P), \qquad (3)$$

which holds whenever $A \nabla B$ is defined.

Let R be a ring and $\mathfrak{M}(R)$ the set of all square matrices over R. We define a *matrix pre-ideal* in R as a subset $\mathfrak{a}$ of $\mathfrak{M}(R)$ satisfying the following three conditions:

M.1. $\mathfrak{a}$ includes all non-full matrices,
M.2. If $A, B \in \mathfrak{a}$ and their determinantal sum C (with respect to some row or column) exists, then $C \in \mathfrak{a}$,
M.3. If $A \in \mathfrak{a}$, then $A \oplus B \in \mathfrak{a}$ for all $B \in \mathfrak{M}(R)$.

If, further, we have

M.4. $A \oplus 1 \in \mathfrak{a}$ implies $A \in \mathfrak{a}$,

we call $\mathfrak{a}$ a *matrix ideal*. A matrix pre-ideal is called *proper* if it is not $\mathfrak{M}(R)$; clearly a matrix ideal is proper if and only if it does not contain the element 1 or equivalently, I_0.

We note the following consequences of the definitions, where $\mathfrak{a}$ denotes any matrix pre-ideal.

(a) A *hollow* matrix, i.e. $n \times n$ with an $r \times s$ block of zeros, where $r + s > n$, is not full (Proposition 3.1.2), and hence is in $\mathfrak{a}$; clearly this still holds if the zero block only arises after certain permutations of rows and of columns. In particular, any square matrix with a zero row or column is hollow and so is in $\mathfrak{a}$.

(b) In an equation $C = A \nabla B$, if B is non-full, then A lies in $\mathfrak{a}$ if and only if C does. For we have $B \in \mathfrak{a}$, so if $A \in \mathfrak{a}$, then $C \in \mathfrak{a}$ by M.2, while if $C \in \mathfrak{a}$, we can bring B over to the other side, changing the sign of the row or column to be added; clearly this leaves B non-full.

(c) Let $A \in \mathfrak{a}$. Then the result of adding any right multiple of one column of A (or any left multiple of a row) to another again lies in $\mathfrak{a}$. Writing $A = (A_1, A_2, \ldots, A_n)$, we have

$$(A_1 + A_2 c, A_2, \ldots, A_n) = (A_1, A_2, \ldots, A_n) \nabla (A_2 c, A_2, \ldots, A_n)$$
$$= A \nabla (A_2, \ldots, A_n) \begin{pmatrix} c & \\ 0 & I_{n-1} \end{pmatrix};$$

on the right we have the determinantal sum (with respect to the first column) of A and a non-full matrix, hence the result is in $\mathfrak{a}$. Similarly for rows.

(d) If $A \in \mathfrak{a}$, then the result of interchanging any two columns (or rows) of A and changing the sign of one of them again lies in $\mathfrak{a}$. In particular, if $A \oplus B \in \mathfrak{a}$, then $B \oplus A \in \mathfrak{a}$. This follows in familiar fashion from (c). Writing only the two columns in question, we have, by repeated application of (c),

$$(A_1, A_2) \to (A_1 + A_2, A_2) \to (A_1 + A_2, -A_1) \to (A_2, -A_1).$$

(e) Let $A \in R_m, B \in R_n, C \in {}^n R^m$. Then

$$\begin{pmatrix} A & 0 \\ C & B \end{pmatrix} \in \mathfrak{a} \Leftrightarrow \begin{pmatrix} A & 0 \\ 0 & B \end{pmatrix} \in \mathfrak{a}. \tag{4}$$

For given A, B, C let A_1, C_1 be the first columns of A, C respectively, and

7.3 Matrix ideals

write $A = (A_1, A'), C = (C_1, C')$. We have

$$\begin{pmatrix} A & 0 \\ C & B \end{pmatrix} = \begin{pmatrix} A_1 & A' & 0 \\ 0 & C' & B \end{pmatrix} \nabla \begin{pmatrix} 0 & A' & 0 \\ C_1 & C' & B \end{pmatrix}, \tag{5}$$

where the determinantal sum is with respect to the first column. By (a) the second matrix on the right is not full, hence by (b),

$$\begin{pmatrix} A & 0 \\ C & B \end{pmatrix} \in \mathcal{a} \Leftrightarrow \begin{pmatrix} A_1 & A' & 0 \\ 0 & C' & B \end{pmatrix} \in \mathcal{a}.$$

In a similar way we can vary the other columns of C and so prove the assertion. An analogous argument, using B, shows that for any $m \times n$ matrix C,

$$\begin{pmatrix} A & C \\ 0 & B \end{pmatrix} \in \mathcal{a} \Leftrightarrow \begin{pmatrix} A & 0 \\ 0 & B \end{pmatrix} \in \mathcal{a}.$$

(f) If $\mathcal{a}$ is actually a matrix ideal, then for any two square matrices A, B of the same order, $AB \in \mathcal{a}$ if and only if $A \oplus B \in \mathcal{a}$. Assume that $A \oplus B \in \mathcal{a}$; using (e) and (c) several times, we obtain in turn

$$\begin{pmatrix} A & 0 \\ 0 & B \end{pmatrix} \to \begin{pmatrix} A & 0 \\ I & B \end{pmatrix} \to \begin{pmatrix} A & -AB \\ I & 0 \end{pmatrix} \to \begin{pmatrix} AB & A \\ 0 & I \end{pmatrix} \to \begin{pmatrix} AB & 0 \\ 0 & I \end{pmatrix}.$$

By applying M.4 we find that $AB \in \mathcal{a}$. The converse follows by reversing our steps.

(g) If A belongs to a matrix ideal $\mathcal{a}$, then the result of permuting the rows or the columns of A in any way again belongs to $\mathcal{a}$. For we can achieve any permutation by multiplying by an appropriate permutation matrix on the left or right.

(h) Let $f : R \to S$ be a ring homomorphism and $\mathcal{a}$ a matrix ideal in S. Then $f^{-1}\mathcal{a} = \{A \in \mathfrak{M}(R) | A^f \in \mathcal{a}\}$ is a matrix ideal in R. This follows because any homomorphism preserves non-full matrices and is compatible with determinantal sums and diagonal sums.

There is an operation on matrix ideals analogous to residual division of ideals. Let $\mathcal{a}$ be a matrix ideal and Σ a subset of $\mathfrak{M}(R)$ that is *directed:* $1 \in \Sigma$ and if $P_1, P_2 \in \Sigma$, then there exists $P \in \Sigma$ that is conjugate by signed permutation matrices, to $P_1 \oplus P_1'$ and $P_2 \oplus P_2'$, for suitable P_i'. We shall define the *residual division* of $\mathcal{a}$ by Σ as

$$\mathcal{a}/\Sigma = \{A \in \mathfrak{M}(R) | A \oplus P \in \mathcal{a} \text{ for some } P \in \Sigma\}. \tag{6}$$

It is easily seen that $\mathcal{a}/\Sigma$ is again a matrix ideal; M.1 is clear and M.2 follows by (3): if $A \oplus X, B \oplus Y \in \mathcal{a}$, then $A \oplus Z, B \oplus Z \in \mathcal{a}$, where Z is conjugate to $X \oplus X', Y \oplus Y'$, for some X', Y', and hence $(A \oplus Z) \nabla (B \oplus Z) \in \mathcal{a}$. Now

M.3 follows by (d), and in the same way we can verify M.4, to show that $\mathfrak{a}/\Sigma$ is a matrix ideal. This is still true if $\mathfrak{a}$ is merely a matrix pre-ideal but all the unit matrices lie in Σ.

We observe that $\mathfrak{a}/\Sigma \supseteq \mathfrak{a}$, by the definition (6). Suppose that $\mathfrak{a}$ is a proper matrix ideal and that $\mathfrak{a} \cap \Sigma = \varnothing$; we claim that $\mathfrak{a}/\Sigma$ is then proper. For if not, then $1 \oplus A \in \mathfrak{a}$ for some $A \in \Sigma$ hence $A \in \mathfrak{a}$, which is a contradiction. Conversely, when $\mathfrak{a} \cap \Sigma \neq \varnothing$, then $\mathfrak{a}/\Sigma$ is improper, as we see by reversing the argument. Summing up, we have

Proposition 7.3.1. *Let R be a ring, $\mathfrak{a}$ a matrix ideal and Σ a directed subset of $\mathfrak{M}(R)$. Then the set $\mathfrak{a}/\Sigma$ defined by (6) is a matrix ideal; this is still true if $\mathfrak{a}$ is a matrix pre-ideal and Σ contains all unit matrices. Further, $\mathfrak{a}/\Sigma$ is proper if and only if $\mathfrak{a} \cap \Sigma = \varnothing$.* ∎

Taking Σ to consist only of unit matrices, we obtain by the same argument

Corollary 7.3.2. *Let $\mathfrak{a}$ be any matrix pre-ideal in a ring R and let $\mathfrak{J}$ be the set of all unit matrices over R. Then $\mathfrak{a}/\mathfrak{J}$ is the least matrix ideal containing $\mathfrak{a}$.* ∎

Let $(\mathfrak{a}_\lambda)$ be any family of matrix ideals. Then it is clear that $\mathfrak{a} = \cap \mathfrak{a}_\lambda$ is again a matrix ideal. We can therefore speak of the 'least' matrix ideal containing a given subset $\mathfrak{X}$ of $\mathfrak{M}(R)$. This least matrix ideal is also called the matrix ideal *generated* by $\mathfrak{X}$. Similarly we can define the matrix pre-ideal generated by $\mathfrak{X}$. Explicitly this is obtained by taking the set of all matrices $X \oplus A (X \in \mathfrak{X}, A \in \mathfrak{M}(R))$ and non-full matrices and taking its closure under determinantal sums. For this set is contained in any matrix pre-ideal containing $\mathfrak{X}$, and it satisfies M.1 and M.2; let us show that it also satisfies M.3. If the set contains C, then (for suitable bracketing)

$$C = B_1 \nabla \ldots \nabla B_r \quad (B_i = X \oplus A \text{ or non-full}), \tag{7}$$

hence for any $P \in \mathfrak{M}(R)$, by (3),

$$C \oplus P = (B_1 \oplus P) \nabla \ldots \nabla (B_r \oplus P),$$

with the same bracketing.

Thus the matrix pre-ideal generated by a set $\mathfrak{X}$ consists precisely of all determinantal sums (7). Together with Corollary 3.2 this shows the truth of

Proposition 7.3.3. *The matrix ideal generated by a set $\mathfrak{X}$ is proper if and only if the unit matrix (of any size) cannot be expressed as a determinantal sum of non-full matrices and matrices of the form $X \oplus A$, where $X \in \mathfrak{X}$ and $A \in \mathfrak{M}(R)$.* ∎

Let z_0 be the matrix pre-ideal generated by the empty set. Clearly this is the least matrix pre-ideal and it consists precisely of all determinantal sums of non-full matrices. By Proposition 3.3 we obtain

Corollary 7.3.4. *A given ring has proper matrix ideals if and only if no unit matrix can be written as a determinantal sum of non-full matrices.* ∎

For example, any ring without UGN (and *a fortiori* any ring without IBN) has no proper matrix ideals, because the unit matrix of a certain size is actually non-full.

We now come to the definition of prime matrix ideals, which stand in the same relation to matrix ideals as do prime ideals to ideals. It will be convenient to begin by defining a multiplication of matrix ideals; this is an associative and commutative operation, like the multiplication of ideals in a commutative ring.

Definition Given two matrix ideals $\mathfrak{a}_1, \mathfrak{a}_2$ in a ring R, their *product*, denoted by $\mathfrak{a}_1 \mathfrak{a}_2$, is defined as the matrix ideal generated by all matrices $A_1 \oplus A_2$ with $A_i \in \mathfrak{a}_i (i = 1, 2)$.

The following lemma is often useful in constructing products. In writing a quotient set $\mathfrak{a}/\Sigma$, if $\Sigma = \{1, A, 1 \oplus A\}$, we shall simply write $\mathfrak{a}/A$.

Lemma 7.3.5. *In any ring R, let $\mathfrak{X}_i \subseteq \mathfrak{M}(R) (i = 1, 2)$, let $\mathfrak{X}$ be the set of matrices $A_1 \oplus A_2 (A_i \in \mathfrak{X}_i)$ and $\mathfrak{a}_1, \mathfrak{a}_2, \mathfrak{a}$, the matrix ideals generated by $\mathfrak{X}_1, \mathfrak{X}_2, \mathfrak{X}$, respectively. Then $\mathfrak{a} = \mathfrak{a}_1 \mathfrak{a}_2$. In particular, writing (A) for the matrix ideal generated by A, we have $(A)(B) = (A \oplus B)$, for any $A, B \in \mathfrak{M}(R)$.*

Proof. Clearly $\mathfrak{X} \subseteq \mathfrak{a}_1 \mathfrak{a}_2$; hence $\mathfrak{a} \subseteq \mathfrak{a}_1 \mathfrak{a}_2$. To establish equality, let $A_i \in \mathfrak{X}_i$; then $A_1 \oplus A_2 \in \mathfrak{a}$ by definition, hence $\mathfrak{X}_1 \subseteq \mathfrak{a}/A_2$ and so $\mathfrak{a}_1 \subseteq \mathfrak{a}/A_2$. It follows that $B_1 \oplus A_2 \in \mathfrak{a}$ for all $B_1 \in \mathfrak{a}_1$, so fixing B_1 we have $\mathfrak{X}_2 \subseteq \mathfrak{a}/B_1$, hence $\mathfrak{a}_2 \subseteq \mathfrak{a}/B_1$ and so $B_1 \oplus B_2 \in \mathfrak{a}$ for all $B_i \in \mathfrak{a}_i$, and it follows that $\mathfrak{a}_1 \mathfrak{a}_2 \subseteq \mathfrak{a}$. ∎

From this lemma it easily follows that the multiplication of matrix ideals is associative. We write $\mathfrak{a}_1 \mathfrak{a}_2 \mathfrak{a}_3$, etc. for repeated products, and abbreviate $\mathfrak{a}\mathfrak{a}, \mathfrak{a}\mathfrak{a}\mathfrak{a}, \ldots$ as $\mathfrak{a}^2, \mathfrak{a}^3, \ldots$. From property (g) it follows that the multiplication is commutative, and by M.3 the product is contained in the intersection of the factors:

$$\mathfrak{a}_1 \mathfrak{a}_2 = \mathfrak{a}_2 \mathfrak{a}_1 \subseteq \mathfrak{a}_1 \cap \mathfrak{a}_2.$$

A matrix ideal $\mathcal{P}$ is said to be *prime* if it is proper and

$$A \oplus B \in \mathcal{P} \Rightarrow A \in \mathcal{P} \text{ or } B \in \mathcal{P}.$$

An alternative description is given in

Proposition 7.3.6. *For any proper matrix ideal $\mathcal{P}$ in a ring R the following three conditions are equivalent:*

(a) $\mathcal{P}$ *is prime,*
(b) *for any matrix ideals $\mathcal{a}_1, \mathcal{a}_2$ we have $\mathcal{a}_1 \mathcal{a}_2 \subseteq \mathcal{P} \Rightarrow \mathcal{a}_1 \subseteq \mathcal{P}$ or $\mathcal{a}_2 \subseteq \mathcal{P}$,*
(c) *for any matrix ideals $\mathcal{a}_i \supseteq \mathcal{P}(i = 1, 2)$ we have $\mathcal{a}_1 \mathcal{a}_2 \subseteq \mathcal{P} \Rightarrow \mathcal{a}_1 = \mathcal{P}$ or $\mathcal{a}_2 = \mathcal{P}$.*

Proof. (a) $\Rightarrow$ (b). Let $\mathcal{P}$ be prime and $\mathcal{a}_1 \mathcal{a}_2 \subseteq \mathcal{P}$ but $\mathcal{a}_i \not\subseteq \mathcal{P}(i = 1, 2)$. Then there exists $A_i \in \mathcal{a}_i$ but $A_i \notin \mathcal{P}$. Since $\mathcal{P}$ is prime, $A_1 \oplus A_2 \notin \mathcal{P}$, but $A_1 \oplus A_2 \in \mathcal{a}_1 \mathcal{a}_2 \subseteq \mathcal{P}$, a contradiction.

(b) $\Rightarrow$ (c) is clear; to prove (c) $\Rightarrow$ (a), suppose that $A_1 \oplus A_2 \in \mathcal{P}$. Consider the matrix ideal $\mathcal{P}_i$ generated by $\mathcal{P}$ and A_i; for any $B_i \in \mathcal{P} \cup \{A_i\}$, $B_1 \oplus B_2 \in \mathcal{P}$, hence $\mathcal{a}_1 \mathcal{a}_2 \subseteq \mathcal{P}$ (by Lemma 3.5). By hypothesis, $\mathcal{a}_1$ or $\mathcal{a}_2$ must equal $\mathcal{P}$, so A_1 or A_2 lies in $\mathcal{P}$, i.e. (a). ∎

For any matrix ideal $\mathcal{a}$ we define its *radical* as the set

$$\sqrt{\mathcal{a}} = \{A \in \mathfrak{M}(R) | \oplus^r A \in \mathcal{a} \text{ for some } r \geq 1\}.$$

This set is again a matrix ideal: M.1, M.3 and M.4 clearly hold, and to prove M.2, we note that (by (3) of 7.3), $\oplus^n (A \nabla B)$ is a determinantal sum of terms $C_1 \oplus \cdots \oplus C_n$, where each C_i is A or B. Hence if $\oplus^r A$ and $\oplus^s B$ lie in $\mathcal{a}$, then $\oplus^{r+s-1} (A \nabla B) \in \mathcal{a}$; therefore $\sqrt{\mathcal{a}}$ also satisfies M.2 and so is a matrix ideal. More generally, if $\mathcal{a}$ is a matrix pre-ideal, then the radical of the corresponding matrix ideal $\mathcal{a}/\mathcal{J}$ is given by

$$\sqrt{(\mathcal{a}/\mathcal{J})}\{A \in \mathfrak{M}(R) | \oplus^r A \oplus I_s \in \mathcal{a} \text{ for some } r, s \geq 1\}.$$

A matrix ideal $\mathcal{a}$ will be called *semiprime* if $\sqrt{\mathcal{a}} = \mathcal{a}$; e.g. $\sqrt{\mathcal{a}}$ is always semiprime, in fact it is the least semiprime matrix ideal containing $\mathcal{a}$, as is easily verified (see Exercise 15).

The usual method of constructing prime ideals also works for prime matrix ideals:

Theorem 7.3.7. *Let R be any ring, Σ a non-empty subset of $\mathfrak{M}(R)$ closed under diagonal sums and $\mathcal{a}$ any matrix ideal such that $\mathcal{a} \cap \Sigma = \emptyset$. Then there exists a matrix ideal $\mathcal{P}$ which is maximal subject to the conditions $\mathcal{P} \supseteq \mathcal{a}, \mathcal{P} \cap \Sigma = \emptyset$, and any such matrix ideal is prime.*

Proof. The collection $\mathcal{C}$ of all matrix ideals containing $\mathcal{a}$ and disjoint from Σ is clearly inductive, so by Zorn's lemma it has a maximal member $\mathcal{P}$, and this satisfies the conditions of the theorem. Any such $\mathcal{P}$ is proper, because

7.3 Matrix ideals

$\Sigma \neq \emptyset$; now let $\mathfrak{a}_i \supseteq \mathcal{P}(i = 1, 2)$ be matrix ideals such that $\mathfrak{a}_1 \mathfrak{a}_2 \subseteq \mathcal{P}$. If $\mathfrak{a}_i \neq \mathcal{P}$ for $i = 1, 2$, then $\mathfrak{a}_i \supset \mathcal{P}$, so by the maximality of $\mathcal{P}$, $\mathfrak{a}_i \cap \Sigma \neq \emptyset$. Take $A_i \in \mathfrak{a}_i \cap \Sigma$; then $A_1 \oplus A_2 \in \mathcal{P} \cap \Sigma$, which is a contradiction. Hence $\mathfrak{a}_1$ or $\mathfrak{a}_2$ equals $\mathcal{P}$, and so $\mathcal{P}$ is prime, by Proposition 3.6. ∎

This theorem shows for example that every maximal proper matrix ideal is prime; we need only take $\Sigma = \mathcal{J} = \{I_n | n = 0, 1, \ldots\}$. Of course, one of its main uses in the commutative case is the representation of semiprime ideals as the intersection of prime ideals; a corresponding result holds in the general case.

Theorem 7.3.8. *If R is any ring and $\mathfrak{a}$ a matrix ideal in R, then*

$$\sqrt{\mathfrak{a}} = \cap\{\mathcal{P} | \mathcal{P} \text{ prime} \supseteq \mathfrak{a}\}. \qquad (8)$$

Thus $\sqrt{\mathfrak{a}}$ is the intersection of all prime matrix ideals containing $\mathfrak{a}$.

Proof. If $\mathcal{P}$ is a prime matrix ideal and $\mathcal{P} \supseteq \mathfrak{a}$, then $\sqrt{\mathfrak{a}} \subseteq \sqrt{\mathcal{P}} = \mathcal{P}$, hence $\sqrt{\mathfrak{a}} \subseteq \cap \mathcal{P}$, where the intersection is taken over all prime matrix ideals containing $\mathfrak{a}$. To establish equality, let $A \notin \sqrt{\mathfrak{a}}$ be given and consider the set Σ_A of all diagonal sums of copies of A. By definition, $\Sigma_A \cap \mathfrak{a} = \emptyset$, hence by Theorem 3.7 there is a maximal matrix ideal $\mathcal{P}_0$ containing $\mathfrak{a}$ and disjoint from Σ_A, and $\mathcal{P}_0$ is prime. Thus $A \notin \cap \mathcal{P}$ and this proves equality in (8). ∎

Taking $\mathfrak{a}$ to be semiprime, we obtain

Corollary 7.3.9. *In any ring a matrix ideal is semiprime if and only if it is an intersection of prime ideals.* ∎

Exercises 7.3

1. Give an example to show that property (f) need not hold for matrix pre-ideals.
2. Let $\mathfrak{a}$ be an ideal in a commutative ring R. Show that the set $\mathfrak{a}^*$ of matrices A such that $\det A \in \mathfrak{a}$ is a matrix ideal. More generally, define $\mathfrak{X}^*$ as the set $(\mathfrak{X})^*$, where $(\mathfrak{X})$ is the ideal generated by $\mathfrak{X} \subseteq \mathfrak{M}(R)$. If $\mathfrak{a}$ is a matrix ideal of R, show that the set $\mathfrak{a}^{\blacklozenge}$ of elements $\det A$, $A \in \mathfrak{a}$, if closed under addition, is an ideal. Verify that $\mathfrak{a}^{*\blacklozenge} = \mathfrak{a}$, $\mathfrak{a}^{\blacklozenge *} \supseteq \mathfrak{a}$, and that equality holds here if R is Euclidean. What is the relation between $\mathfrak{a}^{\blacklozenge}$ and the 1×1 matrices in $\mathfrak{a}$?
3. Show that a ring with a proper matrix ideal has UGN.
4. Show that a field has precisely one proper matrix ideal.
5. A matrix is called *degenerate* if two rows are left (or two columns are right) linearly dependent. Show that if A is non-full, then $A \oplus I$ (for suitable I) is a determinantal sum of hollow and degenerate matrices.
6. Given any ring R, let $\mathfrak{a}$ be a matrix ideal in R and $\mathfrak{X} \subseteq \mathfrak{M}(R)$. Show that the set $\mathfrak{a} : \mathfrak{X} = \{A \in \mathfrak{M}(R) | A \oplus X \in \mathfrak{a} \text{ for all } X \in \mathfrak{X}\}$ is a matrix ideal, the *quotient* of

$\mathfrak{A}$ by $\mathfrak{X}$ (not to be confused with $\mathfrak{A}/\mathfrak{X}$), and that it may be characterized as the largest matrix ideal $\mathfrak{Q}$ satisfying $X \oplus Q \in \mathfrak{A}$ for all $X \in \mathfrak{X}$, $Q \in \mathfrak{Q}$. If the matrix ideal generated by $\mathfrak{X}$ is denoted by $(\mathfrak{X})$, show that $\mathfrak{A} : \mathfrak{X} = \mathfrak{A} : (\mathfrak{X})$.

7. Show that if $\mathfrak{A}, \mathfrak{A}_1$ are matrix ideals and $\mathfrak{X} \subseteq \mathfrak{M}(R)$, then $\mathfrak{A} : \mathfrak{X} = \cap \{\mathfrak{A} : \{X\} | X \in \mathfrak{X}\}$, $(\mathfrak{A}_1 : \mathfrak{A}_2) : \mathfrak{A}_3 = \mathfrak{A}_1 : \mathfrak{A}_2 \mathfrak{A}_3$, $(\cap \mathfrak{A}_\lambda) : \mathfrak{X} = \cap (\mathfrak{A}_\lambda : \mathfrak{X})$.

8. Let $\mathscr{P}$ be a minimal prime matrix ideal in a ring R. Show that for each $A \in \mathscr{P}$ there exists a matrix $C \notin \mathscr{P}$ and a positive integer r such that $C \oplus^r A$ is a determinantal sum of non-full matrices.

9. A matrix ideal $\mathfrak{Q}$ is said to be *primary* if it is proper and $A \oplus B \in \mathfrak{Q}$, $A \notin \mathfrak{Q}$ implies that $\oplus^r B \in \mathfrak{Q}$ for some $r \geq 1$. Show that for any primary matrix ideal $\mathfrak{Q}$, $\sqrt{\mathfrak{Q}}$ is prime (in this case $\mathfrak{Q}$ is also called $\sqrt{\mathfrak{Q}}$-primary). Give an example of a matrix ideal $\mathfrak{Q}$ such that $\sqrt{\mathfrak{Q}}$ is prime but $\mathfrak{Q}$ is not primary. Show that if $\mathfrak{Q}$ is $\mathscr{P}$-primary, then for any $\mathfrak{X} \subseteq \mathfrak{M}(R)$, $\mathfrak{Q} : \mathfrak{X}$ is proper if and only if $\mathfrak{X} \nsubseteq \mathfrak{Q}$ and when this is so, it is $\mathscr{P}$-primary.

10*. Let $\mathfrak{A}$ be a matrix ideal that has a primary decomposition, i.e. $\mathfrak{A} = \mathfrak{Q}_1 \cap \ldots \cap \mathfrak{Q}_r$, where $\mathfrak{Q}_i$ is $\mathscr{P}_i$-primary. Show that $\sqrt{\mathfrak{A}} = \mathscr{P}_1 \cap \ldots \cap \mathscr{P}_r$, and show that $\mathfrak{A}$ also has a primary decomposition in which the $\mathscr{P}_i$ are all different and no $\mathfrak{Q}_i$ contains the intersection of all the others (such a primary decomposition is called *irredundant*). Show that any two irredundant primary decompositions of $\mathfrak{A}$ have the same number of terms, and these can be numbered so that corresponding primes are equal. If $\mathscr{P}_1, \ldots, \mathscr{P}_s$ (for suitable numbering) are those of the $\mathscr{P}_i$ disjoint from a given set $\mathfrak{X}$, show that the corresponding intersection $\mathfrak{Q}_1 \cap \ldots \cap \mathfrak{Q}_s$ is determined entirely by $\mathfrak{A}$ and $\mathfrak{X}$ (independent of the particular decomposition), but this is not necessarily true of each $\mathfrak{Q}_i$. (*Hint*: For the last part localize at the complement of $\mathscr{P}_1 \cup \ldots \cup \mathscr{P}_s$.)

11°. Find conditions on a ring R under which every matrix ideal is a finite intersection of primary matrix ideals (recall that in a commutative Noetherian ring every ideal has a finite primary decomposition; see e.g. BA, Theorem 10.8.8).

12. Show that the union and intersection of any chain of prime matrix ideals are again prime. Show that for any two prime matrix ideals $\mathfrak{A} \subset \mathfrak{B}$ there exist prime matrix ideals $\mathscr{P}_1, \mathscr{P}_2$ such that $\mathfrak{A} \subseteq \mathscr{P}_1 \subset \mathscr{P}_2 \subseteq \mathfrak{B}$, but there is no prime matrix ideal between $\mathscr{P}_1$ and $\mathscr{P}_2$.

13. Find conditions on 2×2 matrices over a field for the determinantal sum of two nilpotent matrices to be nilpotent.

14. If $C = A \nabla B$, where the determinantal sum is with respect to a column, show that $PC = PA \nabla PB$, for any P of appropriate size, but that in general $CP \neq AP \nabla BP$. However, show that $CP = AP \nabla BP$ when P is diagonal (or more generally, if the determinantal sum is with respect to the first column and the first row of P after the first element is zero).

15. Show that for any matrix ideal $\mathfrak{A}$ in a ring R the following are equivalent: (a) $\mathfrak{A}$ is semiprime, (b) for any matrix ideal $\mathfrak{Q}$, $\mathfrak{Q}^2 \subseteq \mathfrak{A} \Rightarrow \mathfrak{Q} \subseteq \mathfrak{A}$ and (c) for any matrix ideal $\mathfrak{Q}$, $\mathfrak{Q} \supseteq \mathfrak{A} \supseteq \mathfrak{Q}^2 \Rightarrow \mathfrak{Q} = \mathfrak{A}$. Verify that for any matrix ideal $\mathfrak{A}$, $\sqrt{\mathfrak{A}}$ is semiprime and that $\sqrt{\sqrt{\mathfrak{A}}} = \sqrt{\mathfrak{A}}$. Deduce that $\sqrt{\mathfrak{A}}$ is the least semiprime matrix ideal $\supseteq \mathfrak{A}$.

16°. Investigate rings in which every matrix ideal is finitely generated (e.g. examine the commutative case). Investigate rings in which every finitely generated matrix ideal can be generated by a single matrix.

17°. Investigate the notion of a 'sum' or 'join' of matrix ideals; is the lattice of matrix ideals modular? For what rings is the lattice of matrix ideals distributive?
18°. Develop a theory of one-sided matrix ideals, admitting determinantal sums only with respect to columns (or only rows), and find an application.

7.4 Constructing the localization

Given a ring R and a set Σ of matrices over R, we shall be interested in constructing the universal localization R_Σ. The direct construction is of course straightforward, as we saw in Section 7.2, but the main problem is to obtain information about the kernel of the natural map $\lambda : R \to R_\Sigma$, in particular, conditions for it to be injective. For this reason we shall take a somewhat less direct route.

Since we are mainly concerned with the embedding of rings in fields, we shall restrict the matrices to be inverted to be square. Thus Σ will be a subset of $\mathfrak{M}(R)$; we further assume that Σ is upper multiplicative and *factor-stable*, by which is meant that if a product of square matrices PQ is in Σ, then $P, Q \in \Sigma$.

We shall construct R_Σ as the set of all displays $a - (x, A, u)$, intended to represent $a - xA^{-1}u$, as defined in Section 7.1, modulo a certain equivalence. Thus we consider the set $B(\Sigma)$ of all displays

$$\alpha = \begin{pmatrix} A & u \\ x & a \end{pmatrix}, \qquad (1)$$

where $a \in R$, x, u are a row and column respectively over R and $A \in \Sigma$. The expression $a - xA^{-1}u$ will be called its *value*. The special case where A, x, u are absent just represents the element a of R; such a display is said to be *scalar*. On each display we define the following *elementary* operations:

F.1. Replace u by $u + Af$ and a by $a + xf$, where f is any column of appropriate size.
F.2. Replace A by AQ and x by xQ, where $Q \in \Sigma$.
F.3. Replace the display (1) by

$$\begin{pmatrix} A & 0 & u \\ 0 & G & 0 \\ x & p & a \end{pmatrix},$$

where $G \in \Sigma$ and p is a row over R.

The process F.3 will be called *inserting a trivial block*. The inverse of F.3, removing a trivial block, also counts as an elementary operation; similarly for

F.2 of course the inverse of F.1 is again of the same form. The application of F.1 yields a display with value $a + xf - xA^{-1}(u + Af) = a - xA^{-1}u$; similarly the result of F.2 is $a - xQ(AQ)^{-1}u = a - xA^{-1}u$, and from F.3 we have

$$a - \begin{pmatrix} x & p \end{pmatrix} \begin{pmatrix} A^{-1} & 0 \\ 0 & G^{-1} \end{pmatrix} \begin{pmatrix} u \\ 0 \end{pmatrix} = a - xA^{-1}u;$$

where the mapping λ is always understood; thus in each case the value is unchanged.

In addition we have the operations F.1'–F.3', where rows and columns are interchanged, thus in F.1' we add a left multiple of the row block containing A to the last row, in F.2' we multiply the first row block by a matrix from Σ on the left and in F.3' the trivial block has a zero row instead of a zero column.

Given two displays α, β, we shall write $\alpha \sim \beta$ to mean that β is obtained from α by a sequence of elementary operations. Clearly this relation is an equivalence on $B(\Sigma)$; the equivalence class containing $\alpha \in B(\Sigma)$ is written $[\alpha]$ and the set of all such classes is denoted by $M(\Sigma)$. We shall define two binary operations on $B(\Sigma)$, to represent addition and multiplication:

$$\begin{pmatrix} A & u \\ x & a \end{pmatrix} [+] \begin{pmatrix} B & v \\ y & b \end{pmatrix} = \begin{pmatrix} A & 0 & u \\ 0 & B & 0 \\ x & y & a \end{pmatrix} \nabla \begin{pmatrix} A & 0 & 0 \\ 0 & B & v \\ x & y & b \end{pmatrix} = \begin{pmatrix} A & 0 & u \\ 0 & B & v \\ x & y & a+b \end{pmatrix}, \quad (2)$$

$$\begin{pmatrix} A & u \\ x & a \end{pmatrix} \otimes \begin{pmatrix} B & v \\ y & b \end{pmatrix} = \begin{pmatrix} A & 0 & u \\ 0 & I & 0 \\ x & 0 & a \end{pmatrix} \begin{pmatrix} I & 0 & 0 \\ 0 & B & v \\ 0 & y & b \end{pmatrix} = \begin{pmatrix} A & uy & ub \\ 0 & B & v \\ x & ay & ab \end{pmatrix}. \quad (3)$$

A small calculation (left to the reader) shows that (2) represents the sum and (3) the product, of $a - xA^{-1}u$ and $b - yB^{-1}v$. If we apply an elementary operation to a display on the left-hand side of either of these equations, this is equivalent to applying a corresponding operation of the same type to the right-hand side, as is easily verified. This allows us to define these operations on the equivalence classes, i.e. the elements of $M(\Sigma)$. The class containing the scalar display (0) is called the *zero element* of $M(\Sigma)$ and is denoted by 0. A condition for a display to represent zero is given by a form of Malcolmson's criterion:

Lemma 7.4.1. *In any ring R, the display α given by (1) represents the zero element of $M(\Sigma)$ if and only if there exist $F, G, P, Q \in \Sigma$, rows f, p and columns*

7.4 Constructing the localization

g, q over R such that

$$\begin{pmatrix} A & 0 & 0 & u \\ 0 & F & 0 & 0 \\ 0 & 0 & G & g \\ x & f & 0 & a \end{pmatrix} = \begin{pmatrix} P \\ p \end{pmatrix} (Q \quad q). \tag{4}$$

Thus the display is non-full in this case.

Proof. We note that the matrix on the left of (4) can be factorized as

$$\begin{pmatrix} I & 0 & 0 & 0 \\ 0 & F & 0 & 0 \\ 0 & 0 & I & 0 \\ 0 & f & 0 & 1 \end{pmatrix} \begin{pmatrix} A & 0 & 0 & u \\ 0 & I & 0 & 0 \\ 0 & 0 & I & 0 \\ x & 0 & 0 & a \end{pmatrix} \begin{pmatrix} I & 0 & 0 & 0 \\ 0 & I & 0 & 0 \\ 0 & 0 & G & g \\ 0 & 0 & 0 & 1 \end{pmatrix}.$$

Suppose that α represents zero; then it can be reduced by elementary operations to the form 0, a non-full display, so if the inverses of F.3, F.3' are omitted, we obtain a non-full display as in (4). Conversely, it is clear that the left-hand side of (4) is equivalent to α given by (1), and for a display of the form (4), we apply F.2, F.2', F.1 and the inverse of F.3 to obtain

$$\begin{pmatrix} PQ & Pq \\ pQ & pq \end{pmatrix} \to \begin{pmatrix} P & Pq \\ p & pq \end{pmatrix} \to \begin{pmatrix} I & q \\ p & pq \end{pmatrix} \to \begin{pmatrix} I & 0 \\ p & 0 \end{pmatrix} \to (0). \blacksquare$$

From the form of the definitions (2), (3) it is clear that both operations are associative. The addition and multiplication formulae with a scalar display have the forms

$$\begin{pmatrix} A & u \\ x & a \end{pmatrix} [+](b) = \begin{pmatrix} A & u \\ x & a+b \end{pmatrix}, \quad \begin{pmatrix} A & u \\ x & a \end{pmatrix} [\times](b) = \begin{pmatrix} A & ub \\ x & ab \end{pmatrix};$$

it follows that 0 is the neutral element for addition and 1 is the neutral for multiplication. Further, if we form the sum of a display and the same display with the sign of the last column changed, we obtain

$$\begin{pmatrix} A & u \\ x & a \end{pmatrix} [+] \begin{pmatrix} A & -u \\ x & -a \end{pmatrix} = \begin{pmatrix} A & 0 & u \\ 0 & A & -u \\ x & x & 0 \end{pmatrix} \to \begin{pmatrix} A & A & 0 \\ 0 & A & -u \\ x & x & 0 \end{pmatrix}$$

$$\to \begin{pmatrix} A & 0 & 0 \\ 0 & A & -u \\ x & 0 & 0 \end{pmatrix} \to \begin{pmatrix} A & 0 \\ 0 & I \\ x & 0 \end{pmatrix} \begin{pmatrix} I & 0 & 0 \\ 0 & A & -u \end{pmatrix}.$$

This shows that each display has a negative and the classes form a group under addition. To prove distributivity, in the form $(\alpha + \beta)\gamma = \alpha\gamma + \beta\gamma$, let

$$\alpha = \begin{pmatrix} A & u \\ x & a \end{pmatrix}, \quad \beta = \begin{pmatrix} B & v \\ y & b \end{pmatrix}, \quad \gamma = \begin{pmatrix} C & w \\ z & c \end{pmatrix}.$$

The display representing $\alpha\gamma + \beta\gamma$ is

$$\begin{pmatrix} A & uz & uc \\ 0 & C & w \\ x & az & ac \end{pmatrix} \oplus \begin{pmatrix} B & vz & vc \\ 0 & C & w \\ y & bz & bw \end{pmatrix} = \begin{pmatrix} A & uz & 0 & 0 & uc \\ 0 & C & 0 & 0 & w \\ 0 & 0 & B & vz & vc \\ 0 & 0 & 0 & C & w \\ x & az & y & bz & sc \end{pmatrix},$$

where $s = a + b$. On applying F.3, F.3' and the inverse of F.3 in turn we obtain

$$\begin{pmatrix} A & uz & 0 & 0 & uc \\ 0 & C & 0 & 0 & w \\ 0 & 0 & B & vz & vc \\ 0 & 0 & 0 & C & w \\ x & az & y & bz & sc \end{pmatrix} \to \begin{pmatrix} A & uz & 0 & uz & uc \\ 0 & C & 0 & C & w \\ 0 & 0 & B & vz & vc \\ 0 & 0 & 0 & C & w \\ x & az & y & sz & sc \end{pmatrix} \to \begin{pmatrix} A & uz & 0 & uz & uc \\ 0 & C & 0 & 0 & 0 \\ 0 & 0 & B & vz & vc \\ 0 & 0 & 0 & C & w \\ x & az & y & sz & sc \end{pmatrix}$$

$$\to \begin{pmatrix} A & 0 & uz & uc \\ 0 & B & vz & vc \\ 0 & 0 & C & w \\ x & y & sz & sc \end{pmatrix},$$

and this represents $(\alpha + \beta)\gamma$. The other distributive law follows similarly. Thus we have shown that $M(\Sigma)$ is a ring under the operations (2) and (3).

If we assume that our ring has UGN (to ensure that the unit matrices of all orders are full), then $M(\Sigma) \neq 0$; for the display consisting of a unit matrix, representing 1, is full and so cannot satisfy an equation (4). Since the matrices in Σ are to be inverted, we shall also assume these matrices to be full, to avoid pathologies. Thus we have a mapping $\varphi : M(\Sigma) \to R_\Sigma$ given by

$$\begin{pmatrix} A & u \\ x & a \end{pmatrix} \mapsto a - xA^{-1}u, \tag{5}$$

which by our earlier remarks depends only on the equivalence class of the display and from (2), (3) is easily verified to be a homomorphism. Further, it is Σ-inverting, for if $A \in \Sigma$, then the display $\begin{pmatrix} A & -e_j^\mathrm{T} \\ e_i & 0 \end{pmatrix}$ represents the (i, j)-entry of A^{-1}, as we see from (5). Now the universality of R_Σ shows that φ has an inverse and so is an isomorphism.

We sum up the result as

Theorem 7.4.2. *Let R be a ring with unbounded generating number and Σ an upper multiplicative factor-stable set of full matrices over R. Then the universal localization R_Σ is a non-zero ring.* ■

We can now achieve our aim of showing that each prime matrix ideal is the singular kernel of a homomorphism to a field:

Theorem 7.4.3. *Let R be any ring.*

(i) *If $\mathcal{P}$ is any prime matrix ideal of R, then the localization $R_\mathcal{P}$ is a local ring and the singular kernel of its residue class field is $\mathcal{P}$.*
(ii) *If K is an epic R-field, with singular kernel $\mathcal{P}$, then $\mathcal{P}$ is a prime matrix ideal and the local ring $R_\mathcal{P}$ has a residue-class field isomorphic to K.*

Proof. (i) Since $\mathcal{P}$ is a prime matrix ideal, R has UGN; further, the complement Σ of $\mathcal{P}$ in $\mathfrak{M}(R)$ is multiplicative and we can form the localization $R_\Sigma = R_\mathcal{P}$. By Theorem 4.2 this is a non-zero ring.

To show that R_Σ is a local ring, let $A \in \Sigma$ and suppose that its $(1, 1)$-minor is in $\mathcal{P}$. Writing $A = (A_1, A')$, we have

$$A - e_{11} = (A_1, A')\nabla(-e_1, A'). \tag{6}$$

The first matrix on the right is just A, while the second lies in $\mathcal{P}$, because the first column after the first entry is 0 and its $(1, 1)$-minor is in $\mathcal{P}$. Hence the left-hand side lies in Σ, and applying Proposition 2.6, we see that $R_\Sigma = R_\mathcal{P}$ is a local ring, as claimed. Now the remaining assertion of (i) follows from the construction of $R_\mathcal{P}$.

To prove (ii) let K be an epic R-field with singular kernel $\mathcal{P}$. Its complement is the set of all square matrices inverted over K, which is clearly multiplicative. To verify M.1–M.4 for $\mathcal{P}$, we take matrices in $\mathcal{P}$ and examine their images in K. Any non-full matrix is singular, hence M.1 holds. Given two singular matrices A, B whose determinantal sum C is defined, say $A = (A_1, A')$, $B = (B_1, A')$, $C = (A_1 + B_1, A')$, where A_1, B_1 are columns and A' is $n \times n - 1$, either the columns of A' are right K-linearly dependent or A_1, B_1 are right K-linear combinations of the columns of A'. In both cases it follows that C is singular, and this proves M.2. Now M.3 is clear. For M.4 suppose that $A \oplus 1$ is a singular matrix over K; then A must be singular, and this proves M.4. Finally, $\mathcal{P}$ is prime because its complement is clearly multiplicative. Thus $\mathcal{P}$ is a prime matrix ideal and by (i), $R_\mathcal{P}$ is a local ring; moreover any square matrix over R lies in $\mathcal{P}$ if and only if its image in K is singular, and hence the residue-class field is isomorphic to K. ■

Over an epic R-field K, any non-zero element has positive stable rank, hence $f = 0$ if and only if its numerator, or also its display, is singular over K, or equivalently, lies in the singular kernel. Thus we have

Corollary 7.4.4. *Let R be any ring with an epic R-field K with singular kernel $\mathcal{P}$. Then an element of K is non-zero if and only if its numerator is not in $\mathcal{P}$, or equivalently, a display representing it is not in $\mathcal{P}$.* ∎

Theorem 4.3 tells us that for any ring R there is a natural bijection between the (isomorphism classes of) epic R-fields and prime matrix ideals of R; moreover, by Theorem 2.7, specialization of fields corresponds to inclusion of their singular kernels. Thus the set $X(R)$ of isomorphism classes of epic R-fields, partially ordered by specialization, may also be thought of as the set of all prime matrix ideals of R, partially ordered by inclusion. This way of looking at $X(R)$ also assures us that it may be regarded as a set.

Each square matrix A over R determines a subset of $X(R)$, called its *singularity support*:

$$\mathcal{D}(A) = \{x \in X(R) | A \notin \mathcal{P}_x\}, \tag{7}$$

where $\mathcal{P}_x$ is the prime matrix ideal corresponding to x. It is easily verified that

$$\mathcal{D}(A \oplus B) = \mathcal{D}(A) \cap \mathcal{D}(B), \quad \mathcal{D}(I) = X(R).$$

Hence the collection of sets $\mathcal{D}(A)$ admits finite intersections, and so may be taken as a base for the open sets of a topology on $X(R)$. The topological space $X(R)$ so defined is called the *field-spectrum* of R. It is analogous to the usual prime spectrum of a commutative ring, to which it reduces when R is taken to be commutative. We shall not pursue this point of view but merely note that $X(R)$ satisfies the conditions for a spectral space, which characterizes the prime spectrum of a commutative ring (see Hochster [69], Cohn [72c], and Exercise 5 below).

We can also describe the behaviour of homomorphisms under localization.

Theorem 7.4.5. *Let R, R' be rings with prime matrix ideals $\mathcal{P}$, $\mathcal{P}'$, their complements Σ, Σ' and corresponding epic R-fields K, K' respectively and let $f : R \to R'$ be a homomorphism. Then*

(i) f extends to a specialization if and only if f maps $\mathcal{P}$ into $\mathcal{P}'$ and
(ii) f extends to a homomorphism $K \to K'$ if and only if f maps $\mathcal{P}$ to $\mathcal{P}'$ and Σ to Σ'.

Proof. (i) The set $\mathcal{P}' f^{-1} = \{A \in \mathfrak{M}(R) | Af \in \mathcal{P}'\}$ is a prime matrix ideal of R and the condition $\mathcal{P} f \subseteq \mathcal{P}'$ of (i) is equivalent to $\mathcal{P} \subseteq \mathcal{P}' f^{-1}$, which is the criterion for a specialization, by Theorem 2.7. This proves (i); for (ii) we have

$\Sigma f \subseteq \Sigma'$, hence the denominators of elements of K remain invertible over K', so we have a homomorphism, and again the condition is clearly necessary. ∎

Exercises 7.4

1. Give an example of a saturated set of matrices (see Exercise 2.12) whose complement is not a union of prime matrix ideals. (*Hint*: Consider the invertible matrices in a general ring.)
2*. (Gerasimov [79]) Let R be a ring, $\Sigma \subseteq R$ and $\mathcal{Z}$ the least matrix ideal (generated by the empty set). Show that the universal Σ-inverting map $\lambda : R \to R_\Sigma$ is injective if and only if $\mathcal{Z}/\bar{\Sigma}$ contains no 1×1 matrices other than 0, where $\bar{\Sigma}$ denotes the saturation of Σ.
3. A topological space is called *irreducible* if it is non-empty and cannot be written as a union of two proper closed subsets. Show that for any ring R, a closed subset of its field-spectrum $X(R)$ is irreducible if and only if it is the closure of a single point.
4. Show that the singularity support $\mathcal{D}(A)$ is compact, for any matrix A. (*Hint*: Assume that $\mathcal{D}(A) = \cup \mathcal{D}(A_i)$ and denote by $\mathcal{C}$ the matrix ideal generated by the A_i. Show that $A \in \sqrt{\mathcal{C}}$ and deduce that $A \in \sqrt{\mathcal{C}_0}$, where $\mathcal{C}_0$ is the matrix ideal generated by a finite subfamily of the A_i).
5*. Let R be a ring and $X = X(R)$ its field-spectrum. Verify that X satisfies Hochster's axioms for a spectral space (Hochster [69]):
   (i) X is a T_0-space, i.e. given $x, y \in X$, $x \neq y$, there is an open subset containing precisely one of x, y;
   (ii) X is compact;
   (iii) the compact open sets admit finite intersections and form a base for the open sets, and
   (iv) any irreducible closed subset is the closure of a single point.
6. Verify that the displays (2) and (3) represent the sum and product, respectively, as stated in the text.
7. Show that a ring homomorphism $f : R \to S$ induces a continuous map $f^* : X(S) \to X(R)$.
8°. Describe the rings whose field-spectrum is a T_1-space (i.e. every prime matrix ideal is maximal).
9°. Which rings have a Noetherian field-spectrum (i.e. ACC on prime matrix ideals)?
10°. For a given ring (e.g. a semifir) find the localizations that are right Ore.
11*. Given $A \in \mathfrak{M}(R)$, define Σ_A as the complement of $\cup \{\mathcal{P} | \mathcal{P} \in \mathcal{D}(A)\}$, write $R(A)$ for the universal localization at Σ_A and R_A for the universal localization at A. Show that there is a canonical homomorphism $R_A \to R(A)$, which is an isomorphism when R is commutative, but not in general. Show also that the canonical homomorphism $\phi : R \to R(A)$ induces a homomorphism $\phi^* : X(R(A)) \to \mathcal{D}(A)$. Given $A, B \in \mathfrak{M}(R)$, verify that $\mathcal{D}(A) \supseteq \mathcal{D}(B) \Rightarrow \Sigma_A \subseteq \Sigma_B$ and hence obtain a homomorphism $\phi_{AB} : R(A) \to R(B)$. Show that $\{R(A), \phi_{AB}\}$ is a presheaf of rings over R (see e.g. Macdonald [68]). Verify that the corresponding sheaf R^* has as stalk over $x \in X(R)$ the universal localization at $\mathcal{P}_x$ (see Cohn [72c, 79]).

12. Let R be a fir and $\mathscr{P}$ a prime matrix ideal of R. Show that $\mathscr{Q} = \cap_n \mathscr{P}^n$ is again prime. Is $\mathscr{Q}$ always the least matrix ideal?
13°. (W. Stephenson) If R is a commutative ring, the clopen (= closed and open) subsets of $X(R)$ correspond to idempotents (see Jacobson [89], p. 406). What corresponds to a clopen subset of $X(R)$ in the general case? Is there a commutative ring R such that $X(k\langle X\rangle) \cong X(R)$?
14. Let R be a ring with UGN, but not weakly finite. Suppose further that the set Φ of all full matrices is upper multiplicative and stable under products. Show that the natural map $R \to R_\Phi$ is not an embedding.
15°. Let R be any ring and $\mathscr{P}$ a prime matrix ideal, so that the localization $R_\mathscr{P}$ is a local ring (by Theorem 4.3). Find the conditions on $\mathscr{P}$ for $R_\mathscr{P}$ to be a valuation ring (i.e. R is the set of all elements of value ≥ 0 in a valuated field).

7.5 Fields of fractions

The results of the last section can be used to answer various questions about the embeddability of rings in fields, or more generally, the existence of R-fields.

Let R be a ring; clearly R-fields exist if and only if there exist epic R-fields, or equivalently, by Theorem 4.3, prime matrix ideals. To find whether R has prime matrix ideals we go back to the method of generating matrix ideals described in Section 7.3. Let us again write $\mathcal{Z}_0$ for the set of all determinantal sums of non-full matrices. Thus $A \in \mathcal{Z}_0$ precisely if $A = C_1 \nabla \ldots \nabla C_r$, where each C_i is non-full and the right-hand side is suitably bracketed. From the description in Section 7.3 it is clear that $\mathcal{Z}_0$ is the least matrix pre-ideal of R. Let $\mathcal{Z} = \mathcal{Z}_0/\mathcal{J}$ be the matrix ideal generated by $\mathcal{Z}_0$ and put $\mathcal{N} = \sqrt{\mathcal{Z}}$. We note that $\mathcal{N}$ has the two properties (which actually characterize it): (i) for any matrix A, if $\oplus^m A \in \mathcal{N}$ for some $m \geq 1$, then $A \in \mathcal{N}$, (ii) if $A \in \mathcal{N}$, then for some $m, r \geq 1$, $\oplus^m A \oplus I_r$ is a determinantal sum of non-full matrices. It is thus the analogue of the nilradical, and we shall call it the *matrix nilradical*. Clearly $\mathcal{N}$ is proper if and only if $\mathcal{Z}$ is proper, and this will be the case precisely when the unit matrix (of any order) cannot be written as a determinantal sum of non-full matrices. By Theorem 3.8, $\mathcal{N}$ is the intersection of all prime matrix ideals in R, so $\mathcal{N}$ is proper if and only if R has prime matrix ideals. By Theorem 4.3 this means that R has epic R-fields if and only if $\mathcal{N}$ is proper. We therefore obtain the following criterion for the existence of R-fields:

Theorem 7.5.1. *Let R be any ring. Then there exists a homomorphism of R into a field if and only if its matrix nilradical is proper; equivalently, no unit matrix over R can be written as a determinantal sum of non-full matrices.* ∎

7.5 Fields of fractions

This includes the well-known necessary condition: if a ring can be mapped into a field, then the unit matrix (of any order) is full, i.e. R has UGN. If the matrix nilradical of R is prime, it is the least prime matrix ideal of R, for any prime matrix ideal $\mathcal{P}$ satisfies $\mathcal{P} \supseteq \mathcal{Z}$, hence $\mathcal{P} = \sqrt{\mathcal{P}} \supseteq \sqrt{\mathcal{Z}} = \mathcal{N}$. This makes it clear that R has a least prime matrix ideal precisely when the matrix nilradical is prime. By the correspondence of epic R-fields and prime matrix ideals described in Section 7.4 we therefore have

Theorem 7.5.2. *Let R be any ring. Then R has a universal R-field if and only if the matrix nilradical of R is prime.* ∎

Next we obtain a criterion for the invertibility of a matrix, which results from the following more general formulation:

Theorem 7.5.3. *Let R be a ring and P, Q any square matrices over R. Then there is a homomorphism from R to a field mapping P to an invertible matrix and Q to a singular matrix if and only if no diagonal sum $I \oplus (\oplus^r P)$ can be written as a determinantal sum of non-full matrices and matrices $Q \oplus B$, where $B \in \mathfrak{M}(R)$.*

Proof. The condition for a homomorphism of the required sort to exist is that there should be a prime matrix ideal containing Q but not P. Let (Q) denote the matrix ideal generated by Q; there is a prime matrix ideal containing Q but not P if and only if $P \notin \sqrt{(Q)}$. So the required condition is that $\oplus^r P \notin (Q)$, i.e. there is no equation

$$I \oplus^r P = C_1 \nabla \ldots \nabla C_s \quad (C_i \text{ non-full or of the form } Q \oplus B_i). \tag{1}$$
∎

In particular, taking $Q = 0$, we obtain

Corollary 7.5.4. *For any square matrix P over a ring R there is an R-field inverting P if and only if no diagonal sum $I \oplus (\oplus^r P)$ can be written as a determinantal sum of non-full matrices.* ∎

Secondly, take $P = I$; then we find

Corollary 7.5.5. *For any square matrix Q over a ring R there is an R-field over which Q becomes singular if and only if no unit matrix I can be written as a determinantal sum of non-full matrices and matrices of the form $Q \oplus B$, where B is any square matrix.* ∎

From these results it is easy to derive a criterion for the embeddability of a ring in a field. We recall that an integral domain R is embeddable in a field if and only if for each $a \in R^\times$ there is an a-inverting homomorphism into a field (see Cohn [71b], SF, p. 13 or Exercise 7 below). By Corollary 5.4, this holds if and only if there is no equation

$$I \oplus aI = C_1 \nabla \ldots \nabla C_t \quad (C_i \text{ non-full}). \tag{2}$$

In particular, aI cannot be expressed in this way. Conversely, if there is an expression (2) with I of order r and aI of order s, we multiply both sides by $aI_r \oplus I_s$ and observe that the determinantal sum is distributive with respect to multiplication by diagonal matrices. Thus we obtain aI as a determinantal sum of non-full matrices. This proves

Corollary 7.5.6. *A ring R can be embedded in a field if and only if it is an integral domain and no non-zero scalar matrix can be written as a determinantal sum of non-full matrices.* ∎

An alternative formulation is given in

Theorem 7.5.7. *A ring R is embeddable in a field if and only if it is non-zero and no diagonal matrix with all elements on the main diagonal non-zero can be written as a determinantal sum of non-full matrices.*

Proof. Clearly the conditions are necessary. Suppose they are satisfied and $a, b \in R^\times$ are such that $ab = 0$. Then

$$\begin{pmatrix} a & 0 \\ 0 & b \end{pmatrix} = \begin{pmatrix} a & 0 \\ 1 & b \end{pmatrix} \nabla \begin{pmatrix} 0 & 0 \\ -1 & b \end{pmatrix} = \begin{pmatrix} a \\ 1 \end{pmatrix}(1 \quad b)\nabla \begin{pmatrix} 0 \\ 1 \end{pmatrix}(-1 \quad b),$$

and here both matrices on the right are non-full. Thus the condition of Theorem 5.7 is sufficient to exclude zero-divisors, and by Corollary 5.6 it is sufficient for embeddability in a field. ∎

It may be of interest to note that the conditions of Corollary 5.6, apart from the absence of zero-divisors, are in the form of quasi-identities, as required by general theory (see e.g. UA, p. 235 and Exercise 6 below).

We have already found a condition for the existence of a universal R-field: it was that the matrix nilradical $\mathcal{N}$ should be prime (Theorem 5.2). Moreover, there will be a universal field of fractions if and only if, further, $\mathcal{N}$ contains no non-zero elements of R. But frequently we are interested in the special case when there is a universal field of fractions in which every full matrix becomes invertible. This is the maximal set that can be inverted, since no non-full matrix can ever be inverted over a field.

We recall that a ring-homomorphism is called *honest* if it keeps all full matrices full. In particular, a homomorphism to a field K is honest if and only if it inverts all full matrices; this is possible only when the singular kernel of

7.5 Fields of fractions

K consists precisely of the non-full matrices. The set of all full matrices over a given ring R will be denoted by $\Phi = \Phi(R)$, and an R-ring or a homomorphism is called *fully inverting* if it is Φ-inverting. Any fully inverting homomorphism $f : R \to S$ to a non-zero ring S must be injective, for every non-zero element of R is full, as 1×1 matrix, and so maps to an invertible element of S. We shall need to assume various conditions on the set of all full matrices; the next lemma simplifies our task:

Lemma 7.5.8. *Let R be any ring. If the set of all full matrices over R is lower multiplicative, then the product of any two full matrices of the same order is again full. The converse holds if every full matrix is stably full.*

Proof. Let S, T be full matrices of the same order. Then we have the following series of elementary transformations:

$$\begin{pmatrix} S & 0 \\ I & T \end{pmatrix} \to \begin{pmatrix} 0 & -ST \\ I & T \end{pmatrix} \to \begin{pmatrix} 0 & -ST \\ I & 0 \end{pmatrix} \to \begin{pmatrix} ST & 0 \\ 0 & I \end{pmatrix},$$

which shows that ST is full, as claimed. For the converse we note that the above argument can be reversed when ST is stably full. ∎

We shall need a condition for a universal localization to be a field, rather than just a local ring (Proposition 2.6). In fact we have two slightly different situations with essentially the same proof.

A multiplicative set Σ of full matrices is said to be *factor-inverting* if for any square matrices A, B such that $AB \in \Sigma$, A is invertible over R_Σ. Clearly every factor-stable multiplicative set of full matrices is factor-inverting (though not conversely).

Proposition 7.5.9. *Let R be a ring.*

(i) *If the set Φ of all full matrices over R is lower multiplicative and $R_\Phi \neq 0$, then R_Φ is a field;*
(ii) *if Σ is a factor-inverting set of matrices over R and $f : R \to S$ is an honest Σ-inverting homomorphism, so that $f = \lambda f'$, where $\lambda : R \to R_\Sigma$, $f' : R_\Sigma \to S$, then f' is injective.*

Proof. In both cases we have to show that certain elements in a localization are zero. In case (i) they are the non-units in R_Φ, in case (ii) the elements of R_Σ in the kernel of f'.

Let p be such an element of R_Φ, R_Σ respectively and let

$$Au = 0 \tag{3}$$

be an admissible system for p. In case (i), when p is a non-unit in R_Φ, its numerator $(A_0 \quad A_\bullet)$ is not invertible over R_Φ and hence is not full over R. In case (ii), $pf' = 0$, hence the numerator $(A_0 \quad A_\bullet)$ is not full over S, by Cramer's rule (since the right-hand factor in (9) of Proposition 1.5 has a zero row), and because f is honest, $(A_0 \quad A_\bullet)$ is not full over R.

Thus in both cases, $(A_0 \quad A_\bullet) = PQ$, where $P \in {}^nR^{n-1}$, $Q \in {}^{n-1}R^n$ and so

$$A = (A_0 \quad A_\bullet \quad A_n) = (PQ \quad A_n) = (P \quad A_n)\begin{pmatrix} Q & 0 \\ 0 & 1 \end{pmatrix}.$$

We claim that (P, A_n) is invertible over R_Φ, R_Σ respectively. For (P, A_n) is a left factor of the denominator $(A_\bullet, A_n)$ and in case (i) the latter is full over R, hence so is (P, A_n) and so it is invertible over R_Φ. In case (ii) the denominator is in Σ and since Σ is factor-inverting, (P, A_n) becomes invertible over R_Σ. We can therefore in both cases cancel the left factor (P, A_n) in (3) and conclude that $p = 0$. ∎

Let us take $\Sigma = \Phi$ in (ii) and note that Φ is necessarily factor-inverting. This yields

Corollary 7.5.10. *Let $f : R \to S$ be a fully inverting homomorphism of rings. If either (i) $S \neq 0$ and $\Phi(R)$ is multiplicative, or (ii) S has UGN, then f is injective and R_Φ is a field.*

Proof. By combining (i) and (ii) of Proposition 5.9 we obtain (i); it remains to prove (ii). By Proposition 5.9 (i), the set Φ' of matrices inverted by f contains Φ. If $\Phi' \neq \Phi$, a non-full matrix becomes invertible over S, so a unit matrix is non-full and S cannot satisfy UGN. ∎

We shall use this result to derive a criterion for the universal localization R_Φ to be a field. The following lemma will be needed in the proof.

Lemma 7.5.11. (Magic Lemma) *Let R be any ring and let $A, B, C \in \mathfrak{M}(R)$ such that B is non-full and $C = A \nabla B$. Then $A = ST$, $C = SUT$, where $S, T \in \mathfrak{M}(R)$ and $U \in E(R)$. Hence C is full if and only if A is full whenever the full matrices of any given order over R admit products, in particular, when the set of all full matrices is lower multiplicative.*

Proof. Suppose that in $C = A \nabla B$, the determinantal sum is with respect to the first column; by hypothesis, B is non-full, say $B = PQ$, where P is $n \times (n-1)$ and Q is $(n-1) \times n$. Write $Q = (Q_1, Q')$, where Q_1 is a column and Q' is square of order $n - 1$. Then

$$A = (A_1 \quad PQ') = (A_1 \quad P)(1 \oplus Q'). \tag{4}$$

Now $C = (A_1 + B_1 \quad PQ') = (A_1 + PQ_1 \quad P)(1 \oplus Q')$, hence
$$C = (A_1 \quad P)\begin{pmatrix} 1 & 0 \\ Q_1 & I \end{pmatrix}\begin{pmatrix} 1 & 0 \\ 0 & Q' \end{pmatrix}.$$
Thus we have $A = ST, C = SUT$, where S, T, U are square matrices over R and $U \in E(R)$.

If we assume that A is full, then S, T are full, hence so is SUT, by the hypothesis and Lemma 5.8; this argument can be reversed to show that when C is full, then so is A. ∎

This result allows us to give conditions for the universal localization to exist.

Theorem 7.5.12. *Let R be a non-zero ring and Φ the set of all its full matrices. Then there is an honest map to an epic R-field, which is necessarily the universal localization R_Φ if and only if Φ is lower multiplicative. In particular this holds for every Sylvester domain.*

Proof. The conclusion requires that the set of all non-full matrices should be the unique least prime matrix ideal. This shows the condition to be necessary. Conversely, when it is satisfied, then every unit matrix lies in Φ, so R has UGN; further, the determinantal sum of any non-full matrices, when defined, is again non-full, by Lemma 5.11 and it follows that the non-full matrices form a matrix pre-ideal, $\mathcal{Z}$ say. In fact this is a matrix ideal, for if $A \oplus I \in \mathcal{Z}$, we know that $I \notin \mathcal{Z}$ by UGN, so $A \in \mathcal{Z}$. The same argument (and the fact that 1 is full) shows that $\mathcal{Z}$ is prime; clearly it is the least prime matrix ideal. Hence there is an honest map to an epic R-field K, the residue-class field of R_Φ. Now Corollary 5.10 (i) shows that $R_\Phi \cong K$. The last part follows by the law of nullity, which shows Φ to be lower multiplicative. ∎

We observe that the condition given here generalizes the condition 'no zero-divisors'. To give an example of a ring with UGN where the condition fails, we use the following construction due to Bergman. In the subring $k[x^2, x^3]$ of $k[x]$ consider the matrices
$$\begin{pmatrix} x^2 & x^5 \\ x^3 & x^6 \end{pmatrix} = \begin{pmatrix} x^2 \\ x^3 \end{pmatrix}(1 \quad x^3), \quad \begin{pmatrix} x^3 & x^5 \\ x^4 & x^6 \end{pmatrix} = \begin{pmatrix} x^3 \\ x^4 \end{pmatrix}(1 \quad x^2). \tag{5}$$
Both are non-full, but their determinantal sum with respect to the first column is the matrix
$$\begin{pmatrix} x^2 + x^3 & x^5 \\ x^3 + x^4 & x^6 \end{pmatrix}, \tag{6}$$
which is easily seen to be full. However, the embedding of $k[x^2, x^3]$ in $k[x]$ is not honest, since the matrix (6) is not full over $k[x]$.

We remark that a non-zero ring is Hermite whenever the set of all full matrices admits diagonal sums. By Proposition 0.4.4 this will follow if we show that the stable rank of any matrix exists and equals its inner rank. Let A be a matrix of rank r; by Theorem 5.4.9 it has an $r \times r$ full submatrix, which by elementary transformations may be taken to be in the top left-hand corner. Now the diagonal sum $A \oplus 1$ includes a full submatrix of order $r + 1$ and so has inner rank $r + 1$; by induction the inner rank of $A \oplus I_s$ is $r + s$, hence A has stable rank r as claimed, and the result follows.

We now return to the situation of Theorem 5.12 and show that the class of rings considered there can be characterized in different ways.

Theorem 7.5.13. *For any non-zero ring R the following conditions are equivalent:*

(a) R is a Sylvester domain,
(b) the set of all full matrices over R is lower multiplicative,
(c) every full matrix over R is stably full and full matrices admit products,
(d) the set of all non-full matrices is a prime matrix ideal, the matrix nilradical,
(e) if Φ denotes the set of all full matrices, then R_Φ is a field, necessarily the universal field of fractions of R,
(f) R has an inner rank preserving homomorphism to the universal field of fractions of R.

Proof. (a) $\Rightarrow$ (b) In a Sylvester domain the set of all full matrices is lower multiplicative, by the law of nullity. Now (b) $\Rightarrow$ (c) follows by Lemma 5.8 and the above remark. (c) $\Rightarrow$ (d): By Theorem 5.12 it will be enough to show that the set of all full matrices is lower multiplicative. Given full matrices A, B the matrices $A \oplus I$, $B \oplus I$ are full because full matrices are stably full, so we may take A, B to have the same order. By hypothesis AB is full, and so is $AB \oplus I$. By elementary transformations as in the proof of Lemma 5.8 we conclude that the set of full matrices is lower multiplicative. It follows as in the proof of Theorem 5.12 that the set of all non-full matrices is a prime matrix ideal. (d) $\Leftrightarrow$ (e) follows from the proof of Theorem 5.2. (d, e) $\Rightarrow$ (f): When (d), (e) hold, then the homomorphism $\lambda : R \to R_\Phi$ is honest and R_Φ is a field, hence by Corollary 5.4.10, λ preserves the inner rank. By (d) the non-full matrices form the least prime ideal, so R_Φ is a universal field of fractions. (f) $\Rightarrow$ (a) follows because any field clearly satisfies Sylvester's law of nullity. ■

This result tells us that Sylvester domains form the precise class of rings that have a universal field of fractions over which every full matrix can be inverted. In particular, since any semifir is a Sylvester domain, we have

7.5 Fields of fractions

Corollary 7.5.14. *Every semifir R has a universal field of fractions K, such that every full matrix over R can be inverted over K.* ∎

We note the following consequences of Theorem 5.13.

Corollary 7.5.15. *Over any Sylvester domain the product of any two right full matrices (if defined) is right full.*

Proof. Since a matrix over R is right full if and only if it is right regular over its universal field of fractions, the result follows. ∎

Any automorphism clearly preserves the set of full matrices; hence we obtain

Corollary 7.5.16. *Any automorphism of a Sylvester domain extends to an automorphism of the universal field of fractions.* ∎

The same result can be proved for derivations:

Theorem 7.5.17. *Any derivation of a Sylvester domain extends to a derivation of its universal field of fractions.*

Proof. Let R be a Sylvester domain with a derivation δ and let U be its universal field of fractions. The derivation can be expressed as a homomorphism from R to R_2:

$$\Delta : a \mapsto \begin{pmatrix} a & a^\delta \\ 0 & a \end{pmatrix}; \tag{7}$$

it induces a homomorphism $\Delta_n : R_n \to R_{2n}$ such that every full matrix over R maps to an invertible matrix over U. For if A is full over R, then it is invertible over U and so

$$\begin{pmatrix} A & A^\delta \\ 0 & A \end{pmatrix} \text{ has the inverse } \begin{pmatrix} A^{-1} & -A^{-1}A^\delta A^{-1} \\ 0 & A^{-1} \end{pmatrix}.$$

Hence Δ can be extended to a unique homomorphism from U to U_2, again denoted by Δ. Clearly this again has the form (7) and the (1, 2)-entry is a derivation of U extending δ, unique because the extension of Δ was unique. ∎

For pseudo-Sylvester domains we have the following analogue of Theorem 5.13. Let us call a homomorphism of rings *stably honest* if it keeps stably full matrices stably full. Clearly every honest homomorphism is stably honest, but the converse need not hold. It is easily verified that Proposition 5.9 (ii) remains true if f is merely stably honest. For in Cramer's rule for an element $p \neq 0$,

$$\begin{pmatrix} A_\bullet & -A_0 \end{pmatrix} = \begin{pmatrix} A_\bullet & A_\infty \end{pmatrix} \begin{pmatrix} I & u_\bullet \\ 0 & p \end{pmatrix},$$

the left-hand side is now stably full, hence so are the factors on the right and it follows that $pf' \neq 0$. We also note that the analogue for Theorem 5.12 for stably honest maps, with the set Φ^* of all stably full matrices instead of Φ holds, with the same proof as before.

Theorem 7.5.18. *For any ring R the following are equivalent:*

(a) R is an S-ring.
(b) the set of all matrices that are not stably full is a prime matrix ideal,
(c) if Φ^* is the set of all stably full matrices, then R_{Φ^*} is a field,
(d) R has a stable rank preserving homomorphism to a field K.

Moreover, the field K is a universal R-field; it is a field of fractions if and only if R is weakly finite. Thus every pseudo-Sylvester domain has a universal field of fractions.

Proof. (a) $\Rightarrow$ (b) follows since the stably full matrices admit diagonal sums. (b) $\Leftrightarrow$ (c) follows from the analogue of Theorem 5.2. (b, c) $\Rightarrow$ (d) follows by the analogue of Corollary 5.4.10 and (d) $\Rightarrow$ (a) follows since S-rings satisfy the law of nullity for the stable rank.

Now K is a universal R-field because the matrices that are not stably full form a prime matrix ideal, clearly the least such. If in addition R is weakly finite, then the elements of stable rank 0 are 0, by Proposition 0.1.3, and conversely, any subring of a field is weakly finite. ∎

We now return to the embedding of Theorem 4.5.3 to show that it extends to the universal fields of fractions, i.e. it is honest:

Theorem 7.5.19. *Let G, F be the free k-algebras of rank 2 and infinite rank respectively. Then there is an honest embedding $F \to G$.*

Proof. As in the proof of Theorem 4.5.3 we take $G = k\langle x, y \rangle$, $F = k\langle Z \rangle$, where $Z = \{z_0, z_1, \ldots\}$ and define the embedding by

$$\beta : z_n \mapsto [\ldots [y, x], \ldots, x] \quad \text{with } n \text{ factors } x. \tag{8}$$

We shall denote the universal fields of fractions of F, G by U, V respectively. On G we have a derivation δ defined by $x^\delta = 0$, $y^\delta = yx - xy$. The restriction of δ to F (via the embedding (8)) is the mapping defined by $\delta : z_i \mapsto z_{i+1}$, and the subalgebra of G generated by x and the image of F has the form $F[x; 1, \delta]$. Clearly β can be extended to a homomorphism mapping x to x, so the subalgebra generated by x and z_0 maps to the subalgebra generated by x, y, i.e. the whole of G. Thus $F[x; 1, \delta] \cong G$ and we have an embedding

$$F[x; 1, \delta] \to V. \tag{9}$$

Now δ extends to a derivation of U, again denoted by δ and we can form the skew polynomial ring $U[x; 1, \delta]$ with $F[x; 1, \delta]$ as subalgebra. Any full matrix over F becomes invertible over U and so remains full over $F[x; 1, \delta]$; thus it maps to an invertible matrix over V and in this way we obtain an embedding $U[x; 1, \delta] \to V$. In particular we thus obtain the desired embedding of U in V. ∎

The results of this section also allow us to give a short proof of a weak form of the inertia theorem:

Theorem 7.5.20. *Let D be a field, E a subfield and X a set. Then the inclusion $D_E\langle X\rangle \to D_E\langle\!\langle X\rangle\!\rangle$ is honest.*

Proof. Suppose first that X is finite. Put $R = D_E\langle X\rangle$, $R' = D_E\langle\!\langle X\rangle\!\rangle$ and let $S = R[[t]]$ be the power series ring in a central indeterminate t. We have D-linear homomorphisms $R \to R' \to S$, where $x \mapsto x \mapsto xt$ any $x \in X$. Any matrix $A = A(x)$ over R becomes $A(xt)$ over S. If A is full it is invertible over $D_E(\langle X\rangle)$, hence $A(xt)$ is invertible over the rational function field $D_E(\langle X\rangle)(t)$ and hence also over the formal Laurent series field $D_E(\langle X\rangle)((t))$. Hence it must be full over $S = D_E\langle X\rangle[[t]]$ and *a fortiori* A is full over $D_E\langle\!\langle X\rangle\!\rangle$, as we had to show. If X is infinite, the same proof still applies, since A can involve only a finite part of X. ∎

Exercises 7.5

1. Let R be a ring with a fully inverting R-field K. Show that every honest endomorphism of R extends to a unique endomorphism of K. Show also that every derivation of R extends to a unique derivation of K.
2. Let $R = k\langle x, y, z, t\rangle$ and define an endomorphism α of R by the rules: $x \mapsto xz, y \mapsto xt, z \mapsto yz, t \mapsto yt$. Show that α is injective, but not honest.
3. Show that a ring R is embeddable in a direct product of fields if and only if no non-zero scalar matrix can be written as a determinantal sum of non-full matrices.
4. Let R be an Ore domain with field of fractions K. Show that $A \in R_n$ is invertible over K if and only if Ac is full over R for all $c \in R^\times$.
5. Let R be a commutative ring. Show that for any $A \in \mathfrak{M}(R)$, $\det A = 0$ if and only if A becomes singular over every R-field. Deduce that $\det A = 0$ if and only if $\oplus^r A$, for some $r \geq 1$, can be written as a determinantal sum of non-full matrices.
6. A condition of the form $A_1 \wedge \ldots \wedge A_r \Rightarrow A$, where A_i, A are atomic formulae or A is $F, =$ false, is called a *quasi-identity*. The class of algebras defined by a set of quasi-identities is called a *quasi-variety*; it may be characterized as a universal class admitting direct products (see UA, p. 235). Verify that the following are quasi-varieties: (i) the class of monoids embeddable in groups, (ii) the class $\mathfrak{F}$ of

subrings of strongly regular rings (i.e. rings such that for each c there exists x such that $c^2 x = c$).

7. Show that any subring of a strongly regular ring that is an integral domain can be embedded in a field. Verify that a direct product of fields is strongly regular and deduce that a subring of a direct product of fields is embeddable in a field if and only if it is an integral domain.

8. Show that any filtered ring with inverse weak algorithm has a universal field of fractions. If U is the universal field of fractions of the free power series ring $k\langle\!\langle X\rangle\!\rangle$, show that the subfield of U generated by $k\langle X\rangle$ is the universal field of fractions of the latter. (*Hint*: Use the inertia theorem 2.8.16.)

9. Let $R = k\langle x, y, z\rangle$, denote its universal field of fractions by U and the subalgebra generated by x, y, z and $t = zx^{-1}y$ by S. Show that the matrix $\begin{pmatrix} x & y \\ z & t \end{pmatrix}$ is full over S but not invertible over U.

10°. Let R be a ring with a 2×2 matrix C satisfying $C^n = 0$ for some $n \geq 1$. Show that for any entry c of C^2, and a suitable unit matrix I, cI is a determinantal sum of non-full matrices. Find an explicit expression for cI when R is generated over k by the entries of C with the relation $C^3 = 0$.

11. Let R be a ring with a universal R-field U. If $\Sigma \subseteq \mathfrak{M}(R)$ is such that R_Σ admits a homomorphism to a field, show that the map $R \to U$ factors through R_Σ and U is also the universal R_Σ-field. Show also that a homomorphism $f : R \to S$ is honest whenever fg, for some homomorphism g from S, is honest. Deduce that for any Sylvester domain R and any set of matrices Σ for which there is an R_Σ-field, the map $R \to R_\Sigma$ is honest.

12. Let F be the free group on a set X. Using Exercise 11, show that the natural mapping $k\langle X\rangle \to kF$ is honest.

13. Let R be a semifir with centre k and with universal field of fractions U. If E is a finite commutative extension field of k such that $R \otimes_k E$ is again a semifir, show that $U \otimes E$ is a field and is the universal field of fractions of $R \otimes E$.

14*. (A. H. Schofield) Let R be any ring with an epic R-field K. Verify that $K(t)$ is an epic $R[t]$-field, and deduce that the embedding $f : R \to R[t]$ induces the surjection $f^* : X(R[t]) \to X(R)$, where $X(R)$ denotes the set of all epic R-fields or equivalently, the set of all prime matrix ideals of R. Show that the inverse image of K under f^* consists of the different $K[t]$-fields and deduce that if R has a universal R-field U, then $U(t)$ is a universal $R[t]$-field.

15°. (Bergman and Dicks [78]). Let R be a ring with a universal field of fractions U and let Σ be the set of all matrices inverted over U. Show that the natural homomorphism $R_\Sigma \to U$ is surjective, and find an example where it is not an isomorphism.

16°. Given $n > 1$, find a criterion for a ring to be embeddable in an $n \times n$ matrix ring over a field (see Schofield [85], Theorem 7.4).

17°. Let A be a square matrix over a free algebra. Verify that if A is invertible, the matrix ideal generated by it is improper. Is the converse true?

18. (Cohn [71a]) Let D be a field with two subfields E, F that are isomorphic, via an isomorphism $f : E \to F$. Show, by taking the universal field of fractions of a suitable HNN-construction (SF, p. 231, or also Exercise 21 of Section 2.5 above) that D can be embedded in a field in which f is realized as an inner automorphism.

(This process allows two elements transcendental over a central subfield k to be made conjugate in an extension field. It can be shown to apply more generally to square matrices; see Cohn [73a,77b], SF, Section 5.5).

19. Let G be a group whose group algebra kG is embeddable in a field K, and let F be the free group on a set X. By considering $K\langle X, X^{-1}\rangle$ show that the group algebra of the direct product $G \times F$ can be embedded in a field.
20. Show that the determinantal sum of the matrices in (5) is non-full over $k[x]$, but full over $k[x^2, x^3]$.
21. (Bergman [74b]). Let R be a k-algebra with 27 generators, arranged as three 3×3 matrices P, U, V with defining relations in matrix form: $UV = VU = I$, $P^2 = P$, $UP = (I - P)U$. Show that there are no R-fields and use Theorem 5.1 to deduce that I can be written as a determinantal sum of non-full matrices. (*Hint*: Observe that R can be realized by 2×2 matrices over a field, and deduce that R has UGN. It follows from Theorem 2.11.2 that R is a 2-fir.)
22. Give an example to show that the set of all full matrices over $k[x, y, z]$ is not multiplicative.
23. Let R be a right Ore domain with field of fractions K. Show that a square matrix A over R is invertible over K if and only if Ac is full over R for all $c \in R^\times$.
24°. Let A be a full matrix over a semifir R. If $c \in R$ and A is totally coprime to c (see Section 6.2), is A still full (mod (c))?
25*. Let R be a commutative ring. Show that for any $A \in \mathfrak{M}(R)$, $\det A = 0$ if and only if A becomes singular in every R-field. Deduce that $\det A = 0$ if and only if $\oplus^r A$ for suitable $r \geq 1$ can be written as a determinantal sum of non-full matrices.

7.6 Numerators and denominators

Given any ring R with an epic R-field K, our object here will be to compare the different admissible systems for the same element or matrix over K. For simplicity we shall only discuss elements of K; the only other case needed, that of a square matrix, is then easily deduced. However, we shall allow the admissible matrix A to have index other than 1, thus A will be $r \times (m + 1)$, say. Then the system $Au = 0$ is said to have *order* m and we put $o(A) = m$; we shall modify the definition given in Section 7.1 by calling the system *admissible* for p if it has a unique normalized solution u (i.e. $u_0 = 1$) and $u_\infty = p$, even though now m need not equal r. We shall see in Proposition 6.3 that in any case $m \leq r$; thus an admissible matrix may have an index ≤ 1, while for index 1 we are in the case considered in Section 7.1. As before, in an admissible matrix of order m, the first m columns represent the numerator and the last m the denominator of the system. If $\mathcal{P}$ is the singular kernel of the R-field K, we shall also call A $\mathcal{P}$-*admissible*, or K-*admissible*, but the reference to $\mathcal{P}$ or K is omitted when there is no risk of confusion.

In order to compare different systems we shall describe some operations which can be carried out on a system without changing its solution. Let A be an admissible matrix for p (over R); if u satisfies $Au = 0$, then it also satisfies $PAu = 0$ for any matrix P for which the product is defined, and the two systems are equivalent provided that P is right regular. On the other hand, the solutions of $Au = 0$ and $AQu = 0$ are not generally the same, but we are only concerned with the ratio of the first and last components of u, and this ratio is preserved when Q is suitably restricted. Let us define a *bordered* matrix as a matrix Q^*, not necessarily square, which agrees with the unit matrix in the first and last row. We shall define the following two operations on an admissible matrix A for p.

λ. Replace A by PA such that PA is again admissible for p,
ρ. Replace A by AQ^*, where Q^* is bordered, such that AQ^* is again admissible for p.

The inverse operations, cancelling a left or a bordered right factor from A so as to obtain again an admissible matrix for p, are denoted by λ', ρ', respectively, and an operation is called *trivial* if the matrix P or Q^* is invertible, and called *full* if P or Q^* is full. We shall say that A admits a certain operation if the result of performing it on A is again admissible for the same element.

Just as matrices may be thought of as defining certain finitely presented modules, so admissible matrices, essentially matrices with distinguished first and last column (and a rank condition) define left *bipointed* modules, i.e. modules with a distinguished pair of elements. We shall need a notion of equivalence for these matrices corresponding to stable association but preserving these two columns.

Two matrices A, A' are said to be *biassociated* (explicitly: associated as maps of bipointed modules) if there exist invertible matrices U, V such that

$$AU = VA', \tag{1}$$

and U is bordered. The matrices $A = (A_0, A_\bullet, A_\infty)$, $A' = (A_0', A_\bullet', A_\infty')$ are *stably biassociated* if they satisfy a relation

$$\begin{pmatrix} A & 0 \\ 0 & I \end{pmatrix} U = V \begin{pmatrix} I & 0 \\ 0 & A' \end{pmatrix}, \tag{2}$$

where U, V are invertible and U is bordered, with the unit rows corresponding to the first and last columns of A. Over a weakly finite ring this condition can again (as in Proposition 0.5.6) be restated in terms of comaximal relations.

Proposition 7.6.1. *Let R be a weakly finite ring and let A, A' be matrices over R of the same index. Then A, A' are stably biassociated if and only if there is a comaximal relation*

$$AB' = BA', \tag{3}$$

where B' is a bordered matrix.

Proof. If A, A' are stably biassociated, then by definition we have a relation (2), where U, V are invertible, and this relation is balanced because $i(A) = i(A')$. Moreover, U is obtained from a bordered matrix by certain row and column permutations and (by formula (10) after Proposition 0.5.6) has the partitioned form

$$U = \begin{pmatrix} D' & -B' \\ -C' & A' \end{pmatrix},$$

where $-B'$ is bordered and D' has first and last rows zero. If we partition V correspondingly and call the $(1, 2)$-block B, then a comparison of $(1, 2)$-blocks in (2) gives the comaximal relation (3) with bordered matrix $-B'$. Conversely, given a comaximal relation (3), then by Proposition 0.5.6 we obtain two mutually inverse matrices

$$V = \begin{pmatrix} A & B \\ C & D \end{pmatrix}, \quad V^{-1} = \begin{pmatrix} D' & -B' \\ -C' & A' \end{pmatrix}. \tag{4}$$

Since the first row of $-B'$ is $e_1 = (1, 0, \ldots, 0)$, we can subtract multiples of the $(r + 1)$th column of V^{-1} from the first r columns so as to reduce the first row of D' to 0, and make the corresponding change in V. This will only affect C and D, but not A or B, in V. Similarly, since the last row of $-B'$ is e_{m+1}, we can reduce the last row of D' to zero (at the expense of further changes to C and D). Since the relation $V^{-1}V = I$ has been preserved, we see that the first and last rows of D are zero, while C is bordered. Now the argument of Proposition 0.5.6 shows that A and A' are stably biassociated. ∎

From the proof we see that when A, A' are stably biassociated, they occur as blocks within pairs of inverse matrices, as in (4), with $-B'$ bordered. If in the expression (4) for V^{-1} we consider the cofactor of the $(1, 1)$-entry of B', we again obtain a pair of mutually inverse matrices, but with the first column of A, A' omitted. This shows that the denominators of A, A' are stably associated. A similar argument, omitting the last row and column of B' in (4), shows that the numerators of A, A' are stably associated, as are the cores, and of course here weak finiteness is not needed. Thus we obtain

Proposition 7.6.2. *Over any ring R, if matrices A, A' are stably biassociated, then the denominators (obtained by omitting the first column), the numerators (obtained by omitting the last column) and the cores (obtained by omitting the first and last columns) are stably associated.* ∎

Let us next record the conditions under which the operations $\lambda, \lambda', \rho, \rho'$ preserve admissibility. For simplicity we shall say that a matrix $A \in {}^m R^n$ is 'left (right) regular over K' to mean 'its image in ${}^m K^n$ is left (right) regular'.

Proposition 7.6.3. *Let R be a ring with epic R-field K, and let $A \in {}^r R^{m+1}$.*

(i) *The matrix A is admissible over K if and only if $(A_\bullet, A_m)$ is right regular over K but A is not; in particular this entails that $r \geq m$.*

(ii) *Suppose that $B = PA$, where $B \in {}^s R^{m+1}$ and $P \in {}^s R^r$. If A is admissible for $p \in K$, then B is admissible for p if and only if $P(A_\bullet, A_m)$ is right regular over K; this is so whenever P is right regular over K. If B is admissible for p, then A is admissible for p if and only if A is not right regular over K.*

(iii) *Suppose that $B = AQ^*$, where $B \in {}^r R^{n+1}$ and $Q^* \in {}^{m+1} R^{n+1}$ is bordered. If A is admissible for p, then B is admissible for p if and only if Q^* is right regular over K but B is not; when this is so, $m \geq n$. If B is admissible for p, then A is admissible for p if and only if its denominator $(A_\bullet, A_m)$ is right regular over K.*

(iv) *Let P be a matrix and Q^* a bordered matrix over R, both right regular over K. Then A is admissible $\Leftrightarrow PA$ is admissible $\Leftrightarrow AQ^*$ is admissible.*

Proof. We have already seen (i), and (ii) follows because A and PA are admissible for the same element p (by uniqueness), if they are admissible at all, while admissibility follows since the rank (over K) is non-increasing under multiplication. Similarly (iii) and (iv) follow because the product of right regular matrices is right regular over K. ∎

We remark that when R is a Sylvester domain and K its universal field of fractions, regularity over K can be replaced by fullness over R in this proposition.

We next examine conditions under which an admissible system can be simplified by operations λ, λ', ρ, ρ'. For any matrix A over R and any R-field K the rank of A over K will be called the *K-rank* of A, written $\rho_K(A)$.

Lemma 7.6.4. *Let R be a ring with epic R-field K, and let $A \in {}^r R^{m+1}$ be a K-admissible matrix. Then the index of A is at most 1, i.e. $o(A) = m \leq r$ and the following are equivalent:*

(a) *the index is exactly 1, i.e. $o(A) = r$,*

(b) *A is left regular over K,*

(c) *the number of rows of A cannot be decreased by an operation λ.*

Further, any K-admissible matrix can be transformed by an operation λ to a K-admissible matrix which is left regular over K.

If R is a Sylvester domain and K its universal field of fractions, then (a)–(c) are equivalent to:

(d) *the number of rows of A cannot be decreased by an operation λ',*

7.6 Numerators and denominators

and moreover, any K-admissible matrix can be transformed by an operation λ' to a K-admissible matrix that is left full, and every operation λ or λ' admitted by a left full matrix that does not increase the number of rows is full.

Proof. Since $Au = 0$ has the unique normalized solution u over K, the K-rank of A is m, whence $r \geq m$, with equality if and only if $\rho_K(A) = r$, i.e. A is left regular over K; thus (a) $\Leftrightarrow$ (b).

Next let us write

$$A'' = EA, \quad E \in {}^m R^r, \quad A'' \in {}^m R^{m+1}, \qquad (5)$$

where E consists of m rows of the $r \times r$ unit matrix, chosen so that the denominator formed from A'' is regular over K. Then A'' is K-admissible; this shows that (a) $\Leftrightarrow$ (c), and that we can reach a matrix satisfying (a) by an operation λ.

When R is a Sylvester domain with universal field of fractions K, we note that all full matrices become invertible over K and ρ_K is the inner rank. Since $\rho A = m$, we can write $A = PA'$, where P is right full $r \times m$ and A' is left full $m \times (m+1)$. Clearly A' is not full, so by Proposition 6.3(ii), A' is admissible; moreover P is full precisely when $r = m$. This shows that (a) $\Leftrightarrow$ (d) and it also proves the last assertion. ∎

From now on we shall restrict attention to the case of a semifir R with universal field of fractions U; by an admissible matrix we shall understand a matrix that is U-admissible. As the homomorphism to U is fully inverting, this will allow us to decide when an admissible matrix is left prime. We recall from Section 3.1 that over a weakly finite ring any left prime admissible matrix is left full and so has index 1. The next two lemmas hold for Sylvester domains (except for the last sentence in each), but they will only be needed for semifirs.

Lemma 7.6.5. *Let R be a semifir and let $A \in {}^r R^{m+1}$ be an admissible matrix. Then the following conditions are equivalent:*

(a) A is left prime (hence the number of rows in A is $o(A)$, i.e. $r = m$),
(b) every non-trivial operation λ' admitted by A increases the number of rows.

Further, if R has left ACC_r then any admissible matrix $A \in {}^r R^{m+1}$ can be transformed by an operation λ' : $A = PA'$ to a left prime admissible matrix $A' \in {}^m R^{m+1}$.

Proof. If (a) holds, A is left prime; then A is left full and no operation λ' can decrease the number of rows. Moreover, in any factorization of A,

$$A = PB, \quad \text{where } P \in R_r, B \in {}^r R^{m+1}, \qquad (6)$$

P is a unit, so any non-trivial operation λ' increases the number of rows, i.e. (b) holds. Conversely, if (b) holds, then by Lemma 6.4, A is left full and in any equation (6), P is a unit and hence A is left prime, i.e. (a) holds.

If A has left ACC_r we can apply Proposition 5.9.1 to obtain an equation (6), where B is left prime. ∎

To establish a corresponding result for the operations ρ, ρ', let us call an admissible matrix A *right* $*$-*prime* if its core $A_\bullet$ is right prime. By the *core* of a bordered matrix we understand the submatrix obtained by omitting the first and last rows and columns.

Lemma 7.6.6. *Let R be a semifir and let A, A' be admissible matrices. If $A = A'Q^*$, where Q^* is bordered (and necessarily right full), then $o(A) \leq o(A')$, with equality if and only if Q^* is full. For any admissible matrix A the following are equivalent:*

(a) A is right $*$-prime,
(b) every bordered square right factor of A is a unit,
(c) all full operations ρ' admitted by A are trivial.

Further, if R has right ACC_{m-1}, then any admissible matrix $A \in {}^r R^{m+1}$ can be reduced by a full operation $\rho' : A = A'Q^*$ to a matrix that is right $*$-prime.

Proof. If A, A' are admissible, where

$$A = A'Q^*, \quad A \in {}^r R^{m+1}, A' \in {}^r R^{m'+1}, Q^* \in {}^{m'+1} R^{m+1}, \tag{7}$$

with a bordered matrix Q^*, then Q^* is right full, by Proposition 6.3 (iii), hence $m \leq m'$, with equality if and only if Q^* is full.

Assume now that $m' = m$ in (7); comparing cores in (7), we find that $A_\bullet = A'_\bullet Q_\bullet$, where $Q_\bullet$ is the core of Q^*, and clearly Q^* is a unit if and only if $Q_\bullet$ is. It follows that $A_\bullet$ is right prime if and only if Q^* is a unit in every factorization (7). Thus (a) ⇔ (b), and (b) ⇔ (c) is clear because any full operation ρ' preserves the order and so has a square matrix Q^*. Conversely, when Q^* is square, ρ' preserves the order and so is full.

Finally, when R has right ACC_{m-1}, we can by Proposition 5.9.1 write

$$A_\bullet = A'_\bullet Q_\bullet, \quad \text{where } A'_\bullet \in {}^r R^{m-1}, Q_\bullet \in R_{m-1},$$

where $A'_\bullet$ is right prime and $Q_\bullet$ is full. Hence on writing $A = A'Q^*$, where

$$A' = (A_0 \quad A_\bullet \quad A_\infty), \quad Q^* = \begin{pmatrix} 1 & 0 & 0 \\ 0 & Q_\bullet & 0 \\ 0 & 0 & 1 \end{pmatrix},$$

we find that Q^* is full, whence A' is admissible and right $*$-prime. ∎

7.6 Numerators and denominators

In the rest of this section our aim is to take a semifir R with universal field of fractions U and describe the relation between the various admissible systems for a given element of U. We begin with a basic result giving conditions under which two admissible systems determine the same element of U. A matrix Q^* is said to be *strictly bordered* if it has the form $Q^* = 1 \oplus Q_\bullet \oplus 1$, and the corresponding operation ρ is said to be *strict*. The equation

$$Q^* = \begin{pmatrix} 1 & 0 & 0 \\ Q_0 & Q_\bullet & Q_\infty \\ 0 & 0 & 1 \end{pmatrix} = \begin{pmatrix} 1 & 0 & 0 \\ Q_0 & I & Q_\infty \\ 0 & 0 & 1 \end{pmatrix} \begin{pmatrix} 1 & 0 & 0 \\ 0 & Q_\bullet & 0 \\ 0 & 0 & 1 \end{pmatrix}$$

shows that every operation ρ consists of a trivial operation followed by a strict operation.

Lemma 7.6.7. *Let R be a semifir and U its universal field of fractions. Then two admissible matrices $A \in {}^rR^{m+1}$ and $B \in {}^sR^{n+1}$ determine the same element $p \in U$ if and only if there exist matrices P, P', strictly bordered matrices Q^*, Q'^* and admissible matrices A', B' such that*

$$A = PA', \quad B = B'Q^*, \tag{8}$$

$$P'A' = B'Q'^*. \tag{9}$$

Moreover, we can always take P to be right full, Q^* to be full and A' to be left full.

Proof. Assume that (8), (9) hold; by (8), A and A' determine the same element p of U, and B, B' determine the same element q. If $A'u = 0$, we apply (9); Q'^*u has the form $(1, v_\bullet, p)^T$, so if $p \neq q$, then $B'Q'^*u \neq 0$ and we have a contradiction. Thus $p = q$.

Conversely, assume that A and B determine the same element p. It follows that

$$\begin{pmatrix} A_0 & A_\bullet & A_\infty & 0 & A_\infty \\ B_0 & 0 & B_\infty & B_\bullet & 0 \end{pmatrix}$$

is an admissible system for $p - p = 0$, so its numerator is not right full: its matrix

$$\begin{pmatrix} A_0 & A_\bullet & A_\infty & 0 \\ B_0 & 0 & B_\infty & B_\bullet \end{pmatrix}$$

has inner rank $k < m + n$. Applying the partition lemma to a rank factorization, we get

$$\begin{pmatrix} A_0 & A_\bullet & A_\infty & 0 \\ B_0 & 0 & B_\infty & B_\bullet \end{pmatrix} = \begin{pmatrix} P & 0 \\ P' & B'_\bullet \end{pmatrix} \begin{pmatrix} A' & 0 \\ -Q'_\bullet & Q_\bullet \end{pmatrix}. \tag{10}$$

Here the corner blocks of zeros on the right are $r \times n - 1$ and $m \times n - 1$.

On equating blocks we find

$$A = (A_0 \quad A_\bullet \quad A_\infty) = PA', \tag{11}$$
$$(B_0 \quad 0 \quad B_\infty) = P'A' - B'_\bullet Q_\bullet, \tag{12}$$
$$B_\bullet = B'_\bullet Q_\bullet. \tag{13}$$

Here (11) is the first equation of (8) and (13) yields the second equation of (8), if we set

$$B' = \begin{pmatrix} B_0 & B'_\bullet & B_\infty \end{pmatrix}, \quad Q^* = \begin{pmatrix} 1 & 0 & 0 \\ 0 & Q_\bullet & 0 \\ 0 & 0 & 1 \end{pmatrix}.$$

Finally, we obtain (9) from (12) by taking Q'^* to be the matrix $Q'_\bullet$ topped by the first row and tailed by the last row of the unit matrix.

It remains to prove that P is right full, A' is left full and Q^* is full. This will also show (by Proposition 6.3) that A', B' are admissible.

We denote by t the number of columns of P, P' or, equivalently, the number of rows of A' (see (10)) and by t' the number of columns of $B'_\bullet$, or equivalently, the number of rows of $Q'_\bullet$, $Q_\bullet$. Since (10) was a rank factorization, we have

$$t + t' = k < m + n. \tag{14}$$

But $t \geq \rho P \geq \rho A = m$, by (11) and $t' \geq \rho Q_\bullet \geq \rho B_\bullet = n - 1$, by (13). A comparison with (14) shows that $t = \rho P = m$ and $t' = \rho Q_\bullet = n - 1$. Hence P is right full, A' has index 1 and so is left full, by Lemma 6.4, and $Q_\bullet$, therefore also Q^*, is full, as claimed. ∎

Note that although we proved P to be right full and Q^* full, we have not proved this for P' and Q'^*. Hence the common value of the two sides of (9) need not itself be an admissible matrix, and we have not obtained a chain of operations λ, λ', ρ, ρ' connecting A and B.

To study the case where $Q'^* = 1 \oplus Q'_\bullet \oplus 1$ may be non-full, we take a rank factorization $Q'_\bullet = S_\bullet T_\bullet$ and write $S^* = 1 \oplus S_\bullet \oplus 1$, $T^* = 1 \oplus T_\bullet \oplus 1$; then

$$Q'^* = S^* T^*, \tag{15}$$

where S^* is strictly bordered and right full and T^* is strictly bordered and left full. By the first part of the above proof, $A'u = 0$; hence on combining this with (9) and (15) we get

$$0 = P'A'u = B'Q'^*u = (B'S^*)(T^*u).$$

Since T^* is bordered, the vector T^*u has first entry 1, therefore $B'S^*$ is not right full. But by Proposition 6.3(iii), $B'S^*$ is admissible, because S^* is right full. Thus $B' \mapsto B'S^*$ is an operation ρ taking B' to an admissible matrix, but this does not lead to a 'path' from A to B unless T^* is full. However, we can ensure that T^* is full by imposing a further condition on A and B.

An admissible matrix B is said to be *m-blocked* for $m \geq 1$, if B cannot be transformed by any full operation ρ', followed by any strict operation ρ, to an admissible matrix of smaller order than m. If B is $o(B)$-blocked, it will be called *minimal admissible*, or simply *minimal*.

Let $B'S^*$ be of order h; if B is m-blocked, then since $B \mapsto B'$ by a full operation ρ' and $B' \mapsto B'S^*$ by a strict operation ρ, it follows that

$$h = o(B'S^*) \geq m.$$

Now T^* is $(h+1) \times (m+1)$ and is left full, so $h \leq m$, and hence $h = m$. It follows that $Q'^* = S^*T^*$ is right full, and so $P'A' = B'Q'^*$ is admissible, and P' is right full. This proves

Corollary 7.6.8. *In the situation of Lemma 6.7, if A is minimal admissible, then the common value of the two sides of (9) is an admissible matrix. Hence we can pass from A to B by a chain of four operations λ', λ, ρ', ρ, in that order. Here ρ is full, and if A is left full, then λ' is full.* ∎

In Lemma 6.7, if A is left prime, P must be a unit, and if B is right *-prime, then Q^* must be a unit. Let us call an admissible matrix A *reduced* if it is left prime and right *-prime. Then we can state the result just proved as

Corollary 7.6.9. *In the situation of Lemma 6.7, if A is left prime and B is right *-prime, then there exist a matrix P and a bordered matrix Q^* such that $PA = BQ^*$. If, moreover, A, B are reduced, then they are stably biassociated.*

Proof. All but the last part has been proved, and the last part follows because under the given conditions the relation $PA = BQ^*$ is coprime, hence comaximal, by Proposition 3.1.4. ∎

We remark that if A is right *-prime, any full operation ρ' must be trivial; hence a reduced matrix A that is not minimal admits a strict operation $\rho : A \mapsto AQ^*$ with a matrix Q^* of negative index.

We now come to our main result, the comparison theorem for numerators and denominators (although they are not mentioned explicitly).

Theorem 7.6.10. *Let R be a semifir, U its universal field of fractions and $p \in U$.*

(i) *We can pass between any two admissible matrices A, B for p by some sequence of operations λ, λ', ρ, ρ' (at most six in all).*

(ii) *The minimal admissible matrices for p are just the admissible matrices of least order for p, and we can pass from one minimal admissible matrix for p to any other by four operations:* λ', λ, ρ', ρ, *in that order, where* ρ, ρ' *are full.*

(iii) *If A and B are left full admissible matrices for p, then we can pass by a full operation* λ' *from A to an admissible matrix* A'' *and by a full operation* ρ' *from B to an admissible matrix* B'' *such that* A'' *and* B'' *are stably biassociated.*

(iv) *Any two reduced admissible matrices A, B for p are stably biassociated, and if there exists a reduced admissible matrix for p, then there exists a reduced minimal admissible matrix for p.*

Proof. We begin with (ii); by Lemma 6.6 and an induction on the order we see that every element has a minimal admissible matrix. If A, B are minimal admissible for p, with $o(A) \leq o(B)$, say, then B is certainly $o(A)$-blocked, so by Corollary 6.8, the two sides of (9) represent an admissible matrix and we have a path from A to B by our four operations. By Lemma 6.7, P can be chosen to be right full and Q^* can be chosen to be full. Since B is minimal, Q'^* must be full. This proves the last part of (ii).

Since all minimal admissible matrices for p have the same order and since from any admissible matrix A for p we can construct a minimal admissible matrix by repeated operations ρ', ρ, λ' that do not increase the order, it follows that the common order of all minimal admissible matrices for p is the least order of any admissible matrix for p; this proves the rest of (ii).

To prove (i) let B be any admissible matrix; we can always pass from B to some minimal admissible matrix B' by two operations. Now A will be $o(B')$-blocked and we can pass from A to B' by four operations ρ', ρ, λ', λ. Hence, we can pass between any two admissible matrices by a sequence of at most six operations ρ', ρ, λ', λ, ρ', ρ, so (i) holds.

Next let A, B be left full admissible matrices for p, as in (iii). By Lemma 6.7, (8), we can transform A by a full operation λ' to A', and B by a full operation ρ' to B' such that

$$P'A' = B'Q'^*, \quad \text{where } Q'^* \text{ is strictly bordered.} \tag{16}$$

Since B is left full and Q^* in (8) is full, B' is left full, hence the matrix (P', B') is also left full. To show that $(A', Q'^*)^T$ is right full, suppose that $(A', Q'^*)^T x = 0$ for some $x \in {}^m U$. Since A' is admissible for p, the vector x is a right multiple of a vector $u = (1, u\bullet, p)^T$, so if $x \neq 0$, its first entry cannot vanish. But $Q'^* x = 0$, so by the bordered form of Q'^*, the first entry of x is 0. This shows that $x = 0$, so $(A', Q'^*)^T$ is right full. Further, A' and B', being left

7.6 Numerators and denominators

full admissible matrices, both have index 1, by Lemma 6.4, so by Proposition 3.1.3 we can cancel full left and right factors from the two sides of (16) so as to obtain a comaximal relation.

It remains to show that these cancellations have the form of admissible operations λ' and ρ'. The first of these conditions is immediate; to get the second we note that any full right factor of Q'^* can be taken to be bordered. To achieve this we first conjugate Q'^* by a permutation matrix to bring the first two rows to the form $(I_2 \quad 0)$ and then apply the partition lemma to transform the factorization to the form

$$Q'^* = \begin{pmatrix} s & 0 \\ S' & S'' \end{pmatrix} \begin{pmatrix} t & 0 \\ T' & T'' \end{pmatrix}, \tag{17}$$

where $st = 1$; we note that s is $2 \times \nu$ and t is $\nu \times 2$, for some $\nu \geq 1$. If Q'^* is $(m+2) \times (m+2)$, then the second factor has a $\nu \times m$ block of zeros, and since it must be full, we conclude by Proposition 3.1.2 that $\nu \leq 2$. Here equality must hold because $st = 1$; hence s is a unit and we can by an inessential modification reduce s, t to 1 in (17). Now we can cancel the bordered full right factor (and undo the conjugation) and this corresponds to an operation ρ'. The resulting comaximal relation

$$P''A'' = B''Q''^*$$

shows that A'' and B'' are stably biassociated.

The first statement of (iv) now follows from (iii), by the definition of 'reduced'. For the second assertion we apply the reduction in (iii) to a minimal admissible matrix A and a reduced matrix B. Of the resulting matrices A'' and B'', the first will again be minimal, because the operations applied to A were full, while the second can be taken equal to B, because B admits no non-trivial operations λ' or ρ'. Hence A'', being stably biassociated to the reduced matrix B, will be a minimal reduced admissible matrix for p. ∎

In Section 7.1 we defined the left depth of p as the least value of the order of any admissible matrix for p. From Theorem 6.10 (ii) we see that it can be obtained as the order of any minimal admissible matrix for p. We note that whereas a minimal admissible matrix for a given $p \in U$ always exists, there may be no reduced matrix (in the absence of ACC). Moreover, a reduced matrix for p need not be minimal, e.g. over $k\langle x, y \rangle$ the matrix $\begin{pmatrix} x & -1 & 0 \\ 0 & 1 & y \end{pmatrix}$ is admissible for $y^{-1}x$ and is reduced, but not minimal. However, when there is a reduced matrix, then there is also a minimal reduced matrix, by Theorem 6.10(iv).

We remark that the analogues of these theorems hold when p is a matrix over U. The simplest case is that of a square matrix, say $v \times v$. This is most easily handled by taking T to be a semifir with universal field of fractions V and writing $R = T_v, U = V_v$. An $m \times n$ matrix over R is now an $mv \times nv$ matrix over T; if its inner rank over T is r, then the *rank* of A over R may be defined as r/v. So the rank is now no longer an integer but a rational number with denominator dividing v. By a *left full* matrix we understand as before an $m \times n$ matrix of rank m; clearly this will also be left full as matrix over T. Right full and full matrices are defined correspondingly. The definitions of left and right prime, maximal and reduced, and of the basic operations $\lambda, \rho, \lambda', \rho'$ are all as before. An admissible system for $p \in U$ can again be defined as a system with right full denominator, of rank equal to the rank of the matrix. If this rank is m and the numerator has rank s, then the rank of p over U is $s - m + 1$, a rational number between 0 and 1.

With these conventions Theorem 6.10 and the lemmas leading up to it hold as stated, but for matrix rings over semifirs. The proofs are similar to those given here and will be left to the reader.

Exercises 7.6

1. Show that the conditions of Lemmas 6.4–6.6 are invariant under stable biassociation; interpret these conditions as conditions on the corresponding module with a pair of distinguished elements.
2. If A is an admissible matrix for p, find an admissible matrix for $p(1 + p)^{-1}$.
3. Show that all full (but no non-full) operations $\lambda, \lambda', \rho, \rho'$ preserve the conditions of Lemma 6.4, all full operations ρ' but not all full operations ρ preserve the conditions of Lemma 6.5, all full operations λ' but not all full operations λ preserve the conditions of Lemma 6.6.
4. Let A be an admissible matrix; if its core has an invertible $v \times v$ submatrix, show that the order of A can be decreased by v.
5. Let U be the universal field of fractions of the free algebra $k\langle x, y, z\rangle$. Show that $A = \begin{pmatrix} x & -z & 0 \\ 0 & z & y \end{pmatrix}$ is an admissible matrix for $y^{-1}zz^{-1}x = y^{-1}x$. Verify that no operation ρ will decrease the order of A, but that $y^{-1}x$ has left depth 1.
6. State and prove the matrix analogues of the results in this section.
7°. Discuss the matrix equation $Au = 0$, where u in normalized form is not unique but its last component is unique.

7.7 The depth

In the depth we have a numerical invariant for the elements of an epic R-field that has no analogue in the commutative case; there we can write every element as

$ab^{-1} = b^{-1}a$ and this has depth 1. By contrast we shall see that in the universal field of fractions of a semifir the depth is generally unbounded. But first we shall examine the case of elements of minimum depth. Given a ring R with epic R-field K, if a matrix $P \in {}^m K^n$ is given by an admissible system (in the sense defined in Section 7.6) with null core (i.e. a core with 0 columns), then the system has the form

$$(A \quad B) \begin{pmatrix} I \\ -P \end{pmatrix} = 0, \quad A \in {}^m R^n, B \in R_m, \tag{1}$$

where B is invertible over K, and hence

$$P = B^{-1}A, \quad B \text{ invertible over } K. \tag{2}$$

Conversely, any matrix P of the form (2) is given by the admissible system (1). Similarly, P has the form

$$P = AB^{-1}, \quad B \text{ invertible over } K, \tag{3}$$

precisely if it is given by a right admissible system $(I \quad -P)\begin{pmatrix} A \\ B \end{pmatrix} = 0$, or equivalently (by the left–right dual of Proposition 1.5), by

$$\begin{pmatrix} 0 & A & -I \\ -I & B & 0 \end{pmatrix} \begin{pmatrix} I \\ B^{-1} \\ P \end{pmatrix} = 0. \tag{4}$$

We remark that the leftmost matrix in (4) has the right annihilator $(B, I, A)^T$ in R, which is right regular over K. The general situation can be described as follows.

Proposition 7.7.1. *Let R be a ring and K an epic R-field. Given $P \in {}^m K^n$:*

(i) *P can be written in the form (2) if and only if it has a (left) admissible representing matrix A whose core is null (i.e. $v \times 0$ for some v), or equivalently, whose core is left annihilated by a matrix which is regular over K.*

(ii) *P can be written in the form (3) if and only if it has an admissible matrix which is right annihilated by a matrix over R which is right regular over K.*

Proof. (i) If P is given by (2), it has the admissible system (1) with null core, and the latter is left annihilated by I_m. Conversely, if A is an admissible matrix for P, where $CA_\bullet = 0$ for a matrix C over R that is regular over K, then CA is again admissible for P, by Proposition 6.3(iv), and its core is $CA_\bullet = 0$; if this had a positive number of columns, the denominator of CA would not be invertible over K, a contradiction; so CA and hence A itself, must have a null core.

(ii) If P has the form (3), it has the admissible system (4) in which the left factor is right-annihilated by $(B, I, A)^T$, which is right regular over K. Conversely, let P have the admissible matrix A that is right-annihilated by C, right regular over K. By the uniqueness of the normalized solution U of (4) over K, if C is a solution, then $C - UC_0$ is a solution with top block zero, hence if it were non-zero, we could add it to U, which would contradict the uniqueness of the normalized solution. ∎

Let us restate the most important special case, where K is a fully inverting field. For simplicity we shall limit ourselves to elements. In that case (2), (3) just mean that the left depth $d_l(p)$, respectively the right depth $d_r(p)$ of p is 1.

Corollary 7.7.2. *Let R be a ring with fully inverting field K. Given $p \in K$, $d_l(p) = 1$ if and only if the core of one (and hence every) admissible matrix for p is a zero-divisor in R, and $d_r(p) = 1$ if and only if some (and hence every reduced) admissible matrix for p is a zero-divisor in R.*

Further, if R is a semifir, then it is left Bezout if and only if the left depth of elements of K is bounded by 1, and correspondingly on the right.

Proof. The main result is clear from Proposition 7.1, while the last part follows because a semifir is left Bezout if and only if it is left Ore, or equivalently, it has a field of fractions in which every element has the form $b^{-1}a$, in which case that is its only field of fractions. ∎

Let R be a semifir and U its universal field of fractions. The various depths of element of U satisfy certain inequalities, which were noted in Exercise 5 of Section 7.1. To find when equality holds we can take admissible matrices A, B for our elements p, q that are reduced minimal and ask whether the matrix

$$A.B = \begin{pmatrix} B_0 & B_\bullet & B_\infty & 0 & 0 \\ 0 & 0 & A_0 & A_\bullet & A_\infty \end{pmatrix}, \tag{5}$$

which is admissible for pq, is reduced minimal. The following result, without giving a complete answer, shows when this matrix is reduced.

Proposition 7.7.3. *Let R be a semifir with a universal field of fractions U. Given $p, q \in U$ with admissible matrices A, B respectively, which are reduced, if $A.B$ is the matrix for pq given by (5), then*

(i) *$A.B$ is left prime whenever the denominator of A and the numerator of B have no non-unit similar left factors;*
(ii) *$A.B$ is right *-prime whenever the numerator of A and the denominator of B have no non-unit similar right factors.*

7.7 The depth

Proof. Let us write N_p, D_p for the numerator and denominator, respectively, of p, so that $A = (N_p, A_\infty) = (A_0, D_p)$ and similarly for q. We have

$$A.B = \begin{pmatrix} B_0 & B_\bullet & B_\infty & 0 & 0 \\ 0 & 0 & A_0 & A_\bullet & A_\infty \end{pmatrix} = \begin{pmatrix} N_q & B_\infty & 0 \\ 0 & A_0 & D_p \end{pmatrix}. \quad (6)$$

To prove (i) we assume that D_p and N_q have no similar non-unit left factors and that $A.B$ is not left prime. We have a factorization that by the partition lemma can be written

$$\begin{pmatrix} N_q & B_\infty & 0 \\ 0 & A_0 & D_p \end{pmatrix} = \begin{pmatrix} P' & 0 \\ P & P'' \end{pmatrix} \begin{pmatrix} Q' & 0 \\ Q^* & Q'' \end{pmatrix}, \quad (7)$$

where the first factor on the right is square. Thus if A, B are of orders m, n, respectively, then P' is $n \times n$, P'' and Q' are $m \times m$, P is $m \times n$, Q' is $n \times (n+1)$ and Q^* is $m \times (n+1)$. Thus $B = P'Q'$; since B is left prime, P' is a unit and so may be taken to be I. It follows that $Q' = B$. Further, we have

$$D_p = P''Q'', \quad (0, A_0) = PB + P''Q^*. \quad (8)$$

Write $Q^* = (Q, q)$, where Q is $m \times n$ and q is $m \times 1$. The second equation (8) shows that

$$A_0 = PB_\infty + P''q, \quad PN_q = -P''Q. \quad (9)$$

Now N_q and P'' are full and since A is left prime, the same is true of (P, P''), hence by the second equation (9), P is similar to a left factor of Q and P'' is similar to a left factor of N_q. But P'' is also a left factor of D_p, by (8), so it must be a unit. Hence the first factor in (7) is a unit and this means that $A.B$ is left prime.

To prove (ii) we assume that N_p and D_q have no similar non-unit right factor and that $A.B$ is not right *-prime. This means that the core of $A.B$ has a non-unit square right factor. Applying the partition lemma in two ways we obtain

$$\begin{pmatrix} B_\bullet & B_\infty & 0 \\ 0 & A_0 & A_\bullet \end{pmatrix} \begin{matrix} n \\ m \end{matrix} = \begin{matrix} n \\ m \end{matrix} \begin{pmatrix} P' & 0 \\ P & P'' \end{pmatrix} \begin{pmatrix} Q' & 0 \\ Q & Q'' \end{pmatrix} \begin{matrix} n \\ m-1 \end{matrix}$$
$$\quad n \quad m-1 \quad n \quad m-1$$

$$= \begin{matrix} n \\ m \end{matrix} \begin{pmatrix} S' & S \\ 0 & S'' \end{pmatrix} \begin{pmatrix} T' & T \\ 0 & T'' \end{pmatrix} \begin{matrix} n-1 \\ m \end{matrix}, \quad (10)$$
$$\quad n-1 \quad m \quad n-1 \quad m$$

where the number of rows and columns is indicated. Now $P''Q'' = A_\bullet$ and since A is right *-prime, Q'' must be a unit and so may be taken to be I. Similarly $B_\bullet = S'T'$ and it follows that T' is a unit and so may be taken to be I. Since the two right factors in the products in (10) are associated, it follows that T''

is similar to Q', but $P'Q' = D_q$, $S''T'' = N_p$; this means that N_p and D_q have similar factors, contradicting our assumption, so the result follows. ∎

Here the condition (on similar shared factors) cannot be omitted, as examples show (see Exercise 8). The conclusion may be expressed symbolically by saying that if $D_p^{-1} N_q$ is in lowest terms, then $A.B$ is left prime and if $N_p D_q^{-1}$ is in lowest terms, then $A.B$ is right *-prime.

If R is a semifir and U its universal field of fractions, then for any matrix A over R its rank over U, $\rho_U A$, equals ρA, as we have seen in Section 7.5. Now let K be any epic R-field; then the rank of A over K, $\rho_K A$, say, is the maximum of the orders of square submatrices of A that are regular over K, i.e. which do not lie in the singular kernel. Since this kernel is least for the universal field of fractions, the rank is then greatest, so we have

$$\rho_K A \leq \rho_U A.$$

This observation can be used to show that for fully atomic semifirs the depth of an element cannot increase on specialization. We begin by proving a theorem on 'universal denominators', which is of independent interest.

Theorem 7.7.4. *Let R be a semifir and U its universal field of fractions. If an element p of U can be defined by a reduced admissible matrix A over R and there is a matrix for p which remains admissible over an epic R-field K, then A is also K-admissible.*

If, moreover, R is a fully atomic semifir, then every element of U can be defined by a reduced admissible matrix.

Proof. Let $p \in U$ and let A be a reduced admissible matrix over R defining p. By hypothesis there is a U-admissible matrix B for p, of index 1, which is also K-admissible. By Theorem 6.10(iii) we can pass by full operations λ', ρ' from A to A'' and from B to B'' such that A'' and B'' are stably biassociated. Since A is left prime, λ' is trivial and we may take $A'' = A$; thus A is stably biassociated to B''. Now B is K-admissible, so the system $Bz = 0$ has a unique normalized solution and $B = B''Q^*$, therefore $B''y = 0$ has a normalized solution $y = Q^*z$ in K. The denominator of B'' is invertible over K, because it is a left factor of the denominator of B, which is known to be invertible over K. Thus B'' is K-admissible, hence so is A.

If R is fully atomic, and $p \in U$ is defined by an admissible matrix A, then A can be transformed to a left prime matrix A' by an operation λ', using Proposition 5.9.1, and A' can be transformed to a minimal reduced matrix by an operation ρ', using the left–right dual of Proposition 5.9.1 and Theorem 6.10 (iv). ∎

We can now prove the result on the behaviour of the depths over different R-fields announced earlier.

Theorem 7.7.5. *Let R be a fully atomic semifir, U its universal field of fractions and K any epic R-field. Then for any element p of U in the minimal domain of the specialization $\phi : U \to K$ we have $d(p) \geq d(p\phi)$, where d is any one of the depths.*

Proof. Consider the left depth, say; p may be defined by a minimal reduced admissible $m \times (m + 1)$ matrix A, which by Theorem 7.4 is still admissible for $p\phi$ over K. Now $d(p) = m$ by definition, and since A is K-admissible, $d(p\phi) \leq m$, as claimed. ∎

For more general rings the depth may increase on specialization, as Exercise 2 shows.

We now turn to compute some examples of depths. We recall from Exercise 1.5 the following inequalities for the left depth, where $f : R \to S$ is a Σ-inverting homomorphism, $a \in \operatorname{im} f$ and $p, q \in R_\Sigma(s)$:

$$d(p - q), d(pq) \leq (p) + d(q), \quad d(pa) \leq d(p) \leq d(ap) \leq d(p) + 1. \quad (11)$$

The examples will show in particular that some of these inequalities are sharp and in the free algebra and the free power series ring (on more than one free generator) the depth assumes every positive integer value.

Let R be a semifir and U its universal field of fractions. First consider an element of the form

$$p = b_n^{-1} a_n b_{n-1}^{-1} a_{n-1} \ldots b_1^{-1} a_1, \quad (12)$$

where no a_i or b_i is 0 or a unit. An admissible matrix for $b^{-1} a$ is $(a, -b)$, hence by (5) of 7.1,

$$A = \begin{pmatrix} a_1 & -b_1 & 0 & 0 & \ldots & & \\ 0 & a_2 & -b_2 & 0 & \ldots & & \\ \ldots & & \ldots & & \ldots & & \\ 0 & 0 & 0 & 0 & \ldots & a_n & -b_n \end{pmatrix} \quad (13)$$

is an admissible matrix for p. We shall need a condition for A to be left prime; let us recall that two elements a, b of a ring are *totally coprime* if no non-unit factor of a is similar to a factor of b.

Lemma 7.7.6. *Let R be a semifir and $a_i, b_i \in R (i = 1, \ldots, n)$ not zero or units. Then the matrix A in (13) is left prime provided that each a_i is totally coprime to each b_j.*

Proof. Any $n \times n$ left factor P of A must also be a left factor of the numerator and denominator of A, namely the square matrices

$$A_1 = \begin{pmatrix} a_1 & -b_1 & & \\ & a_2 & -b_2 & \\ & & \ddots & \\ & & & a_n \end{pmatrix}, \quad A_2 = \begin{pmatrix} -b_1 & & & \\ a_2 & -b_2 & & \\ & \ddots & & \\ & & a_n & -b_n \end{pmatrix},$$

obtained by omitting the last and first column, respectively, from A. Now A_1 has $a_1, \ldots, a_n$ down the main diagonal and 0 below it, hence by unique factorization over R (Theorem 3.2.7) and Proposition 3.2.8, P must be a product of factors similar to factors of $a_1 \ldots a_n$ and therefore totally coprime to $b_1 \ldots b_n$. But A_2 has (apart from sign) $b_1, \ldots, b_n$ down the main diagonal and 0 above it, so P is also a product of factors similar to factors of $b_1 \ldots b_n$. This is possible only when P is a unit, hence A is indeed left prime, as claimed. ∎

We can now give sufficient conditions for the element (12) to have left depth n:

Theorem 7.7.7. *Let R be a semifir and U its universal field of fractions. Consider the element p of U given by (12), where the a_i, b_i are in R and are neither 0 nor units. If each a_i is totally coprime to each b_j and there exists a proper ideal $\mathfrak{c}$ of R containing all the a_i, b_i such that $R/\mathfrak{c}$ is weakly finite, then the left depth of p, as element of U, is n.*

Proof. We know that the matrix A in (13) is admissible for p; if we can show it to be minimal reduced, then it will follow that $d_1(p) = n$. By Lemma 7.6, A is left prime; its core $A_\bullet$ is of the same form as A but transposed, and the same argument shows $A_\bullet$ to be right prime, thus A is right *-prime, and hence reduced. By Lemma 3.3.11, A is not stably associated to a matrix of smaller order, hence A is minimal and the result follows. ∎

To give an example, in the free algebra $R = k\langle x, y \rangle$, x and y are dissimilar atoms, and if $\mathfrak{c}$ is the ideal generated by x, y, then $R/\mathfrak{c} \cong k$, hence $(y^{-1}x)^n$ has depth n. The same holds in the free power series ring $k\langle\!\langle x, y \rangle\!\rangle$. As an example not covered by Theorem 7.7, consider in $k\langle\!\langle x, y, z \rangle\!\rangle$ the element $y^{-1}zx^{-1}yz^{-1}x$. This is determined by the admissible matrix

$$\begin{pmatrix} x & -z & 0 & 0 \\ 0 & y & -x & 0 \\ 0 & 0 & z & -y \end{pmatrix}. \tag{14}$$

7.7 The depth

It can be shown by a direct argument that its depth is 3 (see Exercise 5).

As a second type of element consider

$$q = a_1^{-1} + a_2^{-1} + \cdots + a_n^{-1}, \quad \text{where } a_i \in R^\times. \tag{15}$$

An admissible matrix for q is given by the $n \times (n+1)$ matrix

$$B = \begin{pmatrix} 1 & -a_1 & -a_1 & \cdots & \cdots & -a_1 \\ 1 & a_2 & 0 & \cdots & \cdots & 0 \\ 1 & 0 & a_3 & 0 & \cdots & 0 \\ & \cdots & & & \cdots & \\ 1 & 0 & 0 & \cdots & \cdots & a_n & 0 \end{pmatrix}. \tag{16}$$

By subtracting the last column from the others after the first, we see that B is left prime, provided that any two of the a's are totally coprime. Under this condition it also follows that B is right $*$-prime, and if B is stably biassociated to an $m \times (m+1)$ matrix, then $m \geq n$, provided that there exists a proper ideal $\mathfrak{c}$ containing all the a_i such that $R/\mathfrak{c}$ is weakly finite. Hence we obtain

Proposition 7.7.8. *Let R be a semifir with universal field of fractions U. Given $a_1, \ldots, a_n \in R$, each a_i not 0 or a unit and such that any two are totally coprime and if further, there exists a proper ideal $\mathfrak{c}$ containing all the a_i such that $R/\mathfrak{c}$ is weakly finite, then the element (15) has depth n.* ∎

Exercises 7.7

1. In the universal field of fractions of $k\langle x, y, z\rangle$ compute the left, right, upper and lower depths of xy^{-1}, $y^{-1}z$, and $xy^{-1}z$.
2. (G. M. Bergman) Let R be the k-algebra generated by x, y, z, w with defining relation $wx = yz$ and denote by U, V the universal fields of fractions of the free algebras $k\langle x, y, z\rangle, k\langle x, z\rangle$, respectively. Verify that there is a homomorphism $R \to U$ with $x, y, z, w \mapsto x, y, z, yzx^{-1}$ and a homomorphism $R \to V$ with $x, y, z, w \mapsto x, 0, z, 0$. Show that U, V are epic R-fields with a specialization $U \to V$ defined by $x, y, z \mapsto x, 0, z$. Show also that xz^{-1} has left depth 1 in U but that its left depth in V is 2.
3. Let R be a semifir with universal field of fractions U. Given $p \in U$, let A, B be two admissible matrices for p. By applying Theorem 6.10(iii), find an admissible matrix C for p that is K-admissible, for any epic R-field K, whenever either A or B is.
4. Obtain analogues of Lemma 7.6 and Theorem 7.7 when R is a full matrix ring over a semifir.
5. Prove that the matrix (14) is reduced, by showing that it has no left factor similar to x, y or z. (*Hint*: If the matrix is A, consider a comaximal relation $uA = xA'$, where $u \in R^3$ and use the fact that the eigenring of x is k.) Apply Lemma 3.3.11 to show

that the matrix is not stably biassociated to a matrix of lower order and hence show that $d_l(y^{-1}zx^{-1}yz^{-1}x) = 3$.
6. Use the method of Exercise 5 to show that the left depth of $xy^{-1}x + x^2y^{-2}x^2 + \cdots + x^n y^{-n} x^n$ in $k\langle\!\langle x, y \rangle\!\rangle$ is $n + 1$ (see Cohn [85a]).
7. Find the depth of $xy^{-1}x + x^2y^{-2}x + \cdots + x^n y^{-n} x$ and of $xy^{-1}x + x^2y^{-1}x^2 + \cdots + x^n y^{-1} x^n$ in $k\langle x, y \rangle$.
8*. In a semifir R let a, b be non-zero non-units without a common left factor and let $c, d, e, a', c' \in R$ be such as to satisfy a comaximal relation $c'a = a'c$. Put $A = (a'd - c'b, -a'e)$, $B = (a, -b)$ and suppose that d, e are such that A is left prime. Show that A, B are both minimal reduced but that $A.B$ is not left prime. Find the similar factors of the denominator of A and the numerator of B.
9°. Find a non-Ore domain R admitting a field of fractions in which every element can be written as $ba^{-1}c$ for some $a, b, c \in R$. (For an example of a field construction with an upper depth bounded by 2, see Bergman [83]).

7.8 Free fields and the specialization lemma

Let D be a field with a central subfield k and X a non-empty set. As we have seen in Section 2.4, the tensor D-ring $R = D_k\langle X \rangle$ is a fir and so, by Corollary 5.14, it has a universal field of fractions, which we shall call the *free D-field* on X over k and denote by $D_k(\langle X \rangle)$, or U in what follows. For each element u of U we have the representation $u = d - bA^{-1}c$ introduced in Section 7.1 with the display

$$\begin{pmatrix} A & c \\ b & d \end{pmatrix}. \qquad (1)$$

By Proposition 0.1.3, any non-zero element has positive stable rank, and as we saw in Section 7.1, u is stably associated to the display (1), hence $u = 0$ if and only if a display representing it is singular over U, or equivalently, non-full over R. This holds more generally for every Sylvester domain, and it yields another formulation of Malcolmson's criterion (Lemma 4.1):

Proposition 7.8.1. *Let R be a Sylvester domain and $u \in U$, its universal field of fractions. Then $u \neq 0$ if and only if any display representing it is full over R.* ∎

Although we are calling $D_k(\langle X \rangle)$ the 'free' field, we have not examined whether it has a generating set that is 'free' in the usual sense; in other words, given a rational relation in $D_k(\langle X \rangle)$, it is not clear whether it can be trivialized, using only the algebraic operations and the fact that f^{-1} is the inverse of f for every $f \neq 0$. To answer this question, our first task is to formulate it precisely; this is best put in a more general context. Let R be a ring with a universal field of fractions U, with the natural embedding $\phi : R \to U$. Starting from R we define

7.8 Free fields and the specialization lemma

recursively pairs $(R(n), f(n))$, where $R(n)$ is a ring and $f(n) : R(n) \to U$ a homomorphism, as follows: we put $(R(0), f(0)) = (R, \phi)$, and if $(R(n), f(n))$ has been defined, where $f(n) : R(n) \to U$ is the given homomorphism and $\varphi_n : R(n-1) \to R(n)$ is a homomorphism such that $f(n-1) = \varphi_n f(n)$, then $R(n+1)$ is obtained by adjoining to $R(n)$ formal inverses of those elements that do not lie in ker $f(n)$, and $\varphi_{n+1} : R(n) \to R(n+1)$ is the natural map, while $f(n+1)$ is the extension of $f(n)$ to $R(n+1)$. The direct limit of the $(R(n), f(n))$ will be written $(R(\infty), f(\infty))$ or $(R(\infty), f)$. Here $f = f(\infty)$ is a homomorphism of $R(\infty)$ into U, where every element of $R(\infty)$ not in ker f is a unit and it follows that $R(\infty)$ is a local ring. The non-zero elements of ker f may be regarded as the 'non-trivial' relations holding in the universal field of fractions, and our aim is to show that there are no such relations.

Theorem 7.8.2. *Let R be a Sylvester domain with a universal field of fractions U and let $(R(\infty), f)$ be defined as above. Then the canonical homomorphism $f : R(\infty) \to U$ is an isomorphism.*

Proof. The definition of f shows it to be surjective, so to prove that it is an isomorphism we must verify that for any rational expression u in $R(\infty)$, $uf = 0$ implies $u = 0$. For any $u \in R(\infty)$ we have a representation $u = bA^{-1}c$ and by Proposition 8.1, $uf = 0$ if and only if the matrix

$$\begin{pmatrix} A & -c \\ b & 0 \end{pmatrix} \qquad (2)$$

is non-full. When this is so, u has stable rank 0 and so $u = 0$, by Proposition 0.1.3. ∎

This result shows in particular that the universal field of fractions of the free algebra is really a 'free field' and it answers a question of Bergman [70a], who asks whether each rational identity $f = 0$ in the free field is an algebraic consequence of the fact that g^{-1} is the inverse of g, for each non-zero g of the free field. This fact cannot be proved by using only the elements actually inverted in f; for we have the rational identity $y(xy)^{-1}x = 1$ holding in any field, but not in a ring with elements x, y such that $xy = 1$ but $yx \neq 1$.

However, this and the corresponding matrix expression is the only exception, as we shall now show. In U consider an expression $f(x_1, \ldots, x_r)$ formed from the variables $x_1, \ldots, x_r$ over the ground field by the four operations $+, -, \times, ^{-1}$. When we substitute the values of R for $x_1, \ldots, x_r$, the result may be undefined, e.g. in a field, if $(g - g)^{-1}$ occurs; if the result is either undefined or zero, f is called an *absolute rational identity*. Of course every non-zero expression will have an inverse when we have a universal field of fractions U at our disposal. If we represent f as a component of the normalized solution of a matrix equation $Au = 0$, then to say that f is an absolute rational identity just means that either the numerator or the denominator of A is not full. By a *rational identity* we mean a rational expression that for any values of the variables in the ring is either undefined or zero.

Theorem 7.8.3. *Let D be a field that is a k-algebra and let R be a D-ring that is a k-algebra. Then every absolute rational identity is a rational identity for R if and only if R is weakly finite.*

Proof. Suppose that R is weakly finite. Let f be an absolute rational identity with the admissible matrix $A = A(x)$. Since $A(x)$ has a non-full numerator or a non-full denominator, the same is true of $A(a)$, where $a = (a_i)$, for any map $x_i \mapsto a_i$ from $D_k\langle x_1, \ldots, x_r\rangle$ to R. If the denominator of $A(a)$ is not full, $f(a)$ is undefined, so assume that the denominator of $A(a)$ is full, but its numerator is not. In that case, by Cramer's rule, $f(a) \oplus I$ is not full:

$$\begin{pmatrix} f(a) & 0 \\ 0 & I \end{pmatrix} = \begin{pmatrix} p \\ P \end{pmatrix} (q \quad Q),$$

where p is a row, q a column and P, Q are square. Hence $PQ = I$, $pQ = 0 = Pq$, $pq = f(a)$. Since R is weakly finite, $QP = I$, so $p = 0 = q$ and $f(a) = 0$, as we wished to show.

Conversely, if R is not weakly finite, then there are square matrices P, Q over R such that $PQ = I$, $QP \neq I$. Let P, Q be $n \times n$, say; writing S, T for $n \times n$ matrices with indeterminate entries, consider the matrix equation

$$T(ST)^{-1}S - I = 0.$$

Written out in full, the left-hand side consists of n^2 expressions in the entries of S, T and $(ST)^{-1}$; thus they are rational expressions that are defined and equal to 0 in the free field $D_k(\langle X \rangle)$, and so are absolute rational identities, but not all of them hold when we put $S = P, T = Q$, though all are defined. ∎

We shall now digress to give another proof of the existence of a fully inverting field for a free algebra that is independent of the theory developed in this chapter but makes use of the specialization lemma, a result of independent interest. For the proof we shall need (i) a slight extension of Amitsur's theorem on generalized polynomial identities, (ii) the inertia theorem (Theorem 2.9.16) and (iii) some auxiliary results to be described below, as well as a result on polynomial rings (Lemma 1.4.9). The existence proof of U will also make use of the ultraproduct theorem (Appendix C).

7.8 Free fields and the specialization lemma

We begin by stating the special case of Amitsur's theorem needed here. Let A be a k-algebra; we recall that an element of the free algebra $A\langle X\rangle$ that vanishes for all values of X in A is called a *polynomial identity*. By a *generalized polynomial identity* (GPI) of A one understands a non-zero element p of the free A-ring $A_k\langle X\rangle$ that vanishes under all maps $X \to A$. Thus the indeterminates commute with the elements of k but not with those of A. Amitsur [65] proved that a primitive k-algebra A satisfies a GPI if and only if it is a dense ring of linear transformations of a (possibly infinite-dimensional) vector space over a skew field of finite dimension over its centre, and A contains a non-zero transformation of finite rank. We shall only need the case where A itself is a skew field; in this case the existence of a non-zero transformation of finite rank means that the vector space itself must be finite-dimensional and Amitsur's theorem takes the following form (see FA, theorem 7.8.4):

Theorem 7.8A *A skew field satisfies a generalized polynomial identity if and only if it is of finite dimension over its centre.*

We shall want a generalization of this result in which the ground field, K say, need not be central or even commutative, and the types of identities we consider will be statements that an element p of $A_K\langle X\rangle$ vanishes only for values of the indeterminates in X ranging over the centralizer of K in A. The proof we shall give is similar to that of Theorem 8.A; we shall follow Martindale's proof in Herstein [76], with simplifications due to Bergman. We shall denote the centralizer of a subfield K of a field D by K'. Clearly K' is again a subfield and $K'' \supseteq K$; if equality holds here, K is said to be *bicentral* in D. Given a field D with a subfield K, we shall call an element $f \in D_K\langle X\rangle$ *left K-finite* if the values obtained by replacing the indeterminates in f by elements of K' all lie in a finite-dimensional left K-subspace of D. If this holds when the indeterminates take values in a subfield L of K', f will be called *left (L,K)-finite*. Thus 'K-finite' is short for '(K', K)-finite'. Our aim will be to prove that certain classes of skew fields satisfy no polynomial identities, but one difficulty is that when polynomial identities do occur, the least degree of such an identity may be quite large, which means that inductive arguments cannot be used. However, if instead of demanding that a polynomial of degree greater than 1 be identically zero, we require its values to lie in a finite-dimensional K-subspace, i.e. to be K-finite, then there is another such polynomial of smaller degree. Specifically we have

Theorem 7.8.4. *Let D be a field with a bicentral subfield K and X an infinite set such that some element f of $D_K\langle X\rangle$ of positive degree is left K-finite. If $x \in X$, then $cx \in D_K\langle X\rangle$ is left K-finite for some $c \in D^\times$.*

Proof. Suppose that $cx \in D_K\langle X \rangle$ is not left K-finite, for any $c \in D^\times$; we have to show that the same holds for all polynomials of positive degree. Let $f \in D_K\langle X \rangle$ be of positive degree and suppose that f is left K-finite. By linearization, i.e. replacing $f(\ldots, x_i, \ldots)$ by $f(\ldots, x_i + y, \ldots) - f(\ldots, x_i, \ldots) - f(\ldots, y, \ldots) + f(\ldots, 0, \ldots)$, we may take f to be multilinear of least degree, n say. In each term of f the n indeterminates will occur in some order. We shall group these terms according to which indeterminate occurs last and assume f chosen among left K-finite polynomials of degree n so as to minimize the number m of distinct indeterminates that occur in last position.

Let x_1 be one of the indeterminates occurring in last position and write

$$f = \sum_1^h g_i x_1 b_i + \sum_1^k p_j x_1 q_j, \tag{3}$$

where $g_i, p_j \in D_k\langle X \rangle$, $b_i \in D^\times$ and each q_j is multilinear in some non-empty subset of $\{x_2 \ldots x_n\}$. Further, assume the element f and the expression (3) for it chosen (among all GPIs of degree n in which only m indeterminates occur in the last position, and all possible such expressions) so as to minimize the number $h + k$ of summands. Replacing f by fb_1^{-1} we may assume that $b_1 = 1$. Note that $h \neq 0$, by our assumption that f has terms with x_1 in last position. If $h = 1$ and $k = 0$, then $f = g_1 x_1$ is left K-finite. By induction on n, g_1 is not left K-finite, so we can specialize the indeterminates to give g_1 a value $c \neq 0$; now cx_1 is left K-finite, which contradicts the hypothesis. Thus either $h > 1$ or $k > 0$. For any $y \in X$ we have the left K-finite polynomial

$$f(x_1, \ldots, x_n)y - f(x_1 y, \ldots, x_n) = \sum_2^h g_i x_1 (b_i y - y b_i) + \sum_1^k p_j x_1 (q_j y - y q_j). \tag{4}$$

Since h was chosen minimal in (3), $1, b_2 \ldots b_n$ are left K-linearly independent, in particular $b_i \notin K$ for $i > 1$, and since K is bicentral in D, none of the terms $b_i y - y b_i$ in the first sum vanishes identically for $y \in K'$. Choosing $y = c \in K'$ such that $b_2 c \neq c b_2$, we obtain a left K-finite polynomial with a smaller h and the result follows by induction, unless $h = 1$, when the first sum on the right of (4) is absent. In this case the second sum in (3) cannot be absent, by what has been proved. If for some $j, q_j y - y q_j$ vanishes identically for all $y \in K'$, then

$$q_j \in K'' = K \quad \text{for all values of the } x\text{'s in } K'. \tag{5}$$

We can choose the values of $x_3, \ldots, x_n$ so that the value of q_j is $r_j \neq 0$, by induction on n. If $r_j = \sum c_i x_2 d_i$, then $\sum c_i K' d_i \subseteq K$, by (5), so r_j is left K-finite and the result follows by induction on m.

We may thus assume that the first sum on the right of (4) is absent and that $q_j y - y q_j$ does not vanish identically for any j; for suitable $y = y_0 \in K'$ the left-hand side of (4) is then a left K-finite polynomial f_1, again multilinear in $x_1, \ldots, x_n$ with no term in which x_1 is last. Moreover, each term in f_1 has the x's in the same order as some term in f, so if x_n does not come last in any term of f, then the same is true for f_1. We now apply the same reduction to $x_2, \ldots, x_n$ in turn and finally get a left K-finite polynomial f^* in which no x_ν comes last. This is impossible, so this case cannot occur. ∎

We note the special case of degree 1, which is often useful, with another condition, which is easily verified.

Corollary 7.8.5. *Let D be a field and K a bicentral subfield. Given $a_1, \ldots, a_r \in D$, which are right K-linearly independent and $b_1, \ldots, b_r \in D^\times$ such that $\sum a_i x b_i \in D_K \langle x \rangle$ is left K-finite; then there exists $c \in D^\times$ such that cx is left K-finite, or equivalently, the conjugate subfield $cK'c^{-1}$ spans a finite-dimensional left K-subspace of D.* ∎

The extension of Amitsur's theorem follows as a special case of Theorem 8.4:

Corollary 7.8.6. *Let D be a field and K a bicentral subfield such that cx is not left K-finite for any $c \in D^\times$. Then every non-zero element of $D_K \langle X \rangle$ has a non-zero value for some choice of values for the elements of X in K'.* ∎

We can now state the main result of this section.

Lemma 7.8.7. (Specialization lemma) *Let D be a field whose centre C is infinite. Let H be a subfield of D containing C, with centralizer K such that cx is not left (H, K)-finite for any $c \in D^\times$. Then any full matrix over $D_K \langle X \rangle$ is non-singular for some choice of values of X in H.*

Proof. Let $A = A(X)$ be a full $n \times n$ matrix over $D_K \langle X \rangle$; let r be the supremum of its ranks as its arguments range over H and assume that $r < n$. By a translation $x_i \mapsto x_i + a_i$ ($a_i \in H$) we may assume that the maximum rank is attained at the point $x = 0$, and by elementary transformations we may take the principal $r \times r$ minor of $A(0)$ to be invertible. Thus if

$$A(X) = \begin{pmatrix} B_1(X) & B_2(X) \\ B_3(X) & B_4(X) \end{pmatrix},$$

where B_1 is $r \times r$, then $B_1(0)$ is invertible. Given $a \in H^X$ and any $c \in C$, we have $\rho A(ca) \leq r$, hence by Lemma 1.4.9 (since C is infinite), the rank of $A(ta)$ over the rational function field $D(t)$ is at most r, and the same holds over the Laurent series field $D((t))$. Now $B_1(ta)$ is a polynomial in t with matrix coefficients and constant term $B_1(0)$, a unit, hence $B_1(ta)$ is invertible over the power series field $D[[t]]$. By Proposition 5.4.6, since $r < n$, the equation

$$B_4(ta) = B_3(ta)B_1(ta)^{-1}B_2(ta)$$

holds over $D[[t]]$, for all $a \in H^X$. This then means that the matrix

$$B_4(tX) - B_3(tX)B_1(tX)^{-1}B_2(tX) \qquad (6)$$

vanishes whenever X is replaced by any values in H. Now (6) is a power series in t with coefficients that are matrices over $D_K\langle X \rangle$; thus the coefficients are generalized polynomial identities (or identically 0), so that by Corollary 8.6 the expression (6) vanishes identically on H, as an element in the t-adic completion of $R = D_K\langle X \rangle[t]$. Hence the same equation holds in the (t, X)-adic completion, and also in the X-adic completion; for in each case the matrix $B_1(tX)$ has constant term $B_1(0)$ and so is invertible. Thus we may set $t = 1$ in (6) and find that

$$B_4(X) = B_3(X)B_1(X)^{-1}B_2(X) \quad \text{in } D_K\langle\!\langle X \rangle\!\rangle;$$

in other words, $A(X)$ is non-full over the power series ring unless $r = n$. By Theorem 5.20, or also Theorem 2.9.16, $A(X)$ is non-full over $D_K\langle X \rangle$, which contradicts the hypothesis. So every full matrix $A(X)$ is non-singular for some set of values of X in H. ∎

We note the special case where $H = D$ and K is the centre of D:

Corollary 7.8.8. *Let D be a field with infinite centre C and such that $[D : C] = \infty$. Then any full matrix over $D_C\langle X \rangle$ is non-singular for some set of values of X in D.* ∎

If $[D : C]$ is finite, there are non-trivial identities over D, so this condition cannot be omitted. Whether the hypothesis that C be infinite can be relaxed is not known (see Exercise 4).

We can now prove the existence of a universal field of fractions for the tensor D-ring. The proof uses the ultraproduct theorem, in the form stated in Appendix C.

Theorem 7.8.9. *Let D be a field with infinite centre C and K a bicentral subfield with centralizer K', such that cx is never left K-finite, for any $c \in D^\times$ and $x \in X$. Then $D_K\langle X \rangle$ (for any set X) has a fully inverting field of fractions.*

Proof. Consider the natural mapping

$$D_K\langle X \rangle \to D^{K'^X}, \qquad (7)$$

7.8 Free fields and the specialization lemma

where $p \in D_K\langle X \rangle$ is mapped to (p_f) with $p_f = p(Xf)$ for any $f \in K'^X$. With each square matrix A over $D_K\langle X \rangle$ we associate the subset $\mathcal{D}(A)$ of K'^X consisting of all $f \in K'^X$ such that $A(Xf)$ is non-singular and so invertible at a point of K'^X, the *singularity support* of A. Of course $\mathcal{D}(A) = \emptyset$ unless A is full, but by Lemma 8.7, $\mathcal{D}(A) \neq \emptyset$ whenever A is full. If A, B are invertible at a point of K'^X, then so is $A \oplus B$, hence we have

$$\mathcal{D}(A) \cap \mathcal{D}(B) = \mathcal{D}(A \oplus B).$$

Hence the collection of subsets of K'^X containing some $\mathcal{D}(A)$, where A is full, is a filter $\mathcal{F}$, which is contained in an ultrafilter $\mathcal{C}$ say, and we have a homomorphism into the ultrapower

$$D_K\langle X \rangle \to D^{K'^X}/\mathcal{C}. \tag{8}$$

By definition every full matrix A over $D_K\langle X \rangle$ is invertible on the set $\mathcal{D}(A)$ of $\mathcal{F}$ and is therefore invertible in the ultrapower. Now the subfield of the ultrapower generated by the image of $D_K\langle X \rangle$ is the desired field of fractions inverting all full matrices over $D_K\langle X \rangle$. ∎

The ultrapower on the right of (8) is again a field, V say, and from the construction it is clear that the embedding of $D_K\langle X \rangle$ in V induces an *elementary* embedding of D in V, i.e. a mapping preserving all first-order sentences.

Taking $K = C$, we have $K' = D$ and this yields

Corollary 7.8.10. *Let D be a field with infinite centre C and such that $[D : C] = \infty$. Then $D_C\langle X \rangle$, for any set X, has a fully inverting field of fractions.* ∎

Exercises 7.8

1. Verify the 'if' part of Theorem 8.A, i.e. show that any field finite-dimensional over its centre satisfies a non-trivial GPI.
2. Let D be a field and Y any subset of D. Show that the centralizer Y' of Y is a subfield of D, which is bicentral.
3. Let k be an infinite commutative field and let A be a full matrix over $k\langle X \rangle$. Show that there exists an integer $n = n(A)$ such that for every central division k-algebra D of finite dimension at least n, A is non-singular for some set of values of X in D [*Hint*: Use Kaplansky's theorem on polynomial identities (FA, theorem 8.3.6); alternatively, assume the contrary and form an ultraproduct.]
4°. Let K be an infinite field with finite centre, and let A be a square matrix over K. Does there always exist $\alpha \in K$ such that $A - \alpha I$ is non-singular? (Note that for any finite field F there is a matrix A such that $A - xI$ is singular for all values of x in F.)
5. For any set I consider the Boolean algebra B of all subsets of I. Show that for any proper ideal $\mathfrak{a}$ of B, the set of all complements of members of $\mathfrak{a}$ is a filter on I, and

conversely, the set of complements of the members of any filter on I is an ideal of B. Deduce that every filter is contained in a maximal filter, i.e. an ultrafilter. Verify that an ultrafilter is characterized by the fact that for any subset J of I it contains either J or its complement.

6. Let K be a field with centre C and X a set. Show that every matrix over $K_C\langle X, X^{-1}\rangle$ is stably associated to a linear matrix $B = A_0 + \sum A_i x_i (x_i \in X)$. If $[K : C] = \infty, |C| = \infty$, deduce that for any full matrix A over $K_C\langle X, X^{-1}\rangle$ there exist non-zero values of X in K for which A is non-singular. [Hint: Apply the specialization lemma to $B \oplus \text{diag}(x_1, \ldots, x_r)$.]

7. (G. M. Bergman) Verify that in Theorem 8.9 the field obtained is an extension of D by a generic X-tuple of K'-valued indeterminates, i.e. a family of elements satisfying the generalized rational identities holding for elements of K' in D, and only those.

8°. (G. M. Bergman) Does Theorem 8.9 hold when the hypothesis 'cx is never left K-finite' is weakened to 'D is infinite-dimensional over K' (or even over its centre)'?

9°. (G. M. Bergman) Show that in the homomorphism (8) the rational closure of $D_K\langle X\rangle$ is a field isomorphic to the field constructed in the proof.

10. State and prove a generalization of Theorem 8.9 in which K is replaced by a family of subfields (K_i), one for each $x_i \in X$.

11. (G. M. Bergman) Let D be a field with subfields K and L such that L is contained in a finite-dimensional left K-subspace of D. Are K and L necessarily contained in a common subfield of D of finite left K-dimension? [Hint: In the field $\mathbb{H}$ of real quaternions let $i' = \alpha i + \beta j$, where $\alpha, \beta \in \mathbb{R}$ satisfy $\alpha^2 + \beta^2 = 1$ and are transcendental over $\mathbb{Q}$. Consider $K = \mathbb{Q}(i)$ and $L = \mathbb{Q}(i')$.]

7.9 Centralizers in the universal field of fractions of a fir

It is obvious that the centre of the free algebra $k\langle X\rangle$, where X has more than one element, is just k; that the centre of its universal field of fractions is also k is less evident. This will follow from the results to be proved in this section, in which we determine more generally the centre of the universal field of fractions of a fir. We begin with a technical result on relations between elements of a semifir and of its universal field of fractions.

Lemma 7.9.1. *Let R be a semifir and U its universal field of fractions. Given $p, p' \in U$ and $c, c' \in R$ such that*

$$cp' = pc', \tag{1}$$

suppose further that there are reduced admissible matrices $A \in {}^m R^{m+1}$ for p', $A' \in {}^n R^{n+1}$ for p' (this is so, for example, when R is fully atomic). Then there exist $P \in {}^m R^n$ and $Q^* \in {}^{m+1} R^{n+1}$ such that

$$PA' = AQ^*, \quad \text{where } Q^* = \begin{pmatrix} c' & 0 & 0 \\ Q_0 & Q_\bullet & Q_\infty \\ 0 & 0 & c \end{pmatrix}. \tag{2}$$

Moreover, given any one solution P_0, Q_0^* of (2), *the general solution has the form*

$$P = P_0 + A_{\bullet}N, \quad Q^* = Q_0^* + N^*A', \quad \text{where}$$
$$A_{\bullet} \text{ is the core of } A, N^* = (0 \quad N \quad 0)^T, N \in {}^{m-1}R^n. \quad (3)$$

Proof. Since A is a reduced admissible matrix for p, it follows that AC, where

$$C = \begin{pmatrix} c' & 0 & 0 \\ 0 & I_{m-1} & 0 \\ 0 & 0 & c \end{pmatrix},$$

is an admissible matrix for p'. Now A is reduced, so AC is certainly *-prime and we may write $AC = P_1 A_1$, where $P_1 \in R_m$ is full and A_1 is left prime and hence a reduced admissible matrix for p'. By Theorem 6.10 (iv), A_1 and A' are stably biassociated, so there is a comaximal relation $P_0 A' = A_1 Q_1^*$, where Q_1^* is bordered. Hence $P_1 P_0 A' = ACQ_1^*$; writing $P = P_1 P_0$ and $Q^* = CQ_1^*$, we obtain (2).

Suppose now that P_1, Q_1^* and P_2, Q_2^* both satisfy (2); then on writing $P = P_1 - P_2, Q^* = Q_1^* - Q_2^*$, we have

$$PA' = AQ^*, \quad \text{where } Q^* = (0 \quad Q \quad 0)^T \text{ and } Q \in {}^{m-1}R^{n+1}. \quad (4)$$

From the form of Q^* the right-hand side of (4) does not involve the first or last column of A, so we can write it as

$$PA' = A_{\bullet}Q. \quad (5)$$

Since A is right *-prime, its core $A_{\bullet}$ is right prime, and A' is left full; by applying Proposition 3.1.6, we find that $P = A_{\bullet}N$ and $Q = NA'$ for some $N \in {}^{m-1}R^m$. It follows that $Q^* = N^*A'$, where Q^*, N^* are obtained by topping and tailing Q, N by rows of zeros. This shows that all solutions of (2) can be expressed in terms of a given one as in (3), and it is clear that all the expressions (3) are solutions of (2). ∎

It is clear that the same result holds, with the same proof, when p, p', c, c' are full matrices.

We shall use Lemma 9.1 with Corollary 7.2 to show that under suitable conditions on c, c' any p occurring in an equation $cp = pc'$ has depth at most 1, but two auxiliary results will be needed. In the first place we note a lemma on coprime relations.

Lemma 7.9.2. *In a 2-fir R consider a coprime relation*

$$ab' = ba'. \quad (6)$$

Put $p = b^{-1}a = a'b'^{-1}$; if

$$cp = pc', \quad \text{where } c, c' \in R^\times, \tag{7}$$

then there exist $d, d' \in R$ such that

$$bc = db, ac' = da, \quad ca' = a'd', c'b' = b'd'. \tag{8}$$

Proof. By (7) and the form of p we have $bca' = ac'b'$. Since (6) is an LCRM of a, b, it follows that $ca' = a'd', c'b' = b'd'$, where $d' \in R$; (6) is also an LCLM of a', b', so we conclude that $bc = db, ac' = da$ for some $d \in R$. ∎

We shall also need an elementary result from linear algebra.

Lemma 7.9.3. *Let K be a field and $A \in {}^nK^p, B \in {}^mK^p$. Then there exists $N \in {}^mK^n$ with at most one entry 1 in each row and all the rest 0, such that $\rho(B + NA) \geq \min(m, \rho A)$.*

Proof. We assume that for some r, where $0 \leq r \leq \min(m, \rho A)$ the first r rows of N have been chosen as indicated so that the first r rows of $B + NA$ are left linearly independent. If $r = \min(m, \rho A)$, the conclusion follows by taking the remaining rows of N to be zero, so we assume that $r < \min(m, \rho A)$ and use induction on r. Let U be the space spanned by the first r rows of $B + NA$. If the $(r + 1)$st row of B is not in U, we can take the $(r + 1)$st row of N to be zero to get the inductive step; otherwise there is a row of A after the first r that is not in U, say the ith; then we take the $(r + 1, i)$-entry of N equal to 1 and now $B + NA$ has an $(r + 1)$st row not in U. Now the result follows by induction. ∎

We can now accomplish the objective announced earlier.

Proposition 7.9.4. *Let R be a fully atomic semifir and U its universal field of fractions. Given $p \in U, c, c' \in R$ such that (7) above holds, assume further that (i) c, c' are not zero and not both units, or (ii) R is a persistent semifir over a commutative field k and c, c' are not both algebraic over k. Then p has left and right depth at most 1, with left and right fractions*

$$p = a'b'^{-1} = b^{-1}a, \quad \text{where } ab' = ba' \text{ is coprime}, \tag{9}$$

and for any such representation (9) of p, there exists $d' \in R$ such that

$$ca' = a'd', c'b' = b'd'. \tag{10}$$

Moreover, in case (ii) there exist $e \in R$ and a non-zero polynomial f over k such that

$$p = f(c)^{-1}e = ef(c')^{-1}, \quad ce = ec'. \tag{11}$$

Proof. Since R is fully atomic, there is a reduced admissible matrix A for p, and by Lemma 9.1 there exist square matrices P, Q^* satisfying

$$PA = AQ^*, \quad \text{where } Q^* \text{ is as in (2)}. \tag{12}$$

Let us first take case (i). If A is $m \times (m+1)$, then from (12) we have

$$P(A_0 \quad A_\bullet) = (A_0 \quad A_\bullet)\begin{pmatrix} c' & 0 \\ Q_0 & Q_\bullet \end{pmatrix},$$

$$P(A_\bullet \quad A_\infty) = (A_\bullet \quad A_\infty)\begin{pmatrix} Q_\bullet & Q_\infty \\ 0 & c \end{pmatrix}. \tag{13}$$

The two equations (13) show that c and c' have the same length, hence they are both non-units. Let us assume that p has depth greater than 1; then $A_\bullet$ is regular over R, by Corollary 7.2. Moreover, given one solution P_0, Q_0^* of (2), the general solution has the form

$$P = P_0 + A_\bullet N, \quad Q^* = Q_0^* + N^*A, \quad \text{where } N^* = (0 \quad N \quad 0)^T.$$

Thus we have

$$Q_\bullet = Q_{0\bullet} + NA_\bullet, \quad \text{where } N \in {}^{m-1}R^m. \tag{14}$$

Now $A_\bullet$ has rank $m-1$, so to each row of $Q_{0\bullet}$ we can add a linear combination of the rows of $A_\bullet$ so as to ensure that $Q_\bullet$ has rank $m-1$ over U, by Lemma 9.3 with m, ρA replaced by $m-1$. Thus for suitable choice of N in (14), $Q_\bullet$ has inner rank $m-1$ over R and so is full. Since $c, c' \neq 0$, it follows that Q^* is full, therefore the two sides of each equation in (13) have rank m and it follows that P is full. By unique factorization, P, $\begin{pmatrix} c' & 0 \\ Q_0 & Q_\bullet \end{pmatrix}$, $\begin{pmatrix} Q_\bullet & Q_\infty \\ 0 & c \end{pmatrix}$ each are products of similar factors, in particular they have the same length. Now A is left prime, so (12) is left coprime, hence by Corollary 3.1.4, P is similar to a left factor, P_1 say, of Q^*; thus

$$Q^* = \begin{pmatrix} I & 0 \\ 0 & c \end{pmatrix}\begin{pmatrix} c' & 0 & 0 \\ Q_0 & Q_\bullet & Q_\infty \\ 0 & 0 & 1 \end{pmatrix} = P_1 P_2, \tag{15}$$

where P_1 is similar to P and P_2 is a highest common right factor of A and Q^* (see (12)). It follows from this and (13) that P_1 and $\begin{pmatrix} c' & 0 \\ Q_0 & Q_\bullet \end{pmatrix}$ are products of similar factors; by unique factorization applied to (15), P_2 and c are products of similar factors; in particular, P_2 and c have the same length. Moreover, as we saw, P_2 is a right factor of A; thus A has a full right factor of length $l(c)$. Now $cp = pc'$ implies $c^n p = pc'^n$ for all $n \geq 1$. If we repeat the argument with c, c'

replaced by c^n, c'^n, we find that A has a full right factor of length $l(c^n) = nl(c)$. But the length of full right factors of A is bounded, by the left–right dual of Proposition 5.9.1, so we have a contradiction. This shows the left depth of p to be 1, and the same holds for the right depth, by symmetry.

We have $p = ab^{-1} = b'^{-1}a'$, so d' to satisfy (10) exists by Lemma 9.2 and this completes the proof of case (i).

Turning to case (ii), we have by (7), $f(c)p = pf(c')$ for any polynomial f over k; hence if one of c, c' is algebraic, both are, so in fact neither can be algebraic. If p had depth greater than 1, the matrix A would be regular, hence its eigenring would be algebraic over k, by Theorem 4.6.9. Then by (10) we can find a non-zero polynomial f over k such that $f(Q^*) = SA$, where $S \in {}^{m+1}R^m$. Then the first row of $f(Q^*)$ takes the form

$$(f(c') \quad 0 \quad \ldots \quad 0) = sA$$

for some $s \in R^m$. Thus $s(A_\bullet, A_\infty) = 0$, but $(A_\bullet, A_\infty)$ is full and so is regular, hence $s = 0$ and it follows that $f(c') = 0$. This contradicts the assumption that c' is not algebraic over k and it follows that the right depth of p is at most 1. By symmetry the same is true of the left depth and we can prove (10) as before.

Now in (10), $b' \neq 0$, so by another application of Theorem 4.6.9, we can find $f \in k[x]^\times$ such that

$$f(c') = b'd_0,$$

where $d_0 \neq 0$ because c' is not algebraic over k. Put $e = a'd_0$; then $p = a'b'^{-1} = ef(c')^{-1}$, but by (7), $f(c)p = p(f(c') = e$, hence

$$p = f(c)^{-1}e = ef(c')^{-1}.$$

Moreover, $ce = cf(c)p = f(c)cp = f(c)pc' = ec'$, so (11) holds. ∎

In Proposition 9.4(i), if p has depth greater than 1, then c, c' are 0 or units, hence the elements of R whose conjugates under p again lie in R, form a subfield of R:

Corollary 7.9.5. *Let R be a fully atomic semifir with universal field of fractions U. If $p \in U$ has depth greater than 1, then $pRp^{-1} \cap R$ is a subfield of R.* ∎

As a further application of Proposition 9.4 we can determine the centre of U:

Theorem 7.9.6. *Let R be a fully atomic semifir with universal field of fractions U. Then the centre C of U coincides with the centre of R, unless R is a principal ideal domain. In that case C consists of all elements $ab^{-1} = b^{-1}a (a, b \in R)$*

7.9 Centralizers in the universal field of fractions of a fir

for which an automorphism α of R exists such that $ra = ar^\alpha, rb = br^\alpha$ for all $r \in R$.

Proof. If $U = R$, there is nothing to prove; otherwise we take a non-zero non-unit c in R and $p \in C$; then $cp = pc$, so p has depth 1, by Proposition 9.4, say $p = ab^{-1}$. Now $pb = bp$, hence $ab = ba$ and so $p = b^{-1}a$. Since R is atomic, we can take this left fraction to be reduced, and by Proposition 9.4 we have, for each $r \in R$ an $r' \in R$ such that $ra = ar', rb = br'$. It is clear that the map $\alpha : r \mapsto r'$ is an endomorphism; by symmetry it has an inverse and so is an automorphism of R. If $r' = r$ for all $r \in R$ and a, b are units, then $p \in R$. When this holds for all $p \in C$, then U and R have the same centre. Otherwise R is a two-sided Ore domain, by Theorem 6.4.2, and hence a principal ideal domain. Now the form of the centre of R is given by Proposition 6.4.4. ∎

The exception does in fact occur since, as we have seen in Section 6.3, every Krull domain can occur as the centre of a principal ideal domain. To illustrate these results, let D be a field that is a k-algebra and X a set. If $|X| > 1$, then $D_k\langle X \rangle$ is a non-principal fir with centre k; therefore its universal field of fractions has centre k, by Theorem 9.6. We state the result as

Corollary 7.9.7. *Let D be a field, k a subfield of its centre and let X be any set. If either $|X| > 1$, or $X \neq \varnothing$ and $D \neq k$, then the universal field of fractions of $D_k\langle X \rangle$ has centre k.* ∎

Next we look at centralizers in the universal field of fractions. To put the result in context we recall that if D is any field with centre k and $a \in D$ is algebraic of degree n over k, then the centralizer C_a of $a \in D$ is a subfield such that $[D : C_a] = n$ (see e.g. FA, Section 5.1). This may be expressed loosely by saying that a has a 'large' centralizer; by contrast, in the universal field of fractions of a semifir R, the elements of R have 'small' centralizers, as the next result shows. We shall need to assume persistence; this holds e.g. for free algebras.

Theorem 7.9.8. *Let R be a fully atomic persistent semifir over k with universal field of fractions U. Given $c \in R$, not algebraic over k,*

(i) *if the centralizer of c in R is C, then its centralizer in U is the localization of C at $k[c]^\times$; moreover, C is an Ore domain and*
(ii) *if R contains the field of fractions $k(c)$ and c is conjugate in U to an element c' of R, say $cp = pc'$, then p is a unit in R and R also contains $k(c')$.*

Proof. (i) Let $p \in C$. By Proposition 9.4 there exist $a \in R$ and $f \in k[c]^\times$ such that $p = f(c)^{-1}a = af(c)^{-1}$ and $ac = ca$, therefore $a \in C$ and the form of the centralizer follows. Now let $p, q \in C$, $q \neq 0$; then $q^{-1}p = af(c)^{-1}$ for some $a \in C$, by what has been proved, hence $pf(c) = qa$ and this shows C to be a right Ore domain; by symmetry it is also left Ore.

(ii) Again by Proposition 9.4 we have $p = f(c)^{-1}a$, but now $f(c)^{-1} \in R$, so $p \in R$. Applying the same proposition to p^{-1}, we find that $p^{-1} = bg(c)^{-1}$, hence $p^{-1} \in R$ and this shows p to be a unit in R; further R contains $p^{-1}k(c)p = k(c')$. ∎

Here the condition of persistence cannot be omitted, as is shown by the example $R = F\langle X \rangle$, where F is a commutative transcendental field extension of k and $|X| > 1$. This is a fir, but the centralizer of any $x \in X$ contains $F(x)$ and so is larger than $F[x]_{k[x]^\times}$.

We recall from Section 0.6 that for any element c of a ring R, the *left idealizer* is defined as $I_l(c) = \{x \in R \mid cx \in Rc\}$; similarly $I_r(c) = \{x \in R \mid xc \in cR\}$ is the *right idealizer*. We note that $I_r(c) \cong I_l(c)$ by means of the 'inner' automorphism $\iota(c)$ defined by $x\iota(c) = y$ if $xc = cy$. We also remark that a regular element c is left invariant if and only if $I_l(c) = R$ and right invariant if and only if $I_r(c) = R$. If K is a field with a subring R, then the *right normalizer* of R in K is defined as the set of right R-invariant elements in K:

$$N_r = \{u \in K^\times \mid Ru \subseteq uR\}, \tag{15}$$

and the elements of N_r are said to be *right R-normalizing*. Left normalizing elements and the left normalizer N_l are defined similarly.

Proposition 7.9.9. *Let R be a fully atomic semifir with universal field of fractions U and let N_r be the right normalizer of R in U. Then any $p \in N_r$ can be written as ab^{-1}, where a, b are non-zero right coprime elements of R. Given any two non-zero right coprime elements $a, b \in R$, we have $ab^{-1} \in N_r$ if and only if*

$$I_r(a) = R, \quad I_l(a) \subseteq I_l(b). \tag{16}$$

Moreover, $p^{-1}xp = x\iota(a)\iota(b)^{-1}$ for any $x \in R$.

Proof. In any subring R of a field U, if $a, b \in R^\times$, (16) is sufficient for $p = ab^{-1} \in U$ to right normalize R and conjugation by p will have the form $\iota(a)\iota(b)^{-1}$. It remains to prove the converse when R is a fully atomic semifir with universal field of fractions U.

We may assume that $R \neq U$, since otherwise the result holds trivially. Given $p \in N_r$ and a non-zero non-unit c, we have (7) for some $c' \in R$, hence by Proposition 9.4, p is a reduced right fraction, say $p = ab^{-1}$, where a, b are right coprime elements of R. Now (7) holds for any $c \in R$ and a suitable $c' \in R$; by Proposition 9.4 we get the relations (8), from which (16) follows. ∎

We next examine the relation between $U(R)$, the set of units of R, and the normalizers.

Proposition 7.9.10. *Let R be a fully atomic semifir with universal field of fractions U, normalizers N_r, N_l of R in U, and $T = U(R)$. Then the following conditions are equivalent:*

(a) $N_r \cup N_l \subseteq R$,
(b_r) $N_r = T$,
(b_l) $N_l = T$,
(c) Any left or right invariant element of R is in T.

Proof. We shall prove (a) $\Rightarrow$ (b_r) $\Rightarrow$ (c) $\Rightarrow$ (a); the equivalence with (b_l) will then follow by the symmetry of (a) (and of (c)). The first two implications hold for any subring R of a field U.

(a) $\Rightarrow$ (b_r). Clearly $T \subseteq N_r$. Conversely, if $p \in N_r$, then $p^{-1} \in N_l$; hence p and p^{-1} both lie in R, by (a); so p is a unit of R and hence $p \in T$.

(b_r) $\Rightarrow$ (c). If $a \in R \backslash T$ is right invariant, then $a \in N_r$, but $a \notin T$, in contradiction to (b_r). If $a \in R \backslash T$ is left invariant, then a^{-1} is right invariant and this again contradicts (b_r).

(c) $\Rightarrow$ (a). Let $p \in N_r$. By Proposition 9.9, p has the form $p = ab^{-1}$, where $I_r(a) = R$, $I_l(a) \subseteq I_l(b)$. Thus a is right invariant and by (c) it is a unit in R; hence it is also left invariant, i.e. $I_l(a) = R$. It follows that $I_l(b) = R$, i.e. b is left invariant, so also a unit, and hence $p \in R$. The corresponding assertion for $p \in N_l$ follows by symmetry. ∎

With R and U as before, let $p \in U$ be a right R-normalizing element with the reduced right fraction $p = ab^{-1}$. Then $a' = p^{-1}ap \in R$, and so $ab^{-1} = b^{-1}a'$, which is a left fraction for p, though not necessarily reduced. If b and a' are not left coprime, we can obtain a reduced left fraction by cancelling the highest common left factor (which exists in an atomic semifir). In this case the lengths of the numerator and denominator in the left fraction for p are less than those in the right fraction.

Suppose now that $p \in N_r \cap N_l$. Then we have reduced left and right fractions for p : $p = ab^{-1} = b'^{-1}a'$. By the above argument $l(a) \leq l(a') \leq l(a)$; thus $l(a') = l(a)$ and similarly $l(b') = l(b)$. By the above reduction we can write both reduced fractions with the same denominator: $p = ab^{-1} = b^{-1}a'$. Taking inverses, we obtain reduced fractions for the normalizing element p^{-1} having the same numerator, b. But by Proposition 9.9 the numerator of a right normalizing element is right invariant; by symmetry the numerator of a left normalizing element is left invariant, hence b is an invariant element of R. Now b was the denominator of the reduced right fraction for p; interchanging the roles of p and p^{-1} again we see that the numerator a is also invariant. Thus we have

Proposition 7.9.11. *Let R be a fully atomic semifir and U its universal field of fractions. Then the (two-sided) normalizer N of R in U is the group of all elements ab^{-1} where a, b are invariant elements of R such that $ab = ba$.*

Each member of N can be written as a reduced right fraction ab^{-1}, unique up to right multiplication of a, b by a common unit, and as a reduced left fraction $b^{-1}a$, unique up to left multiplication of a, b by a common unit. ∎

This result shows in particular that U contains R-normalizing elements other than units of R if and only if R contains non-unit invariant elements. When this is so, then R is a principal ideal domain, by Theorem 6.4.2. So we obtain

Corollary 7.9.12. *Let R be a fully atomic semifir and U its universal field of fractions. If R is not a principal ideal domain, every R-normalizing element of U is a unit in R.* ∎

This result shows in particular that for a free algebra R of rank at least 2 the universal field of fractions contains no R-normalizing elements other than the units of R.

Exercises 7.9

1. Let R be a fully atomic semifir that is doubly persistent (i.e. $R \otimes k(t)$ is persistent), with universal field of fractions U. Show that if x is any element of R, not algebraic over k, then the centralizer of x in U is matrix algebraic over k.
2. (G. M. Bergman) Let S be a semifir that is a k-algebra (distinct from k). Show that in the coproduct $R = S_k^* k[x]$ any left or right invariant element is a unit. (*Hint*: Verify that any element of R right comaximal with x has the form $xa + u$, where u is a unit of S, and apply this to the comaximal relation $xa' = ax'$. Deduce that the only elements similar to x are its associates uxv, and that this also holds for $1 + x$ in place of x.)
3. (W. Dicks) Let $R = k\langle X \rangle$, U its universal field of fractions, and let $k(s)$ be a purely transcendental field extension of k. Show that $U \otimes k(s)$ is an integral domain and remains one under all commutative field extensions of $k(s)$. Deduce that $U \otimes k(s)$ is 1-inert in $U \otimes k(s, t)$ and hence that U is not a doubly persistent integral domain.
4. Let $c \in k\langle X \rangle \setminus k$, so that the centralizer of c has the form $k[p]$, by Bergman's centralizer theorem (Theorem 6.7.7). Show that the centralizer of c in the universal field of fractions of $k\langle X \rangle$ is $k(p)$.
5. Let K be a field with a subring R, and let N be the (two-sided) normalizer of R in K. Show that $RN = NR$.

6. Show that the free power series ring $R = k\langle\langle x, y \rangle\rangle$ is a persistent semifir over k; deduce that the normalizer of R in its universal field of fractions is contained in R.
7°. For which semifirs R is it the case that the normalizer of R in its universal field of fractions is contained in R? Are there any 1-atomic non-Ore semifirs for which this statement is not true?
8*. Let $R = k\langle X \rangle$ and U its universal field of fractions. Suppose that $p \in U$ is quadratic over $R : p^2 - ap - b = 0\, (a, b \in R)$. If the companion matrix $Z = \begin{pmatrix} 0 & 1 \\ b & a \end{pmatrix}$ is algebraic over k, show that $p \in R$. (*Hint*: Consider first the case that k is algebraically closed.) When Z is not algebraic over k and p has left and right depth greater than 1, show that the admissible matrix for p has algebraic eigenring and obtain a contradiction. Deduce that again $p \in R$, thus R is 'quadratically closed' in U.
9°. Show that $k\langle X \rangle$ is 'relatively algebraically closed' in its universal field of fractions.
10. (Klein [72b]) Show that for any commutative field k there is a non-commutative field with an involution, and with k as its centre.
11*. Let D be a field of characteristic not 2 and α an involution on D. Show that the fixed set of α need not be a subring of D. (*Hint*: Consider the quaternions.)
12*. (G. M. Bergman) Let D be a field of characteristic not 2, with an involution α whose fixed set is a field k. Show that either $D = k$ or D is a commutative quadratic extension of k, or k is central in D and $[D : k] = 4$. (*Hint*: Show first that every element of $D \setminus k$ must be quadratic over k.)

7.10 Determinants and valuations

The determinant is an important scalar invariant associated with matrices over a commutative ring. For general rings there is no satisfactory analogue, although for skew fields one has the Dieudonné determinant, which in effect provides a homomorphism from $GL_n(K)$ to abelian groups. Our main object in this section is to calculate the Dieudonné determinant for the universal field of fractions of a fir, particularly for the free algebra $k\langle X \rangle$. However, we shall look at this problem in a more general setting, which will also allow us to describe valuations on these fields in terms of the rings.

Let R be any ring, Σ a factor-stable multiplicative set of full matrices over R and G an abelian group, written additively. By a *G-value* on R we understand a mapping $v : \Sigma \to G$ such that

V.1. $v(E) = 0$ for any elementary matrix E,
V.2. $v(A \oplus I) = v(A)$ for $A \in \Sigma$, and
V.3. $v(AB) = v(A) + v(B)$ whenever $AB \in \Sigma$.

If we write $v(-1) = \varepsilon$, then by V.1, V.3, $2\varepsilon = 0$; thus if G is torsion-free, then $\varepsilon = 0$, and so

V.4. $v(-1) = 0$.

However, we shall not assume V.4 in what follows. We note the following consequences of V.1–V.3:

$$V.5.\, v(I \oplus A) = v(A).$$

This follows from the equation

$$\begin{pmatrix} I & 0 \\ 0 & A \end{pmatrix} = \begin{pmatrix} 0 & I \\ I & 0 \end{pmatrix} \begin{pmatrix} A & 0 \\ 0 & I \end{pmatrix} \begin{pmatrix} 0 & I \\ I & 0 \end{pmatrix},$$

by applying v and using V.1–V.3.

$$V.6.\, v\begin{pmatrix} A & C \\ 0 & B \end{pmatrix} = v(A) + v(B) = v\begin{pmatrix} A & 0 \\ D & B \end{pmatrix}.$$

The first equation follows by writing

$$\begin{pmatrix} A & C \\ 0 & B \end{pmatrix} = \begin{pmatrix} I & 0 \\ 0 & B \end{pmatrix} \begin{pmatrix} I & C \\ 0 & I \end{pmatrix} \begin{pmatrix} A & 0 \\ 0 & I \end{pmatrix}.$$

The right-hand side is $v(I \oplus B) + v(A \oplus I)$, by V.1, V.3 and this reduces to $v(A) + v(B)$, by V.2, V.5. The second equation V.6 follows similarly.

By V.1–V.3, $v(A) = v(B)$ whenever A is stably E-associated to B. In particular,

V.7. $v(A)$ is unchanged if we interchange two columns of A and change the sign of one of them.

Further, an application of the magic lemma (Lemma 5.11) shows that two matrices 'differing' by a non-full matrix have the same value:

Lemma 7.10.1. *Let R be a ring, Σ a factor-inverting multiplicative set of square matrices over R and v a G-value on Σ. In any equation*

$$C = A \nabla B,$$

if $A \in \Sigma$ and B is non-full, then $C \in \Sigma$ and we have $v(A) = v(C)$.

Proof. By Lemma 5.11 we see that $A = ST, C = SUT$, where U is a product of elementary matrices. Hence $v(U) = 0$ and so $v(A) = v(S) + v(T) = v(C)$. ∎

In the applications Σ will usually be $\Phi(R)$, the set of all full matrices; moreover, $\Phi(R)$ will need to admit diagonal sums and products (when defined). By Theorem 5.13, R is then a Sylvester domain. For these reasons we shall assume from now on that R is a Sylvester domain; then R has a universal field of fractions U, with a fully inverting homomorphism $\lambda : R \to U$.

7.10 Determinants and valuations

Let v be a G-value on $\Phi(R)$; our object is to extend v to U. If $p \in U^\times$ has the admissible system $Au = 0$, our aim will be to define v by putting

$$v(p) = v(A_0 \quad A_\bullet) - v(-A_\infty \quad A_\bullet), \quad \text{where}$$
$$A \text{ is an admissible matrix for } p, \qquad (1)$$

but to do so we must show that the right-hand side of (1) is independent of the choice of A. We shall first use v to define a function on admissible matrices. Let $A = (A_0 \quad A_\bullet \quad A_\infty)$ be an admissible matrix for a non-zero element of U; then $(A_0, A_\bullet)$ and $(-A_\infty, A_\bullet)$ are both full and we put

$$V(A) = v(A_0, A_\bullet) - v(-A_\infty, A_\bullet). \qquad (2)$$

If A is admissible for p and B is admissible for q, then the matrix

$$A.B = \begin{pmatrix} B_0 & B_\bullet & B_\infty & 0 & 0 \\ 0 & 0 & A_0 & A_\bullet & A_\infty \end{pmatrix} \qquad (3)$$

is admissible for pq, and we have

$$V(A.B) = v\begin{pmatrix} B_0 & B_\bullet & B_\infty & 0 \\ 0 & 0 & A_0 & A_\bullet \end{pmatrix} - v\begin{pmatrix} 0 & B_\bullet & B_\infty & 0 \\ -A_\infty & 0 & A_0 & A_\bullet \end{pmatrix}.$$

Here the second term becomes, by a column interchange (and change of sign)

$$v\begin{pmatrix} -B_\infty & B_\bullet & 0 & 0 \\ -A_0 & 0 & -A_\infty & A_\bullet \end{pmatrix};$$

hence we obtain (using V.6),

$$V(A.B) = v(B_0 \quad B_\bullet) + v(A_0 \quad A_\bullet) - v(-B_\infty \quad B_\bullet) - v(-A_\infty \quad A_\bullet)$$
$$= V(A) + V(B).$$

Further, $(1, -1)$ is an admissible system for 1, and

$$V(1 \quad -1) = v(1) - v(1) = 0.$$

If S is the semigroup of admissible systems with the multiplication (3) (clearly an associative operation), then $V : S \to G$ is a homomorphism, by what has been shown; we still have to check that V has the same value on all admissible matrices defining a given $p \in U$.

Let $A = (A_0, A_\bullet, A_\infty)$ be admissible for p; then $\bar{A} = (-A_\infty, A_\bullet, -A_0)$ is admissible for p^{-1} and hence $A.\bar{A}$ is admissible for $pp^{-1} = 1$. So if A, B are any admissible matrices for the same p, then $A.\bar{B}$ is admissible for $pp^{-1} = 1$, and clearly $V(\bar{B}) = -V(B)$, hence

$$V(A.\bar{B}) = V(A) + V(\bar{B}) = V(A) - V(B).$$

So to prove that v is well-defined we need only show that $V(C) = 0$ for any admissible matrix C for 1.

Suppose then that C is admissible for 1; the condition for this is that $(C_0 + C_\infty, C_\bullet)$ is non-full. Then

$$(C_0 \quad C_\bullet) = (C_0 + C_\infty \quad C_\bullet) \nabla (-C_\infty \quad C_\bullet),$$

and by Lemma 10.1, $V(C) = v(C_0 \quad C_\bullet) - v(-C_\infty \quad C_\bullet) = 0$, as claimed. Thus we have proved

Theorem 7.10.2. *Let R be a Sylvester domain with universal field of fractions U and a G-value v on $\Phi(R)$. Then v can be extended to a homomorphism of $U^\times$ into G by (1), where A is any admissible matrix for p.* ∎

Let us see how this relates to the Dieudonné determinant. If K is any field and $n \geq 1$, then any matrix $A \in GL_n(K)$ can be reduced to diagonal form by elementary transformations; more precisely we have

$$A = (I_{n-1} \oplus \alpha) U, \quad \text{where } \alpha \in K^\times, U \in E_n(K).$$

If $A = (I \oplus \beta) V$ is another such expression, then $\alpha \beta^{-1}$ can be shown to belong to $K^{\times\prime}$, the derived group (commutator subgroup) of $K^\times$. Hence the coset of α in $K^{ab} = K^\times / K^{\times\prime}$ is an invariant of the matrix A, called the *Dieudonné determinant* (for the proof see Dieudonné [43], or also FA, Section 9.2). Let us denote this coset by Det A; it is easily seen that the map $A \mapsto$ Det A is universal for homomorphisms into abelian groups; in fact it may be shown to induce an isomorphism $GL_n(K)^{ab} \to K^{ab}$, with an exception noted below (see Bass [68]). Thus for any field K and any $n \geq 1$, except for $K = \mathbb{F}_2$ and $n = 2$, there is an isomorphism

$$GL_n(K)^{ab} \cong K^{ab}. \tag{4}$$

We can define matrix multiplication generally for square matrices of different orders over any ring R by regarding A as $A \oplus I$, for I of sufficiently high order. In this way $\Phi(R)$ becomes a monoid for any Sylvester domain R and we shall write $\Phi(R)^{ab}$ for the universal abelian group of $\Phi(R)$. The same device allows us to embed $GL_n(R)$ in $GL_{n+1}(R)$ and define the *stable* general linear group $GL(R) = \varinjlim GL_n(R)$. In order to evaluate its abelianization we recall a lemma from linear algebra:

Lemma 7.10.3. *Let R be a weakly 1-finite ring and let $n \geq 3$. Then every elementary matrix of order n lies in the derived group $GL_n(R)'$ and every diagonal matrix of order n in which the product of the diagonal elements is 1 lies in $GL_n(R)'$.*

7.10 Determinants and valuations

Proof. We write $B_{ij}(a) = I + ae_{ij}$; To express it as a commutator we have

$$((B_{ik}(a), B_{kj}(1)) = B_{ik}(-a)B_{kj}(-1)B_{ik}(a)B_{kj}(1) = B_{ij}(a),$$

where i, j, k are any three distinct subscripts. Next consider a diagonal matrix, in which the product of the diagonal entries is 1. It follows that all diagonal entries are units; now a 2×2 matrix of this form can be written as a product of elementary matrices:

$$c \oplus c^{-1} = B_{21}(c^{-1})B_{12}(-c)B_{21}(c^{-1})B_{21}(-1)B_{12}(1)B_{21}(-1),$$

and any diagonal matrix with the product of the diagonal entries equal to 1 can be written as a product of matrices of this form, with the diagonal topped and tailed by 1s, since

$$\text{diag}(a_1, a_2, \ldots, a_n) = \text{diag}(a_1, a_1^{-1}, 1, \ldots, 1).\text{diag}(1, a_1a_2, a_3, \ldots, a_n) = \ldots$$
$$= \text{diag}(a_1, a_1^{-1}, 1, \ldots, 1)\text{diag}(1, a_1a_2, (a_1a_2)^{-1}, 1, \ldots, 1)$$
$$\times \ldots \text{diag}(1, \ldots, 1, a_1 \ldots a_{n-1}, a_n);$$

hence it can be expressed as a product of elementary matrices, and so as a product of commutators. ∎

For $n = 2$ the conclusion still holds, provided that $x^3 = x$ is not an identity for R (see FA, Section 3.5 and Exercise 2 below).

Still taking R to be a Sylvester domain with universal field of fractions U, we have the natural map $\Phi(R) \to GL(U)^{ab}$, which yields a homomorphism

$$f : \Phi(R)^{ab} \to GL(U)^{ab}. \tag{5}$$

We claim that the natural map $\Phi(R) \to \Phi(R)^{ab}$ is a $\Phi(R)^{ab}$-value on $\Phi(R)$; V.2 and V.3 are clear, while V.1 follows by Lemma 10.3. By Theorem 10.2 it can be extended to a map $U^{\times} \to \Phi(R)^{ab}$; combining it with the Dieudonné determinant map we thus obtain a homomorphism from $GL(U)^{ab}$ to $\Phi(R)^{ab}$. Now Cramer's rule shows that p is stably associated to the numerator times the inverse of the denominator and this shows the map to be inverse to (5); hence the latter is an isomorphism. The abelianization $GL(U)^{ab}$ is called the *Whitehead group* of U and is denoted by $K_1(U)$. Thus we obtain

Theorem 7.10.4. *Let R be a Sylvester domain, U its universal field of fractions and $\Phi(R)^{ab}$ the universal abelian group of the monoid $\Phi(R)$ of all full matrices over R. Then there is a natural isomorphism with the Whitehead group of U:*

$$K_1(U) = \Phi(R)^{ab}. \qquad \blacksquare \tag{6}$$

To investigate the structure of this group more closely let us return to Theorem 10.2 and consider the case $G = \mathbb{Z}$. By an $\mathbb{N}$-*value* on R we understand a function v on $\Phi(R)$ with values in $\mathbb{N}$, satisfying V.1–V.3. Since $\mathbb{N} \subseteq \mathbb{Z}$, Theorem 10.2 shows that v can be extended to a $\mathbb{Z}$-value on $U^\times$; here $v(-1) = 0$, because $\mathbb{Z}$ is torsion-free.

Proposition 7.10.5. *Let R be a Sylvester domain and v any $\mathbb{N}$-value on R. Then $v(P) = 0$ for $P \in GL(R)$ and*

$$v(A) = v(A'), \tag{7}$$

whenever A and A' are stably associated.

Proof. If $P \in GL(R)$, then $v(P) \geq 0$, $v(P^{-1}) \geq 0$, but $v(P) + v(P^{-1}) = v(I) = 0$, hence $v(P) = 0$. Now (7) is an immediate consequence. ∎

To give an example of $\mathbb{N}$-values, let R be a fully atomic semifir and let us define a *matrix prime* as an equivalence class of matrix atoms over R under stable association. For each matrix prime p_i we define an $\mathbb{N}$-value v_i as follows. Given $A \in \Phi(R)$, we put $v_i(A) = r$ if p_i occurs just r times in a complete factorization of A. By unique factorization (Theorem 3.2.7), r is independent of the factorization chosen and V.1–V.3 are easily checked. We shall call v_i the *simple* $\mathbb{N}$-*value* associated with p_i. More generally, if for each matrix prime p_i we pick an integer $n_i \geq 0$, then $w = \sum n_i v_i$ is an $\mathbb{N}$-value, for it is defined on each full matrix A : $w(A) = \sum n_i v_i(A)$, where the sum is finite because $v_i(A) = 0$ for almost all i. In fact every $\mathbb{N}$-value arises in this way, for if w is any $\mathbb{N}$-value on R, take an atom P_i in the class p_i and put $n_i = w(P_i)$; then w and $\sum n_i v_i$ have the same value on each atom and hence on all of $\Phi(R)$. This proves

Theorem 7.10.6. *Let R be a fully atomic semifir and let (v_i) be the simple $\mathbb{N}$-values corresponding to the matrix primes of R. Then for any family (n_i) of non-negative integers, $\sum n_i v_i$ is an $\mathbb{N}$-value and conversely, every $\mathbb{N}$-value is of this form.* ∎

We remark that with every full matrix A there is associated an $\mathbb{N}$-value w_A that is simple if and only if A is an atom, namely $w_A = \sum n_i v_i$, where $n_i = v_i(A)$ and v_i runs over the simple $\mathbb{N}$-values. We can also use the $\mathbb{N}$-values to characterize fully atomic semifirs:

Proposition 7.10.7. *Let R be a semifir. Then R is fully atomic if and only if there is an $\mathbb{N}$-value w on R such that $w(A) = 0$ precisely if A is a unit.*

7.10 Determinants and valuations

Proof. If R is a fully atomic semifir and v_i are the simple $\mathbb{N}$-values corresponding to the different matrix primes of R, then $w = \sum v_i$ has the desired property. Conversely, when w exists, take any full matrix A and factorize it into non-units in any way:

$$A = P_1 \ldots P_r. \tag{8}$$

Since $w(P_i) \geq 1$ by hypothesis, we have $w(A) = \sum w(P_i) \geq r$, and this provides a bound for the number of factors in (8). Hence we obtain a complete factorization of A by choosing r in (8) maximal. ∎

This result can be extended to fully atomic Sylvester domains if we define a prime in this case as an equivalence class of matrix atoms, where A, B are considered equivalent if $v(A) = v(B)$ for each $\mathbb{N}$-value v on R. Such primes are of course unions of classes of stably associated atoms (by Proposition 10.5), but in general they may include atoms that are not stably associated.

Consider a fully atomic semifir R and let $p_i (i \in I)$ be the family of all matrix primes. For each matrix prime p_i we have a homomorphism $v_i : \Phi(R)^{ab} \to \mathbb{Z}$, and combining all these maps, we have a homomorphism

$$\Phi(R)^{ab} \to \mathbb{Z}^I.$$

But each full matrix maps to 0 in almost all factors of $\mathbb{Z}^I$, hence the image lies in the weak direct power (direct sum) $\mathbb{Z}^{(I)}$. Let us write $D(R)$ for the free abelian group on the matrix primes p_i, written additively, and call $D(R)$ the *divisor group* of R. We have a homomorphism $\lambda : \Phi(R)^{ab} \to D(R)$ and hence, by Theorem 10.4, a homomorphism

$$\mu : K_1(U) \to D(R). \tag{9}$$

The map λ is surjective, by construction, hence so is μ given by (9). We claim that its kernel is $GL(R)/(GL(R) \cap E(U))$.

Any $A \in GL(R)$ satisfies $v_i(A) = 0$ for all i, by Proposition 10.5, hence $A \in \ker \mu$. Conversely, if $([A] - [B])^\mu = 0$, then $A^\lambda = B^\lambda$, hence A and B have the same atomic factors, up to order and stable association. Let $A = P_1 \ldots P_r$ be a complete factorization of A, and let B be the product (in some order) of $Q_1, \ldots, Q_r$, where Q_i is similar to P_i. Replacing A, B by $A \oplus I, B \oplus I$, for I of large enough order, we may assume Q_i to be associated to P_i, say $P_i = U_i Q_i V_i$, where $U_i, V_i \in GL(R)$. Then on writing $\sim$ for equality up to the order of the factors we have

$$A \sim Q_1 \ldots Q_r U_1 \ldots U_r V_1 \ldots V_r \sim BF,$$

where $F \in GL(R)$; hence in $GL(U)$ we have

$$A \equiv BF (\mathrm{mod}\ GL(U)'),$$

and so $[A] - [B] = [F] \in GL(R).GL(U)'$. Here we may replace $GL(U)'$ by $E(U)$, by Whitehead's lemma (see Bass [68], p. 226; FA, lemma 9.2.1, p. 348 or Exercise 7 below), and find

$$\ker \mu = GL(R).E(U)/E(U) \cong GL(R)/(GL(R) \cap E(U)).$$

Moreover, *since $D(R)$ is free abelian*, μ is split by $D(R)$ over its kernel and so we obtain

Theorem 7.10.8. *Let R be a fully atomic semifir with universal field of fractions U and divisor group $D(R)$. Then*

$$K_1(U) \cong U^{ab} \cong D(R) \times [GL(R)/(GL(R) \cap E(U))]. \quad \blacksquare \quad (10)$$

The divisor group inherits a partial ordering from R, defined by writing $p > 0$ for $p \in D(R)$ whenever p is positive on $\Phi(R)$. However, this condition (of all positive divisors being given non-negative values) though necessary, is not sufficient for an element of U to belong to R within U, as is shown by the quotient of similar elements, for example $(xy + 1)(yx + 1)^{-1}$ in $k\langle x, y\rangle$ (see also Exercise 3).

To illustrate Theorem 10.8, consider the free algebra $R = k\langle X\rangle$. By Proposition 2.5.3 and Theorem 2.4.4 we have $GL(R) = E(R).k^\times$, where $c \in k$ is mapped to $c \oplus I$; hence

$$GL(R).E(U)/E(U) \cong E(U).k^\times/E(U) \cong k^\times/(k^\times \cap E(U)).$$

Now it can be shown (Cohn [82b], Révész [83a]) that $k^\times \cap E(U) = 1$, therefore we have for the universal field of fractions U of the free algebra $R = k\langle X\rangle$,

$$K_1(U) \cong D(R) \times k^\times. \quad (11)$$

Thus we obtain

Theorem 7.10.9. *Let $R = k\langle X\rangle$ be the free algebra and U its universal field of fractions. Then the Whitehead group of U is isomorphic to the direct product of the divisor group of R by the multiplicative group of k.* $\blacksquare$

Let us return to the case of a Sylvester domain R and its universal field of fractions U, and examine the case of valuations on U. We recall that a *valuation* on U is a function v on U with values in $G \cup \{\infty\}$, where G is a totally ordered group, taken abelian and written additively for simplicity, such that

v.1. $v(x) = \infty$ if and only if $x = 0$,
v.2. $v(xy) = v(x) + v(y)$,
v.3. $v(x - y) \geq \min\{v(x), v(y)\}$.

Using v.2 we can weaken v.3 to

$$v(p - 1) \geq \min\{v(p), 0\} \text{ for any } p \in U ; \tag{12}$$

moreover, it is clear that a G-value on R gives rise to a valuation on U, provided that (12) holds. To restate this condition in terms of R, let us take an admissible matrix $A = (A_0, A_\bullet, A_\infty)$ for p; then an admissible matrix for $p - 1$ is $(A_0 + A_\infty, A_\bullet, A_\infty)$, so the condition (12) becomes, after a slight rearrangement (bearing in mind (1)),

$$v(A_0 + A_\infty \quad A_\bullet) \geq \min\{v(A_0 \quad A_\bullet), v(A_\infty \quad A_\bullet)\}. \tag{13}$$

Hence v gives rise to a valuation on U provided that

V.8. $v(A \nabla B) \geq \min\{v(A), \quad v(B)\}$,

for any square matrices A, B whose determinantal sum is defined.

In general this condition need not hold, e.g. in the free algebra $R = k\langle x, y\rangle$ consider the simple $\mathbb{N}$-values associated with x. We have $v(xy) = v(yx) = 1$, but $v(xy - yx) = 0$. Nevertheless, there is a valuation on the universal field of fractions U that is associated with x. To find it we write U as a skew function field $K(x; \alpha)$, where K is the universal field of fractions of $k\langle y_i | i \in \mathbb{Z}\rangle$ and α is the shift automorphism $y_i \mapsto y_{i+1}$; thus y_i is realized as $x^{-i} y x^i$. On $K(x; \alpha)$ the order of an element in x (i.e. the exponent of the least power of x occurring with a non-zero coefficient) is the required valuation. In terms of $\mathbb{N}$-values this valuation may be obtained as the sum of certain simple $\mathbb{N}$-values, but this does not seem a very efficient way of constructing the valuation.

Exercises 7.10

1. Let K be a field with centre k, such that for any $c \in K$ there exists $a \in K$ centralizing c and transcendental over $k(c)$, and any two elements transcendental over k are conjugate (see e.g. Cohn [71a], SF, Corollary 5.5.2 or Exercise 5.18). Show that every element of $K^\times$ is a multiplicative commutator and hence that $K_1(K) = 1$.
2. Prove Lemma 10.3 for $n = 2$ and for any field of more than three elements by calculating the commutator of $\text{diag}(a, a^{-1})$ and $B_{12}(c)$.
3*. Let $R = k\langle x, y, z, t\rangle$ be the free algebra and U its universal field of fractions. Show that $\text{Det}\begin{pmatrix} x & y \\ z & t \end{pmatrix} = x(t - zx^{-1}y)$, but that this element of U^{ab} has no representative in R. (Hint: Observe that $k\langle x, x^{-1}y, z, t\rangle$ is also free on $x, x^{-1}y, z, t - zx^{-1}y$ and has the same field of fractions as $k\langle x, y, z, t\rangle$.)

4°. In a fully atomic Sylvester domain R do all complete factorizations of a full matrix have the same number of factors? In R define the relation $A \sim B$ between matrix atoms to mean: $v(A) = v(B)$ for all $\mathbb{N}$-values v. Find explicit conditions for $A \sim B$.

5. (Mahdavi-Hezavehi [82]) Let G be an ordered abelian group and define a *matrix valuation* V on a ring R as a G-value satisfying V.8, defined on a factor-stable multiplicative set Σ, with $V(A) = \infty$ for $A \notin \Sigma$. Show that any valuation v on an R-field K may be extended to a matrix valuation V by putting $V(A) = v(\text{Det } A)$. If $f : R \to S$ is a ring homomorphism and V a matrix valuation on S, show that V^f, defined for $A \in \mathfrak{M}(R)$ by $V^f(A) = V(A^f)$, is a matrix valuation on R.

6. (Mahdavi-Hezavehi [82]) If R is a ring with a matrix valuation V, show that $V^{-1}(\infty)$ is a prime matrix ideal, and if K is the corresponding epic R-field (with singular kernel $V^{-1}(\infty)$), then V induces a valuation v on K. Verify that all valuations on epic R-fields arise in this way. Show that the correspondence between matrix valuations on R and valuations on epic R-fields is bijective, but that different matrix valuations may well define the same valuation on R. (*Hint*: Consider the trivial valuation on K.)

7. (Whitehead lemma) Show that $A \oplus B \equiv B \oplus A \pmod{E(R)}$ for any ring R. Deduce that $GL(R)' = E(R)$.

8*. Let $R = k\langle x, x^{-1}, y\rangle$ and denote its universal field of fractions by U. Writing $A = k\langle y_i | i \in \mathbb{Z}\rangle$, and denoting the shift automorphism $y_i \mapsto y_{i+1}$ by α, verify that $R \cong A[x; \alpha]_{(x)}$. Hence obtain a $\mathbb{Z}$-valued valuation on U such that $v(x) = 1$, $v(x^{-i} y x^i) = 0 (i \in \mathbb{Z})$. Deduce that in U, x cannot be expressed as a sum of (multiplicative) commutators. (*Hint*: Observe that $v(c) = 0$ for any multiplicative commutator c.)

9. Extend Theorem 10.2 to pseudo-Sylvester domains.

10. Let R be a persistent semifir over k. What is the relation between the matrix primes of R and those of $R \otimes k(t)$?

11. Show that over a weakly finite ring the determinantal sum of an invertible matrix and a non-full matrix (when defined) is again invertible.

12. Let R be a ring with UGN, such that the product of full matrices of the same order is full. Show that the diagonal sum of full matrices is full. Verify (using the magic lemma) that any Hermite ring satisfying this condition is a Sylvester domain (Cohn [2000a]).

7.11 Localization of firs and semifirs

Let R be a ring and Σ a set of square matrices over R. In this section we consider which properties pass from R to the localization R_Σ; in particular we shall consider the property of being a semifir or a fir.

We begin by examining what restrictions need to be placed on the set Σ. Let $R = k\langle x, y\rangle$ and take $\Sigma = \{xy\}$, or even $\{xy\}^*$, its multiplicative closure. In R_Σ, xy will have an inverse u and we have $xyu = uxy = 1$, but $yux \neq 1$ (as can easily be verified), hence the universal xy-inverting ring contains the idempotent $yux \neq 0, 1$ and so is not even an integral domain. We can exclude rings of this sort by assuming Σ to be factor-inverting, as defined

7.11 Localization of firs and semifirs

in Section 7.5. Thus whenever $AB \in \Sigma$ for square matrices A, B, then A is invertible over R_Σ. We remark that B will also be invertible over R_Σ, for AB has an inverse C over R_Σ, hence B has the inverse CA: clearly $CA.B = I$ and $A.BC = I$. Since A^{-1} exists over R_Σ, we must have $A^{-1} = BC$ and so $B.CA = A^{-1}A = I$.

If R_Σ is to be a semifir, there is a further similar condition, obtained by taking A, B to be rectangular. Let $A \in {}^rR^n$, $B \in {}^nR^r$; If $AB \in \Sigma$, then AB is invertible over R_Σ, so if R_Σ is to be a Hermite ring, then $r \leq n$ and A must be completable in R_Σ, i.e. there exists $A' \in {}^{n-r}R_\Sigma^n$ such that $\begin{pmatrix} A \\ A' \end{pmatrix}$ is invertible. A multiplicative set of square matrices satisfying this condition is called *factor-complete*. This condition is again left–right symmetric, by Theorem 0.4.1. Explicitly, if AB has the inverse C over R_Σ and $(A, A')^T$ has the inverse (P, P'), then B can be completed to (B, P'), with inverse $(CA, A' - A'BCA)^T$, as is easily checked. From the definition it is clear that any factor-complete set is factor-inverting, and all the matrices in it are full. Like factor-invertibility, the condition of being factor-complete depends on R_Σ and there is no obvious way of expressing it in terms of R and Σ alone.

We note that if Σ is factor-complete and $ABC \in \Sigma$, where A is square, then B is completable in R_Σ. For by factor-completeness, AB is completable in R_Σ, say $(AB, P)^T \in GL(R_\Sigma)$, and A is invertible, therefore so is $(B, P)^T = (A^{-1} \oplus I)(AB, P)^T$.

Our first result states that these properties are preserved by passing to the multiplicative closure:

Lemma 7.11.1. *Let R be a semifir, Σ a set of full matrices over R and Σ^* the set of all matrices of the form*

$$X = P \begin{pmatrix} C_1 & *** & *** & *** \\ 0 & C_2 & *** & *** \\ \cdots & \cdots & \cdots & *** \\ 0 & 0 & \cdots & C_r \end{pmatrix} Q, \quad P, Q \in GL(R), C_i \in \sum \cup \{I\}. \quad (1)$$

Then Σ^* is multiplicative, and if Σ is factor-inverting or factor-complete then so is Σ^*.

Proof. It is clear that Σ^* is multiplicative. Suppose now that Σ is factor-complete and assume that the matrix X in (1) can be factorized as $X = AB$; we have to show that A or equivalently, B is completable in R_Σ. Thus we have to find B' such that $(B, B') \in GL(R_\Sigma)$. It comes to the same to show that $(BQ^{-1}, B') \in GL(R_\Sigma)$, so we may assume that $P = I = Q$ in (1). We shall use induction on the order r of X; when $r = 1$, then either $AB \in \Sigma$ and the result follows by the hypothesis on Σ, or $AB = I$ and then the result follows because R is an Hermite ring. Now let $r > 1$ and write

$$AB = \begin{pmatrix} X_1 & X_2 \\ 0 & X_4 \end{pmatrix}. \tag{2}$$

By the partition lemma there exists $T \in GL(R)$ such that

$$AT = \begin{pmatrix} A_1 & A_2 \\ 0 & A_4 \end{pmatrix}, \quad T^{-1}B = \begin{pmatrix} B_1 & B_2 \\ 0 & B_4 \end{pmatrix},$$

where $A_1 B_1 = X_1$, $A_4 B_4 = X_4$. By induction on r there exist B_1', B_4' such that (B_1, B_1'), $(B_4, B_4') \in GL(R_\Sigma)$, hence we can find B' such that

$$(B \quad B') = T \begin{pmatrix} B_1 & B_2 & B_1' & 0 \\ 0 & B_4 & 0 & B_4' \end{pmatrix} \in GL(R_\Sigma).$$

This shows Σ^* to be factor-complete. When Σ is factor-inverting, the proof is similar, but simpler, since A, B in (2) are now restricted to be square. ∎

We shall need two auxiliary results, one on relations in the universal field of fractions and one on the preservation of linear independence in passing from R to R_Σ. The first of them actually holds in Sylvester domains.

Lemma 7.11.2. *Let R be a Sylvester domain and U its universal field of fractions. Given a relation*

$$A = BD^{-1}C, \tag{3}$$

over U, where $D \in R_n$ is full, $A \in {}^m R^p$, $B \in {}^m R^n$ and $C \in {}^n R^p$; then there exist $P, Q \in R_n$, $B' \in {}^m R^n$, $C' \in {}^n R^p$ such that

$$D = QP, \quad B = B'P, \quad C = QC', \quad A = B'C'. \tag{4}$$

Proof. We have the following series of elementary transformations, leaving the inner rank unchanged:

$$\begin{pmatrix} A & B \\ C & D \end{pmatrix} \to \begin{pmatrix} A & B \\ D^{-1}C & I \end{pmatrix} \to \begin{pmatrix} 0 & B \\ 0 & I \end{pmatrix} \to \begin{pmatrix} 0 & 0 \\ 0 & I \end{pmatrix},$$

where we have used (3). The inner rank over U is n, by inspection of the last matrix. Since the embedding $R \to U$ is rank preserving, by Corollary 5.4.10, the rank is the same in U as in R, hence

$$\begin{pmatrix} A & B \\ C & D \end{pmatrix} = \begin{pmatrix} B' \\ Q \end{pmatrix} (C' \quad P),$$

where $P, Q \in R_n$, $B' \in {}^m R^n$, $C' \in {}^n R^p$. On multiplying out, we obtain (4). ∎

7.11 Localization of firs and semifirs

Lemma 7.11.3. *Let R be a semifir and Σ a factor-inverting set of matrices over R. Given any matrix $X \in {}^m R^n$, there exists $T \in GL(R_\Sigma)$ such that XT is a matrix over R whose non-zero columns are right R_Σ-linearly independent.*

Proof. Suppose that the columns of X are right R_Σ-linearly dependent, say

$$Xv = 0, \qquad (5)$$

where $0 \neq v \in {}^n R_\Sigma$. Each component v_i of v is a component of the solution of a matrix equation with matrix in Σ. We can combine these equations into one, $Cv^* = b$, where C is an $N \times N$ matrix in Σ for some $N \geq n$, $b \in {}^N R$, $v^* \in {}^N R_\Sigma$ and the components of v occur among those of v^*. By permuting the columns of C so that the rows relating to v come first, we can write this system as

$$C \begin{pmatrix} v \\ v' \end{pmatrix} = b. \qquad (6)$$

The original equation (5) may now be written as

$$(X \quad 0) \begin{pmatrix} v & 0 \\ v' & I \end{pmatrix} = 0, \qquad (7)$$

where I is the unit matrix of order $N - n$. If in (6) we partition C as $C = (C_1, C_2)$, where $C_1 \in {}^N R^n$, we can write (7) over R_Σ as

$$(X \quad 0) C^{-1} (b \quad C_2) = 0.$$

This equation still holds over the universal field of fractions U of R, so by Lemma 11.2 there exist full $N \times N$ matrices Q, C', an $m \times N$ matrix X' and an $N \times (N - n + 1)$ matrix V over R, such that

$$C = QC', \quad (X, 0) = X'C', \quad (b, C_2) = QV, \, X'V = 0. \qquad (8)$$

The third equation of (8) shows that $b = QV_1$, where V_1 is the first column of V, so we can rewrite (6) as $QC'(v, v')^T = QV_1$; since $QC' \in \Sigma$, which is factor-inverting, we can cancel Q from (6). We now have the same situation with Q replaced by I. Thus the equations (8) are replaced by

$$(X, 0) = X'C, \quad X'(b, C_2) = 0. \qquad (9)$$

On replacing X' by $X'T$ and C by $T^{-1}C$ for suitable $T \in GL_N(R)$, we may assume that $X' = (Y, 0)$, where the columns of Y are right linearly independent over R. We have to show that Y has fewer than n columns, so let us assume the contrary. The equation $X'C_2 = 0$ shows the first n rows of C_2 to be zero, so C

takes the form

$$C = \begin{pmatrix} C_1' & 0 \\ C_3' & C_4' \end{pmatrix} = \begin{pmatrix} C_1' & 0 \\ 0 & I \end{pmatrix} \begin{pmatrix} I & 0 \\ C_3' & C_4' \end{pmatrix},$$

where C_1' is $n \times n$, and by the assumption on Σ is invertible over R_Σ. Now the equation $X'b = 0$ shows the first n entries of b to be 0, but by (6) these are just the entries of $C_1'v$. Hence $v = 0$, a contradiction, which proves the result. ∎

We now come to the first main result of this section, showing that localization preserves semifirs.

Theorem 7.11.4. *Let R be a semifir and Σ any set of full matrices over R. Then the universal localization R_Σ is again a semifir.*

Proof. Let Σ' be the set of all matrices over R that are inverted over R_Σ. Clearly Σ' is multiplicative and consists of full matrices. Moreover it is factor-inverting, for if $AB \in \Sigma'$ then A and B are regular, by Corollary 2.3.2. Now over R_Σ there exists C such that $ABC = CAB = I$; hence $BCABC = BC$ and cancelling BC on the right we obtain $BCA = I$. Thus $A^{-1} = BC$ and similarly $B^{-1} = CA$.

Clearly $R_{\Sigma'} = R_\Sigma$ so we may assume Σ to be factor-inverting. To show that R_Σ is a semifir, let $u \in (R_\Sigma)^n$ and consider the right ideal of R_Σ generated by the components of u. If the components are right linerally dependent, there exists $v \in {}^n R_\Sigma$, $v \neq 0$, such that $uv = 0$. As in the proof of Lemma 11.3 we can write

$$C \begin{pmatrix} v \\ v' \end{pmatrix} = b, \tag{10}$$

where $C \in \Sigma$, $b \in {}^N R$. Now the equation $uv = 0$ may be written

$$(u \quad 0) \begin{pmatrix} v & 0 \\ v' & I \end{pmatrix} = 0,$$

and we can again partition C as $C = (C_1, C_2)$, $C_1 \in {}^n R^n$, so that

$$(u \quad 0)C^{-1}(b \quad C_2) = 0. \tag{11}$$

Here (b, C_2) is $N \times (N - n + 1)$. We claim that this matrix is right full; for if not, then $(b, C_2) = C_2''(b', C_2')$ where C_2'' is $N \times (N - n)$, b' is a column and C_2' is square of order $N - n$. Hence

$$C = (C_1 \quad C_2) = (C_1 \quad C_2'' C_2') = (C_1 \quad C_2'') \begin{pmatrix} I & 0 \\ 0 & C_2' \end{pmatrix},$$

7.11 Localization of firs and semifirs

and we can rewrite (10) as

$$(C_1 \quad C_2'') \begin{pmatrix} I & 0 \\ 0 & C_2' \end{pmatrix} \begin{pmatrix} v \\ v' \end{pmatrix} = b = (C_1 \quad C_2'') \begin{pmatrix} 0 \\ b' \end{pmatrix}.$$

Since Σ is a factor-inverting, the first factor is invertible over R_Σ. Cancelling it, we find that $v = 0$, which is a contradiction; this shows (b, C_2) to be right full, i.e. of inner rank $N - n + 1$. By the left–right dual of Lemma 11.3 we can transform (b, C_2) by an invertible matrix over R_Σ to a matrix over R with r rows that are left R_Σ-linearly independent, followed by 0s, say $T(b, C_2)$. Then $(u, 0)C^{-1}T^{-1}$ has at most $N - r$ non-zero columns and since $r \geq \rho(b, C_2) = N - n + 1$, we have at most $N - (N - n + 1) = n - 1$ non-zero columns. Thus we have a generating set of less than n elements for our right ideal. Hence R_Σ is a semifir by Theorem 2.3.1.(c). ∎

When R is not a semifir, it is no longer true that every factor-inverting set is factor-complete. If the set Σ is not factor-complete, we cannot expect an Hermite ring, but as we shall see, the localization is a pseudo-Sylvester domain provided that Σ is factor-inverting. We shall also need to assume Σ multipliative, since Lemma 11.1 is no longer available now.

Theorem 7.11.5. *Let R be an S-ring and let Σ be a multiplicative set of stably full matrices over R. Then the localization R_Σ is again an S-ring. Moreover, (i) if Σ is factor-inverting, then R_Σ is weakly finite, hence a pseudo-Sylvester domain and (ii) if Σ is factor-complete, then R_Σ is a Sylvester domain.*

In both cases (i) and (ii) the canonical map $R \to R_\Sigma$ is injective.

Proof. Let P be a stably full $n \times n$ matrix over R_Σ; by Cramer's rule, applied to P, we have an equation

$$(A_\bullet \quad -A_0) = (A_\bullet \quad A_\infty) \begin{pmatrix} I & Q \\ 0 & P \end{pmatrix}, \tag{12}$$

where Q is over R_Σ, the A's are over R and $(A_\bullet, A_\infty) \in \Sigma$. Since P is stably full, so is the second factor on the right of (12), while the first factor is invertible over R_Σ, so the term on the left is stably full over R_Σ, hence stably full over R and so invertible over U, the universal R-field. It follows that P is invertible over U, hence the map $R_\Sigma \to U$ is stably honest, therefore by Theorem 5.18, R_Σ is an S-ring, as claimed.

Let us now return to (12) and suppose that P is an $n \times n$ matrix over R_Σ of stable rank r. Suppose that $A_\bullet$ has t columns, and so $t + n$ rows, since $(A_\bullet, A_\infty)$ is square. By increasing t if necessary we may enlarge the unit factor on the right of (12) and thus we may assume that the rank of $\begin{pmatrix} I & Q \\ 0 & P \end{pmatrix}$ is stabilized at $t+r$. Hence $(A_\bullet, -A_0)$ has the same inner rank and we have an equation

$$(A_\bullet \quad -A_0) = B(C_\bullet \quad C_0), \tag{13}$$

over R, where B has $t+r$ columns. Therefore

$$(A_\bullet \quad A_\infty) = (BC_\bullet \quad A_\infty) = (B \quad A_\infty)\begin{pmatrix} C_\bullet & 0 \\ 0 & I \end{pmatrix}. \tag{14}$$

Further, $(A_\bullet \quad -A_0) = (B \quad A_\infty)\begin{pmatrix} C_\bullet & C_0 \\ 0 & 0 \end{pmatrix}$, so (12) now becomes

$$(B \quad A_\infty)\begin{pmatrix} C_\bullet & C_0 \\ 0 & 0 \end{pmatrix} = (B \quad A_\infty)\begin{pmatrix} C_\bullet & 0 \\ 0 & I \end{pmatrix}\begin{pmatrix} I & Q \\ 0 & P \end{pmatrix}.$$

We thus obtain

$$(B \quad A_\infty)\begin{pmatrix} C_\bullet Q - C_0 \\ P \end{pmatrix} = 0, \tag{15}$$

where (B, A_∞) is $(t+n) \times (t+r+n)$.

We consider the two cases separately. (i) Σ is factor-inverting. Let P be a square matrix of stable rank 0. Then (B, A_∞) is a square left factor of $(A_\bullet, A_\infty)$, by (14), and hence is invertible over R_Σ. Cancelling this factor from (15), we find that $P = 0$, so R_Σ is weakly finite, by Proposition 0.1.3. Thus R_Σ is a pseudo-Sylvester domain, as claimed.

(ii) Σ is factor-complete. Then the pair of factors on the right of (14) is completable in R_Σ; if

$$\begin{pmatrix} B & A_\infty \\ X & Y \end{pmatrix} \quad \text{and} \quad \begin{pmatrix} S & T \\ U & V \end{pmatrix}$$

are mutually inverse, then by (15) we have $P = VZ$ for some matrix Z over R_Σ. Here the number of columns of V is the number of rows of X, which is the index of (B, A_∞), i.e. $(t+r+n) - (t+n) = r$. Thus $\rho P \leq r = \rho^* P$, and this shows that the stable rank and the inner rank of P over R_Σ are the same, therefore by Propositions 0.4.4 and 5.6.1. R_Σ is a Sylvester domain. Finally it is clear that in both cases the canonical map $R \to R_\Sigma$ is injective. ∎

In order to obtain a localization theorem for firs, we shall use the following result of Bergman and Dicks [78]:

Theorem 7.11.6. *Let R be a left hereditary ring and Σ any set of matrices over R. Then the universal localization R_Σ is again left hereditary.*

7.11 Localization of firs and semifirs

Proof. Let us write $S = R_\Sigma$; given $M \in {}_R\text{Mod}$, $A \in R_n$, we have an abelian group endomorphism θ_A of ${}^n M$ given by

$$\theta_A : u \mapsto Au \ (u \in {}^n M). \tag{16}$$

Let Mod^Σ be the full subcategory of ${}_R\text{Mod}$ consisting of those modules M for which (16) is an automorphism for all $A \in \Sigma$. Then it is clear that Mod^Σ is equivalent to ${}_S\text{Mod}$; by the 5-lemma it follows that Mod^Σ is closed under extensions in ${}_R\text{Mod}$.

Now S is an epic R-ring, hence for $M \in \text{Mod}^\Sigma$, $\text{Hom}_R(S, M) \cong M$ and it follows that $\text{Ext}_R^1(S, -)$ vanishes on Mod^Σ. Given $C \in {}_S\text{Mod}$, we have an injective resolution of C as R-module

$$0 \to C \to I_0 \to I_1 \to 0.$$

This is of length at most 1 because R is left hereditary. We now apply $\text{Hom}_R(S, -)$ and recall that if $\lambda : R \to S$ is the canonical map, then $\text{Hom}_R(S, I) = I^\lambda$ is the coinduced module; by Appendix Lemma B.6, this is S-injective whenever I is R-injective, while $C^\lambda = C$, as we have seen:

$$0 \to C \to I_0^\lambda \to I_1^\lambda \to 0.$$

Thus C has an injective resolution of length at most 1, and so $\text{l.gl.dim}.S \leq 1$ and S is left hereditary, as claimed. ∎

As a consequence we obtain

Theorem 7.11.7. *Let R be a fir and Σ a factor-inverting set of matrices. Then R_Σ is a fir.*

Proof. By Theorem 11.4, R_Σ is a semifir and it is hereditary by Theorem 11.6, therefore R_Σ is a fir, by Corollary 2.3.12. ∎

This shows for example that the ring $k\langle\langle X \rangle\rangle_{\text{rat}}$ discussed in Section 2.9 is a fir, for it has the form R_Σ, where Σ is the set of all matrices with invertible constant term, clearly a factor-inverting matrix set. We also note

Corollary 7.11.8. *The group algebra of a free group over a field is a fir.*

Proof. Let F be the free group on a set X and consider the group algebra KF, where K is a field. It can be obtained from the tensor ring $K\langle X\rangle$ by localizing at the set Σ consisting of all diagonal matrices with diagonal elements in $X \cup \{1\}$. This set is easily seen to be factor-inverting, so the conclusion follows by Theorem 11.7. ∎

508   *Skew fields of fractions*

Let us now examine how the divisor group of a fir, or more generally, a fully atomic semifir, behaves under localization. We shall assume that R is a fully atomic semifir and Σ a factor-complete set of matrices. A matrix atom A over R and also the associated simple $\mathbb{N}$-value is called Σ-*irrelevant* if A becomes a unit in R_Σ and Σ-*relevant* otherwise. We note that a Σ-relevant $\mathbb{N}$-value extends to a $\mathbb{Z}$-value that is not an $\mathbb{N}$-value on R_Σ.

Theorem 7.11.9. *Let R be a fully atomic semifir and Σ a factor-complete set of matrices over R. Then the localization R_Σ is again a fully atomic semifir and every matrix atom over R either becomes a unit or remains an atom over R_Σ.*

Proof. Let P be any full matrix over R_Σ and using Cramer's rule, write

$$U(P \oplus I)V = A, \tag{17}$$

where $A \in \mathfrak{M}(R), U, V \in GL(R_\Sigma)$. Denote by w the sum of all Σ-relevant $\mathbb{N}$-values on R; then w is an $\mathbb{N}$-value on R_Σ and $w(U) = w(V) = 0$, hence $w(A) = w(P)$, so $w(P) = 0$ if and only if P is a unit. By Proposition 10.7 we conclude that R_Σ is fully atomic; it is a semifir by Theorem 11.4.

For the second part let A be a matrix atom over R and suppose that over R_Σ we have $A = B_1 B_2$, where the B_i are square non-units. By Cramer's rule we have $U_i(B_i \oplus I)V_i = C_i$ ($i = 1, 2$), where C_i is a matrix over R and $U_i, V_i \in GL(R_\Sigma)$. Hence

$$A \oplus I = U_1^{-1} C_1 V_1^{-1} U_2^{-1} C_2 V_2^{-1}. \tag{18}$$

Take complete factorizations of C_1, C_2 over R and let w be as before. Then $w(A) = 1$, and $w(C_i) \geq 1$ because C_i, like B_i is a non-unit. By (17) we have

$$1 = w(A) = w(C_1) + w(C_2) \geq 2,$$

a contradiction; this shows A to be an atom or a unit over R_Σ. ∎

The fact that R_Σ is fully atomic may also be proved as follows. Denote by w the sum of all Σ-relevant simple $\mathbb{N}$-values on R; then w is an $\mathbb{N}$-value on R_Σ and by Cramer's rule, for any full matrix A over R_Σ, $w(A) = 0$ if and only if A is invertible over R_Σ. Hence we can apply the criterion of Proposition 10.7 to reach the desired conclusion.

For any fully atomic semifir R, the divisor group may be defined as in Section 7.10, and if Σ is a factor-complete set of matrices over R, then the divisor group can again be defined over R_Σ, by Proposition 11.9. Our next result describes the mapping between these divisor groups:

Proposition 7.11.10. *Let R be a fully atomic semifir and Σ a factor-complete set of matrices over R, so that R_Σ is again fully atomic. Then*

(i) *any two matrix atoms over R that are stably associated over R_Σ are stably associated over R,*

(ii) *every matrix P over R_Σ is stably associated over R_Σ to the image of a matrix P' over R, and if P is an atom, then P' can also be taken to be an atom.*

Proof. (i) Let A, A' be matrix atoms over R and suppose that they are not stably associated over R; we may assume further that A is Σ-relevant. If v is the simple $\mathbb{N}$-value corresponding to A, then v is an $\mathbb{N}$-value on R_Σ and $v(A) = 1, v(A') = 0$; hence A, A' cannot be stably associated over R_Σ.

(ii) Let P be a matrix over R_Σ; then we have an equation (17), hence P is stably associated to a matrix A over R. Now suppose that P is an atom over R_Σ and denote by w the sum of all Σ-relevant $\mathbb{N}$-values on R; then w is an $\mathbb{N}$-value on R_Σ. Since P is an atom, we have $1 = w(P) = w(A)$; hence in a complete factorization of A over R there is only one factor, P' say, which is Σ-relevant, and clearly P is stably associated over R_Σ to P'. ∎

Let A be a matrix atom over R and denote by $[A]_R$ the corresponding matrix prime of R, as defined in Section 7.10. If A is Σ-relevant, it remains an atom over R_Σ and so defines a matrix prime $[A]_{R_\Sigma}$ there. It is clear that stably associated atoms over R remain stably associated over R_Σ, hence the correspondence $[A]_R \mapsto [A]_{R_\Sigma}$ defines a homomorphism

$$\phi : D(R) \to D(R_\Sigma). \qquad (19)$$

Let $D_\Sigma(R)$ be the subgroup of $D(R)$ generated by the matrix primes defined by Σ-relevant atoms; we claim that $D_\Sigma(R) \cong D(R_\Sigma)$. For the restriction of ϕ to $D_\Sigma(R)$ is injective, by Proposition 11.10 (i), and surjective by (ii). Thus we obtain

Theorem 7.11.11. *Let R be a fully atomic semifir, Σ a factor-complete set of matrices over R and denote by $D_\Sigma(R)$ the subgroup of $D(R)$ generated by the matrix primes defined by the Σ-relevant matrix atoms of R. Then the embedding $R \to R_\Sigma$ induces an isomorphism*

$$D_\Sigma(R) \cong D(R_\Sigma).$$

Moreover, if (19) is the induced homomorphism, then

$$D(R) = D_\Sigma(R) \times \ker \phi, \qquad (20)$$

where $\ker \phi$ is the subgroup of $D(R)$ generated by the Σ-irrelevant primes.

Proof. The first part is proved by the above remarks and (20) follows because $D_\Sigma(R)$ is free and so provides a splitting. ∎

Exercises 7.11

1. Let R be a semifir with universal field of fractions U. If $a_i a_0^{-1} = b_i b_0^{-1}$ in U (for $a_i, b_i \in R, i = 1, 2$), find $a, b, c_i \in R$ such that $a_i = c_i a, b_i = c_i b$. Prove a corresponding result for matrices over R.

2*. Show that the canonical non-IBN ring $V_{m,n}$ is hereditary and that $V_{n,n}$ is a fir. (*Hint*: Use Theorem 11.6 for rectangular matrices. It can be shown that $V_{m,n}$ is an r-fir, where $r = \min(m, n) - 1$. See Section 2.11 or SF, Section 5.7.)

3*. (A. H. Schofield) Let R be a semifir and $\Sigma(n)$ the set of all full $n \times n$ matrices, for some $n \geq 1$. Show that $\Sigma(n)$ is factor-complete; deduce that $R_{\Sigma(n)}$ is a semifir. Show that the chain of rings obtained by iterating this process (for a fixed n) has as its union the universal field of fractions of R. Verify that different values of n give cofinal chains.

4. Let R be a k-algebra with $2n(n + 1)$ generators arranged as an $n \times (n + 1)$ matrix A and an $(n + 1) \times n$ matrix B, with defining relation (in matrix form) $AB = I$. It follows by Theorem 2.11.2 that R is an n-fir (see also SF, theorem 5.7.6). Show that the universal localization of R at the set of all full $n \times n$ matrices is not an n-fir. (*Hint*: Note that the localization is not Hermite.)

5. (A. H. Schofield) Let R be a k-algebra on 18 generators arranged as two 3×3 matrices A, B with defining relation $AB = 0$. By Theorem 2.11.2 it follows that R is a 2-fir (see also SF theorem 5.7.6). Show that $R_{R^\times}$ is not an integral domain. (*Hint*: Apply elementary operations to reduce some of the entries to 0.)

6. Let R be a ring and $\Sigma \subseteq \mathfrak{M}(R)$. Show that if Σ is multiplicative and R_Σ is a local ring, then Σ is factor-complete.

7. Let R be a ring with an ideal $\mathfrak{a}$ such that $R/\mathfrak{a}$ is Hermite. Show that the set of all matrices over R invertible (mod $\mathfrak{a}$) is factor-complete.

8. Let R be a semifir and $\Sigma \subseteq \mathfrak{M}(R)$. Show that R_Σ is a semifir if and only if R_Σ is weakly finite and whenever $AB \in \Sigma$, where $A \in {}^m R^n, B \in {}^n R^m, m \leq n$, then $I_n - B(AB)^{-1} A$ has inner rank $n - m$ over R_Σ.

9*. Let F be the free group on a set Y. Show that if X is a set with a bijection $\phi : Y \to X$, then the group algebra kF may be embedded in the free power series ring $k\langle\langle X \rangle\rangle$ by mapping $y \mapsto 1 - y\phi, y^{-1} \mapsto \Sigma(y\phi)^n$. Show further that this embedding is totally inert. (*Hint*: Verify that every matrix over kF is stably associated to a matrix over $k\langle Y \rangle$ and use the inertia theorem.)

10°. For a free algebra of infinite rank let Σ_0 be a finite set of matrices such that the multiplicative set generated by Σ_0 is factor-inverting. Find a finite set Σ_1 containing Σ_0 and such that the multiplicative set generated by Σ_1 is factor-complete.

11°. (G. M. Bergman) Let R be a fully atomic semifir and Σ the union of a set of similarity classes of some matrix atoms over R. What can be said about the set Σ^* of similarity classes of matrix atoms inverted over R_Σ? Examine the cases where R is (i) commutative, (ii) a free algebra.

12. Let R be a Sylvester domain with a homomorphism to an Hermite ring. Show that the set Σ of all matrices inverted is factor-complete and R_Σ is a Sylvester domain. (See Cohn [2000a]. *Hint*: Use Cramer's rule and Theorem 11.5.)

7.12 Reversible rings

We now examine a special class of rings for which some of the conditions for embeddability become much simpler. A non-zero ring will be called *reversible* if $ab = 0$ implies $ba = 0$. Clearly every non-zero commutative ring is reversible, as well as every integral domain and more generally, any reduced ring. As we shall see, reversible rings share some of the properties of commutative rings.

Lemma 7.12.1. *Let R be a reversible ring and suppose that $a_1, \ldots, a_n \in R$ satisfy $a_1 \ldots a_n = 0$. Then*

$$a_i a_{i+1} \ldots a_{i-1} = 0 \text{ for } i = 2, \ldots, n \text{ (where the subscripts are taken mod } n\text{)} \tag{1}$$

and

$$c_0 a_1 c_1 a_2 \ldots c_{n-1} a_n c_n = 0 \text{ for all } c_0, \ldots, c_n \in R. \tag{2}$$

Proof. If $a_1 \ldots a_n = 0$, then (1) follows by writing $x = a_1 \ldots a_{i-1}, y = a_i \ldots a_n$ and applying reversibility. Given (1), we have $a_i a_{i+1} \ldots a_{i-1} c_{i-1} = 0$ for any c_i, hence $a_1 \ldots a_{i-1} c_{i-1} a_i \ldots a_n = 0$, and by repeating this process we obtain (2). ∎

We recall that an ideal $\mathfrak{a}$ is *nilpotent* if $\mathfrak{a}^n = 0$, for some $n \geq 1$. It is called a *nil ideal* if all its elements are nilpotent. A ring without nilpotent ideals (other than 0) is called *semiprime* and in any ring R a *nilradical* is a nil ideal N such that R/N is semiprime. A commutative ring has a unique nilradical, but general rings may have more than one, in fact there is an *upper nilradical*, the sum of all nil ideals, and a *lower nilradical*, the intersection of all prime ideals (see FA, Section 8.5).

Proposition 7.12.2. *In any reversible ring, every nil right ideal generates a two-sided nil ideal and the set of all nilpotent elements is a nil ideal, the unique nilradical.*

Proof. Let R be a reversible ring and take a nilpotent element $x \in R$. If $x^r = 0$, then $x^{r_1} c_1 x^{r_2} c_2 \ldots x^{r_n} c_n = 0$, for any $c_i \in R$ and any r_i such that $r_1 + \cdots + r_n = r$, by Lemma 12.1. Now if $\mathfrak{a}$ is a nil right ideal, then for any $a, b \in \mathfrak{a}, a^r = b^s = 0$ for some $r, s \geq 1$, hence any product of $r + s - 1$ factors a or

b vanishes; more generally, $(xa + yb)^{r+s-1} = 0$ for any $x, y \in R$ and so $R\mathfrak{a}$ is a nil ideal. Next let N be the set of all nilpotent elements of R and suppose that $x, y \in N$, say $x^r = y^s = 0$; the same argument shows that $x + y \in N$. Further, if $x^r = 0$, then $(bxc)^r$ has r factors x and so is 0, and this shows N to be an ideal, a nil ideal, by definition. Now any element of N generates a nilpotent ideal, hence N is the unique nil radical. ∎

To simplify the conditions for embeddability in fields we shall need more than reversibility; in fact one would expect a matrix type of condition. This suggests the following definition. A ring R is said to be *n-reversible* if R is non-zero and for any square matrices A, B of the same order, at most n, AB is full whenever BA is full. Thus 'reversible' is the same as '1-reversible'. If R is n-reversible for all n, it is said to be *fully reversible*. We shall need a couple of lemmas on the behaviour of these rings; the first one is a consequence of the magic lemma:

Lemma 7.12.3. *Over a fully reversible ring R the determinantal sum of any two non-full matrices, when defined, is again non-full.*

Proof. Let $C = A \nabla B$, where A, B are non-full; by Lemma 5.11 we have $A = ST, C = SUT$. Since R is fully reversible and A is non-full, so is TS, TSU and $SUT = C$, as we had to show. ∎

Lemma 7.12.4. *Over a fully reversible ring every full matrix is stably full.*

Proof. Let R be a fully reversible ring and suppose that A is a full $n \times n$ matrix such that $A \oplus 1$ is not full. Then there exist square matrices P, Q of order n, a row $p = (p_1, p_2, \ldots, p_n)$ and a column $q = (q_1, q_2, \ldots, q_n)^T$ such that

$$\begin{pmatrix} A & 0 \\ 0 & 1 \end{pmatrix} = \begin{pmatrix} P \\ p \end{pmatrix} (Q \quad q). \tag{3}$$

Since $Pq = 0$, it follows that PQ' is not full, where $Q' = (q, e_2, \ldots, e_n)$, for PQ' has its first column zero. By elementary transformations we can reduce the first column of Q' to the form $(q_1, 0, \ldots, 0)^T$, hence $P(q_1 \oplus I)$ is not full, so neither is $(q_1 \oplus I)P$, and it follows that $(p_1 q_1 \oplus I)P$ is not full. The same is true of $(p_i q_i \oplus I)P$, for if T is the matrix obtained from the unit matrix by interchanging the first and ith rows, then Tq is the column q with the first and ith component interchanged and $PT.Tq = 0$. By the previous argument it follows that $(p_i q_i \oplus I)PT$ is not full, so this is also true of $(p_i q_i \oplus I)P$. We now form the determinantal sum of these matrices with respect to the first row; bearing in mind that $\sum p_i q_i = 1$ (by (1)), and find by Lemma 12.3 that P is not full. This contradicts the fact that $A = PQ$ is full, and the conclusion follows. ∎

We have seen (Theorem 5.13) that a ring is a Sylvester domain if and only if the set of all non-full matrices is a prime matrix ideal. The next result shows that fully reversible rings are characterized by a slightly weaker condition:

Theorem 7.12.5. *Let R be any ring. Then R is fully reversible if and only if the set of all non-full matrices is a proper matrix ideal.*

Proof. Assume that R is fully reversible. By Lemma 12.3, the set $\mathcal{Z}$ of all non-full square matrices is closed under determinantal sums and it is proper, since R is non-zero. Further, if $A \oplus I \in \mathcal{Z}$ then $A \in \mathcal{Z}$ because full matrices are stably full, by Lemma 12.4. Thus $\mathcal{Z}$ is a proper matrix ideal, clearly the least one. Conversely, suppose that the set $\mathcal{Z}$ of all non-full matrices is a proper matrix ideal; then R is non-zero. Now assume that $AB \in \mathcal{Z}$. Then $A \oplus B \in \mathcal{Z}$ by Section 7.3 (f), but this is equivalent to $B \oplus A \in \mathcal{Z}$ and so $BA \in \mathcal{Z}$; so R is fully reversible, as we had to show. ∎

By combining this result with Theorem 5.13, we obtain

Corollary 7.12.6. *Every Sylvester domain is fully reversible.* ∎

As we saw in Section 7.3, the matrix nilradical $\mathcal{N}$ is the intersection of all prime matrix ideals (Theorem 3.8):

$$\mathcal{N} = \sqrt{\mathcal{Z}} = \cap \mathcal{P}_\lambda, \tag{4}$$

where $\mathcal{P}_\lambda$ runs over all prime matrix ideals. Write $K_\lambda = R/\mathcal{P}_\lambda$ for the epic R-field defined by $\mathcal{P}_\lambda$; the natural maps $\phi_\lambda : R \to K_\lambda$ can be combined to a homomorphism $\phi : R \to \Pi K_\lambda$. Its kernel is given by the equation

$$\ker \phi = \mathcal{N} \cap R = (\mathcal{N})_1, \tag{5}$$

where the subscript 1 indicates the subset of 1×1 matrices. To prove (5), if $x\phi = 0$, then x lies in each $\mathcal{P}_\lambda$, so $x \in \mathcal{N}$. Conversely, if $x \in \mathcal{N}$, then x maps to 0 in all the K_λ and so lies in $\ker \phi$ thus (5) is proved.

For a fully reversible ring R we saw in Proposition 12.2 that the set N of all nilpotent elements is the nilradical of R; we claim that $N = \ker \phi$. Clearly $N \subseteq \ker \phi$; conversely, if $x\phi = 0$, then $xI \in \mathcal{N}$ for a unit matrix of a certain order n, say, which means that xI is non-full, hence (by Section 7.3 (f)), $x^n \in \mathcal{N}$, that is, x is nilpotent and so $x \in N$. The residue-class ring $R^* = R/N$ is reduced, and so is reversible. If $\{\mathcal{P}_\lambda\}$ is the family of all prime matrix ideals of R and $K_\lambda = R/\mathcal{P}_\lambda$ as before, then as we have seen, the homomorphism $\phi : R \to \Pi K_\lambda$ has kernel N, hence R^* is embedded in ΠK_λ. We thus obtain

Theorem 7.12.7. *Let R be a fully reversible ring and N its nilradical. Then R/N is a subring of a direct product of fields; in particular, any fully reversible ring has a homomorphism into a field.* ∎

We recall that a non-zero ring R is called *prime* if $aRb = 0$ implies $a = 0$ or $b = 0$; the notion of semiprime ring introduced earlier can also be characterized by: $aRa = 0 \Rightarrow a = 0$. Now Theorem 12.7 yields

Corollary 7.12.8. *Any fully reversible semiprime ring is a subring of a direct product of fields. In particular, any such ring has a homomorphism to a field.* ∎

In order to find embeddability conditions we can make Theorem 12.7 more precise. Let R be any ring and Σ a set of full matrices over R closed under diagonal sums. If, further, $\Sigma \cap \mathcal{Z} = \emptyset$, then by Theorem 3.7, there is a prime matrix ideal $\mathcal{P}$ disjoint from Σ and over the corresponding R-field $R/\mathcal{P}$ the members of Σ become invertible. So far we have not had to assume full reversibility. But if we now assume that R is a fully reversible prime ring and take Σ to consist of all diagonal matrices with no zero on the main diagonal, then since $\mathcal{Z}$ contains no non-zero element of R, $\Sigma \cap \mathcal{Z} = \emptyset$ by Section 7.3 (f), and we obtain an embedding in a field. This yields

Theorem 7.12.9. *A fully reversible ring is embeddable in a field if and only if it is an integral domain or, equivalently, a prime ring.*

Proof. Clearly the condition is necessary, and any integral domain is prime. Conversely, if R is a prime ring, then, being reversible, it is an integral domain, and by Theorem 12.7 it is a subring of a direct product of fields, say $R \subseteq P = \Pi_\Lambda K_\lambda$. Let ϕ_λ be the projection $R \to K_\lambda$ and for any $x \in R$, put $\Lambda(x) = \{\lambda \in \Lambda | x\phi_\lambda \neq 0\}$. Then for any $x_1, \ldots, x_n \in R^\times$, since R is embedded in P,

$$\Lambda(x_1) \cap \ldots \cap \Lambda(x_n) = \Lambda(x_1 \ldots x_n) \neq \emptyset,$$

so there is a filter on Λ including all the $\Lambda(x)$, and this is contained in an ultrafilter $\mathcal{F}$. Let $\psi : P \to P/\mathcal{F}$ be the natural map, where $P/\mathcal{F}$ is a field, by Appendix Theorem C.1. For each $x \in R$, $x\phi_\lambda \neq 0$ for all λ in a member of $\mathcal{F}$, hence $x\phi\psi \neq 0$ and $\phi\psi$ provides an embedding of R in the field $P/\mathcal{F}$. ∎

Exercises 7.12

1. Show that a ring is an integral domain if and only if it is prime and reversible, and that it is reduced (i.e. $x^2 = 0 \Rightarrow x = 0$) if and only if it is semiprime and reversible.
2. Let R be a ring such that $abc = 0 \Rightarrow cba = 0$. Show that if $a_1 a_2 \ldots a_n = 0$, then the product of the a_i in any order is zero (such a ring is called *symmetric*).

3. Show that every strongly regular ring (see Exercise 5.6) is reduced; deduce that if in addition it is prime it must be a field.
4. Let R be a strongly regular ring and $0 \neq a \in R$; show that any ideal $\mathfrak{p}$ maximal subject to $a \notin \mathfrak{p}$ is prime, and deduce that R is a subdirect product of fields.
5. Show that a strongly regular ring is embeddable in a field if and only if it is an integral domain. [*Hint*: For the sufficiency take $R \subseteq P = \Pi K_\lambda$ (K_λ a field), with projections $\varepsilon_\lambda : P \to K_\lambda$ and for each $x \in P$ define $\Gamma_x = \{\lambda | x\varepsilon_\lambda = 0\}$, $I_x = \Pi\{K_\lambda | \lambda \in \Gamma_x\} = \mathrm{Ann}(x)$. Verify that I_x is an ideal in P, and if I is the ideal generated by all I_x with $x \in R^\times$, show that $R \cap I = 0$. Deduce that the map $P \to P/I$ is $R^\times$-inverting, hence any homomorphism of P/I into a field provides an embedding of R in K.]
6. Let R be a subring of a direct product of fields. Use the proof of Theorem 12.9 to show that R is embeddable in a field if and only if it is an integral domain.
7. (V. O. Ferreira) A ring is said to be *unit-stable* if it is non-trivial and for any square matrices A, B of the same order and any invertible matrix U, AUB is full whenever AB is full. Verify that every fully reversible ring is unit-stable, and conversely, a unit-stable ring is fully reversible if the stable rank equals the inner rank (this follows from Proposition 0.4.3, but give a direct proof).
8. Show that any direct product of fields is fully reversible.
9*. (Cohn [99]) Show that a subring of a field need not be fully reversible. (*Hint*: Find two full 3×3 matrices over the polynomial ring $k[x, y, z]$ whose product is non-full.)
10. Show that a fully reversible Hermite ring need not be embeddable in a field. (*Hint*: Try the direct product of two fields.)

Notes and comments on Chapter 7

It is well known that a commutative ring R can be embedded in a field if and only if R is an integral domain. This condition is clearly still necessary in the non-commutative case, but no longer sufficient. This was first shown by Malcev [37] in answer to van der Waerden ([30], p. 49) who had written: "Die Möglichkeit der Einbettung nicht kommutativer Ringe ohne Nullteiler in einen sie umfassenden Körper bildet ein ungelöstes Problem, außer in ganz speziellen Fällen". Malcev's counter-example was in the form of a monoid ring QS, where S is a monoid with cancellation, but not embeddable in a group (see SF, Section 1.2). The existence of such a monoid also provided a counter-example to the claim by Sushkevich [36], that every cancellation monoid is embeddable in a group. Malcev followed up his example by two papers (Malcev [39, 40]), which gave a set of necessary and sufficient conditions for a monoid to be embeddable in a group, in the form of quasi-identities and showing that these conditions formed an infinite set that could not be replaced by a finite subset.

The following classification is taken from Bokut [81]. Let $\mathfrak{D}_0$ be the class of integral domains, $\mathfrak{D}_1$ the class of rings R such that $R^\times$ is embeddable in a group, $\mathfrak{D}_2$ the class of *invertible* rings, i.e. rings R such that the universal $R^\times$-inverting mapping is injective, and $\mathcal{E}$ the class of rings embeddable in fields. Then it is clear that

$$\mathfrak{D}_0 \supseteq \mathfrak{D}_1 \supseteq \mathfrak{D}_2 \supseteq \mathcal{E}.$$

Here all the inclusions are strict: in answering van der Waerden's question (whether $\mathfrak{D}_0 = \mathcal{E}$), Malcev proved that $\mathfrak{D}_0 \neq \mathfrak{D}_1$ and he raised the question whether $\mathfrak{D}_1 = \mathcal{E}$. This was answered 30 years later by three people, independently and almost simultaneously: Bowtell [67b] and Klein [67] gave examples showing that $\mathfrak{D}_2 \neq \mathcal{E}$, while Bokut [67, 69] gave examples to show that $\mathfrak{D}_1 \neq \mathfrak{D}_2$. The examples of Bowtell and Klein can be obtained quite simply by the methods described in Section 2.11 (see also SF, Section 5.7), which provide n-firs that are not $(n + 1)$-firs, while Bokut's proofs have now been simplified by Valitskas [87]. The examples of n-firs that are not embeddable in (n+1)-firs can also be used to show that the necessary and sufficient conditions of Section 7.5 for embeddability in a field cannot be replaced by a finite subset (they are not equivalent to a finite set of elementary sentences, see Cohn [74b] and UA, p. 344). Dubrovin [87] proves that the group algebra of any right-ordered group over a field is invertible.

Exercise 5.6 shows that the conditions for an integral domain to be embeddable in a field take the form of quasi-identities, as expected from the general theory (UA,VI.4). Thus $\mathcal{E} = \mathcal{J} \cap \mathfrak{D}_0$, where $\mathcal{J}$ is a quasi-variety, namely the class of all subrings of strongly regular rings (see Exercise 5.6 and SF, Section 1.2). The result that a subdirect product of rings in $\mathcal{E}$ is in $\mathcal{E}$ if and only if it is in $\mathfrak{D}_0$ was proved by Burmistrovich [63] (see also Exercise 12.6).

Malcev's solution of the embedding problem for monoids gave no hint for the corresponding problem of embedding rings in fields. Until 1970 the only purely algebraic method of embedding rings in fields was based on Ore's method (Ore [31], see Section 0.7); for a topological method see Cohn [61b] and SF, Section 2.6. L. Schwarz [49] defined a form of quotient ring for a free algebra $R = k\langle X \rangle$ (essentially a localization $R_{R^\times}$), but he did not succeed in constructing a field of fractions for $k\langle X \rangle$ (it may be of interest to note that Schwarz began in 1931 as a student of E. Noether and then became a student of H. Weyl; he submitted a Habilitationsschrift in Halle, but withdrew it later).

The first field of fractions for $R = k\langle x, y \rangle$ had been obtained by R. Moufang [37] by embedding R in kG, where G is the free metabelian group on x, y, and constructing a field of power series for kG (Schwarz seems to have been unaware of her work). In the light of Moufang's construction it is relevant to observe that the free monoid X^* cannot be embedded in a nilpotent group on X, because such a group always satisfies non-trivial monoid identities (see Malcev [53]). Later, Malcev [48] and B. H. Neumann [49] independently showed that $k\langle X \rangle$ has a field of fractions, by embedding it in the group ring kF, where F is the free group on X, and constructing the field of formal power series kF in terms of an ordering on F (see Corollary 1.5.10 above and SF, Section 2.4).

The basic idea underlying this chapter, to invert matrices rather than just elements, was inspired by the rationality criteria of Schützenberger [62] and Nivat [69] for non-commutative power series rings. The observation that these criteria had quite general validity, coupled with the notion of a 'full' matrix (derived from the 'inner rank' defined by Bergman [67]) was exploited in Cohn [71a] to embed firs in fields. These criteria survive in Theorem 1.2 (the forms given by Schützenberger and Nivat correspond to (b) and (c), respectively). The application to firs is based on the following theorem (Cohn [71a]): let R be a ring and M a subset of R. Then the universal localization $R \to R_M$ is injective provided that M consists of regular elements of R and for any $p, q \in M$, $\mathrm{Hom}_R(R/pR, R/qR)$ is 0 or a field according as $p \neq q$ or $p = q$. This shows for example that any atomic 2-fir is in $\mathcal{D}_2$. Other interesting consequences depend on the fact that the coproduct of fields is a fir and so is embeddable in a field (SF, Chapter 5).

The notion of a 'free' field was introduced by Amitsur [66] as a result of studying generalized rational identities (see also Bergman [70a] and SF, Chapter 7). J. Lewin [74b] proved that the subfield of the Malcev–Neumann power series field generated by kF is the universal field of fractions of $k\langle X\rangle$. This result has also been proved recently by Reutenauer [99] by means of graph theory.

The development in Section 7.2 is based on Cohn [72a]; the method of matrix equations, Section 7.1, and matrix ideals, Sections 7.3 and 7.4, was first described in Cohn [72b]. Proposition 2.1 and the discussion of ring epimorphisms is based on Knight [70]. The term 'matrix nilradical' is new here, though the concept already appeared in FR.2. The term emphasizes the analogy with the commutative case, where there is a unique nilradical, the radical of 0, which is the greatest nilideal; by contrast, a general (non-commutative) ring usually has more than one nilradical (see FA, Section 8.5).

In FR.1 the construction of an epic R-field with prescribed singular kernel was based on an axiomatic description of fields as groups with an extra element 0 and an operation $x \mapsto 1 - x$ (see Cohn [61b]; Dicker [68]; Leissner [71] and, for an amusing connexion with Mersenne primes, Hotje [83]). In FR.2 this was replaced by a direct but quite lengthy proof; its place has now been taken by Theorem 4.3. The proof of the latter went through several versions; the present form was greatly helped by comments from G. M. Bergman. The consequences of this result are traced out in Section 7.5; in particular this leads to explicit conditions for a ring to be embeddable in a field (Theorem 5.7) and the existence of a universal field of fractions for any semifir (Cohn [72b]); this result was generalized to Sylvester domains by Dicks and Sontag [78]. The more general condition of Theorem 5.13 is taken from Cohn [2000a].

The proof of Theorem 5.20, that $k\langle X\rangle$ is honestly embedded in $k\langle\langle X\rangle\rangle$, is new. As is well known, the free algebra of countable rank can be embedded in a free algebra of rank 2, but the obvious embedding will not be honest; in Cohn [90] the theory is applied to obtain an honest embedding, which is presented in Theorem 5.19.

The field spectrum $X(R)$ defined in Section 7.4 was introduced by Cohn [72b]; it leads to an 'affine scheme' $(X; \bar{R})$ associated with a general ring (see also Cohn [79]). For any ring one has a sheaf of local rings over X, and a natural homomorphism

$$\gamma : R \to \Gamma(X, \bar{R}),$$

into the ring of global sections, but, of course, one cannot expect an isomorphism (as in the commutative case, see e.g. Macdonald [68]). Thus, for example, if R has no homomorphism into fields, then $X(R) = \emptyset$ and the scheme is trivial. Besides the sections arising from R (the 'integral' sections) one also has the following 'rational' sections. Let $A \in \mathfrak{M}(R)$; if A becomes invertible in every localization, then the entries of A^{-1} define *rational sections* and one may ask: (i) for which rings is every global section rational and (ii) for which rings is every rational section integral? For fully atomic semifirs (i) holds, by Theorem 7.4. There is also a generalization of the field spectrum to take into account homomorphisms into simple Artinian rings (see Cohn [79]; Ringel [79]). An even more general spectrum, the epi-spectrum, has been defined by Bergman [70b]; for commutative rings this agrees with the usual prime spectrum and, in general, it is never empty, but is difficult to determine explicitly (for Noetherian rings it agrees with the previous case, see Cohn [79]).

Sections 7.6 and 7.7 are extracted with only small changes from the trilogy Cohn [82a, 85a, 85b], where further details on semifirs of bounded depth and centralizers in field coproducts can be found. Proposition 7.3, on expressing a fraction in 'lowest terms' is new. Section 7.9 is based on Cohn [85b], but some of the proofs have been streamlined; Propositions 9.9–9.11 and Corollary 9.12 are based on Bergman [87].

Section 7.8 contains the specialization lemma (Lemma 8.7), which in FR.2 formed part of Section 5.9 and whose proof incorporates simplifications due to Bergman. Theorem 8.2, taken from Cohn and Reutenauer [99], answers a question of Bergman by showing that 'free fields' really are free, using a simple case of Malcolmson's criterion (Proposition 8.1). Theorem 8.3 was proved in Cohn [82a].

Section 7.10 is based on Cohn [82b, 83], with some simplifications; for a different approach (and more precise results for free algebras) see Révész [83a]. There is a corresponding development of valuations, briefly mentioned in Exercises 10.5 and 10.6 (Mahdavi-Hezavehi [82]) and a description of orderings of epic R-fields in terms of 'matrix cones' on R, by Révész [83b]; see also SF; Chapter 9).

The localization theorem for semifirs, Theorem 11.4, is taken from Cohn and Dicks [76], which corrects and complements results of Cohn [74b]. The extension to Sylvester domains in Theorem 11.5 was obtained by Cohn and Schofield [82]. Bergman and Dicks [78] first proved Theorem 11.6 on localization of hereditary rings; the very simple proof given here is essentially due to Dlab and Ringel [85], in a formulation by W. Dicks. The corresponding statement for flat epimorphisms is much easier to prove, see Hudry [70]. The behaviour of atoms under localization (Propositions 11.9–11.10 Theorem 11.11) is described in Cohn [82b], where these results are used to prove the following theorem. Let $R = k\langle X \rangle$ be the free algebra on an infinite set X, let X_0 be a subset of X with an infinite complement, and denote by Σ the set of all full matrices totally coprime to the elements of X_0. Then R_Σ is a simple principal ideal domain. This result is used in Cohn and Schofield [85] to construct a simple Bezout domain that is right but not left principal, thus answering exercise 1.2.9° of FR.1.

Section 7.12 is based on Cohn [99], with minor simplifications. We note that Proposition 12.2 provides a positive answer to Köthe's conjecture for reversible rings by showing that a reversible ring with a non-zero nil right ideal has a non-zero nil ideal.

Appendix

This appendix gives a brief summary of facts needed from lattice theory, homological algebra and logic, with references to proofs or sometimes the proofs themselves. In each section some reference books are listed, with an abbreviation which is used in quoting them in the appendix.

A. Lattice theory

LT: G. Birkhoff, Lattice Theory, 3rd Edition. Amer. Math. Soc. Providence RI 1967.
BA: P. M. Cohn, Basic Algebra, Groups, Rings and Fields. Springer, London 2002.
FA: P. M. Cohn, Further Algebra and Applications. Springer, London 2003.
UA: P. M. Cohn, Universal Algebra, 2nd Edition. D. Reidel, Dordrecht 1981.

(i) We recall that a *lattice* is a partially ordered set in which any pair of elements a, b has a supremum (i.e. least upper bound, briefly: sup), also called *join* and written $a \vee b$, and an infimum (i.e. greatest lower bound, briefly: inf), also called *meet* and written $a \wedge b$. It follows that in a lattice L every finite non-empty subset has a sup and an inf; if every subset has a sup and an inf, L is said to be *complete*. A partially ordered set that is a lattice (with respect to the partial ordering) is said to be *lattice-ordered*. It is possible to define lattices as algebras with two binary operations $\vee, \wedge$ satisfying certain identities, so that lattices form a variety of algebras (LT, p. 9, UA, p. 63 or BA, Section 3.1).

If we reverse the ordering in a lattice, we again obtain a lattice, in which meet has been replaced by join and vice versa. This is the basis of the principle of duality in lattice theory, by which we obtain from each theorem about lattices (except the self-dual ones) another one that is its dual.

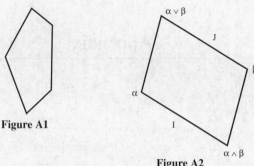

Figure A1

Figure A2

Given any elements $x \leq y$ in a lattice L, we define the *interval* from x to y as the set $[x, y] = \{z \in L | x \leq z \leq y\}$; it is a sublattice of L, i.e. closed under the operations of join and meet in L. If $[x, y]$ contains no elements apart from x and y, it is called a *link*, and y is said to *cover* x. The maximal number of links in a chain is called the *length* of the lattice; thus a lattice of length n has chains of $n + 1$ elements, but no more.

(ii) Nearly all the lattices we deal with will be *modular*, i.e. they satisfy the *modular law* (Dedekind [1900]):

$$(x \vee y) \wedge z = x \vee (y \wedge z) \quad \text{for all } x, y, z \text{ such that } x \leq z. \tag{1}$$

As stated here, this law is not actually an identical relation, since it involves a condition, but we observe that it is equivalent to the following identity:

$$((x \wedge z) \vee y) \wedge z) = (x \wedge z) \vee (y \wedge z).$$

An important example of a modular lattice, which will much occupy us here (and which incidentally is responsible for the name) is the set $\text{Lat}_R(M)$ of all submodules of a module M over a ring R, partially ordered by inclusion. A criterion for modularity is provided by

Proposition A.1 *A lattice is modular if and only if it contains no sublattice isomorphic to the 5-element lattice of length 3* (Fig. A1; see LT, p. 13, BA, Section 3.1). ■

The following well-known consequence of (1) will be used without further reference:

Lemma A.2 *Let M be a module and M' a submodule of M. Then any direct summand of M that is contained in M' is a direct summand of M'.*

Proof. Assume that $M = P \oplus Q$ and $P \subseteq M'$. Then by (1),

$$M' = (P \oplus Q) \cap M' = P \oplus (Q \cap M'). \qquad \blacksquare$$

If a, b are any two elements of a modular lattice L, there is an isomorphism between the intervals $I = [a \wedge b, a]$ and $J = [b, a \vee b]$, given by the mapping $x \mapsto x \vee b$, with inverse $y \mapsto y \wedge a$. This fact is often referred to as the *parallelogram law*; the proof is an easy exercise (see LT, p. 13, BA, p. 35). Two intervals related in this way are said to be *perspective*; more generally, two intervals are said to be *projective* if we can pass from one to the other by a series of perspectivities.

The parallelogram law shows that projective intervals are isomorphic as lattices. This may not tell us much, e.g. when the intervals are links, referring to simple modules, but for modules a stronger assertion can be made: by the Noether isomorphism theorem, perspective intervals give isomorphic module quotients: $(a + b)/b \cong a/(a \cap b)$, hence by induction, so do projective intervals.

Many basic theorems on modules are lattice-theoretic in nature, in the sense that the results can be stated and proved in terms of lattices. This is true of the next group of theorems, which are all used in the text (mainly in Chapter 3):

Theorem A.3 (Schreier refinement theorem) *In a modular lattice, two finite chains between the same end-points have refinements that are isomorphic, in the sense that their intervals can be paired off in such a way that corresponding intervals are projective* (LT, p. 66, BA, p. 37). ∎

This result shows in particular that any two maximal chains between given end-points in a modular lattice have the same length. A lattice of finite length necessarily has a greatest element, usually denoted by 1, and a least element, denoted by 0. The length of the interval $[0, a]$ is called the *height* of a. A modular lattice is of finite length whenever it satisfies the ascending and descending chain conditions, for then all its maximal chains are finite and of equal length. As in group theory, the Schreier refinement theorem has the following consequence:

Theorem A.4 (Jordan–Hölder theorem) *In a modular lattice of finite length, any chain can be refined to a maximal chain, and any two maximal chains are isomorphic* (LT, p. 166, BA, p. 36). ∎

(iii) A lattice L is said to be *distributive*, if

$$(x \vee y) \wedge z = (x \wedge z) \vee (y \wedge z) \quad \text{for all } x, y, z \in L. \tag{2}$$

This is easily seen to be equivalent to its dual (obtained by interchanging $\vee$ and $\wedge$) and to imply (1), so that every distributive lattice is modular. Like modularity, distributivity can be characterized by the non-existence of a certain type of sublattice:

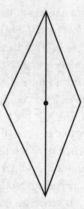

Figure A3

Proposition A.5 *A lattice L is distributive if and only if it is modular and does not contain a 5-element sublattice of length 2* (LT, p. 14, UA, p. 73 or BA, 3.1). ∎

There is precisely one 5-element modular lattice of length 2, up to isomorphism; it is shown in Fig. A3. Given an interval $[a, b]$ and $x \in [a, b]$, an element $x' \in [a, b]$ is called a *relative complement* of x in $[a, b]$ if $x \vee x' = b$, $x \wedge x' = a$. Proposition A.5 shows that a lattice is distributive if and only if relative complements, when they exist, are unique, while Proposition A.1 states that a lattice is modular if and only if relative complements of a given element are incomparable.

In a lattice L of finite length we have unique least and greatest elements 0 and 1, respectively; a relative complement in $[0, 1]$ is called a *complement*. Such complements may not exist in a distributive lattice, but when they do, they are unique, by Propositions A.1 and A.5. A distributive lattice with 0 and 1 in which every element has a complement is called a *Boolean algebra*; the latter can also be defined as an abstract algebra with two binary operations: $\vee$, $\wedge$ and a unary operation $x \mapsto x'$, satisfying the appropriate set of identities (see BA, Section 3.4).

(iv) Distributive lattices are much more special than modular ones, as is clear from Chapters 3 and 4. It is also apparent from the special form taken by the next two theorems in the distributive case. An element a in a lattice with least element 0 is said to be expressed as a *join of independent elements*: $a = a_1 \vee \ldots \vee a_n$, if $a_i \neq 0$ and

$$a_i \wedge (\bigvee_{j \neq i} a_j) = 0 \quad \text{for } i = 1, \ldots, n.$$

If a is not 0 and cannot be expressed as a join of two independent elements, it is said to be *indecomposable*. We can now state another result, well-known for modules, in terms of lattices:

Theorem A.6 (Krull–Schmidt theorem)   *In a modular lattice of finite length, let*

$$c = a_1 \vee \ldots \vee a_m \quad \text{and}$$
$$c = b_1 \vee \ldots \vee b_n$$

be two representations of c as a join of independent indecomposable elements. Then $m = n$ and for some permutation $i \mapsto i'$ of $1, 2, \ldots, n$, $[0, a_i]$ is projective with $[0, b_{i'}]$. In the case of a distributive lattice projectivity may be replaced by equality: $a_i = b_{i'}$ (LT, p. 168, UA, p. 73). ∎

A finite decomposition $a = a_1 \vee \ldots \vee a_n$ is called *irredundant*, if no a_i can be omitted. If $a \neq 0$ and no irredundant decomposition of a with more than one term exists, a is said to be *join-irreducible*. For such elements there is also a decomposition theorem:

Theorem A.7 (Kurosh–Ore theorem)   *In a modular lattice L, let*

$$c = p_1 \vee \ldots \vee p_r \quad \text{and} \quad c = q_1 \vee \ldots \vee q_s$$

be two irredundant decompositions of c into join-irreducible elements. Then $s = r$ and the p's may be exchanged against the q's, i.e. after suitable renumbering of the q's we have

$$c = q_1 \vee \ldots \vee q_i \vee p_{i+1} \vee \ldots \vee p_r \quad (i = 1, \ldots, r-1).$$

If moreover, L is distributive, then the p's and q's are equal except for their order (LT, p. 58, UA, p. 76f.). ∎

It is clear how an algebraic notion like 'homomorphism' is to be interpreted for lattices, namely as a join-and-meet-preserving mapping. An order-preserving mapping of lattices need not be a lattice-homomorphism, but an order-preserving bijection with an order-preserving inverse is a lattice-isomorphism, because the lattice structure can be defined in terms of the ordering.

A homomorphism of a modular lattice L that collapses an interval I (i.e. identifies all its points) will clearly collapse all intervals perspective with I and hence all intervals projective with I. Conversely, if we collapse all intervals projective with a given one, we obtain a homomorphic image of L. Thus each congruence on L (i.e. each collection of inverse image sets of a given homomorphic image) is a union of projectivity classes of intervals. As an illustration consider the 4-point chain 4 and its homomorphic image obtained by collapsing the top and bottom interval; the result is a 2-point chain 2 and we have a homomorphism $4 \to 2$. If we apply the same homomorphism to the square of 4, we obtain $4 \times 4 \to 2 \times 2$. Here each point of 2×2 has as congruence class the vertices of one of the corner squares in 4×4.

(v) There is a useful representation theorem for lattices due to G. Birkhoff, which is actually true for any variety of abstract algebras and it is no more difficult to state and prove it in that form (although this assumes that the reader has met varieties and congruences before, see UA, pp. 57, 162, LT, p. 26 or FA, Chapter 1). In what follows we shall be dealing with algebras defined by a family of operations and subject to certain identities, for example, modular lattices. Given such a family of algebras (A_i), their *direct product* is the Cartesian product ΠA_i, on which the operations can be defined componentwise. A subalgebra of ΠA_i that projects onto each factor A_i is called a *subdirect product* of the A_i. If an algebra C can be written as a subdirect product of the family (A_i) where none of the projections $C \to A_i$ is an isomorphism, C is said to be *subdirectly reducible*, otherwise C is *subdirectly irreducible*.

Theorem A.8 (Birkhoff's representation theorem) *Every algebra A (of a given variety) can be expressed as a subdirect product of a (possibly infinite) family of subdirectly irreducible algebras.*

Proof (sketch). Given any pair of elements $x, y \in A$, $x \neq y$, there is (by Zorn's lemma) a maximal congruence q not identifying x and y. The homomorphism $A \to A/\mathfrak{q}$ separates x and y, whereas every proper homomorphic image of $A/\mathfrak{q}$ identifies them, therefore $A/\mathfrak{q}$ is subdirectly irreducible. By combining these homomorphisms, for all pairs x, y, we obtain the required representation. ∎

For example, as is easily verified, a distributive lattice is subdirectly reducible, unless it consists of a single link 2. As a consequence every distributive lattice can be expressed as a subdirect power of 2 (see Theorem 3.4.4).

B. Categories and homological algebra

H: S. Mac Lane, Homology. Springer, Berlin-Göttingen-Heidelberg 1963.
C: S. Mac Lane, Categories for the working mathematician. Springer, New York-Heidelberg-Berlin 1971.
UA. P. M. Cohn, Universal Algebra 2nd Ed. D. Reidel Dordrecht 1981.

SF: P. M. Cohn, Skew Fields, Theory of General Division Rings, Cambridge University Press 1995.
BA: P. M. Cohn, Basic Algebra, Groups, Rings and Fields. Springer, London 2002.
FA: P. M. Cohn, Further Algebra and Applications. Springer, London 2003.

(i) We shall not give a formal definition of a category here; let us just recall that if one merely looks at morphisms or maps (by identifying objects with their identity maps) a *category* is a class (i.e. a 'big' set) with a set of maps between any pair of objects and a multiplication, not necessarily everywhere defined, but where defined it is associative, with left and right neutrals for multiplication. As in the main text we shall compose maps from left to right, thus fg will mean 'first f, then g'. In diagrams each map is represented by an arrow going from the *source* (or *domain*) to the *target* (or *codomain*). If the collection of all objects is a set, we have a *small* category; a small category in which there is at most one map between any two objects is essentially a partially ordered set. The anti-isomorph of a category $\mathcal{C}$, obtained by reversing all the arrows, is again a category, called the *opposite* of $\mathcal{C}$ and denoted by $\mathcal{C}^o$. Most of the categories in this book are categories of modules and their homomorphisms.

Let $\mathcal{C}$ be any category and write $\mathcal{C}(X, Y)$ for the set of all maps $X \to Y$ with source X and target Y. A *subcategory* $\mathcal{B}$ is defined (as in algebra) as a subclass of $\mathcal{C}$ closed under multiplication when defined, and containing with any maps its left and right neutrals. If for any objects X, Y in $\mathcal{B}$ we have $\mathcal{B}(X, Y) = \mathcal{C}(X, Y)$, $\mathcal{B}$ is said to be *full*; clearly a full subcategory is determined by its objects alone. A subcategory $\mathcal{B}$ of $\mathcal{C}$ is called *dense* if every object of $\mathcal{C}$ is a direct limit of $\mathcal{B}$-objects.

In any category a map is called an *isomorphism* if it has a two-sided inverse, and two objects are *isomorphic* if there is an isomorphism between them. An object X_0 in a category $\mathcal{C}$ is said to be *initial* if for each object Y there is precisely one map $X_0 \to Y$; clearly any two initial objects are isomorphic, by a unique isomorphism. Dually, an initial object in the opposite category $\mathcal{C}^o$ is called a *final* object for $\mathcal{C}$. An object that is both initial and final is called a *zero* object and is written 0.

(ii) In a diagram of maps there may be several ways of passing from a given object to another one; if all ways compose to the same map, the diagram is said to be *commutative*. In any category consider a diagram consisting of two maps f, g with the same target (Fig. B1). The different ways of completing this figure to a commutative square as in Fig. B2 form themselves a category in which the 'maps' are maps between the new objects added to get a 'commutative wedge'. A final object in this category is called the *pullback* of f and g. Thus the pullback

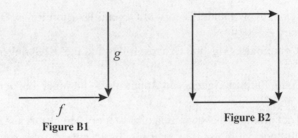

Figure B1   **Figure B2**

consists of a pair of maps f', g' with the same source, such that $f'g = g'f$, and 'universal' with this property; in other words, the pullback of f and g is just their least common left multiple (when it exists). *Pushouts* are defined dually, as the least common right multiple of two maps with the same source (see C, p. 65ff. BA, p. 88). An important example is provided by a pair of rings R, S with a common subring A; the two inclusion maps $A \to R$, $A \to S$ have a pushout, called the *coproduct* of R and S over A and written $R_A^* S$. The coproduct of fields is discussed in chapter 5 of SF.

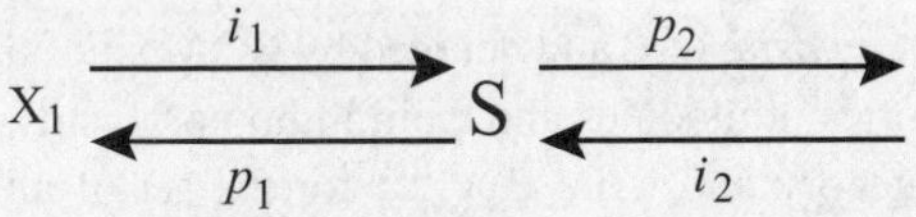

$$X_1 \xrightarrow[p_1]{i_1} S \xrightarrow[i_2]{p_2} X_2 \tag{1}$$

Figure B3

(iii) A category $\mathcal{A}$ is called *additive* if (a) $\mathcal{A}(X, Y)$ is an abelian group such that composition when defined is distributive: $\alpha(\beta + \beta') = \alpha\beta + \alpha\beta'$, $(\alpha + \alpha')\beta = \alpha\beta + \alpha'\beta$, (b) there is a zero object, and (c) to each pair of objects X_1, X_2 there corresponds an object S and maps as shown in Fig. B3, such that $i_1 p_1 = 1$, $i_2 p_2 = 1$, $p_1 i_1 + p_2 i_2 = 1$. The object S is called the *direct sum* of X_1 and X_2 with injections i_ν and projections p_ν. For example, Mod_R is an additive category in which the direct sum has its usual meaning. A category satisfying just (a) is called *preadditive*; a small preadditive category is called a *ringoid*; if there is a single object we have a ring.

Let $\mathcal{A}$ be an additive category; with each map $f : X \to Y$ we can associate a new category whose objects are maps α of $\mathcal{A}$ with target X and satisfying $\alpha f = 0$, and whose maps, between α and α', are maps λ from the source of α to that of α' such that $\alpha = \lambda \alpha'$. A final object μ in this category (when one exists) is called a *kernel map* for f and its source is a *kernel* for f, written $\ker f$; μ is always a *monomorphism* or *monic*, i.e. $\xi \mu = \xi' \mu$ implies $\xi = \xi'$. It follows that its source $\ker f$ is unique up to isomorphism. The dual notion is the *cokernel map* of f, whose target is written $\text{coker} f$. It is an *epimorphism* or *epic*, i.e. it can be cancelled whenever it appears as a left-hand factor.

Given $f : X \to Y$, we define its *image* as $\operatorname{im} f = \ker \operatorname{coker} f$, its *coimage* as $\operatorname{coim} f = \operatorname{coker} \ker f$; of course they may not exist, but when they do, we have the following picture:

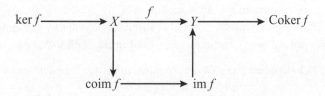

Here the map $\operatorname{coim} f \to \operatorname{im} f$ follows from the definition of im and coim. If the additive category $\mathcal{A}$ is such that every map has a kernel and a cokernel and the natural map $\operatorname{coim} f \to \operatorname{im} f$ is an isomorphism, then $\mathcal{A}$ is said to be *abelian*. We remark that the notion of an abelian category is again self-dual, and also note that in an abelian category a map that is both epic and monic is necessarily an isomorphism. The category of all right modules over a ring R is an abelian category, as is easily seen; of course not every abelian category is isomorphic to one of the form Mod_R; for example, if R is any non-zero ring, then the opposite of Mod_R is not isomorphic to a category of the form Mod_S, for any ring S (FA, p. 47). The following result is an easy consequence of the definitions:

Proposition B.1 *Let $\mathcal{A}$ be an abelian category and $\mathcal{B}$ a full subcategory. Then $\mathcal{B}$ is abelian provided that the direct sums, kernels and cokernels, taken in $\mathcal{A}$, of maps in $\mathcal{B}$ again lie in $\mathcal{B}$.* ∎

We remark that in most 'concrete' categories (where the objects have an underlying set structure), such as rings and modules, monomorphisms are injective; this simplifies the terminology. However, an epimorphism need *not* be surjective, thus in the category of rings the embedding $\mathbb{Z} \to \mathbb{Q}$ is an epimorphism, though in module categories it is true that all epimorphisms are surjective.

(iv) A sequence of maps in an abelian category

$$\cdots \longrightarrow X_{i-1} \xrightarrow{f_{i-1}} X_i \xrightarrow{f_i} X_{i+1} \longrightarrow \cdots$$

is *exact* at X_i if $\ker f_i = \operatorname{im} f_{i-1}$; if it is exact at each object, it is called an *exact sequence*. Verifying the exactness of sequences in diagrams is often called 'diagram chasing'. An exact sequence beginning and ending in 0 cannot have just one non-zero term, and if it has two, the map connecting them must be an isomorphism. Thus the first non-trivial case is that of a three-term exact

sequence:
$$0 \to A' \to A \to A'' \to 0, \tag{2}$$

also called a *short* exact sequence. For example, in the case of modules, (2) represents an extension: A is an extension of A' by A''.

The following two results are easily proved for modules by a diagram chase (and this is all we need), but in fact they hold in any abelian categories:

Lemma B.2 (five lemma)   *Given a commutative diagram with exact rows*

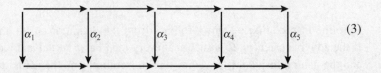
(3)

if α_1 is epic and α_2, α_4 are monic, then α_3 is monic. Dually, if α_5 is monic and α_2, α_4 are epic, then α_3 is epic. In particular, if α_1, α_2, α_4, α_5 are isomorphisms, then so is α_3 (H, p. 14, C, p. 201, FA, p. 71). ∎

Lemma B.3 (three-by-three lemma)   *Given a commutative diagram with exact rows and columns as shown, there is a unique way of filling in the first column so as to keep the diagram commutative and then the new column is exact, too* (H, p. 49, FA, p. 41). ∎

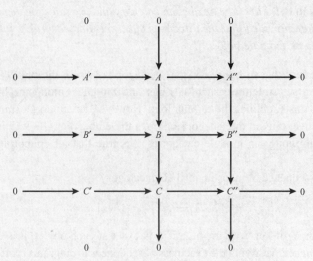

(v) With every type of algebraic system there is associated the notion of a homomorphism, i.e. a structure-preserving mapping. In the case of categories one speaks of functors. Thus a *functor* $F : \mathcal{A} \to \mathcal{B}$ is a mapping from one category, $\mathcal{A}$, to another, $\mathcal{B}$, which preserves neutrals and satisfies

$$F(\alpha\beta) = F\alpha.F\beta. \tag{4}$$

If instead of (4) we have

$$F(\alpha\beta) = F\beta.F\alpha, \tag{5}$$

F is said to be a *contravariant functor*, in contrast to the sort defined by (4), which is called *covariant*. Thus a contravariant functor from $\mathcal{A}$ to $\mathcal{B}$ may also be regarded as a covariant functor from $\mathcal{A}^\circ$ to $\mathcal{B}$, or from $\mathcal{A}$ to $\mathcal{B}^\circ$. If F and G are two covariant functors from $\mathcal{A}$ to $\mathcal{B}$, a *natural transformation* $t : F \to G$ is a function that assigns to each $\mathcal{A}$-object X a $\mathcal{B}$-map $t_X : FX \to GX$ such that for each map $X \to Y$ the square

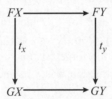

commutes. If each t_X is an isomorphism, t is called a *natural isomorphism*, or in the case of contravariant functors, a *natural anti-isomorphism* or also a *duality*; if $\mathcal{B}$ is a concrete category and t_X is an inclusion map, F is called a *subfunctor* of G.

Given categories $\mathcal{A}$ and $\mathcal{B}$, if there is a pair of functors $F : \mathcal{A} \to \mathcal{B}, G : \mathcal{B} \to \mathcal{A}$ such that the compositions (from left to right) $FG : \mathcal{A} \to \mathcal{A}$ and $GF : \mathcal{B} \to \mathcal{B}$ are naturally isomorphic to the identity functor, $FG \cong 1$, $GF \cong 1$, then $\mathcal{A}$ and $\mathcal{B}$ are said to be *equivalent* and F, G are *equivalences* between them. In the case of contravariant functors we speak of *anti-equivalence* or *duality*, and $\mathcal{A}, \mathcal{B}$ are said to be *dual*.

Let A, B be any rings. If the module categories Mod_A and Mod_B (or equivalently, $_A\text{Mod}$ and $_B\text{Mod}$) are equivalent, then A and B are said to be *Morita-equivalent*. For example, any ring R is Morita-equivalent to the matrix ring $\mathfrak{M}_n(R)$, for any $n \geq 1$, and these are the only such rings when R is projective-free. It can be shown that A and B are Morita-equivalent if and only if there exist bimodules $_A P_B, _B Q_A$ such that $P \otimes_B Q \cong A$, $Q \otimes_A P \cong B$ as bimodules (see FA, p. 151).

(vi) Let $\mathcal{A}$ be a class of algebras and X a set. Then we can form the *comma category* $(X, \mathcal{A})$ whose objects are maps $\alpha : X \to G$ from X to an $\mathcal{A}$-object G

and whose morphisms from $\alpha : X \to G$ to $\beta : X \to H$ are $\mathcal{A}$-homomorphisms $f : G \to H$ such that $\beta = \alpha f$. An initial object in $(X, \mathcal{A})$, if it exists, is called a *universal $\mathcal{A}$-object* for X; as initial object it is unique up to isomorphism. For example, the free group on a set X is a universal object in this sense. We remark that if with each group G we associate its underlying set $\sigma(G)$ by means of the 'forgetful' functor from groups to sets, then we have a natural bijection of morphism sets:

$$\mathrm{Gp}(F_X, G) \cong \mathrm{Ens}(X, \sigma(G)), \tag{6}$$

where Gp, Ens (= Ensemble, French for 'set') denote the categories of groups and sets, respectively. More generally, a pair of functors $T : \mathcal{A} \to \mathcal{B}, S : \mathcal{B} \to \mathcal{A}$ is called an *adjoint pair*, and S is *left adjoint*, T *right adjoint*, if for any $\mathcal{A}$-object X and $\mathcal{B}$-object V,

$$\mathcal{A}(V^S, X) \cong \mathcal{B}(V, X^T). \tag{7}$$

It is not hard to show that each of S, T determines the other up to natural isomorphism by (7) (H, p. 266, FA, p. 45). For example the free group functor is the left adjoint of the forgetful functor σ, by (6). More generally, almost every universal construction arises as the left adjoint of a forgetful functor.

The following lemma is often useful in discussing universal constructions; although we state it for rings, it clearly holds quite generally for all types of algebras:

Lemma B.4 *Let A, B be any rings and f, g two homomorphisms from A to B. If f and g agree on a generating set of A, then $f = g$.*

Proof. The set $\{x \in A \mid xf = xg\}$ is easily seen to be a subring of A; since it contains a generating set it must be the whole of A. ∎

(vii) We shall mainly be concerned with additive (in particular abelian) categories; in that case all functors are assumed to be *additive*, i.e. $F(\alpha + \beta) = F\alpha + F\beta$. For example, $\mathrm{Hom}_R(M, -)$ and $M \otimes -$ are additive functors. We note that an additive functor preserves direct sums when they exist; this is easily verified from the definition. In what follows, all functors are tacitly assumed to be additive.

Any functor F transforms an exact sequence

$$\xrightarrow{\alpha} A \xrightarrow{\beta} \tag{8}$$

into a sequence

$$\xrightarrow{F\alpha} FA \xrightarrow{F\beta} \tag{9}$$

with composition zero: $F\alpha.F\beta = 0$, but (9) need not be exact. However, if (8) is split exact, i.e. if im $\alpha = \ker \beta$ is a direct summand of A, then the same is true of (9). The functor F is said to be *exact* if it transforms any exact sequence into an exact sequence.

Exact functors are rare; thus $\operatorname{Hom}_R(P, -)$ is exact if and only if P is projective. This may be taken as the definition of a projective module, and it can then be proved that P is projective if and only if it is a direct summand of a free module. Similarly, $\operatorname{Hom}_R(-, I)$ is exact if and only if I is injective. There is no simple description of injective modules, as there is for projectives, but in testing for injective modules, Baer's criterion is often useful. This states that a right R-module I is injective if and only if every homomorphism $\mathfrak{a} \to I$ where $\mathfrak{a}$ is a right ideal, can be extended to a homomorphism $R \to I$ (FA, p. 53).

Every module can be embedded in an injective module, and the least injective containing a given module M is unique up to isomorphism and is called the *injective hull* of M; further, M can be shown to be an essential submodule of its injective hull (i.e. it meets every non-zero submodule); the injective hull of M is also called an *essential extension* of M. There is an alternative definition of injective module and injective hull as a sort of algebraic closure, but this will not be needed here (see UA, p. 261, FA, p. 166).

(viii) Although few functors are exact, many have a partial exactness property, which we now describe.

A covariant functor F is called *left exact* if the sequence

$$0 \longrightarrow FA' \longrightarrow FA \longrightarrow FA'' \longrightarrow 0 \qquad (10)$$

obtained by applying F to the short exact sequence (2), is exact except possibly at FA''; if (10) is exact except possibly at FA', then F is called *right exact*. For a contravariant functor $D : \mathcal{A} \to \mathcal{B}$ these terms are defined by applying the definitions just given to the associated covariant functor from $\mathcal{A}°$ to $\mathcal{B}$. A routine verification shows that $\operatorname{Hom}_R(-, -)$ is left exact in each of its arguments.

With each functor F we can associate a series of derived functors $F^i (i \in \mathbb{Z})$ that measure the departure of F from exactness. Assume further that every object is quotient of a projective and subobject of an injective (the category 'has enough projectives and injectives'). When F is exact, $F^i = 0$ for $i \neq 0$ and F^0 is naturally isomorphic to F, in symbols $F^0 \cong F$. For a left exact functor F, $F^i = 0$ for $i < 0$ and (a) $F^0 \cong F$, (b) $F^n I = 0$ for $n > 0$ and I injective, and (c) to each short exact sequence (2) there corresponds a long exact sequence

$$0 \to F^0 A' \to F^0 A \to F^0 A'' \xrightarrow{\Delta} F^1 A' \to F^1 A \to \ldots \qquad (11)$$

with a 'connecting homomorphism' Δ that is a natural transformation. Moreover, F^i is uniquely determined up to isomorphism by (a)–(c).

As an example let us take $FA = \text{Hom}_R(M, A)$. The derived functor in this case is denoted by Ext. Thus $\text{Ext}_R^0(M, N) \cong \text{Hom}_R(M, N)$, $\text{Ext}_R^i(M, I) = 0$ for $i > 0$ and I injective, and we have an exact sequence

$$0 \to \text{Hom}_R(M, A') \to \text{Hom}_R(M, A) \to \text{Hom}_R(M, A'') \to \text{Ext}_R^1(M, A')$$
$$\to \text{Ext}_R^1(M, A) \to \text{Ext}_R^1(M, A'') \to \text{Ext}_R^2(M, A') \to \ldots \quad (12)$$

Next let us take $\text{Hom}_R(M, N)$, regarded as a functor in M. This is contravariant and we have to replace injectives in (b) above by projectives. The derived functor is the same as before, namely Ext. Thus we have $\text{Ext}_R^i(P, N) = 0$ for $i > 0$ and P projective, and from any short exact sequence (2) we obtain

$$0 \to \text{Hom}_R(A'', N) \to \text{Hom}_R(A, N) \to \text{Hom}_R(A', N) \to \text{Ext}_R^1(A'', N)$$
$$\to \text{Ext}_R^1(A, N) \to \text{Ext}_R^1(A', N) \to \text{Ext}_R^2(A'', N) \to \ldots \quad (13)$$

Whenever R is a field or, more generally, when R is a semi-simple Artinian ring, then $\text{Ext}^i = 0$ for all $i \neq 0$, because for such rings all exact sequences split, so Hom is then exact in each argument. Most of the rings considered in this book are hereditary, which amounts to saying that Ext^i vanishes identically for $i > 1$. Therefore (12) and (13) reduce to 6-term sequences in this case. For any module M, the *projective dimension*, also called *homological dimension*, is defined as the least $n \geq 0$ such that $\text{Ext}_R^{n+1}(M, -) = 0$, or ∞ if no such n exists; the *injective* or *cohomological dimension* is the least $n \geq 0$ such that $\text{Ext}_R^{n+1}(-, M) = 0$ or ∞ if no such n exists. They are denoted by pd.M, id.M, respectively. Now the *right global dimension* of R is defined as r.gl.dim.$R = \sup(\text{pd}.M)$, where M ranges over all right R-modules, and the left global dimension is defined similarly in terms of left R-modules. In a Noetherian ring these dimensions are equal, but in general they may be different (see Section 1.6).

Let us briefly consider the connexion of Ext with module extensions. In (12) put $M = A''$ and consider the image of $1 \in \text{Hom}_R(A'', A'')$ under Δ; this is an element θ of $\text{Ext}_R^1(A'', A')$. It can be shown (H, Chapter 3, FA, p. 71) that two short exact sequences (2) and

$$0 \to A' \to B \to A'' \to 0 \quad (14)$$

give rise to the same θ if and only if the extensions A and B are isomorphic in the sense that there is an isomorphism $f : A \to B$ making the diagram

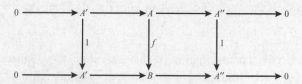

commutative. Moreover, every element of $\operatorname{Ext}_R^1(A'', A')$ gives rise to an extension in this way; so the isomorphism classes of extensions of A' by A'' may be classified by $\operatorname{Ext}_R^1(A'', A')$. We remark that the element θ can also be obtained as the image of 1 under Δ by putting $N = A'$ in (13). This element θ is sometimes called the *obstruction*; it is zero if and only if the extension (2) splits.

(ix) Since every free module is projective, every module can be written as a homomorphic image of a projective module, in other words, for any ring R the category of R-modules has 'enough projectives'. Given a module M, we can write it as a homomorphic image of a projective module P_0, say $f_0 : P_0 \to M$. Similarly we can find a projective module P_1 mapping onto ker f_0 and by continuing in this way we get an exact sequence of projective modules P_i:

$$\ldots \to P_1 \to P_0 \to M \to 0, \tag{15}$$

called a *projective resolution* of M. Dually, M has an *injective resolution*

$$0 \to M \to I_0 \to I_1 \to \ldots, \tag{16}$$

where the I_i are injective. If in (15) we replace P_1 by the kernel of the map $P_1 \to P_0$, we obtain a short exact sequence

$$0 \to Q \to P \to M \to 0, \quad P \text{ projective.} \tag{17}$$

This is called a *presentation* of M. Clearly if P is finitely generated, then so is M; if there is a presentation (17) of M with P, Q both finitely generated, (17) is called a *finite presentation* of M and M is said to be *finitely presented*; if there is a presentation of M with Q finitely generated (but not necessarily P), M is said to be *finitely related*. Different presentations of M are compared in

Lemma B.5 (extended Schanuel lemma) *Given two presentations of a module M, one of them by a projective module, say (17) and*

$$0 \to Q' \to P' \to M \to 0, \tag{18}$$

where P is projective (but not necessarily P'), then there is an exact sequence

$$0 \to Q \to P \oplus Q' \to P' \to 0.$$

In particular, if P' is also projective, then

$$P \oplus Q' \cong P' \oplus Q. \qquad \blacksquare$$

(See FA, p. 58 and Theorem 0.5.3 above; the last assertion is the actual form of Schanuel's lemma.)

This shows in particular that if M is finitely presented then in any presentation (18) of M in which P' is finitely generated, the same is true of Q'. For a finitely presented R-module M we also note that any submodule M' of M is finitely generated if and only if M/M' is finitely presented. For we may take $M = F/K$, where F is free of finite rank and K is a finitely generated submodule, and take M' to be of the form G/K, where $K \subseteq G \subseteq F$. Since K is finitely generated, M' is finitely generated if and only if G is finitely generated, and this is the condition for $M/M' \cong F/G$ to be finitely presented.

(x) The tensor product may be defined as the left adjoint of the Hom-functor. Thus if modules $U_{R,R} V_S$ are given, their tensor product $U \otimes_R V$ is determined up to isomorphism as a right S-module by the equation

$$\mathrm{Hom}_R(U, \mathrm{Hom}_S(V, W)) \cong \mathrm{Hom}_S(U \otimes_R V, W), \quad (19)$$

where W is a right S-module; (19) is also referred to as *adjoint associativity*. From the left exactness of Hom and (19) it is easily proved that the tensor product is right exact in each of its arguments (FA, p. 45).

Let $f : R \to S$ be a ring homomorphism. With each right R-module M we can associate an S-module, the *coinduced extension* $M^f = \mathrm{Hom}_R(S, M)$ (see FA, p. 54).

Lemma B.6 *If I is an injective right R-module, then the coinduced extension I^f is an injective right S-module.*

Proof. If N is any S-module, we have, by (19),

$$\mathrm{Hom}_S(N, I^f) \cong \mathrm{Hom}_R(N \otimes_S S, I) = \mathrm{Hom}_R(N, I),$$

hence $\mathrm{Hom}_S(-, I^f)$ is an exact functor, and the assertion follows. ∎

Suppose that U is a free right R-module with basis (u_λ); in particular, such a basis always exists if R is a field. Then for any left R-module V, each element of $U \otimes V$ can be written in the form

$$f = \sum u_\lambda \otimes a_\lambda \quad (a_\lambda \in V),$$

and moreover, the coefficients a_λ are uniquely determined by f. In general no such convenient normal form exists, but one has the following independence property of the tensor product (see Cohn [59], BA, Proposition 4.8.9; for a special case see also Bourbaki [61], p. 41):

Lemma B.7 *Let R be any ring and $U_{R,R} V$ modules over R, where V is generated by a family $(e_\beta, \beta \in B)$ with defining relations $\sum a_{\alpha\beta} e_\beta = 0$ ($\alpha \in A$). Given any family (x_β) of elements in U indexed by B and almost all 0, if*

$$\sum x_\beta \otimes e_\beta = 0 \text{ in } U \otimes V,$$

then there exist elements $y_\alpha \in U$, almost all 0, such that

$$x_\beta = \sum y_\alpha a_{\alpha\beta}. \tag{20}$$

Proof. By hypothesis V has a presentation

$$0 \to G \xrightarrow{\lambda} F \xrightarrow{\mu} V \to 0,$$

where F is free on a family (f_β) and G is the submodule of F spanned by the elements $\sum a_{\alpha\beta} f_\beta$. Tensoring with U and observing that this operation is right exact, we obtain the exact sequence

$$U \otimes G \xrightarrow{\lambda'} U \otimes F \xrightarrow{\mu'} U \otimes V \to 0.$$

By hypothesis, $(\sum x_\beta \otimes f_\beta)\mu' = 0$, hence exactness shows that

$$\sum x_\beta \otimes f_\beta = \left(\sum y_\alpha \otimes a_{\alpha\beta} f_\beta\right)\lambda',$$

for some elements $y_\alpha \in U$, almost all 0. Now λ' is the homomorphism induced by the inclusion $G \to F$, and F is free on the f's, so on equating coefficients we obtain (20). ∎

A right module U is said to be *flat* if $U \otimes -$ is an exact functor. E.g. any free module, and more generally, any projective module is flat. We note an easy consequence of the definition:

Proposition B.8 *Let R be any ring. If U is a flat right R-module and $\mathfrak{a}$ is a left ideal of R, then the natural mapping*

$$\nu : U \otimes \mathfrak{a} \to U\mathfrak{a} \tag{21}$$

is an isomorphism.

Proof. Only the injectivity requires proof and this follows by tensoring the exact sequence

$$0 \to \mathfrak{a} \to R \to R/\mathfrak{a} \to 0$$

with U and using the fact that the latter is flat. ∎

The following criterion for flatness is easily proved. It suggests that any 'torsion' in a flat module is due to relations in the ring, which accounts for the name.

Proposition B.9 *Let R be any ring. A right R-module U is flat if and only if, whenever*

$$uc = 0, \quad \text{where } u \in U^n, c \in {}^nR,$$

there exist $B \in {}^m R^n$ and $x \in U^m$ such that

$$u = xB, \quad Bc = 0.$$

(For a proof see FA, p. 161). ∎

Proposition B.9 shows that a module U is flat if and only if every finitely generated submodule of U is contained in a flat submodule; in other words, flatness is a local property. The definition shows that U is flat if and only if $\text{Tor}_1(U, -) = 0$, where Tor is the left derived of the tensor product. Here it is enough to show that $\text{Tor}_1(U, -)$ vanishes on all finitely generated modules, by the remark just made.

The *weak dimension* of a module M is defined as

$$\text{wd}.M = \sup\{n | \text{Tor}_n^R(-, M) \neq 0\} \,;$$

thus if $\text{wd}.M = r$ and we take a projective resolution (15) of M, then $\ker(P_r \to P_{r-1})$ is flat. Now the *weak global dimension* of a ring R is defined as

$$\text{w.gl.dim}.R = \sup\{n | \text{Tor}_n^R \neq 0\}.$$

Clearly this is left–right symmetric and is majorized by the left and the right global dimensions of R, while for Noetherian rings all three global dimensions coincide (see FA, p. 77).

(xi) A ring R is said to be *right coherent* if every finitely generated right ideal is finitely related; *left coherent* is defined similarly. It follows easily that when R is right coherent, then every finitely generated submodule of a free right R-module is finitely related. Some equivalent formulations of coherence are given in the next result:

Theorem B.10 *For any ring R the following assertions are equivalent:*

(a) R is right coherent,
(b) the right annihilator of any row vector over R is finitely generated, i.e. given $u \in R^n$, if $uB = 0$ for some $B \in {}^n R^l$, then there exist $C \in {}^n R^r$, $D \in {}^r R^l$ such that $B = CD$, $uC = 0$,
(c) the right annihilator of any matrix over R is finitely generated, i.e. given $A \in {}^m R^n$, if $AB = 0$ for some $B \in {}^n R^l$, then there exist $C \in {}^n R^r$, $D \in {}^r R^l$ such that $B = CD$, $AC = 0$,
(d) the dual of every finitely presented left R-module is finitely presented,
(e) the direct product of any family of flat left R-modules is flat,
(f) every direct power R^I is left flat,
(g) every finitely generated submodule of a free right R-module is finitely related.

Proof. The proof follows the pattern $a \Rightarrow b, c \Rightarrow a \wedge c \Leftrightarrow d, b \Rightarrow e \Rightarrow f \Rightarrow c \wedge f \Rightarrow g \Rightarrow a$. (a) $\Rightarrow$ (b). By definition R is right coherent precisely if every finitely generated right ideal is finitely related. Consider $u = (u_1, \ldots, u_n) \in R^n$ and let the family of all relations be $\sum u_i b_{i\lambda} = 0$. By (a) these relations can be generated by a finite subset; writing the $b_{i\lambda}$ as a matrix B with n rows and possibly infinitely many columns, we see that there exists $C \in {}^nR^r$ and a matrix D with r rows such that $B = CD$ and $uC = 0$.

(a) $\wedge$ (c) $\Leftrightarrow$ (d). Let V be finitely presented, say

$$R^m \xrightarrow{\alpha} R^n \longrightarrow V \to 0,$$

where α is given by $A \in {}^mR^n$. Dualizing, we obtain the exact sequence

$$0 \to V^* \longrightarrow {}^nR \longrightarrow {}^m R,$$

and V^* is the annihilator of A in ${}^nR$. Now (c) is just the condition for V^* to be finitely generated, but then it is finitely presented by (a). Conversely, if V^* is finitely generated, (a) and so also (c) holds.

(b) $\Rightarrow$ (e). Let (V_α) be a family of flat left R-modules and consider $V = \Pi V_\alpha$. If $av = 0$, where $a \in R^n$, $v \in {}^nR$, let $v = (v_\alpha)$, $v_\alpha \in {}^nV_\alpha$; then $av_\alpha = 0$. Since V_α is flat, there exists $B_\alpha \in {}^nR^{m_\alpha}$, $v'_\alpha \in {}^{m_\alpha}V_\alpha$ such that $v_\alpha = B_\alpha v'_\alpha$, $aB_\alpha = 0$. By (b), $B_\alpha = CD_\alpha$ for some C, D_α such that $aC = 0$. Now $v = (CD_\alpha v'_\alpha)$ and this shows V to be flat.

(e) $\Rightarrow$ (f) is clear. To prove (f) $\Rightarrow$ (c), suppose that $AB = 0$, where $A \in {}^mR^n$ and B has possibly infinitely many columns. Since R^I is left flat, there exists $C \in {}^nR^r$ such that $B = CD$ for some $D \in {}^rR^I$ and $AC = 0$. The same argument shows that (c) $\Rightarrow$ (f), so (c) $\Rightarrow$ (c) $\wedge$ (f).

(c) $\wedge$ (f) $\Rightarrow$ (g). Let U be a finitely generated submodule of a free right R-module F. By omitting basis elements of F not involved in the generators of U, we can take F to be finitely generated. If U is generated by the columns of the $m \times n$ matrix A and B is the matrix of all relations between the columns of A, then by (f) there exist $C \in {}^nR^r$ and D with r rows such that $B = CD$, $AC = 0$. Hence U is finitely related, by the columns of C. Finally (g) $\Rightarrow$ (a) is clear. ∎

Corollary B.11 *Over a right coherent ring the intersection of any two finitely generated submodules of a free right R-module is finitely generated.*

Proof. If U, V are submodules of a free module F, generated by $u_1, \ldots, u_m$ and $v_1, \ldots, v_n$, respectively, then the elements of $U \cap V$ are all of the form

$$\sum u_i a_i = \Sigma v_j b_j, \quad \text{where } a_i, b_j \in R,$$

and they thus correspond to the relations $\sum u_i a_i - \sum v_j b_j = 0$. But the module of all these relations is finitely generated, by (c), hence $U \cap V$ is indeed finitely generated. ∎

We also show that over a coherent ring the finitely presented modules form an abelian category.

Theorem B.12 *If R is a right coherent ring, then the category of all finitely presented right R-modules is closed under taking finitely generated submodules, finite direct sums, kernels and cokernels, hence it is an abelian subcategory of the category of all right R-modules.*

Proof. If M is a finitely presented right R-module over a right coherent ring R, then any finitely generated submodule M' of M is finitely presented. For if $M = F/K, M' = G/K$, where F is free, then G, being a finitely generated submodule of F, is finitely presented by Theorem B.10 (g), and now the remark after Lemma B.5, applied to G and $K \subseteq G$ shows $M' = G/K$ to be finitely presented, which proves the first claim. To prove the rest, let $f : M \to N$ be an R-linear map between finitely presented right R-modules; then im f is a finitely generated submodule of N and so is finitely presented. If we apply the remark after Lemma B.5 to im $f \cong M/\ker f$ and coker $f \cong N/\text{im} f$, we see that ker f is a finitely generated submodule of M, hence finitely presented, and coker f is likewise finitely presented. The rest is clear. ∎

C. Ultrafilters and the ultraproduct theorem

The properties of a mathematical structure are described by statements in which the variables refer to elements of the structure and are usually *quantified*, i.e. they are limited by a *universal quantifier* ∀ (for all) or an *existential quantifier* ∃ (there exists). Such statements are called *elementary sentences*. Any set of elementary sentences thus defines a class of structures; this is called a *universal class* if all sentences have only universal quantifiers. In the text we only need one result from logic, the ultraproduct theorem, and only a special case of it. We briefly explain the background.

Let I be a set. A *filter* $\mathcal{C}$ on I is defined as a collection of subsets of I not including $\varnothing$, closed under finite intersections, and such that any subset of I containing a member of $\mathcal{C}$ itself belongs to $\mathcal{C}$. With the help of Zorn's lemma it is easy to prove that every filter is contained in a maximal filter, also called *ultrafilter* and characterized by the fact that for every subset of I it contains either the subset or its complement (but not both). Let $\{R_i | i \in I\}$ be a family of rings and $P = \Pi R_i$ their direct product, with canonical projections $\pi_i : P \to R_i$. For any ultrafilter $\mathcal{F}$ on I, the reduced product $\Pi R_i / \mathcal{F}$, also called *ultraproduct*, is defined as the quotient of ΠR_i given by the rule

$$a\mathcal{F} = 0 \text{ if and only if } a\pi_i = 0 \text{ for all } i \text{ in some member of } \mathcal{F}.$$

C. Ultrafilters and the ultraproduct theorem

The special case where all the factors are equal is called an *ultrapower*. The ultraproduct theorem of logic (Los's theorem) states that in an ultraproduct $\Pi A_i / \mathcal{F}$ (of algebras, say) any elementary sentence holds if and only if it holds in all the factors corresponding to a member of $\mathcal{F}$. We shall only need the following special case:

Theorem C.1 *Any ultraproduct of fields is again a field.*

Proof. Given a family of fields $K_i (i \in I)$, let $D = \Pi K_i / \mathcal{F}$ be an ultraproduct, formed with an ultrafilter $\mathcal{F}$ on I. Clearly D, as a quotient of the product ΠK_i is a ring; we have to show that it is a field. Let $a = (a_i)$ be a non-zero element of D. Then $J = \{i \mid a_i \neq 0\} \in \mathcal{F}$; we define

$$a' = \begin{cases} a_i^{-1} & \text{if } i \in J, \\ 0 & \text{otherwise.} \end{cases}$$

Then $aa' = a'a = 1$, because the components are 1 for all $i \in J$. Thus a has an inverse and the result follows. ∎

Bibliography and author index

Apart from listing works referred to in the text, the bibliography includes papers related to the main topics, although there has been no attempt at complete coverage of such general topics as principal ideal domains. References to some of the author's books are by two-letter abbreviations, which are listed in Notes to the reader.

Items are referred to by author and the year of publication, except that 19 is omitted for items published between 1920 and 1999. The page references at the end of the entries indicate places in the text where the entry is quoted (or implicitly referred to); other references to an author are listed after the name.

Akgül, M.
    83. Derivation in the free associative algebra. *Hacettepe Bull. Nat. Sci. Eng.* **12** (1983), 245–256.

Albrecht, F.
    61. On projective modules over semihereditary rings. *Proc. Amer. Math. Soc.* **12** (1961), 638–639.   183

Almkvist, G., Dicks, W. and Formanek, E.
    85. Hilbert series of fixed free algebras and noncommutative classical invariant theory. *J. Algebra* **93** (1985), 189–214.   386

Amitsur, S. A.   223
    48. On unique factorization in rings (Hebrew). *Riveon Lematematika* **2** (1948), 28–29.
    58. Commutative linear differential operators. *Pacif. J. Math.* **8** (1958), 1–10.   58
    63. Remarks on principal ideal rings. *Osaka J. Math.* **15** (1963), 59–69.   106
    65. Generalized polynomial identities and pivotal monomials. *Trans. Amer. Math. Soc.* **114** (1965), 210–226.   477
    66. Rational identities and applications to algebra and geometry. *J. Algebra* **3** (1966), 304–359.   517

Anderson, D. D. and Camillo, V. 99. Semigroups and rings whose 0-products commute. *Comm. Algebra* **27** (1999), 2847–2852.

Andreeva, T. A. and Kharchenko, V. K. 96. On rings of quotients of free algebras (Russian). *Algebra i Logika* **35** (1996), 655–662 [Engl. transl. in *Algebra and Logic* **35** (1996), 366–370].

Andrunakievich, V. A. and Ryabukhin, Yu. M. 79. *Radicals of Algebras and Structural Theory* (Russian), Nauka, Moscow 1979.   141

Anick, D. J. 398
Antoine, R. 98. On rigid monoids with right ACC_1. *Comm. Algebra* **26** (2)(1998), 507–513.
Antoine, R. and Cedó, F. 2003 Inverting elements in rigid monoids. *Comm. Algebra* **31** (2003), 4179–4194.
Artamonov, V. A. and Cohn, P. M. 99. The skew field of rational functions on the quantum plane. *J. Math. Sci.* **93** (6) (1999), 824–829.
Asano, K. 38. *Nichtkommutative Hauptidealringe.* Act. Sci. Ind. No. 696. Hermann, Paris. 1938. 105
Auslander, M. 329
Barbilian, D. 56. *Teoria Aritmetica a Idealilor (in inele necomutative).* Ed. Acad. Rep. Pop. Romine, Bucuresti 1956. 223
Bass, H. 261f.
   64. Projective modules over free groups are free. *J. Algebra* **1** (1964), 367–373. 183
   68. *Algebraic K-theory.* Benjamin, New York 1968. 58, 498
Bass, H., Connell, E. H. and Wright, D. 82. The Jacobian conjecture: reduction of degree and formal expansion of the inverse. *Bull. Amer. Math. Soc.* **7** (1982), 287–330. 405
Baumslag, B. and Baumslag, G. 71. On ascending chain conditions. *Proc. London Math. Soc.* (3) **22** (1971), 681–704. 183
Baumslag, G. (see Baumslag, B.) 408
Beauregard, R. A. 69. Infinite primes and unique factorization in a principal right ideal domain. *Trans. Amer. Math. Soc.* **141** (1969), 245–253. 223
   73. Overrings of Bezout domains. *Canad. Math. Bull.* **16** (1973), 475–477.
   74. Right bounded factors in an LCM-domain. *Trans. Amer. Math. Soc.* **200** (1974), 251–266. 350
   77. Left and right invariance in an integral domain. *Proc. Amer. Math. Soc.* **67** (1977), 201–205.
   80. Left versus right LCM domains. *Proc. Amer. Math. Soc.* **78** (1980), 464–466. 242
   88. Left Ore principal right ideal domains. *Proc. Amer. Math. Soc.* **102** (1988), 459–462.
Beauregard, R. A. and Johnson, R. E. 70. Primary factorization in a weak Bezout domain. *Proc. Amer. Math. Soc.* **25** (1970), 662–665. 223
Beck, I. 72. Projective and free modules. *Math. Zeits.* **129** (1972), 231–234. 19
Bedoya, H. and Lewin, J. 77. Ranks of matrices over Ore domains. *Proc. Amer. Math. Soc.* **62** (1977), 233–236.
Behrens, E.-A. 65. *Algebren.* Bibliographisches Institut, Mannheim 1965. 226
Bergman, G. M.   12, 37, 57f., 65, 95, 97, 124, 153, 156, 170f., 191, 207, 231, 261f., 268, 303f., 320, 329f., 339, 350f., 374, 379, 386, 396, 408f., 427f., 473, 482, 490f., 510, 517f.
   67. Commuting elements in free algebras and related topics in ring theory. Thesis, Harvard University 1967.   51, 58f., 152, 183f., 223, 237, 260, 268, 281, 289, 325, 329, 378f., 407, 516
   69. Centralizers in free associative algebras. *Trans. Amer. Math. Soc.* **137** (1969), 327–344. 407
   70a. Skew fields of noncommutative rational functions, after Amitsur. Séminaire Schützenberger-Lentin-Nivat, Année 1969/70, No. 16, Paris 1970. 517

70b. Notes on epimorphisms of rings. Preprint, Berkeley 1970. 517
71a. Groups acting on hereditary rings. *Proc. London Math. Soc.* (3) **23** (1971), 70–82 [corr. ibid. **24** (1972), 192]. 156, 408
71b. Hereditary commutative rings and centres of hereditary rings. *Proc. London Math. Soc.* (3) **23** (1971), 214–236.
72a. Boolean rings of projection maps. *J. London Math. Soc.* (2) **4** (1972), 593–598. 110, 330
72b. Hereditarily and cohereditarily projective modules. Proc. Conf. on Ring Theory at Park City, Utah, 1971 (ed. R. Gordon). Academic Press, New York, 1972, pp. 29–62. 183, 330
73. Infinite multiplication of ideals in $\aleph_0$-hereditary rings. *J. Algebra* **24** (1973), 56–70. 330
74a. Modules over coproducts of rings. *Trans. Amer. Math. Soc.* **200** (1974), 1–32. 183, 261
74b. Coproducts and some universal ring constructions. *Trans. Amer. Math. Soc.* **200** (1974), 33–88. 58, 183, 185, 455
76. Rational relations and rational identities in division rings. *J. Algebra* **43** (1976), I. 252–266, II. 267–297.
77. On the existence of subalgebras of direct products with prescribed d-fold projections. *Algebra Universalis* **7** (1977), 341–356.
78a. The diamond lemma for ring theory. *Adv. in Math.* **29** (1978), 178–218. 65, 181
78b. Conjugates and nth roots in Hahn–Laurent group rings. *Bull. Malaysian Math. Soc.* (2) **1** (1978), 29–41 [historical addendum ibid. **2** (1979), 41–42]. 106
83. S fields finitely right generated over subrings. *Comm. in Algebra* **11** (17) (1983), 1893–1902. 474
87. Normalizers of semifirs in their universal fields. *Proc. London Math. Soc.* (3) **54** (1987), 83–87. 518
89. Radicals, tensor products and algebraicity. Isr. Math. Conf. Proc. Ring Theory in Honor of S. A. Amitsur (ed. L. Rowen), 1989, pp. 150–192.
90. Ordering coproducts of groups and semigroups. *J. Algebra* **133** (1990), 313–339.
99. Supports of derivations, free factorizations and ranks of fixed subgroups in free groups. *Trans. Amer. Math. Soc.* **351** (1999), 1531–1550.
2002. Constructing division rings as module-theoretic direct limits. *Trans. Amer. Math. Soc.* **354** (2002), 2079–2114. 278, 329
a. The class of free subalgebras of a free associative algebra is not closed under taking unions of chains, or pairwise intersections. To appear. 374

Bergman, G. M. and Cohn, P. M. 69. Symmetric elements in free powers of rings. *J. London Math. Soc.* (2) **1** (1969), 525–534. 386
71. The centres of 2-firs and hereditary rings. *Proc. London Math. Soc.* (3) **23** (1971), 83–98. 407

Bergman, G. M. and Dicks, W. 75. Universal derivations. *J. Algebra* **36** (1975), 193–211.
78. Universal derivations and universal ring constructions. *Pacif. J. Math.* **79** (1978), 293–337. 66, 454, 506, 518

Bergman, G. M. and Lewin, J. 75. The semigroup of ideals of a fir is (usually) free. *J. London Math. Soc.* (2) **11** (1975), 21–32. 408f.

Berstel, J. and Perrin, D. 85. *Theory of Codes*. Academic Press, Orlando 1985.
360, 407
Berstel, J. and Reutenauer, C. 88. *Rational Series and their Languages*. Springer-Verlag, Berlin 1988. 408
Bessenroth-Timmerscheidt, C., Brungs, H.-H. and Törner, G. 85. Right chain rings. Schriftenreihe d. Fachber. Math. Universität Duisburg Gesamthochschule. 1985.
Birkhoff, G. 67. *Lattice Theory*, 3rd edn. Amer. Math. Soc. New York 1967. 116, 232, 261, 335, 519
Blum, E. K. 65. A note on free semigroups with two generators. *Bull. Am. Math. Soc.* **71** (1965), 678–679.
Bokut, L. A. 59
 64. Some examples of rings without zero-divisors (Russian). *Algebra i Logika* **3** (5–6) (1964), 5–28.
 65. Factorization theorems for certain classes of rings without zero-divisors I, II (Russian). *Algebra i Logika* **4** (1965), I. (4), 25–52; II. (5), 17–46.
 67. The embedding of rings in skew fields (Russian). *Dokl. Akad. Nauk SSSR* **175** (1967), 755–758. 516
 69. On Malcev's problem (Russian). *Sibirsk. Mat. Zh.* **10** (1969), 965–1005. 516
 81. *Associative Rings* 1, 2 (Russian). NGU, Novosibirsk 1981. 515
 87. Embeddings of rings (Russian). *Uspekhi Mat. Nauk* **42**:4 (1987), 87–111 [Engl. transl. *Russian Math. Surveys* (1987), 105–138].
 97. The method of Gröbner and Shirshov bases (Russian), in *Algebra, Geometry, Analysis and Math. Physics* (Russian) (Novosibirsk, 1996), 30–39, 189. Izdat. Ross. Akad. Nauk Sibirsk. Otdel. Inst. Mat. Novosibirsk 1997.
Bonang, F. 89. Noetherian rings whose subidealizer subrings have pan-ACC. *Comm. Alg.* **17** (5) (1989), 1137–1146. 111
Bourbaki, N. 61. *Algèbre Commutative*, Chapters 1 & 2. Act. Sci. et Ind. 1290. Hermann, Paris 1961. 534
 72. *Commutative Algebra*, Chapters 1–7. Addison-Wesley, Reading, MA. 1972.
183, 338
Bowtell, A. J. 67a. The multiplicative semigroup of a ring and the embedding of rings in skew fields. Thesis, London University 1967. 182f., 191, 223
 67b. On a question of Malcev. *J. Algebra* **9** (1967), 126–139. 516
Bowtell, A. J. and Cohn, P. M. 71. Bounded and invariant elements in 2-firs. *Proc. Camb. Phil. Soc.* **69** (1971), 1–12. 407
Bray, U. and Whaples, G. 83. Polynomials with coefficients from a division ring. *Canad. J. Math.* **35** (1983), 509–515.
Brenner, J. L. 55. Quelques groupes libres de matrices. *C. R. Acad. Sci. Paris* **241** (1955), 1689–1691. 152
Brungs, H.-H. (see also Bessenroth-Timmerscheidt, C.) 69a. Ringe mit eindeutiger Faktorzerlegung. *J. reine angew. Math.* **236** (1969), 43–66. 214, 223
 69b. Generalized discrete valuation rings. *Canad. J. Math.* **21** (1969), 1404–1408. 104, 106
 71. Overrings of principal ideal domains. *Proc. Amer. Math. Soc.* **28** (1971), 44–46.
 73. Left Euclidean rings. *Pacif. J. Math.* **45** (1973), 27–33.
 74. Right invariant right hereditary rings. *Canad. J. Math.* **26** (1974), 1186–1191.
 76. Rings with a distributive lattice of right ideals. *J. Algebra* **40** (1976), 392–400.

78. Unique factorization in rings with right ACC_1. *Glasgow Math. J.* **19** (1978), 167–171. 223

86. Bezout domains and rings with a distributive lattice of right ideals. *Canad. J. Math.* **38** (1986), 286–303.

Brungs, H.-H. and Törner, G. 81. Right chain groups and the generalized semigroup of divisibility. *Pacif. J. Math.* **97** (1981), 293–305. 237

84a. Extensions of chain rings. *Math. Zeits.* **185** (1984), 93–104.

84b. Skew power series rings and derivations. *J. Algebra* **87** (1984), 368–379.

Burkov, V. D. 81. Derivations of polynomial rings (Russian, English summary). *Vestnik Moscov. Univ. Ser. I Mat. Mekh.* 1981, No. 2, 51–55, 87.

Burmistrovich, I. E. 63. On the embedding of rings in skew fields (Russian). *Sibirsk. Mat. Zh.* **4** (1963), 1235–1240. 516

Camillo, V. P. (see also Anderson, D. D.) 75. Distributive modules. *J. Algebra* **36** (1975), 16–25. 231, 261

Carcanague, J. 71. Idéaux bilatères d'un anneau de polynômes non commutatifs sur un corps. *J. Algebra* **18** (1971), 1–18. 407

Cartan, H. and Eilenberg, S. 56. *Homological Algebra*. Princeton University Press, Princeton, NJ 1956. 183

Cauchon, G. 92. Séries de Malcev-Neumann sur le groupe libre et questions de rationalité. *Theor. Comput. Sci.* **98** (1992), 79–97.

96. Séries formelles croisées. *J. Pure Appl. Alg.* **107** (1996), 153–169.

Cedó, F. (see also Antoine, R.) 88. A question of Cohn on semifir monoid rings. *Comm. Algebra* **16** (6)(1988), 1187–1189. 175

89. On semifir monoid rings. *Publ. Mat.* **33** (1989), 123–132.

97. Strongly prime power series rings. *Comm. Algebra* **25** (7) (1997), 2237–2242.

Cedó, F. and Herbera, D. 95. The Ore condition for polynomial and power series rings. *Comm. Algebra* **23** (14)(1995), 5131–5159.

Cedó, F. and Pitarch. A. 92. On rigid monoids and 2-fir monoid rings. *Comm. Algebra* **20**(8)(1992), 2295–2303.

94. Construction of left fir monoid rings. *J. Algebra* **165** (1994), 645–660.

Chase, S. U. 61. A generalization of the ring of triangular matrices. *Nagoya Math. J.* **18** (1961), 13–25. 183

62. On direct sums and products of modules. *Pacif. J. Math.* **12** (1962), 847–854. 204, 223

Chatters, A. W. 84. Non-commutative unique factorization domains. *Math. Proc. Camb. Phil Soc.* **95** (1984), 49–59.

Chatters, A. W. and Jordan, D. A. 86. Non-commutative unique factorization rings. *J. London Math. Soc.* (2) **33** (1986), 22–32.

Cheng, C. C. and Wong, R. W. 82. Hereditary monoid rings. *Amer. J. Math.* **104** (1982), 935–942.

Chevalley, C. 408

Chiba, K. 96. Free subgroups and free subsemigroups of division rings. *J. Algebra* **184**, (1996), 570–574.

Choo, K. G. 74. Whitehead groups of twisted free associative algebras. *Pacif. J. Math.* **50** (1974), 399–402.

77. Grothendieck groups of twisted free associative algebras. *Glasgow Math. J.* **18** (1977), 193–196.

Clark, W. E. 373

Claus, H. J. 55. Über die Partialbruchzerlegung in nicht notwendig kommutativen euklidischen Ringen. *J. reine angew. Math.* **194** (1955), 88–100.

Cohn, P. M. (see also Artamonov, V. A., Bergman, G. M., Bowtell, A. J.)
- 59. On the free product of associative rings. *Math. Zeits.* **71** (1959), 380–398.   534
- 60. On the free product of associative rings II. The case of (skew) fields. *Math. Zeits.* **73** (1960), 433–456.   183
- 61a. On a generalization of the Euclidean algorithm. *Proc. Camb. Phil. Soc.* **57** (1961), 18–30.   105, 175, 183f., 366
- 61b. On the embedding of rings in skew fields. *Proc. London Math. Soc.* (3) **11** (1961), 511–530.   516f.
- 61c. Quadratic extensions of skew fields. *Proc. London Math. Soc.* (3) **11** (1961), 531–556.
- 62a. Factorization in non-commutative power series rings. *Proc. Camb. Phil. Soc.* **58** (1962), 452–464.   184, 223, 407
- 62b. On subsemigroups of free semigroups. *Proc. Amer. Math. Soc.* **13** (1962), 347–351.   407f.
- 62c. *Free Rings.* Yale University Lecture Notes. Yale University, New Haven, CT 1962.
- 63a. Noncommutative unique factorization domains. *Trans. Amer. Math. Soc.* **109** (1963), 313–331 [corr. ibid. **119** (1965), 552].   58, 106, 183, 223
- 63b. Rings with a weak algorithm. *Trans. Amer. Math. Soc.* **109** (1963), 332–356.   183f., 407
- 64a. Subalgebras of free associative algebras. *Proc. London Math. Soc.* (3) **14** (1964), 618–632.   408
- 64b. Free ideal rings. *J. Algebra* **1** (1964), 47–69 [corr. ibid. **6** (1967), 410].   183
- 66a. Some remarks on the invariant basis property. *Topology* **5** (1966), 215–228.   3, 58, 182ff., 185
- 66b. On the structure of the GL_2 of a ring. *Publ. Math. IHES* No. 30, Paris 1966.   106, 152, 155, 184
- 66c. A remark on matrix rings over free ideal rings. *Proc. Camb. Phil. Soc.* **62** (1966), 1–4.   58, 183
- 66d. Hereditary local rings. *Nagoya Math. J.* **27**-1 (1966), 223–230.   206, 223
- 67. Torsion modules over free ideal rings. *Proc. London Math. Soc.* (3) **17** (1967), 577–599.   104, 183, 329
- 68. Bezout rings and their subrings. *Proc. Camb. Phil. Soc.* **64** (1968), 251–264.
- 69a. Free associative algebras. *Bull. London Math. Soc.* **1** (1969), 1–39.   223, 260
- 69b. Rings with a transfinite weak algorithm. *Bull. London Math. Soc.* **1** (1969), 55–59.   184f.
- 69c. Dependence in rings II. The dependence number. *Trans. Amer. Math. Soc.* **135** (1969), 267–279.   183, 185
- 70a. Factorization in general rings and strictly cyclic modules. *J. Reine Angew. Math.* **239/40** (1970), 185–200.   58, 214, 223
- 70b. Torsion and protorsion modules over free ideal rings. *J. Austral. Math. Soc.* **11** (1970), 490–498.   276
- 70c. On a class of rings with inverse weak algorithm. *Math. Zeits.* **117** (1970), 1–6.   184, 330
- 71a. The embedding of firs in skew fields. *Proc. London Math. Soc.* (3) **23** (1971), 193–213.   37, 185, 223, 454, 499, 516
- 71b. Rings of fractions. *Amer. Math. Monthly* **78** (1971), 596–615.   517

71c. Un critère d'immersibilité d'un anneau dans un corps gauche. *C. R. Acad. Sci. Paris Ser. A* **272** (1971), 1442–1444.

72a. Universal skew fields of fractions. *Symposia Math.* **VIII** (1972), 135–148.   517

72b. Skew fields of fractions and the prime spectrum of a general ring. *Lectures on Rings and Modules.* Lecture Notes in Math. No. 246. Springer-Verlag, Berlin, 1972, pp. 1–71.   446, 517

72c. *Rings of fractions.* University of Alberta Lecture Notes, Edmonton 1972.   58, 442f.

72d. Generalized rational identities. Proc. Conf. on Ring Theory at Park City, Utah, 1971 (ed. R. Gordon). Academic Press, New York, 1972, pp. 107–115.

73a. The similarity reduction of matrices over a skew field. *Math. Zeits.* **132** (1973), 151–163.   455

73b. Free radical rings. *Colloq. Math. Soc. J. Bolyai,* **6**. Rings, modules and radicals. Keszthely, Hungary, 1971 (1973), pp. 135–145.   184

73c. Unique factorization domains. *Amer. Math. Monthly* **80** (1973), 1–17.   214, 223

73d. Bound modules over hereditary rings. In *Selected Questions on Algebra and Logic* (volume dedicated to the memory of A. I. Malcev). Izd. Nauka, SO, Novosibirsk, 1973, pp. 131–141.   329

74a. Progress in free associative algebras. *Isr. J. Math.* **19** (1974), 109–151.

74b. Localization in semifirs. *Bull. London Math. Soc.* **6** (1974), 13–20.   329, 516, 518

74c. The class of rings embeddable in skew fields. *Bull. London Math. Soc.* **6** (1974), 147–148.

76a. Morita equivalence and duality. Queen Mary College Lecture Notes, London. 1966, 1976.

76b. The Cayley–Hamilton theorem in skew fields. *Houston J. Math.* **2** (1976), 49–55.   320, 329

77a. A construction of simple principal right ideal domains. *Proc. Amer. Math. Soc.* **66** (1977), 217—222 [corr. ibid. **77** (1979), 40].   77, 329

77b. Full modules over semifirs. *Publ. Math. Debrecen* **24** (1977), 305–310.   455

78. Centralisateurs dans les corps libres, Ecole de Printemps d'Informatique Théorique. Séries formelles en variables non-commutatives et applications (ed. J. Berstel), 1978, pp. 45–54.

79. The affine scheme of a general ring. *Applications of Sheaves: Durham Research Symposium 1977.* Lecture Notes in Math. No. 753. Springer-Verlag, Berlin (1979), pp. 197–211.   58, 443, 517f.

80. On semifir constructions. *Word Problems II, the Oxford Book* (eds. S. I. Adian, W. W. Boone and G. Higman). North-Holland, Amsterdam 1980, pp. 73–80.

82a. The universal field of fractions of a semifir I. Numerators and denominators. *Proc. London Math. Soc.* (3) **44** (1982), 1–32.   58, 183, 223, 325, 329, 518

82b. The divisor group of a fir. *Publ. Sec. Mat. Univ. Autonom. Barcelona* **26** (1982), 131–163.   498, 518

82c. Ringe mit distributivem Faktorverband. *Abh. Braunschweig. Wiss. Ges.* **33** (1982), 35–40.   262, 300

82d. Torsion modules and the factorization of matrices. *Advances in Non-commutative Ring Theory.* Proc. 12th George Hudson Symposium (ed. P. J. Fleury). Lecture Notes in Math. No. 951. Springer-Verlag, Berlin (1982), pp. 1–11.   300, 329

83. Determinants on free fields. *Contemp. Math.* **13** (1983), 99–108. 578
84a. Fractions. *Bull. London Math. Soc.* **16** (1984), 561–574.
84b. Embedding problems for rings and semigroups. Proc. 25. Arbeitstagung über allgemeine Algebra, Darmstadt, 1983 (ed. P. Burmeister *et al.*). N. Heldermann Verlag, Berlin, 1984, pp. 115–126.
85a. The universal field of fractions of a semifir II. The depth. *Proc. London Math. Soc.* (3) **50** (1985), 69–94. 223, 474, 518
85b. The universal field of fractions of a semifir III. Centralizers and normalizers. *Proc. London Math. Soc.* (3) **50** (1985), 95–113. 262, 518
85c. On coproducts of ordered systems. *Algebra and Order.* Proc. First International Symp. Ordered Algebraic Structures. (ed. S. Wolfenstein), Luminy-Marseille, 1984. N. Heldermann Verlag, Berlin 1986, pp. 3–12.
87a. Right principal Bezout domains. *J. London Math. Soc.* (2) **35** (1987), 251–262. 77, 97, 106
87b. The rational hull of a semifir. *J. London Math. Soc.* (2) **35** (1987), 263–275.
89a. Distributive factor lattices for free rings. *Proc. Amer. Math. Soc.* **105** (1989), 34–41. 261
89b. An algebraist looks at coding theory, or how to parse without punctuation. *Bull. Inst. Math. Applns.* **25** (1989), 192–195.
89c. Around Sylvester's law of nullity. *The Math. Scientist,* **14** (1989), 73–83.
89d. Generalized polynomial identities and the specialization lemma. In *Ring Theory*, (ed. L. H. Rowen, in Honor of S. A. Amitsur). *Israel Math. Conf. Proc.* **1** (1989), 242–246.
90. An embedding theorem for free associative algebras. *Math. Pannon.* **1**/1 (1990), 49–56. 261f., 517
92a. Modules over hereditary rings. *Contemp. Math.* **130** (1992), 111–119. 183
92b. A remark on power series rings. *Publ. Secc. Mat.* **36** (1992), 481–484. 184
97a. The universal skew field of fractions of a tensor product of free rings. *Colloq. Math.* **72** (1) (1997), 1–8 [corr. ibid. **73** (2)(1998), 1–9]. 300
97b. Cyclic Artinian modules without a composition series. *J. London Math. Soc.* (2) **55** (1997), 231–235. 175
99. Reversible rings. *Bull. London Math. Soc.* **31** (1999), 641–648. 515, 518
2000a. From Hermite rings to Sylvester domains. *Proc. Amer. Math. Soc.* **128** (2000), 1899–1904. 58, 511
2000b. On the construction of Sylvester domains. *J. Algebra* **234** (2000), 423–432.
2002. The automorphism group of the free algebra of rank two. *Serdica Math. J.* **28** (3)(2002), 255–266. 409
2003. Some remarks on projective-free rings. *Algebra Universalis,* **49** (2)(2003), 159–164.
2006. Localization in general rings; a historical survey. In Noncommutative Localization in Algebra and Topology (Ed. A. A. Ranicki) LMS Lecture Note Series 330. Cambridge University Press 2006.
Cohn, P. M. and Dicks, W. 76. Localization in semifirs II. *J. London Math. Soc.* (2) **13** (1976), 411–418. 184, 518
80. On central extensions of skew fields. *J. Algebra* **63** (1980), 143–151.
Cohn, P. M. and Gerritzen, L. 2001. On the group of symplectic matrices over a free associative algebra. *J. London Math. Soc.* (2) **63** (2001), 353–363.

Cohn, P. M. and Mahdavi-Hezavehi, M. 80. Extensions of valuations on skew fields. *Proc. Ring Theory Week, Antwerp* (ed. F. Van Oystaeyen). Lecture Notes in Math. No. 825. Springer-Verlag, Berlin 1980, pp. 28–41.

Cohn, P. M. and Reutenauer, C. 94. A normal form in free fields. *Canad. J. Math.* **46** (3) (1994), 517–531.

   99. On the construction of the free field. *Internat. J. of Alg. and Comp.* **9** (3&4) (1999), 307–323.   518

Cohn, P. M. and Schofield, A. H. 82. On the law of nullity. *Math. Proc. Camb. Phil. Soc.* **91** (1982), 357–374.   58f., 303, 329, 518

   85. Two examples of principal ideal domains. *Bull. London Math. Soc.* **17** (1985), 25–28.   518

Connell, E. H. (see Bass, H.)

Corner, A. L. S. 69. Additive categories and a theorem of W. G. Leavitt. *Bull. Amer. Math. Soc.* **75** (1969) 78–82.   182

Coxeter, H. S. M.   408

Cozzens, J. H. 72. Simple principal left ideal domains. *J. Algebra* **23** (1972), 66–75.

Cozzens, J. H. and Faith, C. 75. *Simple Noetherian Rings.* Cambridge Tracts in Math. No. 69. Cambridge University Press, Cambridge 1975.   224

Curtis, C. W. 52. A note on non-commutative polynomial rings. *Proc. Amer. Math. Soc.* **3** (1952), 965–969.   105

Czerniakiewicz, A. J. 71–72. Automorphisms of a free associative algebra of rank 2. *Trans. Amer. Math. Soc.* I. **160** (1971), 393–401; II. **171** (1972), 309–315.   405, 409

Dauns, J. 70a. Embeddings in division rings. *Trans. Amer. Math. Soc.* **150** (1970), 287–299.

   70b. Integral domains that are not embeddable in division rings. *Pacif. J. Math.* **34** (1970), 27–31.

   82. *A Concrete Approach to Division Rings.* N. Heldermann Verlag, Berlin 1982.

Dedekind, J. W. R. 1900. Über die von drei Moduln erzeugte Dualgruppe. *Math. Ann.* **53** (1900), 371–403 [*Ges. Werke* **II**, 236–271].   520

Dicker, R. M. 68. A set of independent axioms for a field and a condition for a group to be the multiplicative group of a field. *Proc. London Math. Soc.* (3) **18** (1968), 114–124.   517

Dicks, W. (see also Almkvist, G., Bergman, G. M., Cohn, P. M.)   373f., 408, 490, 518

   72. On one-relator associative algebras. *J. London Math. Soc.* (2) **5** (1972), 249–252.

   74. Idealizers in free algebras. Thesis, London University 1974.   184, 379

   77. Mayer–Vietoris presentations over colimits of rings. *Proc. London Math. Soc.* (3) **34** (1977), 557–576.   141

   79. Hereditary group rings. *J. London Math. Soc.* (2) **20** (1979), 27–38.

   81. An exact sequence for rings of polynomials in partly commuting indeterminates. *J. Pure Appl. Algebra* **22** (1981), 215–228.

   83a. The HNN construction for rings. *J. Algebra* **81** (1983), 434–487.

   83b. Free algebras over Bezout domains are Sylvester domains. *J. Pure Appl. Algebra* **27** (1983), 15–28.

   83c. A free algebra can be free as a module over a non-free subalgebra. *Bull. London Math. Soc.* **15** (1983), 373–377.

   85a. Homogeneous elements of free algebras have free idealizers. *Math. Proc. Camb. Phil. Soc.* **97** (1985), 7–26.

85b. On the cohomology of one-relator associative algebras. *J. Algebra* **97** (1985), 79–100.
Dicks, W. and Formanek, E. 82. Poincaré series and a problem of S. Montgomery. *Lin. Multilin. Algebra* **12** (1982), 21–30. 386, 408
Dicks, W. and Menal, P. 79. The group rings that are semifirs. *J. London Math. Soc.* (2) **19** (1979), 288–290. 185
Dicks, W. and Schofield, A. H. 88. On semihereditary rings. *Comm. Algebra* **16** (6) (1988), 1243–1274.
Dicks, W. and Sontag, E. D. 78. Sylvester domains. *J. Pure Appl. Algebra* **13** (1978), 243–275. 290, 293, 299, 329, 517
Dieudonné, J. 43. Les déterminants sur un corps non-commutatif. *Bull. Soc. Math. France* **71** (1943), 27–45. 494
73. *Introduction to the Theory of Formal Groups*. M. Dekker, New York, 1973. 223
Dixmier, J. 68. Sur les algèbres de Weyl. *Bull. Soc. Math. France* **96** (1968), 209–242.
Dlab, V. and Ringel, C. M. 85. A class of bounded hereditary Noetherian domains. *J. Algebra* **92** (1985), 311–321. 518
Donkin, S. 408
Doss, R. 48. Sur l'immersion d'un semi-groupe dans un groupe. *Bull. Sci. Math.* **72** (1948), 139–150. 59
Drensky, V. 373
2000. *Free algebras and PI-algebras*. Springer-Verlag, Singapore 2000.
Dress, F. 71. Stathmes euclidiens et séries formelles. *Acta Arithm.* **19** (1971), 261–265. 97
Drogomizhska, M. M. 70. Augmentation of a matrix to an invertible one in an Ore domain (Ukrainian, Russian summary). *Visnik L'viv Politekhn. Inst.* **44** (1970), 50–60, 210. 58
Dubois, D. W. 66. Modules of sequences of elements of a ring. *J. London Math. Soc.* **41** (1966), 177–180. 329
Dubreil-Jacotin, M.-L. 47. Sur l'immersion d'un semigroupe dans un groupe. *C. R. Acad. Sci. Paris* **225** (1947), 787–788. 407
Dubrovin, N. I. 82. Non-commutative valuation rings (Russian). *Trudy Mat. Obshch.* **45** (1982), 265–280.
86. On elementary divisor rings. *Izv. Vyssh. Uchebn. Zaved. Mat.* **11** (1986), 14–20, 85.
87. Invertibility of the group ring of a right-ordered group over a division ring (Russian). *Mat. Zametki* **42** (1987), 508–518, 622. 516
Dumas, F. 86a. Hautes dérivations et anneaux de séries formelles non-commutatifs en caractéristique nulle. *C. R. Acad. Sci. Paris I. Math.* **303** (1986), 383–385.
86b. Hautes dérivations équivalentes en caractéristique nulle. *C. R. Acad. Sci. Paris I. Math.* **303** (1986), 841–843.
88. Hautes dérivations équivalentes dans les corps gauches en caractéristique nulle. *C. R. Acad. Sci. Paris I. Math.* **306** (1988), 519–521.
96. Sous-corps complets de corps gauches de séries de Laurent á hautes dérivations. Anneaux et modules (Colmar, 1991) 55–71. Travaux en cours 51. Hermann, Paris 1996.
Eilenberg, S. (see also Cartan, H.) 66
Elizarov, V. P. 69. Rings of fractions (Russian). *Algebra i Logika* **8**(4) (1969), 381–424. 59

Euclid (-300). *Elements*. 105
Evans, M. J. 95. Some rings with many non-free stably free modules. *Quart. J. Math. Oxf.* **2**(46), (1995), 291–297.
Faith, C. (see also Cozzens, J. H.) 73. *Algebra I, Rings, Modules and Categories*. Springer-Verlag, Berlin 1973 (repr. 1981). 223
Faizov, S. K. 81. Free ideal categories (Russian). *Ukrainskii Mat Zh.* **33** (1981), 626–630. 185
Farber, M. and Vogel, P. 92. The Cohn localization of the free group ring. *Math. Proc. Camb. Phil. Soc.* **111** (1992), 433–443.
Farbman, P. 95. Non-free two-generator subgroups of $SL_2(\mathbb{Q})$. *Publ. Sec. Mat. Univ. Aut. Barcelona* **39** (1995), 379–391. 152
   99. Amalgamation of domains. *J. Algebra* **221** (1999), 90–101.
Feller, E. H. 60. Intersection irreducible ideals of a non-commutative principal ideal domain. *Canad. J. Math.* **12** (1960), 592–596. 223
   61. Factorization and lattice theorems for a bounded principal ideal domain. *Duke Math. J.* **28** (1961), 579–583.
Ferreira, V. O. 515
   2000. Tensor rings under field extensions. *J. Algebra*, **231** (2000), 342–363.
   2001 Commutative monoid amalgams with natural core. *Comm. Algebra* **29** (2001), 757–767.
   2002 Constants of derivations on free associative algebras. *Glasgow Math. J.* **44** (2002), 177–183.
Ferreira, V. O., Murakami, L. S. I. and Paques, A. 2004. A Hopf-Galois correspondence for free algebras. *J. Algebra* **276** (2004), 407–416.
Figueiredo, L. M. V., Gonçalves, J. Z. and Shirvani, M. 96. Free group algebras in certain division rings. *J. Algebra* **185** (1996), 298–313.
Fisher, J. L. 71. Embedding free algebras in skew fields. *Proc. Amer. Math. Soc.* **30** (1971), 453–458. 427
   74a. The poset of skew fields generated by a free algebra. *Proc. Amer. Math. Soc.* **42** (1974), 33–35.
   74b. The category of epic R-fields. *J. Algebra* **28** (1974), 283–290.
Fitting, H. 35. Primärkomponentenzerlegung in nichtkommutativen Ringen. *Math. Ann.* **111** (1935), 19–41. 58
   36. Über den Zusammenhang zwischen dem Begriff der Gleichartigkeit zweier Ideale und dem Äquivalenzbegriff der Elementarteilertheorie. *Math. Ann.* **112** (1936), 572–582. 28, 58, 191, 223
Fliess, M. 184
   70a. Inertie et rigidité des séries rationnelles et algébriques. *C. R. Acad. Sci. Paris Ser. A* **270** (1970), 221–223. 184
   70b. Transductions algébriques. *R.I.R.O.* **R**-1 (1970), 109–125.
   70c. Sur le plongement de l'algèbre des séries rationnelles non commutatives dans un corps gauche. *C. R. Acad. Sci. Paris Ser. A* **271** (1970), 926–927.
   71. Deux applications de la représentation matricielle d'une série rationnelle non-commutative. *J. Algebra* **19** (1971), 344–353.
Formanek, E. (see Almkvist, G., Dicks, W.)
Förster, A. 88. Zur Theorie einseitig Euklidischer Ringe (English and Russian summaries). *Wiss. Z. Pädagog. Hochsch. 'Karl Liebknecht', Potsdam* **32**(1) (1988), 143–147.

Fox, R. H. 53. Free differential calculus I. Derivation in the free group ring. *Ann. Math.* **57** (1953), 547–560. 86
Fröberg, R. 97. *An Introduction to Gröbner Bases.* John Wiley & Sons, Chichester 1997. 184
Gabel, M. R. 25
Gatalevich, A. I. and Zabavskii, B. V. 97. Non-commutative elementary divisor rings (Ukrainian; English, Russian and Ukrainian summaries). *Mat. Metodi Fiz-Mekh. Polya* **40** (1997), 86–90.
Gentile, E. R. 60. On rings with a one-sided field of quotients. *Proc. Amer. Math. Soc.* **11** (1960), 380–384. 51, 59
Gerasimov, V. N. 73. Rings that are nearly free (Russian). In *Rings* II (Russian), pp. 9–19. Inst. Mat. Sibirsk. Otdel. Akad. Nauk SSSR, Novosibirsk 1973.
76. Distributive lattices of subspaces and the word problem for one-relator algebras (Russian). *Algebra i Logika* **15**(4) (1976), 384–435, 487.
79. Inverting homomorphisms of rings (Russian). *Algebra i Logika* **18** (1979), 648–663. 443
82. Localization in associative rings (Russian). *Sibirsk. Mat. Zh.* **23** (1982), 36–54. 183
93. Free associative algebras and inverting homomorphisms of rings. Three papers on algebras and their representations. 1–76. *Amer. Math. Soc. Transl. Ser.* 2, **156** (1993).
Gerritzen, L. (see also Cohn, P. M.) 98. On infinite Gröbner bases in free algebras. *Indag. Math. (N.S.)* **9** (1998), 491–501.
Gerritzen, L. and Holtkamp, R. 98. On Gröbner bases of non-commutative power series. *Indag. Math.(N.S.)* **9** (1998), 503–519.
Gersten, S. M. 65. Whitehead groups of free associative algebras. *Bull. Amer. Math. Soc.* **71** (1965), 157–159.
74. K-theory of free rings. *Comm. in Algebra* **1** (1974), 39–64.
Goldie, A. W. 58. The structure of prime rings under ascending chain conditions. *Proc. London Math. Soc.* (3) **8** (1958), 589–608. 59
Goldschmidt, D. 85
Gonçalves, J. Z. and Shirvani, M. (see also Figueiredo, L. M. V.) 96. On free group algebras in division rings with uncountable center. *Proc. Amer. Math. Soc.* **124** (1996), 685–687.
98. Free group algebras in the field of fractions of differential polynomial rings and enveloping algebras. *J. Algebra* **204** (1998), 372–385.
Goodearl, K. R. 79. *von Neumann Regular Rings.* Pitman, London 1979. (2nd edn, Krieger Publ. Co. Malabar, Florida, 1991). 58
Goodearl, K. R. and Warfield Jr., R. B. 81. State spaces of K_0 of Noetherian rings. *J. Algebra* **71** (1981), 322–378. 58
Gorbachuk, O. L. and Komarnickii, M. Y. 77. Radical filters in a principal ideal domain (Russian, English summary). *Doklady Akad. Nauk Ukrain. SSR Ser. A*, 2(1977), 103–104, 191.
Gordon, B. and Motzkin, T. S. 65. On the zeros of polynomials over division rings. *Trans. Amer. Math. Soc.* **116** (1965), 218–226 [corr. ibid. **122** (1966), 547].
Goursaud, J. M. and Valette, J. 75. Anneaux de groupe héréditaires et semihéréditaires. *J. Algebra* **34** (1975), 205–212. 357

Govorov, V. E. 72. Graded algebras (Russian). *Mat. Zametki* **12** (1972), 197–204.
   73. The global dimension of algebras (Russian). *Mat. Zametki* **14** (1973), 399–406.
   81. Algebras of homological dimension 1 (Russian). *Mat. Sb. (N.S.)* **116**(158) (1981), 111–119.
   96. On Poincaré series for codes. *J. Math. Sci.* **80**(5) (1996), 2161–2173.
Grams, A. 74. Atomic rings and the ascending chain conditions for principal ideals. *Proc. Camb. Phil. Soc.* **75** (1974), 321–329.   55
Grätzer, G. 78. *General Lattice Theory*. Math. Reihe Band 52. Birkhäuser Verlag, Basel. 1978.   245, 262
Green, E. L., Mora, T. and Ufnarovski, V. 98. The non-commutative Gröbner freaks. *Symbolic Rewriting Techniques* (Ascona, 1995), 93–104. Progr. Comput. Sci. Appl. Logic 15. Birkhäuser, Basel. 1998.
Guazzone, S. 62. Sui Λ-moduli liberi i alcuni teoremi di C. J. Everett, *Rend. Sem. Mat. Univ. Padova* **32** (1962), 304–312.   183
Guralnick, R. M., Levy, L. S. and Odenthal, C. 88. Elementary divisor theorem for non-commutative principal ideal domains. *Proc. Amer. Math. Soc.* **103** (1988), 1003–1011.   106
Hahn, H. 1907. Über die nichtarchimedischen Größensysteme. *S.-B. Akad. Wiss. Wien IIa* **116** (1907), 601–655.   106
Hall Jr., M. 59. *The Theory of Groups*. Macmillan, New York 1959.   95
Hasse, H. 28. Über die eindeutige Zerlegung in Primelemente oder in Primhauptideale in Integritätsbereichen. *J. Reine Angew. Math.* **159** (1928), 3–12.   77
Hausknecht, A. O.   77
Hedges, M. C. 87. The Freiheitssatz for graded algebras. *J. London Math. Soc.* (2) **35** (1987), 395–405.
Helm, P. R. 83. Generators and relations for certain linear groups over rings of linear operators. *Comm. in Algebra* **11** (5) (1983), 551–565.   152
Hensel, K. 27. Über eindeutige Zerlegung in Primelemente. *J. Reine Angew. Math.* **158** (1927), 195–198.
   29. Über Systeme in einfachen Körpern. *J. Reine Angew. Math.* **160** (1929), 131–142.
Herbera, D. (see Cedó, F.)
Herstein, I. N. 76. *Rings with Involution*, Chicago Lectures in Math. Chicago University Press, Chicago 1976.   477
Hiblot, J.-J. 75. Des anneaux euclidiens dont le plus petit algorithme n'est pas à valeurs finies. *C. R. Acad. Sci. Paris Ser. A* **281** (1975), 411–414.   105
   76. Sur les anneaux euclidiens. *Bull. Soc. Math. France* **104** (1976), 33–50.
Higman, G. 40. The units of group rings. *Proc. London Math. Soc.* (2) **46** (1940), 231–248.   311
   52. Ordering by divisibility in abstract algebras. *Proc. London Math. Soc.* (3) **2** (1952), 326–336.   106
Higman, G., Neumann, B. H. and Neumann, H. 49. Embedding theorems for groups. *J. London Math. Soc.* **24** (1949), 247–254.   185, 407
Hilbert, D. 1899. *Grundlagen der Geometrie*. Festschrift zur Feier der Enthüllung des Gauss–Weber Denkmals in Göttingen. B. G. Teubner, Leipzig 1899, 2. Auflage 1903.   106
Hochster, M. 69. Prime ideal structure in commutative rings. *Trans. Amer. Math. Soc.* **142** (1969), 43–60.   442f.

Holtkamp, R. (see Gerritzen, L.)
Hotje, H. 83. On Cohn functions. *Annals of Discrete Math.* **18** (1983), 467–468. 517
Hudry, A. 70. Quelques applications de la localisation au sens de Gabriel. *C. R. Acad. Sci. Paris Ser. A-B* **270** (1970), A8–A10. 518
Hughes, I. 70. Division rings of fractions for group rings. *Comm. Pure Appl. Math.* **23** (1970), 81–88.
Ikeda, M. 69. Über die maximalen Ideale einer freien assoziativen Algebra. *Hamb. Abh.* **33** (1969), 59–66. 260
Ince, E. L. 27. *Ordinary Differential Equations.* Longmans, London 1927. 66
Isbell, J. 81. Submodules of free modules. *Amer. Math. Monthly* **88** (1981), 53.
Ischebeck, F. 71. Die Zentren hereditärer Ringe. *Math. Ann.* **193** (1971), 83–88. 407
Iyudu, N. K. 2000. Subalgebra membership problem and subalgebra standard basis. Algebra, Proceedings of the International Algebraic Conference on the Occasion of the 90th Birthday of A.G. Kurosh, Moscow, Russia, May 25–30, 1998. Walter de Gruyter, Berlin (2000), pp. 137–143.
Jacobson, N. 34. A note on non-commutative polynomials. *Ann. Math.* **35** (1934), 209–210. 105
    37. Pseudo-linear transformations. *Ann. Math.* **38** (1937), 484–507. 106
    43. *Theory of Rings.* American Mathematical Society, New York 1943. 105f., 350, 407
    50. Some remarks on one-sided inverses. *Proc. Amer. Math. Soc.* **1** (1950), 352–355. 12
    64. *Structure of Rings* (rev. edn). American Mathematical Society, Providence 1964, RI. 408
    89. *Basic Algebra II*, 2nd edn. Freeman, San Francisco 1989. 444
Jaffard, P. 60. *Les Systèmes d'Idéaux.* Dunod, Paris 1960. 183
James, G. D. and Kerber, A. 81. *The Representation Theory of the Symmetric Group.* Encyclopedia of Mathematics and its Applications, Vol. 16. Addison-Wesley, Reading, MA. 1981. 382
Jategaonkar, A. V. 69a. A counter-example in ring theory and homological algebra. *J. Algebra* **12** (1969), 418–440. 98f., 104ff., 350
    69b. Rings with a transfinite left division algorithm. *Bull. Amer. Math. Soc.* **75** (1969), 559–561.
    69c. Ore domains and free algebras. *Bull. London Math. Soc.* **1** (1969), 45–46. 140
    71. Skew polynomial rings over semisimple rings. *J. Algebra* **19** (1971), 315–328.
Jensen, C. U. 63. On characterizations of Prüfer rings. *Math. Scand.* **13** (1963), 90–98. 225, 231
    66. A remark on semihereditary local rings. *J. London Math. Soc.* **41** (1966), 479–482.
    69. Some cardinality questions for flat modules and coherence. *J. Algebra* **12** (1969), 231–241. 183
Johnson, R. E. (see also Beauregard, R. A.) 63. Principal right ideal rings. *Canad. J. Math.* **15** (1963), 297–301.
    65. Unique factorization in a principal right ideal domain. *Proc. Amer. Math. Soc.* **16** (1965), 526–528. 223
    67. The quotient domain of a weak Bezout domain. *J. Math. Sci.* **2** (1967), 21–22.
Jøndrup, S. 70. On finitely generated flat modules. *Math. Scand.* **26** (1970), 233–240.
    77. The centre of a right hereditary ring. *J. London Math. Soc.* (2) **15** (1977), 211–212.

De Jonquières, J. P. E. de F.
Jooste, T. de W. 71. Derivations in free power series rings and free associative algebras. Thesis, London University 1971. 170
  78. Primitive derivations in free associative algebras. *Math. Zeits.* **164** (1978), 15–23.
Jordan, D. A. (see also Chatters, A. W.) 65
Jung, H. W. E. 42. Über ganze birationale Transformationen der Ebene. *J. reine u. angew. Math.* **184** (1942), 161–174. 405
Kalajdzić, G. 95. On Euclidean algorithms with some particular properties. *Publ. Inst. Math. N.S.* **57** (71) (1995), 124–134.
Kaplansky, I. 49. Elementary divisors and modules. *Trans. Amer. Math. Soc.* **66** (1949), 464–491. 58, 85, 106
  58. Projective modules. *Ann. Math.* **68** (1958), 372–377. 19, 110, 183
Kazimirskii, P. S. 62. A theorem on elementary divisors for the ring of differential operators (Ukrainian, Russian summary). *Dopovidi Akad. Nauk Ukraïn. RSR* (1962), 1275–1278.
Kazimirskii, P. S. and Lunik, F. P. 72. Completion of a rectangular matrix over an associative ring to an invertible one (Ukrainian, Russian summary). *Dopovidi Akad. Nauk Ukraïn. RSR, Ser. A* (1972), 505–506. 25, 58
Kemer, A. R. 91. Identities of associative algebras. In *Algebra and Analysis* (Kemerevo, 1988), 65–71. Amer. Math. Soc. Transl. Ser. 2, 148. Providence, RI. 1991.
Kerber, A. (see James, G. D.)
Kerr, J. W. 82. The power series ring over an Ore domain need not be Ore. *J. Algebra* **75** (1982), 175–177. 105
Kharchenko, V. K. (see also Andreeva, T. A.) 77. Galois theory of semiprime rings (Russian). *Algebra i Logika* **16** (1977), 313–363. 388
  78. On algebras of invariants of free algebras (Russian). *Algebra i Logika* **17** (1978), 478–487. 381, 408f.
  80. The actions of groups and Lie algebras on non-commutative rings (Russian). *Uspekhi Mat. Nauk* **35**(2) (1980), 67–90 [Engl. transl. *Russian Math Surveys* **35** (2) (1980), 77–104].
  81. Constants of derivations on prime rings (Russian). *Izv. Akad. Nauk Ser. Mat.* **45** (1981), 435–461 [Engl. transl. *Math. USSR Izv.* **18** (1982), 381–410].
  84. Noncommutative invariants of finite groups and Noetherian varieties. *J. Pure Appl. Algebra* **31** (1984), 83–90. 408
  91. *Automorphisms and Derivations of Associative Rings*. Mathematics and its Applications (Soviet Series), Vol. 69. Kluwer Academic, Dordrecht 1991. 408
Kirezci, M. 7, 183
Klein, A. A. 67. Rings nonembeddable in fields with multiplicative semigroups embeddable in groups. *J. Algebra* **7** (1967), 100–125. 516
  69. Necessary conditions for embedding rings into fields. *Trans. Amer. Math. Soc.* **137** (1969), 141–151. 182f.
  70. Three sets of conditions on rings. *Proc. Amer. Math. Soc.* **25** (1970), 393–398.
  72a. A remark concerning the embeddability of rings into fields. *J. Algebra* **21** (1972), 271–274. 288
  72b. Involutorial division rings with arbitrary centers. *Proc. Amer. Math. Soc.* **34** (1972), 38–42. 491
  80. Some ring-theoretic properties implied by embeddability in fields. *J. Algebra* **66** (1980), 147–155.

92. Free subsemigroups of domains. *Proc. Amer. Math. Soc.* **116** (1992), 339–341.
Knight, J. T. 70. On epimorphisms of non-commutative rings. *Proc. Camb. Phil. Soc.*
68 (1970), 589–600. 517
Knus, M. A. 68. Homology and homomorphisms of rings. *J. Algebra* **9** (1968),
274–284. 408
Kolotov, A. T. 78. On free subalgebras of free associative algebras (Russian). *Sibirsk.
Mat. Zh.* **19** (1978), 328–335. 374, 379, 408
Komarnickii, M. Ya. (see Gorbachuk, O. L.)
Korotkov, M. V. 76. Description of rings with single-valued division with remainder
(Russian). *Uspekhi Mat. Nauk* **31** (1976), No. 1 (187), 253–254. 106
Koshevoi, E. G. 66. On the multiplicative semigroup of a class of rings without zero-
divisors (Russian). *Algebra i Logika* **5**(5) (1966), 49–54. 206, 223
70. On certain associative algebras with transcendental relations (Russian). *Algebra
i Logika* **9**(5) (1970), 520–529. 140
71. Pure subalgebras of free associative algebras (Russian). *Algebra i Logika* **10**
(1971) 183–187. 379
Köthe, G. 518
Kozhukhov, I. B. 82. Free left ideal semigroup rings (Russian). *Algebra i Logika* **21**(1)
(1982), 37–59. 185
Kraft, L. G. 49. A device for quantizing, grouping and coding amplitude modu-
lated pulses. M.S. Thesis, Electrical Engineering Dept., MIT, Cambridge, MA
1949. 408
Krasilnikov, A. N. and Vovsi, S. M. 96. On fully invariant ideals in the free group
algebra. *Proc. Amer. Math. Soc.* **124** (1996), 2613–2618.
Krob, D. 91. Some examples of formal series used in non-commutative algebra.
Theoret. Comput. Sci. **79** (1991), 111–135.
Krull, W. 28. Zur Theorie der zweiseitigen Ideale in nichtkommutativen Bereichen.
Math. Zeits. **28** (1928), 481–503.
54. Zur Theorie der kommutativen Integritätsbereiche. *J. Reine Angew. Math.* **192**
(1954), 230–252.
Kryazhovskikh, G. V. and Kukin, G. P. 89. On subrings of free rings (Russian). *Sibirsk.
Mat. Zh.* **30** (1989), 87–97.
Kukin, G. P. (see Kryazhovskikh, G. V.)
Lallement, G. 79. *Semigroups and Combinatorial Applications.* John Wiley & Sons,
New York, 1979. 360
84. Some problems on rational power series in non-commuting variables. Proc. 1984
Marquette Conf. on semigroups.
Lam, T. Y. 76. Series summation of stably free modules. *Quart. J. Math. Oxford Ser.*
(2) **27** (1976), 37–46. 25
78. *Serre's Conjecture.* Lecture Notes in Math. No. 635. Springer-Verlag, Berlin,
1978. 25, 58, 297
Lam, T. Y. and Leroy, A. 88. Algebraic conjugacy classes and skew polynomial rings.
In *Perspectives in Ring Theory* (eds F. van Oystaeyen and L. Le Bruyn), Proc.
Antwerp Conf. in Ring Theory. Kluwer Academic, Dordrecht 1988, pp. 153–203.
Lambek, J. 73. Noncommutative localization. *Bull. Amer. Math. Soc.* **79** (1973),
857–872.
Landau, E. 1902. Ein Satz über die Zerlegung homogener linearer Differentialausdrücke
in irreduzible Faktoren. *J. Reine Angew. Math.* **124** (1902), 115–120. 105

Lane, D. R. 184, 373, 386
75. Fixed points of affine Cremona transformations of the plane over an algebraically closed field. *Amer. J. Math.* **97** (1975), 707–732.
76. Free algebras of rank two and their automorphisms. Thesis, London University 1976. 381, 408

Lazard, M. 184

Leavitt, W. G. 57. Rings without invariant basis number. *Proc. Amer. Math. Soc.* **8** (1957), 322–328. 7, 58
60. The module type of a ring. *Trans. Amer. Math. Soc.* **103** (1960), 113–130.

Leissner, W. 71. Eine Charakterisierung der multiplikativen Gruppe eines Körpers. *Jber. Deutsche Math.-Verein.* **73** (1971), 92–100. 517

Lemmlein, V. 54. On Euclidean rings and principal ideal rings (Russian). *Dokl. Akad. Nauk SSSR* **97** (1954), 585–587. 73

Lenstra Jr., H. W. 74. Lectures on Euclidean rings. Universität Bielefeld 1974. 73, 98, 105f.

Leroy, A. (see also Lam, T. Y.) 85. Dérivées logarithmiques pour une S-dérivation algébrique. *Comm. Algebra* **13** (1985), 85–99.

Lesieur, L. 65
78. Conditions Noethériennes dans l'anneau de polynômes de Ore A[X, σ, δ]. Sém. d'Algèbre Paul Dubreil, 30ème année (Paris 1976–77), pp. 220–234. Lecture Notes in Math. No. 641. Springer-Verlag, Berlin 1978.

Leutbecher, A. 78. Euklidischer Algorithmus und die Gruppe GL_2. *Math. Ann.* **231** (1978), 269–285. 105

Levi, F. W. 44. On semigroups. *Bull. Calcutta Math. Soc.* **36** (1944), 141–146. 407

Levi-Civita, T. 1892. Sugli infiniti ed infinitesimali attuali quali elementi analitici. *Atti Ist. Veneto di Sci. Lett. ed Arti, Ser. 7a* **4**(1892–93), 1765–1816. (= Op. mat. vol. primo, pp. 1–39, Bologna 1954). 106

Levitzki, J. 58

Levy, L. S. and Robson, J. C. (see also Guralnick, R. M.) 74. Matrices and pairs of modules. *J. Algebra* **29** (1974), 427–454.

Lewin, J. (see also Bedoya, H., Bergman, G. M.) 263
68. On Schreier varieties of linear algebras. *Trans. Amer. Math. Soc.* **132** (1968), 553–562.
69. Free modules over free algebras and free group algebras: the Schreier technique. *Trans. Amer. Math. Soc.* **145** (1969), 455–465. 143, 184, 408
72. A note on zero-divisors in group rings. *Proc. Amer. Math. Soc.* **31** (1972), 357–359.
73. On some infinitely presented associative algebras. *J. Austral. Math. Soc.* **16** (1973), 290–293.
74a. A matrix representation for associative algebras. *Trans. Amer. Math. Soc.* **188** (1974), I. 293–308, II. 309–317.
74b. Fields of fractions for group algebras of free groups. *Trans. Amer. Math. Soc.* **192** (1974), 339–346. 517
77. Ranks of matrices over Ore domains. *Proc. Amer. Math. Soc.* **62** (1977), 233–236.
88. The symmetric ring of quotients of a 2-fir. *Comm. in Algebra* **16** (8) (1988), 1727–1732.

Lewin, J. and Lewin, T. 68. On ideals of free associative algebras generated by a single element. *J. Algebra* **8** (1968), 248–255.

78. An embedding of the group algebra of a torsion-free one-relator group in a field. *J. Algebra* **52** (1978), 39–74.

Lewin, T. (see Lewin, J.)

Lissner, D. 65. Outer product rings. *Trans. Amer. Math. Soc.* **116** (1965), 526–535.   58

Loewy, A. 1903. Über reduzible lineare homogene Differentialausdrücke. *Math. Ann.* **56** (1903), 549–584.   105

1920. Begleitmatrizen und lineare homogene Differentialausdrücke. *Math. Zeits.* **7** (1920), 58–128.

Lopatinskii, Y. B. 45. Linear differential operators (Russian). *Mat. Sb.* **17** (59) (1945), 267–288.

46. A theorem on bases (Russian). *Trudy Sek. Mat. Akad. Nauk Azerb. SSR* **2** (1946), 32–34.

47. On some properties of rings of linear differential operators (Russian). *Nauch. Zap. L'vov Univ. I Ser. fiz.-mat.* **2** (1947), 101–107.

Lorenz, M. 86. On free subalgebras of certain division algebras. *Proc. Amer. Math. Soc.* **98** (1986), 401–405.

Lorimer, J. W. 92. The classification of compact right chain rings. *Math. Forum* **4** (1992), 335–347.   19

Lothaire, M. 97. *Combinatorics on Words.* Cambridge Math. Library, Cambridge University Press, 1997, Cambridge.   360, 366, 407

Lunik, F. P. (see also Kazimirskii, P. S.) 70. The intersection of principal ideals in a certain non-commutative ring. (Ukrainian, Russian summary). *Visnik L'viv Politekhn. Inst.* **44** (1970), 24–26, 209.   183

Lyndon, R. C. and Schupp, P. E. 77. *Combinatorial Group Theory.* Springer-Verlag, Berlin 1977.   409

Ma, W. X. 93. A decomposition of elements of the free algebra. *Proc. Amer. Math. Soc.* **118** (1993), 37–45.

McAdam, S. and Rush, D. E. 78. Schreier rings. *Bull. London Math. Soc.* **10** (1978), 77–80.   290

Macdonald, I. G. 68. *Algebraic Geometry: Introduction to Schemes.* Benjamin, New York 1968.   443, 517

Macintyre, A. 79. Combinatorial problems for skew fields. I. Analogue of Britton's lemma and results of Adjan–Rabin type. *Proc. London Math. Soc.* (3) **39** (1979), 211–236.   185

Mac Lane, S. 63. *Homology.* Springer-Verlag, Berlin.   27f., 524

71. *Categories for the Working Mathematician.* Springer-Verlag, Berlin 1971.   524

McLeod, J. B. 58. On the commutator subring. *Quart. J. Math Oxford Ser.* (2) **9** (1958), 207–209.   141

McMillan, B. 56. Two inequalities implied by unique decipherability. *IRE Trans. Inform. Theory* **IT**-2 (1956), 115–116.   408

Magnus, W. 40. Über Gruppen und zugeordnete Liesche Ringe. *J. Reine Angew. Math.* **182** (1940), 142–149.

Mahdavi-Hezavehi, M. (see also Cohn, P. M.) 79. Matrix valuations on rings. Antwerp Conf. on Ring Theory. M. Dekker, New York 1979, pp. 691–703.

82. Matrix valuations and their associated skew fields. *Resultate d. Math.* **5** (1982), 149–156.   500, 518

Makar-Limanov, L. G.   47, 97

70a. On automorphisms of certain algebras (Russian). Thesis, Moscow University, 1970. 405, 409
70b. The automorphisms of the free algebras with two generators (Russian). *Funkcional. Anal. i Prilozhen.* **4** (1970), 107–108.
75. On algebras with one relation. *Uspekhi Mat. Nauk* **30**, No. 2 (182)(1975), 217.
83. The skew field of fractions of the Weyl algebra contains a free non-commutative subalgebra. *Comm. in Algebra* **11**(17) (1983), 2003–2006.
84a. On free subsemigroups. *Semigroup Forum* **29** (1984), 253–254.
84b. On free subsemigroups of skew fields. *Proc. Amer. Math. Soc.* **91** (1984), 189–191.
85. Algebraically closed skew fields. *J. Algebra* **93** (1985), 117–135.
Makar-Limanov, L. G. and Malcolmson, P. 91. Free subalgebras of enveloping fields. *Proc. Amer. Math. Soc.* **111** (1991), 315–322.
Malcev, A. I. 37. On the immersion of an algebraic ring into a field. *Math. Ann.* **113** (1937), 686–691. 37, 182, 515
39. Über die Einbettung von assoziativen Systemen in Gruppen I (Russian, German summary). *Mat. Sb.* **6**(48) (1939), 331–336. 37, 515
40. Über die Einbettung von assoziativen Systemen in Gruppen II (Russian, German summary). *Mat. Sb.* **8**(50) (1940), 251–264. 37, 515
48. On the embedding of group algebras in division algebras (Russian). *Dokl. Akad. Nauk SSSR* **60** (1948), 1499–1501. 106, 516
53. Nilpotent semigroups (Russian). *Uchen. Zap. Ivanovsk. Ped. In-ta* **4** (1953), 107–111. 516
Malcolmson, P. (see also Makar-Limanov, L. G.) 78. A prime matrix ideal yields a skew field. *J. London Math. Soc.* (2) **18** (1978), 221–233.
80a. On making rings weakly finite. *Proc. Amer. Math. Soc.* **80** (1980), 215–218. 58
80b. Determining homomorphisms to skew fields. *J. Algebra* **64** (1980), 399–413.
82. Construction of universal matrix localizations. Advances in Non-commutative. Ring Theory. Proc. 12th G. H. Hudson Symp., Plattsburgh, 1981. Lecture Notes in Math. No. 951. Springer-Verlag, Berlin 1982, pp. 117–131.
84. Matrix localizations of n-firs. *Trans. Amer. Math. Soc.* **282** (1984), I. 503–518, II. 519–527.
93. Weakly finite matrix localizations. *J. London Math. Soc.* (2) **48** (1993), 31–38.
Martindale III, W. S. 82. The extended center of coproducts. *Canad. Math. Bull.* **25**(2) (1982), 245–248.
Martindale III, W. S. and Montgomery, M. S. 83. The normal closure of coproducts of domains. *J. Algebra* **82** (1983), 1–17. 409
May, M. 92. Universal localization and triangular rings. *Comm. in Algebra* **20** (1992), 1243–1257.
Melançon, G. 93. Constructions des bases standard des K⟨A⟩-modules à droite. *Theor. Comp. Sci.* **117** (1993), 255–272.
Menal, P. (see also Dicks, W.) 79. Remarks on the GL_2 of a ring. *J. Algebra* **61** (1979), 335–359. 184
81. The monoid rings that are Bezout domains. *Arch. Math.* **37** (1981), 43–47. 185
Mez, H.-C. 84. HNN-extensions of algebras and applications. *Bull. Austral. Math. Soc.* **29** (1984), 215–229.

Mikhalev, A. A., Shpilrain, V. and Yu, J.-T. 2004. *Combinatorial Methods, Free Groups, Polynomials and Free Algebras.* Canadian Math. Soc. Books in Mathematics 19. Springer-Verlag, New York 2004.
Mikhalev, A. A. and Zolotykh, A. A. 98. Standard Gröbner–Shirshov bases of free algebras over rings I. Free associative algebras. *Internat. J. Algebra Comput.* **8** (1998), 689–726.
Milnor, J. W. 71. *Introduction to Algebraic K-theory.* Ann. Math. Studies No. 72. Princeton University Press, Princeton, NJ 1971. 14
Minc, H. 78. *Permanents.* Encyclopedia of Maths. and its Applns, Vol. 6. Addison-Wesley, Reading, MA, 1978. 191
Minkowski, H. 153
Mitchell, B. 72. Rings with several objects. *Adv. Math.* **8** (1972), 1–161. 185
Molien, T. E. 1897. Über die Invarianten der linearen Substitutionsgruppen. *Sitzungsber. Akad. Wiss. Berlin* **52** (1897), 1152–1156. 408
Montgomery, M. S. (see also Martindale III, W. S.) 83. von Neumann finiteness of tensor products of algebras. *Comm. in Algebra* **11** (6) (1983), 595–610. 7, 182
Montgomery, M. S. and Passman, D. S. 84. Galois theory of prime rings. *J. Pure Appl. Algebra* **31** (1984), 139–184. 388, 408
87. Prime ideals in fixed rings of free algebras. *Comm. Algebra* **15** (1987), 2209–2234.
Mora, T. (see Green, E. L.)
Moran, S. 73. Some subalgebra theorems. *J. Algebra* **27** (1973), 366–371. 408
Motzkin, T. S. (see also Gordon, B.) 49. The euclidean algorithm. *Bull. Amer. Math. Soc.* **55** (1949), 1142–1146. 105
Moufang, R. 37. Einige Untersuchungen über geordnete Schiefkörper. *J. Reine Angew. Math.* **176** (1937), 203–223. 516
Murakami, L. S. I. (see Ferreira, V. O.)
Nagao, H. 59. On GL(2,K[x]). *J. Inst. Polytech. Osaka City Univ., Ser. A* **10** (1959), 117–121. 401
Nagata, M. 398
57. A remark on the unique factorization theorem. *J. Math. Soc. Japan* **9** (1957), 143–145. 56, 59
Nakajima, H. 83. Regular rings of invariants of unipotent groups. *J. Algebra* **85** (1983), 253–286. 408
Nakayama, T. 38. A note on the elementary divisor theory in non-commutative domains. *Bull. Amer. Math. Soc.* **44** (1938), 719–723. 106
Neumann, B. H. (see also Higman, G.) 49. On ordered division rings. *Trans. Amer. Math. Soc.* **66** (1949), 202–252. 106, 516
Neumann, H. (see Higman, G.)
Nivat, M. 68. Transduction des langages de Chomsky. *Ann. Inst Fourier* **18** (1) (1968), 339–456. 184
69. Séries rationnelles et algébriques en variables non commutatives. Cours DEA 1969/70. 516
Nöbeling, G. 68. Verallgemeinerung eines Satzes von Herrn Specker. *Invent. Math.* **6** (1968), 41–55. 330
Noether, E. 516
1916. Der Endlichkeitssatz der Invarianten endlicher Gruppen. *Math. Ann.* **77** (1916) 89–92. 408

Noether, E. and Schmeidler, W. 1920. Moduln in nichtkommutativen Bereichen, insbesondere aus Differential- und Differenzenausdrücken. *Math. Zeits.* **8** (1920), 1–35. 262

Novikov, B. V. 84. On a question of P. Cohn (Russian). *Teor. Polugrupp Prilozh.* **7** (1984), 50–51. 57

Nuñez, P. 1567. *Libro de Algebra*, Antwerp 1567. 105

Odenthal, C. (see Guralnick, R. M.)

Ojanguren, M. and Sridharan, R. 71. Cancellation in Azumaya algebras. *J. Algebra* **18** (1971), 501–505. 25, 58

O'Neill, J. D. 91. An unusual ring. *J. London Math. Soc.* (2) **44** (1991), 95–101. 58

Ore, O. 31. Linear equations in non-commutative fields. *Ann. Math.* **32** (1931), 463–477. 37, 58, 516

　32. Formale Theorie der linearen Differentialgleichungen. *J. Reine Angew. Math.* **167** (1932), 221–234, II. **168** (1932), 233–252. 58, 65, 105

　33a. Theory of non-commutative polynomials. *Ann. Math.* **34** (1933), 480–508. 105, 223

　33b. On a special class of polynomials. *Trans. Amer. Math. Soc.* **35** (1933), 559–584.

Osofsky, B. L. 71. On twisted polynomial rings. *J. Algebra* **18** (1971), 597–607.

Ostmann, H.-H. 50. Euklidische Ringe mit eindeutiger Partialbruchzerlegung. *J. Reine Angew. Math.* **188** (1950), 150–161.

Palmer, Th. W. 94. Banach algebras and general theory of *-algebras. Cambridge University Press, Cambridge 1994. 11

Paques, A. (see Ferreira, V. O.)

Parimala, S. and Sridharan, R. 75. Projective modules over polynomial rings over division rings. *J. Math. Kyoto Univ.* **15** (1975), 129–148.

Passman, D. S. (see Montgomery, M. S.)

Paul, Y. 73. Unique factorization in a 2-fir with right ACC_1. *Riv. Mat. Univ. Parma* (3)**2** (1973) 115–119. 223

Perrin, D. (see Berstel, J.)

Perron, O. 54. *Die Lehre von den Kettenbrüchen I*, 3rd edn. B. G. Teubner, Stuttgart 1954.

Pierce, R. S. 82. *Associative Algebras*. Springer-Verlag, New York 1982. 226

Pikhtilkov, S. A. 74. Primitivity of a free associative algebra with a finite number of generators (Russian), *Uspekhi Mat. Nauk* **29** (1974), No. 1(175), 183–184.

Pinchuk, S. I. 94. A counterexample to the strong real Jacobian conjecture. *Math. Z.* **217** (1994), 1–4. 405

Piontkowskii, D. I. 98. On Hilbert series and relations in algebras (Russian). *Uspekhi Mat. Nauk* **53** (6) (1998), 257–258 [Engl. transl. *Russian Math. Surveys* **53** (1998), 1360–1361].

　99. On free products in varieties of associative algebras (Russian). *Mat. Zametki* **65** (1999), 693–702 [Engl. transl. *Math. Notes* **65** (1999), 582–589].

Pitarch, A. (see also Cedó, F.) 90. Monoid rings that are firs. *Publ. Mat.* **34** (1990), 217–221.

Poincaré, H. 105

Pontryagin, L. 39. *Topological Groups*. Princeton University Press, Princeton, NJ 1939. 329

Prest, M. Y. 83. Existentially complete prime rings. *J. London Math. Soc.* (2) **28** (1983), 238–246. 183

Ranicki, A. A. (ed.) (a) Noncommutative localization in algebra and topology (to appear).

Renshaw, J. 99. On subrings of an amalgamated product of rings. *Colloq. Math.* **79** (1999), 241–248.
Reutenauer, C. (see also Berstel, J., Cohn, P. M.) 156, 366
  83. Sulla fattorizzazione dei codici. *Ricerche di Mat.* **XXXII** (1983), 115–130.
  96. Inversion height in free fields. *Selecta Math. (N.S.)* **2** (1996), 93–109.
  99. Malcev–Neumann series and the free field. *Expo. Math.* **17** (1999), 469–478.
517
Révész, G. 81. Universal fields of fractions, their orderings and determinants. Thesis, London University 1981.
  83a. On the abelianized multiplicative group of a universal field of fractions. *J. Pure Appl. Algebra* **27** (1983), 277–297. 498, 518
  83b. Ordering epic R-fields. *Manu. Math.* **44** (1983), 109–130. 518
  84. On the construction of the universal field of fractions of a free algebra. *Mathematika* **31** (1984), 227–233.
Richard, J. 70. Représentations matricielles des séries rationnelles en variables non-commutatives. *C. R. Acad. Sci. Paris Ser. A-B* **270** (1970), A224–A227.
Ringel, C. M. (see also Dlab, V.) 79. The spectrum of a finite-dimensional algebra. In *Ring Theory* (ed. F. Van Oystaeyen), Proc. 1978 Antwerp Conf. M. Dekker, New York 1979, pp. 535–597. 517
Roberts, M. L. 250, 263, 320
  82. Normal forms, factorizations and eigenrings in free algebras. Thesis, London University, 1982. 260, 329
  84. Endomorphism rings of finitely presented modules over free algebras. *J. London Math. Soc.* (2) **30** (1984), 197–209. 329
  89. The centre of a 2-fir. *Bull. London Math. Soc.* **21** (1989), 541–543. 407
Robson, J. C. (see also Levy, L. S.) 67. Rings in which finitely generated right ideals are principal. *Proc. London Math. Soc.* (3) **17** (1967), 617–628.
  72. Idealizers and hereditary Noetherian prime rings. *J. Algebra* **22** (1972), 45–81. 37
Rodosski, K. I. 80. On Euclidean rings (Russian). *Dokl. Akad. Nauk SSSR* **253** (1980), 819–822. 105
Roitman, M. 77. Completing unimodular rows to invertible matrices. *J. Algebra* **49** (1977), 206–211.
Roos, J.-E. 67. Locally distributive spectral categories and strongly regular rings. *Reports of the Mid-West Category Seminar.* Lecture Notes in Math. No. 47, Springer-Verlag, Berlin 1967. 261
  81. Homology of loop spaces and of local rings. *Proc. 18th Scand. Cong. Math.* Progress in Math. Birkhäuser, Basel 1981, pp. 441–468. 184
Rosenmann, A. 93. Essentiality of fractal ideals. *Internat. J. Alg. Comput.* **3** (1993), 425–445.
Rosenmann, A. and Rosset, S. 91. Essential and inessential fractal ideals in the group ring of a free group. *Bull. London Math. Soc.* **23** (1991), 437–442. 262
  94. Ideals of finite codimension in free algebras and the fc-localization. *Pacif. J. Math.* **162** (1994), 351–371. 184
Rosset, S. (see also Rosenmann, A.) 47
Rush, D. E. (see McAdam, S.)
Russell, P. 80. Hamburger–Noether expansions and approximate roots of polynomials. *Manu. Math.* **31** (1980), 29–95. 407
Ryabukhin, Yu. M. (see Andrunakievich, V. A.)

Samuel, P.   72, 97
   68. Unique factorization. *Amer. Math. Monthly* **75** (1968), 945–952.   175
   71. About Euclidean rings. *J. Algebra* **19** (1971), 282–301.   105
Sanov, I. N. 67. Euclidean algorithm and onesided prime factorization for matrix rings (Russian). *Sibirsk. Mat. Zh.* **8** (1967), 846–852.   72
Schlesinger, L. 1897. *Handbuch der Theorie der Differentialgleichungen.* B. G. Teubner, Leipzig 1897.   105
Schmeidler, W. (see Noether, E.)
Schofield, A. H. (see also Cohn, P. M., Dicks, W.)
                       112, 213, 272, 329, 387, 427, 454, 510
   85. Representations of rings over skew fields. London Math. Soc. Lecture Notes, No. 92, Cambridge University Press, Cambridge 1985.   454
Schupp, P. E. (see Lyndon, R. C.)
Schur, I. 1904. Über vertauschbare lineare Differentialausdrücke. *Berl. Math. Ges. Sitzber.* **3** (Archiv d. Math. Beilage (3) **8**) (1904), 2–8.   106, 378
Schützenberger, M.-P. 59. Sur certains sous-demi-groupes qui interviennent dans un problème de mathématiques appliquées. *Publ. Sci. Univ. d'Alger. Ser A* **6** (1959), 85–90.   407
   62. On a theorem of R. Jungen. *Proc. Amer. Math. Soc.* **13** (1962), 885–890.   516
Schwarz, L. 47. Zur Theorie der nichtkommutativen rationalen Funktionen. *Berl. Math. Tagung, Tübingen* **1946** (1947) 134–136.
   49. Zur Theorie des nichtkommutativen Polynombereichs und Quotientenrings. *Math. Ann.* **120** (1947/49), 275–296.   516
Serbin, H. 38. Factorization in principal ideal rings. *Duke Math. J.* **4** (1938), 656–663.
Serre, J.-P.   408
Sheiham, D. 2001. Non-commutative characteristic polynomials and Cohn localization. *J. London Math. Soc.* (2) **64**(2001), 13–28.
Shephard, G. C.   408
Shepherdson, J. C. 51. Inverses and zero-divisors in matrix rings. *Proc. London Math. Soc.* (3) **1** (1951), 71–85.   58
Shirvani, M. (see Figueiredo, L. M. V., Gonçalves, J. Z.)
Shpilrain, V. and Yu, J.-T. (see also Mikhalev, A. A.) 2003 Factor algebras of free algebras: on a problem of G. Bergman *Bull. London Math. Soc.* **35** (2003), 706–710.
Sierpin'ska, A. 73. Radicals of rings of polynomials in non-commutative indeterminates (Russian summary). *Bull. Acad. Polon. Sci. Ser. Math. Astronom. Phys.* **21** (1973), 805–808.
Silvester, J. R. 73. On the K_2 of a free associative algebra. *Proc. London Math. Soc.* (3) **26** (1973), 35–56.
Singh, S.   74
Skornyakov, L. A. 65. On Cohn rings (Russian). *Algebra i Logika* **4**(3) (1965), 5–30.   185
   66. On left chain rings (Russian). *Izv. vyss. uchebn. zaved. Mat.* **53** (1966), 114–117.
   67. The homological classification of rings (Russian). *Mat. Vesnik* **4** (19) (1967), 415–434.
Smits, T. H. M. 67. Skew polynomial rings and nilpotent derivations. Proefschrift, Technische Hogeschool, Delft. Uitgeverij Waltman, Delft 1967.   91
   68a. Nilpotent S-derivations. *Indag. Math.* **30** (1968), 72–86.   106

68b. Skew polynomial rings. *Indag. Math.* **30** (1968), 209–224.
Snider, R. L. 82. The division ring of fractions of a group ring. P. Dubreil and M.-P. Dubreil-Jacotin Alg. Sém. 35th Année (Paris, 1982), 325–339. Lecture Notes in Math. No. 1029. Springer-Verlag, Berlin 1983.
Sontag, E. D. (see Dicks, W.)
Specker, E. 50. Additive Gruppen von Folgen ganzer Zahlen. *Portugal. Math.* **9** (1950), 131–140. 329
Springer, T. A. 77. *Invariant Theory.* Lecture Notes in Math. No. 585, Springer-Verlag, Berlin, 1977. 408
Sridharan, R. (see Ojanguren, M., Parimala, S.)
Stafford, J. T. 77. Weyl algebras are stably free. *J. Algebra* **48** (1977), 297–304. 88
85. Stably free projective right ideals. *Compositio Math.* **54** (1985), 63–78. 106
Stallings, J. R. 68. On torsion-free groups with infinitely many ends. *Ann. of Math.* **88** (1968), 312–333. 409
Stanley, R. P. 79. Invariants of finite groups and their applications to combinatorics. *Bull. Amer. Math. Soc.* **1** (1979), 475–511. 408
Stenström, B. 75. *Rings of Quotients.* Grundl. Math. Wiss. 217. Springer-Verlag, Berlin 1975. 329
Stephenson, W. 261, 444
66. Characterizations of rings and modules by means of lattices. Thesis, London University 1966. 58
74. Modules whose lattice of submodules is distributive. *Proc. London Math. Soc.* (3) **28** (1974), 291–310. 231, 261
Stevin, S. 1585. *Arithmétique*, Antwerp (Vol. II of collected works, 1958). 105
Sushkevich, A. K. 36. On the embedding of semigroups in groups of fractions (Ukrainian). *Zap. Khark. Mat. O-va* **4**(12) (1936), 81–88. 515
Suslin, A. A. 76. On a theorem of Cohn (Russian). Rings and Modules. *Zap. Nauchn. Sem. Leningrad Otdel. Mat. Inst. Steklov (LOMI)* **64** (1976), 121–130.
Sylvester, J. J. 1884. On involutants and other allied species of invariants to matrix systems. *Johns Hopkins Univ. Circ.* **III** (1884), 9–12, 34–35. 329
Tarasov, G. V. 67. On free associative algebras (Russian). *Algebra i Logika* **6**(4) (1967), 93–105. 184
Taylor, J. L. 73. Functions of several noncommuting variables. *Bull. Amer. Math. Soc.* **79** (1973), 1–34.
Tazawa, M. 33. Einige Bemerkungen über den Elementarteilersatz. *Proc. Imper. Acad. Japan* **9** (1933), 468–471.
Teichmüller, O. 37. Der Elementarteilersatz für nichtkommutative Ringe. *S.-Ber. Preuss. Akad. Wiss. phys.-math.* **Kl.** (1937), 169–177. 106
Todd, J. A. 408
Tong, W. T. 89. On rings in which finitely generated projective modules are free. *J. Math. Res. Exposition* **9** (1989), 319–323.
94. IBN rings and orderings on Grothendieck groups. *Acta Math. Sinica (N.S.)* **10** (1994), 225–230.
Törner, G. (see Bessenroth-Timmerscheidt, C., Brungs, H.-H.)
Tronin, S. N. 98. Retracts and retractions of free algebras (Russian). *Izv. Vyssh. Uchebn. Zaved. Mat.* (1998) No. 1, 67–78 [Engl. transl. *R. Math. (VUZ)* **42** (1998), 65–77].

Tuganbaev, A. A. 84. Distributive modules and rings (Russian). *Uspekhi Mat. Nauk* **39** (1984), 157–158.
- 87. Hereditary rings (Russian). *Mat. Zametki* **41** (1987), 303–312, 456.
- 90. Distributive rings and modules (Russian). *Mat. Zametki* **47**(2) (1990), 115–123.
- 98. Semidistributive modules and rings. *Mathematics and its Applications* 449. Kluwer Academic, Dordrecht 1998.
- 99. Semidistributive modules. *J. Math. Sci. (New York)* **94** (1999), 1809–1887.

Tyler, R. 75. On the lower central factors of a free associative ring. *Canad. J. Math.* **27** (1975), 434–438.

Ufnarovski, V. (see Green, E. L.)

Umirbaev, U. U. 92. On the membership problem for free associative algebras (Russian). 11th International Workshop on Mathematical Logic, Kazan 1992, p. 145.
- 93a. Universal derivations and subalgebras of free algebras. Algebra, Proc. 3rd Internat. Conf. on Algebra (in memory of M. I. Kargapolov) (eds Y. Ershov *et al.*), Krasnoyarsk, Russia, August 23–28, 1993. W. de Gruyter, Berlin.
- 93b. Some algorithmic questions of associative algebras (Russian). *Algebra i Logika* **32**(4) (1993), 450–470.
- 96. Universal derivations and subalgebras of free algebras. *Algebra* (Krasnoyarsk, 1993), 255–271. W. de Gruyter, Berlin.

Valette, J. (see Goursaud, J. M.)

Valitskas, A. I. 82. The absence of a finite basis for the quasi-identities of rings embeddable in radical rings (Russian). *Algebra i Logika* **21**(1) (1982), 13–36.
- 84. The representation of finite-dimensional Lie algebras in radical rings (Russian). *Doklady Akad. Nauk SSSR* **279** (1984), 1297–1300 [Engl. transl. *Soviet Math. Dokl* **30** (1984), 794–798].
- 87. Examples of non-invertible rings imbedded into groups (Russian). *Sibirsk. Mat. Zh.* **28** (1987), 35–49, 223. 516
- 88. Examples of radical rings not embeddable in simple radical rings (Russian). *Doklady Akad. Nauk SSSR* **299** (1988) [Engl. transl. *Soviet Math. Dokl* **37** (1988), 449–451].

Vamos, P. 76. Test modules and cogenerators. *Proc. Amer. Math. Soc.* **56** (1976), 8–10. 224
- 78. Finitely generated artinian and distributive modules are cyclic. *Bull. London Math. Soc.* **10** (1978), 287–288. 231

Vechtomov, E. M. 90. On the general theory of Euclidean rings (Russian). Abelian groups and modules, No. 9 (Russian), 3–7. 155 Tomsk Gos. University, Tomsk 1990.

Verevkin, A. B. 96. A condition for graded freeness (Russian). *Izv. Vyssh. Uchebn. Zaved. Mat.* (1996) No. 9, 14–15.

Vogel, P. (see Farber, M.)

Voskoglou, M. G. 86. Extending derivations and endomorphisms to skew polynomial rings. *Publ. Inst. Math. (Beograd) N.S.* **39**(53) (1986), 79–82.

Vovsi, S. M. (see also Krasilnikov, A. N.) 93. On the semigroups of fully invariant ideals of the free group algebra and the free associative algebra. *Proc. Amer. Math. Soc.* **119** (1993), 1029–1037.

Warfield Jr., R. B. (see Goodearl, K. R.)

van de Water, A. 70. A property of torsion-free modules over left Ore domains. *Proc. Amer. Math. Soc.* **25** (1970), 199–201.

van der Waerden, B. L. 30. *Moderne Algebra*. I. Springer-Verlag, Leipzig 1930.   515
Webb. P.   386
Webber, D. B. 70. Ideals and modules in simple Noetherian hereditary rings. *J. Algebra* **16** (1970), 239–242.   11
Wedderburn, J. H. M. 21. On division algebras. *Trans. Amer. Math. Soc.* **22** (1921), 129–135.
   32. Noncommutative domains of integrity. *J. Reine Angew. Math.* **167** (1932), 129–141.   106, 184
Westreich, S. 91. On Sylvester rank functions. *J. London Math. Soc.* (2) **43** (1991), 199–214.
Weyl, H.   516
Whaples, G. (see Bray, U.)
Williams, R. E.   373
   68a. A note on weak Bezout rings. *Proc. Amer. Math. Soc.* **19** (1968), 951–952.   183
   68b. Sur une question de P. M. Cohn. *C. R. Acad. Sci. Paris Ser. A* **267** (1968), 79–80.   191
   69a. A note on rings with weak algorithm. *Bull. Amer. Math. Soc.* **75** (1969), 959–961 [see correction in *Math. Rev.* **40** (1970), No. 7308].
   69b. On the free product of rings with weak algorithm. *Proc. Amer. Math. Soc.* **23** (1969), 596–597.   184
Wolf, M. C. 36. Symmetric functions of noncommuting elements. *Duke Math. J.* **2** (1936), 626–637.
Wong, R. W. (see also Cheng, C. C.) 78. Free ideal monoid rings. *J. Algebra* **53** (1978), 21–35.   185
Wood, R. M. W. 85. Quaternionic eigenvalues. *Bull. London Math. Soc.* **17** (1985), 137–138.
Wright, D. (see Bass, H.)
Yu, J.-T. (see Mikhalev, A. A., Shpilrain, V.)
Zabavskii, B. V. (see also Gatalevich, A. I.) 87. Non-commutative rings of elementary divisors (Russian). *Ukrain. Mat. Zh.* **39** (1987), 440–444, 541.
Zaks, A. 70. Hereditary local rings. *Michigan Math. J.* **17** (1970), 267–272.
   72. Restricted left principal ideal rings. *Isr. J. Math.* **11** (1972), 190–215.
   82. Atomic rings without ACC on principal ideals. *J. Algebra* **74** (1982), 223–231.   55
Zolotykh, A. A. (see Mikhalev, A. A.)

Subject index

Any terms not found here may be in the list of notations and conventions p. xvi.
Left and right or upper and lower properties are usually not listed separately, thus 'left depth' and 'upper depth' are listed under 'depth'.

abelian category 527
absolute property 238
absolute rational identity 475f.
abstract atomic factor 249
ACC = ascending chain condition xvii
ACC_{dense} 321f.
additive category 526, 530
additive functor 530
a-adic filtration 163
adjoint associativity 534
adjoint functor, pair 530
admissible subcategory 225
admissible system, matrix 414, 455
affine automorphism 397
affine scheme 517
-algebra xviii
algebraic algebra, matrix 251f.
algebraic power series 167, 184, 323
algebra of invariants 379
Amitsur's theorem 477f.
antichain 370
anti-ideal 358f.
associated elements, matrices, maps xviii, 28, 74
atom, (n-)atomic xvii, 55, 74
augmentation ideal 136
augmentation-preserving automorphism 397

Baer's criterion 531
balanced relation 188, 304f.
BDT bounded decomposition type 6
Bergman's centralizer theorem 378, 407

Bezout domain 73, 109, 115f., 121, 183, 351f.
Bezout relation 54, 183
biassociated matrices 456
bicentral subfield 477
bipointed module 456
bipolar structure 399
Birkhoff's representation theorem 210, 524
block, -factorization 220
-blocked matrix 463
Boolean algebra 245, 522
bordered matrix 456
bound component, module 264
bound of an element 80, 341
bounded element, module 253, 341
bridge category 185

cancellable ring 23
canonical non-IBN ring 7, 510
capacity 17
cardinality of I, ($|I|$) xvi
category 525
centred automorphism 397
chain ring 200
characteristic of a module 26, 120, 144
cleavage, cleft 220f.
closed submodule, closure 282
code 361ff., 407
codomain 525
cofactor 133, 169
cofinal sequence, subset 110, 322
cogenerated 264
coherent ring 298, 536

Subject index

coimage, cokernel 526f.
coinduced extension 534
column rank 80, 283
comaximal pair, relation xviii, 31, 149, 188
comaximal transposition 197, 309
comma category 529f.
comma-free code 407
commensurable elements xviii
commutative diagram 525
companion matrix 253
comparison theorem 463
complement (in a lattice) 522
completable matrix 20
complete direct decomposition 216
complete factorization xvii, 196
complete (inversely filtered) ring 158, 375
complete prefix code 374
completely primary ring 217, 251, 346
completion 158
complex-skew polynomial ring 64, 209f., 213
conductor 332
conical monoid xvii, 14, 52, 172f., 358
conjugate idempotents 13
connected inversely filtered K-ring 161
connecting homomorphism 531
continuant polynomial 148
contravariant functor 529
convex 93
coprime pair, relation xviii, 138
coproduct 135, 526
core of admissble system 414, 460
core of bipolar structure 399
covariant functor 529
cover (in a lattice) 520
Cramer's rule 416
Czerniakiewicz-Makar-Limanov theorem 402, 409

decomposable element 214ff.
Dedekind's lemma 388
defect theorem 366
degenerate matrix 435
degree (-function) 60f., 123, 380
denominator 38, 414
dense subcategory 342, 525
dense submodule 282
dependence number 127, 157, 184
depth 413, 467ff.
derivation 41, 61
derived functor 513f.
derived set 67, 105

determinantal sum 429
DFL = distributive factor lattice 232
diagonal matrix sum 428
diagram chasing 527
Dieudonné determinant 494
differential operator ring 64, 105, 339, 378
dimension of a display 416
dimension, (co-)homological 532
direct limit 123
direct power, sum 1, 525
direct product 524
display 416
distributive lattice 225ff., 521f.
distributive module 116, 226f.
divisibility preordering 52
division algorithm (DA) 66ff.
divisor group $D(-)$ 497
domain 420, 525
dual module xvi, 2
duality 529
duality for modules 193f., 269ff.
dyad 58

E-related 150
E-ring 117
eigenring $E(-)$ 33, 58, 105, 251ff.
elementary divisor ring 85, 106
elementary divisors 84
elementary embedding 481
elementary matrix xvii, 196f.
elementary operation 80, 437
elementary sentence 538
epimorphism, epic 419, 526
equidivisible 43
equivalence of categories 529
essential extension 270, 531
essential left factor 280
essentially distinct factorizations 207
Euclidean algorithm 68, 105
Euclidean ring 67
Euler's theorem 140
exact functor 531
exact sequence 527
exchange principle 157, 184
extended elementary group $E_2^*(R)$ xvii, 147
Ext, extension of modules 532f.

factor-complete matrix set 501
factor-inverting matrix set 447, 500
factor-stable matrix set 437
factorial duality 195

568 Subject index

fastest algorithm 68
Fibonacci numbers 148
field of fractions 39, 419
field spectrum $X(-)$ 442
filter 538
filtered ring, filtration 125ff., 134
final object 525
(L, K)-finite 477
finite free resolution 26, 79
finitely presented, related 26, 533f.
fir = free ideal ring 110, 136, 183, 371
firoid 185
Fitting's lemma 217, 230, 276
five-lemma 528
flat module 50, 122, 369, 535f.
forgetful functor 530
formal degree 131f.
formal Laurent series 88
formal power series 88
formula xix
Fox derivative 86
fraction 413
Frame–Robinson–Thrall formula 382
free associative algebra 135
free D-field on a set 474f.
free ideal ring, see fir
free K-ring 135
free monoid 357ff.
free product 183, 398
free subset 360, 376
Frobenius' inequality 299
Frobenius–König theorem 191
full matrix 3, 186, 428
full relation 188
full subcategory 525
fully atomic semifir 196
fully invariant submodule 229
fully inverting homomorphism 447
fully reducible matrix 220
fully reversible ring 512
functor 529

G-value 491
Galois theory 388, 408
Gauss's lemma 140
generalized polynomial identity (GPI) 477
GE-related 150
GL-related 31, 150, 189
global dimension 107, 532
graded ring 141
graph of a mapping 226

Gröbner basis 184
Grothendieck category 276
Grothendieck group 14, 98

Hasse's criterion 77, 131
HCF, HCLF, HCRF highest common (left, right) factor 54, 154
height (of lattice element) 521
hereditary ring 108, 183
Hermite ring 19, 58, 87, 115
higher derivation 91
Higman's trick 311
Hilbert basis theorem 63
Hilbert series 142, 184, 382
HNN-construction 141, 185, 399, 407, 454
hollow matrix 187, 430
homogeneous subalgebra 368f.
homological dimension 532
honest ring homomorphism 287, 446

IBN = invariant basis number 2, 58
idealizer $I(-)$ 33, 58, 488
idempotent matrix 12ff.
image 527
indecomposable 523
independence property of tensor product 534
index of a matrix xviii
inert subring 165
inertia lemma 255f., 292
inertia theorem 165, 184, 453
inessential modification 113, 222, 305
initial object 525
injective hull 531
injective resolution 533
inner derivation 42, 64
inner rank 3, 58
integral closure 332
integral element, extension 332, 338
integral section 517
intersection theorem 328
interval 210, 520
Inv-atom 334
Inv-(in)decomposable element 345
invariant 379
invariant element, monoid, ring 53, 231, 236
invariant factors 84, 254
invariant matrix 335
invariant principal ideal 333
inverse filtration 157
inverse weak algorithm 157, 161, 184
invertible ring 515

-inverting 37, 411
involution 326
irredundant decomposition 215, 523
isomorphic idempotents 13
isomorphism of factorizations 75, 208
isotone (= order-preserving) map 243
iterated skew polynomial ring 98

J-(skew polynomial) ring 98, 106
Jacobian matrix 405
Jacobson radical 15, 102, 201, 230
join 519
join-irreducible 244, 523
de Jonquières automorphism 397
Jordan–Hölder theorem 75, 192, 246, 521

Kaplansky's theorem 109f.
kernel (map) 526
Koszul resolution 297
Kraft–McMillan inequality 362, 408
Kronecker delta 8
Kronecker function ring 353
Krull domain 337, 351
Krull–Schmidt theorem 83, 217, 276, 523
K-theory 58
Kurosh–Ore theorem 216, 523

large element (in a ring) 40
lattice 519ff.
lattice isomorphism 10
Laurent polynomial ring 86
law of nullity 189, 290f.
LCM, LCRM, LCLM least common (right, left) multiple 54, 117
leading term 94, 173
leapfrog construction 148
least matrix (pre-) ideal 444f.
left (right) full matrix, relation 186ff.
left prime matrix 187f.
Leibniz's formula 47
length of a chain or lattice 520
length of a monoid element 56, 75, 141, 358
level of a matrix 203
linearization by enlargement 311
linear dependence xviii, 118
link in a lattice 210, 520
local homomorphism 16
local rank 293
local ring 15
localization 39, 110f., 500ff.
Łos's theorem 539

lower segment 370
Lüroth's theorem 367

magic lemma 448, 492
Magnus' theorem 328
Malcev conditions 37
Malcev–Neumann power series 94
Malcolmson's criterion 438f., 474
matrix algebraic k-algebra 251
matrix atom 195
matrix ideal 430
matrix local ring 17
matrix nilradical 444f.
matrix pre-ideal 429
matrix prime 496
matrix reduction functor 11, 58, 180
matrix units xvii, 8
matrix valuation 500
maximal code 362
meet 519
meta-Artinian, -Noetherian module 228
meta(semi)fir 112, 183
minimal admissible matrix 463
minimal bound module 278, 308
minimal domain 420
modular lattice, law 520
monic matrix 312
monic normal form 313
monic polynomial xviii
monomial right K-basis 132
monomorphism, monic 526
monoid xvi
Morita equivalence 9, 119, 219, 529
multiplicative (matrix-)set 38, 411f.

Nagata's theorem 57
Nakayama's lemma 16f.
natural filtration 134
natural transformation 529
negative module 273
neutral 525
nilradical 511
normalizer, normalizing element 488
null matrix 9
nullity, law of 189, 290f.
numerator 38, 414

obstruction 533
one-sided fir 175, 185, 204f.
opposite category 525
opposite ring xvi, 10

570 Subject index

order of an admissible system 414
order-function 88, 137, 157
ordered series ring 94
Ore domain 39, 111, 128, 200, 338
Ore set 38, 111

palindrome 436
pan-ACC xviii, 6, 111, 183
parallelogram law 209, 521
partition lemma 119, 183
permutation matrix xvii
persistent property 238
perspective intervals 521
PID = principal ideal domain 73ff.
pivot matrix 416
place-permutation 382
pointed module 279
polynomial identity 477
positive module 272
power series ring 88, 137, 242
preadditive category 526
prefix (set), code 361
presentation 25f., 533
primal ring 289
primary decomposition 214ff., 223
primary matrix ideal 436
prime element 52, 334
prime matrix (left, right) 187
*-prime matrix (right) 460
prime matrix ideal 433
prime ring 514
principal valuation ring 162, 202, 375
product of matrix ideals 433
projective intervals 521
projective resolution 533
projective-free ring 2, 24, 58, 115
projective-trivial ring 24
proper factorization xvii, 307
proper specialization 426
protorsion module 276
pseudo-Sylvester domain 300ff., 452, 505
pullback 525f.
pure display 416
pure subalgebra 379
pushout 526
PVR = principal valuation ring 194

quasi-Euclidean ring 105
quasi-Frobenius ring 339
quasi-identity, -variety 453
quaternions 64, 72, 77, 261

quotient 67
quotient of matrix ideals 435f.

radical of a matrix ideal 434f.
rank (row, column) 80, 284
rank factorization 3, 285
rank of a free algebra 137
rank of a free module 2
rank of a module 49, 288
rational identity 475f.
rational power series 167, 184
rational section 517
rationality criterion 90
reduced admissible matrix 463
reduced fraction 413
reduced ring 104, 511, 515
refinement of a factorization 208
regular element, matrix xvii, 187
regular factorization 307
regular field extension 238
regularly embedded 368
relative invariant 379
remainder 67
residual division of matrix ideals 431
retract of a ring 123, 256
reversible ring 511ff.
right *-prime matrix 460
rigid domain 116, 200, 202
rigid element, monoid 43, 185, 199, 358
-ring xviii
ring of fractions 39
ringoid 185, 526
root of a monoid element 360, 377
row rank 80, 284

S-inverting ring 37f.
Σ-(ir)relevant 508
S-ring 300ff., 452, 505
saturated matrix set 427
scalar (term of) display 416, 437
Schanuel's lemma 26ff., 30, 223, 533
Schreier refinement theorem 208, 521
Schreier set 370
Schreier–Lewin formula 144, 184, 372
Schur's lemma 195, 278, 343
see-saw lemma 252
semi-Euclidean ring 85
semifir 113, 183
semifree module 49, 122
semihereditary ring 108, 183
semimaximal left ideal 37

Subject index

semiprime matrix ideal 434f.
semiprime ring 511
Serre's conjecture 297
shear 397
short exact sequence 528
signed permutation matrix xvii, 159
similar elements, matrices 27, 189
similar right ideals 76f., 186
simple $\mathbb{N}$-value 496
singular kernel 423
singularity support 442, 481
skew Laurent polynomial 86
skew polynomial 61f., 105f.
skew power series ring 248
skew rational function field 63
Smith normal form 80
source 525
spatial module 293
special module 269
specialization 420
specialization lemma 479f.
spectral space 443
split idempotent 24, 288
square-free module 233
stabilized matrix 300
stable matrix atom 23
stable general linear group 494
stable rank 5, 300
stably associated 28, 189
stably biassociated 456
stably free module 15, 19
stably full 5, 301
stably honest homomorphism 451
standard basis 10
standard form for GE_2 146, 401
strict operation 461
strictly bordered matrix 461
strictly positive (negative) module 273
strong DFL-property 234, 332
strong E_2-ring 117, 155
strong G-ring 117
strong prime ideal 104, 206
strong v-dependence 126, 184
strongly bound module 270
strongly regular ring 454, 515
subcategory 525
subdirectly (ir)reducible 524
subfunctor 529
subhomomorphism 420
suffix (set) 361
support 93

Sylvester domain 291, 450
symmetric ring 514

tame automorphism 398
target 525
TC-ring 224
tensor K-ring 134
tensor product 534
tertiary radical 350
three-by-three lemma 528
topological fir 160
Tor-functor 536
Tor-simple 195
torsion class, torsion-free class 264
torsion element 48, 77f.
torsion-free module 48
torsion module 192, 273
torsion theory 264
total divisor 79, 106
total inertia 165, 184
totally coprime elements 344, 471
totally unbounded 253, 344
totally uncleft 221
trace form, map 387
transcendental matrix 252
transduction 133, 136, 184
transfinite degree-function 171
transfinite weak algorithm 172, 185
transfinitely Euclidean 68, 105
translation 396
translation ring 65
transpose of a module 269, 329
triangular automorphism 397
trivial filtration 125
trivial(izable) relation 113f.
truncated filtered ring 137
type of a non-IBN ring 4

UF-monoid 52f., 334
UFD unique factorization domain 52, 75, 192
UGN unbounded generating number 2, 58
ultrafilter, -power, -product 538
unbound (n-unbound) module 264
uncleft 220
unfactorable matrix 307
uniform module 51
unimodular row xviii, 20
unique factorization of invariant elements 334f.
unique remainder algorithm 69, 99, 106

unit-closed, -closure 373, 395, 404
unit-linear dependence 118
unitriangular matrix 159
universal denominator 470
universal derivation 66
universal derivation bimodule 145
universal field of fractions 421
universal group 38
universal localization 422ff.
universal object 530
universal R-field 421
universal Σ-inverting homomorphism, ring 422
upper segment 243, 370

v-dependence 125
v-generator 130
V-ring 224
valuation (ring) 337, 498f.
value of a display 437

weak algorithm (n-term) 126, 183, 380f.
weak (global) dimension 536
weak v-basis 129, 160
weakly (n-)finite 2, 58
weakly semihereditary 109, 183
weight of a subset 361
Weyl algebra 64, 87, 152, 199
WF = weakly finite 2, 58
Whitehead group 495
Whitehead's lemma 498, 500
wild automorphism 398

X-inner, -outer 387

Young tableau 382

zero-delay code 361
zero-divison xvii
zero object 325
Zero ring xvi